D1207498

Methods in Enzymology

Volume 278

FLUORESCENCE SPECTROSCOPY

METHODS IN ENZYMOLOGY

EDITORS-IN-CHIEF

John N. Abelson   Melvin I. Simon

DIVISION OF BIOLOGY
CALIFORNIA INSTITUTE OF TECHNOLOGY
PASADENA, CALIFORNIA

FOUNDING EDITORS

Sidney P. Colowick and Nathan O. Kaplan

Methods in Enzymology

Volume 278

Fluorescence Spectroscopy

EDITED BY

Ludwig Brand

DEPARTMENTS OF BIOLOGY AND BIOPHYSICS
McCOLLUM–PRATT INSTITUTE
JOHNS HOPKINS UNIVERSITY
BALTIMORE, MARYLAND

Michael L. Johnson

DEPARTMENT OF PHARMACOLOGY
UNIVERSITY OF VIRGINIA
HEALTH SCIENCES CENTER
CHARLOTTESVILLE, VIRGINIA

ACADEMIC PRESS
San Diego London Boston New York Sydney Tokyo Toronto

This book is printed on acid-free paper. 

Academic Press
15 East 26th Street, 15th floor, New York, New York 10010
http://www.apnet.com

Academic Press Limited
24-28 Oval Road, London NW1 7DX, UK
http://www.hbuk.co.uk/ap/

International Standard Book Number: 0-12-182179-X

PRINTED IN THE UNITED STATES OF AMERICA
97 98 99 00 01 02 EB 9 8 7 6 5 4 3 2 1

Table of Contents

Contributors to Volume 278

Article numbers are in parentheses following the names of contributors.
Affiliations listed are current.

MARCEL AMELOOT (6), *Limburgs Universitair Centrum, B-3590 Diepenbeek, Belgium*

PAUL J. ANGIOLILLO (5), *Johnson Research Foundation, Department of Biochemistry and Biophysics, School of Medicine, University of Pennsylvania, Philadelphia, Pennsylvania 19104*

MARY D. BARKLEY (9), *Department of Chemistry, Case Western Reserve University, Cleveland, Ohio 44106*

JOSEPH M. BEECHEM (3), *Department of Molecular Physiology and Biophysics, Vanderbilt University School of Medicine, Nashville, Tennessee 37232*

NOËL BOENS (6), *Department of Chemistry, Katholieke Universiteit Leuven, B-3001 Heverlee, Belgium*

LUDWIG BRAND (2, 15), *Departments of Biology and Biophysics, McCollum–Pratt Institute, Johns Hopkins University, Baltimore, Maryland 21218*

ENRICO BUCCI (28), *Department of Biochemistry, University of Maryland Medical School, Baltimore, Maryland 21201*

PATRIK R. CALLIS (7), *Department of Chemistry and Biochemistry, Montana State University, Bozeman, Montana 59717*

TANYA E. S. DAHMS (10), *Department of Biochemistry, University of Ottawa, Ottawa, Ontario, Canada K1H 8M5*

LESLEY DAVENPORT (24), *Department of Chemistry, Brooklyn College of the City University of New York, Brooklyn, New York 11210*

JOHN F. ECCLESTON (18), *Division of Physical Biochemistry, National Institute for Medical Research, London, NW7 1AA, United Kingdom*

MICHAEL EDIDIN (21), *Department of Biology, Johns Hopkins University, Baltimore, Maryland 21218*

MAURICE R. EFTINK (11, 12), *Department of Chemistry, University of Mississippi, University, Mississippi 38677*

PEGGY S. EIS (16), *Third Wave Technologies, Inc., Madison, Wisconsin 53711*

ARI GAFNI (4), *Institute of Gerontology, Department of Biological Chemistry, University of Michigan, Ann Arbor, Michigan 48109*

ZYGMUNT GRYCZYNSKI (28), *Department of Biochemistry, University of Maryland Medical School, Baltimore, Maryland 21201*

MYUN K. HAN (17), *Department of Biochemistry, Georgetown University Medical Center, Washington, DC 20007*

JOHN J. HILL (19), *School of Pharmacy, University of Wisconsin–Madison, Madison, Wisconsin 53706*

CHRISTOPHER W. V. HOGUE (8), *National Center for Biotechnology Information, National Library of Medicine, National Institutes of Health, Bethesda, Maryland 20894*

KALINA HRISTOVA (23), *Department of Physiology and Biophysics, College of Medicine, University of California, Irvine, California 92697*

DAVID M. JAMESON (18), *Department of Biochemistry and Biophysics, University of Hawaii at Manoa, Honolulu, Hawaii 96822*

MICHAEL L. JOHNSON (29), *Department of Pharmacology, University of Virginia, Health Sciences Center, Charlottesville, Virginia 22908*

AKIRA KOMORIYA (2), *OncoImmunin, Inc., College Park, Maryland 20742*

ANDRZEJ KOWALCZYK (6), *Institute of Physics, Nicholas Copernicus University, 87-100 Toruń, Poland*

MONIQUE LABERGE (5), *Johnson Research Foundation, Department of Biochemistry and Biophysics, School of Medicine, University of Pennsylvania, Philadelphia, Pennsylvania 19104*

ALEXEY S. LADOKHIN (22, 23), *Palladin Institute of Biochemistry, National Academy of Sciences of Ukraine, Kiev, 252030, Ukraine*

JOSEPH R. LAKOWICZ (14), *Center for Fluorescence Spectroscopy, Department of Biochemistry and Molecular Biology, University of Maryland at Baltimore, Baltimore, Maryland 21201*

KYUNG BOK LEE (25), *Department of Biology, Johns Hopkins University, Baltimore, Maryland 21218*

S. PAUL LEE (17), *Department of Biochemistry, Georgetown University Medical Center, Washington, DC 20007*

YUAN CHUAN LEE (25, 26), *Department of Biology, Johns Hopkins University, Baltimore, Maryland 21218*

SHERWIN S. LEHRER (13), *Boston Biomedical Research Institute, Boston, Massachusetts 02114; Department of Neurology, Harvard Medical School, Boston, Massachusetts 02115*

JACEK LUBKOWSKI (28), *Macromolecular Structure Laboratory, NCI-FCRDC, ABL-Basic Research Program, Frederick, Maryland 21702*

JANOS MATKO (21), *Department of Biophysics, University School of Medicine, 4012 Debrecen, Hungary*

KOJI MATSUOKA (26), *Faculty of Engineering, Saitama University, Urawa, Saitama 338, Japan*

RICHARD E. MCCARTY (27), *Department of Biology, Johns Hopkins University, Baltimore, Maryland 21218*

MARK L. MCLAUGHLIN (9), *Department of Chemistry, Louisiana State University, Baton Rouge, Louisiana 70803*

DAVID P. MILLAR (20), *Department of Molecular Biology, The Scripps Research Institute, La Jolla, California 92037*

SHIN-ICHIRO NISHIMURA (26), *Division of Biological Sciences, Graduate School of Science, Hokkaido University, Hokkaido 060, Japan*

BEVERLY Z. PACKARD (2), *OncoImmunin, Inc., College Park, Maryland 20742*

J. B. ALEXANDER ROSS (8), *Department of Biochemistry, Mount Sinai School of Medicine, New York, New York 10029*

CATHERINE A. ROYER (19), *School of Pharmacy, University of Wisconsin–Madison, Madison, Wisconsin 53706*

JOSEPH A. SCHAUERTE (4), *Institute of Gerontology, Department of Biological Chemistry, University of Michigan, Ann Arbor, Michigan 48109*

M. C. R. SHASTRY (12), *Department of Chemistry, University of Mississippi, University, Mississippi 38677*

DUNCAN G. STEEL (4), *Institute of Gerontology, Department of Biological Chemistry, University of Michigan, Ann Arbor, Michigan 48109*

ARTHUR G. SZABO (8, 10), *Department of Chemistry and Biochemistry, University of Windsor, Windsor, Ontario, Canada N9B 3P4*

HENRYK SZMACINSKI (14), *Center for Fluorescence Spectroscopy, Department of Biochemistry and Molecular Biology, University of Maryland at Baltimore, Baltimore, Maryland 21201*

EWALD TERPETSCHNIG (14), *Institute for Analytical Chemistry, Bio and Chemosensors, University of Regensburg, 93040 Regensburg, Germany*

DMITRI D. TOPTYGIN (2), *Department of Biology, Johns Hopkins University, Baltimore, Maryland 21218*

JANE M. VANDERKOOI (5), *Johnson Research Foundation, Department of Biochemistry and Biophysics, School of Medicine, University of Pennsylvania, Philadelphia, Pennsylvania 19104*

GREGORIO WEBER (1), *School of Chemical Sciences, University of Illinois, Urbana, Illinois 61801*

STEPHEN H. WHITE (23), *Department of Physiology and Biophysics, College of Medicine, University of California, Irvine, California 92697*

WILLIAM C. WIMLEY (23), *Department of Physiology and Biophysics, College of Medicine, University of California, Irvine, California 92697*

PENGGUANG WU (15), *Tularik Inc., South San Francisco, California 94080*

MENGSU YANG (20), *Department of Molecular Biology, The Scripps Research Institute, La Jolla, California 92037*

Preface

Fluorescence spectroscopy stands out among the optical techniques used to investigate biomolecules in that the lifetime of the excited states is sufficiently long for a variety of chemical and physical interactions to take place prior to emission. These include solvent reorientation, complex formation, proton transfer, and transfer of excited-state energy from one chromophore to another. Studies of these processes provide information on the chemical and physical nature of the environment. In this way information regarding specific sites on proteins, nucleic acids, and biological membranes can be obtained and related to their function. Excitation of electronic transitions in organic molecules involves orientational photoselection. Fluorescence anisotropy measurements provide information about rotational diffusive motions and thus tell us about the size and shape of macromolecules and their interaction with small molecules and with other macromolecules.

There have been significant advances in instrumentation, particularly in regard to time-resolved measurements and procedures for data analysis. More information on intrinsic fluorescent probes, such as tryptophan, have become available and new types of probes and ways of attaching them to macromolecules have been described. It is now possible to do nanosecond time-resolved measurements through a microscope. In this way many of the approaches initially used in solution with purified molecules can now be carried over directly to work with organelles and cells.

Topics covered in this volume include discussions of the fluorescence of tryptophan, the origin of the excited states, and their sensitivity to environmental perturbations. Studies of constrained tryptophan residues have been helpful in this regard as have been studies of tryptophan in crystals. The fluorescence of proteins can be utilized to study the equilibria and kinetics of protein–protein and protein–nucleic acid interactions. Incorporation of tryptophan derivatives into proteins has led to spectrally enhanced proteins which have been helpful in these studies. New technology is described which allows fluorescence lifetimes to be used to study slower (millisecond to second) kinetic processes. Applications of fluorescence to studies of nucleic acids and protein–nucleic acid interactions are moving forward at a rapid pace. New probes have been developed and energy transfer and emission anisotropy are being used to advantage. Fluorescence provides excellent ways to measure DNA cleavage.

Phosphorescence occurs on slower time scales than fluorescence. The finding that many proteins exhibit room temperature phosphorescence provided that oxygen is excluded has made time-dependent protein phosphorescence an important new technique for studying conformational changes and protein folding. The development of metal–ligand complexes with very long fluorescence decay times represents a major breakthrough for studies of events on a long time scale, in particular rotational motion of macromolecular complexes.

Several chapters describe applications of fluorescence to studies of bilayer membranes as well as of cell surfaces. Fluorescence spectroscopy continues to be of value in studies of complex systems such as chloroplast ATP synthase and hemoglobin.

From the breadth of these topics it is clear that fluorescence techniques have wide application for studies of biological problems. This volume touches on some of these important areas.

Ludwig Brand
Michael L. Johnson

METHODS IN ENZYMOLOGY

VOLUME I. Preparation and Assay of Enzymes
Edited by SIDNEY P. COLOWICK AND NATHAN O. KAPLAN

VOLUME II. Preparation and Assay of Enzymes
Edited by SIDNEY P. COLOWICK AND NATHAN O. KAPLAN

VOLUME III. Preparation and Assay of Substrates
Edited by SIDNEY P. COLOWICK AND NATHAN O. KAPLAN

VOLUME IV. Special Techniques for the Enzymologist
Edited by SIDNEY P. COLOWICK AND NATHAN O. KAPLAN

VOLUME V. Preparation and Assay of Enzymes
Edited by SIDNEY P. COLOWICK AND NATHAN O. KAPLAN

VOLUME VI. Preparation and Assay of Enzymes (*Continued*)
Preparation and Assay of Substrates
Special Techniques
Edited by SIDNEY P. COLOWICK AND NATHAN O. KAPLAN

VOLUME VII. Cumulative Subject Index
Edited by SIDNEY P. COLOWICK AND NATHAN O. KAPLAN

VOLUME VIII. Complex Carbohydrates
Edited by ELIZABETH F. NEUFELD AND VICTOR GINSBURG

VOLUME IX. Carbohydrate Metabolism
Edited by WILLIS A. WOOD

VOLUME X. Oxidation and Phosphorylation
Edited by RONALD W. ESTABROOK AND MAYNARD E. PULLMAN

VOLUME XI. Enzyme Structure
Edited by C. H. W. HIRS

VOLUME XII. Nucleic Acids (Parts A and B)
Edited by LAWRENCE GROSSMAN AND KIVIE MOLDAVE

VOLUME XIII. Citric Acid Cycle
Edited by J. M. LOWENSTEIN

VOLUME XIV. Lipids
Edited by J. M. LOWENSTEIN

VOLUME XV. Steroids and Terpenoids
Edited by RAYMOND B. CLAYTON

VOLUME XVI. Fast Reactions
Edited by KENNETH KUSTIN

Volume XXXVI. Hormone Action (Part A: Steroid Hormones)
Edited by Bert W. O'Malley and Joel G. Hardman

Volume XXXVII. Hormone Action (Part B: Peptide Hormones)
Edited by Bert W. O'Malley and Joel G. Hardman

Volume XXXVIII. Hormone Action (Part C: Cyclic Nucleotides)
Edited by Joel G. Hardman and Bert W. O'Malley

Volume XXXIX. Hormone Action (Part D: Isolated Cells, Tissues, and Organ Systems)
Edited by Joel G. Hardman and Bert W. O'Malley

Volume XL. Hormone Action (Part E: Nuclear Structure and Function)
Edited by Bert W. O'Malley and Joel G. Hardman

Volume XLI. Carbohydrate Metabolism (Part B)
Edited by W. A. Wood

Volume XLII. Carbohydrate Metabolism (Part C)
Edited by W. A. Wood

Volume XLIII. Antibiotics
Edited by John H. Hash

Volume XLIV. Immobilized Enzymes
Edited by Klaus Mosbach

Volume XLV. Proteolytic Enzymes (Part B)
Edited by Laszlo Lorand

Volume XLVI. Affinity Labeling
Edited by William B. Jakoby and Meir Wilchek

Volume XLVII. Enzyme Structure (Part E)
Edited by C. H. W. Hirs and Serge N. Timasheff

Volume XLVIII. Enzyme Structure (Part F)
Edited by C. H. W. Hirs and Serge N. Timasheff

Volume XLIX. Enzyme Structure (Part G)
Edited by C. H. W. Hirs and Serge N. Timasheff

Volume L. Complex Carbohydrates (Part C)
Edited by Victor Ginsburg

Volume LI. Purine and Pyrimidine Nucleotide Metabolism
Edited by Patricia A. Hoffee and Mary Ellen Jones

Volume LII. Biomembranes (Part C: Biological Oxidations)
Edited by Sidney Fleischer and Lester Packer

Volume LIII. Biomembranes (Part D: Biological Oxidations)
Edited by Sidney Fleischer and Lester Packer

Volume LIV. Biomembranes (Part E: Biological Oxidations)
Edited by Sidney Fleischer and Lester Packer

VOLUME 108. Immunochemical Techniques (Part G: Separation and Characterization of Lymphoid Cells)
Edited by GIOVANNI DI SABATO, JOHN J. LANGONE, AND HELEN VAN VUNAKIS

VOLUME 109. Hormone Action (Part I: Peptide Hormones)
Edited by LUTZ BIRNBAUMER AND BERT W. O'MALLEY

VOLUME 110. Steroids and Isoprenoids (Part A)
Edited by JOHN H. LAW AND HANS C. RILLING

VOLUME 111. Steroids and Isoprenoids (Part B)
Edited by JOHN H. LAW AND HANS C. RILLING

VOLUME 112. Drug and Enzyme Targeting (Part A)
Edited by KENNETH J. WIDDER AND RALPH GREEN

VOLUME 113. Glutamate, Glutamine, Glutathione, and Related Compounds
Edited by ALTON MEISTER

VOLUME 114. Diffraction Methods for Biological Macromolecules (Part A)
Edited by HAROLD W. WYCKOFF, C. H. W. HIRS, AND SERGE N. TIMASHEFF

VOLUME 115. Diffraction Methods for Biological Macromolecules (Part B)
Edited by HAROLD W. WYCKOFF, C. H. W. HIRS, AND SERGE N. TIMASHEFF

VOLUME 116. Immunochemical Techniques (Part H: Effectors and Mediators of Lymphoid Cell Functions)
Edited by GIOVANNI DI SABATO, JOHN J. LANGONE, AND HELEN VAN VUNAKIS

VOLUME 117. Enzyme Structure (Part J)
Edited by C. H. W. HIRS AND SERGE N. TIMASHEFF

VOLUME 118. Plant Molecular Biology
Edited by ARTHUR WEISSBACH AND HERBERT WEISSBACH

VOLUME 119. Interferons (Part C)
Edited by SIDNEY PESTKA

VOLUME 120. Cumulative Subject Index Volumes 81–94, 96–101

VOLUME 121. Immunochemical Techniques (Part I: Hybridoma Technology and Monoclonal Antibodies)
Edited by JOHN J. LANGONE AND HELEN VAN VUNAKIS

VOLUME 122. Vitamins and Coenzymes (Part G)
Edited by FRANK CHYTIL AND DONALD B. MCCORMICK

VOLUME 123. Vitamins and Coenzymes (Part H)
Edited by FRANK CHYTIL AND DONALD B. MCCORMICK

VOLUME 124. Hormone Action (Part J: Neuroendocrine Peptides)
Edited by P. MICHAEL CONN

VOLUME 125. Biomembranes (Part M: Transport in Bacteria, Mitochondria, and Chloroplasts: General Approaches and Transport Systems)
Edited by SIDNEY FLEISCHER AND BECCA FLEISCHER

VOLUME 126. Biomembranes (Part N: Transport in Bacteria, Mitochondria, and Chloroplasts: Protonmotive Force)
Edited by SIDNEY FLEISCHER AND BECCA FLEISCHER

VOLUME 127. Biomembranes (Part O: Protons and Water: Structure and Translocation)
Edited by LESTER PACKER

VOLUME 128. Plasma Lipoproteins (Part A: Preparation, Structure, and Molecular Biology)
Edited by JERE P. SEGREST AND JOHN J. ALBERS

VOLUME 129. Plasma Lipoproteins (Part B: Characterization, Cell Biology, and Metabolism)
Edited by JOHN J. ALBERS AND JERE P. SEGREST

VOLUME 130. Enzyme Structure (Part K)
Edited by C. H. W. HIRS AND SERGE N. TIMASHEFF

VOLUME 131. Enzyme Structure (Part L)
Edited by C. H. W. HIRS AND SERGE N. TIMASHEFF

VOLUME 132. Immunochemical Techniques (Part J: Phagocytosis and Cell-Mediated Cytotoxicity)
Edited by GIOVANNI DI SABATO AND JOHANNES EVERSE

VOLUME 133. Bioluminescence and Chemiluminescence (Part B)
Edited by MARLENE DELUCA AND WILLIAM D. MCELROY

VOLUME 134. Structural and Contractile Proteins (Part C: The Contractile Apparatus and the Cytoskeleton)
Edited by RICHARD B. VALLEE

VOLUME 135. Immobilized Enzymes and Cells (Part B)
Edited by KLAUS MOSBACH

VOLUME 136. Immobilized Enzymes and Cells (Part C)
Edited by KLAUS MOSBACH

VOLUME 137. Immobilized Enzymes and Cells (Part D)
Edited by KLAUS MOSBACH

VOLUME 138. Complex Carbohydrates (Part E)
Edited by VICTOR GINSBURG

VOLUME 139. Cellular Regulators (Part A: Calcium- and Calmodulin-Binding Proteins)
Edited by ANTHONY R. MEANS AND P. MICHAEL CONN

VOLUME 140. Cumulative Subject Index Volumes 102–119, 121–134

VOLUME 141. Cellular Regulators (Part B: Calcium and Lipids)
Edited by P. MICHAEL CONN AND ANTHONY R. MEANS

VOLUME 142. Metabolism of Aromatic Amino Acids and Amines
Edited by SEYMOUR KAUFMAN

VOLUME 143. Sulfur and Sulfur Amino Acids
Edited by WILLIAM B. JAKOBY AND OWEN GRIFFITH

VOLUME 144. Structural and Contractile Proteins (Part D: Extracellular Matrix)
Edited by LEON W. CUNNINGHAM

VOLUME 145. Structural and Contractile Proteins (Part E: Extracellular Matrix)
Edited by LEON W. CUNNINGHAM

VOLUME 146. Peptide Growth Factors (Part A)
Edited by DAVID BARNES AND DAVID A. SIRBASKU

VOLUME 147. Peptide Growth Factors (Part B)
Edited by DAVID BARNES AND DAVID A. SIRBASKU

VOLUME 148. Plant Cell Membranes
Edited by LESTER PACKER AND ROLAND DOUCE

VOLUME 149. Drug and Enzyme Targeting (Part B)
Edited by RALPH GREEN AND KENNETH J. WIDDER

VOLUME 150. Immunochemical Techniques (Part K: *In Vitro* Models of B and T Cell Functions and Lymphoid Cell Receptors)
Edited by GIOVANNI DI SABATO

VOLUME 151. Molecular Genetics of Mammalian Cells
Edited by MICHAEL M. GOTTESMAN

VOLUME 152. Guide to Molecular Cloning Techniques
Edited by SHELBY L. BERGER AND ALAN R. KIMMEL

VOLUME 153. Recombinant DNA (Part D)
Edited by RAY WU AND LAWRENCE GROSSMAN

VOLUME 154. Recombinant DNA (Part E)
Edited by RAY WU AND LAWRENCE GROSSMAN

VOLUME 155. Recombinant DNA (Part F)
Edited by RAY WU

VOLUME 156. Biomembranes (Part P: ATP-Driven Pumps and Related Transport: The Na,K-Pump)
Edited by SIDNEY FLEISCHER AND BECCA FLEISCHER

VOLUME 157. Biomembranes (Part Q: ATP-Driven Pumps and Related Transport: Calcium, Proton, and Potassium Pumps)
Edited by SIDNEY FLEISCHER AND BECCA FLEISCHER

VOLUME 158. Metalloproteins (Part A)
Edited by JAMES F. RIORDAN AND BERT L. VALLEE

VOLUME 159. Initiation and Termination of Cyclic Nucleotide Action
Edited by JACKIE D. CORBIN AND ROGER A. JOHNSON

VOLUME 160. Biomass (Part A: Cellulose and Hemicellulose)
Edited by WILLIS A. WOOD AND SCOTT T. KELLOGG

VOLUME 161. Biomass (Part B: Lignin, Pectin, and Chitin)
Edited by WILLIS A. WOOD AND SCOTT T. KELLOGG

VOLUME 162. Immunochemical Techniques (Part L: Chemotaxis and Inflammation)
Edited by GIOVANNI DI SABATO

VOLUME 163. Immunochemical Techniques (Part M: Chemotaxis and Inflammation)
Edited by GIOVANNI DI SABATO

VOLUME 164. Ribosomes
Edited by HARRY F. NOLLER, JR., AND KIVIE MOLDAVE

VOLUME 165. Microbial Toxins: Tools for Enzymology
Edited by SIDNEY HARSHMAN

VOLUME 166. Branched-Chain Amino Acids
Edited by ROBERT HARRIS AND JOHN R. SOKATCH

VOLUME 167. Cyanobacteria
Edited by LESTER PACKER AND ALEXANDER N. GLAZER

VOLUME 168. Hormone Action (Part K: Neuroendocrine Peptides)
Edited by P. MICHAEL CONN

VOLUME 169. Platelets: Receptors, Adhesion, Secretion (Part A)
Edited by JACEK HAWIGER

VOLUME 170. Nucleosomes
Edited by PAUL M. WASSARMAN AND ROGER D. KORNBERG

VOLUME 171. Biomembranes (Part R: Transport Theory: Cells and Model Membranes)
Edited by SIDNEY FLEISCHER AND BECCA FLEISCHER

VOLUME 172. Biomembranes (Part S: Transport: Membrane Isolation and Characterization)
Edited by SIDNEY FLEISCHER AND BECCA FLEISCHER

VOLUME 173. Biomembranes [Part T: Cellular and Subcellular Transport: Eukaryotic (Nonepithelial) Cells]
Edited by SIDNEY FLEISCHER AND BECCA FLEISCHER

VOLUME 174. Biomembranes [Part U: Cellular and Subcellular Transport: Eukaryotic (Nonepithelial) Cells]
Edited by SIDNEY FLEISCHER AND BECCA FLEISCHER

VOLUME 175. Cumulative Subject Index Volumes 135–139, 141–167

VOLUME 176. Nuclear Magnetic Resonance (Part A: Spectral Techniques and Dynamics)
Edited by NORMAN J. OPPENHEIMER AND THOMAS L. JAMES

VOLUME 177. Nuclear Magnetic Resonance (Part B: Structure and Mechanism)
Edited by NORMAN J. OPPENHEIMER AND THOMAS L. JAMES

VOLUME 248. Proteolytic Enzymes: Aspartic and Metallo Peptidases
Edited by ALAN J. BARRETT

VOLUME 249. Enzyme Kinetics and Mechanism (Part D: Developments in Enzyme Dynamics)
Edited by DANIEL L. PURICH

VOLUME 250. Lipid Modifications of Proteins
Edited by PATRICK J. CASEY AND JANICE E. BUSS

VOLUME 251. Biothiols (Part A: Monothiols and Dithiols, Protein Thiols, and Thiyl Radicals)
Edited by LESTER PACKER

VOLUME 252. Biothiols (Part B: Glutathione and Thioredoxin; Thiols in Signal Transduction and Gene Regulation)
Edited by LESTER PACKER

VOLUME 253. Adhesion of Microbial Pathogens
Edited by RON J. DOYLE AND ITZHAK OFEK

VOLUME 254. Oncogene Techniques
Edited by PETER K. VOGT AND INDER M. VERMA

VOLUME 255. Small GTPases and Their Regulators (Part A: Ras Family)
Edited by W. E. BALCH, CHANNING J. DER, AND ALAN HALL

VOLUME 256. Small GTPases and Their Regulators (Part B: Rho Family)
Edited by W. E. BALCH, CHANNING J. DER, AND ALAN HALL

VOLUME 257. Small GTPases and Their Regulators (Part C: Proteins Involved in Transport)
Edited by W. E. BALCH, CHANNING J. DER, AND ALAN HALL

VOLUME 258. Redox-Active Amino Acids in Biology
Edited by JUDITH P. KLINMAN

VOLUME 259. Energetics of Biological Macromolecules
Edited by MICHAEL L. JOHNSON AND GARY K. ACKERS

VOLUME 260. Mitochondrial Biogenesis and Genetics (Part A)
Edited by GIUSEPPE M. ATTARDI AND ANNE CHOMYN

VOLUME 261. Nuclear Magnetic Resonance and Nucleic Acids
Edited by THOMAS L. JAMES

VOLUME 262. DNA Replication
Edited by JUDITH L. CAMPBELL

VOLUME 263. Plasma Lipoproteins (Part C: Quantitation)
Edited by WILLIAM A. BRADLEY, SANDRA H. GIANTURCO, AND JERE P. SEGREST

VOLUME 264. Mitochondrial Biogenesis and Genetics (Part B)
Edited by GIUSEPPE M. ATTARDI AND ANNE CHOMYN

VOLUME 284. Lipases (Part A: Biotechnology) (in preparation)
Edited by BYRON RUBIN AND EDWARD A. DENNIS

VOLUME 285. Cumulative Subject Index Volumes 263, 264, 266–268 (in preparation)

VOLUME 286. Lipases (Part B: Enzyme Characterization and Utilization) (in preparation)
Edited by BYRON RUBIN AND EDWARD A. DENNIS

VOLUME 287. Chemokines (in preparation)
Edited by RICHARD HORUK

VOLUME 288. Chemokine Receptors (in preparation)
Edited by RICHARD HORUK

[1] Fluorescence in Biophysics: Accomplishments and Deficiencies

By GREGORIO WEBER

To characterize the photoexcited emission from molecules in a system of unknown complexity, we should determine the spectral distribution, photon yield, lifetime of the excited state, and polarization of the fluorescence emission, as a function of the wavelength of excitation. The techniques for the determination of each of these quantities have undergone remarkable advances since they were first measured, in the decade from 1920 to 1930, when the quantum theory revealed the origin of the fluorescence emission and provided the first explanation of each of these characteristic properties of the emission. The measurements of fluorescence intensity have been made absolute following the introduction of photon-counting techniques in the 1950s, and what in those days was a complex procedure has been simplified and made much more reliable. By elimination of several sources of uncertainty in the procedure, the precision of the fluorescence intensity as measured by photon counting is now practically dependent on the number counted alone. It thus provides for the achievement of the physical limit of resolution in the many cases in which the emission is heterogeneous as regards its molecular origin. By the application of either of the conjugate methods of determination of the fluorescence lifetime, single-photon delay or phase-modulation techniques, the analysis of complex fluorescence decay has been improved to the point that it provides a benchmark for the ultimate precision and resolvability of numerical measurements in general.

The first measurements of intensity, yield, lifetime of the excitation, and polarization of the fluorescence were all performed by methods that involved the human eye. Today we rely on photoelectric devices with sensitivities comparable to that of the dark-adapted eye, with the additional advantage that this sensitivity does not decrease appreciably as the light intensity increases, as is the case with the eye. The most important advantage of the photoelectric methods is the improved spectral range in which fluorescence can be detected and measured. The ultraviolet fluorescence of amino acids and proteins could not be detected, let alone measured, before the advent of reliable photoelectric methods, although the fluorescence of phenol, to give one example, was detected by the photographic plate before it was measured by means of photomultipliers. Measurements of the fluorescence of bacteriochlorophyll became possible only when photocathodes sensitive to the near-infrared were developed.

0076-6879/97 $25

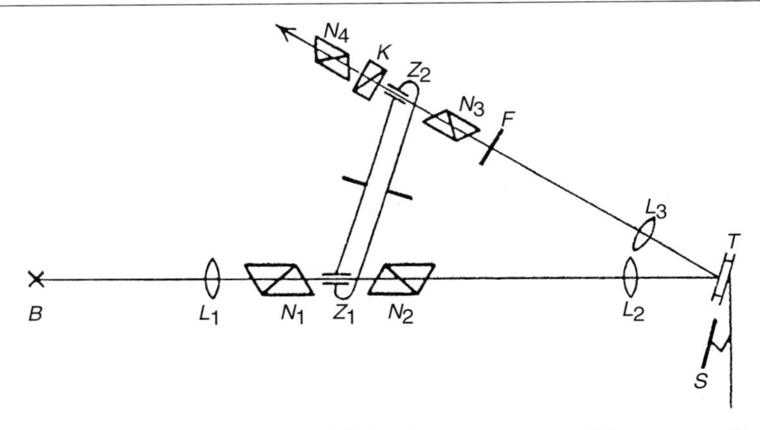

FIG. 1. Original apparatus of Gaviola[1] for the measurement of fluorescence lifetimes, described in text. *B*, Source of exciting light; *T*, cuvette containing the fluorescent solution; *S*, mirror.

The measurements of fluorescence lifetime were, and are still, the more demanding of all those that characterize the fluorescent properties, yet a practically satisfactory precision was achieved by the first direct measurements (done by E. Gaviola in Berlin,[1] employing a visual method) as early as 1926. The apparatus is worth describing, as it relies on the same physical principles that are in use today. However, while in the fluorometer used by Gaviola all the dispositions of the instrument, and the reasons for them, are clearly evident, today we employ black boxes of whose functioning and characteristics we are too often ignorant. Figure 1 shows schematically the apparatus of Gaviola: The exciting light from a carbon arc was modulated by passage through a Kerr cell, Z_1, placed between crossed polarizers N_1 and N_2. A sinusoidal voltage with a frequency of 5 MHz was applied to the Kerr electrodes, and the periodic birefringence of the nitrobenzene, owing to the orientation of the molecules in the electric field, resulted in the creation of a beam of visible light sinusoidally modulated at 10 MHz that emerges from N_2. This light is focused at T, where either a mirror or a cuvette containing a fluorophore solution is placed. The fluorescence, or alternatively the reflected light, reaches the second Kerr cell, Z_2, after going through the polarizer N_3 oriented so as to let through light polarized parallel to that emerging from N_2. Moving the mirror away from Z_1 by an additional length l results in an additional delay in the arrival of the reflected or fluorescent light by a time $t = l/c$, where c is the velocity of light. The time

[1] E. Gaviola, *Z. Physik* **35,** 748 (1926).

between two maxima of transmission of light at Z_2 is $l/(2\pi f)$, where f is the sinusoidal frequency of the voltage applied to the Kerr cells. For f = 5 MHz the Kerr cell transmission period is 3×10^{-8} sec, corresponding to a distance between Z_2 and mirror of some 10 m. Equalizing the arrival of fluorescence at Z_2 with a corresponding increase in mirror–detector distance l with respect to the reflected light permitted Gaviola to measure fluorescence emission delays of 1 to 13 nsec. To establish a fiducial point dependent on the time of arrival at Z_2 in relation to its transmission period, Gaviola introduced a birefringent plate K between Z_2 and N_4, thus creating two images of the light emerging from K_2 polarized in orthogonal directions. The angular displacement of N_4 required to equalize the polarized images is proportional to l. In this way delay curves for the reflected and fluorescence light were established and the difference in l between the minima of the curves gave directly the average delay time between absorption of exciting light and emission of fluorescence. If one remembers that the theory of the fluorescence delay was established by Dushinsky[2] 7 years after these measurements, so that the relation $\tan \theta = \omega\tau$, of the phase delay θ with light excitation frequency $\omega = 2\pi f$ and fluorescence lifetime τ did not yet exist, the coincidence of the fluorescein lifetime, 4.5 ± 0.5 nsec as determined by Gaviola, with present-day observations appears more remarkable as a demonstration of what original thinking and ingenuity of instrumental design can achieve. No less ingenious are the methods used by Wavilov in 1924[3] to estimate the quantum yield of fluorescein in alkaline solution as better than 0.8, the measurements by Weigert and Kaeppler[4] and Perrin[5] of the dependence of the polarization of fluorescence on the viscosity, and the demonstration by Gaviola and Pringsheim[6] of nonradiative energy transfer between excited fluorescein molecules, all done at about the same time. However, all these methods could be applied only to the brightest fluorescence with spectral maxima not too distant from the maximum sensitivity of the human retina.

Today we can easily make the same observations on a much extended spectral range, on emissions 1000 times weaker, and with a precision several hundred times better. In fact we are probably close to reaching the phenomenological limit of resolvability in each of these areas of experimentation. This has been particularly important in biophysics, where natural emissions are often weak and practically always heterogeneous. The improved preci-

[2] F. Dushinsky, *Z. Physik* **81,** 7 (1933).
[3] S. I. Wavilov, *Z. Physik* **42,** 311 (1927).
[4] F. Weigert and G. Kaeppler, *Z. Physik* **25,** 99 (1924).
[5] F. Perrin, *Ann. Physique* **12,** 169 (1929).
[6] E. Gaviola and P. Pringsheim, *Z. Physik* **43,** 384 (1927).

sion of present-day methods is indeed desirable, but can often deviate our attention from the explanation of the larger differences, which is most necessary, toward consideration of the much smaller ones, which is not nearly as necessary. Often I am called to review papers on fluorescence in which an author has spent a great deal of effort in characterizing a third component of the emission that forms perhaps 1–2% of the total, while not paying similar attention to the origin of the difference between the two major components of the emission.

Present Shortcomings

Fluorophore–Solvent Interactions

While the technological accomplishments involved in the detection and measurement of fluorescence properties are illustrated every day in the communications of a number of scientific journals, mention of the large gaps in our knowledge of the relation of those many details to the actual molecular properties that are their origin is far less common. The original conception of these relations goes back to the decade from 1950 to 1960, more specifically to the work of Bayliss and McRae,[7] Lippert,[8] Bakshiev,[9] and others. The common approach has been the assumption that the molecular interactions of solvent and fluorophore result in changes of the ground and electronic excited states and that these are revealed by the shifts in absorption and emission in the particular cases with respect to the same chromophore properties in the gas phase, or in a solvent medium in which all molecular interactions are expected to be reduced to the minimum. The differences between absorption or emission in the gas phase and an "indifferent" solvent, such as hexane, are already appreciable. They usually involve a "universal red shift" of the spectra toward longer wavelengths on the order of $kT/2$, or 100 cm^{-1} at room temperature, which is attributed to the increased polarizability of chromophores in the excited state as compared to the ground state. Of more importance is the disappearance of recognizable vibrational levels through broadening that makes virtually impossible a detailed characterization of the vibrational effects that result from fluorophore–solvent interactions. In interpreting the absorption and fluorescence of chromophores in a condensed medium, we are thus obliged to compare absorption and fluorescence emissions in various environments to those observed in solvents with the lowest polarizability such as hexane

[7] N. S. Bayliss and E. G. McRae, *J. Phys. Chem.* 1002 (1954).
[8] E. Lippert, *Z. Elektrochem.* **61,** 962 (1957).
[9] N. G. Bakshiev, *Opt. Spektrok.* (Engl. trans.) **12,** 350 (1962).

or cyclohexane. From the initial observations it became evident that chromophores can be broadly divided into those that possess a strongly polar excited state, in which separation of charge is much larger than in the ground state,[8] and apolar fluorophores, in which there is a much more regular electronic distribution in both ground and excited states. The best examples of these latter cases are the unsubstituted aromatic hydrocarbons dissolved in apolar solvents. In these hydrocarbons the vibrational levels are resolved in the absorption and to a lesser degree in the emission. Because the emissions from these apolar fluorophores vary little from one solvent to another, they cannot provide us with much information about the surroundings and about the possible changes in the medium associated with physiological states, which is the main aim of the biophysicist. In other words they are uninteresting environmental probes. However, the spectral changes of the polar fluorophores exhibit exquisite sensitivity to the environment, that varies with the electronic state to which they are excited. In these cases both electronic and vibrational levels are affected to different degrees, making the analysis that much more difficult. Two important ingredients add to the difficulties. First, the ground-state interactions between the solvent molecules and the chromophore involve energies on the order of 0.5–2 kcal that are too small to resist the thermal fluctuations and to determine a unique energetic state. They produce a collection of states and, under monochromatic excitation, some of the solvent–chromophore dispositions are much more likely to be excited than others. Second, the excited-state interactions selected by monochromatic excitation evolve in a time-dependent manner toward an equilibrium on their interaction with the solvent. The fluorescence lifetime lasts picoseconds to nanoseconds, and during this time the equilibrium in the interaction between the fluorophore in its newly acquired excited electronic distribution and the surrounding solvent molecules may or may not be reached. If the characteristic time for this equilibration is ρ and the lifetime of the excited state is τ, then the ratio $\rho/(\rho + \tau)$[9] determines the spectral distribution of the emission. If $\rho \gg \tau$, then the emission reflects the excited-state characteristics selected at the time of the excitation, often designated as the *Franck–Condon excited state* and, if $\rho \ll \tau$, then the emission is uniform for any wavelength of excitation and reflects the equilibrium excited state attained after excitation. Because ρ is practically determined by the reorientation of the solvent molecules, ρ lasts only a few picoseconds in a fluid solvent but in a semirigid solvent, such as glycerol at room temperature, it can readily become much longer than τ and it then gives a direct demonstration of the character of the Franck–Condon excited state reached on excitation. Although the spectral relaxation effects of fluorescent probes like ANS or PRODAN that indicate the interaction of fluorophore and solvent in the excited state have received a great deal of attention and are often used as indicators of the polarity of the environment, much less use has been made of the dependence of the emission spectrum on the excitation wavelength. Each partial charge, or electric pole, of a polar chromophore may be taken as an independent site of interaction with the molecules of the solvent or the macromolecule to which they are attached. In the simplest case of a molecule such as PRODAN with only two equal identifiable charges of opposite sign, the energy of interaction u of each individual solvent dipole with the field of a fluorophore monopole depends on the angle θ made by the solvent dipole direction, with the direction determined by the monopole and the dipole center.[10] Maximum attraction $-u_0$, or repulsion, $+u_0$, respectively, correspond to angles of 0 and 180° between these two directions and zero interaction to a 90° angle between them. For intermediate directions

$$u(\theta) = u_0 \cos \theta \tag{1}$$

u_0 depends on the strength of the solvent dipole μ_s and the monopole charge $q = l/\mu$, where μ is the dipole moment determined by the two opposite fluorophore charges in the ground state, and l is the distance between them. If a designates the distance between monopole and dipole center, then

$$u_0 = (\mu_s \mu)/(la^2) \tag{2}$$

The angular interactions follow the Langevin (1905) distribution,[10] the average of which is given by

$$\mathscr{L}(u_0/kT) = c \tanh(u_0/kT) - (kT/u_0) \tag{3}$$

The level from which absorption takes place, S_0, is related to the level corresponding to the absence of interaction with the surroundings, S_0^0, is

$$S_0 = S_0^0 + 2u_0 \mathscr{L}(u_0/kT) \tag{4}$$

On excitation, the dipole moment of the chromophore is changed from μ to μ^* and the interaction with the solvent dipoles involves an energy of interaction in the excited state $u_0^* = u_0\mu^*/\mu$ and the energy of the excited state S_1 is related to its unperturbed energy S_1^0 by the relation

$$S_1 = S_1^0 + 2u_0^* \mathscr{L}(u_0/kT) \tag{5}$$

Equation (5) takes account of the Franck–Condon principle in that the nuclear motions of the solvent necessary to establish the new Langevin distribution characteristic of the excited state cannot follow the instantaneous change in the electronic distribution within the fluorophore so that

[10] R. B. Macgregor and G. Weber, *Ann. N.Y. Acad. Sci.* **366,** 140 (1981).

the Langevin distribution of the ground state remains undisturbed during the absorption process. The energy E_A of the absorption is

$$E_A = S_1 - S_0 = S_1^0 - S_0^0 + (u_0^* - u_0)\mathscr{L}(u_0/kT) \tag{6}$$

If the conditions in the excited state are such as to favor the attainment of the equilibrium of the dipoles in the solvent cage with the monopoles of the fluorophore during the excited state lifetime, the transition $\mathscr{L}(u_0/kT) \Rightarrow \mathscr{L}(u_0^*/kT)$ will take place. Its effect is a further change in the energy S_1^0 to the level S_1' so that

$$S_1' = S_1^0 + 2u_0^*\mathscr{L}(u_0^*/kT) \tag{7}$$

corresponding to a much stronger interaction with the solvent dipoles in the proportion μ^*/μ. The final step is the emission of radiation reaching a level that in accordance with the Franck–Condon principle equals

$$S_0' = S_0^0 + 2u_0\mathscr{L}(u_0^*/kT) \tag{8}$$

The energy of the emission is therefore

$$E_F = S_1' - S_0' = S_1^0 - S_0^0 + 2(u_0^* - u_0)\mathscr{L}(u_0^*/kT) \tag{9}$$

The relations between the various levels in Eqs. (3)–(8) are shown in Fig. 2. The importance of the Langevin distribution cannot be overestimated, as

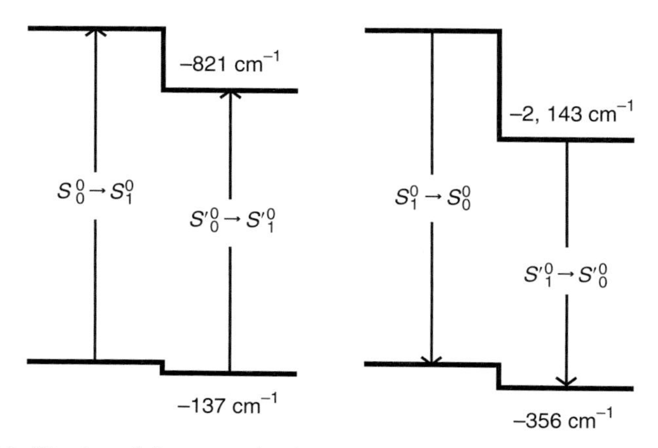

FIG. 2. Modification of the energy levels of the ground and fluorescence states owing to interaction of the solvent dipoles with the fluorophore monopoles under conditions that permit the transition $\mathscr{L}(u_0/kT) \Rightarrow \mathscr{L}(u_0^*/kT)$ to occur. Parameters used: $\mu_{sol} = \mu = 2$ D, $\mu^* = 12$ D, $a = 3.4$ Å, $l = 8.2$ Å. Temperature of observation, 293.1 K. S and S', respectively, are the ground and excited singlet states, with subscripts and superscripts as shown in Eqs. (2)–(7). The shift of the absorption (*left*) with respect to an indifferent medium is 743 cm^{-1}. The shift of the emission (*right*) is 1779 cm^{-1}.

it permits one immediately to see that the fluorophores in the ground state that constitute the red edge of the absorption correspond to the geometry of maximal interaction between fluorophore and solvent dipole and that the emission from these fluorophores must be close to that observed for the whole population when the equilibrium of fluorophore with solvent is reached in cases in which $\mu^* \gg \mu$. An example is given in Fig. 3, taken from Macgregor and Weber.[10] It shows the fluorescence emission of PRODAN in glycerol at $-41°$ when excited at 325 nm (close to the center of absorption) and at 425 nm (red edge of the absorption). The mean wave numbers of emission are, respectively, 23,056 and 21,108 cm^{-1} and the difference, 1948 cm^{-1}, represents an energy loss of 6 kcal in the excited state. To obtain the maximum information, the absorption and emission spectra must be related.

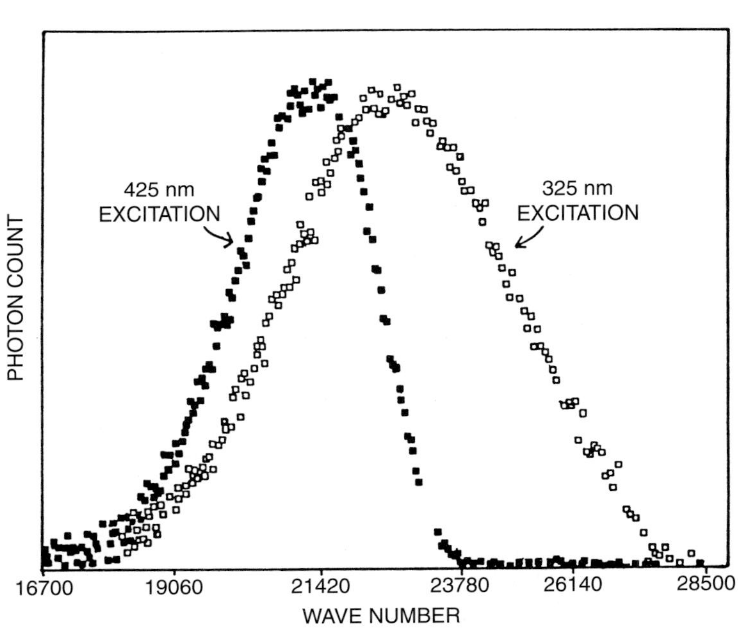

FIG. 3. Dependence of the fluorescence emission of PRODAN in glycerol at $-41°$ on the excitation wavelength. (Adapted from Macgregor and Weber.[10])

Absorption observations are particularly difficult in biological molecules *in situ* owing to their low concentration as well as to the variable scattering and the complex background absorption spectra of the samples. These difficulties can be largely circumvented by relating the emission spectrum to the excitation spectrum rather than to the absorption spectrum of the samples. The necessary variation of the wavelength of excitation requires a source of exciting light with a continuum of wavelengths, like the xenon arc, and is therefore not readily feasible with monochromatic lasers. The extensive use of lasers is perhaps the explanation of the concentration on the fluorescence properties of the samples with neglect of the important relations of excitation and emission. Synchrotron radiation, which consists of a true continuum extending from the infrared to the vacuum ultraviolet and beyond, offers distinct advantages for these studies of the dependence of the fluorescence emission on the exciting wavelengths. Be as it may, we are currently far from having a reasonable general theory of the relations of the changes on the absorption and emission to the molecular interactions that generate them, a theory that should help us in the interpretation of events in biological membranes and proteins. As noted previously, the difficulty arises from the statistical nature of the ground-state interactions of fluorophore and solvent molecules. Although the difference in the emission from PRODAN in cyclohexane and water may reach as much as 5 kcal, the corresponding differences in the mean energies of absorption are several times smaller as shown in Fig. 2. If an equilibrium excited state is reached, the fluorescence may be looked on as being that from a single molecular species, but the absorption must of necessity be considered as arising in a plurality of species. The distribution of charges in the fluorophore in ground and excited states can make the difference between cases that are sufficiently simple to permit calculation of the effects of environment and those that are not so favorable. A polar fluorophore placed in a solvent has a distribution of centers of charge that depend essentially on the fluorophore, but the intensity of the effects depends on the polarity of the solvent. Generally, the distribution of charges in the fluorophore can be reasonably assigned from its chemical constitution, and the cases most favorable for analysis are those with unique centers of positive and negative charge at a fixed distance, as is the case for PRODAN and other truly dipolar chromophores. Less favorable are the cases in which the molecule has potential charges at more than two places. An example of such multipolar chromophores are the phthalimides, much studied by Russian spectroscopists from the point of view of their solvent interactions, and also fluorescein. It is clear that the general effects depend on the Franck–Condon principle and the separation of charges in the chromophore in both ground and excited states. The latter characterizes the electronic properties of the fluorophore and will not be influenced appreciably by the characteristics of the medium. In contrast, the surroundings of the fluorophore equilibrate according to its electrostatic or electrodynamic properties as well as to its distribution of charges in the chromophore molecule. This equilibrium is always reached in the ground state, which predates the excitation by a time that is always long enough to permit equilibration. The excited state, however, may not last long enough to permit equilibrium to be reached. In general, we can distinguish several cases.

1. Fluorophore and medium are apolar and all interactions thus depend on mutual polarizabilities. The excited state is always much more polarizable than the ground state, and the result is a "universal" red shift of both absorption and emission of approximate magnitude $kT/2$ with respect to the gas-phase levels of the chromophore. Changing from an apolar to a polar environment produces only small changes.

2. The fluorophore is polar and to a first approximation it can be described as having a dipole moment in the excited state larger than that of the ground state. If the medium is apolar, the red shift in the absorption owing to the lowering of S_0 is evidently smaller than the shift in the emission so that the stabilization of the excited state is larger than that of the ground state. The reason for this difference is found in the excess induction produced by the increased separation of charges in the excited state, acting on a solvent made up of molecules without a permanent distribution of charges.

3. The largest effects appear in the case of a polar fluorophore in a solvent containing permanent dipoles, a case that gives rise to a Langevin distribution of interactions as examined previously. Calculations of the effects in media without permanent dipoles but with molecules of different polarizabilities, such as saturated hydrocarbons and unsaturated or aromatic hydrocarbons, are more difficult because of uncertainties regarding local polarizabilities. In fact, many calculations of molecular interactions, including those inside proteins, not only of chromophores but also among the amino acid residues themselves, are hampered by ignorance of the local polarizabilities involved. An example of the importance of atomic polarizabilities is given by the pK of the chloroacetic acids that decreases progressively with chlorination owing to the larger polarizability of the C–Cl bonds as compared to the C–H bonds. The spectral shifts of prosthetic groups bound to proteins are often much more pronounced than those that can be obtained by dissolving the chromophore in solvents of variable polarity. Perhaps the most remarkable example is that of the opsins, which give rise to three distinct pigments with maxima at 410, 532, and 563 nm[11] by

[11] A. B. Asenjo, J. Rim, and D. D. Oprian, *Neuron* **12,** 1131.

combination of the specific apoproteins with the same chromophore, 11-*cis*-retinal, which has an absorption maximum at 440 nm in methanol solution. Equally demonstrative are the rabbit antibodies to fluorescein described by Voss and collaborators,[12] which show displacements of the fluorescein absorption maximum as large as 28 nm. These displacements correspond to a change of 1058 cm^{-1}, or 2.5 kcal, at room temperature with respect to the absorption of fluorescein in water. The shifts of the fluorescein absorption in different solvents are five or more times smaller. Yagi *et al.*[13] have observed displacement of the flavin absorption and appearance of vibrational resolution of the flavin absorption spectrum in the ternary complexes of D-amino-acid oxidase with coenzyme and benzoate, another indication of the complex effects that follow the existence of a fixed environment.

Two main causes are of importance in the generation of such large protein effects:

a. The intensity of the interaction of fluorophore and solvent is influenced by the thermal agitation in inverse proportion to u_0/kT. If we limit the monopole charges in the chromophore to those corresponding to a dipole moment of 2 D units, characteristic of many aromatic hydrocarbons with OH and NH_2 substituents, we find that we have broad Langevin distributions, even in solvents with the largest dipole moment. If we could suspend or restrict the effect of the thermal agitation and thus obtain a unique maximal interaction of solvent dipole and chromophore, excluding as well all possible unfavorable interactions between the solvent dipoles themselves, we would expect to increase by a factor of 2 or 3 the spectral displacements. Evidently this single cause, although significant, cannot contain the whole explanation.

b. To understand the further causes present here, we must refer to the structure of the protein surrounding the chromophore. In a solvent in which the free motions minimize the energy of fluorophore–solvent interactions, we cannot expect to find a positive solvent charge in proximity to the positive monopole of the chromophore. However, we can expect to find such a situation in a protein where an amino acid residue is placed so as to generate a repulsive electrostatic interaction with the chromophore monopole; that is, if stronger attractive interactions at other, spectrally uninvolved places are sufficiently large to overcome the local electrostatic repulsion that determines the spectral changes. If this situation obtains in the ground state, but disappears in the excited state owing to the new

[12] E. W. Voss, personal communication (1996).

[13] K. Yagi, T. Ozawa, M. Naoi, and A. Kotaki, *in* "Flavins and Flavoproteins" (K. Yagi, ed.), p. 237. University of Tokyo Press, Tokyo, 1968.

distribution of charges in the chromophore, we have a greatly destabilized ground state and a neutral excited state. Further, we expect a red shift by a cause that cannot be experimentally duplicated in solvents. If the electrostatic repulsion takes place in the excited state while the ground state is neutral, we have a displacement to the blue, which could reach a large effect. Unlike the former effect, this latter effect can be observed in solvents because the subsequent nuclear rearrangement responsible for the effects cannot modify, by the Franck–Condon principle, the virtually instantaneous absorption. A blue displacement of this type is seen in 1-aminoethylaminonaphthalene. At neutral pH, absorption and emission are displaced to the blue in relation to the spectra in alkaline pH when the aliphatic amino group is deprotonated.

Perhaps the main reason for our deficiencies in the knowledge of the effect of molecular interactions on the absorption and emission spectra derives from the small energies involved and the uncertainties of the statistics that demand a complete enumeration of all probable causes. The relatively fixed atomic structure of the proteins ought to permit us to carry out realistic calculations in which all relevant interactions are taken into account. However, such calculations will be anything but trivial. In cases in which the protein as a whole imposes repulsive interactions on the chromophore, one does not know how far from the chromophore it will be necessary to go to take into account all possible contributions that permit the repulsive effect to occur. The necessity to compute both the ground and excited state may mean that the problem will involve considerations of quantum mechanics as well as electrostatics, thus adding to the difficulties of a straightforward interpretation. Presently, one can only expect to establish semiempirical correlations between the protein environment and the spectral displacement.[14] Regardless, an important new way of addressing these questions must be mentioned: The methods of molecular biology permit us to replace amino acids interacting with the chromophore with others of different polarity so that the conclusions as to the relative importance of the elementary interactions can be checked in a fashion that is not at all possible by observations on solvents of different characteristics. Important work in this direction in the most promising case, that of the visual pigments, is in progress and it has been already shown that the difference in the absorption spectra of the visual pigments are determined by a small number of polar amino acids (perhaps only seven).[11] On the basis of these observations one can be optimistic about the possibility of the future interpretation of spectroscopic perturba-

[14] R. B. Macgregor and G. Weber, *Nature* **319,** 70 (1986).

tion by the environment, and it seems that if we acquire a high degree of certainty regarding the origin of the molecular effects on the absorption and emission of chromophores, it will be due to the fixed models presented by the proteins and the additional possibilities introduced by the methods of genetic replacement of amino acids. Looking further into the importance of these studies, one realizes that they will be able to give us a direct glimpse of the coupling of electrostatic properties of the environment and the electronic properties, an aspect that we believe to be of importance in the determination of the properties and the time evolution of enzyme–substrate complexes.

Red-Edge Effects

Another area in which the interpretation of the data of fluorescence in terms of molecular properties is lacking is that of the red-edge effects, that is, those effects that involve the modification of the fluorescence emission when excitation falls at the long-wave absorption edge. Investigation of this spectral region is often important in biological samples because it offers the best possibilities of detecting compositional heterogeneities. Investigation of red-edge effects has been particularly difficult for technical reasons: As one approaches the edge of the absorption the need for purity of the preparations becomes more stringent. Also, the decreasing absorption with increasing wavelength results in a parallel decrease of the signal-to-noise (S/N) ratio and increasing difficulties in the separation of fluorescence from exciting light. I do not know of studies in which the absolute quantum yield has been reliably determined as a function of the exciting wavelength in this region. When the temperature is decreased by an appropriate amount, for example, 60 to 100° below room temperature in fluid apolar solvents such as cyclohexane or dodecane, the red edge of the absorption spectrum shifts to the blue, and this displacement is to be attributed to the decrease in thermally excited molecules, which would absorb from levels above that of the lowest level of the ground state. At the higher temperature the red-edge absorption is primarily due to these hot molecules, and if the quantum yield is unity when excited at the absorption maximum, and continues to be unity with conservation of the fluorescence spectrum throughout the edge of the absorption, excitation with wavelengths longer than the maximum of emission must derive part of the radiated energy from the thermal energy of the sample, which will then be cooled by the emission. This cooling by fluorescence has been demonstrated[15] for a sample of small

[15] R. I. Epstein, M. I. Buchwald, B. C. Edwards, T. R. Gosnell, and C. E. Mungan, *Nature (London)* **377,** 500 (1995).

dimensions and negligible absorption of the exciting radiation, and indeed the cooling effect could be observed only under conditions in which reabsorption of the emitted fluorescence is negligible. As the reabsorption of the emitted fluorescence is virtually confined to those wavelengths of absorption shorter than the maximum of emission, the resulting secondary fluorescence must produce heating of the sample that more than compensates for the cooling due to the primary fluorescence owing to direct excitation, unless the secondary fluorescence is negligible.

Another red-edge phenomenon that is incompletely understood is the disappearance of electronic energy transfer among identical fluorophores on excitation at the red edge of the absorption.[16] In substances such as indole and naphthalene derivatives, in which there are two distinct, largely overlapping absorption bands L_a and L_b[17] responsible for the longest wavelength absorption and the fluorescence, it is difficult or impossible to decide whether the electronic energy transfer is due to resonance between identical oscillators (homotransfer) or represents the transfer resulting from overlap of the fluorescence from one transition with the absorption of the other, as is the case for two different fluorophores (heterotransfer). A similar difficulty is present when interactions with the environment give rise to homogeneous broadening of the absorption, and a case of this nature has been clearly demonstrated by the experiments of Valeur and co-workers.[18] Indeed, if homogeneous broadening were the complete explanation of the failure of energy transfer at the red edge of the absorption it would be necessary to conclude that homotransfer is only a particular but trivial case of heterotransfer. However, before accepting this important conclusion, it appears necessary to study in more detail those cases in which the last absorption band is not degenerate, so that it does not result either from superposition of L_a and L_b transitions, or from multiple interactions with the solvent. Naphthacene and other elongated aromatic systems in which the L_a and L_b transitions are well separated, and also nonaromatic fluorophores dissolved in apolar solvents with lowest polarizability, such as aliphatic hydrocarbons, appear to be the only systems that could in principle permit the demonstration of true resonance homotransfer free from heterotransfer arising from homogeneous broadening.

Thermodynamics of Fluorescence

Fluorescence is important as a paradigm of natural processes that follow the laws of thermodynamics. The appearance of an excited state by light

[16] M. Shinitzky and G. Weber, *Proc. Natl. Acad. Sci. U.S.A.* **65,** 823 (1970).

[17] B. Valeur and G. Weber, *Photochem. Photobiol.* **25,** 441 (1977).

[18] M. N. Barberan Santos, J. Canceill, L. Julien, J.-M. Lehn, J. Pouget, and B. Valeur, *J. Phys. Chem.* **97,** 11376 (1993).

absorption introduces a perturbing free energy in the system and the subsequent processes of emission, internal conversion without radiation, transfer of the excitation to another molecule, conversion to the triplet state, and photochemical reaction are alternative paths by which reestablishment of the equilibrium takes place. According to the second law of thermodynamics, all of these processes must occur with an increase in entropy, that is, with production of excess thermal energy that is dissipated either within the system itself or its surroundings. In accordance with the second law, even in the case of fluorescence with unit quantum yield, some of the absorbed energy is lost as heat on account of the Stokes shift. In other cases we expect the heat loss to be larger. If we inquire about the reasons that determine one path rather than another, we encounter concepts of quantum mechanics that are wholly unrelated to entropy: They are symmetry properties of ground and excited state, differences in spin, and coupling of the vibrational modes of the solvent with the excited molecule that can lead to internal conversion. Thus the competing states of fluorescence emission and other radiationless processes present us with proof that the laws of thermodynamics limit, but do not completely determine, the evolution of the system toward equilibrium. At the same time they permit us to deduce that in general there must be many cases in which the final outcome of natural processes is determined, not simply by the heat dissipation, but by probabilities that escape a thermodynamic interpretation.

[2] Design of Profluorescent Protease Substrates Guided by Exciton Theory

By Beverly Z. Packard, Dmitri D. Toptygin, Akira Komoriya, and Ludwig Brand

Owing to the essential roles played by proteases in both normal and pathophysiological processes,[1–15] the development of methods for detecting

[1] S. P. Colowick and N. O. Kaplan, eds. *Methods Enzymol.* **1,** (1955).
[2] S. P. Colowick and N. O. Kaplan, eds. *Methods Enzymol.* **2,** (1955).
[3] S. P. Colowick and N. O. Kaplan, eds. *Methods Enzymol.* **3,** (1957).
[4] S. P. Colowick and N. O. Kaplan, eds. *Methods Enzymol.* **4,** (1957).
[5] S. P. Colowick and N. O. Kaplan, eds. *Methods Enzymol.* **5,** (1962).
[6] S. P. Colowick and N. O. Kaplan, eds. *Methods Enzymol.* **6,** (1963).
[7] G. E. Perlmann and L. Lorand, eds. *Methods Enzymol.* **19,** (1970).
[8] L. Lorand, ed. *Methods Enzymol.* **45,** (1976).
[9] L. Lorand, ed. *Methods Enzymol.* **80,** (1981).

0076-6879/97 $25

proteolytic activities has received the attention of a multitude of investigators. Moreover, current interest in protease inhibitors as therapeutic agents for the treatment of metastatic cancer, inflammation, and acquired immunodeficiency syndrome (AIDS) underscores the importance of having sensitive and selective assays. Although the basic techniques for measuring proteolytic activities are both extensive and diverse, improvements in sensitivity, specificity, and potential for automation represent important problems in methods in enzymology. Assays for protease activities should be applicable both for solution work and for microscopy.

In this chapter we provide a brief history of the design of spectroscopic assays for protease activity. We then describe the application of exciton theory to dye dimers and show how the latter can be used in the design of a new class of protease substrates.

Chromophoric Substrates for Protease Activity Determination

The principle behind spectrophotometric and fluorescence assays for protease activity is the generation of a change in an optical signal on hydrolysis of a polypeptide containing an amino acid sequence that is recognizable by a specific protease. The amino acid residues in a protease substrate are designated as P_n and P'_n, where n is the distance from the scissile bond and the absence or presence of a prime indicates whether the residue is on the amino or the carboxyl side, respectively:

$$P_n \ldots P_3P_2P_1 \Uparrow P'_1P'_2P'_3 \ldots P'_n$$

where $\Uparrow$ represents the cleavage site.

A major element of substrate design is the type and position of the chromophore or fluorophore. The signal that represents protease activity can be a change in intensity, wavelength of absorption or emission peak, or, in the case of fluorescence, a change in emission anisotropy.[16–18]

[10] L. Lorand and K. G. Mann, eds. *Methods Enzymol.* **222,** (1993).
[11] L. Lorand and K. G. Mann, eds. *Methods Enzymol.* **223,** (1993).
[12] J. F. Riordan and B. L. Vallee, eds. *Methods Enzymol.* **226,** (1993).
[13] K. C. Kuo and J. A. Shafer, eds. *Methods Enzymol.* **241,** (1994).
[14] A. J. Barrett, ed. *Methods Enzymol.* **244,** (1994).
[15] A. J. Barrett, ed. *Methods Enzymol.* **248,** (1995).
[16] C. G. Knight, *Methods Enzymol.* **248,** 18 (1995).
[17] G. A. Krafft and G. A. Wang, *Methods Enzymol.* **241,** 70 (1994).
[18] R. Bolger and W. Checovich, *Biotechniques* **17,** 585 (1994).

Absorbance Measurements

Early workers in the field noted that the absorbance of some chromophores such as *p*-nitroaniline[19] or *p*-nitrophenylalanine[20] was sensitive to amide linkage. Exploiting this observation, a family of protease substrates was developed in which absorbance changes could be observed on cleavage of an amide bond between one of these chromophores and an amino acid. Numerous short peptides containing sequences recognized by proteases of interest were synthesized and assayed in this way. For example, in a series of peptides with *p*-nitrophenylalanine in the P_1' position, the decrease in absorbance at 300 nm owing to deacylation of the dye was used to detail subsite interactions for aspartic proteinase substrate design.[21]

Fluorescence Measurements Using Single Fluorophore

The next generation of optical protease detection reagents made use of fluorescent dyes and resulted in a significant enhancement of sensitivity. Fluorophores in this group, which include naphthylamines,[22] coumarins,[23,24] and aminoquinolines,[25] were placed in amide linkages, usually at the P_1' site. One of the most sensitive families of reagents that follow this design are the rhodamine-110 substrates, with which a large increase in fluorescence is observed on protease addition.[26] Water-soluble macromolecular fluorogenic substrates in which fluoresceinated peptide moieties are released from a precipitated polymer have also shown some utility,[27] as have chemiluminescent- and bioluminescent-containing substrates.[28]

Fluorescence Measurement Using Two Fluorophores

Unfortunately, substrates in which only one side of the scissile bond contains a recognition sequence are limited to proteases such as trypsin, which do not require specificity on the carboxyl side of the cleavage site. Thus, addition of sequence information on both sides can provide a signifi-

[19] B. F. Erlanger, N. Kokowsky, and W. Cohen, *Arch. Biochem. Biophys.* **95,** 271 (1961).
[20] T. Hofmann and R. S. Hodges, *Biochem. J.* **203,** 603 (1982).
[21] B. M. Dunn, M. Jimenez, B. F. Parten, M. J. Valler, C. E. Rolph, and J. Kay, *Biochem. J.* **237,** 899 (1986).
[22] A. J. Barrett and H. Kirschke, *Methods Enzymol.* **80,** 535 (1981).
[23] M. Zimmerman, E. Yurewicz, and G. Patel, *Anal. Biochem.* **70,** 258 (1976).
[24] M. Zimmerman, E. Yurewicz, and G. Patel, *Anal. Biochem.* **78,** 47 (1977).
[25] P. J. Brynes, P. Bevilacqua, and A. Green, *Anal. Biochem.* **116,** 408 (1981).
[26] S. P. Leytus, L. L. Melhado, and W. F. Mangel, *Biochem. J.* **209,** 299 (1983).
[27] F. Anjuere, M. Monsigny, and R. Mayer, *Anal. Biochem.* **198,** 342 (1991).
[28] J. Monsees, W. Miska, and R. Geiger, *Anal. Biochem.* **221,** 329 (1994).

cant advantage in specificity, as can be reflected in both K_m and k_{cat} values.[29] The effectiveness of this approach was initially realized by addition of a second chromophore to quench the fluorescence of the first. In a study in which quenching as a function of distance between the two chromophores was investigated, a decrease in quenching from 27- to 18-fold was found on increasing the interprobe distance from a di- to a tripeptide.[30] With up to four amino acids between the two chromophores, collisional quenching was assumed to be the operative mechanism. Interestingly and somewhat prophetically, the data indicated that a further increase in distance between the fluorophore and the quencher might still yield useful information. It was suggested that long-range quenching might be due to an increased collisional frequency between the two ends of the polypeptide when the substrate exceeded a certain length. Support for this notion was found by later studies in which the conformational effects of specific amino acid sequences were shown to impact on the degree of quenching.[31–33] It is shown in the next section that there are alternative explanations for the quenching of some substrates containing two fluorophores.

Fluorescence Resonance Energy Transfer

Limitations inherent in the design of protease substrates on the basis of collisional quenching led several investigators to utilize the principles of Förster-type resonance energy transfer (FRET).[34–44] In this approach two spectroscopically complementary probes are placed on opposite sides

[29] J. C. Powers, A. D. Harley, and D. V. Myers, *in* "Acid Proteases, Structure, Function, and Biology" (J. Tang, ed.), pp. 141–157. Plenum, New York, 1977.

[30] A. Carmel, E. Kessler, and A. Yaron, *Eur. J. Biochem.* **73,** 617 (1977).

[31] C. F. Vencill, D. Rasnick, K. V. Crumley, N. Nishino, and J. C. Powers, *Biochemistry* **24,** 3149 (1985).

[32] M. J. Castillo, K. Kurachi, N. Nishino, I. Ohkubo, and J. C. Powers, *Biochemistry* **22,** 1021 (1983).

[33] D. A. Jewell, W. Swietnicki, B. M. Dunn, and B. A. Malcolm, *Biochemistry* **31,** 7862 (1992).

[34] Th. Förster, *Ann. Phys.* **2,** 55 (1948).

[35] S. A. Latt, H. T. Cheung, and E. R. Blout, *J. Am. Chem. Soc.* **87,** 995 (19965).

[36] L. Stryer, *Annu. Rev. Biochem.* **47,** 819 (1978).

[37] L. Stryer and R. P. Haugland, *Proc. Natl. Acad. Sci. U.S.A.* **58,** 710 (1967).

[38] H. Fairclough and C. R. Cantor, *Methods Enzymol.* **48,** 347 (1978).

[39] B. W. VanDer Meer, G. Coker III, S.-Y. S. Chem, "Resonance Energy Transfer, Theory, and Data." VCH, New York, 1994.

[40] P. Selvin, *Methods Enzymol.* **246,** 300 (1995).

[41] R. M. Clegg, *Curr. Opin. Biotech.* **6,** 103 (1995).

[42] P. Wu and L. Brand, *Anal. Biochem.* **218,** 1 (1994).

[43] S. A. Latt, D. S. Auld, and B. L. Vallee, *Anal. Biochem.* **50,** 56 (1972).

[44] A. Carmel, M. Zur, A. Yaron, and E. Katchalski, *FEBS Lett.* **30,** 11 (1973).

of the cleavage site. The fluorescence emission spectrum of one chromophore (the donor) must overlap the absorption spectrum of the second chromophore (the acceptor). The latter does not necessarily have to fluoresce[45] but from the experimental point of view it can be advantageous if it does. The Förster equation indicates the important parameters in this process:

$$k_T = \frac{1}{\tau_D}\left(\frac{R_0}{R}\right)^6, \qquad R_0^6 = 8.785 \times 10^{-5}\frac{(\kappa^2 \phi_D J)}{n^4} \tag{1}$$

where k_T is the rate constant, τ_D is the donor lifetime in the absence of the acceptor, R is the donor–acceptor distance, and R_0 is the Förster or critical distance at which the energy transfer rate is equal to the decay rate. In Eq. (1) the Förster distance is measured in angstroms, and it is expressed in terms of the orientational factor κ^2 (dimensionless), as well as the spectroscopic properties of the two chromophores: ϕ_D is the quantum yield of the donor in the absence of the acceptor, n is the index of refraction (dimensionless), and J is the spectral overlap integral between the donor and the acceptor (in nm^4 cm^{-1} liter/mol).

Protease activity can be monitored using resonance energy transfer by measuring the increase in fluorescence of the donor on hydrolysis or by following the decrease in fluorescence of the acceptor. Energy transfer is lost on proteolysis since the distance of the dyes becomes too great. As is discussed as follows, invocation of a Förster-type mechanism implies spectral independence of the donor and the acceptor. Although resonance energy transfer has been used to characterize a diversity of proteases,[46] spectral independence has rarely been demonstrated.

Substrates Based on Exciton Theory

Kasha[47,48] has characterized various examples of resonance interactions between two fluorophores. The Förster mechanism discussed previously is defined as a weak dipole–dipole interaction; at the other end of the scale[49,50] is the strong dipole–dipole interaction described by the Simpson–Peterson

[45] G. T. Wang, E. Matayoshi, H. JanHuffaker, and G. A. Krafft, *Tetrahedron Lett.* **31,** 6493 (1990).

[46] C. G. Knight, *Methods Enzymol.* **248,** 18 (1995).

[47] M. Kasha, "Physical and Chemical Mechanisms in Molecular Radiation Biology" (W. A. Glass and M. N. Varma, eds.), pp. 231–255. Plenum, New York, 1991.

[48] M. Kasha, *Radiat. Res.* **20,** 55–71 (1963).

[49] M. Kasha, H. R. Rawls, and M. Ashraf El-Bayoumi, *Pure Appl. Chem.* **2,** 371–392 (1965).

[50] R. M. Hochstrasser and M. Kasha, *Photochem. Photobiol.* **3,** 317–331 (1964).

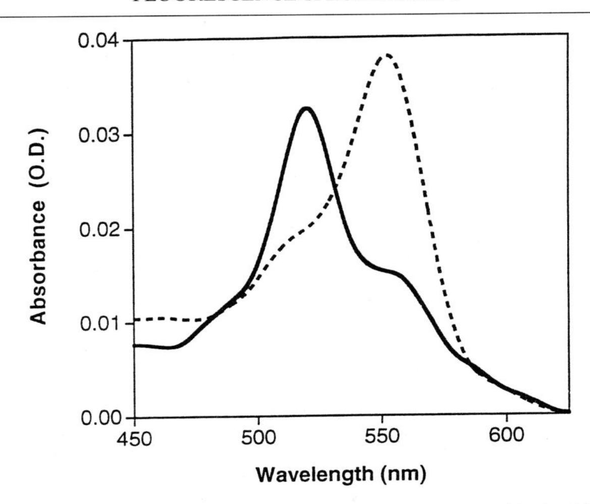

FIG. 1. Absorption spectra of tetramethylrhodamine monomer (dashed line) and dimer (solid line).

model.[51] The latter is associated with significant changes in absorption spectra that should not be found in the case of Förster-type energy transfer.

Changes in absorption spectra are observed in solutions of rhodamines at concentrations greater than 1×10^{-5} M[52–56] or when two rhodamines are held in close proximity by virtue of being conjugated to small[57,58] or large[59–61] molecules. Under these conditions, interaction between two rhodamines leads to a blue shift in the absorption maximum from 552 to 518 nm (Fig. 1); the latter has been interpreted in the framework of exciton theory in terms of an H-type dimer, which is defined by an alignment between the two dyes such that a radius vector connecting the two chromo-

[51] W. T. Simpson and D. L. Peterson, *J. Chem. Phys.* **26,** 588 (1957).

[52] O. Valdes-Aguilera and D. C. Neckers, *Acc. Chem. Res.* **22,** 171 (1989).

[53] Th. Förster and E. König, *Ber. Bunsenes. Physik* **61,** 344 (1957).

[54] J. E. Selwyn and J. I. Steinfeld, *J. Phys. Chem.* **76,** 762 (1972).

[55] I. L. Arbeloa and P. R. Ojeda, *Chem. Phys. Lett.* **87,** 556 (1982).

[56] T. P. Burghardt, J. E. Lyke, and K. Ajtai, *Biophys. Chem.* **59,** 119 (1996).

[57] D. K. Luttrull, O. Valdes-Aguilera, S. M. Linden, J. Paczkowski, and D. C. Neckers, *Photochem. Photobiol.* **47,** 551 (1988).

[58] B. Z. Packard, D. D. Toptygin, A. Komoriya, and L. Brand, *Proc. Natl. Acad. Sci. U.S.A.* **93,** 11640 (1996).

[59] P. Ravdin and D. Axelrod, *Anal. Biochem.* **80,** 585 (1977).

[60] K. Ajtai, P. J. K. Ilich, A. Ringler, S. S. Sedarous, D. J. Toft, and T. P. Burghardt, *Biochemistry* **31,** 12431 (1992).

[61] B. D. Hamman, A. V. Oleinikov, G. G. Jokhadze, D. E. Bochkariov, R. R. Traut, and D. M. Jameson, *J. Biol. Chem.* **271,** 7568 (1996).

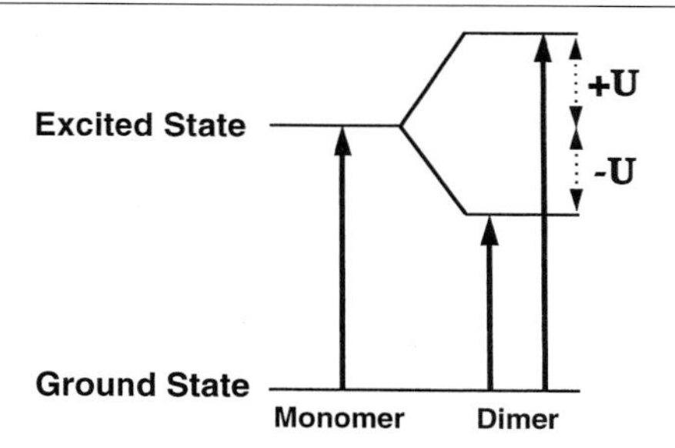

FIG. 2. Exciton theory.

phores is perpendicular to their transition dipoles. Although previous depictions of rhodamine dimers have indicated face-to-face structures,[62–64] more recent data point toward an edge-to-edge alignment between the two dyes.[58]

In the exciton model (Fig. 2), absorption and emission of light are by the dimeric unit. This cooperativity is conceptually quite different from the resonance energy transfer described by Förster, where a spectral separation of the two chromophores gives rise to absorption of light by the donor chromophore, resonance energy transfer to the acceptor chromophore, and emission with the spectral characteristics of the acceptor exclusively.

According to exciton theory, in the ground-state dimer the excited energy level splits into two levels, each shifted by $\pm U$ compared to the position of the excited energy level in the monomer. The value of U is given by Eq. (2),

$$U = \frac{|d|^2}{n^2 R^3} |\kappa| \tag{2}$$

where $|d|^2$ is the squared module of the transition dipole, n is the solvent refractive index ($n = 1.333$ in water), and κ is the orientation factor [$\kappa = \cos\,\theta_{12} - 3(\cos\,\theta_{1R})(\cos_{2R})$]. θ_{12} is the angle between the transition dipoles in dyes 1 and 2, and θ_{1R} and θ_{2R} represent the angles between each dipole and the radius vector R connecting the two dipoles, respectively.

In the case of H-dimers, $\theta_{12} = 0$ and $\theta_{1R} = \theta_{2R} = 90°$, which yields $\kappa = 1$. Another characteristic of this dimer is a twofold increase in the probability of radiative transitions between the ground level and the upper excited level, whereas the radiative transitions between the ground level and the lower excited level are forbidden. This results in a blue shift in the

[62] K. K. Rohatgi, *J. Mol. Spectrosc.* **27,** 545 (1968).

[63] K. Kemnitz and K. Yoshihara, *J. Phys. Chem.* **95,** 6095.

[64] T. Kajiwara, R. W. Chambers, and D. R. Kearns, *Chem. Phys. Lett.* **22,** 37 (1973).

absorption spectrum and in a strong decrease in the fluorescence quantum yield (fluorescence has its origin in the lowest excited level, the radiative transition from which to the ground state is forbidden in H-dimers).

In contrast, in a J-dimer $\theta_{12} = \theta_{1R} = \theta_{2R} = 0$, which yields $\kappa = -2$, and here the radiative transitions between the ground state and the upper excited state are forbidden, whereas transitions between the ground state and the lower excited state are enhanced twofold. Thus, one observes a red shift in the absorption spectrum and an increased fluorescence quantum yield.

A significant element that derives from incorporation of exciton theory into protease substrate design concerns the potential enhancement in the amount of amino acid sequence information that can be incorporated. This can be critical since the inclusion of distal substrate recognition residues, e.g., P_n and P'_n where $n > 2$, may contribute significantly toward defining the protease substrate specificity as expressed in K_m and k_{cat} values.

This increase in allowable sequence information stems from the intrinsic affinity of fluorophores for each other. The potential of a dye for dimer formation is a function of several properties; these include its charge, symmetry, and dipole strength. For example, rhodamines, which have a +1 charge on the xanthene component of their structures, i.e., the three fused rings, form dimers starting at concentrations in the 10^{-5} M range. Fluoresceins, in contrast, lack charge on their xanthenes (lower than pH 10) and with a reported K_d of 24 M^{-1} [65] do not form dimers until much higher concentrations are reached. (Fluorescein and rhodamine dimer structures are likely to differ.) Thus, the ~100-Å limit of Förster theory does not necessarily apply for substrates in which exciton bands are seen in the absorption spectra and, therefore, the size of the peptide sequence becomes less of a limiting factor. The complete extent of this approach awaits further definition since conformational integrity at the cleavage site must be maintained, but the thrust of a significant increase in the permissible amount of sequence data in a substrate could be quite profound.

Another impact that the introduction of exciton theory has for protease detection derives both from the presence of only a single type of fluorophore being necessary, and the radiative emission from H-dimers being quenched. The former represents a significant advantage over a Förster-type mechanism, where the signal is a chromatic (as well as an intensity) change and, as such, detection is a function of filter utilization. In contrast, use of homodoubly labeled substrates not only results in an intrinsically more favorable signal-to-noise ratio but the signal itself is truly fluorogenic owing to the protease-induced dequenching, which is against a nearly black back-

[65] Th. Förster and E. Konig, *Z. Elektrochem.* **61,** 344 (1957).

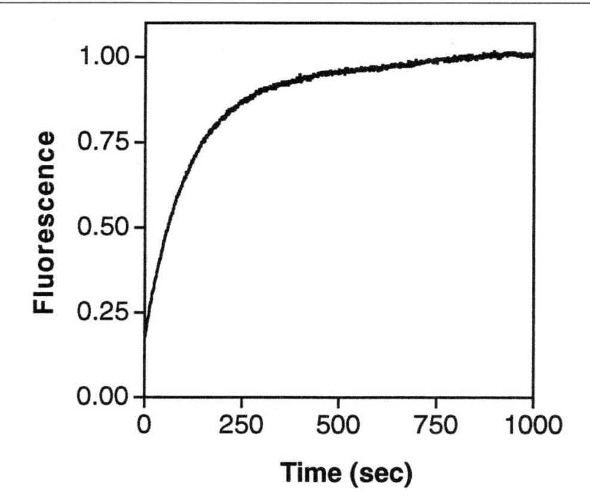

FIG. 3. Fluorescence of a protease substrate shown as a function of time.

ground. An example of this is shown in Fig. 3, where the increase in fluorescence of a protease substrate homodoubly labeled with two tetramethylrhodamines is shown as a function of time after addition of the serine protease elastase. Here, in the intact peptide, 96% of the fluorescence is quenched.

One application that originates from the capability of homodoubly labeling a substrate rather than heterolabeling, which is still possible but somewhat counterproductive, is that much simpler optics can be used. For example, one can measure protease activities associated with cells by using a fluorescence microscope equipped with standard filters. In contrast, a heterodoubly labeled substrate would require replacement of the barrier filter on the emission side with a bandpass filter if both the donor and acceptor were fluorophores.

In conclusion, chromogenic/fluorogenic assays for the detection of protease activities have shown great versatility as tools for characterizing physiological functions in which proteolysis plays essential roles. Addition of substrates whose designs are guided by exciton theory has the potential for uncovering arrays of new proteases as well as providing characterization not previously available, e.g., more favorable k_{cat}/K_m ratios. This should result from the increase in sequence information allowable in substrates with fluorescence characteristics that can be described by exciton theory, particularly in comparison with previous spectroscopic substrates on which considerable limitations have been placed. Furthermore, use of fluorophores with both excitation and emission in the visible range will allow detection of protease activities in biological systems where autofluorescence has been a formidable problem in the past.

[3] Picosecond Fluorescence Decay Curves Collected on Millisecond Time Scale: Direct Measurement of Hydrodynamic Radii, Local/Global Mobility, and Intramolecular Distances during Protein-Folding Reactions

By JOSEPH M. BEECHEM

Introduction

The use of fluorescence spectroscopy to study protein-folding kinetics has had a long and successful history. Fluorescence spectroscopy is a sensitive technique with high intrinsic timing resolution (integrated photons can be easily quantitated in milliseconds). With this extreme sensitivity, fluorescence folding studies have generated extensive, high-quality data with sufficient signal-to-noise (S/N) ratios to resolve complex multiphasic folding kinetics. One serious problem, however, has plagued all of these studies. How was one to interpret this complex change in (typically) an integrated total-intensity signal in terms of fundamental physical properties associated with folding intermediates? Given a simple steady state fluorescence signal, no fundamental interpretation is really possible. Therefore, application of fluorescence spectroscopy to the protein-folding problem was in a fundamental "Catch-22" position: What use are high-quality measurements, if what is being measured cannot be understood?

What we have attempted to do is to take fluorescence spectroscopy out of this Catch-22 position. Our strategy has been simple: If one wishes to get more information out of the fluorescence signals, one must be able to put more fluorescence signals into the detectors. The only question is: How many fluorescence signals can one get into detectors with millisecond timing resolution, simultaneously? The answer, which it is hoped will become apparent from the data to be presented, is, "all of them!" Of all of the fluorescence signals, the most important in terms of providing detailed physical information is the picosecond/nanosecond time-resolved total-intensity and anisotropy decay. The term *double kinetics* is utilized to describe the simultaneous measurement of changes in fluorescence on both picosecond/nanosecond and millisecond time scales simultaneously. Previous to the development of this instrumentation, typical picosecond/nanosecond time-resolved decay data acquisitions required a few minutes. The double-kinetic fluorometer described in this chapter reveals that high signal-to-noise picosecond anisotropy and total-intensity decays can now be obtained in milliseconds.

0076-6879/97 $25

Development of New Class of Fluorescence-Based Kinetic Instrumentation: Two Double-Kinetic *Gedanken* Experiments

In 1986 this author began work on designing a time-resolved fluorescence spectrometer that would be capable of performing the two *gedanken* experiments shown in Figs. 1 and 2. Figure 1 represents a double-kinetic study that measures the instantaneous change in psec/nsec mobility of a single intrinsic tryptophan residue during a refolding experiment. Figure 2 represents a double-kinetic study that measures instantaneous changes in intramolecular distance(s) during a refolding experiment. It is important to recognize the two time scales associated with these experiments. The refolding reaction is being initiated utilizing standard stopped-flow technology with millisecond timing resolution. The light source utilized to obtain the fluorescence signal, however, instead of being continuous, is pulsed, at a repetition rate of approximately 4 million times per second (4 MHz, ~1 psec/pulse). Therefore, within any single 1-msec period, the laser will have excited the sample 4000 times. During these 4000 laser excitations, one can collect "mini" picosecond/nanosecond time-resolved fluorescence signals using time-correlated single-photon counting. Only a low signal-to-noise time-correlated fluorescence decay curve will be obtained during any single millisecond slice, with only hundreds of photons collected during any individual reaction. These mini decay curves are transferred in real time to computer memory, and during the course of a single stopped-flow reaction 100 to 1000 (or so) minidecay curves will be obtained. The stopped-

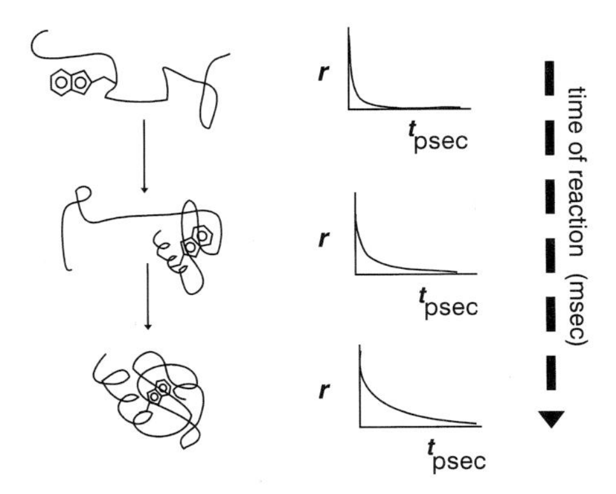

FIG. 1. Schematic of the double-kinetic intrinsic tryptophan fluorescence anisotropy experiment. *Left:* A diagram of a refolding process, wherein the single tryptophan residue is first "captured" into a localized structure followed by native state formation. *Right:* The measured quantity r is the psec/nsec anisotropy (i.e., rotational mobility) of the tryptophan residue, which is collected multiple times on the millisecond refolding time scale.

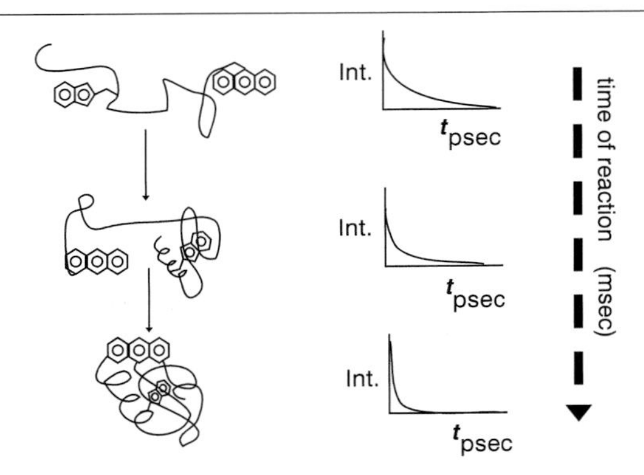

FIG. 2. Schematic of the double-kinetic fluorescence energy-transfer refolding experiment. *Left:* A diagram of a refolding process, wherein an intermediate is formed with a particular distance between two domains in a protein. In a final step, these two domains collapse into a native structure. *Right:* The measured quantity Int is the psec/nsec total-intensity decay of the donor, acceptor, or both (here donor decay is schematized), which shows increased energy transfer during each step. These fluorescence Int vs psec/nsec at each slice along the refolding reaction can be analyzed in terms of the intramolecular distance(s) separating the donor–acceptor. Multiple donor–acceptor probes located at different sites on the protein can be simultaneously analyzed in terms of an internally consistent millisecond motion picture of a refolding reaction.

flow reaction will be "fired again," and additional minidecay curves will be collected and summed into the appropriate millisecond slice obtained from the previous stopped-flow shot(s). In this manner, high-resolution picosecond/nanosecond fluorescence decay curves are collected on the millisecond time scale (the instrumentation and description of these events is described in more detail in the next section).

It is important to realize the additional information content of these double-kinetic experiments. In the tryptophan double-kinetic experiments, the picosecond rotational mobility of the intrinsic tryptophan is measured directly during the refolding experiment. One can determine the time scale associated with tryptophan "packing" into localized structure(s) and (eventually) into the final native state. Measurements of this type (described later in the chapter) have revealed the surprising result that the initially collapsed protein state can exist wherein the tryptophan residues are not packed, and are still rotating as if free in solution.

The double-kinetic fluorescence energy-transfer experiment (Fig. 2) is a measurement that is capable of making millisecond "motion pictures" of changes in intramolecular distances during a protein-folding reaction.

By utilizing cysteine-insertion mutagenesis techniques,[1] one can perform studies of this type using multiple sets of donor–acceptor pairs, and in this manner quantitate a low-resolution structure associated with transiently formed intermediates on the refolding pathway. These experiments are exactly analogous to measuring multiple nuclear magnetic resonance nuclear Overhauser effects (NMR NOEs) in a millisecond. At present, such NMR measurements are impossible. The optical distance measurements are sensitive to change over a large distance scale (5–100 Å) as compared to the local information provided by the NMR NOE. In this manner, a direct experimental approach to measuring long-range structure formation during protein-folding reactions is provided (e.g., domain–domain assembly, high-order α-helix packing, α-helix/β-sheet assembly, and multimer assembly).

Development of Picosecond Time-Resolved Fluorescence Instrument That Can Obtain Multiple Polarized Decay Curves on Millisecond Time Scale

First-Generation Double-Kinetic Instrument

The first nonpolarized (millisecond-based) time-resolved double-kinetic instrument was built in 1989 and can be found in Beechem *et al.*[2] A T-format polarized double-kinetic instrument was then developed, and a brief description of this instrument can be found in Beechem.[3] This initial T-format double-kinetic fluorometer is shown in Fig. 3. The operation of this instrument can be summarized as follows.

1. A pulsed 4-MHz repetition rate laser is focused onto the observation window of a stopped-flow cuvette.
2. A stopped-flow folding reaction is initiated.
3. Time-correlated single-photon counting is performed using 2 multianode (10 linear) microchannel plate detectors.
4. Individual decay curves are collected at parallel and perpendicular polarizations for brief millisecond time periods and transferred from the histogramming memory of the pulse-height analysis analog-to-

[1] M. P. Lillo, J. M. Beechem, B. K. Szpikowska, M. Sherman, and M. T. Mas, *Biochemistry* Submitted (1997).

[2] J. M. Beechem, L. James, and L. Brand, *S.P.I.E. Proc.* **1204,** 686 (1990).

[3] J. M. Beechem, *S.P.I.E. Proc.* **1640,** 676 (1992).

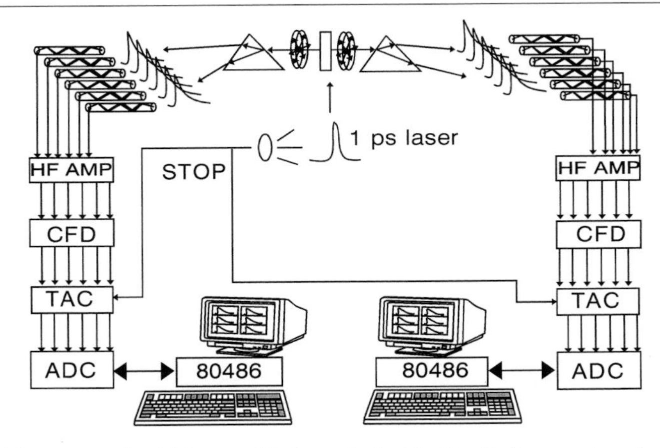

FIG. 3. First-generation T-format, 12-channel, double-kinetic fluorometer. A 1-psec laser pulse at 4 MHz is focused onto the 50-μl observation cuvette of a stopped-flow device (model SFM3; Molecular Kinetics, Pullman, WA). A photodiode monitors the incoming pulse and is fanned out to provide the stop pulse for 12 time-to-amplitude converters (TACs) (model TC862; Tennelec-Nucleus, Oak Ridge, TN). The emitted fluorescence from the observation cuvette is passed through Glan–Thompson polarizers (a complete lens/filter/polarizer combination unit) mounted to both sides of the SFM3 stopped-flow apparatus. The fluorescence through the polarizer units is then focused onto the entrance slits of a polychromator (schematically drawn as a prism) (model HR320; Instruments SA, Edison, NJ). In the exit focal plane of the polychromator, a Hamamatsu (model R3839U-07) 10 linear multianode microchannel plate detector (MAMCP) is mounted. The output of the MAMCP is amplified using high-frequency amplifiers (HF AMPs) (model 774, 2.5 GHz; Phillips Electronic, Mahwah, NJ). The amplified pulses are sent to Tennelec-Nucleus constant-fraction discriminators (CFDs, model 454, modified for MCP usage) and sent to the various TACs (TC 862). The TAC output is digitized by multiple PHA-ADC boards (Tennelec-Nucleus model PCA-II). Custom instrument controller software provides the stop/start/store of each individual millisecond time slice of picosecond/nanosecond data into the computer memory of a controlling 80486 computer (see text for additional details).

digital converters (PHA-ADCs) to the Intel 80486 instrument controlling computer. The multiple millisecond slices of picosecond/nanosecond data are transferred in real time to memory using an in house-designed instrument controller.

5. If the signal-to-noise ratio is sufficient, the experiment is complete and data are written to hard disk. If additional signal is required, repeat stopped-flow reactions are performed, and the time-correlated single photons from each time slice are added together until sufficient signal to noise is obtained (a more detailed flow chart emphasizing hardware details is described in the next section).

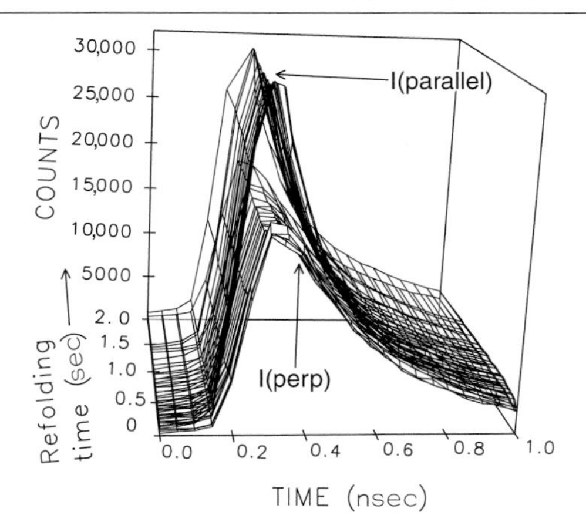

FIG. 4. Typical double-kinetic intrinsic tryptophan polarized time-resolved fluorescence data collected on 10 μM DHFR during the millisecond time scale of refolding. The z axis represents the actual number of photons collected parallel and perpendicular to the vertically polarized laser excitation. The impulse response for this system is currently 32 psec FWHM (full width half-maximum). Note the density of time-resolved data obtained along the refolding time axis.

The only data of this type that had been collected were from the laboratory of L. Brand[4,5] (The Johns Hopkins University, Baltimore, MD). These pioneering works, clearly set the stage for how double-kinetic experiments of this type could be performed. Unfortunately, the excitation sources available at the time were only flashlamp-based systems with low-kilohertz repetition rates. Hence only relatively slow minute time scale double-kinetic studies could be performed. We have improved on the data density obtained in these pioneering experiments in the following manner: the time required to collect each time slice has been reduced from ~60 sec to 1 msec per data set (60,000-fold improvement); the number of photons obtained in the peak of each time-sliced data set was enhanced from ~300–400 to 20,000–30,000 counts (~100-fold improvement); the timing resolution was decreased from 102 psec/channel to 1 to 6 psec/channel (~100-fold improvement); Glan–Thompson polarizers were inserted into the detection channel(s) such that double-kinetic anisotropy decays could be measured; and simultaneous multiple-emission wavelength detection was added (up to 12 time-resolved channels).

Figure 4 shows the quality of data typically obtained with the current

[4] D. G. Walbridge, J. R. Knutson, and L. Brand, *Anal. Biochem.* **161,** 467 (1987).

[5] M. K. Han, D. G. Walbridge, J. R. Knutson, L. Brand, and S. Roseman, *Anal. Biochem.* **161,** 479 (1987).

instrumentation [single tryptophan in 10 μM unfolded phosphoglycerate kinase (PGK) sample]. For clarity only the first 2 sec of data are shown. Data are actually collected in this manner for almost 300 sec, densely collected along the refolding time axis. This instrument represents the only existing functional picosecond/nanosecond polarized time-correlated single photon-counting stopped-flow (with millisecond resolution).

Since this initial design, we have obtained considerable experience in how to best perform double-kinetic experiments. What is now described is the development of a second-generation instrument that is far easier to run, and obtains higher quality data.

Second-Generation Double-Kinetic Instrument

All of the optical aspects of laser input–output were not modified. We found, however, that keeping all 12 parallel time-correlated detection channels operational required simply too much down time. Even more importantly, the T-format data collection lacked any instant-feedback signals, concerning how the experiments are progressing. Since double kinetics collects thousands of time-resolved data points every few milliseconds, and requires nonlinear data analysis for interpretation, there is no way to provide an instant view of what the data really look like. It is difficult, however, not to have instant feedback concerning how well the stopped-flow is performing (e.g., concerning whether both mixers are still perfectly clean and unclogged, whether the protein-folding reaction is still proceeding correctly, and whether there are some slow-time mixing artifacts). For this reason we have gone back to L-format time-resolved detection and assembled two types of instant-feedback total-intensity detectors on the other arm of the stopped-flow instrument (see Fig. 5). In L-format double-kinetic data collection, a stepper motor-controlled polarizer on the stopped-flow apparatus is rotated 90° between each stopped-flow reaction. Since many stopped-flow reactions (hundreds to thousands) are averaged for each complete double-kinetics run, more than sufficient averaging is ensured.

The other arm of the stopped-flow apparatus has been adapted to be both a steady state L-format photomultiplier-based anisotropy detector and a full-emission-wavelength kinetic spectral detector [photomultiplier tube (PMT) and optical multichannel analyzer (OMA) arm of Fig. 5, respectively]. In this manner, both integrated steady state data and double-kinetic time-resolved parallel and perpendicular intensities at multiple-emission wavelengths are simultaneously collected. The fiber optically coupled OMA can be utilized with or without polarizers and allows for the simultaneous collection of 512 emission wavelengths in as little as 1 msec. In this manner,

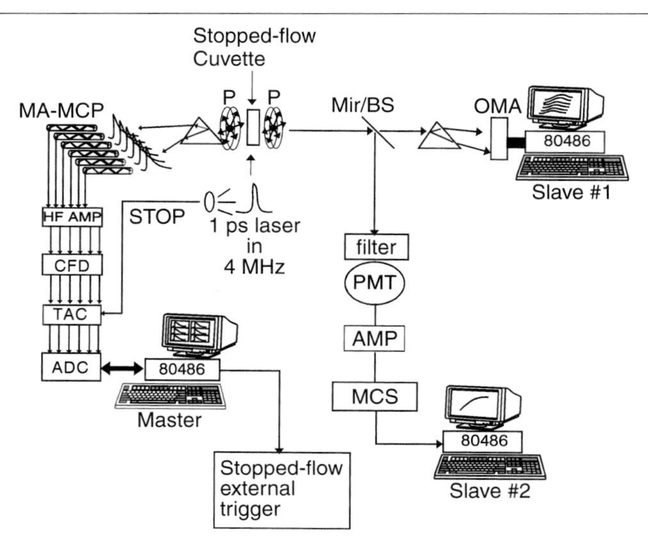

FIG. 5. Second-generation double-kinetic fluorometer, which represents the instrument presently used in the Beechem laboratory. The entire left-hand side has exactly the same hardware as described in Fig. 3. If multiple-emission wavelengths are not absolutely required, the MAMCP is replaced with the new Hamamatsu single-channel MCP (R3809), which together with the new EGG/Ortec (Oak Ridge, TN) pico-timing discriminator yields instrument response functions as low as 30 psec FWHM. The other arm of the stopped-flow detection has now been replaced with a combination steady state wavelength-resolved optical multichannel analyzer (OMA) (model ST1000 controller; Princeton Instruments, Princeton, NJ) and single-channel multichannel scaler (Tennelec-Nucleus model MCS2). (For laboratories just developing these types of measurements, this OMA should be replaced by any of the current high signal-to-noise low light level CCDs that can be obtained.) In this manner the entire data surface, which is simultaneously obtained with each stopped-flow "shot," is as follows: (1) Master computer collects: fluorescence (psec/nsec, msec, polarization, 1 to 6 emission wavelengths); (2) slave 1 collects: fluorescence (msec, polarization, 512 emission wavelengths); (3) slave 2 collects: fluorescence (msec, polarization, integrated over entire emission band).

the full spectrally resolved steady state transitions can be collected simultaneous with the time-resolved detection at a few selected wavelengths. This type of data collection format is the most useful for regular use.

There are two main hardware aspects of double-kinetic data collection that are being improved in this second-generation instrument: (1) improved pulse-height analysis ADC design, with explicit additions specifically for "dead-timeless" double-kinetic time-resolved data collection and (2) higher efficiency processing and real-time storage of time-correlated photons.

Dead-Timeless Double-Kinetic Data Collection

All of the pulse-height analysis analog-to-digital converters (PHA-ADCs) that we utilize are commercial products whose design origin arose from the nuclear physics industry in the 1960s and 1970s. In the current double-kinetic instrument, commercial PHA-ADCs from Tennelec-Nucleus (Oak Ridge, TN) are utilized, primarily owing to the open architecture of their board design. This allows me to totally rewrite the normal data-acquisition controllers with my own protected mode assembly language routines. With this software as many as eight parallel PHA-ADCs may be simultaneously working at any given time within the instrument controller (an expanded bus, 21-slot 80486 microcomputer). Although one could have assembled this multichannel instrument in an alternative laboratory-based bus system (e.g., VME, FAST bus, CAMAC, and fast FERAbus), the cost associated with PHA-ADCs in these configurations was much higher.

Step-by-Step Flow Chart of Double-Kinetic Data Acquisition Sequence

The protected-mode 32-bit assembly language instrument controller, which is the master computer for the double-kinetic experiments, performs the following type of data manipulations in real time during data acquisition (see Fig. 5).

1. The multichannel analyzer (MCA) memory (i.e., where the values from the PHA-ADC are histogrammed into minidecay curves) is split into two equal portions (region 1 and region 2) of 2000 to 4000 channels each. These two regions of MCA memory will be utilized as a double-buffer system. The Tennelec-Nucleus PHA-ADC can be operated in a manner wherein data can be collected in region 1 at the same time that data are read out of region 2. This readout mode is extremely useful for double-kinetic data collection. A variety of assembly language "tricks" is utilized to speed up this data transfer. For instance, in any small millisecond slice, often less than 256 counts ($\leq 2^8$) will be recorded in any single picosecond/nanosecond channel. Hence, one can jump through PHA-ADC memory reading only the lower 8 bits of the ADC. This simple operation decreases the read time by ~50%. Similarly, zeroing out the PHA-ADC between the various millisecond slices also uses up data collection time unnecessarily. This step can be eliminated by collecting all of the individual slices without repeated zeroing by performing a successive-slice subtraction only at the very end of the entire data acquisition sequence. The properly subtracted individual slices can then be obtained using non-real-time operations by starting from the final Nth data slice

and subtracting the signal in the $(N - 1)$ slice. Progressive subtractions, down to slice $N = 1$, generates the same data as if the PHA-ADC were zeroed between each individual data slice. A much higher performance hardware solution to these operations is described following this list.

2. The stopped-flow apparatus is triggered by the master computer by a simple TTL pulse sent out of the parallel port.
3. The master computer then initiates a data collection loop from $i =$ 1 to n, where n equals the total number of time slices desired for this stopped-flow reaction.
4. Time-resolved single photon-counting data are collected for a user-specified time t_n, where t_n is the time period (n). Typically, t_n varies from a few milliseconds to as much as 1 sec in duration. The actual vector time period is specified by the user at program startup and may be spaced (in time) in any manner desired.
5. Since the real-time clocks on typical 80486 and Pentium-based microcomputers are not reliable (on the millisecond time scale), the read and write cycles of the instrument controller are electrically connected to the outside world and logged by a multichannel scaler timer that keeps track (in an absolute sense) of the exact relationship between the individual time-correlated data slices and the stopped-flow time zero (when the reaction was initiated). This time-stamping operation can be performed by connecting a multichannel scaler to any bit of the parallel port of the master computer that toggles this bit high/low as each time-correlated data slice is transferred from PHA-ADC memory to computer main memory. This time stamping can be performed prior to the real stopped-flow experiments, allowing the user to fine-tune the vector t_n such that the high-density picosecond/nanosecond data sets can be obtained at exactly the regions of the refolding reaction that are most important (e.g., closely spaced near time zero of the stopped-flow reaction). Although one can purchase time-stamping PHA-ADCs, this is not necessary. Most time-correlated and stopped-flow fluorescence laboratories will have access to multichannel scalers of some type, which can time stamp and display these results in real time using the above protocol.

Problem: A dead-time problem occurs when t_n is less than the read time associated with getting all of the data out of region 2 of the MCA. The least amount of time in which we can get 4000 channels of data out of the (up to) 8 MCAs (and still maintain all of the other duties of the instrument controller) is approximately 5–10 msec. This means that as we collect data on the sub-10-msec time scale, the time axis that is sampled contains dead time between the points (a solution to this problem is given in point 6).

6. The data for each slice as it is read are maintained in a large matrix with the following structure: Fluorescence counts (100 to 1000 time slices, 2000 time channels parallel, 2000 time channels perpendicular, 1 to 8 MCAs). This amounts to approximately 10 to 100 MB of data obtained by the master computer (Fig. 5) during each single stopped-flow reaction (1 min or so of data collection). An equal density of data is often obtained by the time-synchronized emission wavelength-resolved slave 1 computer (Fig. 5). The data have the simple dimension: Fluorescence counts (512 emission wavelengths, millisecond time). Data densities obtained by the multichannel scaler computer (slave 2) are small (<16,000 points). These data have the structure: Fluorescence counts (parallel or perpendicular, millisecond time). Therefore, with each shot of the stopped flow, 20 to 200 Mbytes of data are collected, on three synchronized computers, representing essentially every possible fluorescence collection axis that is measurable.
7. Complete the loop over all time slices (t_n).
8. Refire the stopped-flow apparatus and signal average for as many reaction sequences as specified. For many of these double-kinetic studies, as many as 100 to 1000 reactions are repeated sequentially. The controlling software does real-time summing during data acquisition. The only user intervention required while this instrument is operating is to refill the stopped-flow syringes every 200 (or so) reactions. This syringe refilling is the rate-limiting aspect of the entire data collection procedure.

To solve the dead-time problem associated with double-kinetic data collection (described previously) a new PHA-ADC board is being designed. This PHA-ADC board will have ultralow digitization dead times and a real-time addressable histogramming memory of ~10 to 100 MB. An on-board arithmetic logic unit (or microprocessor) will perform the histogramming operations with a base page-adjustable addressing register that is controlled by a real-time clock. In this manner, data would not have to be transferred over any bus during data collection. Instead, each data set would be stored on the board, with multiple slices collected naturally by simply incrementing the base-page address of the ALU that is performing the PHA-ADC histogramming. The overall flow chart for this board design in shown in Fig. 6.

This board design offers two double-kinetic enhancements simultaneously. Each time-to-amplitude converter (TAC) is connected to two PHA-

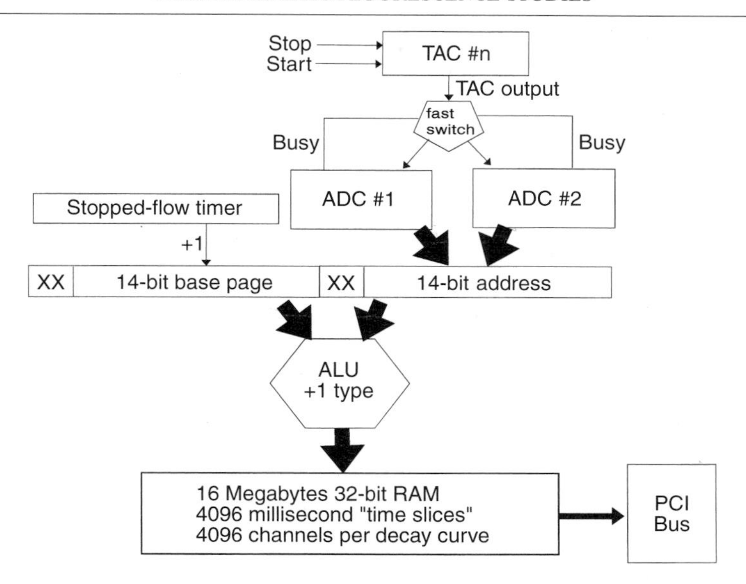

FIG. 6. Schematic of the current PHA-ADC/memory board designed exclusively for double-kinetic studies. The time-to-amplitude converter (TAC) output is shared between two high-speed Wilkinson-style analog-to-digital converters (ADCs) by a fast-switch/amplifier. The channel address corresponding to this TAC output is indexed by a base-page addresser, which is connected to the stopped-flow timer unit. The ALU provides the "photon increment by one operation" into 16 MB of memory on the board. This provides enough room for ~4000 millisecond time slices of decay data, which are each 4000 psec/nsec channels.

ADCs. In a typical time-correlated data collection system, each TAC is connected to a single PHA-ADC. The problem with the 1 TAC → 1 ADC configuration has to do with the finite processing time required by the ADC to digitize the output of the TAC. If the ADC is busy converting a TAC output and an additional TAC output occurs, this photon will be "lost." At relatively low counting rates [1000 to 10,000 counts per second (cps)] the number of lost photons is not significant (<10–20%). However, between 20,000 and 100,000 cps (typical counting rates in a double-kinetic experiment) a standard PHA-ADC system is losing from 50 to 90% of all of the photons recorded by the TAC. In double-kinetic experiments, recorded photon events are at an absolute premium owing to the limited time for sampling each decay curve (milliseconds per decay curve). Therefore, a design that allows both more efficient photon digitization and decay curve addressing would be beneficial.

There are strict requirements for the type of ADC utilized for time-correlated single-photon counting. There are currently only two types of

PHA-ADC that have good enough differential nonlinearity specifications to be utilized for high-resolution measurements: Wilkinson-style ADCs and successive approximation ADCs with sliding scale linearization. Typical data conversion times for the Wilkinson PHA-ADCS are roughly 5 to 50 μsec/photon, dependent on the magnitude of the incoming TAC output. The successive approximation ADC with sliding scale linearization has the advantage of a fixed conversion time (1.5–5.0 μsec).

A prototype board encompassing all aspects of the design features in Fig. 6 (except the stopped-flow timer addresser) has now been constructed. An early prototype of this board is described in Ref. 6. Standard (relatively slow) 150-MHz Wilkinson-type ADCs were utilized to allow for direct comparison with single Wilkinson ADCs already present in the laboratory (PCA-II; Tennelec-Nucleus). Design concerns for this section of the board centered on the ability to match absolutely the timing properties of the two parallel ADC channels. To fine-tune the two ADCs, the fast-switch part of the board was equipped with high-accuracy tunable amplifiers (both offset and gain) that could operate on the TAC output before delivering the signal to the Wilkinson ADC. The ADC matching is so important because the channel number assigned to a TAC output must be absolutely independent of the path (which ADC) by which it was digitized. We have found that the two PHA-ADCs could be matched to <0.3 fsec/channel in a configuration that is stable for >1 year.

Having shown that the parallel Wilkinsons can be matched, it was then necessary to show that they processed photons more efficiently than a single ADC configuration. These data are shown in Fig. 7. Note the large improvement in photon-counting efficiencies for the parallel ADC design over a conventional single-ADC unit. This parallel Wilkinson design is actually as fast (if not faster) than the highest speed commercial PHA-ADC [e.g., the EGG/Ortec (Oak Ridge, TN) model 921 Spectrum master high-rate multichannel buffer]. No commercial multiple parallel Wilkinson unit has ever been produced; commercial manufacturers have attempted to increase ADC throughput simply by speeding up a single ADC. One can show from simple queuing theory, however, that for time-resolved fluorescence applications (which have sums of exponentials-type time-lag histograms) a parallel two-ADC design is more efficient at collecting photons than a single ADC (Wilkinson type) operating at twice the conversion frequency.

It should be emphasized that a full working version of the board described in Fig. 6 is not yet operational. The stopped-flow timer unit and base page indexer are yet to be implemented. Similarly, we plan to replace

[6] G. T. Ying, D. Piston, and J. M. Beechem, *Biophys. J.* **66,** A161 (1993).

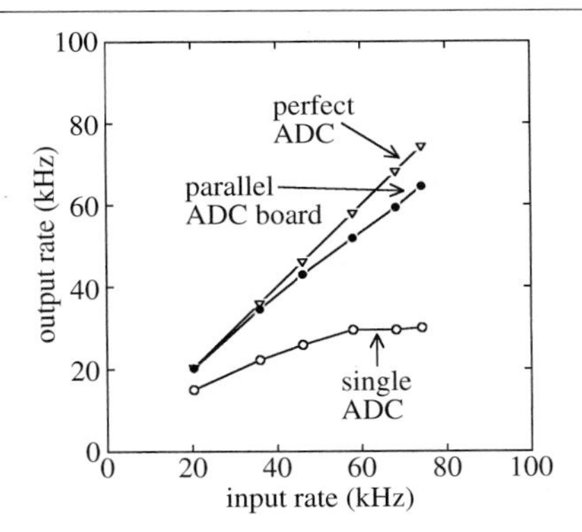

FIG. 7. Input vs output characteristics of the parallel ADC board designed in the Beechem laboratory. The perfect ADC is one that processes 100% of the signals it receives. The parallel ADC is nearly perfect up to approximately 45 kHz. This specification will be greatly improved with the board designed in Fig. 6 by using a much faster timer clock (700 vs 150 MHz). As one approaches the 70-kHz rate, the dead time associated with the TAC (TC 862) begins to become rate limiting.

both 150-MHz Wilkinsons with our own Wilkinson ADCs designed around a common 700-MHz Motorola fast counter. Without going into too many hardware details, by sharing a common counter, both ADCs should be nearly perfectly matched for digitization. Similarly, the memory management for this board still needs to be designed. We are confident, however, that a completely working board performing all of the functions described in Fig. 6 can be assembled within approximately 1 to 2 years. Its performance features will surely exceed any commercial unit, with little cost investment. At this time photon throughput limitations will move from the PHA-ADC to the TAC. We are currently working on a parallel TAC design that will interface with this board also. Once these elements are combined we will have achieved the maximum throughput possible with MCP detectors (~250,000 time-correlated cps; estimates only).

Although a description of this board has been given within the context of data collected in double-kinetic mode, two additional applications must be pointed out. Both the fields of time-resolved fluorescence microscopy and photon-migration imaging examine data of the form: photons (time, spatial x axis, spatial y axis). This board will be immediately applicable for this data type as the 14-bit base page register, instead of incrementing in units of millisecond time, now increment by a spatial axis. Using the full 16-bit base page indexer, 256×256 or 512×512 arrays of time-resolved data could be collected at laser-scanning speeds with this board.

Intrinsic Tryptophan Double-Kinetic Results: Lifetimes/Amplitudes/Total Intensities/Anisotropy

Escherichia coli dihydrofolate reductase (DHFR) is a small, 17,680-dalton protein with five tryptophan residues. Intrinsic tryptophan double-kinetic time-resolved total-intensity and anisotropy experiments were performed during the refolding of this protein.[7] These experiments were performed with essentially one goal in mind. Within the dead time of standard stopped-flow mixing, this protein undergoes significant folding and collapse.[8,9] It was expected, therefore, that examination of the tryptophan lifetimes and rotations (which will be averaged over all five residues) would reveal significant changes and approach native-like rotational and lifetime properties within the dead time of the stopped flow. This result, however, makes the assumption that the tryptophan residues within DHFR would pack into this rapidly collapsed protein state.

The double-kinetic results are shown in Fig. 8, and reveal that this assumption is incorrect. Examination of the time-resolved anisotropy decay of DHFR 20 msec into the folding reaction (Fig. 8, top) reveals a rotational mobility that is essentially identical to that observed in the unfolded state. Similarly, the overall total-intensity decay profile (Fig. 8, bottom) at 20 msec into the refolding reaction is also identical to that observed in the unfolded state. Figure 9 shows in detail the time-resolved double-kinetic decomposition of the observed steady state intrinsic tryptophan fluorescence change (Fig. 9, top) into its primary time-resolved components. Analysis of the total-intensity decay curves at each individual time slice reveals that the initial rise phase in the tryptophan total-intensity curve is due completely to a lifetime change associated with the longest lifetime component (of the three lifetimes that are resolved) (Fig. 9, middle). This rise phase lasts approximately 0.5 sec, in which the amplitude associated with the long lifetime is constant (Fig. 9, bottom). The overall tryptophan total intensity then begins to decrease. One can see that the photophysical origin of this decreasing total intensity is due to a decrease in the amplitude associated with this long lifetime term (Fig. 9, bottom), and not due to any significant change in lifetime (Fig. 9, middle). Hence, the mechanism associated with the change in steady state fluorescence intensity "changes" 0.5 sec into the refolding reaction. Initial changes are completely due to altered dynamic quenching, whereas the decreasing total-intensity phase is almost completely an increased static-quenching mechanism. The time-resolved anisotropy data reveal that at early times, the rotational mobility

[7] B. E. Jones, J. M. Beechem, and C. R. Matthews, *Biochemistry* **34,** 1867 (1995).
[8] P. A. Jennings, B. E. Finn, B. E. Jones, and C. R. Matthews, *Biochemistry* **32,** 3783 (1993).
[9] B. E. Jones, P. A. Jennings, R. A. Pierre, and C. R. Matthews, *Biochemistry* **33,** 15250 (1994).

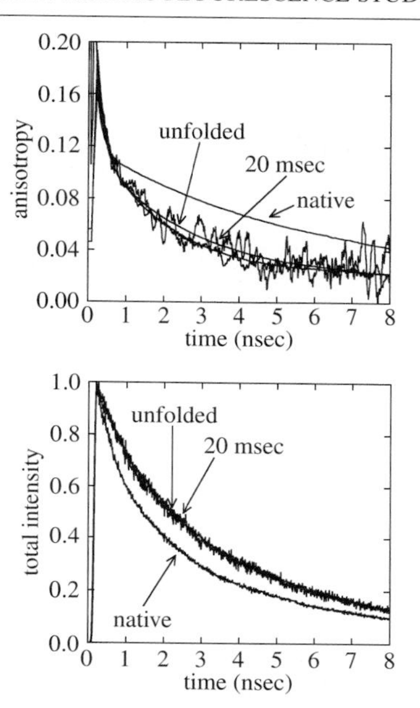

FIG. 8. *Top:* Time-resolved anisotropy of the five tryptophans in *E. coli* DHFR in the unfolded state, 20 msec into a refolding reaction, and native state. Note how the time-resolved rotations of the 20-msec slice nearly superimpose with the completely unfolded rotation pattern. *Bottom:* Time-resolved total intensity of the five tryptophans in *E. coli* DHFR in the unfolded state, 20 msec into a refolding reaction, and native state. Note how the total-intensity function of the unfolded state and the 20-msec slice also superimpose. These data suggest that the majority (all?) of the emitting five tryptophans in DHFR do not pack into the rapidly collapsed state of DHFR.

of the tryptophan residues in DHFR are completely unchanged. Given the fact that other studies (primarily NMR) have revealed a significant amount of secondary structure formation within the dead time of the stopped flow, one must conclude that the tryptophan residues (at least in this protein) do not pack into this initially collapsed structure. This was a surprising, and completely unexpected, result. The time-resolved anisotropy data at 20 msec into the reaction are certainly convincing evidence that these residues are not packing into this initially collapsed structure.

If only steady state signals had been monitored, much weaker conclusions would have been reached, as there are many different photophysical effects (other than changes in rotational mobility) that can alter steady state anisotropy signals. In any steady state study, one can never be certain that the predicted increase in steady state anisotropy (due to a postulated protein collapse) may not be observed simply owing to an increase in the

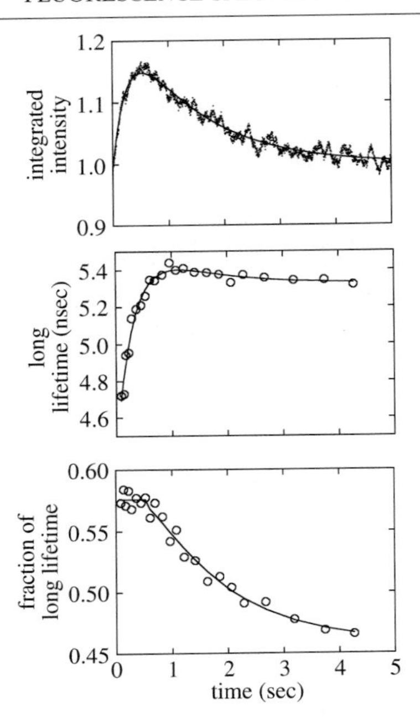

FIG. 9. *Top:* Integrated steady state total intensity of the five tryptophans in *E. coli* DHFR during a refolding reaction collected by multichannel scaling computer slave 2 (Fig. 5). *Middle:* Double-kinetic total intensity lifetime changes of the longest lifetime component of DHFR during the refolding experiment [data collected simultaneous with slave 2 by the master computer (Fig. 5)]. Note the rapid increase in this lifetime component during the first second of refolding, which then stabilizes at 5.3 nsec, whereas the total intensity function (*top*) is decreasing. *Bottom:* Double-kinetic amplitudes associated with the long-lifetime component of DHFR during the refolding reaction. Note the constant-amplitude term during the first 0.5 sec, followed by a decreasing amplitude at later times. These data indicate that the total-intensity "rising" and "falling" phases (*top*) can be decomposed into a dynamic quenching release (up phase) followed by a static quenching term (falling phase).

mean fluorescence lifetime associated with the refolded state (i.e., compensatory changes). Mathematically, the steady state anisotropy that is measured (r_{ss}) is complicated by changes that can originate from either altered fluorescence lifetime(s) (τ) or altered rotational correlation times (ϕ): $r_{ss} = r_0/[1 + (\langle\tau\rangle/\phi)]$. A shortening lifetime during a refolding reaction will yield an increase in r_{ss} independent of any changes in the rotational mobility of the fluorescence probe. Similarly, a lifetime increase during a refolding reaction will work to decrease r_{ss}. The time-resolved anisotropy signal shown in Fig. 8 does not suffer from this type of ambiguity.

Extrinsic Probe Time-Resolved Anisotropy Double-Kinetic Results: Determination of Hydrodynamic Radii of Rapidly Collapsed Protein States

Our initial tryptophan double-kinetic anisotropy studies with DHFR (described previously) revealed that it would be useful to have an additional methodology to quantitate the hydrodynamics associated with rapidly collapsed protein states. To make these measurements, we have concentrated on using the probe 1-anilinonaphthalene 8-sulfonate (ANS) fluorescence. ANS fluorescence has a long history of use as a dye that readily binds to partially folded protein intermediates. Molten globule states from several protein systems have been examined and shown by light-scattering and viscosity measurements to have collapsed structures. These collapsed structures have been shown to interact well with ANS, leading to a large increase in fluorescence when bound.[10,11] This author was initially hesitant to perform detailed work with ANS as a probe of protein-folding reactions owing to its possible adverse effects on refolding kinetics.[12] However, the preliminary data that we have obtained with DHFR and the α subunit of tryptophan synthase (denoted α-TS) have shown that ANS provides an important set of complementary information that cannot be directly obtained with tryptophan fluorescence. The ANS studies allow an investigation of the hydrodynamic properties associated with partially collapsed states that form in the submillisecond and hundreds of millisecond time scales. Our use of this probe will be entirely devoted to characterization of hydrodynamic radii and will not be used (per se) for general kinetic characterization.

The large fluorescence lifetime difference between free ANS (~0.2 nsec) and bound ANS (~10 nsec) makes this probe an ideal fluorophore for double-kinetic studies. Since the percentage of ANS bound during the refolding reaction is small, an integrated steady state signal is composed of both free ANS and bound ANS signals. The fluorescence lifetime difference associated with the bound ANS is so large that even this small percentage of bound molecules can yield significant steady state signal changes that can be easily monitored. In the double-kinetic experiment, however, the recovered fluorescence is doubly time resolved, allowing a magnification of the photons associated with the bound state by another factor of $\geq$1000-fold. For instance, using the measured 0.2 and 10 nsec for the lifetimes of the free and bound ANS, respectively, and moving 2 nsec after the laser pulse (within any single double-kinetic slice) causes the photons from the

[10] S. Tandon and P. M. Horowitz, *J. Biol. Chem.* **264,** 9859 (1989).

[11] G. Semisotnov, N. Rodionova, O. Razgulyaev, V. Uversky, A. Gripas, and R. Gilmanshin, *Biopolymers* **31,** 119 (1991).

[12] M. Engelhard and P. A. Evans, *Proteins* **4,** 1553 (1995).

free ANS to decay to exp(−2 nsec/0.2 nsec) = 4.5×10^{-5} of their initial amplitude. The photons from the bound ANS, however, have decayed to only exp(−2 nsec/10 nsec) = 0.819 (of their initial amplitude). The free ANS-to-bound ratio is approximately 100 : 1 (under our experimental conditions), which yields a signal from bound ANS over free ANS of 0.819/ $(100 \times 4.5 \times 10^{-5})$ = 182 : 1. As one "moves farther out" in nanosecond time in any given millisecond time slice, one can magnify the photons from the bound state even further. For instance, moving out to 5 nsec after the laser pulse yields a signal of bound ANS-to-free ANS ratio of $4.4 \times 10^8 : 1$ (i.e., completely dark-count limited). This double-kinetic spectroscopic enhancement allowed us to determine the rotational correlation time associated with the rapidly collapsed bound-ANS protein state. These double-kinetic anisotropy results are shown in Fig. 10.

These results clearly reveal how combining the ANS rotational correlation time data with the tryptophan rotational correlation time data is superior than either experiment done in isolation. The refolding of DHFR in Fig. 10 can be directly compared with the time scale of events viewed from the tryptophan fluorescence (with the normal caveats concerning the ANS-dependent changes in kinetics) in Fig. 9 as both represent urea jumps from

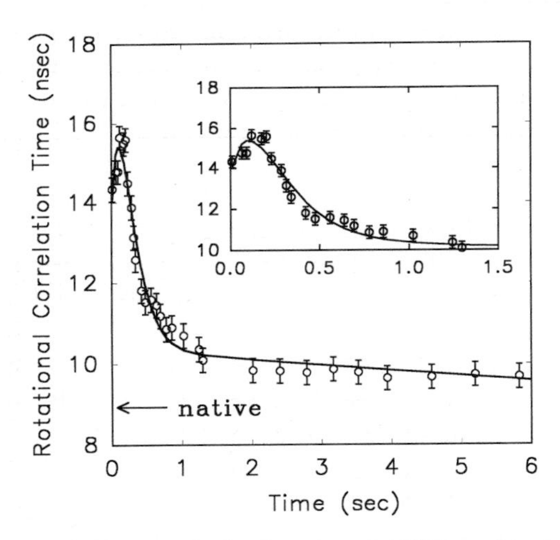

FIG. 10. Double-kinetic time-resolved anisotropy of ANS bound to the rapidly collapsed state of *E. coli* DHFR. Note that within the dead time of the stopped-flow reaction, the rotational correlation time associated with the bound ANS is approximately 1.4 times larger than the native state. Within 100 msec, this correlation time increases to 1.6 times that of the native state, and then rapidly collapses to the native state value of approximately 10 nsec. The changes in hydrodynamic radii of DHFR (corresponding to these rotational correlation times) are shown in Fig. 11 (inset).

5.4 to 0.6 *M*. The ANS rotational correlation time (ϕ) clearly senses this rapidly collapsed protein state. The unexpected result that was obtained from this probe is that within the dead time of the stopped flow, the rapidly collapsed protein state developed a rotational correlation time that was approximately 1.4 times slower than the native state. Within the first 50 msec, this correlation time further lengthens to a maximum of about 1.6 times slower than native. This overexpansion is followed by a rapid contraction to a native state that takes about 2 sec to occur.

Our major initial concern with this result is the possibility that this overexpansion was actually a transiently formed aggregated state (e.g., dimer formation). Therefore, we repeated these original experiments (performed at 30 μM) at 10 μM and more recently at 1 μM. Each of these experiments yielded an identical overexpansion and identical kinetics. If the observed rise phase was associated with dimer formation, this concentration range should have altered the rate by 900 times.

Using the relationship that

$$\phi = \frac{\eta V}{r_{\text{gas}} T} \approx \frac{4\eta\pi r_{\text{hydro}}^3}{3 r_{\text{gas}} T}$$

where η is the solution viscosity, V is the volume of the rotating protein, and T is temperature, one can solve for the effective hydrodynamic radius associated with the rapidly collapsed state (r_{hydro}). This factor of 1.6 overexpansion of the rotational correlation time of DHFR during the first 100 msec of folding translates into an effective hydrodynamic radius increase of $1.6^{1/3} = 17\%$ larger than the native state. The approximation involved in translating the measured ϕ into a hydrodynamic radius involves the implicit assumption of a spherical collapsed state. All of our high signal-to-noise time-resolved anisotropies for this system (50,000 to 100,000 counts at the peak of the vertical emission in the first time slice; ~10 msec) fit extremely well with only a single correlation time for the bound ANS, indicating the collapsed state is reasonably spherical.

Measurements of hydrodynamic radii during protein-refolding reactions are relatively rare. Stable models of molten globules, however, have been obtained for a number of proteins (at low pH values, low salt, high temperature). Viscosity and X ray-scattering studies on these stable molten globules predict an increase in hydrodynamic radii of approximately 10–20%.[13] Since what is measured in a time-resolved fluorescence anisotropy experiment is related to the r_{hydro}^3, small changes in hydrated radii can be obtained. Although we were initially surprised by the large overshoot of the measured correlation time, once put into the perspective of changes in hydrodynamic

[13] I. Nishii, M. Katoaka, F. Tokunaga, Y. Goto, *Biochemistry* **33,** 4903 (1994).

radius, these results are exactly what is to be expected for formation of a molten globule or other expanded rapidly collapsed protein state.

A similar double-kinetic anisotropy study is just being completed with the α subunit of tryptophan synthase (α-TS). This protein also overexpands during the refolding reaction. In the dead time of the stopped flow this protein has a hydrodynamic radius 7% larger than native. Within 2 sec this collapsed state overexpands to a maximum of 16% larger than native, a value close to that obtained with DHFR overexpansion. A plot of the measured hydrated radii with time for both α-TS and DHFR is shown in Fig. 11. Additional experiments with more proteins will be required to determine whether rapid collapse to a spherical radius 16–17% larger than the native state is a general property of refolding proteins.

These results are just one example of the type of data that can be obtained with the time-resolved fluorescence anisotropy double-kinetic instrumentation. This instrumentation is unique in that it (1) makes direct measurements of hydrodynamic radii to high accuracy ($\pm$0.4 Å), (2) uses low protein concentrations (approximately micromolar), and (3) is capable of millisecond timing. It should be emphasized that the interpretation of the changes in the double-kinetic total intensities and rotational correlation times (given previously) certainly represent a great oversimplification of actual protein-folding processes. These simplistic interpretations of the double-kinetic data reflect the majority of the laboratory effort, which, up to this point, has been concentrating on establishing the instrumentation to obtain high signal-to-noise data. We believe that the instrumentation is developed to the stage at which more sophisticated experimental protocols

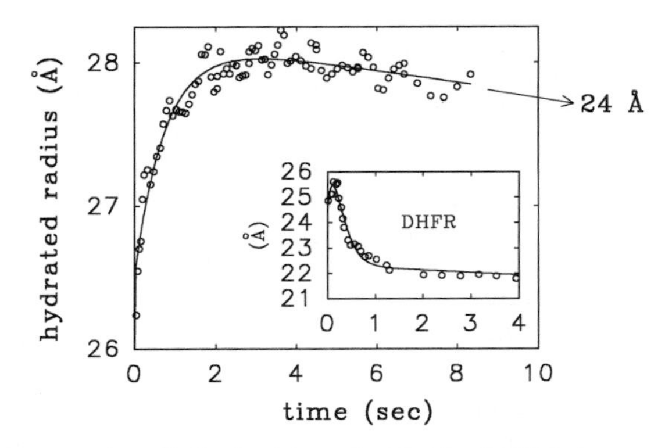

FIG. 11. Hydrodynamic radii obtained from double-kinetic ANS time-resolved anisotropy data for the α subunit of tryptophan synthase and *E. coli* DHFR (inset). In both proteins, the maximum hydrodynamic radius associated with the rapidly collapsed refolded state is 16–17% greater than that of the native state enzyme.

(e.g., double-jump experiments) as well as more sophisticated analysis methodologies (described in the next section) can now be implemented.

New Approaches to Double-Kinetic Data Analysis

A deliberate attempt has been made to perform as little preprocessing analysis of the double-kinetic data as possible. Our efforts in instrumentation development have been to increase the signal-to-noise ratios on the data to the extent that smooth transitions in the time-resolved fluorescence decay parameters could be obtained, even when performing completely unlinked data analysis (note how smoothly the recovered lifetimes, amplitudes, and rotational correlation times change in Figs. 9–11). By far the hardest parameters to determine in achieving smooth double-kinetic transitions are rotational correlation times. The errors in the recovered fluorescence lifetime terms (and associated amplitudes) are generally a factor of 5–10 times less than the errors on the correlation times (depending on the model). Given the measured smooth transitions in the double-kinetic lifetimes and correlation times the question arises: What is the best way to analyze this type of data?

We are currently performing nonlinear least-squares iterative reconvolution analysis to recover the picosecond/nanosecond parameters at a given time slice. These parameters can be represented as F(msec, slice N) = $\tau(N)$, $\phi(N)$, $\alpha(N)$, $\beta(N)$ (step 1), where τ and ϕ represent the fluorescence lifetime(s) and correlation time(s); α and β are their corresponding weights, respectively. These recovered parameters then become "data" that are input to a second set of nonlinear data analysis routines that recover the rate constants linking the various folded states of the proteins being examined. This process can be represented as ϕ (msec time, all slices) analyzed in terms of k_1 (rate for formation of collapsed intermediate), k_2 (rate for formation of overexpanded intermediate), and k_3 (rate for formation of native state size) (step 2). Similar analyses are performed on the recovered fluorescence lifetimes, amplitudes, etc. This multistep nonlinear analysis approach is certainly a suboptimal manner in which to extract the inherent information content from double-kinetic experiments. We have now developed analysis routines that analyze the fluorescence data obtained on both time scales simultaneously, in terms of internally consistent kinetic models.

To perform this analysis, the double-kinetic data are initially analyzed as described previously, just to obtain some "feeling" that the experiments were successful, and to obtain a rough estimation of the time scales associated with the changes and approximate amplitudes. A combined global-type analysis would then be performed, wherein the entire data matrix of

F(psec/nsec, msec, MCA No.), would be simultaneously minimized in terms of internally consistent folding models as follows.

From an empirical examination of the initial double kinetic data, approximate values for many of these rate constants and hydrodynamic radius terms can be obtained as initial guesses. For example, examination of the data for the α-TS data in Fig. 11, reveals that within the dead time of the stopped flow the protein had collapsed to a hydrodynamic radius of ~26.2 Å, yielding an initial estimate of k(collapse) of $\geq 500\ \text{sec}^{-1}$. The protein continues to overexpand for the first 1–2 sec, providing an initial guess for k(expand) of $\sim 3\ \text{sec}^{-1}$ and expands to at least 28 Å. This protein takes a long time to contract completely (data not shown), with a k(contract) $\sim 0.01\ \text{sec}^{-1}$. From the time-resolved anisotropy data of the native state, the hydrodynamic radius is known (24 Å). The key to the global analysis of these data is to combine simultaneously the millisecond and nanosecond models into a single analysis. The linking between these two time scales is made through the population terms of the millisecond time-scale rate constants. Using the initial guess rate constants described previously, the populations of the intermediate states and native state as a function of the millisecond time slices can be calculated. The picosecond/nanosecond data at double-kinetic slices (e.g., 0.1, 0.5, and 4.5 sec) could then be simultaneously analyzed, using

$$
\begin{aligned}
F(t, 0.1\ \text{sec}) &= \underline{0.741} \times F_{I1}(\alpha, \tau, \beta, \phi) + \underline{0.259} \\
&\quad \times F_{I2}(\alpha, \tau, \beta, \phi) + \underline{0.000} F_{I3}(\alpha, \tau, \beta, \phi) \\
F(t, 0.5\ \text{sec}) &= \underline{0.223} \times F_{I1}(\alpha, \tau, \beta, \phi) + \underline{0.774} \\
&\quad \times F_{I2}(\alpha, \tau, \beta, \phi) + \underline{0.002} F_{I3}(\alpha, \tau, \beta, \phi) \\
F(t, 4.5\ \text{sec}) &= \underline{0.000} \times F_{I1}(\alpha, \tau, \beta, \phi) + \underline{0.959} \\
&\quad \times F_{I2}(\alpha, \tau, \beta, \phi) + \underline{0.040} F_{I3}(\alpha, \tau, \beta, \phi)
\end{aligned}
$$

where $F_{IJ}(\alpha, \tau, \beta, \phi)$ is the picosecond/nanosecond decay function (as complex as needed) that describes the physical species IJ. In this case, F_{I1} is the rapidly collapsed state that is disappearing in milllisecond time, F_{I2} is the overexpanded intermediate (population rises then falls), and F_{I3} is the final native state.

These are the proper basis functions to link the millisecond time-scale data together. The underlined numerical values (above) are the millisecond-changing weighting factors that effectively "stitch together" the picosecond/nanosecond decay functions along the stopped-flow reaction. It should be emphasized that the F_{IJ} values do not vary as a function of millisecond time and are linked over the entire double-kinetic run. During global nonlinear minimization as the slow-time millisecond rates are fit, they alter the population parameters associated with the weighting of the picosecond/nanosecond decay functions that are also being minimized. In this manner, there is a natural transfer of information from one time scale to another during the analysis, and both time scales will be simultaneously optimized. This is different than the empirical exponential analyses performed, for example, in Fig. 10. When unlinked double-kinetic analysis is performed only effective ϕ values are actually recovered, representing an averaging of all of the protein species present at that time slice. When the double-kinetics analyses are performed as described in this section, the true nonaveraged ϕ values for each state will be recovered. This simultaneous analysis of both picosecond/nanosecond and millisecond time scales results in an extreme reduction in the number of fitting parameters utilized to describe the entire data surface.

This analysis will also be just as appropriate for the stopped-flow energy-transfer double-kinetic data as it is for the anisotropy data. In the double-kinetic energy-transfer data, each F_{IJ} would represent the distance probability function describing state IJ labeled at sites X and Y. Within this analysis framework, all of the complex sets of picosecond/nanosecond linkages (multiple emission wavelengths, multiple quencher concentrations, etc.) are immediately incorporated.

Double-Kinetic Software/Hardware Information

It is apparent that any time-resolved fluorescence laboratory with access to a stopped-flow instrument could adapt their instrument for double-kinetic studies. Key to the double-kinetic instrumentation is the software that synchronizes and collects all of the time-resolved picosecond/nanosecond data with millisecond stop/start/store capability. Currently, the time-critical aspects of these operations are written in protected-mode 32-bit assembly (80486/Pentium) with the Borland (Scotts Valley, CA) Turbo-Assembler version 4.0. This software can simultaneously operate multiple (up to 16) Tennelec-Nucleus PHA-ADCs (MIS3 system). These assembly language routines are interfaced with a protected-mode Fortran (F77L-EM/32 version 5.1; Lahey, Incline Village, NV) master program that provides the data space and some real-time mathematical operations. Any laboratory that utilizes this hardware and is interested in developing a double-kinetic fluorometer can contact this laboratory (BEECHEM@LHMRBA.HH.VANDERBILT.EDU) for the source code for the master computer. Successful double-kinetic experiments can be performed without any additional hardware.

The software solution is supplemented with two hardware items that allow for higher signal-to-noise data collection. First, the two-PHA/ADC per TAC configuration allows for more efficient digitization of the TAC outputs. The NIM-based unit which we have developed to combine two PHA-ADCs with a single TAC output will work with any Tennelec-Nucleus TAC (e.g., TC862) and two Tennelec-Nucleus PCA-II PHA-ADCs. Any laboratory interested in the detailed schematics for this circuitry can also contact this laboratory at the e-mail address noted above.

The development of the double-kinetic PHA-ADC board (Fig. 6) will clearly provide the optimal methodology with which to perform these types of studies. This board will allow double-kinetic studies to be performed with the same ease as simple stopped-flow fluorescence studies. In fact, laboratories that know that a 1- to 2-year lag will occur before they will be ready to perform double-kinetic studies may wish to wait for this hardware item, instead of using the (stop-gap) solutions previously suggested. Depending on the number of requests, the final implementation of this board design may be altered. If the volume is high enough, certain fabrication techniques could be utilized that would not be practical for the $N = 1$ or 2 required by our laboratory.

The software for the global analysis of double-kinetic data (both time scales simultaneously) is written in protected-mode Fortran (Lahey). Information regarding the availability of this program should be addressed to the Globals Unlimited WWW page at: *http://www.physics.uiuc.edu/groups/fluorescence/globals/introglo.html.* More recent developments on double-kinetic hardware/software from the Beechem laboratory can also be accessed from the WWW at: *http://www.beechemlab.mc.vanderbilt.edu/.*

No attempt has been made to summarize other recently developed protocols that allow for folding data to be collected on picosecond/nanosecond time scales. The interested reader should search for, among others, papers from the laboratories of P. Wright (millisecond stopped-flow NMR), W. Woodruff, R. Hochstrasser, H. Roeder, and W. Eaton [laser T-jump via infrared (IR) absorption, flash photolysis, etc.] for some of these developments.

Acknowledgments

The development of the double-kinetic project has involved a collaborative interaction between many individuals. Initially, the pioneering efforts of the Ludwig Brand laboratory (Ludwig Brand, Dana Walbridge, Jay R. Knutson, Lesley Davenport, Myun Han) are acknowledged. The initial set-up with the multiple Tennelec/Nucleus PHA-ADCs benefited from discussion with Dr. Gary Holtom (Batelle Corp.) and Bob House (Tennelec/Nucleus). The development of the first-generation two-PHA-ADC/TAC circuit was designed/implemented together with the D. W. Piston laboratory (Vanderbilt University, with Dr. G. T. Ying). The design of the parallel ADC/histogramming memory unit (Fig. 6) was carried out with Jason D. B. Sutin (graduate student in the Beechem laboratory).

The experimental double-kinetic work on *E. coli* DHFR and the α subunit of tryptophan synthase was carried out with the assistance of the laboratory of C. R. (Bob) Matthews (Penn State University, with Dr. Bryan Jones and Dr. Osman Bilsen). Experimental efforts of both Jones and Bilsen contributed significantly in establishing the working-type instrument shown in Fig. 5. Bilsen also independently developed software that can simultaneously optimize double-kinetic data on two time scales and can be contacted at: OSMAN@PROTEIN.CHEM.PSU.EDU

The Beechem laboratory also acknowledges an extensive collaboration with the Maria T. Mas laboratory (Beckman Research Institute, City of Hope, Duarte, CA). Steady-state and double-kinetic fluorescence energy-transfer experiments with this laboratory are nearly complete, and papers from this collaboration can be found in *Biochemistry* (M. Pilar Lillo). These experiments were motivated by the pioneering work in this area by Prof. Elisha Haas (Bar-Ilan University, Ramat-Gan, Israel).

The funding for this work was obtained from Vanderbilt University, The Lucille P. Markey Charitable Trust, and the NIH (GM45990, RR5823).

[4] Time-Resolved Room Temperature Tryptophan Phosphorescence in Proteins

By JOSEPH A. SCHAUERTE, DUNCAN G. STEEL, and ARI GAFNI

Introduction

Optical spectroscopy techniques have played an important role in studies of protein structure, dynamics, and interactions. These techniques are relatively nondestructive, require small amounts (and low concentrations) of material, and are rapid and usually easy to use. Fluorescence-based methodologies are particularly suitable for biomolecular studies owing to the high sensitivity of this luminescence to the environment of the emitting chromophore and to its ability to detect an exceedingly small number of molecules. Not surprisingly, fluorescence has found numerous applications in the assay of enzymatic activities, detection of minute amounts of labeled proteins, imaging of single molecules, and in the studies of biomolecular dynamics.

The development of methodologies that utilize the phosphorescence of proteins in fluid solution, an emission that arises from the intrinsic lumiphore, tryptophan, has presented researchers with a range of new tools for studies of protein structure and dynamics. As shown in Fig. 1, tryptophan (Trp) phosphorescence, like fluorescence, involves light emission by an electronically excited state of the molecule; however, while fluorescence connects two singlet energy levels of the indole chromophore, phosphores-

0076-6879/97 $25

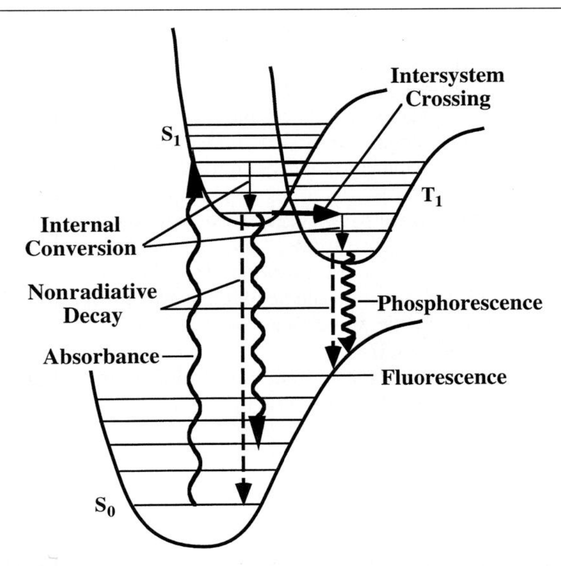

FIG. 1. Jablonski energy diagram schematically depicting the transitions among the ground singlet state of a chromophore (S_0), and the excited states within the singlet ($S_1 \rightarrow S_n$) and triplet ($T_1 \rightarrow T_n$) manifolds. The absorbance of light is quantum mechanically allowed only for the $S_0 \rightarrow S_1$ transition. However, efficient transfer to the triplet manifold occurs from $S_1 \rightarrow T_1$, presumably through excited vibrational states in T_1. Nonradiative processes originating from S_1 and T_1 compete with fluorescence and phosphorescence, respectively, to depopulate the singlet and triplet excited states.

cence involves an electronic transition from the lowest triplet state to the ground singlet state, a transition forbidden by spin selection rules. Its different origin lends Trp phosphorescence several important and useful properties: (1) This emission yields information on the chromophore structure and interactions in the triplet state, information that is additional (rather than redundant) to that obtained from fluorescence. (2) The phosphorescence is red-shifted relative to fluorescence, reflecting the fact that the triplet state is lower in energy than the corresponding singlet state. This allows for easy resolution between the two emissions. (3) Because the transition is dipole forbidden, the phosphorescence decay time is long, typically on the msec–sec time scale, compared to the nsec time scale of fluorescence decay. One important consequence of this slow radiative decay is that the phosphorescence lifetime is markedly more sensitive to quenching by a variety of short- and long-range interactions than the corresponding fluorescence lifetime. Time-resolved room temperature phosphorescence (TR-RTP) can therefore serve as a sensitive probe for the environment of the phosphorescent moiety within the protein.

Early reports on long-lived luminescence from proteins in frozen glasses (e.g., Refs. 1 and 2) served to demonstrate that all three aromatic amino acids—tyrosine, phenylalanine, and tryptophan—can phosphoresce within a frozen protein. While RTP has also been observed from solid (lyophilized) samples of proteins[3] at room temperature, the work of Shah and Ludescher demonstrated that gradual solvation of lyophilized proteins[4] resulted in a concomitant decrease of phosphorescence yield and lifetime. The reduction of phosphorescence efficiency is believed to reflect the decreased internal rigidity of the protein that results from the exchange of solvent H bonds for internal H bonding. Studies show that Trp residues only are responsible for all RTP from proteins in aqueous solutions.

In view of the great sensitivity of Trp phosphorescence to quenching, the first observations of RTP from proteins in fluid solution[5,6] were remarkable in demonstrating that the local rigidity of the protein interior in solution provided the necessary environment. Specifically, Saviotti and Galley in 1974[6] demonstrated that in the absence of oxygen, one of the two tryptophans in horse liver alcohol dehydrogenase (LADH) and in alkaline phosphatase (AP) emitted a long-lived luminescence. Comparison of the emission spectra of the proteins at room temperature with the emission from model aromatics and from proteins at cryogenic temperatures confirmed that the luminescence originates from the triplet state of tryptophan. Extensive deoxygenation was found to be essential in these experiments since in solution the presence of oxygen, which freely diffuses both in the bulk solvent and through the protein, strongly quenches the triplet state of tryptophan, thus reducing the phosphorescence lifetime and intensity. Under deoxygenated conditions, however, RTP is routinely observed in the majority of proteins and displays lifetimes ranging from milliseconds to several seconds. As discussed briefly in the next section, the long lifetime of phosphorescence makes TR-RTP spectroscopy an especially powerful noninvasive probe of changes in protein structure and dynamics in solution.

This chapter addresses the practical aspects of TR-RTP spectroscopy and its applications for protein studies. We focus on intrinsic (Trp) protein phosphorescence; however, the experimental approaches described can be applied to extrinsic phosphorescent chromophores and long-lived lumiphores in general. Several reviews on room temperature phosphorescence

[1] S. V. Konev, "Fluorescence and Phosphorescence of Proteins and Nucleic Acids." Plenum, New York, 1967.

[2] P. Debye and J. O. Edwards, *Science* **116,** 143 (1952).

[3] G. B. Strambini and E. Gabellieri, *Photochem. Photobiol.* **39**(6), 725 (1984).

[4] N. K. Shah and R. D. Ludescher, *Photochem. Photobiol.* **58**(2), 169 (1993).

[5] J. W. Hastings and Q. H. Gibson, *J. Biol. Chem.* **242,** 720 (1967).

[6] M. L. Saviotti and W. C. Galley, *Proc. Natl. Acad. Sci. U.S.A.* **71**(10), 4154 (1974).

include Papp and Vanderkooi,[7] Vanderkooi *et al.*,[8] Vanderkooi and Berger,[9] Vanderkooi,[10] and Sokolovsky and Daniel.[11] In addition, the triplet state of biological macromolecules has been addressed in a review by Geacintov and Brenner.[12]

Physical Basis and Applications of Time-Resolved Room Temperature Phosphorescence Spectroscopy

Although the Trp fluorescence spectrum is sensitive to the polarity of the solvent and shifts in this spectrum can be used to provide significant biological insight, the phosphorescence spectrum of proteins is remarkably insensitive to changes in the solvent environment. In contrast, the rapid radiative decay of the Trp excited singlet state makes the fluorescence lifetime (nsec time scale) relatively insensitive to competitive dynamic quenching processes when compared to phosphorescence where the radiative lifetime is estimated to be on the order of 10 sec.[13] Hence, the power of Trp RTP spectroscopy is manifested in time-resolved experiments. For comparison we note that the temperature-dependent component of the β factors of X-ray crystallography, which report on the structural rigidity, may vary by a factor of four from the core to the surface of a protein, whereas the RTP lifetime, which is sensitive to similar interactions, varies by three to four orders of magnitude. In this section we review the physical issues that control the RTP lifetime of a Trp in a protein in solution and provide illustrative examples of how the TR-RTP data can be used to obtain information on protein structure and dynamics.

Factors That Influence Triplet State Lifetime

Because the electronic transition from the ground (singlet) state to the triplet state is forbidden, direct excitation of the triplet state by light absorption is an extremely weak process. In fact, as depicted in Fig. 1, efficient population of the triplet state is due almost entirely to intersystem crossing where energy absorbed in direct electronic excitation of the first excited singlet state is transferred to the triplet state ($S_1 \rightarrow T_1$), a process

[7] S. Papp and J. M. Vanderkooi, *Photochem. Photobiol.* **49**(6), 775 (1989).
[8] J. M. Vanderkooi, D. B. Calhoun, and S. W. Englander, *Science* **236**(4801), 568 (1987).
[9] J. M. Vanderkooi and J. W. Berger, *Biochim. Biophys. Acta* **976**(1), 1 (1989).
[10] J. M. Vanderkooi, Tryptophan phosphorescence from proteins at room temperature. *In* "Topics in Fluorescence Spectroscopy" (J. R. Lakowicz, ed.), pp. 113–136. Plenum Press, New York, 1992.
[11] M. Sokolovsky and E. Daniel, *Methods Enzymol.* **49,** 236 (1978).
[12] N. E. Geacintov and H. C. Brenner, *Photochem. Photobiol.* **50**(6), 841 (1989).
[13] W. C. Galley and L. Stryer, *Biochemistry* **8**(5), 1831 (1969).

that competes effectively with fluorescence and with internal conversion (where electronic excitation energy is converted to heat) from the singlet state. Following intersystem crossing, radiationless decay into the ground vibrational state of T_1 results in larger energy separation and less efficient mixing of T_1 with S_1 (and hence also with S_0), reflected in the observed long radiative lifetime and weak optical absorption. We note from this discussion that in contrast to fluorescence, where the quantum yield is determined solely by the efficiency of the $S_1 \rightarrow S_0$ transition (given by the ratio of the actual fluorescence lifetime to the radiative lifetime), the intensity of phosphorescence is determined both by the intrinsic quantum efficiency of this luminescence and by the efficiency of intersystem crossing, a process whose rate varies among different tryptophans depending on their conformational environment. Hence, additional information is contained in this normalized intensity.

The observed phosphorescence lifetime combines the intrinsic radiative decay rate with terms due to nonradiative decay mechanisms and to quenching reactions:

$$1/\tau_P = k_r + k_{nr} + k_q[Q] \tag{1}$$

where τ_P is the observed phosphorescence lifetime, k_r is the intrinsic radiative decay rate, k_{nr} is the sum of the rates for the nonradiative decay mechanisms, and k_q is the quenching rate constant for any quencher with concentration $[Q]$. The term k_{nr} is a source of information on protein dynamics because the major nonradiative decay mechanism for Trp triplets is believed to be out-of-plane distortions of the excited indole ring[14] which arise from vibrational modes of the tryptophan residue. On this basis, Strambini and Gonnelli have correlated the tryptophan environment with the observed phosphorescence lifetime[15,16] and demonstrated that in the absence of quenching by specific amino acid residues (e.g., His, Tyr, Trp, cystine/cysteine[17]) the RTP lifetime can be used as a measure of the local flexibility of the tryptophan domain in proteins and that changes in the lifetime reflect changes in this rigidity and hence report on structural changes. It is this sensitivity to local rigidity that is believed to be the origin of the several orders of magnitude decrease in the RTP lifetime between Trp residues deeply buried in a protein core and those on the surface (see

[14] S. K. Lower and M. A. El-Sayed, *Chem. Rev.* **66,** 199 (1966).
[15] G. B. Strambini and M. Gonnelli, *Chem. Phys. Lett.* **115,** 196 (1985).
[16] G. B. Strambini and M. Gonnelli, *J. Am. Chem. Soc.* **117,** 7646 (1995).
[17] M. Gonnelli and G. B. Strambini, *Biochemistry* **34,** 13847 (1995).

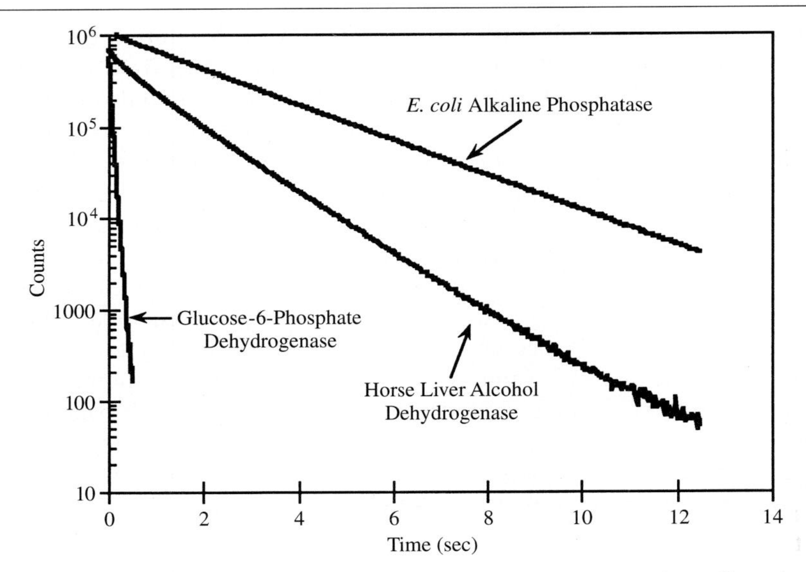

FIG. 2. Demonstration of representative phosphorescence decays of horse liver alcohol dehydrogenase, *E. coli* alkaline phosphatase, and glucose-6-phosphate dehydrogenase (*Leuconostoc mesenteroides*) at room temperature.

Fig. 2), in stark contrast to the less than an order of magnitude change in the temperature factors in X-ray crystallography that also report on rigidity, as discussed above.

Reduction of the triplet state lifetime of tryptophan in proteins can also be caused by quenching by external or internal groups as reflected in the last term of Eq. (1). Studies on the efficiency of quenching by acrylamide[18] and oxygen[19] of the triplet states of model indole compounds or of protein Trp have been conducted to ascertain the mechanism of phosphorescence quenching by an external molecule. Both electron exchange and long-range electron transfer were concluded to play significant roles.[20–23] Accessibility

[18] C. A. Ghiron, M. Bazin, and R. Santus, *Photochem. Photobiol.* **48**(4), 539 (1988).

[19] C. Ghiron, M. Bazin, and R. Santus, *Biochim. Biophys. Acta* **957,** 207 (1988).

[20] J. M. Vanderkooi, S. W. Englander, S. Papp, W. W. Wright, and C. S. Owen, *Proc. Natl. Acad. Sci. U.S.A.* **87**(13), 5099 (1990). [Published erratum appears in *Proc. Natl. Acad. Sci. U.S.A.* **87**(22), 9072 (1990)].

[21] J. M. Vanderkooi, D. B. Calhoun, C. S. Owen, S. Papp, W. W. Wright, and S. W. Englander, *Mol. Cryst. Liq. Cryst.* **194,** 209 (1991).

[22] V. Dadak, J. M. Vanderkooi, and W. W. Wright, *Biochim. Biophys. Acta* **1100**(1), 33 (1992).

[23] F. K. Klemens and D. R. McMillin, *Photochem. Photobiol.* **55**(5), 671 (1992).

of phosphorescent tryptophans to external quenchers[24–32] was employed to assign the observed quenching rates of individual decay components in a protein phosphorescence to the depth of tryptophans below the protein surface, as shown later in the chapter. Theories of phosphorescence quenching by a gated model,[33] or by other models,[34] have been developed to explain the movement of quenching molecules through a densely packed protein interior. These experiments have suggested that proteins can undergo extensive breathing motions to allow access of quenchers into protein interiors that would not be expected from the static picture derived from an X-ray crystal structure. In addition, disulfide groups may also accept electrons from the excited triplet state of tryptophan at room temperature[35,36] as has been observed in studies at cryogenic temperature.[37]

Quenching by heavy atoms[38–40] requires van der Waals contact between the quencher and the phosphorescing chromophore. Heavy atoms impact spin–orbit coupling and reduce both the radiative and observed lifetimes of the chromophore. This will result in an enhanced emission intensity because most nonradiative, competing processes are not similarly affected by the heavy atom.

As we show in the next section, competitive quenching either by dynamic quenchers or by nonradiative processes can be used to obtain information on changes in protein structure. Moreover, the use of rapid data acquisition allows one to obtain this information as a function of time, and thus to follow slow reactions such as the later states of protein folding.

[24] S. Papp, T. E. King, and J. M. Vanderkooi, *FEBS Lett.* **283**(1), 113 (1991).
[25] G. B. Strambini, *Biophys. J.* **52**(1), 23 (1987).
[26] D. B. Calhoun, J. M. Vanderkooi, and S. W. Englander, *Biochemistry* **22**(7), 1533 (1983).
[27] D. B. Calhoun, J. M. Vanderkooi, G. V. D. Woodrow, and S. W. Englander, *Biochemistry* **22**(7), 1526 (1983).
[28] D. B. Calhoun, S. W. Englander, W. W. Wright, and J. M. Vanderkooi, *Biochemistry* **27**(22), 8466 (1988).
[29] S. Papp, J. M. Vanderkooi, C. S. Owen, G. R. Holtom, and C. M. Phillips, *Biophys. J.* **58,** 177 (1990).
[30] G. B. Strambini, *Biophys. J.* **43**(1), 127 (1983).
[31] J. M. Vanderkooi, W. W. Wright, and M. Erecinska, *Biochim. Biophys. Acta* **1207**(2), 249 (1994).
[32] W. W. Wright, C. S. Owen, and J. M. Vanderkooi, *Biochemistry* **31**(28), 6538 (1992).
[33] B. Somogyi, J. A. Norman, and A. Rosenberg, *Biophys. J.* **50**(1), 55 (1986).
[34] J. M. Vanderkooi, C. S. Owen, and W. W. Wright, *Proc. S.P.I.E. Int. Soc. Opt. Eng.* **1640,** 473 (1992).
[35] Z. Li, W. E. Lee, and W. C. Galley, *Biophys. J.* **56**(2), 361 (1989).
[36] Z. Li, A. Bruce, and W. C. Galley, *Biophys. J.* **61,** 1364 (1992).
[37] B. D. Schlyer, E. Lau, and A. H. Maki, *Biochemistry* **31**(18), 4375 (1992).
[38] M. L. Meyers and P. G. Seybold, *Anal. Chem.* **51,** 1609 (1979).
[39] M. Monsigny, F. Delmotte, and C. Helene, *Proc. Natl. Acad. Sci. U.S.A.* **75**(3), 1324 (1978).
[40] G. B. Srambini and E. Gabellieri, *J. Phys. Chem.* **95,** 4352 (1991).

Applications to Studies of Protein Structure and Dynamics

Heterogeneity and Slow Conformational Changes of Proteins. Numerous studies have shown the fluorescence from individual tryptophans in proteins to decay multiexponentially.[41] Explanations for this behavior include the existence of tryptophan rotamers[42,43] and of excited state interactions of the indole chromophore with its environment.[44] Conformational heterogeneity of the host protein has also been inferred from fluorescence studies,[45] which can be interpreted in terms of the energy landscape model of Frauenfelder *et al.*[46] This model predicts the occurrence of numerous shallow local free energy minima in the conformational space of a protein. The rate of interconversion of these conformations depends on the temperature and on the height of the activation energy barriers separating them, which can be derived from an Arrhenius plot.

On the basis of earlier studies, it was assumed that the time scale for interconversion among the different protein conformers in fluid solutions at room temperature would be short compared to the RTP lifetime, hence suggesting a monoexponential decay for the RTP. Indeed, early studies of RTP from proteins containing single emitting tryptophans reported single exponential decays; however, as the quality of the data improved, with better acquisition techniques, it became clear that as a rule RTP decays are multiexponential. Extensively studied examples include liver alcohol dehydrogenase (LADH) and alkaline phosphatase (AP),[47,48] whose RTP decays have been analyzed using either a sum of discrete exponentials or a distribution of decay components (Fig. 3). The observation that both proteins show clear nonexponential decay demonstrates unexpectedly long-lived conformational heterogeneity in either or both the excited and the ground state. One criterion on which to assess the occurrence of ground-state conformational heterogeneity is by observing a correlation between

[41] S.-J. Kim, F. N. Chowdhury, W. Stryjewski, E. S. Younathan, P. S. Russo, and M. D. Barkley, *Biophys. J.* **65,** 215 (1993).

[42] R. F. Chen, J. R. Knutson, H. Ziffer, and D. Porter, *Biochemistry* **30**(21), 5184 (1991).

[43] J. B. Ross, H. R. Wyssbrod, R. A. Porter, G. P. Schwartz, C. A. Michaels, and W. R. Laws, *Biochemistry* **31**(6), 1585 (1992).

[44] Z. Bajzer and F. G. Prendergast, *Biophys. J.* **65,** 2313 (1993).

[45] I. Gryczynski, M. Eftink, and J. R. Lakowicz, *Biochim. Biophys. Acta* **954,** 244 (1988).

[46] H. Frauenfelder, S. G. Sligar, and P. G. Wolynes, *Science* **254,** 1598 (1991).

[47] B. D. Schlyer, J. A. Schauerte, D. G. Steel, and A. Gafni, The nonexponential decay of room temperature phosphorescence: Evidence for several slowly interconverting or static protein conformers. *In* "Time-Resolved Laser Spectroscopy in Biochemistry," Vol. IV. SPIE, Los Angeles, 1994.

[48] B. D. Schlyer, J. A. Schauerte, D. G. Steel, and A. Gafni, *Biophys. J.* **67**(3), 1192 (1994).

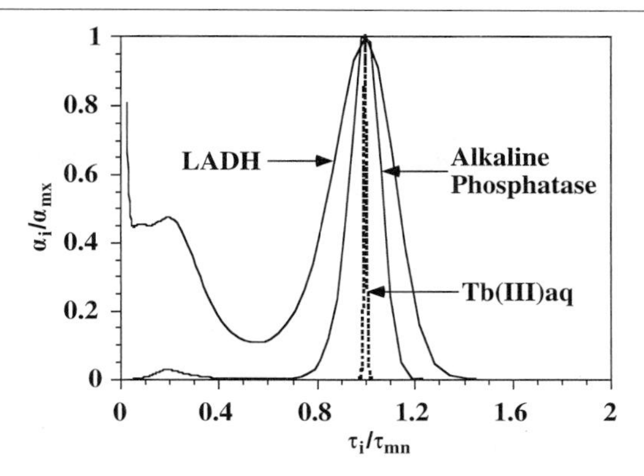

FIG. 3. Comparison of maximum entropy method (MEM) distribution fits of the phosphorescence decays of LADH, AP, and the luminescence of aqueous Tb(III) (excitation at 290 nm; emission observed at 545 nm). Distribution profiles are normalized to the lifetime of the decay [LADH mean lifetime, 712 msec; AP mean lifetime, 1920 msec; Tb(III)aq mean lifetime, 440 μsec]. The normalized distribution width (full width at half-maximum) for LADH (main peak) is 30%, while AP has a width of 15% and Tb(III)aq has a width of 2%. (See Schlyer *et al.*[48])

the excitation wavelength and the phosphorescence lifetime as was done for LADH.[48] These experiments are similar to hole-burning experiments in establishing inhomogeneous line broadening.

Unfolding and Refolding of Proteins: Detection of Transient Intermediate States. The high sensitivity of RTP to protein conformational states makes this technique especially useful in investigating the relationship between structure and biological activity. Enzymatic activity requires the precise interaction between the substrate and the active site of an enzyme. In addition, reaching the transition state of the enzymatic reaction may be facilitated by concerted dynamic movements of the binding and enzymatic domains of the proteins to achieve the reduction in the activation barrier for the reaction being catalyzed. Therefore the efficiency of enzymatic activity may be due to both structural and dynamic terms, information reflected in RTP. Indeed, studies using TR-RTP to follow protein unfolding[49] and refolding[50] showed the presence of active intermediate states. In the latter case, biological activity was recovered while TR-RTP suggested a less rigid core environment that slowly returned to the native state, a result supported by chemical lability studies.

[49] J. V. Mersol, D. G. Steel, and A. Gafni, *Biophys. Chem.* **48**(2), 281 (1993).

[50] V. Subramaniam, N. C. Bergenhem, A. Gafni, and D. G. Steel, *Biochemistry* **34**(4), 1133 (1995).

Detection of Changes in Protein Conformation. Room temperature phosphorescence has been used to investigate such diverse phenomena as the effect of pressure on proteins[51]; to study structural aspects of proteins in micells[52]; to derive information about the effect of solvent microviscosity on protein structure, protein flexibility, and dynamics[7,53–61]; as well as to establish a relationship between protein dynamics and enzyme function.[62,63] Time-resolved RTP was also used to determine the nature of a pH-dependent conformational change in azurin owing to the protonation of a specific His residue.[64] The TR-RTP revealed that the more active conformation was associated with a less rigid form of the protein involved in electron transport.

Phosphorescence Quenching Experiments: Application to Distance Determination. Energy transfer from the triplet state of tryptophan to Tb^{3+} bound to a metal-binding site on the protein[65] can be studied through sensitized Tb^{3+} luminescence.[66] This energy transfer (either by Dexter or Förster mechanism depending on distance) may provide extremely sensitive information on changes in donor–acceptor distances, potentially in real time, to allow detailed structural changes in solution to be followed during slow structural changes.

In another approach to distance determination, energy transfer can be monitored by measuring the change in the RTP lifetime of a buried phosphorescing Trp owing to quenching by freely diffusing acceptors in solution. This interaction becomes effective and relatively straightforward to interpret when the energy transfer is by dipole–dipole coupling and the RTP lifetime is long compared to the characteristic diffusion time of the acceptors. Energy transfer in this so-called rapid diffusion limit has been

[51] P. Cioni and G. B. Strambini, *J. Mol. Biol.* **242,** 291 (1994).
[52] M. Gonnelli and G. B. Strambini, *J. Phys. Chem.* **92,** 2854 (1988).
[53] A. Ansari, C. M. Jones, E. R. Henry, J. Hofrichter, and W. A. Eaton, *Science* **256,** 1796 (1992).
[54] D. B. Calhoun and S. W. Englander, *Biochemistry* **24,** 2095 (1985).
[55] M. Gonnelli, A. Puntoni, and G. B. Strambini, *J. Fluoresc.* **2,** 157 (1992).
[56] M. Gonnelli and G. B. Strambini, *Biophys. J.* **65,** 131 (1993).
[57] S. Kishner, E. Trepman, and W. C. Galley, *Can. J. Biochem.* **57**(11), 1299 (1979).
[58] J. V. Mersol, A. Gershenson, D. G. Steel, and A. Gafni, *Proc. S.P.I.E. Int. Soc. Opt. Eng.* **1640,** 462 (1992).
[59] G. B. Strambini and M. Gonelli, *Biochemistry* **25,** 2471 (1986).
[60] G. B. Strambini, *J. Mol. Liq.* **42,** 155 (1989).
[61] E. Gabellieri and G. B. Strambini, *Eur. J. Biochem.* **221**(1), 77 (1994).
[62] B. Gavish and M. M. Werber, *Biochemistry* **18**(7), 1269 (1979).
[63] B. Gavish, Molecular dynamics and the transient strain model of enzyme catalysis. *In* "The Fluctuating Enzyme" (G. R. Welch, ed.), pp. 263–340. Wiley Interscience, New York, 1986.
[64] J. Hansen, D. G. Steel, and A. Gafni, *Biophys. J.* **71**(4), 2138–2143 (1996).
[65] P. Cioni and G. B. Strambini, *J. Photochem. Photobiol. B: Biol.* **13,** 289 (1992).
[66] B. D. Schlyer, D. G. Steel, and A. Gafni, *J. Biol. Chem.* **270**(39), 22890 (1995).

used to obtain the distance of closest approach between external acceptors (quenchers) and donor chromophores in proteins (i.e., the distance from the surface of the protein to the donor chromophore).[67–70] Hence, the long lifetime of RTP facilitates this measurement since it enables the use of an intrinsic lumiphore, eliminating the complications and ambiguities that arise from the use of extrinsic labels. Complications arise if the quenching is not by Förster energy transfer but, for example, by electron transfer from the donor to the acceptor molecule, an interaction whose dependence on donor–acceptor distance is different.

Use of Time-Resolved Phosphorescence in in Vivo Studies

The long-lived RTP of some proteins provides a unique noninvasive approach allowing one to detect and study these proteins *in vivo,* even in the presence of strong (but short-lived) background luminescence from other cellular components. Horie and Vanderkooi[71] first reported that the long-lived phosphorescence of *Escherichia coli* AP could be easily monitored in a deoxygenated suspension of lyophilized *E. coli* cells by gating out the luminescence from other cellular proteins. Current work in our laboratory uses alkaline phosphatase RTP to study the expression and folding of the enzyme in live *E. coli* following the initiation of its transcription. This study is expected to provide unique information about factors that regulate folding *in vivo.*

Experimental Considerations

Light Sources and Phosphorescence Detection

Since the excitation of the triplet state, from which phosphorescence is emitted, involves intersystem crossing from the excited singlet state, the induction of tryptophan phosphorescence requires the same excitation wavelengths as that of fluorescence, namely in the range of 280–300 nm.

The relatively low intensity of tryptophan phosphorescence relative to fluorescence, owing to the low intersystem crossing yield and the existence of effective triplet state-quenching processes, makes high intensity of excitation light an important consideration. Steady state light sources with extended wavelengths that reach ca. 290 nm include xenon/xenon–mercury

[67] J. V. Mersol, H. Wang, D. G. Steel, and A. Gafni, *Biophys. J.* **61,** 1647 (1992).
[68] J. V. Mersol, D. G. Steel, and A. Gafni, *Biochemistry* **30**(3), 668 (1991).
[69] A. Gafni, J. V. Mersol, and D. G. Steel, *Proc. S.P.I.E. Int. Soc. Opt. Eng.* **1204,** 753 (1990).
[70] U. Gösele, M. Hansen, U. K. A. Klein, and R. Frey, *Chem. Phys. Lett.* **34,** 519 (1975).
[71] T. Horie and J. M. Vanderkooi, *Biochim. Biophys. Acta* **670**(2), 294 (1981).

arc lamps, which are commercially available from Oriel (Stratford, CT), Newport (Irvine, CA), and other sources. Depending on the volume of the sample to be measured, there is a trade-off between total intensity available from high-energy sources such as a 450-W xenon lamp with a large arc size, and the lower output from weaker, but usually brighter, sources such as a 100-W xenon lamp with a small arc size.

An ultraviolet (UV) lens system can be used to focus the excitation light onto the sample. In this case, the excitation wavelength can be controlled by bandpass filters or by UV interference filters (Schott, Duryea, PA; Corion, Holliston, MA; Oriel). One convenient feature of an interference filter is that small rotations of its surface relative to the excitation direction allow wavelengths around the maximum to be selected in a predictable fashion. These filters can be purchased with the bandpass specified for 5 nm, 10 nm, or larger. If a monochromator is to be used to select the excitation wavelength from an extended light source, then attenuation of excitation intensity by one to two orders of magnitude should be expected depending on alignment.

The phosphorescence spectrum of Trp is red-shifted relative to its fluorescence, and has a clear vibronic structure (see Fig. 4); however, because it is also considerably weaker its detection is hindered by the background fluorescence. For effective collection of phosphorescence data, therefore, the excitation source must be pulsed and a mechanical, electrooptical, or electronic gating system (see following discussion) should be used to eliminate the short-lived (tens of nanoseconds) fluorescence contribution, thus allowing for the observation of pure phosphorescence signal. Pulsed excita-

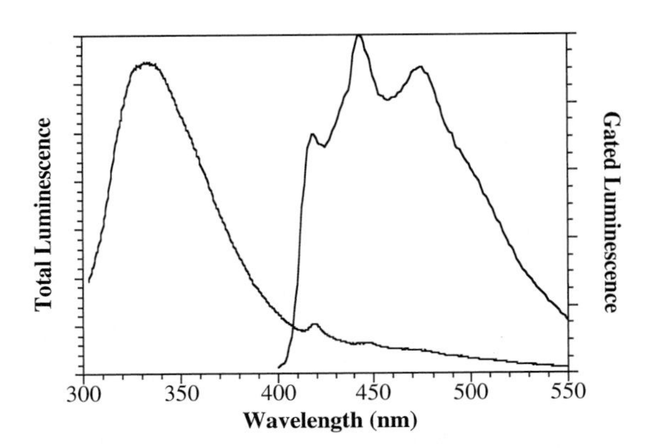

FIG. 4. The corrected steady state total luminescence spectra of alkaline phosphatase (AP) before and after gating away the fluorescence emission (>100 nsec following excitation). Excitation was at 290 nm, and the spectral bandpass was 8 nm. The fluorescence emission was obtained at 10°, where the phosphorescence lifetime of AP is approximately 3.0 sec. The relative contribution of AP phosphorescence is attenuated at higher temperatures.

tion is also used to excite RTP for time-resolved phosphorescence spectroscopy.

The long decay time of Trp phosphorescence (tens of microseconds or longer) makes the demand on pulse excitation duration less critical than in fluorescence decay measurements, and a variety of pulsed sources including pulsed lasers can be used. The elimination of fluorescence from steady state emission spectra is accomplished by gating the emission throughput or the detection system, usually by using a gated photomultiplier tube (PMT) base (model PR1400RF; Products for Research, Danvers, MA), or an optical chopper and a steady state light source. The gated window can be created with a delay gate generator (model 535; Stanford Research, Sunnyvale, CA) and a fast shutter (model SD-10/US14S1T0; Uniblitz, Rochester, NY). The excitation pulse triggers the delay gate generator to open the shutter after a preset dwell time (to remove the excitation pulse and any spurious luminescence). The preferred technique is to use a fast gated preamplifier or discriminator to gate out the fluorescence emission. This is appropriate for photon-counting methods because even if the fluorescence is three to four orders of magnitude stronger than the phosphorescence, the bandwidth limitation for phosphorescence detection (see "Potential Artifacts") requires that the total counts (and therefore the current) be substantially less than the maximum current limitation of most PMTs. Therefore the recovery of the PMT from the fluorescence pulse should occur on the submicrosecond time scale.

While commercial systems are available for measuring steady state phosphorescence spectra (PTI, South Brunswick, NJ; Spex Fluorolog, Edison, NJ), one can assemble the necessary components modularly and achieve good performance. Figure 5 shows an instrument that can be used for either photon counting or analog measurements of time-resolved as well as of steady state phosphorescence spectra (gating out the fluorescence emission). One does not need to reproduce the entire setup, as is discussed, if a *Q*-switched laser source is used instead of a steady state source. To eliminate the fluorescence emission, one takes advantage of its short duration when excited by *Q*-switched lasers (10-nsec pulses, 10–20 nsec of fluorescence emission). In the case of steady state sources, an optical chopper (Stanford Research Systems, Sunnyvale, CA) can modulate the excitation and be followed by a fast shutter (Uniblitz) to gate the fluorescence. There is a limit to the time resolution that these mechanical gates provide (nominally milliseconds).

Laser light sources for UV excitation provide both improved focusing ability for small samples and more intense, monochromatic light, but at higher expense. Typical sources used to provide UV excitation are based on *Q*-switched Nd : YAG lasers. Lasers currently utilized [such as the Spectra-Physics (Mountain View, CA) DCR-11 and Continuum (Santa Clara, CA)

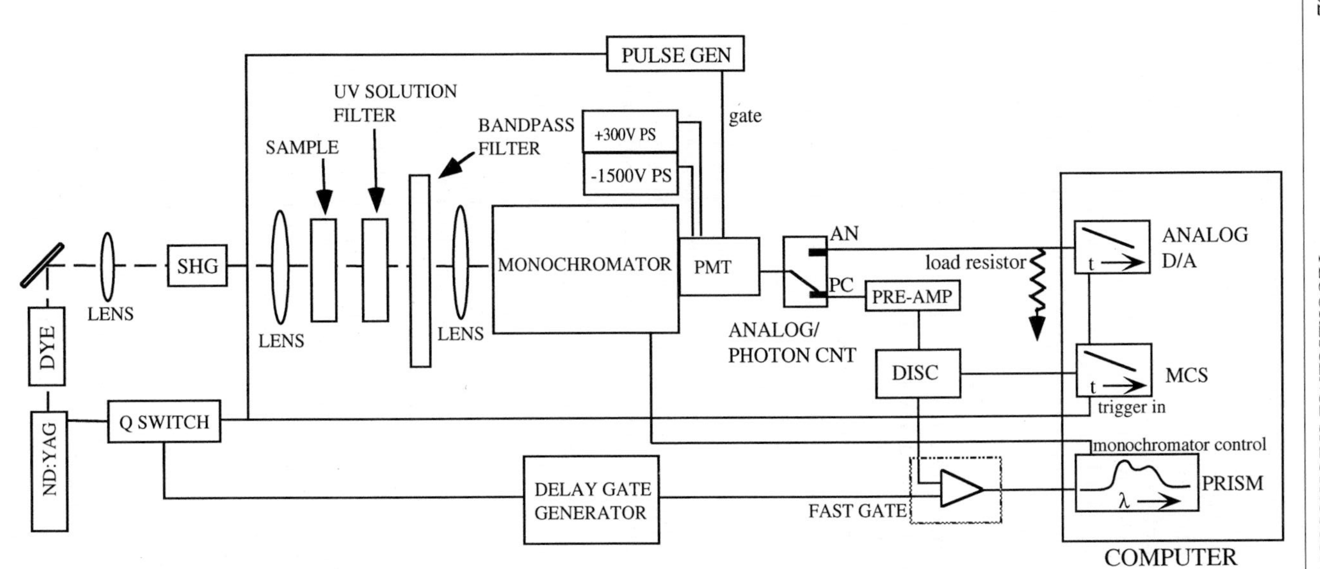

FIG. 5. Instrument for the measurement of room temperature phosphorescence lifetime and emission spectra. Both analog and photon-counting acquisition capability are shown, although photon counting is advantageous for weakly emitting protein samples ($<1\ \mu M$). For dilute systems, the monochromator is replaced by the bandpass filter (Schott KV400) to select the appropriate wavelength region. Phosphorescence spectra can be derived by gating a PMT (Products for Research model PR1400RF) and scanning a monochromator (Instruments SA HR320) with Prism software. A fast preamplifier located directly on the PMT reduces noise (EG&G Ortec VT120), and leads to a discriminator (EG&G Ortec 436, 100-MHz discriminator), which provides the input to the multichannel scaler (EG&G Otec ACEMCS, 2-μsec minimum dwell time) with acquisition software. For decays from dilute samples, there is no gating of the PMT because we have observed that the fluorescence from the sample does not cause the PMT to exceed its rated current maximum.

Surelite] employ low repetition rates (0–10 Hz) with high infrared output power (at 1064 nm) that is frequency doubled to 532 nm (150 mJ/pulse with 10-nsec pulse width). The 532-nm output is used to drive a dye laser (e.g., the Spectra-Physics PDL-3 or the Continuum dye laser) using Rhodamine-6G dye. The output of the dye laser at 570–600 nm (wavelength range depending on dye used) is compressed to a smaller, more intense beam using a two-lens inverted telescope, and is again frequency doubled [Inrad SHG (Northvale, NJ), KDP (potassium dihydrogen phosphate) crystal model R6G] to provide UV excitation in the region of 280–300 nm. The output of the dye laser at 580 nm is approximately 30 mJ/pulse and the UV intensity is approximately 5 mJ/pulse. The beam focusing previously described improves frequency-doubling efficiency; however, because the focused output of these lasers is capable of exceeding the damage threshold of KDP crystals (at 450 MW/cm^2), the SHG crystal should not be at the focus of a lens. Alternatively, one can use crystals such as BBO (β-barium borate) with a damage threshold nominally 20 times larger and a SHG efficiency approximately 4 times larger.

Considerable care must be taken to reduce contributions to the signal from spurious sources of light. Long-lived stray luminescence from quartz cuvettes, lenses, and painted/anodized surfaces can interfere with phosphorescence measurements. Since triplet state emission yields improve dramatically in the solid state, paints and dyes on surfaces and tapes can provide a significant background to protein luminescence studies. Also, filters used in selecting excitation and emission wavelengths often produce luminescence with lifetimes on the order of hundreds of microseconds. Special low-luminescence filters can be purchased (Schott). Also, solution filters suggested by Parker,[72] such as aqueous potassium nitrite, can be used to absorb the UV excitation light once it has passed through the sample to avoid excitation of other components within the sample chamber. It should also be noted that high-excitation intensity of focused laser beams can induce color centers in quartz that may in turn produce luminescence. Depending on the positioning of the cuvette within the excitation beam, such color centers can provide sporadic and irreproducible luminescence background.

Analog versus Photon-Counting Detection. Measurements of luminescence decay in the microsecond to millisecond regime and longer allow for a choice of analog or photon-counting procedures that differ in a number of aspects. For systems with weak luminescence intensity, photon counting is preferable because it allows one to detect a lower level of light, down to

[72] C. A. Parker, "Photoluminescence of Solutions: With Applications to Photochemistry and Analytical Chemistry." Elsevier, Amsterdam, 1968.

individual photons. Moreover, the analysis of luminescence decays derived from photon-counting procedures can benefit from the well-established statistical procedures (see "Data Analysis") based on Poisson statistics. However, for higher levels of light, photon counting is prone to bandwidth artifact (bandwidth distortion) where the data are distorted owing to the inability of the detection electronics to record all photons. Analog detection, which reduces this artifact as detection is limited only by the current output of the PMT (usually 100 μA for short periods of time), is therefore preferred when the light intensity is high. However, the linearity of the PMT response to light intensity must be assured, as its lack may result in distortion of the decay.

When the luminescence decay time is above ca. 100 nsec, photon counting is typically done with a multichannel scaler (MCS), which records the number of photons detected by the PMT as a function of time. Using the excitation source as a trigger, the MCS counts photons for a prespecified dwell time, subsequently inputting the total counts observed during that time into a channel. Photons counted in sequential dwell periods following excitation are stored in sequential channels and form the luminescence decay profile. Multichannel scaler cards are currently available with either 5-nsec or 2-μsec minimum dwell times, where the dwell time can be adjusted up to many seconds (Stanford Research; EG&G Ortec, Oak Ridge, TN). Photon-counting systems can typically be rated at 50–200 MHz. However, it is necessary to consider the limiting bandwidth of all the individual components in the system, and the bandwidth of the combination of components. The preamplifier following the PMT, the discriminator, and the MCS will all combine to provide a photon-counting throughput that is lower than the rated levels if connections between the electronic elements provide stray capacitance. It is necessary to measure the characteristic bandwidth of the entire setup by directly exposing the system to a constant flux of light at an intensity that does not harm the PMT and determine the maximum counting rate.

Deoxygenation Techniques. The excited triplet state of tryptophan is efficiently deactivated by molecular oxygen, whose ground electronic state is a triplet state, hence deoxygenation of the sample is crucial to achieve consistent phosphorescence lifetimes. The importance of oxygen removal can be demonstrated by calculating its effect on the phosphorescence lifetime of alkaline phosphatase, which is approximately 2 sec under deoxygenated conditions, and where the oxygen quenching rate constant has been reported to be $1.2 \times 10^6\ M^{-1}\ \text{sec}^{-1}$.[25] Using this value in Eq. (1) shows that at a concentration of 1 μM oxygen will attenuate the phosphorescence lifetime of AP from 2 to 0.6 sec. Because the concentration of oxygen in air-equilibrated water is approximately 250 μM, it must be reduced by four

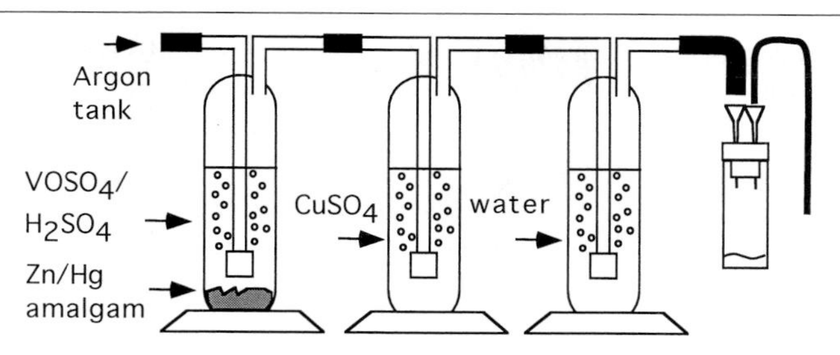

FIG. 6. Further deoxygenation of purified argon is done according to the procedure described by Englander *et al.*[73] Short segments of butyl-rubber tubing connect flasks containing $VOSO_4$ (0.1 *M*) with several milliliters of H_2SO_4 covering 2 cm of amalgamated zinc. The vanadium reacts with the oxygen and is subsequently replenished by the zinc. The vanadyl sulfate is a deep purple color in the appropriate redox state. It turns a light blue when the vanadium is oxidized (adding additional H_2SO_4 will regenerate the original color until the zinc is used up). Minimal contact with tubing is necessary to ensure oxygen-free argon.

orders of magnitude for alkaline phosphatase to be within 5% of its true lifetime. This is commonly done by purging the sample with an inert gas, usually argon. However, even prepurified argon (99.99%) must be further purified to remove traces of oxygen. This may be achieved with the commercial Oxyclear (Labclear, Oakland, CA) or by using the vanadyl sulfate–zinc amalgam system described by Englander *et al.*[73] (see Fig. 6). The color changes from purple to blue in the first flask, indicating when the reagents need to be replenished. The copper sulfate in the second flask precipitates any H_2S generated in the first flask and serves to protect the protein sample from this efficient phosphorescence-quenching agent. The last flask contains water to ensure the argon used for purging is water saturated to avoid evaporating the sample.

Either a slow exchange of argon by continuous purging for several hours, or a more rapid exchange by cycling between high-pressure argon and partial vacuum, can be used for sample deoxygenation. An automated system using sequential cycling between a partial vacuum and high-purity argon, at the rate of approximately 15 sec/cycle, can facilitate sample deoxygenation for phosphorescence measurements. Complete deoxygenation by this rapid exchange can be achieved within 15 min. Three-way gas valves (Cole-Parmer, Niles, IL) controllable by a timer (GraLab; Dimco-Gray, Centerville, OH) are used to provide a highly reproducible deoxygenation procedure. Also, one must be careful when introducing a vacuum to a protein sample to ensure that gas bubbles do not form in the solution, as this will denature the protein. A safer, but slower, technique is to purge the sample with argon over a period of 2–3 hr or longer.

[73] S. W. Englander, D. B. Calhoun, and J. J. Englander, *Anal. Biochem.* **161,** 300 (1987).

Additional deoxygenation techniques that have been described include the use of dithionite, or of enzymatic systems such as a mixture of glucose oxidase and catalase.[73] However, dithionite frequently damages the protein, while glucose oxidase is itself phosphorescent (120-msec lifetime) and its contribution to the decay data must be subtracted from the sample of interest. In our experience, an argon purge is the most consistent technique for deoxygenation.

One centimeter square fluorescence cuvettes from Uvonics (Plainview, NY) or Precision (Framingdale, NY) with blood-serum stoppers are used to contain the sample. Because a tight seal is necessary, a small amount of silicon grease can be used to facilitate insertion of the rubber stoppers. Hypodermic needles (18 to 24 gauge) inserted through the septum provide access for the deoxygenation of the sample. The use of split-cell cuvettes allows the simultaneous deoxygenation of two solutions. Refolding studies of denatured proteins are often studied in our laboratory by simultaneous deoxygenation of both the denatured protein solution and the refolding buffer, with the mixing step, to initiate refolding, being done under deoxygenated conditions.

It is well known that UV irradiation can photochemically damage protein samples (through oxidation of aromatic amino acids and other residues). This is another important reason to deoxygenate the sample fully before data acquisition. In addition, it is important to assess the effect of UV irradiation on the lifetime of the phosphorescence. If the sample is not fully deoxygenated, then its phosphorescence lifetime will gradually increase with excitation time (as oxygen is being depleted by reacting with the protein) and result in a broader distribution of luminescence lifetimes. If prolonged data acquisition from a phosphorescent sample is performed, then the lifetime derived from first few excitation pulses should be compared with that derived from the final shots to ensure lack of change. Ultraviolet excitation that is more defocused is less likely to deoxygenate a defined volume to result in distortion of the data.

It is possible to assess the efficiency of the deoxygenation procedure by determining the lifetime of a standard as suggested by Vanderkooi *et al.*,[74] using the phosphorescence lifetime standards porphine and coproporphyrin derivatives.

Potential Artifacts. It is important to understand the sources of instrumental distortions when photon-counting techniques are used for acquisition of luminescence decay data. Fluorescence decay data gathered by time-correlated single photon-counting (TCSPC) techniques as used in

[74] J. M. Vanderkooi, G. Maniara, T. J. Green, and D. F. Wilson, *J. Biol. Chem.* **262**(12), 5476 (1987).

nanosecond time-resolved fluorescence measurements can be distorted by pulse pile-up, which is the result of failure to record a fraction of the fluorescence photons owing to a high data acquisition rate relative to the rate of excitation pulses of the sample (see Ware[75]). This distortion introduces a false fast decaying component.

Multichannel scalers used in phosphorescence decay acquisitions are capable of distorting data in another manner. Since the MCS requires a certain amount of time to process each event (1/bandwidth = dead time), it will ignore events (photons) that are registered at the PMT during that time. If these events are equally spaced in time, the MCS will faithfully register all photons until their arrival rate is increased beyond the bandwidth of the MCS. However, if the events are uncorrelated in time, as are the photons from a phosphorescent sample, then some counts will be ignored even when the average acquisition rate is below the bandwidth of the system. This distortion results in a downward curvature of the luminescence decay curve, such that analysis of this luminescence assuming a sum of exponential decays will result in the introduction of a false-negative preexponential term into the results.

Correction of this distortion is based on accounting for the probability that an event is missed during the processing of the pulses generated by the PMT, the probability of which depends on the photon flux. Statistically, for any given photon flux, the distribution of the time intervals between uncorrelated photons, is logarithmic in nature.[76] This results in a predictable number of events that will occur during the dead time of the system subsequent to a previously recorded photon.

$$\text{Correction} = 1/(1 + rd) \tag{2}$$

where r is the real count rate and d is the dead time of the counting device. For uncorrelated emission, this correction (see Loudon[76]) already becomes significant when the counting rates are far below the maximum rated bandwidth; hence the maximum count rate must be adjusted to avoid a distortion of the decay.

The correction of the observed count rate at any point in the decay depends on the intensity of light at that time point. For example, if a single laser pulse produces 10,000 counts in the peak channel with a dwell time of 10 msec, the average count rate during that time is 10,000/0.01 = 106 counts/sec. The correction factor for a 100-MHz counting system would be 2%. Once the luminescence has decayed by an order of magnitude (1000

[75] W. R. Ware, "Time-Resolved Fluorescence Spectroscopy in Biochemistry and Biology" (R. B. Cundall and R. E. Dale, eds.), pp. 23–58. Plenum Press, New York, 1983.

[76] R. Loudon, "The Quantum Theory of Light," 2nd Ed. Clarendon Press, Oxford, 1983.

counts in a 10-msec channel), the correction is well below the Poisson noise. Therefore, early channels may need a much larger correction than do subsequent channels.

Artifacts may also occur in the decay data from photochemical modification of the sample owing to the reactivity of excited-state chromophores generated under high-energy UV irradiation. Tryptophan residues exposed to UV irradiation can develop photoproducts with new emission properties.[77–82] This process is generally inhibited under deoxygenated conditions; however, it has been observed to occur anaerobically with irradiated α-crystallin.[83]

Another artifact that affects the derived phosphorescence decay constants, and that is not a problem with fluorescence decay measurements, is due to the long lifetime of the triplet state, which makes it possible to excite a significant fraction of the molecules out of the ground state. The intensity and duration of the excitation pulse will impact the characteristics of the excited-state population in the following way: an excitation pulse that is short compared with the phosphoresce decay time will excite all representative components in the ground-state population (depending on differences in absorption coefficient within the ensemble). However, excitation of the sample by a long pulse (relative to the decay time) followed by a shuttered observation of the phosphorescence will distort the excited-state population toward the longest lived components. Therefore, the lifetime recorded from a heterogeneous sample will depend in part on the nature of the excitation (intensity and length of time of irradiation). Incomplete sample deoxygenation can compound this effect as previously discussed.

Another point of practical significance is that the concentrations of buffers and salts should be limited to the lowest practical levels to avoid the possibility of contamination of the sample by triplet state quenchers. For example, in a 100 mM salt solution, a 1-ppm contamination will yield a quencher concentration of 0.1 μM, and assuming a quenching rate constant of 10^9 M^{-1} sec^{-1} (as shown by iodide ions against buried tryptophans of LADH[26]), this will quench a 1-sec phosphorescence lifetime to 10 msec. Therefore, the purity of buffers and salts is of critical importance in phosphorescence experiments, especially for tryptophan residues that are more exposed to solvent.

[77] K. M. L'Vov, S. V. Kuznetsov, and A. P. Kostikov, *Biofizika* **38,** 574 (1993).
[78] K. M. L'Vov, S. V. Kuznetsov, and A. P. Kostikov, *Biofizika* **38,** 568 (1993).
[79] S. Lerman, *Am. J. Optom. Physiol. Opt.* **64**(1), 11 (1987).
[80] J. L. Redpath, R. Santus, J. Ovadia, and L. I. Grossweiner, *Int. J. Radiat. Biol.* **27**(2), 201 (1975).
[81] E. I. Rivas, A. J. Paladini, and G. Cilento, *Photochem. Photobiol.* **40,** 565 (1984).
[82] T. Thu Ba, *J. Chem. Phys.* **70,** 3544 (1979).
[83] J. W. Berger and J. M. Vanderkooi, *Photochem. Photobiol.* **52**(4), 855 (1990).

Optical densities of samples in luminescence experiments are generally kept low to avoid the inner filter effect. Also, it is important that the protein solutions are clear, as protein aggregation can produce greatly modified phosphorescence signals. Rapid deoxygenation techniques previously described can result in protein aggregation in cases where the pressure changes applied to the protein solution are so rapid as to result in bubbling of the solution. The high surface tension of bubbles can denature proteins and lead to their aggregation.

Data Analysis

Data acquired in the photon-counting mode is well defined by Poisson statistics and allows one to assess the certainty in the parameters derived. This has been well covered in a number of reviews on data analysis including Johnson and Faunt.[84] Several techniques exist for recovering the parameters from luminescence decays, and approaches that have been developed for fluorescence decay analysis can be utilized for phosphorescence decays. The simplest modeling of the data assumes the minimum number of independent first-order decay components to fit time-resolved luminescence and should therefore be the first choice in data analysis. This assumes that the individual emitters in the sample are independently radiating, an assumption that is not true in cases of solvent reorientation (more important for fluorescence), excited-state reactions, or energy transfer between chromophores.

If a decay is truly monoexponential, then regardless of the number of counts collected in the experiment, a single exponential term will always provide a good fit to the data by the criteria of χ^2 (see Bevington[85]). However, if as the number of counts increases the analysis requires more components to achieve an acceptable fit (as judged by the χ^2), then either the sample does not decay truly monoexponentially or there is a systematic error in the acquisition electronics that is distorting the luminescence decay curve. Vix and Lami have discussed the issue of compound Poisson decay statistics for narrow lifetime distributions.[86] If it is important to reliably recover narrow lifetime distributions from decays with a high number of counts, high-purity standards with monoexponential decays (i.e., terbium dipicolinate complex, with a 3.5-msec lifetime) should be utilized to ensure that the instrument is capable of faithfully recording a monoexponential decay.

[84] M. L. Johnson and L. M. Faunt, *Methods Enzymol.* **210,** 1 (1992).

[85] P. R. Bevington, "Data Reduction and Error Analysis for the Physical Sciences." McGraw-Hill, New York, 1969.

[86] A. Vix and H. Lami, *Biophys. J.* **68,** 1145 (1995).

Both the exponential series method (ESM) and the maximum entropy method (MEM) utilize a finite lifetime distribution profile to fit the luminescence decay data. A detailed description of these techniques is beyond the scope of this chapter; however, the reader is referred to previous reviews for a detailed description.[87] It is best to use these techniques in conjunction with the simpler, independent first-order decay model. However, the number of independent first-order decay components necessary to fit the data will depend on the number of counts in the luminescence decay, and will increase in complex systems, as the number of counts in the decay increases. Moreover, each new component will alter the magnitude and location of the previous components recovered by the analysis, distorting the correlation between the decay components and the species emitting the luminescence. However, distribution methods provide a more continuous envelope of lifetimes that is less sensitive to the number of counts in the decay curve.

In attempting to recover lifetime distributions by either of these techniques, it is important to realize that when the distribution narrows the slope in χ^2 space begins to broaden. Therefore, the uncertainty in the width of a narrow distribution should be assessed by techniques such as the Monte Carlo procedure discussed by Johnson and Faunt.[84] The most complete assessment of errors is provided by generating a large number of decays, each with randomly generated standard Poisson statistical error on the given luminescence decay solution. Each decay is analyzed and the distribution results are assessed.

Global analysis procedures (see Knutson *et al.*[88]) are also applicable to phosphorescence decay analysis. However, linkage of decays observed at different emission wavelengths does not provide much new information as the phosphorescence decays are not significantly wavelength dependent. Linkage with excitation wavelength, in contrast, may provide the necessary independent information to improve resolution. Extensive discussion of various aspects of numerical analysis of luminescence decays can be found in Ref. 89.

Summary

The application of luminescence, primarily fluorescence, to the study of protein structure and dynamics has been extensively exploited to facilitate the understanding of complex biological problems. The interest in the

[87] J. H. Justice (Ed.), "Maximum Entropy and Bayesian Methods in Applied Statistics." Cambridge University Press, Cambridge, 1986.

[88] J. R. Knutson, J. M. Beechem, and L. Brand, *Chem. Phys. Lett.* **102**(6), 501 (1983).

[89] L. Brand and M. L. Johnson, *Methods Enzymol.* **210,** (1992).

application of phosphorescence, however, shows that new and complementary information can be had by careful optical studies of the phosphorescence lifetime. As in the early days of fluorescence spectroscopy in proteins, a complete and rigorous interpretation of the room temperature phosphorescence remains to be developed; nevertheless, it is clear that time-resolved phosphorescence yields new information on proteins in solution, for example, the detection of subtle conformational changes during protein folding, which is outside the sensitivity of earlier techniques. In addition, the great sensitivity of the phosphorescence lifetime to structural changes associated with rigidity and of nearby quenchers suggests that detailed structural information can be obtained when this approach is combined with the power of site-directed mutagenesis or other more biophysical techniques such as energy transfer to attached acceptors.

We have presented basic aspects of time-resolved room temperature phosphorescence spectroscopy and demonstrated some useful features of the spectroscopic signals as well as the general approach to data analysis. However, it should be understood that extensions of this approach will easily allow faster and improved time resolution with greater sensitivity to highly quenched phosphorescing states. In addition, many extensions of this approach that are common to fluorescence spectroscopy have yet to be developed. For example, combining time-resolved phosphorescence with anaerobic stopped-flow techniques and more rapid data acquisition electronics will enable studies of conformational dynamics with considerably shortened dead times. Other possibilities include extending the preliminary studies of *in vivo*-based spectroscopy, such as to microscopy.

In conclusion, time-resolved phosphorescence presents a new dimension to biophysical methodologies for the study of proteins, and it is likely that this area will continue to grow in capability as the fundamental understanding improves.

[5] Fluorescence Line Narrowing Spectroscopy: A Tool for Studying Proteins

By JANE M. VANDERKOOI, PAUL J. ANGIOLILLO, and MONIQUE LABERGE

Absorption and emission spectroscopy is used more than any other spectroscopic technique for both structural and analytical applications in biochemistry. In the past, chromophores contained in proteins showed only broad featureless spectra both in absorption and emission even at

METHODS IN ENZYMOLOGY, VOL. 278
0076-6879/97 $25

temperatures approaching absolute zero. Such broad spectra are usually devoid of vibrational information and at most reveal gross electronic transition information (such as λ_{max}). The poor resolution in optical spectroscopy is not a problem particular to biological molecules and can be attributed in part to microscopic disorder of the chromophore–matrix (host–lattice) environment. This leads to what has commonly been called *inhomogeneous line broadening.* The loss of resolution is unfortunate because the more detailed the spectra, the more versatile the spectroscopy becomes in addressing important questions. These questions include the influence of particular groups of atoms on physical, chemical, and biological properties; the paths of photochemical reactions, in particular the possible identification of individual molecules at various stages of photolysis; the distribution of electronic and vibronic states within the envelope of the inhomogeneously broadened line; and, specifically, the correlation of this distribution with biochemical properties.

Obtaining resolved spectra is, therefore, an experimental quest. High-resolution absorption and fluorescence spectra for molecules in a condensed phase were first reported by Shpol'skii,[1,2] who observed that fluorescent aromatic hydrocarbons in *n*-paraffin matrices (*n*-hexane, *n*-heptane, *n*-octane) gave quasiline spectra of widths on the order of 2–20 cm^{-1} at low temperatures. Attempts were then made to obtain similarly resolved spectra for hydrocarbons located in the hydrophobic interior of biological membranes.[3] However, for most chromophores located in biological structures such as proteins, membranes, or nucleic acid polymers, conventional optical spectroscopic techniques resulted only in broad absorption and fluorescence spectra.

With the development of monochromatic laser light sources, it became possible to obtain resolved spectra in the noncrystalline condensed phase. Soviet research groups pioneered the technique of "energy selection" spectroscopy to obtain spectra that have vibrational resolution.[4,5] The basic concept of this method is simple. For many situations, the observed broad absorption band arises because the sample is inhomogeneously broadened. In these cases, narrow-bandwidth light (≤ 1 cm^{-1}) from a laser source is used to excite a subpopulation within an ensemble of molecules. The sample is held at low temperature for reasons discussed in this chapter. Because

[1] E. V. Shpol'skii, *Sov. Phys. Uspekhi* **3,** 372 (1960).

[2] E. V. Shpol'skii, *Sov. Phys. Uspekhi* **5,** 522 (1962).

[3] J. M. Vanderkooi, J. Callis, and B. Chance, *Histochem. J.* **6,** 301 (1974).

[4] A. A. Gorokhovskii, R. Kaarli, and L. A. Rebane, *Opt. Commun.* **16,** 282 (1976).

[5] R. I. Personov, *in* "Spectroscopy and Excitation Dynamics of Condensed Molecular Systems" (V. M. Agranovich and R. M. Hochstrasser, eds.), p. 555. Elsevier North-Holland, Amsterdam, 1983.

only a select group of molecules is interrogated, narrowed spectra may result. The spectra have two main contributions: (1) zero-phonon lines (ZPLs), which are purely electronic or vibronic transitions, and (2) the phonon wings (PWs), which describe unresolved features in the spectrum that arise through the creation or annihilation of matrix (protein) phonons caused by the collective motion of many atoms in response to the dipole change in going from the ground to the excited state.[6] To conform to the description of Austin and Erramilli,[7] localized higher energy vibrational couplings to the electronic transition are termed *vibrons,* whereas the low-frequency collective modes due to matrix couplings (protein or solvent) are *phonons.*

There are several variants of the energy selection technique. The one that we discuss is fluorescence line narrowing (FLN) spectroscopy. Fluorescence spectra with lines of $<10\ cm^{-1}$ width from protein samples were first seen for porphyrin in cytochromes[8–10] and for chlorophyll within an intact leaf.[11] A related technique is hole-burning spectroscopy, which also uses a laser to select a subpopulation of molecules, but looks for the disappearance in the absorption spectra of the irradiated sample. Hole-burning spectroscopy used for biological samples has been described in this series.[12]

In this chapter we describe the technique of FLN as applied to biological systems. We emphasize the unique information that can be obtained, including the following.

1. Vibrationally resolved spectra of the ground-state molecule can be obtained by resonance Raman techniques for nonfluorescent molecules. Fluorescence and resonance Raman spectroscopy can be considered as essentially variants of the same process.[13] Like resonance Raman spectroscopy, FLN can give vibrationally resolved spectra even for molecules in low concentrations, and therefore information on atomic positions and bonds can be determined for a chromophore that is part of a larger complex.

2. The vibrational levels of the excited state molecule can also be acquired. With a lifetime of only a few nanoseconds, vibrational information for the excited-state molecule cannot be obtained by other techniques, except by technically difficult fast kinetic experiments. The vibrational

[6] M. Orrit, J. Bernard, and R. I. Personov, *J. Phys. Chem.* **97,** 10256 (1993).
[7] R. H. Austin and S. Erramilli, *Methods Enzymol.* **246,** 131 (1995).
[8] P. J. Angiolillo, J. S. Leigh, Jr., and J. M. Vanderkooi, *Photochem. Photobiol.* **36,** 133 (1982).
[9] T. Horie, J. M. Vanderkooi, and K. G. Paul, *Biochemistry* **24,** 7935 (1985).
[10] J. M. Vanderkooi, V. T. Moy, G. Maniara, H. Koloczek, and K. G. Paul, *Biochemistry* **24,** 7931 (1985).
[11] R. Avarmaa, I. Renge, and K. Mauring, *FEBS Lett.* **167,** 186 (1984).
[12] J. Friedrich, *Methods Enzymol.* **246,** 226 (1995).
[13] R. M. Hochstrasser and C. A. Nyi, *J. Chem. Phys.* **70,** 1112 (1979).

levels depend on the structure of the excited state molecule, important data for understanding photoreactions.

3. One can obtain access to the electron–phonon coupling. This value tells how strongly the surrounding atoms are influenced by the electrons of the fluorescent molecule.

4. Since the frequency of the transition is separated from phonon interactions, one can obtain the difference in energy between the ground and excited electronic states. For chromophores in proteins, there is some disorder around the chromophores, and this generates a distribution of energy gaps. The distribution of 0,0 energies is quantitatively related to the uniqueness of the environment.

A disadvantage of FLN is that it must be carried out at low temperature, but this need not be a totally undesirable aspect. Temperature itself an be a variable in the experiment and low temperature, besides significantly reducing vibrational broadening, also allows for the examination of low barriers in protein reactions.[7]

Basic Principles of Fluorescence Line Narrowing

Zero Phonon Lines and Phonon Wings

Figure 1 shows an energy diagram used to describe the absorption and emission of light by the molecule. The transition from the lowest electronic level to the next highest electronic level ($S_{0,0} \rightarrow S_{1,0}$) should have a width that reflects the lifetime of the S_1 state. If the lifetime is 10 nsec, the width should be $\sim 3 \times 10^{-4}$ cm^{-1} or 10^{-7} nm at 500 nm. The vibrational modes (vibrons) of the molecule couple with the electronic transition, giving rise to vibronic bands. The relaxation from the vibrational bands are typically tens of picoseconds, but may be as long as or longer than 100 psec.[14] If the excited vibrational state ($S_{1,1}$) relaxes to ($S_{1,0}$) within 10 psec, then the linewidth of the absorption from ($S_{0,0}$) to ($S_{1,1}$) would be 0.3 cm^{-1}. The vibronic transitions (i.e., transitions to or from a vibrational level and/or electronic levels) give rise in the spectrum to the ZPLs.

The energy diagram predicts sharp lines for absorption and emission spectra, which can provide information on the respective vibrational levels of the excited and ground state molecules; from the relative intensity of the bands, an assessment of the Franck–Condon overlap may also be obtained. In reality, such a situation does not occur in condensed matter, owing to interactions with the surrounding atoms. The lattice vibrations

[14] J. R. Ambroseo and R. M. Hochstrasser, *J. Chem. Phys.* **89,** 5956 (1988).

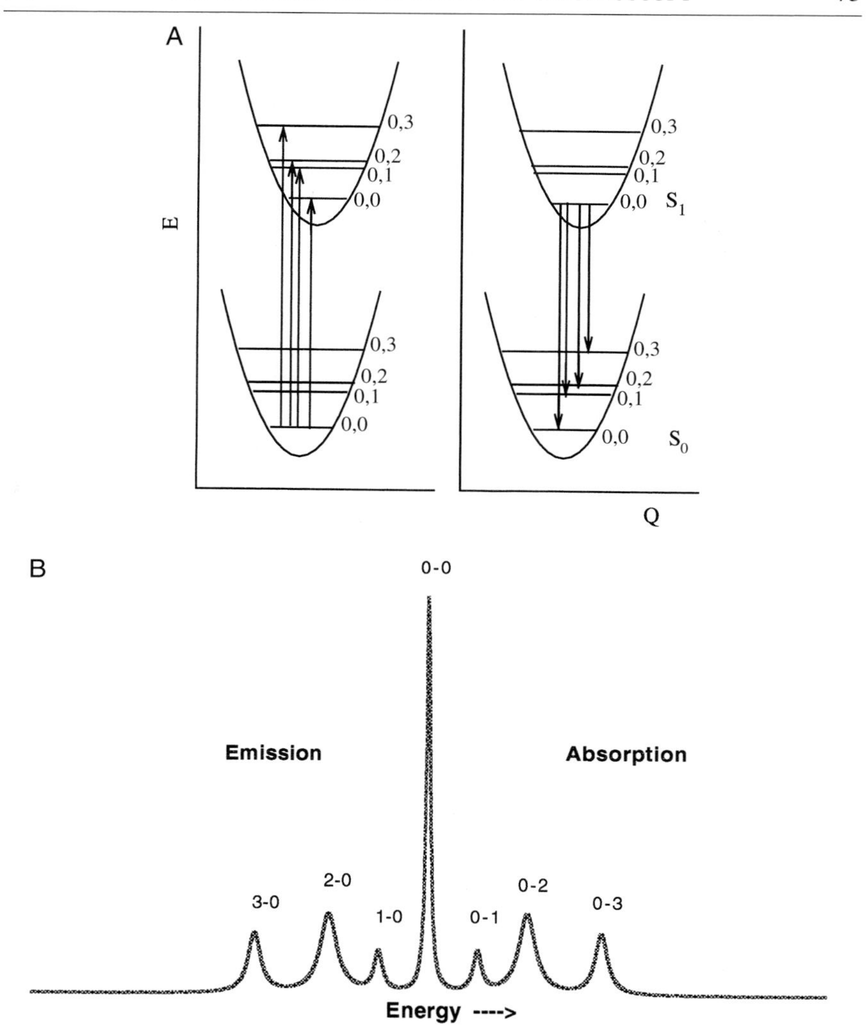

FIG. 1. Schematic potential energy diagram along one normal coordinate (A) and idealized absorption and emission spectra (B) for a chromophore embedded in a matrix. The lowest vibrational level for the ground state is denoted as $S_{0,0}$ and for the excited state as $S_{1,0}$. Arrows represent vertical transitions between two Born–Oppenheimer states. A 0,0 transition is one that arises from a transition between $S_{0,0}$ and $S_{1,0}$. Three vibrational levels are shown; a typical large aromatic chromophore would have several hundred vibrational levels. The 0,0 absorption and emission transitions are resonant.

(phonons) of the surroundings (which in our case is the protein polypeptide chain and the solvent) couple with the electronic transitions; they arise because the charge distribution of the excited-state molecule is different from that of the ground-state molecule, causing the lattice atoms to adjust their positions. This process gives rise to a broad band on the high-energy side of the absorption ZPLs and on the low energy of the emission lines. The broad bands observed are on the order of 100 cm^{-1} in width and are called *phonon wings.* An illustration of a single vibronic ZPL and the PW is shown in Fig. 2.

The experimentally measured ratio of the integrated intensities that describes the coupling of the optical transition to the lattice vibrations (and also the probability of ZPL transitions) is defined as

$$\alpha(T) \equiv \text{(integrated intensity of ZPL)}/[\text{integrated intensity of (ZPL + PW)}] = \exp - [S(T)] \tag{1}$$

This ratio has its origin in X-ray and neutron scattering theory; it is related to the probability of elastic scattering and is called the Debye–Waller factor. It has been retained in these optical phenomena to describe the analogous physics of FLN. The Debye–Waller factor is related to the Huang–Rhys factor, $\exp\ -[S(T)]$, which is a measure of the electron–phonon coupling. α can also be expressed as follows:

$$\alpha(T) = \exp[-S(T)] = \exp - \left\{\int_0^\infty d\omega\ \rho(\omega)^2/[e^{-(\hbar\omega/kT^{-1})}]\right\} \tag{2}$$

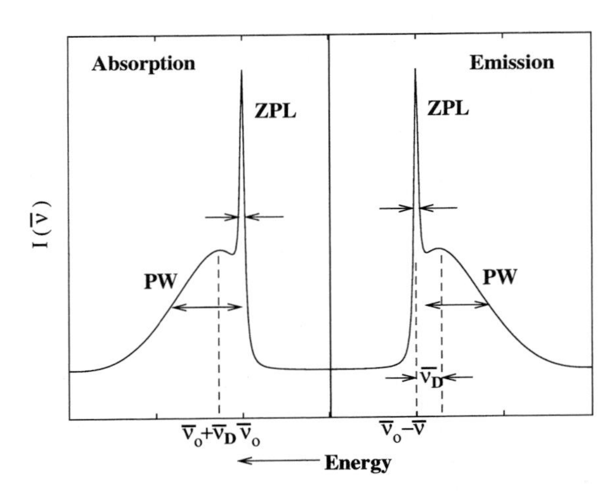

FIG. 2. Zero-phonon line (ZPL) and phonon wing (PW) for one vibronic transition in absorption and emission. The laser excitation is indicated by ν_o, the Debye frequency as ν_D. The ZPL was obtained using a Lorentzian function and the PW functional form was obtained by adapting the equations of Osad'ko.[22]

where $\rho(\omega)$ is the weighted density of matrix phonon states at temperature T, k is the Boltzmann constant, and $\hbar$ is Planck's constant divided by 2π.

When the Huang–Rhys factor approaches 1, the electron–phonon coupling between the optical center and the matrix phonons is weak and only the intense ZPL is observed in the spectrum (as with crystals). When the Huang–Rhys factor is closer to ~0.1, the coupling is intermediate and both ZPL and PW can be observed in the spectrum. In the case of strong coupling, $\exp(-S) < 0.1$ and the PW dominates the spectrum.[15]

The PW is temperature dependent. At temperatures approaching 0 K, the emission intensity will be solely due to the ZPL, but as temperature increases, the value of α decreases. The temperature dependence of the coupling between the chromophore and the solvent may have linear and quadratic terms.[16,17] The linear terms predict that the relative intensity of the ZPL relative to the PW decreases with increasing temperature. The quadratic term predicts that, in addition, there will be a broadening of the ZPL as the temperature is raised. For many systems, the intensity of the ZPL is close to zero at above 50 K. This explains one reason why FLN spectra are seen only at low temperature.

Inhomogeneous Broadening: Shifting of Energy Gap of Zero-Phonon Lines

If all chromophores are in identical environments, then one would expect to observe spectra that are governed by lifetime and phonon-coupling considerations. However, the surrounding atoms can raise and lower the energy levels of both the ground- and excited-state molecules, and when the atoms surrounding the molecule are not identical, then the energy for the $S_{0,0} \rightarrow S_{1,0}$ transition will be different for individual molecules within the ensemble. This gives rise to inhomogeneous broadening, and the situation is diagrammed in Fig. 3. In this case, we have a distribution of energies for the $S_{0,0} \rightarrow S_{1,0}$ transition and the entire absorption line shape is composed of ZPL and PW. For a glass the distribution of energies of the electronic transition (i.e., the ZPLs) can be described by a Gaussian function.[18]

Formally, the entire shape of the spectral band can be described as the sum of ZPL and PW contributions from each chromophore.[5,19,20] Each

[15] V. I. Rakhovski, M. N. Sapozhnikov, and A. L. Shubin, *J. Luminesc.* **28,** 301 (1983).
[16] I. S. Osad'ko, *Sov. Phys. Uspekhi* **22,** 311 (1979).
[17] I. S. Osad'ko and A. A. Shtygashev, *J. Luminesc.* **36,** 315 (1987).
[18] B. B. Laird and J. L. Skinner, *J. Chem. Phys.* **90,** 3274 (1989).
[19] W. C. McColgin, A. P. Marchetti, and J. H. Eberly, *J. Am. Chem. Soc.* **100,** 5622 (1978).
[20] I. I. Abram, R. A. Auerbach, R. R. Birfe, B. E. Kohler, and J. M. Stvenson, *J. Chem. Phys.* **63,** 2473 (1975).

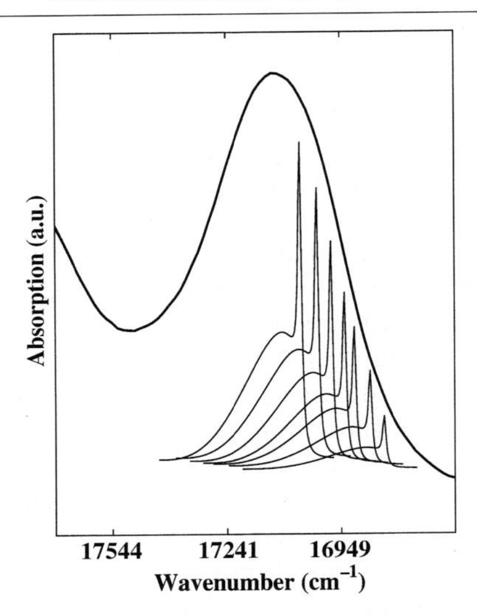

FIG. 3. ZPLs and PWs within an inhomogeneously broadened envelope. The broad absorption is the $Q(0,0)$ visible band of zinc cytochrome c, measured conventionally at room temperature. (The nature of the ZPL and PW line shapes is schematic and is not intended to reflect quantitatively the true nature of the composition of the inhomogeneously broadened absorption or the degree of electron–phonon coupling.)

chromophore can be identified by its 0,0 transition, ν_0, giving rise to ZPL and associated PW with a peak at $\nu_0 + \nu_D$, where ν_D is known as the Debye frequency. In a noncrystalline matrix, the positions of the 0,0 electronic transitions form a distribution, PDF(ν). Each chromophore has its own absorption and emission spectrum, which can be expressed as follows:

$$I_{abs}(\nu - \nu_0) = \alpha(T)\mathrm{ZPL}_{abs}(\nu - \nu_0) + [1 - \alpha(T)]\mathrm{PW}_{abs}[\nu - (\nu_0 + \nu_D)] \quad (3)$$

$$I_{em}(\nu_0 - \nu) = \alpha(T)\mathrm{ZPL}_{em}(\nu_0 - \nu) + [1 - \alpha(T)]\mathrm{PW}_{em}[\nu_0 - (\nu - \nu_D)] \quad (4)$$

where $\alpha(T)$ is the Debye–Waller factor as described previously and ZPL_i and PW_i (i = abs, em) are the ZPL and PW contributions to the line shape (Fig. 2). The total absorption line shape is thus given by the convolution of the distribution of electronic 0,0 transitions and the line-shape function for an individual chromophore (Fig. 3).

$$A(\nu) = \int_{-\infty}^{\infty} d\nu_0 \mathrm{PDF}(\nu_0) I_{abs}(\nu - \nu_0) \quad (5)$$

Similarly, the emission line shape is given by

$$E(\nu) = \int_{-\infty}^{\infty} d\nu_0 \mathrm{PDF}(\nu_0) I_{em}(\nu_0 - \nu) \quad (6)$$

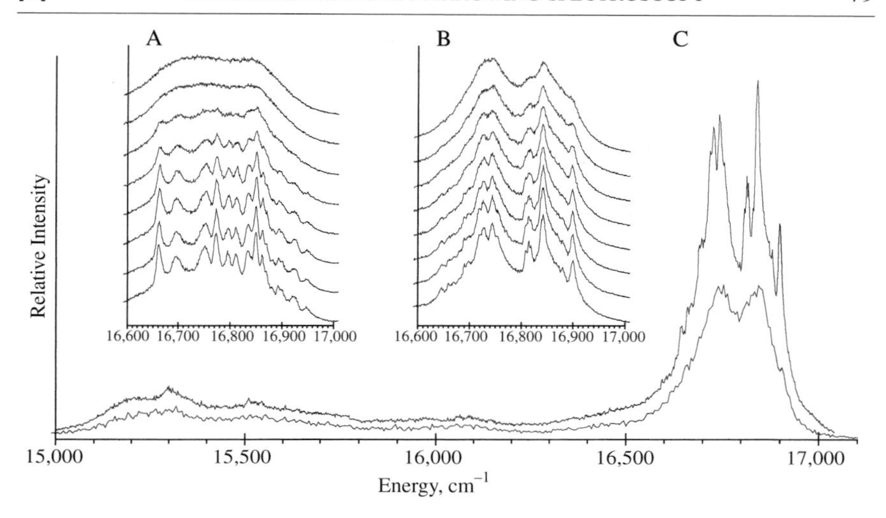

FIG. 4. Effect of temperature of FLN spectra of Mg-myoglobin. (A and B) FLN spectra at different temperatures: the upper spectrum is at 80 K, the lowest is at 10 K. The temperature difference between the intervening spectra was 10 K. Excitation frequency was at 18,000 cm^{-1} (A) and 17,800 cm^{-1} (B). (C) FLN spectrum at 4.2 K (upper curve) and at 77 K (lower curve). Excitation frequency was at 17,800 cm^{-1}. (Unpublished data from A. Kaposi and J. M. Vanderkooi, 1992.)

For $S_{0,0} \rightarrow S_{1,0}$ excitation and emission the functional nature of the ZPL is that of a Lorentzian centered at ν_0 with a linewidth governed by the lifetime of the excited state (in the limit of negligible width to the excitation source). The shape of the PW is determined by the density of states and the magnitude of the electron–phonon coupling and is beyond the scope of this chapter.[16,17,21–23]

When we consider the inhomogeneous broadening, we realize that there is another reason for the experimental need for cryogenic temperature. If the chromophore or environment changes during the fluorescent lifetime, then the energy gap between the ground- and excited-state molecules will change. A change can occur following a chemical reaction (such as proton or electron transfer), a modification in hydrogen bonding, or a physical change in the surroundings that alters the field at the chromophore. In any case, resolution in the spectrum will be lost.

The effect of temperature on FLN spectra is demonstrated in Fig. 4 for Mg-myoglobin. The temperature required to obtain resolved spectra is a

[21] I. S. Osad'ko, E. I. Al'Shits, and R. I. Personov, *Sov. Phys. Solid State* **16,** 1286 (1975).
[22] I. S. Osad'ko, *Sov. Phys. Solid State* **17,** 2098 (1976).
[23] I. S. Osad'ko, *Sov. Phys. JETP* **45,** 827 (1977).

function of the system under study and of the specific experiment being run but, generally speaking, protein systems begin to show resolution below 60–50 K and vibrational features become sharp below ~20 K. Some other features of these spectra are discussed later in this chapter.

Experimental Conditions

Instrumentation

Other than conventional fluorescence equipment (photomultiplier, data acquisition, etc.), FLN requires a means to keep the sample cold and a laser for narrow-band excitation. A high-resolution monochromator or spectrograph is usually used to isolate the emission light, although interference Fourier transform techniques are also suitable.[24]

With respect to low-temperature equipment, refrigerating units or cryostats can be used. The former have the advantage of cooling without requiring a liquid helium transfer step and are more cost effective because no liquid helium needs to be purchased. On the other hand, cryostats run with liquid helium and their main advantage is that they can easily cool down to the 2–4 K range as compared with 10 K for refrigerating units. An attractive alternative is the use of an open flow Dewar such as is conventionally used for electron paramagnetic resonance instrumentation with the corresponding sample tubes. These cooling devices reach down to 4.2 K and sample changeover only necessitates insertion/removal of the sample tube into or from the holding chamber without loss of time while waiting for the Dewar to lower or rise in temperature.

The narrow-line excitation is used to "select" one subpopulation of molecules within the whole ensemble. When fluorescence from aromatic molecules is considered, one is usually examining the nonresonant emission since the lifetimes are too short (nanoseconds) for resonant excitation/emission experiments using conventional mechanical choppers to remove scatter from the exciting light. That is, one excites into a higher vibronic band; this is followed by relaxation to the lowest electronic level; or one excites into the 0,0 band, and observes the transitions to the 0,1 levels. In either case, the lines are 1–10 cm^{-1} wide, the width being due to uncertainty broadening from the vibrational relaxation and possible inhomogeneity in the vibrational transitions. Consequently, the bandwidth of the light used for excitation and the resolution of the monochromator must both be on the order of 1 cm^{-1}. Because fluorescence can be detected at low concentrations the power of the exciting light can be low (<100 mW) and the exciting

[24] J. G. Radziszewski, J. Waluk, M. Nepras, and J. Michl, *J. Phys. Chem.* **95,** 1963 (1991).

beam does not have to be tightly focused. These factors also help to reduce the heating of the sample by the exciting light.

The excitation frequency also needs to be considered. The resolution of FLN spectra is highest when excitation is in the low-energy region of the spectrum. When excitation is resonant with a purely electronic ZPL line the resulting spectrum will show sharp ZPLs, each with an associated PW (see Fig. 2). When the excitation frequency is at higher energy it will be resonant with many PWs and the emission spectrum will consist of an envelope of ZPLs and the convolution of their PWs. (The overlap of many phonon wings at the higher energy side of an absorption spectrum is illustrated in Fig. 3.) In addition to phonon wing considerations, resolved spectra have only been obtained when excitation and emission occurred in the same electronic manifold, i.e., S_0–S_1. Excitation directly into the lowest triplet level (S_0–T_1) also gives resolved phosphorescence spectra for chromophores in glasses.[25] Energy selection for protein samples has not been seen in fluorescence when excitation is into a higher electronic manifold (S_2) or in phosphorescence when the triplet state is populated by intersystem crossing from S_1.

Much of the FLN work in the literature has used visible light, owing to the interest in the colored hemes and chlorophylls involved in electron and energy transfer, but also due to the availability of lasers in the visible range. However, there is no reason why chromophores absorbing in the ultraviolet (UV) should not also show resolved spectra under energy selection conditions, and indeed enhanced resolution was seen for tryptophan in a protein at cryogenic temperatures.[26]

The concentration required for FLN spectroscopy is the same as for conventional fluorescence measurements, usually so that the optical density at the wavelength of excitation is <0.1. Because the sample should be optically clear, a cryosolvent compatible with proteins such as glycerol or ethylene glycol is introduced to give a 20 to 60% by volume solution. This fraction of solvent in water gives a glass or cracked glass that is satisfactory for undistorted optical measurements.

Data Analysis

Vibrational Frequencies of Molecule in S_0 and S_1 States. Unlike conventional fluorescence, the emission spectra of FLN vary with the excitation frequency and therefore the spectra must be examined with reference to the exciting frequency. When the sample is excited into the lowest level

[25] R. L. Williamson and A. L. Kwiram, *J. Phys. Chem.* **83,** 3393 (1979).
[26] T. W. Scott, B. F. Campbell, R. L. Cone, and J. M. Friedman, *Chem. Phys.* **131,** 63 (1989).

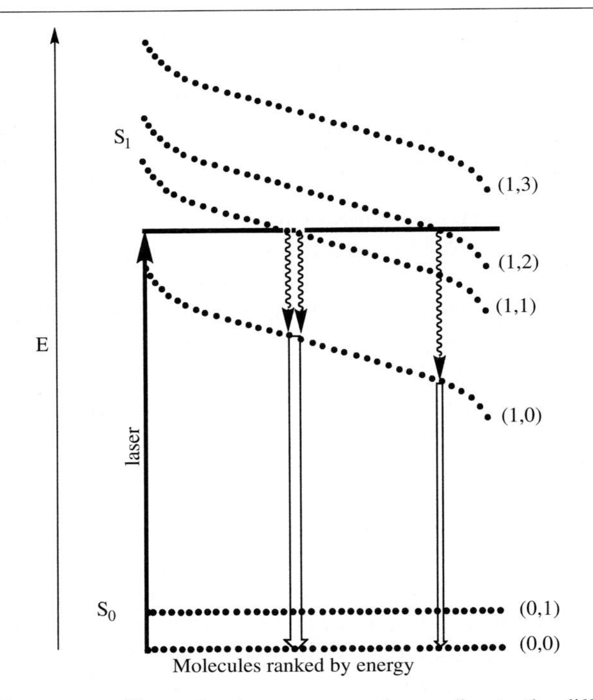

FIG. 5. Energy map. The molecules are arranged according to the difference in energy from the lowest vibrational level of the S_0 and S_1 states, with the ground-state molecules being placed at zero energy, by convention. Open arrows indicate vibrational relaxation; closed arrows show electronic transitions. In the situation shown, two subpopulations of molecules are excited, yielding two 0,0 emission lines. The width of the arrow represents the number of molecules in the individual subset, and illustrates that more molecules are near the mean value of the electronic energy gap than at the extremes.

of the excited state then the observed emission lines arise from the ground-state vibrations. This is equivalent to resonance Raman scattering, and the ground-state vibrational frequencies are determined by subtracting $\nu_{\text{excitation}} - \nu_{\text{emission}}$.

When the sample is excited into the congested higher vibrational levels and the inhomogeneous broadening is bigger than the energy difference of the vibrational lines of the excited state, then it will be possible to excite more than one subpopulation of molecules. For such cases it is useful to consider an *energy map*.[27,28] In Fig. 5 such a diagram is shown. Following

[27] A. D. Kaposi, V. Logovinsky, and J. M. Vanderkooi, *Proc. Soc. Photo-Opt. Instr. Engl.* **1640,** 792 (1992).

[28] A. D. Kaposi and J. M. Vanderkooi, *Proc. Natl. Acad. Sci. U.S.A.* **89,** 11371 (1992).

convention we put the energy levels of all the ground-state molecules to the same level, although it is recognized that both ground- and excited-state molecules are distributed in energy. In addition, we order the molecules in terms of their energy gaps. The major effect of the inhomogeneous environment is that the electronic levels are distributed. Hence the separations of the vibrational levels are shown to be constant in Fig. 5.

The energy map predicts that one excitation frequency can excite more than one population of molecules. Each will relax to the lowest level of the excited state, and then emit. It is important to recognize that these lines will be separated by the energy difference of the vibrational levels of the excited-state molecule. An example is presented in Fig. 6, in which essentially two populations were excited, leading to two main emission spectra. The paired arrows indicate two $S_{1,0} \rightarrow S_{0,0}$ transitions and the accompanying 0,*n* transitions. In other cases, the excitation of a laser will excite more than two vibrational levels; this was seen in Fig. 4, in

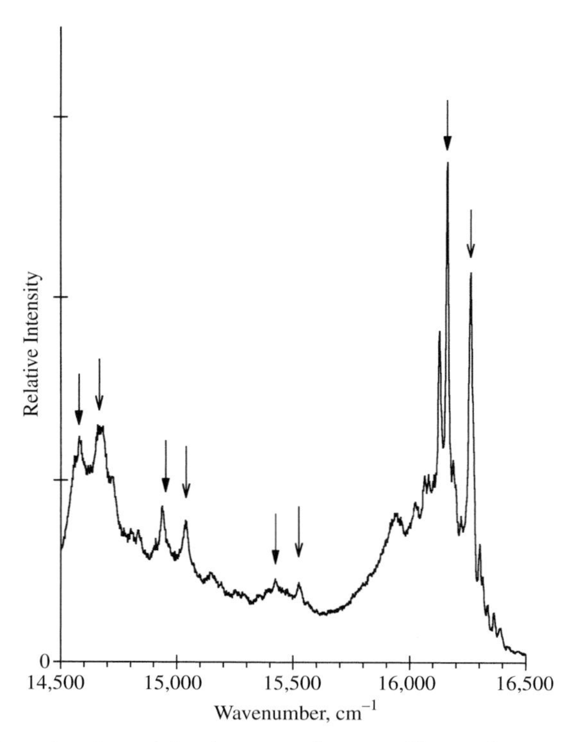

FIG. 6. FLN spectra of metal-free horse cytochrome *c*. The two large peaks indicated by arrows and the smaller peaks between 16,100 and 16,400 cm^{-1} are 0,0 emission lines. The peaks at lower frequency represent vibrons of the ground state.

which the region between 16,600 and 17,000 cm^{-1} shows multiple 0,0 lines and the lower energy region is so congested that sharp lines are not seen.

A point to be emphasized when using the energy map is that if we identify the 0,0 lines, and we subtract $\nu_{excitation} - \nu_{0,0\ emission}$, then we obtain the frequency value of the vibrational level of the molecule in its excited state. If we step through the excitation frequencies then we can construct the vibrational spectrum of the excited-state molecule.

Population Distribution Function. At the same time that vibronic levels are determined by stepping through the excitation frequencies, the distribution of 0,0 transition energies, that is, the number of molecules with 0,0 emission energies in an E, $E + dE$ interval (dN/dE), are obtained.[28–30] This distribution we call the population distribution function (PDF) and a definition is

$$I_m = I_{exc}(dN/dE)A_k V_m F_m K \tag{7}$$

where I_{exc} is the excitation intensity, A_k is the $0,0 \rightarrow 1,k$ absorption transition probability, V_k is related to the $1,k \rightarrow 1,0$ vibrational transition probability, F_m is the fluorescence $1,0 \rightarrow 0,m$ emission probability, and K is a proportionality constant.

In Fig. 7, we illustrate the process of determining the PDF. Figure 7 presents spectra of the fluorescence of zinc-substituted hemoglobin, showing the 0,0 region where the emission lines shift with excitation. If we follow the intensity of one peak (e.g., the peak at 590 cm^{-1} from the excitation), we see that its intensity increases and decreases as the frequency is changed. The inset shows the resultant PDF, which gives the inhomogeneous broadening, undistorted by the phonon wings.

Determination of Zero-Phonon Linewidth of 0,0 Transition. Most protein FLN work involves the study of nonresonant transitions. Ideally, one would like to record the resonant emission of the ZPL because that should give a narrow line with a width determined by the uncertainty broadening by the lifetime of the excited state. (It should be noted that FLN is the optical analog of Mössbauer spectroscopy in that they both gave the ultimate in resolution and that the underlying physics of both spectroscopies are essentially the same.[31] As stated previously, a technical problem in doing this experiment is that the short (nanosecond) decay times prevent the use of conventional choppers to eliminate the exciting light. One approach that seems promising is the use of time-correlated single photon-counting

[29] J. Fuenfschilling and I. Zschokke-Graenacher, *Chem. Phys. Lett.* **91,** 122 (1982).
[30] J. Fidy, J. M. Vanderkooi, J. Zollfrank, and J. Friedrich, *Biophys. J.* **61,** 381 (1992).
[31] R. H. Silsbee and D. B. Fitchen, *Rev. Mod. Phys.* **36,** 432 (1964).

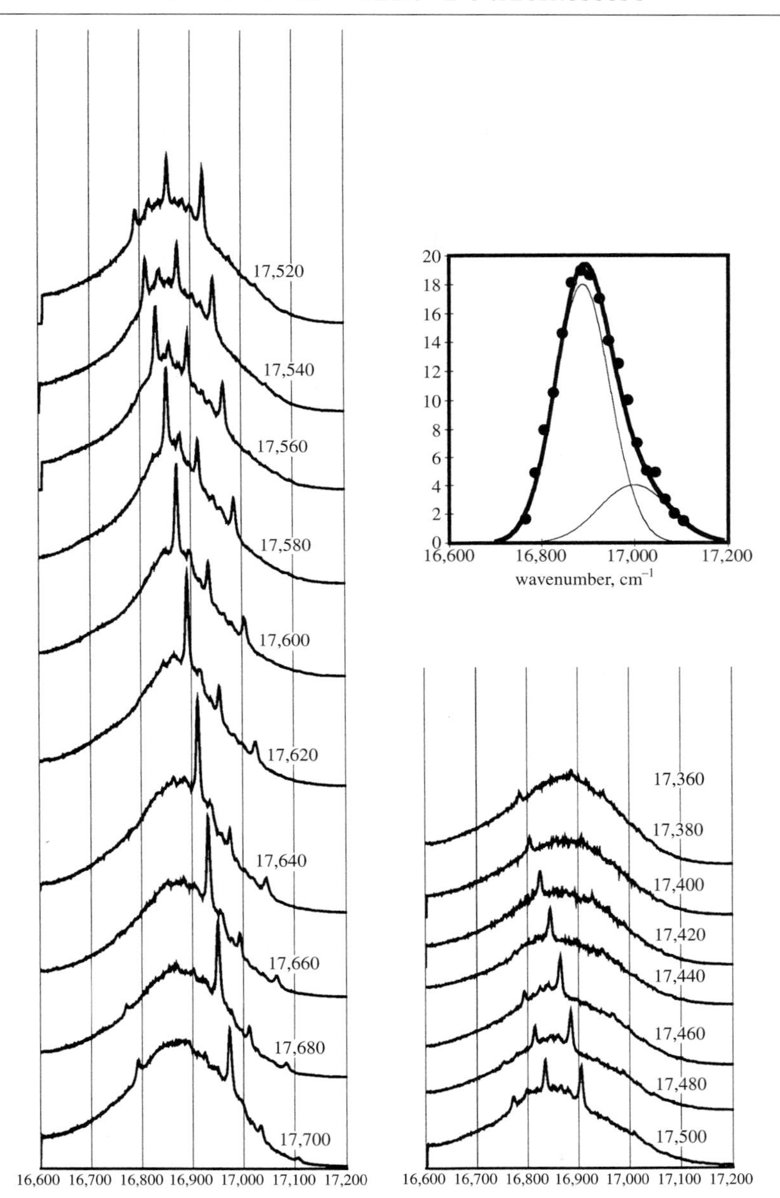

FIG. 7. FLN spectra of α-Zn hemoglobin (zinc has replaced iron in the α chain) excited in the 17,540–17,700 cm^{-1} region. *Inset:* PDF calculated from the data. Temperature: 4.5 K. (K. Sudhakar and J. M. Vanderkooi, unpublished data, 1994.)

methods with time gating.[32] An advantage of this method is that photon counting compensates for the (usually) weak fluorescence resulting from the low exciting powers used to minimize burning effects. Excitation is provided by short laser pulses and resonant spectra are obtained by rejecting the scattered laser light with a gate set after termination of the pulse. Zinc-substituted myoglobin FLN spectra have been reported using this technique.[33] Among the techniques used to measure the homogeneous linewidth of an optical center, such as hole burning, photon echoes, or FLN, the most accurate result will be the FLN measurements, mainly because of the high sensitivity of fluorescence and because the linearity of the fluorescence process produces undistorted spectra.

Fluorescence Line Narrowing as Applied to Proteins

Ground- and Excited-State Vibrational Frequencies

Perhaps one of the most important uses of FLN for protein systems is the extraction of vibrational information from the recorded spectra. The methods of choice in vibrational spectroscopy are usually infrared and Raman spectroscopy, but the first is severely limited by the narrowness of infrared "spectral windows" for aqueous samples and fluorescent materials are not good candidates for Raman work. Fluorescence line narrowing has none of these drawbacks.

As explained in the section on data analysis, FLN spectra can give the vibrational frequencies of both the ground- and excited-state molecules. In Fig. 8A, FLN spectra in the 0,0 region of Mg-myoglobin (myoglobin in which the porphyrin has been made fluorescent by substituting magnesium for iron) are shown at various excitation frequencies that promoted the porphyrin to higher vibronic levels. In Fig. 8B, the spectrum constructed from the individual spectra of Fig. 8A is shown. Figure 8B represents the resolved excitation spectrum of the molecule. Table I lists some of the vibrational frequencies thus extracted from FLN spectra acquired for Mg-myoglobin in its excited state. There is a general downshift in the frequency of the excited-state molecule relative to the ground-state molecule, which is consistent with the general increase in size for an excited-state molecule.

Similar small shifts in frequency were seen for the excited state of zinc cytochrome *c* relative to its ground state[34] and for lumiflavin in *n*-decane

[32] J. S. Ahn, Y. Kanematsu, and T. Kushida, *Phys. Rev. B* **48,** 9058 (1993).
[33] J. S. Ahn, Y. Kanematsu, M. Enomoto, and T. Kushida, *Chem. Phys. Lett.* **215,** 336 (1993).
[34] M. Laberge and J. M. Vanderkooi, *Biospectroscopy* **1,** 421 (1995).

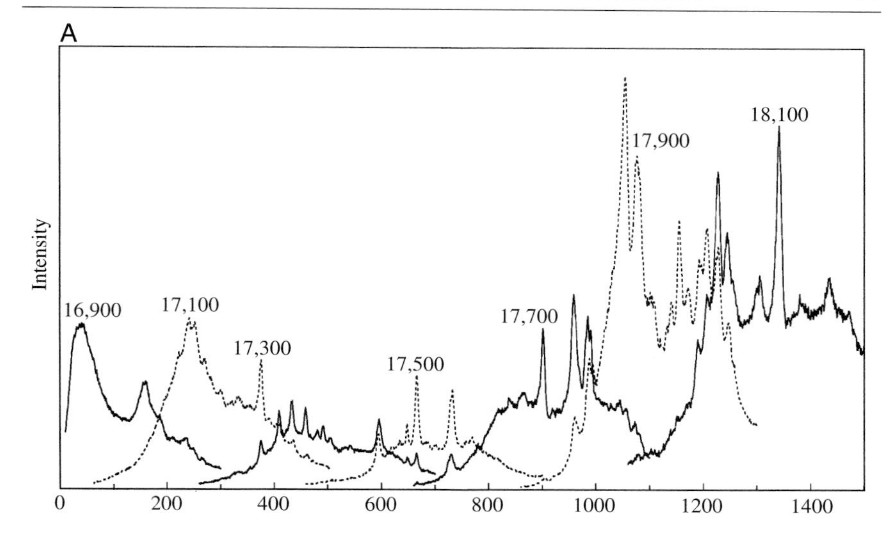

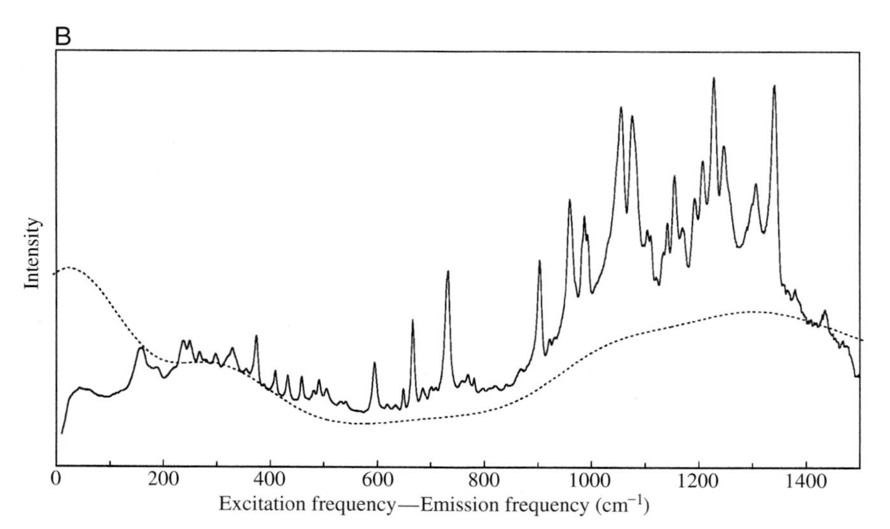

FIG. 8. (A) FLN spectra of MgPP-myoglobin showing in the 1,0 → 0,0 region at 4.2 K; (B) summed emission spectrum (solid line) and broadband excitation spectrum (dotted line). (Data from Kaposi and Vanderlooi,[28] adapted with permission.)

TABLE I
VIBRATIONAL FREQUENCIES OF GROUND-STATE AND SINGLET EXCITED-STATE Mg-MYOGLOBIN[a]

Ground state (cm^{-1})	Excited state (cm^{-1})
670	667
752	734
990	988
1072	1077
1134	1141
1206	1207
1524	1530

[a] Adapted with permission from A. D. Kaposi, J. Fidy, S. S. Stavrov, and J. M. Vanderkooi, *J. Phys. Chem.* **97,** 6317 (1993). Copyright 1993 American Chemical Society.

matrix.[35] The vibrational frequencies of chlorophyll systems in the ground and excited state have also been investigated. Fluorescence line narrowing experiments performed on these proteins yielded ground-state vibrational frequencies and although poor agreement was found between these measurements and infrared (IR)/Raman data, the authors rightly pointed out that IR and Raman studies had covered only some 12% of more than 400 vibrational modes.[36] Excited-state vibrational frequencies have also been determined for chlorophyll *a* and model systems.[37,38]

Fluorescence Line Narrowing Used Analytically

Vibrationally resolved spectra can allow for the identification of molecules and because weak fluorescence signals can be detected, FLN can be used to detect molecules even at low concentration in the presence of many other absorbing molecules. This property was exploited for the identification of the metabolic products of carcinogenic aromatic hydrocarbons.[39,40] The power of the technique is demonstrated by the detection of adducts of benzo[*a*]pyrene diol epoxide with DNA by FLN.[41] In combination with

[35] R. J. Platenkamp, A. J. W. G. Visser, and J. Koziol, *in* "Flavins and Flavoproteins" (V. Massey and C. H. Williams, eds.), p. 35. Elsevier North-Holland, Amsterdam, 1982.
[36] J. Fuenfschilling and D. F. Williams, *Photochem. Photobiol.* **26,** 109 (1977).
[37] K. K. Rebane and R. A. Avarmaa, *Chem. Phys.* **68,** 191 (1982).
[38] S. L. S. Kwa, S. Voelker, N. T. Tilly, R. Van Grondelle, and J. P. Dekker, *Photochem. Photobiol.* **59,** 219 (1994).
[39] R. Jankowiak, P. Lu, and G. J. Small, *J. Pharm. Biomed. Anal.* **8,** 113 (1990).
[40] R. Jankowiak and G. J. Small, *Anal. Chem.* **61,** 1023 (1989).
[41] G. A. Marsch, R. Jankowiak, M. Suh, and G. J. Small, *Chem. Res. Toxicol.* **7,** 98 (1994).

electrophoresis, the spectroscopy could distinguish adducts that were of the internal type from those in which the hydrocarbon adhered only to the surface. In another application, FLN was used to distinguish the binding site of an antibody to a pyrene derivative.[42]

Because FLN has the intrinsic sensitivity of fluorescence, it can detect species at low concentration within tissue. Vibrational fine details obtained by fluorescence have led to the unequivocal identification of porphyrins in muscle slices[43] and in parasites.[44]

Distribution of 0,0 Energy and Relationship to Protein Fluctuations

It has become increasingly evident that the physics describing the "state of matter" of a protein is not as straightforward as once thought. Proteins possess properties reminiscent of both crystalline solids, as shown by their highly regular X-ray diffraction patterns, and amorphous or glassy structures, as documented by low-temperature specific heat measurements (linear in temperature less than 1 K), anomalous thermal conductivity and dielectric response,[45] and large inhomogeneously broadened optical transitions (see as follows). At room temperature, proteins undergo fluctuations on the time scale of vibrations (picoseconds) to large-scale local unfolding of the tertiary structure on the time scale of minutes to days, which means that, at a given instant, there will be a distribution of molecular subconformations. Thus, the range of time scales for relaxation spans some 10 orders of magnitude. Even at temperatures of 2 K and less, it has been demonstrated—using spectral hole burning and photon echo spectroscopies—that proteins undergo relaxation processes that are on the order of days.[46,47] In this respect, proteins exhibit behavior consistent with glasses. These apparent glasslike properties of proteins can be successfully explained using the two-level system (TLS) model.[48,49] In the model of glasses, a TLS comprises a group of atoms or molecules that can be in either of two quantum-mechanical potential wells along a conformational coordinate separated by an energy barrier. The subconformations or conformational substrates show

[42] K. Singh, P. L. Skipper, S. R. Tannenbaum, and R. R. Dasari, *Photochem. Photobiol.* **58,** 637 (1993).

[43] A. V. Novikov, *Photochem. Photobiol.* **59,** 12 (1994).

[44] C. Larralde, S. Sassa, J. M. Vanderkooi, H. Koloczek, J. P. Laclette, F. Goodsaid, E. Sciutto, and C. S. Owen, *Mol. Biochem. Parasitol.* **22,** 203 (1987).

[45] L. R. Narashimhan, K. A. Littau, D. W. Pack, Y. S. Bai, A. Elscher, and M. D. Fayer, *Chem. Rev.* **90,** 439 (1990).

[46] J. Gafert, J. Friedrich, J. M. Vanderkooi, and J. Fidy, *J. Phys. Chem.* **99,** 5223 (1995).

[47] D. T. Leeson, O. Berg, and D. A. Wiersma, *J. Phys. Chem.* **98,** 3913 (1994).

[48] P. W. Anderson, B. I. Halperin, and C. M. Varma, *Philos. Mag.* **25,** 1 (1972).

[49] W. A. Phillips, *J. Low Temp. Phys.* **7,** 351 (1972).

strong similarity with the TLS, the only difference being that, in "pure" glassy materials, the energy barriers are assumed to have a smooth distribution whereas in proteins the conformational substrates have a punctuated barrier distribution and thus a tiered hierarchy is proposed to exist.[50] Each subconformation is more or less closely related to the time-averaged structure revealed by high-resolution techniques such as X-ray or nuclear magnetic resonance (NMR). A variety of approaches show that each subconformation of the protein has fundamentally different chemical and physical properties.[51–54]

Fluorescence line narrowing is an especially powerful technique in studying the distribution of conformations within the population. As the positions of the neighboring atoms change, the chromophore experiences different electric fields and torsional constraints; these change the 0,0 transition energy because conformational changes and fluctuations have large effects on electrostatics.[55–57] Fast cooling of the sample allows the subconformations to be "frozen in," either by being energetically trapped into local minima[58,59] or by being kinetically trapped by the high (approaching infinity) viscosity of the solvent.[60]

It follows, then, that the observed spectral dispersion in the transition frequency should be sensitive to the protein electrostatic properties and the protein flexibility. Indeed, when charge on residues near the chromophore is altered by changing pH, the distribution of energies shifts.[61] The inhomogeneous broadening can also change when substrate is bound to the protein,[62,63] or broaden and shift when the protein is made to unfold by addition

[50] A. Ansari, J. Berendzen, S. F. Bowne, H. Frauenfelder, I. E. T. Iben, T. B. Sauke, E. Shyamsunder, and R. D. Young, *Proc. Natl. Acad. Sci. U.S.A.* **82,** 5000 (1985).

[51] M. Karplus and J. A. McCammon, *Sci. Am.* **254,** 42 (1986).

[52] H. Frauenfelder, F. Parak, and R. D. Young, *Annu. Rev. Biophys. Biophys. Chem.* **17,** 451 (1988).

[53] V. Goldanskii and Y. Krupyanskii, *Q. Rev. Biophys.* **22,** 39 (1989).

[54] R. H. Austin, K. W. Beeson, L. Eisenstein, H. Frauenfelder, and I. C. Gunsalus, *Biochemistry* **14,** 5355 (1975).

[55] J. Wendoloski and J. B. Matthew, *Proteins* **5,** 313 (1989).

[56] S. Northrup, T. G. Wensel, C. F. Meares, J. J. Wendoloski, and J. B. Matthew, *Proc. Natl. Acad. Sci. U.S.A.* **87,** 9503 (1990).

[57] K. Langsetmo, J. A. Fuch, C. Woodward, and K. A. Sharp, *Biochemistry* **30,** 7609 (1991).

[58] H. Frauenfelder, S. G. Sligar, and P. Wolynes, *Science* **254,** 1598 (1991).

[59] R. Elber and M. Karplus, *Science* **235,** 318 (1987).

[60] A. Ansari, C. M. Jones, E. R. Henry, J. Hofrichter, and W. Eaton, *Science* **256,** 1196 (1992).

[61] H. Anni, J. M. Vanderkooi, K. A. Sharp, T. Yonetani, S. C. Hopkins, L. Herenyi, and J. Fidy, *Biochemistry* **33,** 3475 (1994).

[62] J. Fidy, K.-G. Paul, and J. M. Vanderkooi, *Biochemistry* **28,** 7531 (1989).

[63] J. Fidy, G. R. Holtom, K.-G. Paul, and J. M. Vanderkooi, *J. Phys. Chem.* **95,** 4364 (1991).

TABLE II
INHOMOGENEOUS BROADENING: WIDTHS OF 0,0 TRANSITION

Sample	2σ (μ) (cm^{-1})[a]
Chromophore in single crystal	≤ 1[b]
Chromophore in polycrystalline host	1–10[b]
Disordered solids (glasses, polymers)	100–1,000[b]
MP-horseradish peroxidase	52 (16,000), 52 (16,100)[c]
MP-horseradish peroxidase-NHA	43 (16,175)[c]
Zn(II) horse cytochrome *c* (native)	65 (16,989)[d]
Sn(IV) horse cytochrome *c* (native)	65 (18,270)[e]
Zn(II) horse cytochrome *c* (denatured)	361 (17,227)[d]
Porphyrin cytochrome *c* (horse)	48 (16,271), 76 (16,181)[f]
MgPP-myoglobin (horse)	50 (16,022), 76 (16,127)[g]
PP-myoglobin (horse)	25 (16,022), 38 (16,127)[g]

[a] Data obtained from fits to the Gaussian distribution function $I(\bar{\nu}) = I_{max}(\bar{\nu})e^{-[(\bar{\nu}-\mu)^{2}/2\sigma^{2}]}$. All the data refer to distributions obtained at ~10 K or lower.

[b] K. K. Rebane, *Chem. Phys.* **189,** 139 (1994).

[c] J. Fidy, K.-G. Paul, and J. M. Vanderkooi, *Biochemistry* **28,** 7531 (1989). MP is mesoporphyrin IX; the name of horseradish peroxidase was substituted with MP. PP is protoporphyrin IX, and the heme of myoglobin was substituted with it. NHA is naphthohydroxamic acid, a substrate of horseradish peroxidase.

[d] V. Logovinsky, A. D. Kaposi, and J. M. Vanderkooi, *Biochim. Biophys. Acta* **1161,** 149 (1993). Iron was removed from cytochrome *c*, and replaced with zinc or tin, or was used without metal (porphyrin cytochrome *c*).

[e] M. Laberge and J. M. Vanderkooi, *Biospectroscopy* **1,** 413 (1995).

[f] V. Logovinsky, A. D. Kaposi, and J. M. Vanderkooi, *J. Fluoresc.* **1,** 79 (1991).

[g] A. D. Kaposi, J. Fidy, S. S. Stavrov, and J. M. Vanderkooi, *J. Phys. Chem.* **97,** 6317 (1993). The heme was replaced with Mg-protoporphyrin IX (MgPP) or protoporphyrin IX (PP).

of a denaturant.[64] The values for the inhomogeneous width of the optical transition for various chromophores in crystalline, amorphous, and protein matrices are given in Table II. In general, the distribution of inhomogeneity for porphyrins in heme proteins cannot be fit by a single Gaussian function, consistent with the existence of subconformations. The distribution functions are continuous, a result expected because there are many neighboring

[64] V. Logovinsky, A. D. Kaposi, and J. M. Vanderkooi, *Biochim. Biophys. Acta* **1161,** 149 (1993).

atoms, and hence many atomic contacts. This supports the glasslike model of proteins. It is striking, however, that proteins have thus far consistently yielded distributions of ground-state electronic levels that straddle distributions found in crystalline and completely amorphous materials (Table II). Moreover, the distributions found in many proteins can rarely be fit using a mono-Gaussian and in most cases require two or more Gaussians—unlike what would be expected for an amorphous material. The structure of the components that give rise to the multiple distributions remains obscure (although when the free-base porphyrin is used, the different distributions may represent different tautomers; see as follows). The PDF data suggest that, although glasslike by some experimental criteria, proteins are different and thus cannot be completely described by the physics used to model the glassy state. Hole-burning studies are in agreement with this notion and seem to indicate that, in addition to possessing some glasslike properties at low temperature, proteins possess properties indicative of a more structured nature. Thermally induced spectral diffusion broadening in horseradish peroxidase substituted with mesoporphyrin IX demonstrates a discrete step around 12 K suggesting a correlated structural change of the protein.[65] These data suggest conformational substrates with low barriers ($\sim$8 cm^{-1}). It would prove interesting to probe the distribution function as a function of temperature especially in light of the possibility of concerted conformational changes that seem to occur at low temperature. Also, the width of the distributions may prove relevant to the maximal excursions in protein fluctuations occurring at physiological temperatures.

Energy Transfer

For monomeric dilute protein systems, the chromophore, by virtue of being buried within interior hydrophobic portions of the protein, avoids the complication of chromophore–chromophore energy transfer. This need not be the case for proteins when complexes are formed, or when there are multiple binding sites for the chromophore in one protein. Energy transfer between fluorescing donor and acceptors will decrease the contribution of the ZPL.[66] Because energy transfer requires energy matching, and the sample is homogeneously broadened, the possibility of transfer from one ZPL to another is unlikely. More likely, the transfer will occur through the phonon wing. In the latter case the energy selection is lost because the broad phonon wing will excite many molecules in the energy transfer process. The effect of energy transfer is illustrated by the work of Avarmaa

[65] J. Zollfrank, J. Friedrich, J. M. Vanderkooi, and J. Fidy, *Biophys. J.* **59,** 305 (1991).
[66] R. Avarmaa, R. Jaaniso, K. Mauring, I. Renge, and R. Tamkivi, *Mol. Phys.* **57,** 605 (1986).

et al. on the spectral characteristics of an intact leaf.[11,67] Resolution was achieved only when the chlorophyll content was lowered. By growing the plant in darkened conditions, less chlorophyll was synthesized; from the "etiolated" leaf sharp emission lines attributable to chlorophyll could be detected.

Reactions in Excited State

Taking the view that chromophores in proteins are inhomogeneously broadened owing to differences in the positions of the neighboring atoms, one can use energy selection to examine the reactivity of subpopulations. One type of reaction that has been studied is phototautomerization. In free base porphyrin the central hydrogens can reside at the two opposite pyrrole positions and, thus, two tautomers are possible. In addition, the hydrogen atoms do not lie in the plane of the porphyrin ring and so, for each tautomer, there are two up or down positions. Finally, the protein can also bend or twist the porphyrin. The work of Volkers and van der Waals has shown that the absorption and emission of the tautomers of porphine in a hexane crystal are shifted and that one tautomer can convert to the other with light irradiation at 4 K.[68] Similarly, light irradiation of free base porphyrin bound to the heme site in horseradish peroxidase results in conversion of one form into another.[30,69,70] The existence of various tautomers with different transition energies immediately indicates that the heme pocket imposes an asymmetric electric field on the porphyrin.[46] Because the various tautomer/conformers can interconvert in the excited state, irradiation can prepare one or the other form, which can permit further investigation of the asymmetric field imposed on the porphyrin by the protein by using added external electric fields, altered temperature, or pressure.[65,71–73]

Another promising technique is single-molecule fluorescence spectroscopy.[74–76] The emission of a single molecule can be recorded with the assistance of scanning or imaging techniques but a narrow laser, such as a ring dye, could also be used: if the laser width is narrower than the homoge-

[67] I. Renge, K. Mauring, and R. Avarmaa, *Biochim. Biophys. Acta* **766,** 501 (1984).
[68] S. Volkers and J. H. van der Waals, *Mol. Phys.* **32,** 5589 (1976).
[69] J. Fidy, H. Koloczek, K. G. Paul, and J. M. Vanderkooi, *Phys. Chem. Lett.* **142,** 562 (1987).
[70] J. Fidy, K. G. Paul, and J. M. Vanderkooi, *J. Phys. Chem.* **93,** 2253 (1989).
[71] J. Friedrich, J. Gafert, J. Zollfrank, J. Vanderkooi, and J. Fidy, *Proc. Natl. Acad. Sci. U.S.A.* **91,** 1029 (1994).
[72] J. Gafert, J. Friedrich, J. M. Vanderkooi, and J. Fidy, *J. Phys. Chem.* **98,** 2210 (1994).
[73] J. Zollfrank, J. Friedrich, J. Fidy, and J. M. Vanderkooi, *J. Chem. Phys.* **94,** 8600 (1991).
[74] W. E. Moerner and T. Basche, *Angew. Chemie-Int. Ed. Engl.* **32,** 457 (1993).
[75] A. B. Myers, P. Tchenio, M. Z. Zgierski, and W. E. Moerner, *J. Phys. Chem.* **98,** 10377 (1994).
[76] E. Betzig, *Science* **262,** 1422 (1993).

neous linewidth, excitation at any frequency within the inhomogeneous absorption band will excite a small fraction of the total molecules in the inhomogeneous envelope. As the laser is moved further away from the absorption maximum, the density of absorbing centers becomes sufficiently low so that at any excitation frequency, there should be one and only one molecule in resonance. So far, the technique has been used with small crystalline and polymeric systems and awaits application to protein systems.

Summary

Perhaps the most important contribution of FLN is that it provides an experimental approach to relate physical changes in the protein to predicted dynamical behavior. It is clear that the sample is inhomogeneously broadened in a continuous manner, consistent with the damped motion of proteins. At the same time configurational substates can be selected, suggesting that there is indeed a hierarchy of protein motion and structure. As yet, identification of the structure, and relating it to the spectra, has not been achieved. It is clear that the electric field exerted by neighboring atoms shifts the electronic transition, and the inhomogeneity is greater when the surrounding disorder is greater. The inhomogeneity for the chromophore in the protein is dependent on the protein conformation and is intermediate between that of a crystal and a glass. The phonon coupling also depends on the chromophore and the protein. Fluorescence line narrowing provides in addition ground- and excited-state vibrational frequencies, thereby allowing for structural differences between the excited-state and the ground-state molecule to be detected.

Acknowledgment

This work was supported by NIH Grant PO1 GM48130.

[6] Determination of Ground-State Dissociation Constant by Fluorescence Spectroscopy

By ANDRZEJ KOWALCZYK, NOËL BOENS, and MARCEL AMELOOT

Introduction

Fluorescent indicators for the nondestructive determination of concentrations of intracellular calcium, sodium, potassium, and magnesium ions

0076-6879/97 $25

and of intracellular pH have both facilitated and stimulated extensive new research in a variety of areas.[1–4] These indicators change their absorption (excitation) and/or emission properties on binding the ion of interest. The measurements leading to the determination of the ground-state dissociation constant of the formed complexes and subsequently the determination of the unknown ion concentration are usually performed by steady-state fluorimetry. Modern commercial fluorimeters are generally equipped with sophisticated hardware and software to monitor the ion concentrations automatically.

Knowledge of the dissociation constant is important in many fields, e.g., the study of protolytic reactions,[5] ligand–protein interactions,[6] inclusion complexes with cyclodextrins,[7] partitioning between two phases (e.g., between aqueous and micellar phases[8]), and the environment.[9]

Spectrophotometry and fluorimetry are powerful tools for the determination of ground-state dissociation constants. Fluorescence spetroscopy can be used to evaluate dissociation constants provided that the association–dissociation leads to a change in the measured steady-state fluorescence signal. The advantages of fluorescence over absorption measurements comprise higher sensitivity (because fluorescence is detected vs. a dark background) and selectivity (one may avoid the signal from other absorbing molecules). Less probe is required in fluorimetry than in spectrophotometry to attain a similar sensitivity. Fluorimetry generally results in less disturbance of the system and allows less soluble probes to be used. Fluorescence can also be used with samples of high turbidity. Furthermore, the geometric requirements for fluorescence detection are less stringent than for absorption.

However, as fluorescence depends on both ground- and excited-state

[1] R. P. Haughland, "Handbook of Fluorescent Probes and Research Chemicals" (K. D. Larison, ed.), 5th Ed. Molecular Probes, Eugene, Oregon, 1992.

[2] R. Tsien, *Methods Cell Biol.* **30,** 127 (1989).

[3] A. Waggoner, Fluorescent Probes for Analysis of Cell Structure, Function, and Health by Flow and Imaging Cytometry *in* "Applications of the Fluorescence in the Biomedical Sciences" (D. L. Taylor, A. S. Waggoner, R. F. Murphy, F. Lanni, and R. R. Birge, eds.), p. 3. Alan R. Liss, New York, 1986.

[4] A. W. Czarnik, ed., "Fluorescent Chemosensors for Ion and Molecule Recognition," ACS Symposium Series No. 538. ACS, Washington DC, 1992.

[5] A. Weller, *Prog. React. Kinet.* **1,** 189 (1961).

[6] L. D. Ward, *Methods Enzymol.* **117,** 400 (1985).

[7] H. Mwakibete, R. Cristantino, D. M. Bloor, E. Wyn-Jones, and J. F. Holzwarth, *Langmuir* **11,** 57 (1995).

[8] M. Wilhelm, C. L. Zhao, Y. Wang, R. Xu, M. A. Winnik, J. L. Mura, G. Riess, and M. D. Croucher, *Macromolecules* **24,** 1033 (1991).

[9] M. J. Melo, E. Melo, and F. Pina, *Arch. Environ. Contam. Toxicol.* **26,** 510 (1994).

parameters, the fluorimetric determination of the ground-state dissociation constant is not as straightforward as when using spectrophotometry. Whether the ground-state dissociation constant will be correctly determined by fluorimetry depends on the values of the excited-state rate constants and/or on the chosen excitation and emission wavelengths.

The problem of interference of the excited-state reaction in the correct determination of the ground-state dissociation constant was first addressed by Rosenberg and co-workers,[10,11] who indicated that an appropriate selection of instrumental settings could eliminate this interference. Unfortunately, their recommendations appear to be either incorrect or inapplicable.[12]

In this chapter we derive the functional dependence of the measured steady-state fluorescence signal on the concentration of the titrant in the presence of an excited-state reaction. We limit our study to models with a 1:1 stoichiometry between the titrant and the free form of the fluorescent indicator. Only models characterized by a single dissociation constant are considered. We analyze the conditions under which the fluorimetric titration can be used to measure this dissociation constant correctly. It is shown that the interference of the excited-state reaction is avoided when the fluorescence signal is monitored at the isoemissive point. In the absence of an isoemissive point a criterion based on the titrant concentration dependence of the fluorescence decay times provides information on whether the dissociation constant can be correctly determined.

Fluorimetric Titration: Direct Method

General Case

Consider a system consisting of two distinct types of ground-state species and two corresponding types of excited-state species. The kinetic model for such a system is presented in Scheme I. In the ground-state species **1** can undergo a reversible reaction with X to form species **2** described by the dissociation constant K_d [Eq. (1)],

$$K_d = \frac{[1][X]}{[2]} \tag{1}$$

where brackets denote molarities. Species **1** represents the free form of a fluorescent indicator, while species **2** the corresponding bound form. If

[10] L. S. Rosenberg, J. Simons, and S. G. Schulman, *Talanta* **26,** 867 (1979).

[11] L. S. Rosenberg, G. Lam, C. Groh, and S. G. Schulman, *Anal. Chim. Acta* **106,** 81 (1979).

[12] A. Kowalczyk, N. Boens, V. Van den Bergh, and F. C. De Schryver, *J. Phys. Chem.* **98,** 8585 (1994).

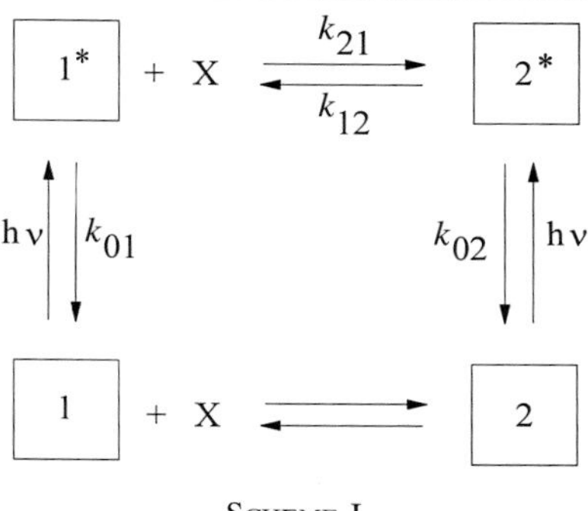

SCHEME I

Scheme I depicts the ionization of a weak acid (species **2**), species **1** is the conjugated base of the weak acid and X is the solvated hydrogen ion. It is assumed that only species **1** and **2**, which are in chemical equilibrium with each other, absorb light at the excitation wavelength λ^{ex}. The model depicted in Scheme I assumes a 1:1 stoichiometry between species **1** and X. Excitation by light creates the excited-state species **1*** and **2***, which can decay by fluorescence (F) and via nonradiative (NR) processes. The composite rate constants for those processes are denoted by k_{01} ($= k_{F1} + k_{NR1}$) and k_{02} ($= k_{F2} + k_{NR2}$). The second-order rate constant describing the association $\mathbf{1}^* + X \rightarrow \mathbf{2}^*$ is represented by k_{21}. The first-order rate constant for the dissociation of **2*** into **1*** and X is denoted by k_{12}.

If the system shown in Scheme I is excited with a δ pulse that does not significantly alter the concentrations of the ground-state species (i.e., in the low excitation limit), the fluorescence decays as a sum of two exponentials[13]

$$f(\lambda^{ex}, \lambda^{em}, t) = \alpha_1 \exp(-t/\tau_1) + \alpha_2 \exp(-t/\tau_2) \qquad t \geq 0 \tag{2}$$

where the amplitudes α_i depend on all the rate constants k_{ij}, the ground-state dissociation constant K_d, the concentration of X, and in addition on λ^{em} and λ^{ex}. The decay times τ_i [Eq. (3)], however, depend exclusively on k_{ij} and [X].

$$(\tau_{1,2})^{-1} = \tfrac{1}{2}\{A_1 + A_2 \mp [(A_1 - A_2)^2 + 4k_{21}k_{12}[X]]^{1/2}\} \tag{3a}$$

with

$$A_1 = k_{01} + k_{21}[X] \tag{3b}$$
$$A_2 = k_{02} + k_{12} \tag{3c}$$

[13] J. B. Birks, "Photophysics of Aromatic Molecules." John Wiley & Sons, New York, 1970.

If the system depicted in Scheme I is excited with light of constant intensity, and if the absorbance of the sample is less than 0.1, the measured steady-state fluorescence signal $F(\lambda^{ex}, \lambda^{em}, [X])$ observed at λ^{em} due to excitation at λ^{ex} is given by

$$F(\lambda^{ex}, \lambda^{em}, [X]) = \{\varepsilon_1(\lambda^{ex})a_1(\lambda^{em}, [X])[1] + \varepsilon_2(\lambda^{ex})a_2(\lambda^{em}, [X])[2]\}\psi(\lambda^{ex}, \lambda^{em}) \quad (4)$$

where $\varepsilon_i(\lambda^{ex})$ represents the molar extinction coefficient of species **i** at λ^{ex}. The coefficients a_i are given by[12]

$$a_1(\lambda^{em}, [X]) = \frac{(k_{02} + k_{12})c_1(\lambda^{em}) + k_{21}[X]c_2(\lambda^{em})}{k_{01}(k_{02} + k_{12}) + k_{02}k_{21}[X]} \quad (5a)$$

$$a_2(\lambda^{em}, [X]) = \frac{k_{12}c_1(\lambda^{em}) + (k_{01} + k_{21}[X])c_2(\lambda^{em})}{k_{01}(k_{02} + k_{12}) + k_{02}k_{21}[X]} \quad (5b)$$

The emission weighting factors $c_i(\lambda^{em})$ are defined as

$$c_i(\lambda^{em}) = k_{Fi} \int_{\Delta\lambda^{em}} \rho_i(\lambda^{em})d\lambda^{em} \quad (6)$$

k_{Fi} is the fluorescence rate constant of species **i***; $\Delta\lambda^{em}$ is the emission wavelength interval where the fluorescence signal is monitored; $\rho_i(\lambda^{em})$ is the steady-state fluorescence spectrum of species **i*** normalized to unity. The ratio $c_i(\lambda^{em})/k_{0i}$ plotted vs λ^{em} represents the emission spectrum of species **i*** normalized to its quantum yield.

The factor $\psi(\lambda^{ex}, \lambda^{em})$ includes instrumental parameters,

$$\psi(\lambda^{ex}, \lambda^{em}) = 2.3dI_0(\lambda^{ex})\xi(\lambda^{em}) \quad (7)$$

where d denotes the excitation light path, $I_0(\lambda^{ex})$ represents the intensity of the exciting light at λ^{ex} impinging on the sample, and $\xi(\lambda^{em})$ is a factor taking into account the efficiency of both the optics and the detector. A change in I_0 and detector sensitivity is automatically taken into account in most modern fluorimeters. Therefore, if in addition the geometry of the experimental set-up is preserved, the factor $\psi(\lambda^{ex}, \lambda^{em})$ can be treated as a constant.

Equation (4) can be rewritten in terms of K_d [Eq. (1)] and the total analytical concentration, C_T, of the fluorescent indicator.

$$C_T = [1] + [2]$$

$$F(\lambda^{ex}, \lambda^{em}, [X]) = \frac{\varepsilon_1(\lambda^{ex})a_1(\lambda^{em}, [X])K_d + \varepsilon_2(\lambda^{ex})a_2(\lambda^{em}, [X])[X]}{K_d + [X]} \times C_T\psi(\lambda^{ex}, \lambda^{em}) \quad (9)$$

If the coefficients a_i are independent of [X], $F(\lambda^{ex}, \lambda^{em}, [X])$ plotted vs. $-\log[X]$ displays a sigmoidal shape with a unique inflection at $-\log[X] = -\log K_d \equiv pK_d$. In fact, the coefficients a_i generally depend on [X]. The functional dependence of these coefficients on $-\log[X]$ is also sigmoidal with a unique inflection at[9,12]

$$-\log[X] = -\log\left[\frac{k_{01}(k_{02} + k_{12})}{k_{02}k_{21}}\right] \equiv pK^{\#} \tag{10}$$

The fast variation of the coefficients a_i in the region of $-\log[X]$ around $pK^{\#}$ may result in additional inflections of $F(\lambda^{ex}, \lambda^{em}, [X])$. Some representative examples of possible titration curves are presented in Figs. 1a and b and 2a. Experimental examples are shown in Figs. 3a, 4a, and 5a.

It is noteworthy that $pK^{\#}$ does not coincide with $pK_d^* = -\log k_{12}/k_{21}$. Although for certain sets of rate constants, both values can be close (see Table I), in principle one cannot use the position of any inflection point to obtain information about pK_d^*.

Usually, the nonsigmoidal shape of some of the titration curves shown in Figs. 1b and 2a excludes the unambiguous determination of pK_d from fluorimetric titrations only. For some combinations of the rate constants and instrumental settings, the excited-state reaction may result only in a shift of the apparently undisturbed sigmoidal titration curve (Fig. 1a and b). This is the most misleading situation because the fluorimetric titration provides no clue that the excited-state reaction interferes with the determination of the correct pK_d.

Special Cases

It follows from Eq. (9) that the independence of the coefficients a_i with respect to [X] is a prerequisite to obtain sigmoidal titration curves F with a unique inflection at $-\log[X] = pK_d$.

In the next section we show that the coefficients a_i become independent of [X] if the fluorescence is observed at the isoemissive point. The subsequent section provides an independent method to determine the range of [X] where the a_i coefficients are practically constant. This test is useful if there is no isoemissive point.

Fluorescence Collected at Isoemissive Point. Assume that the steady-state fluorescence spectra of the individual species **1*** and **2*** are normalized in such a way that the area under each plot is equal to their respective quantum yield. Such plots are described by the previously [Eq. (6)] introduced functions $c_i(\lambda^{em})/k_{0i}$. If these emission spectra are shifted but overlap partially, the wavelength(s) where they cross is (are) called the isoemissive point(s) λ^{isoem}. If these emission spectra are identical, each point of the

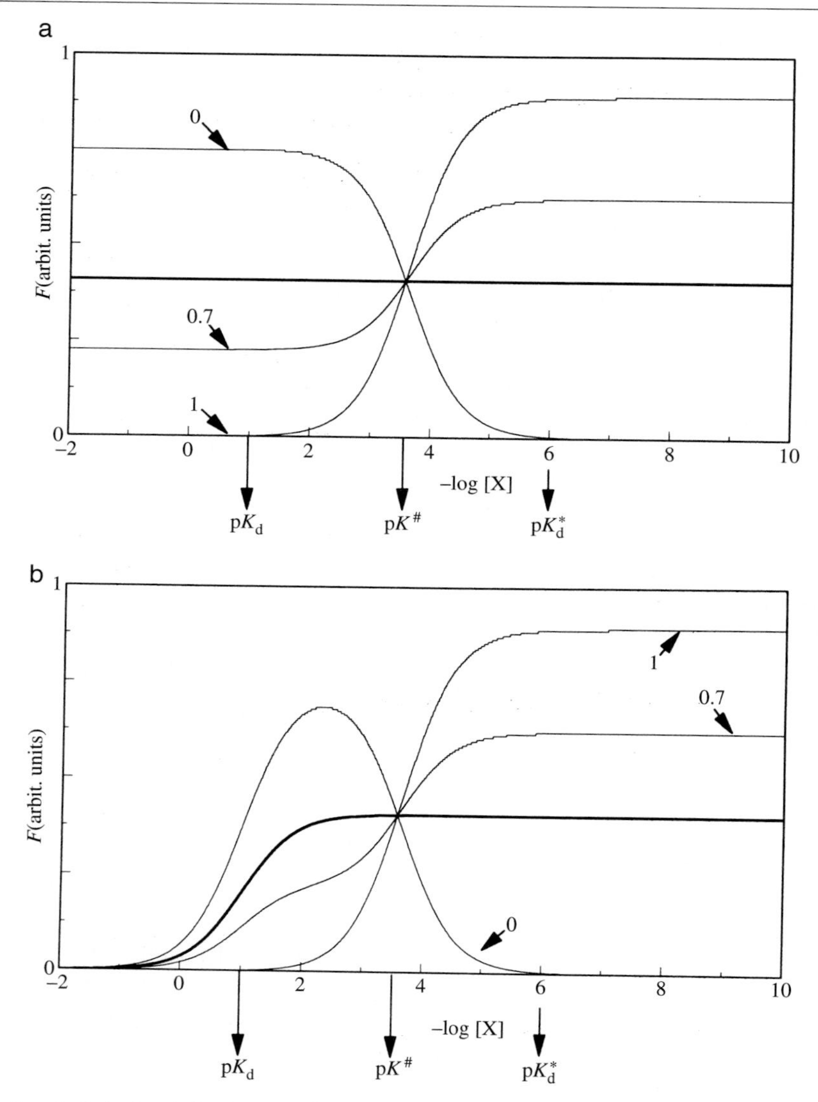

FIG. 1. Calculated fluorimetric titration curves for the system with $k_{01} = 0.112$ nsec^{-1}, $k_{02} = 0.133$ nsec^{-1}, $k_{12} = 5 \times 10^{-4}$ nsec^{-1}, $k_{21} = 500$ (nsec $M)^{-1}$. (a) Excitation at isosbestic point $\varepsilon_1 = \varepsilon_2 = 20{,}000\ M^{-1}$ cm^{-1}. (b) Excitation at λ^{em}, where $\varepsilon_2 = 0\ M^{-1}$ cm^{-1} and $\varepsilon_1 = 20{,}000\ M^{-1}$ cm^{-1}. The numbers on the curves are the c_1 values used. The sum of c_1 and c_2 is normalized to unity. The curves corresponding to the isoemissive point are drawn in boldface. (c) Corresponding decay times as a function of $-\log[X]$ calculated according to Eq. (3). [Adapted with permission from A. Kowalczyk, N. Boens, V. Van den Bergh, and F. C. De Schryver, *J. Phys. Chem.* **98,** 8585 (1994). Copyright 1994 *American Chemical Society.*]

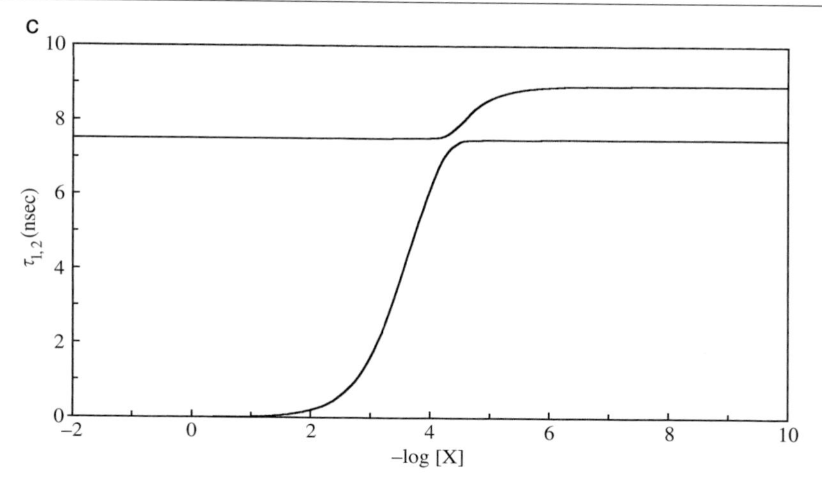

FIG. 1. (*continued*)

emission spectra is an isoemissive point. In all other cases no isoemissive point will be observed. At an isoemissive point one has[14]

$$c_1(\lambda^{\text{isoem}})/k_{01} = c_2(\lambda^{\text{isoem}})/k_{02} \tag{11}$$

The fluorescence quantum yield limited to the wavelength interval around λ^{isoem} is the same for **1*** and **2***. Therefore, the number of photons emitted at an isoemissive point is proportional to the total concentration of **1*** and **2*** and does not depend on the distribution between **1*** and **2***. As a consequence, the fluorescence signal at λ^{isoem} does not depend on the reaction in the excited state. This can be verified by substituting Eq. (11) and Eq. (5). For an isoemissive point the coefficients a_i become equal and independent of [X]:

$$a_1(\lambda^{\text{isoem}}) = a_2(\lambda^{\text{isoem}}) = c_1(\lambda^{\text{isoem}})/k_{01} = c_2(\lambda^{\text{isoem}})/k_{02} \tag{12}$$

As a_1 and a_2 are equal they can be factored out as a common factor in Eqs. (4) and (9). The fluorescence signal observed at an isoemissive point will be directly proportional to the total absorbance.

$$F(\lambda^{\text{ex}}, \lambda^{\text{isoem}}, [\text{X}]) = a_1(\lambda^{\text{isoem}})[\varepsilon_1(\lambda^{\text{ex}})[1] + \varepsilon_2(\lambda^{\text{ex}})[2]]\psi(\lambda^{\text{ex}}, \lambda^{\text{isoem}}) \tag{13}$$

[14] M. Ameloot, N. Boens, R. Andriessen, V. Van den Bergh, and F. C. De Schryver, *J. Phys. Chem.* **95,** 2041 (1991).

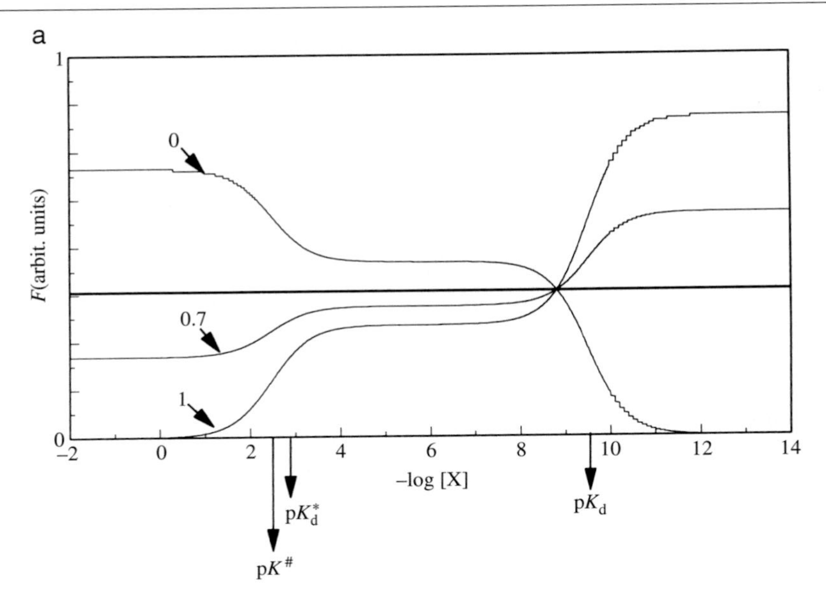

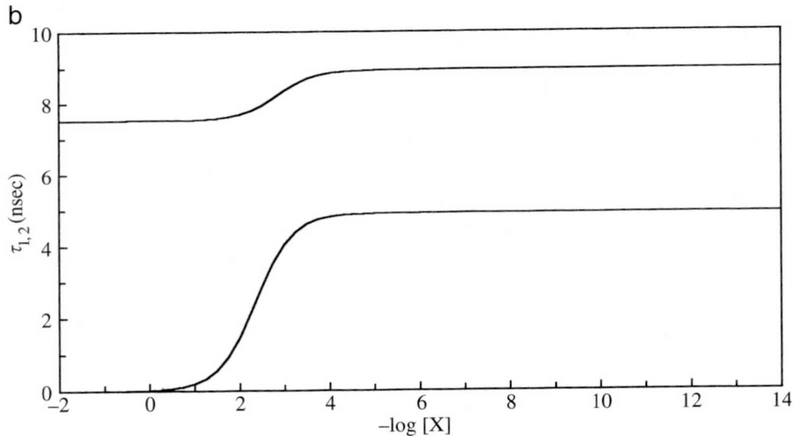

Fig. 2. 2-Naphthol–2-naphtholate system. (a) Calculated fluorimetric titration curves for excitation at the isosbestic point $\varepsilon_1 = \varepsilon_2 = 1500\ M^{-1}\ \text{cm}^{-1}$. The curves corresponding to the isoemissive point are drawn in boldface. (b) Decay times of 2-naphthol as a function of $-\log[X]$ calculated according to Eq. (3). The rate constants used are those of Table I. [Adapted with permission from A. Kowalczyk, N. Boens, V. Van den Bergh, and F. C. De Schryver, *J. Phys. Chem.* **98,** 8585 (1994). Copyright 1994 *American Chemical Society.*]

Equation (9) at an isoemissive point becomes

$$F(\lambda^{\mathrm{ex}}, \lambda^{\mathrm{isoem}}, [\mathrm{X}]) = a_1(\lambda^{\mathrm{isoem}}) \frac{\varepsilon_1(\lambda^{\mathrm{es}})K_\mathrm{d} + \varepsilon_2(\lambda^{\mathrm{ex}})[\mathrm{X}]}{K_\mathrm{d} + [\mathrm{X}]} C_\mathrm{T}\psi(\lambda^{\mathrm{ex}}, \lambda^{\mathrm{isoem}}) \quad (14)$$

The plot of F vs $-\log[\mathrm{X}]$ starts at $F_{\mathrm{min}} \equiv F([\mathrm{X}] \to 0) = \varepsilon_1(\lambda^{\mathrm{ex}})a_1(\lambda^{\mathrm{isoem}})$ $C_\mathrm{T}\psi(\lambda^{\mathrm{ex}}, \lambda^{\mathrm{isoem}})$ for low [X]. It changes rapidly around the inflection at $-\log[\mathrm{X}] = \mathrm{p}K_\mathrm{d}$. The fluorescence then asymptotically approaches $F_{\mathrm{max}} \equiv$ $F([\mathrm{X}] \to \infty) = \varepsilon_2(\lambda^{\mathrm{ex}})a_1(\lambda^{\mathrm{isoem}})C_\mathrm{T}\psi(\lambda^{\mathrm{ex}}, \lambda^{\mathrm{isoem}})$.

In that case the fluorimetric titration curve yields a unique inflection point at $-\log[\mathrm{X}] = \mathrm{p}K_\mathrm{d}$ that corresponds to the midpoint of titration. These cases are illustrated in Figs. 1a and b and 2a by lines drawn in boldface.

Equation (14) is often rearranged in the form of a Hill plot:

$$\log[(F - F_{\mathrm{min}})/(F_{\mathrm{max}} - F)] = \log[\mathrm{X}] - \log K_\mathrm{d} \quad (15)$$

The expression of the left side of Eq. (15) plotted vs log[X] should give a straight line that intersects the abscissa at the value corresponding to $-\log K_\mathrm{d}$. Most often this plot is used to determine K_d. Conversely, if K_d is known, the Hill plot may be used as a calibration curve to determine the unknown concentration of X, provided (1) the impinging light intensity does not vary, and (2) the total concentration C_T of the fluorescent indicator is the same as used to construct the calibration curve. Note that in principle only three fluorescence signals (F, F_{min}, and F_{max}) are sufficient together with the knowledge of K_d to determine an unknown [X].

It must be emphasized that an unfortunate choice of excitation at an isosbestic point and observation at an isoemissive point makes F invariant with [X], thus precluding the determination of K_d (Figs. 1a and 2a).

If an isoemissive point exists, all artifacts of excited-state reaction cancel out automatically. Therefore, it is recommended that fluorimetric titrations be performed at an isoemissive point. There are several methods to determine an isoemissive point.

PROCEDURE a. If the steady-state fluorescence spectra and (relative) quantum yields of the two species can be measured separately, one can normalize these spectra such that the area under each is proportional to their corresponding fluorescence quantum yield. A wavelength where both spectra intersects defines an isoemissive point [Eq. (12)]. If the (relative) quantum yields are not known, the crossing wavelength of the emission spectra of the separate excited-state species normalized to the same absorbance provides the isoemissive point. Because procedure a requires the correct shape of the spectra, they should be corrected for the detection channel sensitivity.

Procedure b. If there is no possibility of obtaining emission spectra of the separate species, the total emission spectra at different [X] can be monitored and normalized to the same number of absorbed photons (i.e., the same absorbance). The emission wavelength where the normalized emission spectra cross defines the isoemissive point. In the special case of excitation at the wavelength corresponding to the true isosbestic point with constant C_T, the emission spectra do not need to be normalized to determine the isoemissive point. Procedure b assumes that $\psi(\lambda^{ex}, \lambda^{em})$ is constant between different fluorescence measurements.

Procedure c. A third method to determine the isoemissive point is provided by global compartmental analysis of time-resolved experiments.[14] Indeed, global compartmental analysis yields estimates of $c_1(\lambda^{em})/c_2(\lambda^{em})$ as a function of emission wavelength, together with values of k_{01} and k_{02}. The emission wavelength where Eq. (11) holds defines an isoemissive point.

Titration in Region of Constant Decay Times. When the fluorescence signal is not monitored at the isoemissive point, the association reaction in the excited state must be slow enough to avoid distortions of the fluorimetric titrations. In this section we provide a method to judge whether interference caused by the excited-state reaction is indeed negligible for K_d determination.

If the rate k_{21}[X] describing the association of **1*** with X is negligible within the range of used X concentrations, the coefficients a_i become practically independent of [X]. The inflection of the fluorimetric titration curve within this [X] range corresponds to K_d. In the limiting case when $k_{21} = 0$, both coefficients are completely independent of [X] and are given by

$$a_1(\lambda^{em}) = \frac{c_1(\lambda^{em})}{k_{01}} \tag{16a}$$

$$a_2(\lambda^{em}) = \frac{k_{12}c_1(\lambda^{em}) + k_{01}c_2(\lambda^{em})}{k_{01}(k_{02} + k_{12})} \tag{16b}$$

In such a case the fluorimetric titration curve shows a unique inflection at $-\log[X] = pK_d$. It should be emphasized that this case represents not merely a theoretical exercise but has important useful applications. A relatively simple experimental test based on time-resolved fluorescence measurements can be used to assess the interference of the excited-state reaction on the fluorimetric determination of K_d. As can be seen from Eq. (3), for the limiting case of $k_{21} = 0$ the decay times become independent of [X]. It can be inferred that, if the decay times are constant over the used concentration range of X, the contribution of the term k_{21}[X] is small enough to make the coefficients a_i practically independent of [X]. Therefore, collecting fluorescence decay traces of the same samples as used in the fluorimetric titration and analyzing them simultaneously[15] as biexponentials with linked decay times allows one to judge whether the infection point of the apparently undisturbed titration curve really corresponds to pK_d. If an inflection occurs in the [X] range of the fluorimetric titration curve where the decay times are invariant, this inflection point can be correctly associated with pK_d. In contrast, the inflection point(s) in the [X] range where the decay times vary cannot be attributed to ground-state dissociation.

It must be noted that the described test based on the simultaneous biexponential analysis of fluorescence decay traces collected over a limited [X] range is much less demanding experimentally than the rigorous global compartmental analysis to estimate numerical values of k_{ij} and K_d. Indeed, there is no need to extend the range of X concentrations in global analysis beyond that used in the fluorimetric titration.

Figure 1a and b presents fluorimetric titration curves obtained for the same photophysical system but under different experimental conditions of excitation and emission. The curves of Fig. 1a show an inflection at $-\log[X] = 3.5$, whereas some curves of Fig. 1b exhibit an additional inflection at $-\log[X] = 1$. Plots of the decay times corresponding to the cases depicted in Fig. 1a and b are shown in Fig. 1c. The variability of the decay times in the region $2 < -\log[X] < 5$ indicates that the inflections shown in Fig. 1a and b in this [X] range are not related to the ground-state dissociation. The constancy of the decay times in the [X] domain of the additional inflection of the curves of Fig. 1b allows one to assign this inflection point to pK_d (=1).

Fluorimetric Titration: Ratiometric Method

Because of bleaching and leakage of a fluorescent indicator from the investigated system, it is often difficult to keep C_T constant during a whole cycle of measurements. In such a case the ratiometric method is the method of choice.[16] The ratiometric method can be performed in two ways. For each concentration of X, one can measure the ratio of the fluorescence signals either at two different excitation wavelengths and at a common emission wavelength, or at two different emission wavelengths and a single excitation wavelength.

The ratio R obtained from dual excitation wavelengths at a common observation wavelength is given[17] by Eq. (17),

[15] J. R. Knutson, J. M. Beechem, and L. Brand, *Chem. Phys. Lett.* **102,** 501 (1983).

[16] G. Grynkiewicz, M. Poenie, and R. Y. Tsien, *J. Biol. Chem.* **260,** 3440 (1985).

[17] V. Van den Bergh, N. Boens, F. C. De Schryver, M. Ameloot, P. Steels, J. Gallay, M. Vincent, and A. Kowalczyk, *Biophys. J.* **68,** 1110 (1995).

$$R(\lambda_1^{ex}/\lambda_2^{ex}, \lambda^{em}, [X]) = \left[\frac{S_1(\lambda_1^{ex}, \lambda^{em}, [X])K_d + S_2(\lambda_1^{ex}, \lambda^{em}, [X])[X]}{S_1(\lambda_2^{ex}, \lambda^{em}, [X])K_d + S_2(\lambda_2^{ex}, \lambda^{em}, [X])[X]}\right]\left[\frac{I_0(\lambda_1^{ex})}{I_0(\lambda_2^{ex})}\right] \tag{17}$$

with $S_i(\lambda_j^{ex}, \lambda^{em}, [X])$ given by

$$S_i(\lambda^{ex}, \lambda^{em}, [X]) = \varepsilon_i(\lambda^{ex})a_i(\lambda^{em}, [X]) \tag{18}$$

The dependence on I_0 can be eliminated by measuring directly F/I_0.

The plot of R as a function of $-\log[X]$ may have a complicated shape unless the coefficients S_i are independent of [X]. This condition can be verified by the constancy of the decay times within the range of [X] used or can be achieved by selection of the isoemissive point as the observation wavelength. In that case the plot of R as a function of $-\log[X]$ has a unique inflection point at $-\log[X] = pK_d - \log S_1(\lambda_2^{ex}, \lambda^{em})/S_2(\lambda_2^{ex}, \lambda^{em})$. The corresponding Hill plot [Eq. (19)] crosses the abscissa in the same point.

$$\log\left(\frac{R - R_{min}}{R_{max} - R}\right) = \log[X] - \log\left[K_d\frac{S_1(\lambda_2^{ex}, \lambda^{em})}{S_2(\lambda_2^{ex}, \lambda^{em})}\right] \tag{19}$$

In Eq. (19), R_{min} and R_{max} represent the ratio R [Eq. (17)] measured at extreme values of [X].

It must be emphasized that the ratio $S_1(\lambda_2^{ex}, \lambda^{em})/S_2(\lambda_2^{ex}, \lambda^{em})$ must be known to determine K_d. This ratio is determined by fluorescence measurements of the same system at the indicated wavelengths using extreme values of [X]: $F_{min}(\lambda_2^{ex}, \lambda^{em}, [X] \to 0) = S_1(\lambda_2^{ex}, \lambda^{em})C_T\psi(\lambda_2^{ex}, \lambda^{em})$ and $F_{max}(\lambda_2^{ex}, \lambda^{em}, [X] \to \infty) = S_2(\lambda_2^{ex}, \lambda^{em})C_{T1}\psi(\lambda_2^{ex}, \lambda^{em})$. Note that these measurements must be performed at the same indicator concentration to ensure that $F_{min}/F_{max} = S_1/S_2$. However, if the observation wavelength is set at the isoemissive point, the ratio $S_1(\lambda^{em}, \lambda_2^{ex})/S_2(\lambda^{em}, \lambda_2^{ex})$ simplifies to $\varepsilon_1(\lambda_2^{ex})/\varepsilon_2(\lambda_2^{ex})$. Furthermore if λ_2^{ex} is set at the isosbestic point, no correction is needed and the titration will give directly the value of pK_d.

An inappropriate choice of excitation and observation wavelengths can make R invariant with [X], thus inhibiting the determination of K_d. The ratio $R(\lambda_1^{ex}/\lambda_2^{ex}, \lambda^{em}, [X])$ is independent of [X] if

$$a_1(\lambda^{em})a_2(\lambda^{em})[\varepsilon_1(\lambda_2^{ex})\varepsilon_2(\lambda_1^{ex}) - \varepsilon_1(\lambda_1^{ex})\varepsilon_2(\lambda_2^{ex})] = 0 \tag{20}$$

This is the case when (1) one of the excited-state species does not emit at the observation wavelength $\lambda^{em}[a_1(\lambda^{em}) = 0$ or $a_2(\lambda^{em}) = 0]$ or when (2) $\varepsilon_1(\lambda_1^{ex})/\varepsilon_2(\lambda_1^{ex}) = \varepsilon_1(\lambda_2^{ex})/\varepsilon_2(\lambda_2^{ex})$. The second condition is fulfilled when the absorption spectra of ground-state species **1** and **2** have identical shapes.

The ratio R at dual emission wavelengths due to a common excitation wavelength is expressed by

$$R(\lambda^{ex}, \lambda_1^{em}/\lambda_2^{em}, [X]) = \left[\frac{S_1(\lambda^{ex}, \lambda_1^{em}, [X])K_d + S_2(\lambda^{ex}, \lambda_1^{em}, [X])[X]}{S_1(\lambda^{ex}, \lambda_2^{em}, [X])K_d + S_2(\lambda^{ex}, \lambda_2^{em}, [X])[X]}\right]\left[\frac{\xi(\lambda_1^{em})}{\xi(\lambda_2^{em})}\right] \quad (21)$$

Again, when S_i is independent of [X] the plot of R vs $-\log[X]$ has an inflection point at $-\log[X] = pK_d - \log S_1(\lambda^{ex}, \lambda_2^{em})/S_2(\lambda^{ex}, \lambda_2^{em})$. The corresponding Hill plot crosses the abscissa in the same point.

The ratio $S_1(\lambda^{ex}, \lambda_2^{em}/S_2(\lambda^{ex}, \lambda_2^{em})$ must be known to determine K_d. This ratio is experimentally determined by fluorescence measurements of the same system at the indicated wavelengths using extreme values of [X]: $F_{min}(\lambda^{ex}, \lambda_2^{em}, [X] \rightarrow 0) = S_1(\lambda^{ex}, \lambda_2^{em})C_T\psi(\lambda^{ex}, \lambda_2^{em})$ and $F_{max}(\lambda^{ex}, \lambda_2^{em}, [X] \rightarrow \infty) = S_2(\lambda^{ex}, \lambda_2^{em})C_T\psi(\lambda^{ex}, \lambda_2^{em})$. As before, the correction factor must be determined for constant C_T. If the excitation wavelength is set at the isoemissive point, the ratio $S_1(\lambda^{ex}, \lambda_2^{em})/S_2(\lambda^{ex}, \lambda_2^{em})$ simplifies to $a_1(\lambda_2^{em})/a_2(\lambda_2^{em})$. Furthermore, if λ_2^{em} is set at an isoemissive point, no correction is needed and the titration will give directly the value of pK_d.

An unsuitable choice of excitation and observation wavelengths can make R invariant with [X], thus precluding the determination of K_d. The ratio $R(\lambda^{ex}, \lambda_1^{em}/\lambda_2^{em}, [X])$ becomes independent of [X] when

$$\varepsilon_1(\lambda^{ex})\varepsilon_2(\lambda^{ex})[a_1(\lambda_2^{em})a_2(\lambda_1^{em}) - a_1(\lambda_1^{em})a_2(\lambda_2^{em})] = 0 \quad (22)$$

This last situation occurs when (1) one of the ground-state species does not absorb at the excitation wavelength λ^{ex} [$\varepsilon_1(\lambda^{ex}) = 0$, $\varepsilon_2(\lambda^{ex}) = 0$] or when (2) $c_1(\lambda_1^{em})/c_2(\lambda_1^{em}) = c_1(\lambda_2^{em})/c_2(\lambda_2^{em})$. This occurs when the fluorescence spectra of excited-state species **3*** and **2*** have identical shapes.

Case Studies

In this section we discuss some fluorescent ion indicator studied in detail by the present authors using time-resolved fluorescence experiments.[17–20] The global compartmental analysis yields numerical values of all excited-

[18] V. Van den Bergh, N. Boens, F. C. De Schryver, M. Ameloot, J. Gallay, and A. Kowalczyk, *Chem. Phys.* **166,** 249 (1992).

[19] Van den Bergh, N. Boens, F. C. De Schryver, J. Gallay, and M. Vincent, *Photochem. Photobiol.* **61,** 442 (1995).

[20] K. Meuwis, N. Boens, F. C. De Schryver, J. Gallay, and M. Vincent, *Biophys. J.* **68,** 2469 (1995).

state rate constants as well as the ground-state dissociation constant. These values are collected in Table I together with the $pK^{\#}$ and pK^{*} values calculated from the estimated k_{ij} values.

Once numerical values of all rate constants are determined, steady-state fluorimetric titration curves [Eq. (9)] and decay times [Eq. (3)] can be calculated. It is then possible to investigate the concentration dependence of the decay times and to assess the interference of the excited-state reaction on the fluorimetric determination of the ground-state dissociation constant.

2-Naphthol + H^+

The absorption and emission spectra of 2-naphthol shift on binding H^+. Values of the ground- and excited-state photophysical parameters are used to simulate titration curves (Fig. 2a) that in most cases display two inflections. The decay times calculated according to Eq. (3) are displayed in Fig. 2b. The decay times are invariant over the $[H^+]$ region of the inflection at pH 9.5 but change around pH 2–3, the region of the second inflection. The constancy of the decay times indicates that $k_{21}[H^+]$ is negligible around pH 9.5 and that as a consequence the pK_d value will be correctly recovered from steady-state fluorescence measurements. The inflection in the pH 2–3 range cannot be associated with the ground-state equilibrium. If the value of the middle plateau instead of the value of F or R at extremely low pH

TABLE I

RATE CONSTANT VALUES ESTIMATED BY GLOBAL COMPARTMENTAL ANALYSIS FOR SOME FLUORESCENT INDICATORS OF BIOLOGICALLY IMPORTANT IONS IN AQUEOUS SOLUTION

Indicator	k_{01} ($nsec^{-1}$)	k_{21} ($M^{-1}nsec^{-1}$)	k_{02} ($nsec^{-1}$)	k_{12} ($nsec^{-1}$)	pK_d	pK^{*a}	$pK^{\#b}$	Ref.
2-Naphthol/H^+	0.112	50	0.133	0.071	9.52	2.85	2.46	*c*
Fura-2/Ca^{2+}	1.424	50	0.549	0.032	6.89	3.19	1.52	*d*
Quin-2/Ca^{2+}	0.863	114	0.088	0.00004	7.15	6.45	2.12	*e*
PBFI/K^+	1.102	0.269	1.8	1.369	2.18	−0.71	−0.86	*f*

[a] $pK^{*} = -\log(k_{12}/k_{21})$.

[b] $pK^{\#}$ is given by Eq. (10).

[c] V. Van den Bergh, N. Boens, F. C. De Schryver, M. Ameloot, J. Gallay, and A. Kowalczyk, *Chem. Phys.* **166,** 249 (1992).

[d] V. Van den Bergh, N. Boens, F. C. De Schryver, M. Ameloot, P. Steels, J. Gallay, M. Vincent, and A. Kowalczyk, *Biophys. J.* **68,** 1110 (1995). The rate constant values are for EGTA as Ca^{2+} buffer.

[e] V. Van den Bergh, N. Boens, F. C. De Schryver, J. Gallay, and M. Vincent, *Photochem. Photobiol.* **61,** 442 (1995).

[f] K. Meuwis, N. Boens, F. C. De Schryver, J. Gallay, and M. Vincent, *Biophys. J.* **68,** 2469 (1995).

is taken as F_{max} and R_{max} in the Hill plot, the ground-state dissociation constant will be properly recovered. For this system an isoemissive point at 387 nm was determined by time-resolved fluorescence measurements.[18]

Fura-2 + Ca^{2+}

The excitation spectra of Fura-2 shift on binding Ca^{2+}. The fluorimetric titration curve reconstructed using data of Table I are displayed in Fig. 3a. The main inflection around $[Ca^{2+}] = 0.1\ \mu M$ is followed by a small one at much higher concentrations ($[Ca^{2+}] = 0.1\ M$). The decay times (Fig. 3b) are constant in the region where the titration curve exhibits the first inflection and start to vary in the region of the small inflection.[17] It may therefore be concluded that (1) the inflection at $[Ca^{2+}] = 132$ nM corresponds to the ground-state dissociation, and (2) because of the small difference between the two left plateaus it does not matter whether F_{max} and R_{max} are determined from the middle or the leftmost plateau.

Quin-2 + Ca^{2+}

This calcium indicator Quin-2 increases sixfold fluorescence intensity on binding Ca^{2+} without visible shift of the fluorescence spectrum. No isoemissive point exists for this system. The titration curve (Fig. 4a) is virtually undisturbed, showing only one inflection around $[Ca^{2+}] = 71$ mM. The comparison with the decay times (Fig. 4b) shows[19] that the titration curve might be distorted for Ca^{2+} concentrations above 1 mM. Again F_{max} and R_{max} can be taken safely at any $[Ca^{2+}]$ from the range 10 μM–1 mM.

PBFI + K^+

On binding K^+ the indicator PBFI increases its fluorescence intensity 2.5-fold and the excitation spectrum shifts. The reconstructed and measured titration curves are displayed in Fig. 5a. The main inflection is around $[K^+] = 6.6$ mM. The plateau for the higher concentrations is, however, slightly distorted owing to the excited-state reaction. The decay times (Fig. 5b) start to change just before the titration curve reaches a plateau.[20] Although most of the titration curve is not distorted, F_{max} and R_{max} are not so well defined. Therefore, the Hill plot might yield slightly different pK_d values.

Conclusion

We have demonstrated the pivotal role of the isoemissive point in deriving the correct K_d from fluorimetric titrations when an excited-state

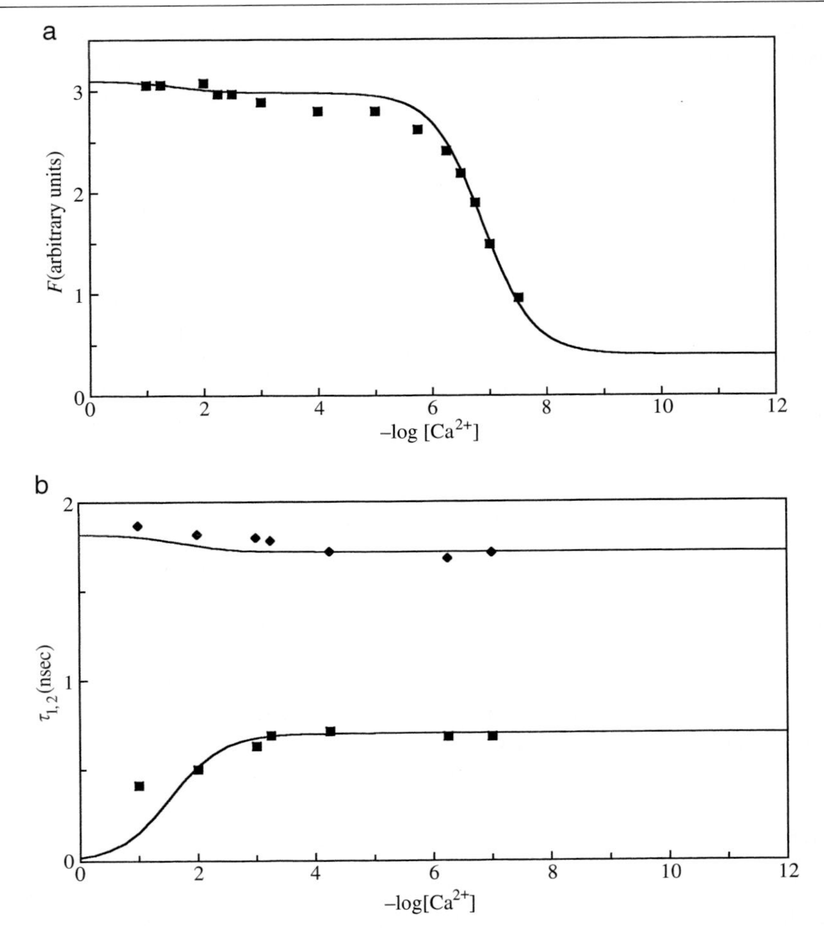

FIG. 3. Fura-2 at pH 7.2 with EGTA as Ca^{2+} buffer in aqueous solution at room temperature. (a) Experimental (■) and calculated (—) fluorimetric titration curve F(340 nm, 510 nm). The F(340 nm, 510 nm) values are calculated with K_d = 130 nM, ε_1 = 13,300 M^{-1} cm^{-1}, and ε_2 = 29,900 M^{-1} cm^{-1} at λ^{ex} = 340 nm; c_1/c_2 = 0.43/0.57 at λ^{em} = 510 nm. (b) Corresponding decay times as a function of $-\log[Ca^{2+}]$ calculated according to Eq. (3). (■ and ◆) Decay times estimated by global biexponential analysis. The rate constants used are those of Table I. [Adapted from V. Van den Bergh, N. Boens, F. C. De Schryver, M. Ameloot, P. Steels, J. Gallay, M. Vincent, and A. Kowalczyk, *Biophys. J.* **68,** 1110 (1995), by permission from *The Biophysical Society.*]

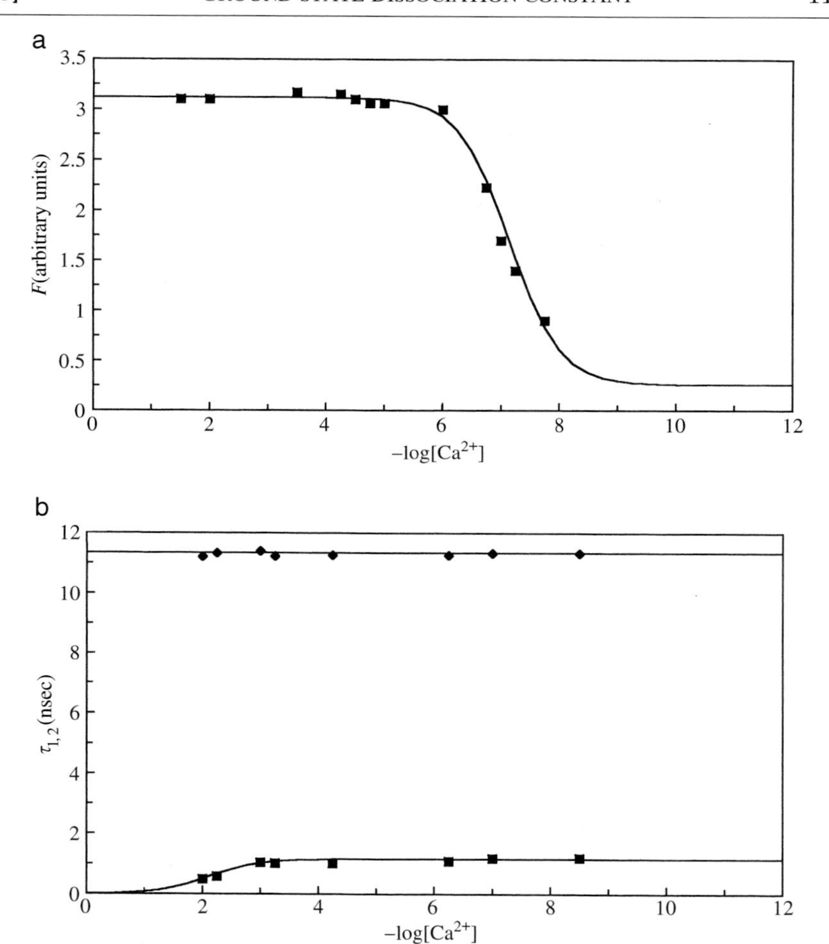

FIG. 4. Quin-2 at pH 7.2 with EGTA as Ca^{2+} buffer in aqueous solution at room temperature. (a) Experimental (■) and calculated (—) fluorimetric titration curve F(339 nm, 490 nm). The F(339 nm, 490 nm) values are calculated with $K_d = 71$ nM, $\varepsilon_1(339\ \text{nm}) = \varepsilon_2(339\ \text{nm}) = 5000\ M^{-1}\ \text{cm}^{-1}$; $c_1/c_2 = 0.45/0.55$ at 490 nm. (b) Corresponding decay times as a function of $-\log[Ca^{2+}]$ calculated according to Eq. (3). (■ and ◆) Decay times estimated by global biexponential analysis. The rate constants used are those of Table I. [Adapted with permission from V. Van den Bergh, N. Boens, F. C. De Schryver, J. Gallay, and M. Vincent, *Photochem. Photobiol.* **61,** 442 (1995).]

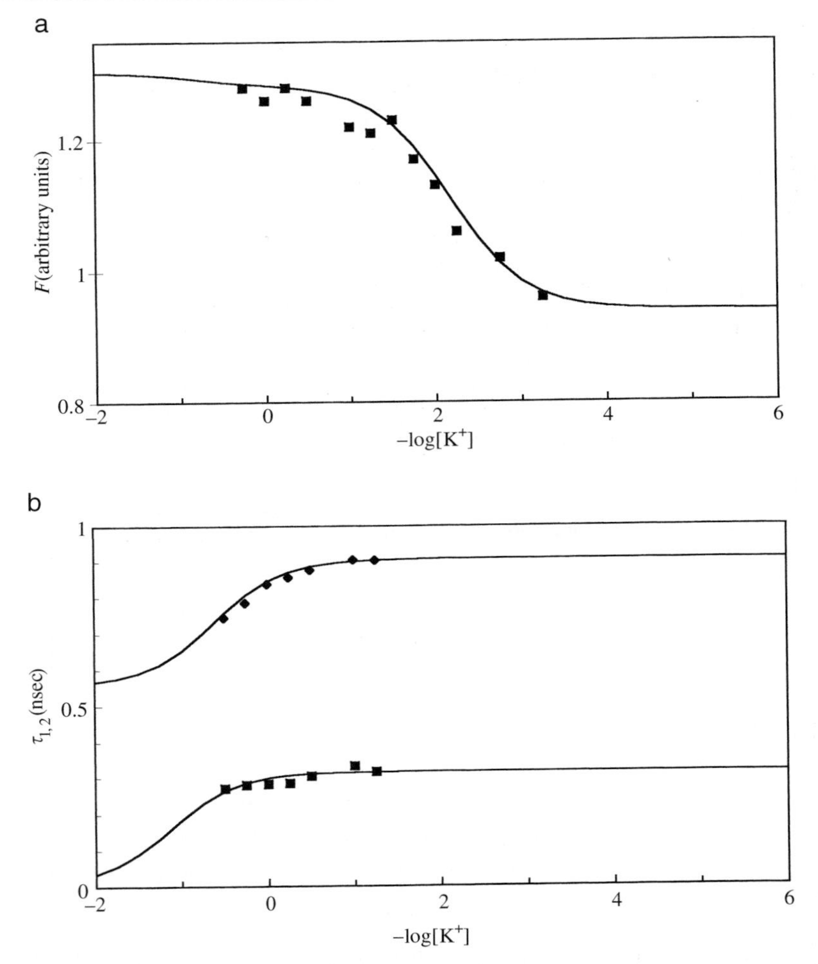

FIG. 5. PBFI at pH 7.2 in aqueous solution at room temperature. (a) Experimental (■) and calculated (—) fluorimetric titration curve F(345 nm, 510 nm). The F(345 nm, 510 nm) values calculated with K_d = 6.62 mM, ε_1 = 28,000 M^{-1} cm^{-1}, and ε_2 = 37,300 M^{-1} cm^{-1} at λ^{ex} = 345 nm; c_1/c_2 = 0.37/0.63 at 510 nm. (b) Corresponding decay times (—) as a function of $-\log[K^+]$ calculated according to Eq. (3). (■ and ◆) Decay times estimated by global biexponential analysis. The rate constant values used are those of Table I. [Adapted from K. Meuwis, N. Boens, F. C. De Schryver, J. Gallay, and M. Vincent. *Biophys. J.* **68,** 2469 (1995), by permission from *The Biophysical Society.*]

reaction is present. The observation at the isoemissive point is the method of choice during fluorimetric titration to avoid interference of the excited-state reaction. If an isoemissive point does not exist, it is still possible to check by time-resolved fluorescence experiments whether the association reaction in the excited state interferes with the determination of K_d. If an inflection occurs in the [X] range of the fluorimetric titration curve where the decay times are invariant, this inflection point can be associated with the correct pK_d. In contrast, the inflection point(s) in the [X] range where the decay times vary cannot be attributed to ground-state dissociation.

When the free and bound forms of a new fluorescent indicator do not have an isoemissive point, time-resolved fluorescence measurements should be performed to estimate the influence of the excited-state reaction on the fluorimetric titration.

Acknowledgments

A.K. thanks the University Research Fund for a Senior Fellowship to stay at the K. U. Leuven. N.B. is an *Onderzoeksdirecteur* of the Belgian *Fonds voor Geneeskundig Wetenschappelijk Onderzoek*. The authors thank Dr. Viviane Van den Bergh and Mrs. Katrien Meuwis for their contributions to the case studies.

[7] ^1L_a and ^1L_b Transitions of Tryptophan: Applications of Theory and Experimental Observations to Fluorescence of Proteins

By PATRIK R. CALLIS

Introduction

More than 1000 papers describing the use of tryptophan (Trp) fluorescence in proteins have appeared in the scientific literature. The vast majority of these have involved enzymes and have exploited changes in the fluorescence intensity, wavelength maximum, fluorescence lifetimes, energy transfer, and anisotropy as a function of folding/unfolding, substrate binding, external quencher accessibility, and so on. In most cases the use of Trp fluorescence has not relied on detailed microscopic information about the Trp electronic states during the analysis of results. A substantial amount of detail concerning the two lowest excited electronic states of Trp, ^1L_a and ^1L_b, has been accumulated, largely from two-photon excitation, jet-cooled and argon matrix spectra, hybrid quantum mechanical–molecular

0076-6879/97 $25

dynamics simulations, and *ab initio* quantum chemical computations. Most of the new results confirm older ideas and results that have been in a prolonged hypothesis stage.

The main objective of this chapter is to raise awareness of detailed characteristics of the ^1L_a and ^1L_b electronic states, including a frank assessment of what may be taken as "fact" and what is speculative, so that a more aggressive interpretation of fluorescence results might be made with confidence when the opportunity arises. The second objective is to present examples of some studies that exploit Trp fluorescence, both as it is commonly used and to point out cases in which the newer information presented here might be used to extract more detail. A tutorial section is first offered for those who might benefit.

General Background

The fluorescence of Trp has been exploited more for the study of proteins than for other amino acids for several reasons. Compared to other amino acids the ultraviolet (UV) absorption and fluorescence extend to longer wavelengths, the absorptivity is the strongest, the fluorescence yield is the largest, and, most important, the fluorescence spectrum and intensity are the most sensitive to the local environment. Since the discovery of these facts by Weber[1,2] and Konev[3] and co-workers, thousands of studies have been conducted.[4–7]

Interpretation of the Trp fluorescence experiments is often clouded by one seemingly unfortunate circumstance: the 280-nm absorption band of Trp is not the result of a single excited state, but of two. They were given the names ^1L_b and ^1L_a by early workers by phenomenological analogy with similar states of substituted benzenes and naphthalenes. (For convenience, transitions involving these states and the ground state are referred to here as ^1L_a and ^1L_b transitions.) The properties of the two states differ considerably. It is the ^1L_a state that has a large dipole compared to the ground state, and thus the high sensitivity to environment. The mere presence

[1] G. Weber, *Biochem J.* **75,** 335 (1960).
[2] G. Weber and F. W. J. Teale, *Trans. Faraday Soc.* **53,** 646 (1957).
[3] S. V. Konev, "Fluorescence and Phosphorescence of Proteins and Nucleic Acids." Plenum, New York, 1967.
[4] J. W. Longworth, *in* "Excited States of Proteins and Nucleic Acids" (R. F. Steiner and I. Weinryb, eds.), p. 319. Plenum, New York, 1971.
[5] J. M. Beechem and L. Brand, *Annu. Rev. Biochem.* **54,** 43 (1985).
[6] A. P. Demchenko, "Ultraviolet Spectroscopy of Proteins." Springer-Verlag, New York, 1986.
[7] M. R. Eftink, *Methods Biochem. Anal.* **35,** 127 (1991).

of the ^1L_b state creates opportunity for confusion and obfuscation, often needlessly. A goal of this chapter is to clarify the issue of when it is possible that ^1L_b is the emitting state in certain proteins.

Much progress has been made toward understanding what has been mentioned here, as becomes evident later in this chapter. However, one of the most striking and potentially useful aspects is the wide variation in intensity (quantum yield) of the fluorescence by Trp in different protein and solvent environments. Not much progress has been made toward understanding what causes such variations, at least not in the sense of being able to make other than empirical predictions.

A variety of phenomena occur within a molecule and its surroundings immediately following the absorption of a photon, and they occur over a large range of time scales. Figure 1 provides an overview. For our purposes, the electronic distribution changes "instantly," or in about 10^{-15} sec (1 fsec), when a photon is absorbed. In such a short time the much heavier nuclei are effectively stationary, the essential content of the Franck–Condon principle. The sudden change in electron distribution has two immediate consequences: redistribution of the electronic charge on the atoms primarily causes a change in the permanent dipole moment of the molecule; redistribution of electron density in the bonds causes the bond strengths (and therefore their lengths) to change.

Thus, the electron redistribution caused by the photon absorption means the molecule is suddenly out of equilibrium, in both an intra- and intermolecular sense. Internally, the atoms may move in response to the changed forces; we say the molecule begins to vibrate. The changed permanent dipole causes the surrounding molecular distribution, on average, to shift in response to the new intermolecular forces. For this latter motion we say there is solvent–solute relaxation or dielectric relaxation. The different names stem from the separation of time scales and location of the action. Most of this chapter is devoted to the interaction with the surroundings, but the internal geometry change is responsible for the overall shape of an absorption or emission band, and is addressed next.

Internal Relaxations: Band Shapes and Dynamics

The unbalanced intramolecular forces caused by excitation means that one or more quanta of a molecular vibration may be simultaneously excited along with the electrons, provided that the photon energy (frequency) sufficiently exceeds the so-called 0-0 frequency (that required to excite the electrons only, often called the "origin") and provided that the vibrational mode is Franck–Condon active. The latter term means that the motions of the vibration coincide, to some extent at least, with the motion of the

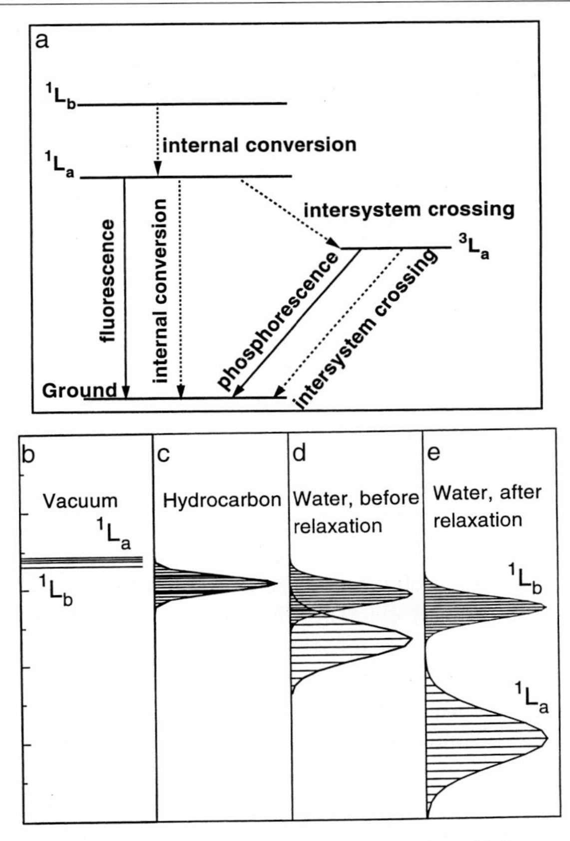

FIG. 1. (a) Jablonski diagram; (b–e) schematic representation of inhomogeneous broadening of tryptophan ^1L_a and ^1L_b states and their relative energies under conditions of (b) vacuum, (c) hydrocarbon solvent, (d) water solvent at the instant of absorption, and (e) water after 1 psec.

nuclei required to change the equilibrium geometry to that of the excited state. The greater the geometry change, the more quanta of vibrational energy can be added, and the broader will be the absorption bandwidth. That is, a larger number of so-called vibronic lines will make up the absorption band envelope. The relative intensities and spacings of these lines constitute a fingerprint for the ^1L_b and ^1L_a spectra, providing they can be resolved. This is because several different normal modes are involved to different extents.

However, in most solutions and proteins these details are blurred by what is termed *inhomogeneous broadening,* a consequence of there being a distribution of local environments experienced by the absorbing molecules. The largest contribution to this effect comes from the variation of the relative orientation of the local electric field and the change in permanent dipole associated with the excitation (not to be confused with the transition dipole).

In condensed phase, the excess vibrational energy is typically lost to the surroundings quickly. After the vibrational relaxation the next event happening internally in the quest by the molecule to reach the ground state is an electronic transition. Return to the ground state by emission of a photon does not always happen. Instead, the molecule has some probability for undergoing a radiationless transition, wherein the electrons change to a lower electronic state, with the energy difference appearing not as a photon, but as vibrational energy (heat). Radiationless transitions were divided into two general classes by Kasha[8]: (1) *internal conversion,* wherein the molecule does not change its electron spin state (e.g., singlet → singlet or triplet → triplet), and (2) *intersystem crossing,* wherein the spin does change (e.g., singlet ↔ triplet). The patterns observed for such changes in condensed phase are so ubiquitous that the behavior is embodied in two rules: *Kasha's rule* and *Vavilov's law.* Kasha's rule states that if a molecule emits, it will always be from the lowest excited state of a given spin multiplicity. For the vast majority of cases this means either from the lowest excited singlet state (spin = 0) or from the lowest excited triplet state (spin = 1). The latter is termed *phosphorescence* and the former is termed *fluorescence.* Kasha's rule arises because typically there is a large energy gap between the lowest excited state and the ground state, compared to spacing between excited states of a molecule. For this reason internal conversion between two excited states is fast compared to the radiative process. Internal conversion to the ground state tends to be slow, because of the large energy gap. Thus, >99.9% of fluorescence is from the lowest excited singlet. Intersystem crossing is slow because it requires a change of spin, and has a rate similar to the fluorescence rate.

Vavilov's law states that the fluorescence quantum yield is essentially independent of the excitation wavelength. This follows because of the rapid descent of the molecule to the zero point of the lowest excited singlet; there is little time for the molecule to cross to the triplet or to the ground state. Excitation of the 280-nm band of Trp involves two electronic states, ^1L_a or ^1L_b, and both rules just introduced appear to apply well to them, although excitation to higher states has been found to yield less fluorescence.[9]

[8] M. Kasha, *Discuss. Farad. Soc.* **9,** 14 (1995).
[9] H. B. Steen, *J. Chem. Phys.* **61,** 3997 (1974).

The fluorescence quantum yield (Φ_f)[9a] of a molecule is the fraction of excitations that result in emission of a photon of fluorescence. For most molecules the quantum yield is not unity even if the molecule is completely isolated following excitation. However, for Trp, one of the most widely exploited properties in proteins is the sensitivity of its fluorescence yield to the microenvironment; the yield varies from about 0.35 to unmeasurably low, for reasons that are not understood in a general way at this time. It is widely believed that low quantum yields are due to the quenching action of amino acid side chains and the peptide chain, and therefore could perhaps be classified as an external relaxation.

External Relaxations

The unbalanced intermolecular forces caused by the electronic excitation are responsible for the environmental effects of great interest in Trp spectroscopy; for example, the fluorescence red shifts that occur to varying degrees in different proteins. Experiments and quantum mechanical calculations of all varieties (see following section) indicate that the ^1L_a transition has a sizeable charge transfer component compared to ^1L_b, with the electron density primarily decreasing on atoms 1 and 3 while increasing on atoms 4, 7, and 9. The resulting increase in permanent dipole moment is usually computed to be 3–5 debyes (D). The response of the environment to this sudden change in dipole can once again be divided into two types: the response of the electrons in the environment and the response of the nuclei in the environment (primarily in the form of molecular rotations and translations). In either case the response of the environment is to create a dipolar *reaction field,* which stabilizes the solute dipole. The electronic response can be considered instantaneous. It will always create a small shift in the absorption and emission proportional to the difference in the squares of the excited and ground-state dipole moments.

The *orientation* of the molecular dipoles of the surroundings is the more important effect. It also depends on the difference in ground- and excited-state dipoles. If there is already a significant dipole in the ground state, there will already exist a reaction field from the partially ordered environment, stabilizing the dipole. If, as is the case for Trp, excitation increases the magnitude of the dipole with little change in direction, then the excited dipole is instantly stabilized from this preexisting field. However, because the new dipole is larger, the mobile components of the environment will

[9a] The same definition applies to phosphorescence. Because phosphorescence is extremely slow, on the order of seconds for aromatic systems, molecules typically have extremely low phosphorescence yields in fluid solutions. However, room temperature phosphorescence from Trp has been observed in proteins when oxygen is rigorously excluded.

be further ordered, creating a larger and more stabilizing reaction field. The extent and time scale of this response depend on the detailed properties of the surroundings. The extent is large and fast for fluid solvents with large dipoles, with the time scale ranging from a few femtoseconds, in the case of the fastest component for fluid water, to times exceeding the excited-state lifetime (several nanoseconds) for the longest components in a glassy solvent or protein. This is the microscopic basis for the large red shift of fluorescence from Trp. A complex spectrum of response times can be expected and, in the case of proteins, can reasonably be expected to span the entire range from 20 fsec to several nanoseconds. This is because there may be OH groups near the Trp that could reorient their large dipoles by torsion about the O–C bond, provided the orientation was favorable. The slow times could, in principle, originate from large-scale cooperative changes in conformation. It is likely that each Trp in each protein will have a unique spectrum of relaxation times. At present this is a largely unexplored arena.

Photoselection, Fluorescence Anisotropy, and Energy Transfer

Somewhat independent of the foregoing is the phenomenon of photoselection, which leads to anisotropy of the emitted fluorescence. When polarized light is absorbed, the rate at which it is absorbed is proportional to the projection squared of the transition dipole moment and polarization direction of the light (see the next section for more comments on the significance of the transition dipole). Thus, the molecules that are most often excited will be the ones oriented such that their transition dipoles are parallel to the polarization direction of the light. This process of photoselection creates a partially ordered population of excited molecules, which in turn means the fluorescence will be anisotropic. The rate at which the anisotropy decays is a measure of the rotational diffusion possible for the molecule and is an invaluable tool for probing the local rigidity, a property that varies greatly in proteins at room temperature. In a protein, the Trp Brownian motion will normally be constrained, and the extent and rate of anisotropy decay are controlled by the directions of the absorbing and emitting moments, which are now known fairly well.

A somewhat related phenomenon is resonance energy transfer, wherein the excitation energy of a Trp can be transferred to another Trp or other absorbing molecule at a rate proportional to the square of the mutual projection of the two transition dipoles, but varying inversely with the sixth power of their separation. There have been few studies that have separated the orientation factor from the distance factor.

Quantum Mechanical Descriptions of ^1L_a and ^1L_b

Considerable progress has been made in bolstering the credibility of quantum mechanical descriptions of the ^1L_a, ^1L_b, and ^3L_a states of Trp, an essential step to microscopic understanding. It is encouraging that quantum mechanical results[10–15] of several types and quality are in substantial qualitative agreement with those from 25 years ago.[16–19] More to the point, they have withstood ever more stringent comparisons with experiment. *Ab initio* results have been presented,[11,13,15] which are certainly more reliable for the ground state and seem to be better in some senses for the excited states.[14,15] The *ab initio* results strongly reinforce most of the previous semiempirical results. In this author's opinion, there seems little doubt concerning the directions of transition dipoles, permanent dipoles, and general electron density changes (within experimental errors).

Molecular Orbitals and Density Matrices

Most quantum mechanical treatments are based on molecular orbital (MO) theory. Figure 2 shows a comparison of the canonical π-electron molecular orbitals of indole obtained from a high-quality *ab initio* procedure (MP2/6-31G*)[13] with those obtained by a typical semiempirical method.[10] In the semiempirical results, the black and white circles (Fig. 2) depict relative amplitudes of π-atomic orbitals, with black and white being opposite in sign. The *ab initio* result is given as contours of orbital amplitude in a plane 1.6 Å above the molecular plane. It is seen that the basic patterns are similar.

Until recently, it has been necessary to use semiempirical methods to describe the excited states of molecules the size of indole, but now at least four *ab initio* studies on the excited states of indole have been accomplished.[11,13,15,20] Excited states in the MO method are described as linear combinations of excitations from the filled MOs to the virtual (unoccupied) MOs, determined by a process known as configuration interaction (CI). As

[10] P. R. Callis, *J. Chem. Phys.* **95,** 4230 (1991).
[11] C. F. Chabalowski, D. R. Garmer, J. O. Jensen, and M. Krauss, *J. Phys. Chem.* **97,** 4608 (1993).
[12] P. R. Callis, *J. Chem. Phys.* **99,** 27 (1993).
[13] L. S. Slater and P. R. Callis, *J. Phys. Chem.* **99,** 8572 (1995).
[14] P. R. Callis, J. T. Vivian, and L. S. Slater, *Chem. Phys. Lett.* **244,** 53 (1995).
[15] L. Serrano-Andrés and B. O. Roos, *J. Am. Chem. Soc.* **118,** 185 (1996).
[16] P. S. Song and W. E. Kurtin, *J. Am. Chem. Soc.* **91,** 4892 (1969).
[17] Y. Yamamoto and J. Tanaka, *Bull. Chem. Soc. Jpn.* **45,** 1362 (1972).
[18] M. Sun and P. Song, *Photochem. Photobiol.* **25,** 3 (1977).
[19] J. Catalan, P. Perez, and A. U. Acuna, *J. Mol. Struct.* **142,** 179 (1986).
[20] D. Hahn and P. R. Callis, *J. Phys. Chem.* (in press) (1997).

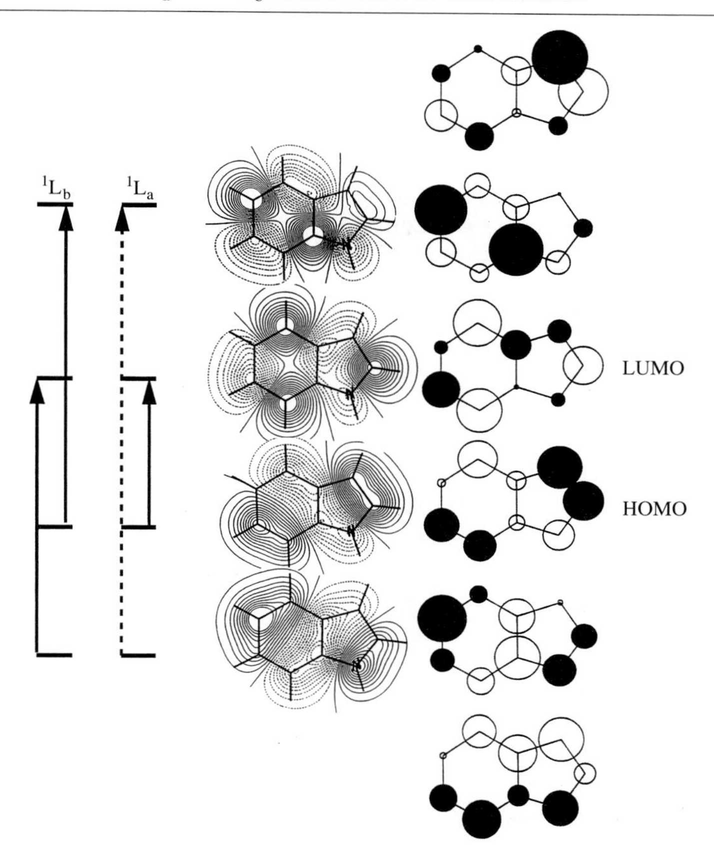

FIG. 2. Molecular orbitals for indole in vacuum from the INDO/S-CI method [P. R. Callis, *J. Chem. Phys.* **95,** 4230 (1991)] compared with those fron an *ab initio* method (MP2/6-31G*) [L. S. Slater and P. R. Callis, *J. Phys. Chem.* **99,** 8572 (1995)]. The main configurations composing the ^1L_a and ^1L_b transition are also shown.

indicated in Fig. 2, the ^1L_a transition is mainly an excitation from the highest occupied MO (HOMO) to the lowest unoccupied MO (LUMO), with a smaller amount of the HOMO − 1 to LUMO + 1 transition. In contrast, the ^1L_b transition is described as a nearly equal superposition of the HOMO → LUMO + 1 and HOMO − 1 → LUMO excitations.

The simplicity of the ^1L_a description allows one to deduce easily the major changes associated with the transition. The charge transfer aspect of the ^1L_a transition is evident from the changes of the squares of the coefficients on the atoms. Thus, it is seen that much electron density is lost from atoms 1 and 3 and deposited at atoms 4, 7, and 9. The result is a large dipole change, with the pyrrole ring becoming more positive. As already mentioned, this is perhaps the single most used property of the Trp molecule because it is the source of the sensitivity to solvent.

Changes in bond lengths determine the details of the absorption and emission band envelopes and are directly related to changes in π-bond orders. When an MO has a node in a bond it contributes negatively to the bond order, and therefore weakens and lengthens the bond. The opposite is true when there is no node. The HOMO is seen to be strongly bonding at the 2–3, 9–4, and 6–7 positions but antibonding at these positions in the LUMO. These bonds are expected to become considerably longer during the ^1L_a transition. In a similar manner, one can see that the 3–9 and 1–2 bonds are expected to become significantly shorter.

The ^1L_b transition properties are quite different and can also be qualitatively deduced from looking at the differences of the MOs involved in the two configurations. After averaging the two excitations, one concludes that there is much less charge transfer than for ^1L_a and that all the bonds in the benzene ring are weakened.

The previous analysis is made more quantitatively by examining the π-electron density difference matrices, depicted in Fig. 3, which incorporate

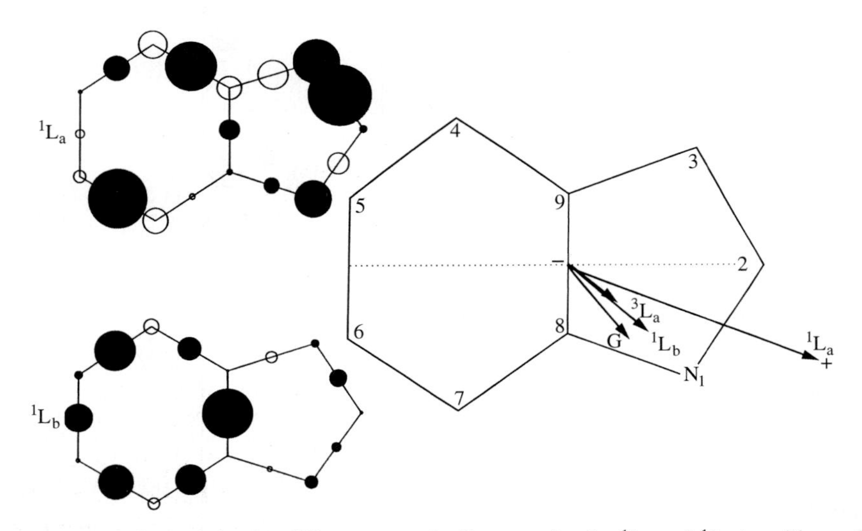

FIG. 3. π-Electron density difference matrix diagrams for the ^1L_a and ^1L_b transitions of indole in vacuum from INDO/S-CI. The circle diameters are proportional to the matrix elements, with black meaning a decrease and white meaning an increase in electron density during the transition. Also shown are the consensus permanent dipole moments computed from *ab initio* methods for the ground, ^1L_b, ^1L_a, and ^3L_a states. The arrow is the positive end. [Adapted from P. R. Callis, *J. Chem. Phys.* **95,** 4230 (1991).]

contributions from about 200 configurations.[10] In Fig. 3, black means a loss of electron density or bond order and white means a gain. It is seen that the behavior is not so different from that deduced by looking at the two HOMOs and LUMOs. A great deal of information is condensed into the density difference figures. The direction and magnitude of the electron transfer are primarily responsible for the detailed solvent sensitivity of the spectra. Thus, if the local electric field points from the six- to the five-membered ring, the ^1L_a transition will be red shifted and vice versa for the opposite direction. The directions and magnitudes of the dipole moments for the ground, ^1L_b, ^1L_a, and ^3L_a states are tabulated in Table I. There is a striking consensus despite a wide range of sophistication in the methods used. The consensus directions and lengths are shown in Fig. 3. There is good agreement with the experimental magnitude in the ground state. The Stark effect experiments of Pierce and Boxer[21] indicate a dipole increase of about 6 debye in glycerol–water glass at 77 K. An experiment on yohimbine[22] provides a confirmation of the direction of the dipole change, wherein a positive charge located near the pyrrole ring caused a pronounced blue shift of the ^1L_a absorption band. The pattern of bond order changes correlates well with the pattern of bond length changes,[13] which, in turn, determine the details of the absorption and emission band shapes, as is seen.

Transition Properties and Transition Densities

Table I also lists the experimental and calculated transition energies (both vertical and 0-0), oscillator strengths, and electric dipole transition moment directions for indole. Again, there is little scatter in the results. (The large deviation seen for the CIS/3-21G case is because of an accidental degeneracy of ^1L_a and ^1L_b using this basis.) Figure 4 (page 126) shows the ^1L_b and ^1L_a transition moment directions for indole as given experimentally.[23] The vertical transition energies are calculated at the ground-state geometries and should be compared to the experimental absorption maxima. The 0-0 transition energy is more difficult to calculate because it corresponds to the difference in the energies calculated for the minimum energy geometries of both states. This is a crucial distinction for indoles because the fluorescent state is the one with the lowest 0-0 transition, often the ^1L_a state, even though the vertical transition lies higher. This is because the ^1L_a transition causes a larger geometry change. Typically, the *ab initio* results are not different than the semiempirical results. It is noted that the *ab initio* CIS 0-0 transition energies for indole in vacuum are much too

[21] D. W. Pierce and S. G. Boxer, *Biophys. J.* **68,** 1583 (1995).
[22] L. J. Andrews and L. S. Forster, *Biochemistry* **11,** 1875 (1972).
[23] B. Albinsson and B. Norden, *J. Phys. Chem.* **96,** 6204 (1992).

TABLE I
GROUND-STATE AND EXCITED-STATE PROPERTIES OF INDOLE

Method	Property: Ground	^1L_b	^1L_a	^3L_a	Ref.
		Permanent dipole (direction)[a]			
CASSCF	1.86 (−50)	0.85 (−41)	5.69 (−12)	1.47 (−34)	*b*
FOCI	2.01 (−48.5)	2.11 (−24.9)	5.84 (−18.0)		*c*
STO-3G	1.80 (−48.8)				*d*
3-21G	2.02 (−45.3)	2.15 (−41.3)	3.22 (−27.3)	1.29 (−37.2)[e]	*d*
6-31G	1.93 (−45.9)[d]	2.11 (−41.2)[e]			
6-31G**	2.05 (−45.3)		3.19 (−27.9)	1.41 (−37.3)	*e*
MP2/6-31G**//MP2	2.19 (−47.4)				*e*
INDO/S-SCI	2.02 (−48)	2.83 (−32)	5.87 (−16)		*f*
INDO/S-SDCI	2.22 (−51)	2.63 (−37)	5.09 (−25)		*f*
PPP	2.18 (−63)	2.24 (−52)	5.65 (−27)	2.11	*g*
Experimental	2.13,[h] 2.09[i]	2.11,[j] 3.45[k]	5.44,[k] ~8[l]		
		Transition energy, cm^{-1}/1000 (nm)			
Vertical					
CASPT2		35.7 (280)	38.2 (262)		*b*
FOCI		42.3 (236)	46.8 (214)		*c*
3-21G		50.1 (200)	50.2 (199)	27.0 (370)	*e*
4-31G		49.8 (201)	49.7 (201)	26.4 (379)	*e*
6-31G**		48.3 (207)	48.0 (208)	26.5 (377)	*e*
6-311G**		47.5 (211)	47.1 (212)		*e*
INDO/S-SCI		34.1 (293)	38.0 (263)		*f*
INDO/S-SDCI		36.3 (275)	42.9 (233)		*f*
Experimental		35.2 (284)	38.5 (260)		*m*
0-0					
CASPT2		35.1 (285)	37.6 (266)		*b*
FOCI		41.2 (243)	46.2 (216)		*c*
3-21G		48.6 (296)	47.6 (219)	23.9 (418)	*e*
4-31G		48.3 (207)	47.4 (211)	23.5 (426)	*e*
6-31G		47.8 (209)			*e*
6-31G**			45.7 (219)	23.4 (427)	*e*
Experimental		35.2 (284)[m]	36.4 (275)[n]		

high and do not give the correct order compared to the more extensive CASPT2 method.[15]

The oscillator strength is a measure of the area under the absorptivity curve and also is a measure of the rate at which the transition emits photons, the inverse of the radiative lifetime. It is proportional to the square of the transition dipole length. The transition dipole directions are the directions of light polarization giving maximal absorbance. The transition dipole is

TABLE I (*continued*)

Method	Property: Ground	$^{1}L_b$	$^{1}L_a$	$^{3}L_a$	Ref.
	Oscillator strength (transition moment direction)				
Vertical absorption					
CASSCF		0.05 (+37)	0.081 (−36)		*b*
FOCI		0.010 (+43)	0.20 (−23)		*c*
3-21G		0.088 (+72)	0.13 (−10)		*e*
4-31G		0.057 (+36)	0.16 (−38)		*e*
6-31G		0.056 (+35)	0.16 (−38)		*e*
6-31G**		0.050 (+48)	0.15 (−30)		*e*
6-31++G**		0.058 (+38)	0.17 (−31)		*e*
6-311G**		0.054 (+40)	0.15 (−32)		*e*
INDO/S-SCI		0.013 (+49)	0.21 (−37)		*f*
INDO/S-SDCI		0.010 (+59)	0.08 (−50)		*f*
CNDO/S		0.006 (+38)	0.34 (−10)		*o*
PPP		0.024 (+54)	0.13 (−41)		*p*
Experimental		0.01,[p] 0.019[q] (+42,[r] ±45,[s] +54[p])	0.11,[p] 0.13[q] (−46,[r] −38[p])		
Vertical emission					
FOCI		0.01 ( )	0.21 (−48)		*c*
3-21G		0.10 (+37)	0.35 (−31)		*e*
4-31G		0.095 (+28)	0.35 (−30)		*e*
6-31G		0.092 (30)			*e*
6-31G**			0.32 (−29)		*e*

[a] Dipole is in debyes; direction is measured counterclockwise from long axis (see Figs. 3 and 4).
[b] L. Serrano-Andrés and B. O. Roos, *J. Am. Chem. Soc.* **118,** 185 (1996).
[c] C. F. Chabalowski, D. R. Garmer, J. O. Jensen, and M. Krauss, *J. Phys. Chem.* **97,** 4608 (1993).
[d] L. S. Slater and P. R. Callis, *J. Phys. Chem.* **99,** 8572 (1995).
[e] D. Hahn and P. R. Callis, unpublished results (1996).
[f] P. R. Callis, *J. Chem. Phys.* **95,** 4230 (1991).
[g] M. Sun and P. Song, *Photochem. Photobiol.* **25,** 3 (1977).
[h] A. L. McClellan, "Tables of Experimental Dipole Moments." Freeman, London, 1963.
[i] W. Caminati and S. De Bernardo, *J. Mol. Struct.* **240,** 253 (1990).
[j] C.-T. Chang, C.-Y. Wu, A. R. Muirhead, and J. R. Lombardi, *Photochem. Photobiol.* **19,** 347 (1974).
[k] H. Lami and N. Glasser, *J. Chem. Phys.* **84,** 597 (1986).
[l] D. W. Pierce and S. G. Boxer, *Biophys. J.* **68,** 1583 (1995).
[m] E. H. Strickland, J. Horwitz, and C. Billups, *Biochemistry* **9,** 4914 (1970).
[n] B. J. Fender, D. M. Sammeth, and P. R. Callis, *Chem. Phys. Lett.* **239,** 31 (1995).
[o] J. Catalan, P. Perez, and A. U. Acuna, *J. Mol. Struct.* **142,** 179 (1986).
[p] Y. Yamamoto and J. Tanaka, *Bull. Chem. Soc. Jpn.* **45,** 1362 (1972).
[q] E. H. Strickland, C. Billups, and E. Kay, *Biochemistry* **11,** 3657 (1972).
[r] B. Albinsson and B. Nordén, *J. Chem. Phys.* **96,** 6204 (1992).
[s] L. A. Phillips and D. H. Levy, *J. Chem. Phys.* **85,** 1327 (1986).

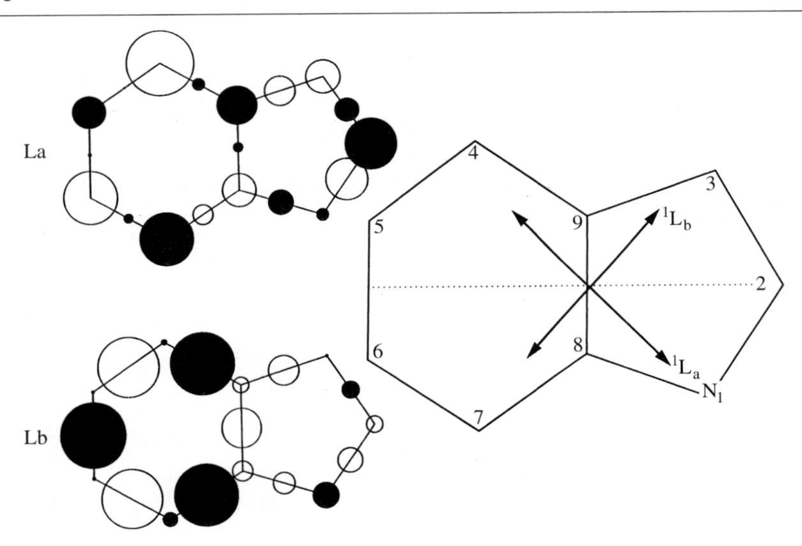

FIG. 4. First-order transition density matrix diagrams for the ^1L_a and ^1L_b transitions of indole in vacuum. Black and white are the relative phases (+ and − are arbitrary). The diameters on atoms are proportional to the amount of the square of that atomic orbital contributed to the product of the ground- and excited-state wavefunctions. In the bonds, the diameter gives the contribution by the product of the neighboring atomic orbitals. The elements on the atom give the transition dipoles. [Adapted from P. R. Callis, *J. Chem. Phys.* **95,** 4230 (1991).]

not a measure of the change in dipole accompanying the change in state; its direction is the direction that electron charge is displaced by mixing the ground and excited states during the transition and involves the product of the ground- and excited-state wavefunctions (transition density). In contrast, the dipole change is the difference of the permanent dipoles of the pure excited and ground state (from the differences of the densities). The two quantities are not generally related.

The first-order transition density matrix diagrams for indole are also shown in Fig. 4. These representations of the product of the wavefunctions of two different states are a complementary and graphic way of describing electronic transitions. The patterns in the benzene ring exhibit major characteristics of the counterpart ^1L_a and ^1L_b transitions of benzene: the ^1L_a transition density is peaked on the *a*toms and that of ^1L_b is on the *b*onds in the benzene ring. The ^1L_b transition is seen to be rather benzene-like, but the ^1L_a transition has a large ethylene-like component in the 2–3 bond in addition to a benzene-like component. The diagonal elements of the transition density reveal the transition dipole direction. The dipoles seen in the 2–3, 4–9, and 6–7 bonds are all pointing the same direction and are in phase, thereby creating a large transition dipole pointing in that direction, a robust result that is insensitive to calculational parameters, substituents, and environmental influences. (Note that the sign of the transition dipole direction, i.e., the phase of the transition density, is not physically significant when considering a single molecule because the absorption and emission rates depend on its square.) In contrast to ^1L_a, the ^1L_b transition density has small diagonal elements. The two opposite elements on atoms 7 and 9 are the main contributors defining the transition dipole direction and magnitude, making it polarized approximately perpendicular to ^1L_a and considerably weaker. Unlike the ^1L_a transition, the ^1L_b transition dipole is quite sensitive to substitution, especially at the 4-, 6-, and 7-positions.[24]

Without the explicit use of transition densities it is often difficult to distinguish ^1L_a and ^1L_b when solvent waters and/or protein environment is included in a calculation, owing to mixing of the two states.

Relevance of Indole Properties to Tryptophan

Unfortunately, no *ab initio* calculations have been done for excited states of Trp or 3-methylindole (3MI). A number of experiments indicate that the main effect of alkyl substitution at the 3-position is to shift the ^1L_b and ^1L_a transitions to the red by about 400 and 1400 cm^{-1}, respectively, thus narrowing the gap by about 1000 cm^{-1}. Semiempirical calculations are in agreement with this,[10] and indicate that transition dipole directions do not change more than about 10°. They also predict a somewhat larger dipole change, in agreement with the larger fluorescence Stokes shift seen for 3MI compared to indole.[25]

Predicted ^1L_b (and ^1L_a) Vibronic Structure and Effect of Broadening

The notion that ^1L_b spectra are structured and ^1L_a spectra are unstructured is so firmly entrenched that any fluorescence band from a protein with a hint of vibronic structure evokes suspicion of being ^1L_b fluorescence. This is because of the high degree of inhomogeneous broadening exhibited by the ^1L_a transition owing to its large dipole change, which usually obscures the structure. In this section the groundwork is laid for identification of ^1L_a emission on the basis of its own characteristic vibronic structure. It is important to eliminate confusion concerning ^1L_b versus ^1L_a fluorescence from proteins because ^1L_b fluorescence would signify an extremely unusual local environment.

[24] M. R. Eftink, L. A. Selvidge, P. R. Callis, and A. A. Rehms, *J. Phys. Chem.* **94,** 3469 (1990).
[25] H. Lami and N. Glasser, *J. Chem. Phys.* **84,** 597 (1986).

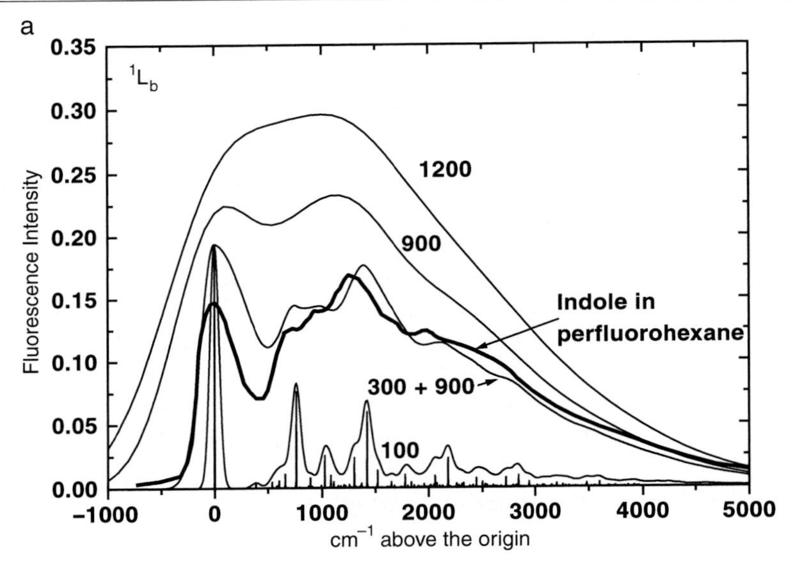

FIG. 5. (a) Comparison of calculated fluorescence band shapes for the ^1L_b transition of indole with those of indole in perfluorohexane at room temperature. The calculated spectra are from the geometry differences and frequencies given by *ab initio* calculations [P. R. Callis, J. T. Vivian, and L. S. Slater, *Chem. Phys. Lett.* **244,** 53 (1995)] and are the thin solid lines, broadened to fwhm (full width half-maximum) widths of 3, 100, 300 + 900, 900, and 1200 cm^{-1}. The notation 300 + 900 means that the two halves of the line have fwhm widths of 300 and 900 cm^{-1}, respectively. The thick solid line is the experimental spectrum. (b) Comparison of calculated fluorescence band shapes for the ^1L_a transition of indole (thin solid lines) with the experimental phosphorescence spectrum of indole (thick solid line), the experimental fluorescence curves from Trp in RNase T_1 at 77 K (dashed line) [J. W. Longworth, *Photochem. Photobiol.* **7,** 587 (1968)] and in ethanol glass at 2 K (dotted line) [T. W. Scott, B. F. Campbell, R. L. Cone, and J. M. Friedman, *Chem. Phys.* **131,** 63 (1989)]. The fwhm broadening (in cm^{-1}) is given for the calculated curves.

One of the most compelling results regarding the credibility of the *ab initio* descriptions of the ^1L_b and ^1L_a states of indole is that the fluorescence spectra predicted from the geometry changes at a relatively inexpensive level of calculation[14] bear a striking resemblance to experimental spectra, including many of the subtle details of the ^1L_b fluorescence of indole under jet-cooled conditions[26] and in solid argon.[27] This is particularly true if the geometry difference is computed from singly excited CI (SCI) and simple Hartree–Fock (HF) calculations on the ^1L_b and ground state, respectively,

[26] G. A. Bickel, D. R. Demmer, E. A. Outhouse, and S. C. Wallace, *J. Chem. Phys.* **91,** 6013 (1989).

[27] B. J. Fender, D. M. Sammeth, and P. R. Callis, *Chem. Phys. Lett.* **239,** 31 (1995).

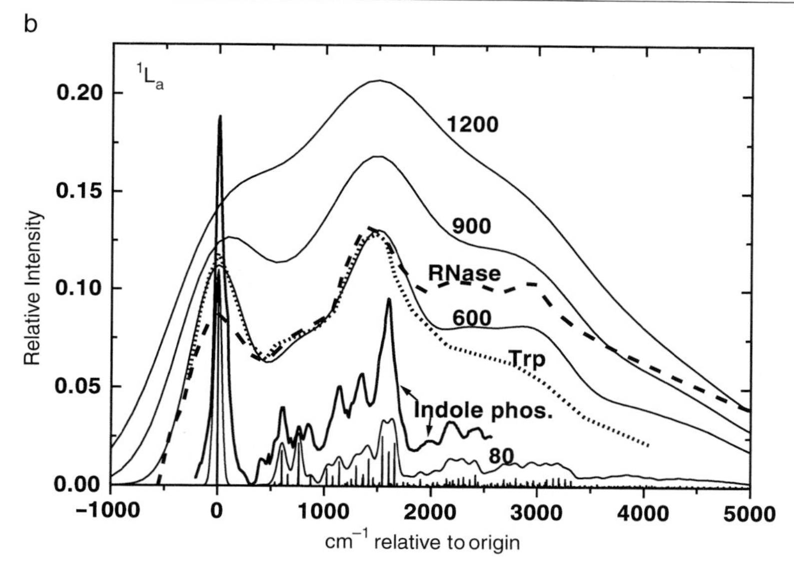

FIG. 5. (*continued*)

using 3-21G or 4-31G basis sets, while the vibrational modes and frequencies are from a high-quality ground-state computation (MP2/6-31G*). Figure 5a shows computed ^1L_b fluorescence spectra using a variety of Gaussian broadening factors, including a comparison to indole fluorescence in perfluorohexane at 20°.

Figure 5b shows a similar computation of the fluorescence from the ^1L_a state of indole, with comparison to two low-temperature experimental fluorescence curves from Trp showing some vibronic structure under conditions where ^1L_a fluorescence is plausible. Because it is difficult to find sharp experimental ^1L_a fluorescence, an experimental phosphorescence spectrum of indole in argon is shown to match with the computed ^1L_a fluorescence with 80 cm^{-1} broadening. As such, it should help decide whether dispersed fluorescence from complexed indoles in cold jets is ^1L_a or ^1L_b. In particular, it should be noted that, just as for ^1L_b, the origin is expected to be the strongest line in a jet spectrum, and that the spectrum will be sparse and structured below 1000 cm^{-1}.

The computed ^1L_a spectrum in Fig. 5b is distinct from that of ^1L_b in the number of significantly Franck–Condon active modes and their relative intensities. The largest difference is that the most active modes are the three C–C stretches ν_8, ν_9, and ν_{10}, predicted to be near 1600 cm^{-1}. The sum of these three Franck–Condon factors is nearly the same as that of the origin, and in the broadened spectra they lead to the characteristic maximum seen in experimental ^1L_a fluorescence and phosphorescence spectra near 1600 cm^{-1}.[3,4] Another difference is that mode 27 (610 cm^{-1}) has intensity comparable to mode 26 (769 cm^{-1}), and both are much weaker than the origin. Together, these differences combine to provide tangible differences between the two types of emission, which are observed in Fig. 5. The difference in the intensity ratio at 610 and 760 cm^{-1} causes distinctive differences in the 500- to 800-cm^{-1} region; the strongly active 1600-cm^{-1} vibrations in the ^1L_a spectrum cause the central maximum to be red shifted about 200 cm^{-1} from that of ^1L_b; and their overtones and combinations create the pronounced shoulder that always extends past 3000 cm^{-1} in ^1L_a emission, but is unseen past 2500 cm^{-1} for ^1L_b emission.

The effect of broadening the spectra is quite interesting and important. The fact that so many vibrations are Franck–Condon active in the ^1L_a spectrum leads to the somewhat surprising result that the origin is the strongest line predicted in high-resolution spectra, yet the maximum absorption appears far from the origin for all observed ^1L_a spectra (which have always been at least modestly broadened). This behavior is because there are more than 5000 lines having Franck–Condon factors $>10^{-4}$ times that of the origin, most of them lying more than 2000 cm^{-1} above the origin. In a highly resolved spectrum they are distinct but too weak to show on a linear scale that shows the full origin line. However, if the lines are broadened either by solvent interaction or simply by a low-resolution spectrometer, many lines coalesce within the bandwidth and are collectively more intense in comparison to the single origin line spread over the same interval. This effect means that the ^1L_a band will appear quite different in different environments, as is demonstrated in Fig. 5b.

Figure 6 compares the well-known Valeur–Weber[28] resolution of the indole absorption into ^1L_b and ^1L_a components with the shapes predicted from *ab initio* geometry changes and vibrational frequencies. Here, the positions of the spectra and their relative intensities were varied to obtain the best fit, but the shapes were unaltered. Although the agreement is not perfect, the theory captures the essence of the vibronic band shapes, again lending credence to the quantum mechanical descriptions of these excited states.

Triplet State

The lowest triplet state of Trp and other indoles has long been designated ^3L_a,[3] implying that the molecular orbital configurations involved are

[28] B. Valeur and G. Weber, *Photochem. Photobiol.* **25,** 441 (1977).

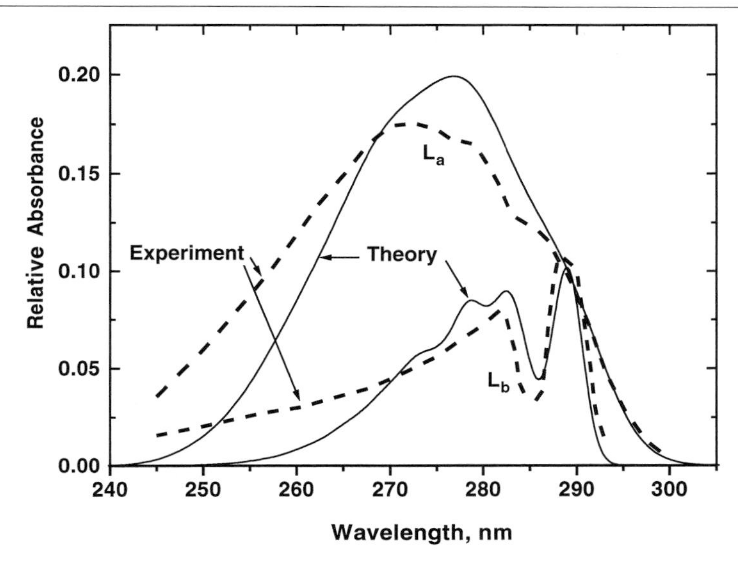

FIG. 6. Comparison of computed [P. R. Callis, J. T. Vivian, and L. S. Slater, *Chem. Phys. Lett.* **244,** 53 (1995)] (CIS/4-31G − HF/4-31G) absorption of indole with those resolved using fluorescence anisotropy [B. Valeur and G. Weber, *Photochem. Photobiol.* **25,** 441 (1977)]. The computed curves were translated and broadened to obtain a reasonable fit and the vibrational frequencies were scaled by the common 0.90 factor.

the same as those for ^1L_a. This result is strongly reinforced by *ab initio* results[11,13,15,20] and by their prediction of strong Franck–Condon activity in the 1600-cm^{-1} C–C stretching vibrations giving rise to a strong peak near 1500–1600 cm^{-1} in every published phosphorescence spectrum of Trp and other simple indoles. What has been somewhat puzzling is that the phosphorescence spectra invariably exhibit much higher resolution of the vibronic structure than the corresponding ^1L_a fluorescence, even in polar solvents. This includes room temperature phosphorescence from certain buried Trps that are protected from quenching by O_2.[29]

The answer to this seeming paradox is provided in quantum mechanical computations,[15,20] which invariably show the ^3L_a state to be described mainly by the same two configurations as ^1L_a, but that have significant differences in the minor configurations. The minor configurations contribute to the dipole with opposite sign in the triplet and singlet states, in such a way that many of the larger terms are canceled,[20] leaving a dipole for the ^3L_a state that is usually predicted to be smaller than that of the ground state.

[29] M. L. Saviotti and W. C. Galley, *Proc. Natl. Acad. Sci. U.S.A.* **71,** 4154 (1974).

This result has been found in semiempirical computations of the INDO/S[30] and PPP[18] types as well as for all *ab initio* computations varying in quality up to CIS/6-31G[20] and CASSCF.[15]

Thus, quantum mechanics predicts the obvious conclusion one would draw from the experimental facts: because there is so little dipole change between the ground and excited state, the energy level spacing is quite insensitive to the wide variations in local electric field found in polar solvents and proteins.

Solvent Effects in Fluid Environment: Exciplexes

The detailed mechanism of the fluorescence shifts of Trp in polar solvents has been a source of fascination and controversy. Discussions have typically involved three subjects: (1) the generic reaction field due to solvent dipoles orienting in response to the increased solute dipole, (2) inversion of the ^1L_a and ^1L_b levels, and (3) excited-state complexes (exciplexes). The first is universally important in pure polar solvent and is unspecific. However, the fluorescence of indole in hydrocarbon solutions containing only 0.1 *M* ethanol was found to be considerably broadened and shifted compared to the corresponding changes in absorption and the bulk dielectric properties.[3,31,32] Mataga *et al.*[31] believed that inversion of ^1L_a and ^1L_b was a major factor in this behavior. Konev,[3] in addition to recognizing the importance of the inversion, also attributed some of the shifting to local changes in solute–solvent interaction caused by the increased excited-state dipole. They also attributed the broadness to a distribution of geometries possible for the emitting solute–solvent combinations. They placed emphasis on hydrogen-bonding involving the indole NH as donor as a contributor to the ^1L_a–^1L_b inversion.

Lumry and co-workers[33–37] attempted to explain the entire phenomenon with specific indole–alcohol exciplexes, which were originally guessed to be charge-transfer complexes involving the indole π electrons, and were formed and stable only in the excited state.[36] This position was encouraged by the ability to decompose the broadened fluorescence into a structured

[30] P. R. Callis, unpublished results (1996).

[31] N. Mataga, Y. Torihashi, and K. Ezumi, *Theor. Chim. Acta* (*Berl.*) **2,** 158 (1964).

[32] M. S. Walker, T. W. Bednar, and R. Lumry, *J. Chem. Phys.* **45,** 3455 (1966).

[33] M. V. Hershberger, R. Lumry, and R. Verrall, *Photochem. Photobiol.* **33,** 609 (1981).

[34] N. Lasser, J. Feitelson, and R. Lumry, *Isr. J. Chem.* **16,** 330 (1977).

[35] M. V. Hershberger and R. Lumry, *Photochem. Photobiol.* **23,** 391 (1976).

[36] M. S. Walker, T. W. Bednar, and R. Lumry, *J. Chem. Phys.* **47,** 1020 (1967).

[37] M. S. Walker, T. W. Bednar, and R. Lumry, *in* "Molecular Luminescence" (E. C. Lim, ed.), pp. 135–152. W. A. Benjamin, New York, 1969.

monomer and unstructured exciplex component, from which rather clear-cut stoichiometries were deduced [2:1 for indole, 1-methylindole (1MI), and 1,3-dimethylindole, but 1:1 for 3MI]. In this mechanism, the broadness of the exciplex emission was attributed to the repulsive nature of the ground state of the complex. However, Hershberger *et al.*[33] now believe that exciplexes are stabilized primarily by dipole–dipole interaction, although at the same time they identify two specific binding sites, N-1 and C-3, designated as nucleophilic and electrophilic attack sites, respectively.

However, quantum mechanical descriptions of the ^1L_a state say that the large dipole arises from loss of π-electron density from the 3- and 1-positions, making C-3 an unlikely candidate for a hydrogen bond acceptor in the excited state. Another problem with this description is that dipole–dipole interactions are, by definition, not specific, formally having maximum stabilization at either of the end-to-end geometries, and also for all possible side-by-side geometries with the dipoles opposed. Aside from these objections, the lack of recognition of both the difference between ^1L_a and ^1L_b fluorescence and the almost obvious importance of the NH donor hydrogen bond for indole and 3MI in comparison to 1MI have hindered complete acceptance of the Lumry picture as an all-encompassing model, whatever elements of truth it may have.

Turning now to the mechanism of the much larger red shift seen in pure water, the relevance of the previous discussion is not clear. Exciplexes due to complexation with water have apparently not been seen. Furthermore, the driving force to produce a specifically bound exciplex with water would seem to be greatly reduced if the water were to come from an environment of pure water as opposed to a nonpolar solvent.

The almost purely electrostatic nature of the exciplex model led at least three groups to merge the fairly successful semiempirical quantum mechanical description of indole excited states with one of the powerful molecular dynamics descriptions of liquid water that have become available within the last several years. Three independent studies were applied to indole or 3MI in water,[38–42] each with similar results within their respective scopes. The study by Muiño and Callis[39,41,42] was the most extensive and included the effect of excited-state geometry changes as well as charge changes in predicting the fluorescence shift and dynamics due to solute–

[38] P. Ilich, C. Haydock, and F. G. Prendergast, *Chem. Phys. Lett.* **158,** 129 (1989).

[39] P. L. Muiño, D. Harris, J. Berryhill, B. Hudson, and P. R. Callis, *Proc. S.P.I.E. Int. Soc. Opt. Eng.* **1640,** 240 (1992).

[40] J. D. Westbrook, R. M. Levy, and K. Krogh-Jespersen, *Proc. S.P.I.E. Int. Soc. Opt. Eng.* **1640,** 10 (1992).

[41] P. L. Muiño and P. R. Callis, *J. Chem. Phys.* **100,** 4093 (1994).

[42] P. L. Muiño and P. R. Callis, *Proc. S.P.I.E. Int. Soc. Opt. Eng.* **2137,** 362 (1994).

solvent relaxation. A simple model is applied wherein the solute is entirely fixed and treated by the INDO/S-SCI method, with the effect of the solvent included as the electrostatic potential given by the Coulomb sum over the point charges (three per water molecule). About 250 water molecules were used in a 20-Å cubic box. The charges used for the indole in the classic molecular dynamics (usually Discover; BioSym, Inc.) were from the quantum mechanics, updated every 5 fsec. Typical Lennard–Jones atom–atom potentials were used and were not modified for the excited state.

Figure 7a shows the vertical ^1L_b and ^1L_a transition frequencies as a function of time during a simulated 30-psec trajectory, with an INDO/S-SCI computation every 5 fsec using the ground-state geometry and charge

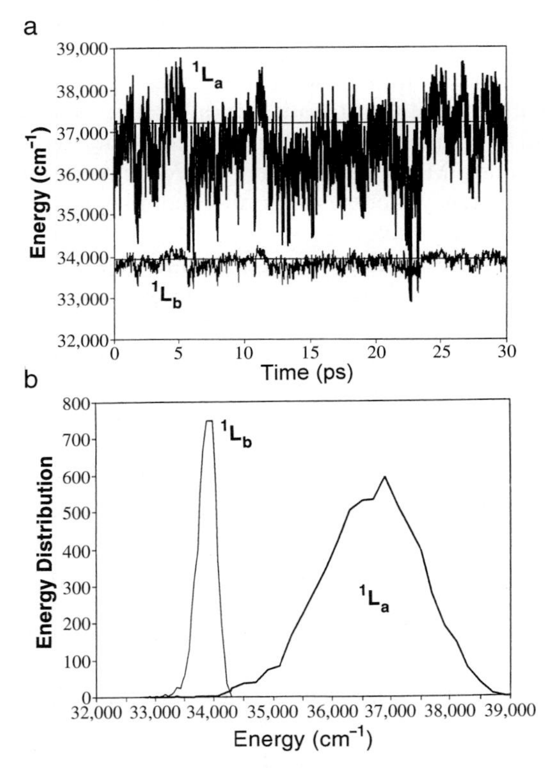

FIG. 7. Results of hybrid quantum mechanical (INDO/S-CI) and classic molecular dynamics simulation of the ^1L_a and ^1L_b electronic transitions of 3-methylindole in water at 300 K [P. L. Muiño and P. R. Callis, *J. Chem. Phys.* **100,** 4093 (1994)]. (a) Thirty-picosescond trajectory with the indole atom charges updated every 5 fsec. The horizontal lines are the vacuum transition energies. (b) Histogram of the 6000 transition energies for ^1L_a and ^1L_b from the trajectory in (a). The widths correspond to the inhomogeneous broadening in water.

distribution on the 3MI. The horizontal lines give the energies in vacuum, so it is seen that there is about a 500-cm^{-1} decrease in the ^1L_a–^1L_b gap caused by the water reaction field due to the ground-state dipole of about 2 D. This means that the ^1L_a 0-0 transition lies below that of ^1L_b about 50% of the time, because it is the vertical transition energy that is being computed and it is known from experiment that the ^1L_a 0-0 lies close to the ^1L_b 0-0 for 3MI in vacuum.[43] The ^1L_a transition shows large fluctuations because of the fluctuations of the local electric field caused by the effectively random motion of the water molecules. The ^1L_b transition shows parallel excursions of much smaller magnitude. This is because the dipole change is much smaller than for ^1L_a, but is in the same direction. The fluctuations indicate the extent of inhomogeneous broadening expected for the two transitions, as shown in Fig. 7b and indicated schematically in Fig. 1d.

The time-resolved fluorescence shift was simulated by changing the indole atomic charges instantly to those of the ^1L_a state. The large new dipole was found to cause the water to create a larger reaction field, which shifts the ^1L_a fluorescence to the red by several thousand cm^{-1}. The relaxation in water at 298 K is extremely fast, about 50% occurring within the first 20 fsec with a Gaussian profile (this is a property of the water[44]). The remainder of the relaxation is finished after about 0.5 psec. The extent of the shift agrees fairly well with what is seen experimentally, if one estimates the shift from the difference of the fluorescence maximum for 3MI in glycol–water at 77 K and that in water at 298 K. This experimental value is 3800 cm^{-1} and is between the simulated values of 2700 cm^{-1} when the excited-state charges are frozen at their $t = 0$ values, and 4800 cm^{-1} when the excited-state charges were allowed to change in response to the solvent field. The larger shift in the latter case is because of the large polarizability of the ^1L_a state; the dipole was found to vary from 4 to 12 D, owing to the fluctuations in the local solvent environment, and was strongly correlated with the transition energy. A cooperative effect in which the reaction field and the solute dipole mutually reinforce each other leads to the increased shift. The large calculated dipole in the solvent is consistent with Stark effect experiments on Trp,[21] which put the change of dipole at about 6 D for the ^1L_a transition.

Thus, a general electrostatic solvent–solute interaction scheme accounts for the large fluorescence Stokes shift in water. Analysis of the source of the red shift in terms of contributions from individual water molecules

[43] D. M. Sammeth, S. S. Siewert, L. H. Spangler, and P. R. Callis, *Chem. Phys. Lett.* **193,** 532 (1992).

[44] R. Jimenez, G. R. Fleming, P. V. Kumar, and M. Maroncelli, *Nature* (*London*) **369,** 471 (1994).

shows that the first 2000 cm^{-1} of the shift, which happens within 20 fsec, has significant contributions from about 20 water molecules within 6 Å of the indole ring center. The contributions were on the order of 100 cm^{-1}/molecule, with none contributing to a blue shift. The contributions were defined as the difference between the indole–water molecule electrostatic interaction that exists 20 fsec after excitation and that which would have existed had the indole remained in the ground state. The contributions are in direct proportion to rotational response to the differential torque on a water molecule created by the excited state dipole, a purely inertial response.

After 150 fsec the picture is different, because Brownian diffusion becomes a factor. It is found that certain molecules contribute as much as 2000 cm^{-1} to a red shift, while others contribute offsetting blue shifts. There was, however, no specific interaction point on the ring that correlated with the larger red-shifting interactions, such as might be envisioned in the site-specific exciplex model. This is not surprising because the major interactions in this model are dipole–dipole. From 10 different trajectories, the 13 water molecules that contribute more than 1500 cm^{-1} shifts were distributed almost randomly about the indole ring, although with orientation such that the dipole–dipole interaction with the indole excited-state dipole was stabilizing.

Solvent Effects in Rigid Environment: Red-Edge Effects

A phenomenon closely related to the inhomogeneous broadening is called the *red-edge effect.*[6,45–47] In the present context, it is the observation that the fluorescence may be seen to shift continuously to longer wavelengths as the excitation wavelength is changed to longer wavelengths in the vicinity of the long-wavelength edge of the absorption spectrum. It is observed only in highly viscous solvents (glasses) or in certain proteins, and it is understood on the basis of there being a distribution of environments about the solute, each providing a different local electric field. This corresponds to the thinking presented in Fig. 7a, which describes the various microenvironment states of one molecule at many times, as the statistically equivalent distribution of environments of many molecules at one time. Thus, the lowest energy absorption will be by the origin transitions of those molecules experiencing the greatest red shift. Providing there is no equilibration of the local environment during the excited-state lifetime, these molecules will also provide the most red-shifted fluorescence. Obvi-

[45] W. C. Galley and R. M. Purkey, *Proc. Natl. Acad. Sci. U.S.A.* **67,** 1116 (1970).
[46] A. P. Demchenko and A. I. Sytnik, *J. Phys. Chem.* **95,** 10518 (1991).
[47] A. P. Demchenko and A. S. Ladokhin, *Eur. Biophys. J.* **15,** 369 (1988).

ously, the extent of this effect is primarily restricted to the inhomogeneous linewidth for the origin transition. For Trp the effect will be much larger if the ^1L_a origin is below that of ^1L_b, the usual situation in polar solvent, as depicted in Fig. 1d.

Experimental Observations

In the following section are detailed experimental results that have been particularly helpful in resolving the ^1L_a–^1L_b manifold of Trp, primarily through studies of model compounds.

Two-Photon Spectroscopy

Two-photon spectroscopy has played a major role in identifying the ^1L_a state of indoles and Trp,[12,43,48–52] ultimately leading to the identification of the ^1L_a origin of indole and 3MI. An intense visible laser pulse with wavelength twice that of the intended one-photon transition can directly excite the ^1L_a and ^1L_b transitions in a weak process closely related to Raman light scattering. The difference from light scattering is that the second photon is absorbed from the laser beam instead of being emitted. As with scattering, the selection rules are complementary to those of one-photon absorption. Detection is normally by fluorescence or some other equally sensitive technique, because typical two-photon absorbance is only about 10^{-6}/cm. In addition to deducing electronic structure, two-photon excitation is finding increased use in fluorescence microscopy[53] and fluorescence anisotropy studies.[54] The aspect of two-photon excitation most exploited for studying ^1L_a and ^1L_b states of Trp is a type of photoselection based on the necessity of the two photons to have a relative polarization dictated by the nature of the final electronic state. It is found experimentally and theoretically that the ^1L_b state requires that the two photons have almost mutually perpendicular polarization, while ^1L_a requires them to be more nearly parallel. These requirements are manifest experimentally in the ratio of two-photon absorptivities for circularly and linearly polarized light, $\Omega = \delta_{cir}/\delta_{lin}$.[55] Ω was found, both experimentally and theoretically[49] for a number

[48] B. E. Anderson, R. D. Jones, A. A. Rehms, P. Ilich, and P. R. Callis, *Chem. Phys. Lett.* **125,** 106 (1986).

[49] A. A. Rehms and P. R. Callis, *Chem. Phys. Lett.* **140,** 83 (1987).

[50] P. R. Callis and A. A. Rehms, *Chem. Phys. Lett.* **208,** 276 (1993).

[51] D. M. Sammeth, S. Yan, L. H. Spangler, and P. R. Callis, *J. Phys. Chem.* **94,** 7340 (1990).

[52] P. L. Muiño and P. R. Callis, *Proc. S.P.I.E. Int. Soc. Opt. Eng.* **2137,** 278 (1994).

[53] W. Denk, J. H. Strickler, and W. W. Webb, *Science* **248,** 73 (1990).

[54] J. R. Lakowicz and I. Gryczynski, *Biophys. Chem.* **45,** 1 (1992).

[55] W. M. McClain, *J. Chem. Phys.* **57,** 2264 (1972).

of indoles in different solvents, to be ~1.4 for ^1L_b and ~0.5 for ^1L_a when the emitted fluorescence is completely depolarized. Because the two photons are absorbed "simultaneously" (within about 1 fsec), this type of photoselection can be used in fluid solutions. Figure 8a shows Ω plotted with the one- and two-photon excitation spectra of Trp in water at 25°.[50] The near superposition of the one- and two-photon excitation spectra (plotted on the basis of equivalent final state energy) in the 280-nm band is the unexpected consequence of the ratio of $^1L_a : {}^1L_b$ absorptivities being nearly the same in the one- and two-photon process when linearly polarized light is used. It can be viewed as a coincidence because it is not predicted by any symmetry considerations and it is not true for circularly polarized light.

It is seen that Ω levels off at about 300 nm with a value of about 0.55, indicating that red-edge excitation at wavelengths >300 nm excites only the ^1L_a transition. Although a similar conclusion has been inferred from one-photon fluorescence anisotropy behavior,[24,28] the latter conclusion is slightly clouded from having been obtained in a glassy solution containing large amounts of glycol, glycerol, etc., often at reduced temperature. Figure 8b shows the resolution of the aqueous Trp excitation spectrum into its

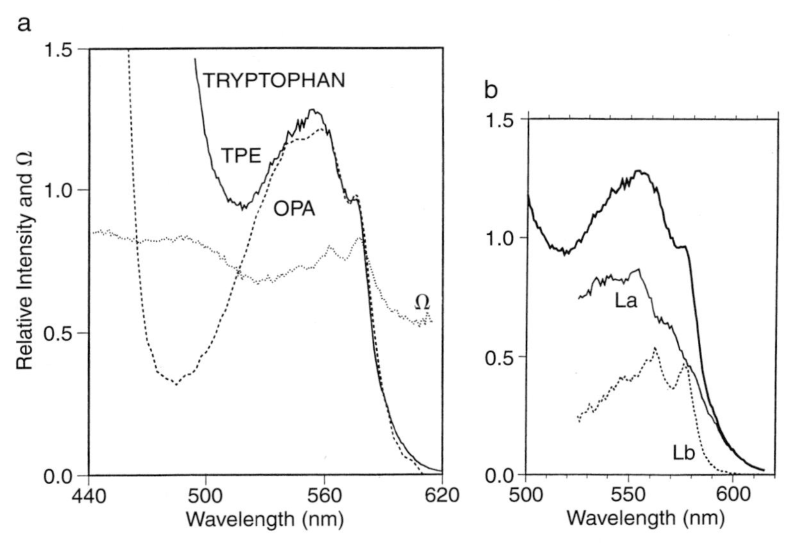

FIG. 8. (a) Two-photon induced fluorescence excitation spectrum of Trp in water at pH 7 (solid line) plotted with its one-photon absorption at one-half the indicated wavelength (dashed line) and the two-photon polarization, Ω (dotted line). (b) Resolution into ^1L_a and ^1L_b components on the basis of $\Omega = 0.52$ for ^1L_a and $\Omega = 1.15$ for ^1L_b. (Adapted from *Chem. Phys. Lett.* **208,** P. R. Callis and A. A. Rehms, 276, Copyright 1993 with kind permission of Elsevier Science–NL, Sara Burgerhartstraat 25, 1055 KV Amsterdam, The Netherlands.)

^1L_a and ^1L_b components, on the basis of the Ω values. The result is similar to that found by one-photon fluorescence anisotropy.[21,24,28]

In proteins it is usually the case that the Trp emission is not depolarized completely, meaning the diagnostic Ω values will be distorted by the fluorescence anisotropy. In such cases one may use the fact that $\Omega = (2I_{CV} + I_{CH})/(I_{VV} + 2I_{VH})$,[12] where the first subscript indicates circular or vertical linear excitation polarization and the second subscript indicates the orientation of the fluorescence polarizer (vertical or horizontal). Thus, one may know which state is absorbing under any conditions, without uncertainties or assumptions concerning the emitting state—an important complement to conventional time-resolved fluorescence anisotropy studies because solvent relaxation can be over in less than 1 psec.

Jet-Cooled and Solid Argon Spectra

A number of experimental observations on indoles in homogeneous environments have revealed the details of the vibronic structure of the ^1L_a and ^1L_b transitions and have definitively located and mapped the ^1L_a states. These studies were done in cold molecular beams[43,51,52] and in argon crystals.[27] Figure 9 compares the one-photon fluorescence excitation spectra of indole in jet and in solid argon. ^1L_a lines found by two-photon Ω are marked with asterisks, and Fig. 9d gives the one-photon fluorescence excitation anisotropy. Two-photon excitation in the jet has revealed transitions of ^1L_a character, not all of which are actually ^1L_a transitions. Although the pair of lines near 470 cm^{-1} above the ^1L_b origin have ^1L_a character, as revealed by both the two-photon polarization and the one-photon anisotropy, they have been shown to be a single ^1L_b transition (split by a Fermi resonance in the jet spectrum) that gains most of its absorptivity by Herzberg–Teller vibronic coupling to the ^1L_a state, the mechanism by which the ^1L_b transition in benzene and naphthalene gains its intensity. The evidence comes from its failure to shift relative to the ^1L_b origin when going from vapor to condensed phase[27] or when complexed with water,[52] with additional proof coming from the low anisotropy of this transition in the fluorescence.[27] The transition at 781 cm^{-1} also falls into this category. In contrast, the cluster of lines extending from 1200 to 1450 cm^{-1} is seen to red shift by about 200 cm^{-1} relative to the ^1L_b lines, as expected for the more polar ^1L_a state owing to the dipole-induced dipole interaction with the argon matrix. This group of lines compose the ^1L_a origin, split because of coupling with a number of ^1L_b vibronic states with nearly the same energy. Such splitting of the origin of a higher excited state is common. Thus, it turns out that the assignment by Strickland *et al.*[56] in 1970 putting the ^1L_a origin 1450 cm^{-1} above ^1L_b was essentially correct.

[56] E. H. Strickland, J. Horwitz, and C. Billups, *Biochemistry* **9,** 4914 (1970).

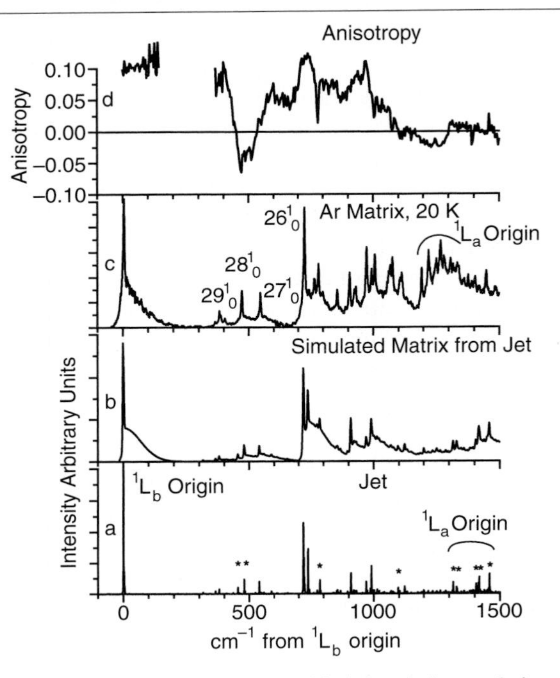

FIG. 9. Jet-cooled and argon-matrix spectra of indole relative to their respective origins at 35,250 cm^{-1} (283.7 nm) and 35,000 cm^{-1} (285.7 nm). (Adapted from *Chem. Phys. Lett.* **239,** B. J. Fender, D. M. Sammeth, and P. R. Callis, 31, Copyright 1995 with kind permission of Elsevier Science–NL, Sara Burgerhartstraat 25, 1055 KV Amsterdam, The Netherlands.) (a) Jet-cooled one-photon fluorescence excitation spectrum of indole seeded in a helium expansion jet. The rotational temperature is about 2 K. The asterisks indicate ^1L_a character as found from two-photon Ω. The cluster of lines from 1200 to 1450 cm^{-1} is the ^1L_a origin; (b) simulated argon matrix fluorescence excitation spectrum by broadening the spectrum in (a); (c) fluorescence excitation spectrum of indole in solid argon at 20 K. The ^1L_a origin lines can be seen to have shifted about 200 cm^{-1} to the red relative to ^1L_b lines; (d) fluorescence anisotropy for the spectrum in (c) with the emission wavelength set to 760 cm^{-1} red of the origin. The strong negative anisotropy near 500 cm^{-1} is due to an ^1L_b line corresponding to one quantum of a vibration that promotes stealing intensity from ^1L_a.

Jet experiments on 3MI reveal that the ^1L_a origin is red shifted to about 500 cm^{-1} above the ^1L_b origin, but remains split into several components.[43] In solid argon the ^1L_a origin of 3MI is further red shifted to within 260 cm^{-1} above the ^1L_b origin and has coalesced into a single line,[57] similar to what happens in perfluorohexane.[56,58] The fluorescence spectrum retains the characteristic pattern of ^1L_b fluorescence seen for indole and 5-methylin-

[57] B. J. Fender and P. R. Callis, *Chem. Phys. Lett.* **262,** 343 (1996).

[58] D. M. Sammeth, Ph.D. dissertation, Montana State University, Bozeman, Montana, 1992.

dole.[57] However, for 2,3-dimethylindole, where ^1L_a is shifted further to the red, the two origins are overlapping in solid argon, with ^1L_a being slightly lower in energy, as indicated by its fluorescence, which is different than that of 3MI and has the characteristic features of ^1L_a fluorescence.[57]

That the ^1L_a origin can be identified distinctly only 260 cm^{-1} above ^1L_b for 3MI in solid argon means that, in principle, ^1L_b can be the fluorescent state in a protein environment, provided the immediate surroundings of the Trp provide a local electric field that causes a sufficient blue shift of ^1L_a.

^1L_a or ^1L_b Fluorescence

Given how precisely the ^1L_a–^1L_b gap is known for 3MI, the question of whether structured, blue-shifted fluorescence from a buried Trp is ^1L_a or ^1L_b has practical significance. If the fluorescence is actually from ^1L_b, a definite statement is made regarding the local electric field and the orientation of the Trp ring relative to the field. It is difficult to find experimental examples of narrow line ^1L_a fluorescence and broadened ^1L_b fluorescence because ^1L_a lies above ^1L_b for indole and 3-substituted indoles under conditions leading to sharp lines. Likewise, solvents that would significantly broaden ^1L_b cause ^1L_a to be the fluorescing state. In this section definitive assignments are made for three cases in which structured fluorescence is suspected of being ^1L_b because it is structured and for a fourth in which fluorescence is assigned ^1L_a because it is not sharp: (1) Trp-48 of azurin,[7,59] (2) ribonuclease T_1,[4,7,60] (3) Trp in ethanol and Trp-37 of hemoglobin with red-edge excitation at 2 K,[61] and (4) the fluorescence of jet-cooled indole plus water π complex.[62,63]

Trp-48 of azurin has a structured fluorescence with maximum at 308 nm, which is almost identical to that of 3MI in hydrocarbon solution.[64] That being established, what then is the character of 3MI fluorescence in hydrocarbon? The spectra of Strickland *et al.*[56] strongly suggest that the ^1L_a and ^1L_b origins are superimposed for 3MI in hexane. Two-photon excitation experiments directly confirm this, with $\Omega = 0.75$ throughout the red edge of the sharp origin line, considerably higher than the value of 0.50 found on the red edge of the 2,3-dimethylindole origin in cyclohexane.[49] This result is strong evidence that absorption is to both types of states or

[59] G. Gilardi, G. Mei, N. Rosato, G. W. Canters, and A. Finazzi-Agro, *Biochemistry* **33,** 1425 (1994).

[60] J. W. Longworth, *Photochem. Photobiol.* **7,** 587 (1968).

[61] T. W. Scott, B. F. Campbell, R. L. Cone, and J. M. Friedman, *Chem. Phys.* **131,** 63 (1989).

[62] M. J. Tubergen and D. H. Levy, *J. Phys. Chem.* **95,** 2175 (1991).

[63] S. Arnold and M. Sulkes, *J. Phys. Chem.* **96,** 4768 (1992).

[64] A. G. Szabo, T. M. Stepanik, D. M. Wayner, and N. M. Young, *Biophys. J.* **41,** 233 (1983).

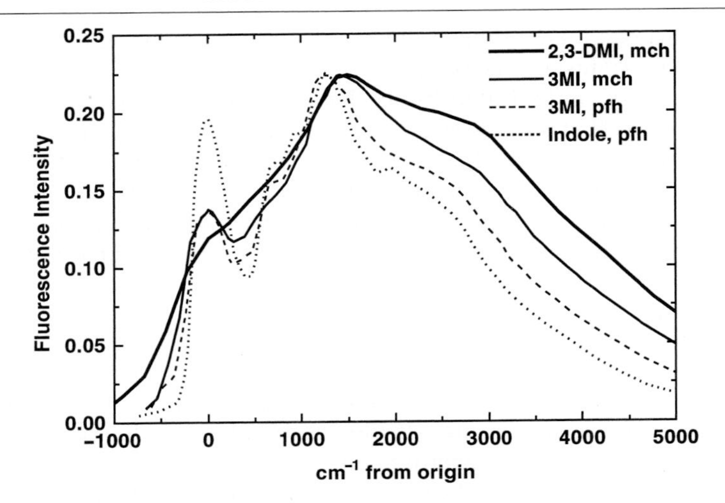

FIG. 10. Comparison of the room temperature fluorescence spectra of indole in perfluorohexane (pfh), 3-methylindole in pfh and methylcyclohexane (mch), and 2,3-dimethylindole (2,3-DMI) in mch (ca. 10^{-4} *M*) at 0.8-nm (100 cm^{-1}) resolution. The latter is known to be purely ^1L_a emission and the indole fluorescence is purely ^1L_b.

to states of hybrid character. The question of the character of the fluorescence is not so clear because, even in hydrocarbon, the more polar ^1L_a state may be preferentially stabilized after excitation, so that the dominant fluorescing state could well be ^1L_a. The local electric field and/or the ability for relaxation could presumably tip the scales either way.

Figure 10 compares the fluorescence spectra of indole and 3MI in perfluorohexane (pfh), 3MI in methylcyclohexane (mch), and 2,3-DMI in mch at room temperature. The types of fluorescence for indole in pfh and 2,3-DMI in mch are pure ^1L_b and pure ^1L_a, respectively. For 3MI in pfh, two-photon studies indicate "contamination" by some ^1L_a by two mechanisms. the Ω value for the ^1L_b origin is only about 1.14 in the jet-cooled spectrum[43] and 1.1 in pfh,[58] whereas it is 1.4 for indole. In addition, there is a possibility of some ^1L_a emission because the ^1L_a origin is only about 400 cm^{-1} above ^1L_b in pfh,[58] and its width is about 400 cm^{-1}. Both factors may contribute to the fluorescence of 3MI in pfh being slightly broader than that of indole in pfh. In any case, the 3MI spectrum in mch appears intermediate between ^1L_a than ^1L_b. Note that the computed spectra in Fig. 5 also support this conclusion. We conclude that Trp-48 in azurin emits from both ^1L_b and ^1L_a at room temperature. However, the spectrum of Burstein *et al.*[65] for Trp-

[65] E. A. Burstein, E. A. Permyakov, V. A. Yashin, S. A. Burkhanov, and A. Finazzi-Agro, *Biochim. Biophys. Acta* **491,** 155 (1977).

48 at 77 K appears to be closer to the spectrum of indole and 3MI in pfh, suggesting that at 77 K Trp-48 emits mainly ^1L_b fluorescence.

Ribonuclease T_1 (RNase T_1) also shows structured, blue-shifted fluorescence at room temperature, and even more so at 77 K.[60] Two lines of reasoning lead to a definite ^1L_a assignment for this fluorescence. For one, the RNase T_1 spectrum in Fig. 5b is similar to that computed for ^1L_a from an *ab initio* procedure. In addition, it can be seen from Fig. 5a that the vibronic structure is not at all what is predicted and seen for ^1L_b. The second approach is to compare this spectrum with broadened phosphorescence spectra (known to be ^3L_a), for which there is quite good agreement. It can be seen from Fig. 5b that the theoretical spectrum broadened to 80 cm^{-1} agrees well with the sharp phorphorescence from indole in argon at 10 K.

A particularly important case in which ^1L_b fluorescence was suspected on the basis of sharp vibrational structure comes from Scott *et al.*,[61] who have observed the red-edge effect and line narrowing in the fluorescence from Trp in ethanol glass and hemoglobin (Hb) in ice at 2 K. For Trp, excitation above 295 nm produces fluorescence with somewhat resolved vibronic structure (~500 cm^{-1}), and which shifts in concert with the excitation wavelength. For excitation with wavelengths $<$294 nm a broader, diffuse emission is seen, with maximum at 313 nm, which does not shift with the excitation wavelength. Scott *et al.*[61] tentatively assigned the structured emission as ^1L_b and unstructured as ^1L_a, following conventional wisdom. This assignment required the assumption that direct excitation to the ^1L_a state is Franck–Condon forbidden, but could be populated by tunneling from higher vibronic levels. However, data presented in this chapter indicate that the ^1L_a origin is quite strong and, in fact, the vibronic structure exhibited fits nicely with that of RNase T_1 at 77 K and the predictions of theory for ^1L_a, but not at all well for ^1L_b (see Fig. 5a). In addition, and quite conclusively, the Stark spectroscopy experiments by Pierce and Boxer[21] on N-acetyltryptophanamide (NATA), melittin, and 5-hydroxytryptophan make it clear that the red-edge absorption in ethanol is ^1L_a. Their associated fluorescence anisotropy experiments on NATA in ethanol at 77 K actually reveal the ^1L_a origin as a shoulder about 500 cm^{-1} to the red of the ^1L_b origin. The fluorescence anisotropy profile is similar to that found by Valeur and Weber,[28] and the red-edge region showed by far the largest Stark effect (corresponding to a dipole change of 6 D) compared to the sharper peak 500 cm^{-1} to the blue (about a 1-D change). In fact, the onset of the broader spectrum at 296 nm is just where Pierce and Boxer[21] find the onset of ^1L_b absorption.

The behavior is what is expected from the red-edge effect, and the broadening seen when exciting at shorter wavelengths was correctly explained by Scott *et al.*[61] Assuming the inhomogeneous broadening of ^1L_a

is about 1000 cm^{-1}, the reasonably strong 0-1 lines at 600 and 700 cm^{-1} above the origin will significantly overlap the origin envelope near its center and higher. At this point, the red edges of the 0-1 lines are being excited along with the blue edge of the origin. The resulting emission will be a mixture of two or more sharp emissions separated by about 600 cm^{-1}. Because the emitted lines already have a 500-cm^{-1} width, there will be little structure left, especially if ^1L_b is also excited. This complexity multiplies as the excitation energy is increased.

Why is the "sharp" emission not as sharp as the laser? One would at first think that line-narrowed emission should have the linewidth of the laser, and this would be true if there were absolutely no relaxation following excitation. In addition to the completely elastic quantum mechanical explanation provided by Scott *et al.*,[61] the following explanation based on overdamped motion may also have merit. Simulations of relaxation by 3MI in butanol at a temperature of 0 K[42] suggest that considerable relaxation may occur in an alcohol glass, even when there is no solvent motion, provided there is a large dipole change in the solute. This arises from torsional responses from nearby O–H dipoles, moving in the free volume of the glass. The population that the laser excites is homogeneous in the transition energy, but is inhomogeneous in the ability to relax, thereby giving a broader distribution of emission.

It is debatable whether jet-cooled ^1L_a fluorescence has been observed, except for 3MI complexed to ND_3[66] and 2,3-DMI complexed to methanol and trimethylamine.[67] For example, spectra by Tubergen and Levy[62] and by Arnold and Sulkes[63] for the indole–water π complex were assigned ^1L_a character on the basis of broadening. However, it was demonstrated that the broadening is quite plausibly due only to the low-frequency intermolecular vibration.[68] Furthermore, the prominence of mode 26 and the relative weakness of mode 27 and modes 8–10 at 1600 cm^{-1} in those spectra suggest that the fluorescence is ^1L_b on the basis of the calculations presented here in Fig. 5. Finally, the two-photon Ω value has been observed to be 1.3, virtually removing any doubt that S_1 is ^1L_b.[69]

Ultrashort Time-Resolved Anisotropy

In a pioneering experiment, which also features ^1L_b fluorescence, Rugierro *et al.*[70] and Hansen *et al.*[71] have observed the time-resolved fluo-

[66] D. R. Demmer, G. W. Leach, and S. C. Wallace, *J. Phys. Chem.* **98,** 12834 (1994).

[67] D. R. Demmer, G. W. Leach, E. A. Outhouse, J. W. Hager, and S. C. Wallace, *J. Phys. Chem.* **94,** 582 (1990).

[68] P. L. Muiño and P. R. Callis, *Chem. Phys. Lett.* **222,** 156 (1994).

[69] K. Short and P. R. Callis, unpublished results (1996).

[70] A. J. Ruggiero, D. C. Todd, and G. R. Fleming, *J. Am. Chem. Soc.* **112,** 1003 (1990).

[71] J. E. Hansen, S. J. Rosenthal, and G. R. Fleming, *J. Phys. Chem.* **96,** 3034 (1992).

rescence and its anisotropy for Trp, azurin, and melittin in water with 500-fsec resolution. For Trp and melittin the anisotropy was found to have a fast decaying component with $\tau = 1.2$ psec, much too fast for depolarization from rotational diffusion. This component is observed only for excitation below 300 nm and fluorescence below 360 nm. It is attributed to internal conversion between ^1L_b and ^1L_a states. At short times after excitation, whichever state is excited will be emitting, leading to high anisotropy. Thus, even though ^1L_b lies above ^1L_a, on average, it is not unreasonable to expect ^1L_b emission prior to internal conversion to the ^1L_a state. Thus, the early emission would have higher anisotropy than that occurring after internal conversion. Similar experiments on Trp-48[71] of azurin, however, did not show the short time component, even though excitation was clearly to a mixture of ^1L_b and ^1L_a. The authors suggest that the close proximity of the two states in the hydrophobic environment leads to much faster internal conversion kinetics, which were not observable in their experiment.

Some Applications to Proteins

One might be concerned that Trp fluorescence might be a limited tool because so few proteins contain a single, strategically located Trp. Modern site-directed mutagenesis techniques have reduced this concern. A variety of strategies have proven effective in which Trp is replaced with tyrosine (Tyr) or phenylalanine (Phe). Below, three examples are briefly presented. In the first, all but one of three native Trps are replaced with Tyr or Phe. Additional mutations are made to assess the effect on the particular Trp. In a second approach, which is useful when so many Trps are present that replacing all but one would seriously compromise the protein function, Trps are replaced one at a time, and their properties determined through difference spectra. The third technique is less passive and potentially powerful: one inserts a Trp into a protein that normally contains none. By finding points that do not alter function seriously and are located in strategic positions, considerable insight may be gained about conformation changes accompanying substrate binding, etc.

T4 Lysozyme

As an example wherein natural Trps are used, one of the most detailed and extensive studies exploiting Trp fluorescence has been that of Harris and Hudson,[72,73] in collaboration with the Matthews group, on T4 lysozyme. The properties of the three native Trps were determined by observing the

[72] D. L. Harris and B. S. Hudson, *Biochemistry* **29,** 5276 (1990).
[73] D. L. Harris and B. S. Hudson, *Chem. Phys.* **158,** 353 (1991).

three single Trp mutants wherein two of the Trps are replaced by Tyr. All Trps are relatively buried according to their similar λ_{max} of 330 nm, yet they exhibit fairly different accessibility to solvent, as judged from crystal structures and external quenching. Trp-158 has a low quantum yield of 0.013, rationalized by its proximity to a cysteine, which is known to be an efficient quencher. Most of the work focuses on Trp-138, which is almost completely buried. It also has a low quantum yield (0.044), apparently because it is hydrogen bonded to a glutamine (Q105). This hydrogen bond is credited for helping to hold Trp-138 rigidly. Several mutants were studied, in which Q105 and A146 (another residue that contacts Trp-138) were replaced. Replacing Q105 with an alanine, which cannot hydrogen bond and is a poor quencher, results in a threefold increase in fluorescence yield, a 5-nm red shift, and a reduced fluorescence anisotropy. In addition, the lifetime changes from being essentially biexponential, dominated by a 1.1-nsec component, to being a single exponential, 5.1-nsec decay. Substitution of bulkier groups for A146 always increased mobility and induced up to 10-nm red shifts in the fluorescence. These effects were found to be largely reversed if a disulfide bridge was introduced locally.

The elongated T4 lysozyme protein rotationally diffuses more slowly about its short axes than about the long axis. This was evident in the anisotropy decay times of the different Trps, which corresponded well with what was expected from the direction of their respective ^1L_a transition dipoles.

In a similar type of study, Gilardi *et al.*[59] prepared mutants wherein I7 and F110 of azurin, two of the hydrophobic residues lining the hydrophobic pocket of Trp-48, were changed to serine. Both absorption and fluorescence were red shifted, and a pronounced red-edge effect was observed, where there was none for the wild type. For excitation at the extreme red edge, the fluorescence from the mutants peaked at 350 nm, i.e., about where fully exposed Trp emits. It is not clear if this represents extremely red-shifted emission from a buried Trp-48 in a fortuitous interaction with the nearby serine (exciplex?), or if the Trp-48 is actually solvent exposed for some conformations of the mutants.

Human Tissue Factor

One of the four Trps of the extracellular domain was replaced with Tyr or Phe with less than 50% loss in activity.[74] The absorption and fluorescence

[74] C. A. Hasselbacher, E. Rusinova, E. Waxman, R. Rusinova, R. A. Kohanski, W. Lam, A. Guha, J. Du, T. C. Lin, I. Polikarpov, C. W. G. Boys, Y. Nemerson, W. H. Konigsberg, and J. B. A. Ross, *Biophys. J.* **69,** 20 (1995).

properties of the four mutants were assessed from difference spectra and were found to agree well with expectations from the crystal structure.

Calmodulin

An example of learning important information from Trp fluorescence from a protein containing no natural Trp is provided by Chabbert *et al.*[75] who have made five isofunctional mutants of calmodulin (CaM), with one Trp either at one of the four calcium-binding loops or in the central helix. They observed the effect on the fluorescence on binding the active-binding peptides from nonmuscle myosin light chain kinase (nmMLCK) and from CaM-dependent protein kinase II (CaMPK-II). A Trp in either loop I or IV exhibited red-shifted fluorescence high yields, long lifetimes, and responded only slightly to the binding. In contrast, fluorescence from a Trp in either loop II or III showed a large decrease in intensity, was blue shifted, and the short-lifetime component seemed to be enhanced on binding. Because loops II and III are in different lobes, this constitutes new evidence that both lobes of calmodulin are involved in peptide binding. Flexibility of the central helix allows the two domains to be significantly closer than in the crystal structure, and the hydrophobic clefts of both lobes can be involved simultaneously in the binding of α-helical amphiphilic peptides. The Trp in the central helix responded differently to binding the two different peptides, as revealed by quenching and fluorescence shifts, becoming more solvent exposed when binding the nmMLCK and less exposed when binding the CaMPK-II. These results are consistent with the hypothesis that the central helix can play a differential role in the recognition of, or response to, CaM-binding structures.

General Comments

How does the fundamental knowledge recounted earlier in this chapter relate to the use of Trp fluorescence? Following are a few observations.

The wavelength of maximum fluorescence intensity (λ_{max}) is universally and unquestioningly used as an indicator of exposure to water, i.e., as an indicator of how deeply the Trp is buried in the protein. This correlates well with the effectiveness of external quenchers, and virtually no exceptions appear to be documented. This would seem to make a strong statement against the idea of exciplex formation, for surely in certain cases exciplexes could form with a buried Trp if the proper group were situated fortuitously.

[75] M. Chabbert, T. J. Lukas, D. M. Watterson, P. H. Axelsen, and F. G. Pendergast, *Biochemistry* **30,** 7615 (1991).

It appears therefore that in the vast majority of cases the large red shift of Trp fluorescence requires the cooperative effect of many water molecules.

It is sometimes stated that red-shifted fluorescence indicates a polar environment, but this alone is incorrect. If the word *polar* is to mean a region containing polar residues and no mobile water, the fluorescence could be strongly blue shifted if the local electric field were oriented against the Trp dipole change, e.g., if a positive charge were near the pyrrole ring or if a negative charge were near the benzene ring. A red shift is guaranteed in a polar environment only if there is sufficient mobility of the polar groups to relax about the large ^1L_a dipole.

In this regard the class of protein fluorescing maximally at about 340 nm[76] is interesting because these are usually totally accessible to external quenchers, yet have fluorescence blue shifted by 15 nm and tend to have restricted motion.[7] To model and understand this class it will be necessary to use fully the details of excited-state charge distribution, dipole direction, and transition dipole directions.

Axelsen *et al.*[77] have undertaken an initial foray at modeling the motion for the Trp in phospholipase A_2, whose fluorescence has λ_{max} at 340 nm, but is maximally exposed to external quenchers and has high anisotropy (cone angle is about 25°). They were able to reproduce this in molecular dynamics modeling, but only if explicit hydrogens were used. The polar hydrogen case gave a poor Trp dipole for the ground state and no attempt was made to use excited-state charges. The considerable difference in charge distribution in the ^1L_a state is likely to play a role in reducing mobility, and the information cited in this chapter should aid such studies. Another issue raised is the direction of the ^1L_a transition dipole during emission, which is apparently $\pm 25°$ relative to the absorbing moment, as deduced from the limiting anisotropy of ~ 0.3.[24,28] There is some theoretical evidence for $\sim 5°$ changes,[12] but this is an area requiring a more thorough study. Table I indicates that differences of a few degrees may be expected, but no clear pattern has emerged. A proper treatment should include averaging over vibrational motions.

With only a few exceptions, Trp fluorescence from proteins exhibits nonexponential decay, which can be resolved into two and sometimes three exponentials.[5,7] The source of the nonexponential decay is generally unknown and is under debate. A long-standing, yet elusive, proposal is the rotomer model, wherein Trp residues may be found in more than one orientation about the bonds connecting the indole ring to the peptide

[76] E. A. Burstein, N. S. Bedenkina, and M. N. Ivkova, *Photochem. Photobiol.* **18,** 263 (1973).

[77] P. H. Axelsen, E. Gratton, and F. G. Prendergast, *Biochemistry* **30,** 1173 (1991).

chain.[78,79] A more recent proposal invokes electron transfer to produce a contact radical ion pair (CRIP) followed by back transfer and subsequent luminescence.[80] The latter is an example of an excited-state reaction; it has been pointed out that any excited-state reaction will lead to at least biexponential decay of fluorescence.[81]

Nevertheless, the lifetimes and weighting coefficients provide additional means of characterizing Trp fluorescence, which can be followed during the course of a transformation. A particularly intriguing pattern is that during the course of substrate binding that causes a large change in quantum yield, it is often found that the decay times do not change significantly; only the relative proportions of the decay components changes.[82,83] Even more intriguing is the typical pattern of results when decay-associated spectra (DAS) are obtained from measuring the lifetime as a function of fluorescence wavelength.[84] With considerable regularity it is found that the longer lifetime component peaks at the longer wavelength. It would seem fruitful to investigate new models for the time-resolved fluorescence in which allowance is made for continuously red-shifting emission due to relaxations.

The preceding statement concerning the correlation of DAS and lifetime is surprising, given that there does not seem to be a correlation between quantum yield and λ_{max}. A survey of most of the known single-Trp proteins reveals that the fluorescence wavelengths are distributed over a considerable range: from about 308 nm to about 355 nm.[7] The observed fluorescence quantum yields are also distributed over a wide range: from about 0.35 to 0.02. It is widely believed that the large variation in yields stems from the proximity of quenching groups in the vicinity of the Trp.[6] One of the most puzzling observations has been that the radiative lifetime, τ_r, varies by about fivefold in this sampling. This behavior is not limited to Trps buried within proteins. Eftink *et al.*[85] have surveyed a large number of Trp analogs. For those that are 3-substituted, the τ_r values range from about 16 to 27 nsec, a 60% range. There is a pronounced correlation with the fluorescence maximum, with the smaller rate at the longer wavelength. Only about a

[78] B. Donzel, P. Gauduchon, and P. Wahl, *J. Am. Chem. Soc.* **96,** 801 (1974).

[79] H. L. Gordon, H. C. Jarrell, A. G. Szabo, K. J. Willis, and R. L. Somorjai, *J. Phys. Chem.* **96,** 1915 (1992).

[80] M. Van Gilst, C. Tang, A. Roth, and B. Hudson, *J. Fluoresc.* **4,** 203 (1994).

[81] L. Van Dommelen, N. Boens, and F. C. De Schryver, *J. Phys. Chem.* **99,** 8959 (1995).

[82] M. P. Brown, N. Shaikh, M. Brenowitz, and L. Brand, *J. Biol. Chem.* **269,** 12600 (1994).

[83] B. K. Szpikowska, J. M. Beechem, M. A. Sherman, and M. T. Mas, *Biochemistry* **33,** 2217 (1994).

[84] C. Hogue, Ph.D. dissertation, University of Windsor, Windsor, Ontario, 1995.

[85] M. R. Eftink, Y. Jia, D. Hu, and C. A. Giron, *J. Phys. Chem.* **99,** 5713 (1995).

20% change is expected on account of the λ^3 dependence of radiative lifetime. The longer wavelength is always with the higher pH. This result is similar to that found by Meech and Phillips,[86] who find a variation from 11 to 33 ns for 3MI as the fluorescence becomes more red shifted due to solvent polarity. Again, the three-fold change in τ_r found by Meech and Phillips[86] is only partly accounted for by the λ^3 dependence, which changes by about 1.8 over the range. The rest of the effect must come from a decrease in the transition dipole length. This was found to be the case in the simulation results described previously.[41]

Conclusions and Outlook

What might be called the adiabatic aspect of Trp spectroscopy is rapidly becoming understood at a fundamental level. That is, we can be optimistic about being able to predict spectra accurately from a given environment. From this we may reasonably expect that spectral results can at least impose constraints on the possible types of environment seen by a Trp in more detail than at present. For this to become a practical reality will no doubt require a period of active modeling and testing.

It is unlikely, however, that Trp fluorescence will reach its full potential without a better understanding of the large variation in quantum yields and the interesting lifetime patterns mentioned. At a basic level, this may require modeling relaxation of the protein environment and will involve the detailed and specific prediction of rates for nonradiative electronic transitions, something that has eluded molecular quantum physicists and chemists in spite of three decades of quite successful generic results. With continued experimental investigation it is possible that at least an empirical grasp will emerge. It is hoped that workers in this field will keep in mind the information presented in this chapter, and exploit it when possible.

Acknowledgments

Assistance from students Bruce Fender, David Hahn, Kurt Short, and Berit Burgess in preparing the manuscript is greatly appreciated. Drs. Maurice Eftink, Bruce Hudson, Art Szabo, Chris Hogue, and Dmitri Toptygin are thanked for helpful discussions. The work contributed from this laboratory was supported by NIH Grant GM31824.

[86] S. R. Meech and D. Phillips, *Chem. Phys.* **80,** 317 (1983).

[8] Enhancement of Protein Spectra with Tryptophan Analogs: Fluorescence Spectroscopy of Protein–Protein and Protein–Nucleic Acid Interactions

By J. B. ALEXANDER ROSS, ARTHUR G. SZABO, and CHRISTOPHER W. V. HOGUE

Fluorescence probes are used extensively to study structures, dynamics, interactions, and environments in a wide variety of biological systems. This chapter focuses on tryptophan analogs as intrinsic fluorescence probes in proteins. The first section provides background and a historical perspective. The next section discusses the principles for spectral enhancement of proteins. The third section describes useful analogs and their spectral properties. The fourth section compares protein production *in vivo* and *in vitro,* discusses factors affecting incorporation, and provides generic methods for analog incorporation and characterization of analog-containing proteins. The last section discusses important spectral features of analog-containing proteins.

Background

The broad utility of fluorescence for studying protein structure, dynamics, and functional interactions is due partly to the response of the excited singlet state to the chemistry of the local environment and partly to the inherent sensitivity of fluorescence spectroscopy. Depending on the nature of the local environment, certain interactions of the excited state may affect its photophysical behavior and measured parameters. Nearby groups may promote proton transfer, hydrogen bonding, exciplex formation, formation of charge transfer complexes, intersystem crossing to the triplet state, or resonance energy transfer. Molecular dynamics that occur during the singlet state lifetime also are important. For example, the excited-state dipole moment of a probe often differs in magnitude and direction from its ground-state dipole moment. Consequently, reorientation of solvent, or other neighboring dipoles, can occur following excitation of the probe. The conformation of the probe itself may change while in the excited state. The impact of these processes on fluorescence parameters, including the characteristic wavelengths and contours of excitation and emission spectra, quantum yield, anisotropy, and lifetime, are discussed in an earlier volume of this series, in an excellent chapter on time-resolved fluorescence measure-

0076-6879/97 $25

ments by Badea and Brand.[1] Also, there is the informative textbook on basic principles of steady-state and time-resolved fluorescence by Lakowicz.[2] Other recommended sources are the series *Topics in Fluorescence Spectroscopy,*[3] edited by Lakowicz, and *Photoluminescence of Solutions* by Parker.[4]

The sensitivity and selectivity of fluorescence are particularly important factors in its role as a tool in modern biology. Extrinsic fluorophores, such as fluorescein and its various derivatives, often are used as probes because their spectral signals can be easily separated from those of intrinsic fluorophores, such as tryptophan (Trp), or pyridine and flavin nucleotide coenzymes. A major issue to be considered in biological applications of extrinsic probes concerns how the probes may affect the system under study. Extrinsic probes frequently are introduced by random labeling procedures because the methods are relatively simple and straightforward. The chemical modification, however, often affects physical structure and functional properties. Also, random labeling often results in a distribution of modified sites, thereby increasing sample heterogeneity. Where possible, site-specific labeling is preferred because changes in structure and function then can be correlated directly with the site(s) of modification and with the ground-state and excited-state chemistries of the probe.

Intrinsic fluorescence probes allow a system to be observed without chemical perturbation. Historically, Trp has been the intrinsic probe of choice for studies of dynamics and function of individual proteins in solution. The absorption of Trp extends to lower energies than that of other natural aromatic amino acids, tyrosine (Tyr) and phenylalanine (Phe),[5,6] and its absorption at wavelengths longer than 293 nm allows Trp fluorescence to be excited independently of Tyr and Phe. The fluorescence emission spectrum and quantum yield of Trp are influenced strongly by the local environment. The emission maximum of a buried Trp residue may be shifted to higher energy by 20–40 nm compared to that in water. In addition, a direct relationship has been demonstrated that links the multiexponential intensity decay kinetics of single Trp residues in peptides with the conformational populations of the indole side chain and local secondary structure of the

[1] M. G. Badea and L. Brand, *Methods Enzymol.* **61,** 378 (1979).

[2] J. R. Lakowicz, "Principles of Fluorescence Spectroscopy," Chap. 2. Plenum, New York, 1983.

[3] J. R. Lakowicz, (ed.), "Topics in Fluorescence Spectroscopy," Vols. 1–3. Plenum, New York, 1991–1992.

[4] C. A. Parker, "Photoluminescence of Solutions." Elsevier, New York, 1968.

[5] J. M. Beechem and L. Brand, *Annu. Rev. Biochem.* **54,** 43 (1985).

[6] M. R. Eftink, *Methods Biochem. Anal.* **35,** 127 (1991).

polypeptide chain.[7–10] Thus, fluorescence lifetime measurements of single Trp residues can provide a unique source of information about structural heterogeneity and dynamics of individual polypeptides and proteins in dilute solutions.

While Trp has unique advantages as an intrinsic probe, it is difficult to use for studying protein–nucleic acid and protein–protein interactions. First, the absorption spectra of nucleic acids completely overlap that of Trp, making it essentially impossible to resolve the absorption due to Trp. Moreover, the strong extinction of nucleic acids in the region used for excitation of Trp fluorescence contributes a substantial inner filter effect.[2,4] When not taken into account, quenching due to this inner filter effect can be incorrectly interpreted as quenching resulting from intermolecular interaction. Second, most proteins contain Trp, and the similar absorption features of the individual proteins make it impossible to assign and interpret absorption or fluorescence changes that result from intermolecular association. In certain cases, to "simplify" the absorption of a protein and its fluorescence spectra, by using site-directed mutagenesis,[11–14] Trp residues can be eliminated or replaced with other residues that have noninterfering spectral properties. However, this procedure may alter structure and/or function. Thus, an important motivation for using Trp analogs in place of Trp as a spectroscopic probe is to address issues regarding spectral overlap and site specificity while minimizing possible deleterious effects on structure or function. Certain Trp residues, nevertheless, may be essential for structure and/or function.[15] As a result, one might expect that certain analogs will not be tolerated as substitute residues.

Sensitivity at low probe concentrations is essential for investigating high-affinity interactions. For example, many gene regulatory proteins tend to self-associate and have specific associations with other proteins as well as with DNA. Often the equilibrium dissociation constants for these interac-

[7] J. B. A. Ross, H. R. Wyssbrod, R. A. Porter, G. P. Schwartz, C. A. Michaels, and W. R. Laws, *Biochemistry* **31,** 1585 (1992).

[8] K. J. Willis, W. Neugebauer, M. Sikorska, and A. G. Szabo, *Biophys. J.* **66,** 1623 (1994).

[9] T. E. S. Dahms and A. G. Szabo, *Biophys. J.* **69,** 569 (1995).

[10] T. E. S. Dahms, K. J. Willis, and A. G. Szabo, *J. Am. Chem. Soc.* **117,** 2321 (1995).

[11] C. M. L. Hutnik, J. P. MacManus, D. Banville, and A. G. Szabo, *Biochemistry* **39,** 7652 (1991).

[12] C. A. Royer, J. A. Gardner, J. M. Beechem, J.-C. Brochon, and K. S. Matthews, *Biophys. J.* **58,** 363 (1990).

[13] D. L. Harris and B. S. Hudson, *Biochemistry* **29,** 5276 (1990).

[14] C. A. Hasselbacher, R. Rusinova, E. Rusinova, and J. B. A. Ross, Spectral enhancement of recombinant proteins with tryptophan analogs: The soluble domain of human tissue factor. *In* "Techniques in Protein Chemistry VI" (John W. Crabb, ed.), pp. 349–356. Academic Press, New York, 1995.

[15] K. C. Chow, H. Xue, W. Shi, and J. T.-F. Wong, *J. Biol. Chem.* **267,** 9146 (1992).

tions are in the nanomolar range. The sensitivity of a fluorescence probe depends mainly on the extinction coefficient and quantum yield. High quantum yield and a high extinction coefficient are practically essential for measurement of sample concentrations in the nanomolar range when using conventional steady-state laboratory instruments. Probe detection depends also on the selectivity provided by the available excitation and emission wavelengths of the probe; the lower concentration limit for detection is determined by the relative background interference from light scatter and fluorescent solution contaminants.

Weak, nonspecific binding interactions become a significant concern at concentrations that extend into the millimolar range. While fluorescence measurements at high concentrations require different optical conditions, the upper concentration limit is determined mainly by the solubility and specific volume of the system being studied. Information regarding necessary precautions, corrections, and appropriate choice of optical conditions for making accurate measurements of low-concentration (optically thin) and high-concentration samples (optically thick) are provided in the texts by Lakowicz[2] and Parker.[4]

The fact that fluorescence spectroscopy is effective over the nanomolar-to-millimolar concentration range allows experimentalists to bridge the information gap between what needs to be learned about structure and functional interactions that occur at *in vivo* concentrations and what is learned from higher resolution structural techniques, such as nuclear magnetic resonance (NMR) spectroscopy or X-ray diffraction, which require sample concentrations in the millimolar range. At these high concentrations, nonspecific interactions must be accounted for in the interpretation of data. In addition, the packing interactions in crystals introduce local perturbations of surface residues. As a consequence, packing interactions may affect certain secondary structural elements and regions of the tertiary structure. For example, the structure of insulin differs in the solution and crystalline states, with the monomers adopting nonequivalent conformations when associated as dimers.[16] In addition, the conformational and ligation states occupied at the high concentrations required for high-resolution techniques may not be representative of those that tend to be occupied at lower, *in vivo* concentrations. An example is catabolite repressor protein (CAP), which is a homodimeric positive regulator of RNA polymerase in *Escherichia coli.* Catabolite repressor protein binds cAMP in a negatively cooperative manner and this ligand modulates the conformation of the protein and its affinity for DNA. The form of CAP that has maximum DNA-

[16] T. L. Blundell, G. Dodson, D. D. Hodgkin, and E. Mercola, *Adv. Protein Chem.* **26,** 279 (1972).

binding activity and activates transcription occurs when CAP is in the singly liganded state, with cAMP bound to only one subunit of the dimer.[17] By contrast, the X-ray crystal structure model of CAP complexed to DNA has cAMP bound to both subunits.[18] This raises the question whether the functionally important conformation is observed from the crystal structure. Although fluorescence does not provide atomic-level resolution, its sensitivity over the nanomolar-to-millimolar concentration range makes possible a thorough physical characterization and thermodynamic analysis of individual proteins and their higher order oligomers, in unliganded, partially liganded, and fully liganded forms. This information facilitates developing a comprehensive understanding of the biological role of macromolecular assemblies and their different ligation states.

Tryptophan Analogs

Tryptophan analogs first were used in the 1950s in phenomenological studies designed to elucidate metabolic pathways and mechanisms for protein synthesis. In these experiments organisms were fed Trp or other amino acid analogs.[19] The first report[20] of analog incorporation into proteins, in 1956, described the growth of a Trp-deficient auxotroph of *E. coli* in media containing 7-azatryptophan (7-ATrp), 2-azatryptophan (tryptazan; 2-ATrp), or 5-methyltryptophan (5-MeTrp) (see structures in Fig. 1). Only 2-ATrp and 7-ATrp were capable of sustaining bacterial growth in these experiments. Several reports examined the role of tryptophanyl-tRNA synthetase (TrpRS) in the incorporation of Trp analogs into protein[20–22]; TrpRS is the key enzyme required for the incorporation of Trp or its analogs into proteins.[23,24] Tryptophan analogs were used in the effort to understand the mechanisms of Trp biosynthesis. While no proteins were purified or studied in these early experiments, one study reported incorporation of radiolabeled 5-hydroxytryptophan (5-OH-Trp), 5-methoxytryptophan, and 7-ATrp into *Bacillus subtilis* proteins.[25] Although there were some premature conclusions regarding nonincorporation of toxic Trp analogs, it became clear from these early experiments that biosynthetic analog incorporation was possible.

[17] T. Heyduk and J. C. Lee, *Biochemistry* **28,** 6914 (1989).
[18] S. C. Schultz, G. C. Schields, and T. A. Steitz, *Science* **253,** 1001 (1991).
[19] H. Halvorson, S. Spiegelman, and R. L. Hinman, *Arch. Biochim. Biophys.* **55,** 512 (1955).
[20] A. B. Pardee, V. G. Shore, and L. S. Prestridge, *Biochim. Biophys. Acta* **21,** 406 (1956).
[21] E. W. Davie, V. V. Koningsberger, and F. Lipmann, *Arch. Biochem. Biophys.* **65,** 21 (1956).
[22] N. Sharon and F. Lipmann, *Arch. Biochim. Biophys.* **69,** 219 (1957).
[23] C. W. V. Hogue, Ph.D. dissertation, University of Ottawa, Ottawa, Ontario, Canada, 1994.
[24] C. W. V. Hogue and A. G. Szabo, *Biophys. Chem.* **48,** 159 (1993).
[25] S. Barlati and O. Ciferri, *J. Bacteriol.* **101,** 166 (1970).

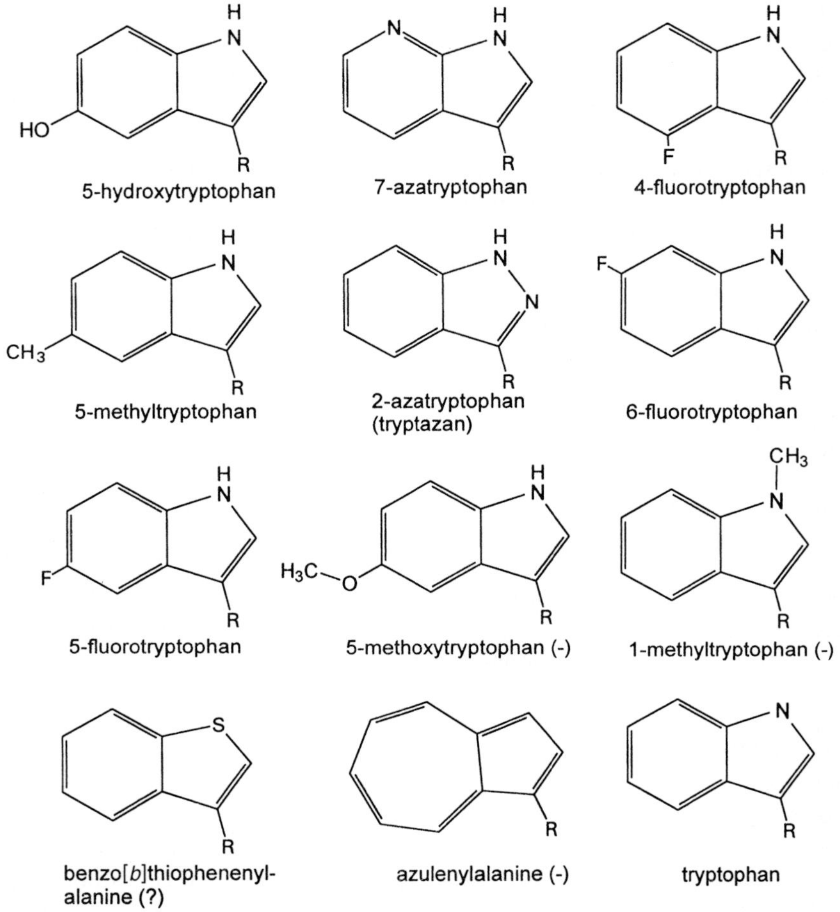

5-hydroxytryptophan
7-azatryptophan
4-fluorotryptophan
5-methyltryptophan
2-azatryptophan (tryptazan)
6-fluorotryptophan
5-fluorotryptophan
5-methoxytryptophan (-)
1-methyltryptophan (-)
benzo[*b*]thiophenenyl-alanine (?)
azulenylalanine (-)
tryptophan

FIG. 1. Structures of tryptophan analogs: (-) indicates compounds that fail at biosynthetic incorporation; (?) indicates untested compounds.

In 1968, Schlesinger[26] was the first to report a study of a purified enzyme in which the Trp residues were replaced by Trp analogs. Schlesinger was investigating the function of alkaline phosphatase, and concluded that incorporation of either 7-ATrp or 2-ATrp had little effect on enzymatic activity. However, incorporation of the analogs produced significant changes in the absorbance and fluorescence properties of the enzyme. The effects of 7-ATrp incorporation on aspartate transcarbamylase (aspartate carbamoyltransferase; ATCase) kinetics were reported by Foote and co-workers.[27]

[26] S. Schlesinger, *J. Biol. Chem.* **243,** 3877 (1968).
[27] J. Foote, D. M. Ikeda, and E. R. Kantrowitz, *J. Biol. Chem.* **255,** 5154 (1980).

They found that allosteric modulation was enhanced by 7-ATrp incorporation. It was suggested that 7-ATrp-199, a part of the catalytic chain, was interacting with the carbamyl phosphate-binding site via a hydrogen bond through the aza nitrogen. These experiments are an example of "analog mutagenesis" that predate the use of site-directed mutagenesis.[28]

By the 1970s, it was recognized that fluorinated amino acids offered spectroscopic utility for use as ^{19}F NMR probes when incorporated into proteins.[29] To this end, Pratt and Ho[30] studied the effect of 4-FTrp, 5-FTrp, and 6-FTrp incorporated into the *E. coli* enzymes lactose permease, β-galactosidase, and D-lactate dehydrogenase. 4-Fluorotryptophan caused the least effects on enzymatic activity and growth rate, increasing the activity of D-lactate dehydrogenase about twofold. The effects of these analogs on individual enzymes were variable. As summarized by Lian and co-workers[31] and also by Luck,[32] ^{19}F NMR has been used to study fluoro analogs of Trp and Phe in many other enzymes.

In 1986, Hudson and co-workers[33] speculated anew that certain Trp analogs would be useful probes for fluorescence spectroscopy. They suggested that azulene and benzo[*b*]thiophene (Fig. 1) amino acid derivatives might be useful fluorophores to substitute for Trp. Koide and co-workers coined the term *alloprotein* to describe, in general, proteins with nonnatural amino acids incorporated in the polypeptide chain.[34]

Petrich and co-workers have supported the potential use of 7-ATrp as a probe in biological fluorescence studies of proteins.[35–45] They have

[28] M. J. Zoller and M. Smith, *Nucleic Acids Res.* **10,** 6487 (1982).

[29] B. D. Sykes, H. I. Weingarte, and M. J. Schlesinger, *Proc. Natl. Acad. Sci. U.S.A.* **71,** 469 (1974).

[30] F. A. Pratt and C. Ho, *Biochemistry* **14,** 3035 (1975).

[31] C. Lian, H. Le, B. Montez, J. Patterson, S. Harrell, D. Laws, I. Matsumura, J. Pearson, and E. Oldfield, *Biochemistry* **33,** 5238 (1994).

[32] L. Luck, *Tech. Protein Chem.* **6,** 487 (1995).

[33] B. S. Hudson, D. L. Harris, R. D. Ludescher, A. Ruggiero, A. Cooney-Freed, and S. Cavalier, Fluorescence probe studies of proteins and membranes. *In* "Applications of Fluorescence in the Biomedical Sciences" (D. L. Taylor, *et al.,* eds.), pp. 159–202. A. R. Liss, New York, 1986.

[34] H. Koide, S. Yokoyama, G. Kawai, J.-M. Ha, T. Oka, S. Kawai, T. Miyake, T. Fuwa, and T. Miyazawa, *Proc. Natl. Acad. Sci. U.S.A.* **85,** 6237 (1988).

[35] M. Négrerie, S. M. Bellefeuille, S. Whitham, J. W. Petrich, and R. W. Thornburg, *J. Am. Chem. Soc.* **112,** 7419 (1990).

[36] M. Négrerie, F. Gai, S. M. Bellefeuille, and J. W. Petrich, *J. Phys. Chem.* **95,** 8663 (1991).

[37] F. Gai, and J. W. Petrich, *J. Am. Chem. Soc.* **114,** 8343 (1992).

[38] M. Négrerie, F. Gai, J.-C. Lambry, J.-L. Martin, and J. W. Petrich, *J. Phys. Chem.* **97,** 5045 (1993).

[39] Y. Chen, R. L. Rich, F. Gai, and J. W. Petrich, *J. Phys. Chem.* **97,** 1770 (1993).

[40] R. L. Rich, Y. Chen, D. Neven, M. Négrerie, F. Gai, and J. W. Petrich, *J. Phys. Chem.* **97,** 1781 (1993).

concentrated on model studies of 7-ATrp fluorescence, examining the properties of 7-azaindole (7-AI), 7-ATrp, and 7-ATrp incorporated into a synthetic peptide. Studies of model compounds under model conditions are clearly essential for fully understanding spectroscopic phenomena. It appears, however, that certain predictions regarding fluorescence behavior of 7-ATrp in proteins, which were based on these model studies, have not been borne out by the results of others[23,24] on 7-ATrp fluorescence when incorporated in proteins.

The unpredictability of the degree of analog incorporation *in vivo* was probably one reason why this method for introduction of fluorescence probes into proteins has been overlooked until recently. In retrospect, biosynthetic analog incorporation was successful in the earlier studies with systems that had natural inducible promoters, such as alkaline phosphatase[26,29] and aspartate transcarbamylase.[27] In these systems, it was possible to uncouple the toxicity of an analog amino acid by first growing cells in media containing the usual amino acid, but while maintaining expression of these proteins in a repressed state. After accumulating sufficient cell mass, analogs were added and the expression was derepressed, resulting in high levels of incorporation (i.e., >85%). Analog incorporation into proteins without natural inducible promoters did not have this advantage, and often required mixtures of analog and native amino acids to sustain cell growth. This situation resulted in lower levels or no observable level of incorporation.

The development in the early 1980s of expression vectors with artificial inducible promoters made possible the construction of highly efficient expression systems for most proteins.[46] The availability of these expression vectors changed the way proteins are produced for study in the laboratory. As demonstrated independently in 1992 by the authors' respective laboratories,[47,48] this advance in molecular biology also made it possible to achieve highly efficient incorporation of Trp analogs *in vivo,* thereby creating the opportunity to use Trp analogs as intrinsic probes in the fluorescence spec-

---

[41] R. L. Rich, M. Négrerie, J. Li, S. Elliott, R. W. Thornburg, and J. W. Petrich, *Photochem. Photobiol.* **58,** 28 (1993).

[42] Y. Chen, F. Gai, and J. W. Petrich, *J. Am. Chem. Soc.* **115,** 10158 (1993).

[43] Y. Chen, F. Gai, and J. W. Petrich, *J. Phys. Chem.* **98,** 2203 (1994).

[44] F. Gai, R. L. Rich, and J. W. Petrich, *J. Am. Chem. Soc.* **116,** 735 (1994).

[45] R. L. Rich, F. Gai, Y. Chen, and J. W. Petrich, *Proc. S.P.I.E.* **2137,** 435 (1994).

[46] E. Amman, J. Brosius, and M. Ptashne, *Gene* **25,** 167 (1983).

[47] C. W. V. Hogue, I. Rasquinha, A. G. Szabo, and J. P. MacManus, *FEBS Lett.* **310,** 269 (1992).

[48] J. B. A. Ross, D. F. Senear, E. Waxman, B. B. Kombo, E. Rusinova, Y. T. Huang, W. R. Laws, and C. A. Hasselbacher, *Proc. Natl. Acad. Sci. U.S.A.* **89,** 12023 (1992).

troscopy of protein–protein and protein–nucleic acid interactions. Using variations of standard methods for expressing recombinant proteins in bacterial Trp auxotrophs, the strategy was to replace Trp residues with analogs that have absorption spectra that are red shifted with respect to that of Trp, for example 5-OHTrp or 7-ATrp (Fig. 2). The red-shifted absorption assures selective excitation of the analog-containing protein when in the presence of other Trp-containing proteins or DNA; the red absorption wavelengths provide a unique spectral window for following the fate of a particular protein in complex mixtures. Nonfluorescent analogs also are of interest, for example, 4-fluorotryptophan (4-FTrp). Proteins with Trp uniformly replaced by a nonfluorescent analog, such as 4-FTrp, are predicted to be "silent" in multiprotein assemblies. In the initial experiments, 5-OHTrp was incorporated into the Y57W mutant of oncomodulin[47] and into λcI repressor.[48] Both spectrally altered proteins retained wild-type functional properties. The utility of the red-edge absorption of 5-OHTrp for studying protein–protein interactions was demonstrated by selective excitation of the fluorescence of the oncomodulin mutant when bound to a specific antibody containing several Trp residues.[47] Likewise, the utility

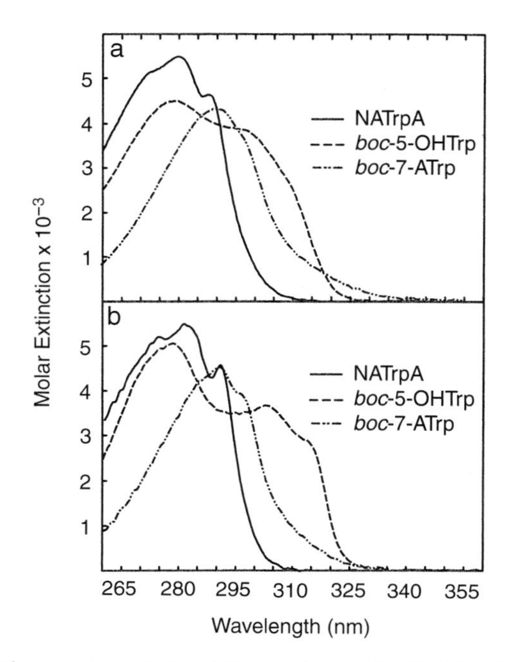

FIG. 2. Absorption spectra of *N*-acetyltryptophanamide (NATrpA), (*t*-Boc)(α-amino) 5-OHTrp, and (*t*-Boc)(α-amino)7-ATrp in neutral pH buffer (a) and in dioxane (b).

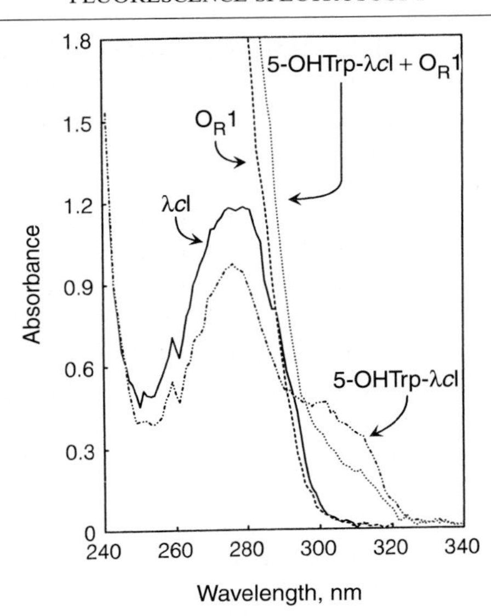

FIG. 3. Absorption spectra obtained using the Beckman Optima XL-A Analytical. Ultracentrifuge of wild-type and 5-OHTrp-containing λ *c*I repressor in the absence and presence of 21-mer operator DNA. (From Ref. 43.)

of 5-OHTrp for studying protein–nucleic acid interactions was demonstrated in analytical ultracentrifugation studies on the effects of DNA binding on the thermodynamics of λ*c*I repressor self-assembly (Fig. 3).[49,50] Spectrally enhanced proteins have been employed subsequently by other groups for fluorescence studies on diverse systems involving protein–protein and protein–DNA interactions.

Another promising technology for introducing Trp analogs into proteins is *in vitro* transcription/translation using a nonsense suppressor tRNA that is chemically acylated with a nonprotein amino acid.[51–55] This provides a

[49] T. M. Laue, D. F. Senear, S. Eaton, and J. B. A. Ross, *Biochemistry* **32,** 2469 (1993).

[50] D. F. Senear, T. M. Laue, J. B. A. Ross, E. Waxman, S. Eaton, and E. Rusinova, *Biochemistry* **32,** 6179 (1993).

[51] J. D. Bain, E. S. Diala, C. G. Glabe, T. A. Dix, and A. R. Chamberlin, *J. Am. Chem. Soc.* **111,** 8013 (1989).

[52] J. D. Bain, D. A. Wacker, E. E. Kuo, M. H. Lyttle, and A. R. Chamberlin, *J. Org. Chem.* **56,** 4615 (1991).

[53] C. J. Noren, S. J. Anthony-Cahill, M. C. Griffith, and P. G. Schultz, *Science* **244,** 182 (1989).

[54] D. M. Mendel, J. A. Ellman, Z. Chang, D. L. Veenstra, P. A. Kollman, and P. G. Schultz, *Science* **256,** 1798 (1992).

[55] J. K. Judice, T. R. Gamble, E. C. Murphy, A. M. de Vos, and P. G. Schultz, *Science* **261,** 1578 (1993).

way to introduce chemically diverse amino acids for the study of protein structure and function. It remains to be shown, however, whether *in vitro* technology will be cost effective when scaled up for the quantities of protein necessary for NMR and X-ray crystallography.[56] Two significant advantages of *in vivo* over *in vitro* incorporation are the high level of protein production—tens to hundreds of milligrams from a few liters of bacterial cell culture—and the simplicity of the methodology.[23] Probably the most significant advantage of *in vitro* over *in vivo* incorporation is that a Trp analog can be placed at any single, unique site in a single or multi-Trp protein. Traditional direct chemical synthetic methods are an alternative way to achieve the latter goal.[57] Finally, engineered peptide ligases[58] might be useful for semisynthesis of larger proteins that contain Trp analogs.

Principles for Fluorescence Enhancement

Brief Overview of Tryptophan Fluorescence

The absorbance of the indole side chain of tryptophan in the wavelength region above 260 nm is due to two overlapping, nearly perpendicular $\pi \rightarrow \pi^*$ transitions, termed ^1L_a and ^1L_b transitions according to Platt notation.[59–62] The relative energies and oscillator strengths of these two electronic transitions depend on the solvent as well as on the nature and position of substituents on the indole ring.[60–64] In vapor phase and nonpolar solvents, such as methylcyclohexane, the more highly structured ^1L_b state is the lower energy state. The broader, less-structured ^1L_a transition is more sensitive to changes in the local environment, being preferentially stabilized by polar solvents. Experimental and theoretical results[60,63–67] show that the oscillator strength of the ^1L_a state is greater than that of the ^1L_b state. Thus, the

[56] W. Kudlicki, G. Kramer, and B. Hardesty, *Anal. Biochem.* **206,** 389 (1992).

[57] W. R. Laws, G. P. Schwartz, E. Rusinova, G. T. Burke, Y.-C. Chu, P. G. Katsoyannis, and J. B. A. Ross, *J. Protein Chem.* **14,** 225 (1995).

[58] D. Y. Jackson, J. Burnier, C. Quan, M. Stanle, J. Tom, and J. A. Wells, *Science* **266,** 243 (1994).

[59] J. F. Platt, *J. Chem. Phys.* **19,** 101 (1951).

[60] S. V. Konev, "Fluorescence and Phosphorescence of Proteins and Nucleic Acids." Plenum, New York, 1967.

[61] D. Creed, *Photochem. Photobiol.* **39,** 537 (1984).

[62] B. Valeur and G. Weber, *Photochem. Photobiol.* **25,** 441 (1977).

[63] E. H. Strickland, J. Horwitz, and C. Billups, *Biochemistry* **9,** 4914 (1970).

[64] H. Lami and N. Glasser, *J. Chem. Phys.* **84,** 597 (1986).

[65] A. A. Rehms and P. R. Callis, *Chem. Phys. Lett.* **140,** 83 (1987).

[66] M. R. Eftink, L. A. Selvidge, P. R. Callis, and A. A. Rehms, *J. Phys. Chem.* **94,** 3469 (1990).

[67] D. W. Pierce and S. G. Boxer, *Biophys. J.* **68,** 1583 (1995).

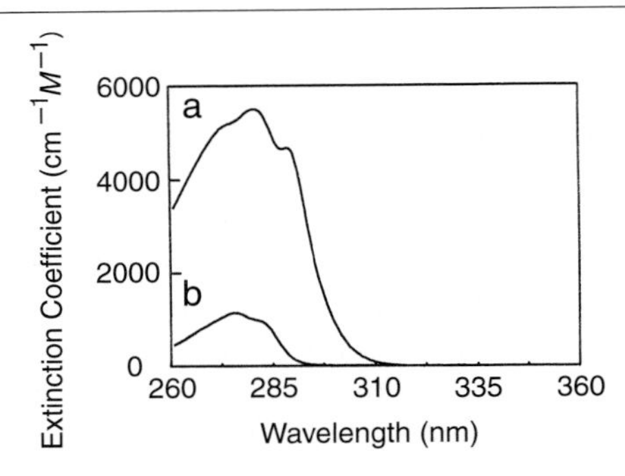

FIG. 4. Absorption spectra of NATrpA and NATyrA in neutral pH buffer.

stronger interaction of the ^1L_a state with polar solvents is attributed to the larger change in its permanent dipole moment following excitation. These factors account for the large Stokes shift of Trp fluorescence when in water and the characteristic "blue shift" that is observed for the emission when the indole side chain is buried in the protein matrix.

Spectral Edge Excitation

The characteristic absorption of proteins and polypeptides in the wavelength region beween 260 and 300 nm is due to Trp, Tyr, and Phe, which are also the fluorescent residues.[68] Most proteins and polypeptides contain Trp, and the fluorescence of this residue generally dominates that of Tyr or Phe. At wavelengths longer than 275 nm, only Tyr and Trp contribute significantly to absorption; Trp has the stronger extinction, and its absorption spectrum extends to lower energies (Fig. 4). Thus, Trp fluorescence can be excited selectively at the "red edge" of its absorbance, at wavelengths longer than 295 nm.

The absorption of the model compounds *t*-Boc(α-amino)5-hydroxytryptophan (*t*-Boc(α-amino)5-OHTrp) and *t*-Boc(α-amino)7-azatryptophan (*t*-Boc(α-amino)7-ATrp) in water, used to model the solvent-exposed aromatic side chains, extend past 320 nm. This is 10–20 nm further to the red than the absorption band of *N*-acetyltryptophanamide (NATrpA) (Fig. 2a and b). This provides a spectral window with significant extinction for selective excitation of either 5-OHTrp or 7-ATrp. In dioxane, which has been used to model the interior of proteins, the absorption spectra of all three model compounds show increased vibrational structure. Whereas the overall absorption envelopes of NATrpA and *t*-Boc(α-amino)5-OHTrp

[68] D. B. Wetlaufer, *Adv. Protein Chem.* **17,** 303 (1962).

exhibit red shifts of a few nanometers, that of *t*-Boc(α-amino)7-ATrp exhibits a blue shift of a few nanometers.

Fluorometric measurements often are carried out with excitation at the absorption maximum of a chromophore.[69] While this practice helps maximize the emission signal, in the case of protein and peptide fluorescence it can lead to ambiguities and misinterpretation of results. For example, Teale and Weber[70] suggested that tyrosine fluorescence is not expected in Trp-containing proteins, even with excitation at 280 nm, which is close to the maximum of Tyr and Trp absorption, because Trp has much greater extinction, its fluorescence spectrum completely overlaps that of Tyr, and it can quench Tyr fluorescence efficiently through energy transfer. (Phenylalanine does not have significant absorption at 280 nm.) Longworth,[71] however, has documented cases in which using 280-nm excitation Tyr fluorescence can be observed in a Trp-containing protein. Usually Tyr has an emission maximum near 303 nm, and generally does not emit significantly at wavelengths greater than 380 nm.[71,72] Therefore, to avoid the fluorescence of Tyr, the better practice is to excite the red edge of Trp absorption at wavelengths greater than 295 nm. It should be noted, however, that Willis and Szabo[73] have observed Tyr fluorescence for a Trp-containing protein even with excitation at 300 nm. Nevertheless, by comparing the spectral contours of emission bands obtained using progressively lower energy excitation wavelengths and then normalized at 380 nm, it is possible to identify an optimal range of excitation wavelengths at which tyrosine fluorescence can be minimized or in most cases essentially eliminated. A decrease in relative intensity on the higher energy side of the emission following excitation toward lower energies is diagnostic of Tyr emission contributing to the fluorescence band. By contrast, buried Trp residues absorb at lower energies and fluoresce at higher energies than exposed Trp residues. In this case, excitation toward lower energies will result in an increase in relative intensity on the higher energy, blue side of the fluorescence band. In this way, emission from Tyr and buried Trp residues can be distinguished, and the principle of red-edge excitation has been used to isolate emission from Trp residues (Fig. 4).

[69] D. Freifelder, "Physical Biochemistry: Applications to Biochemistry and Molecular Biology," 2nd Ed., pp. 542–543. W. H. Freeman, New York, 1982.

[70] F. W. J. Teale and G. Weber, *Biochem. J.* **65,** 476 (1957).

[71] J. W. Longworth, Luminescence of polypeptides and proteins. *In* "Excited States of Aromatic Amino Acids, Peptides, and Proteins" (I. Weinryb and R. F. Steiner, eds.), pp. 319–484. Plenum, New York, 1971.

[72] J. B. A. Ross, W. R. Laws, K. W. Rousslang, and H. R. Wyssbrod, Tyrosine fluorescence and phosphorescence from proteins and polypeptides. *In* "Topics in Fluorescence Spectroscopy 3" (J. R. Lakowicz, ed.), Chap. 1, pp. 1–63. Plenum, New York, 1992.

[73] K. J. Willis and A. G. Szabo, *Biochemistry* **28,** 4902 (1989).

Red-edge excitation also is the key to using 5-OHTrp or 7-ATrp as probe (Fig. 2a and b). For example, others have observed correctly that when 7-ATrp[74] is excited at 288 nm, the analog fluorescence is contaminated by Trp fluorescence. This contamination by Trp fluorescence was the result of using inappropriate excitation wavelengths. As a result, these investigators concluded erroneously that 7-ATrp cannot be particularly useful as an intrinsic spectroscopic probe. By using red-edge excitation at 315 nm, or longer wavelengths, with appropriately narrow bandpasses (~4 nm), we have shown that 5-OHTrp[47,48] and 7-ATrp[24] both can be distinguished from Trp. With red-edge excitation and with detection at the emission wavelength maximum of 5-OHTrp (335 nm) or 7-ATrp (360–400 nm, depending on solvent exposure of this analog), either fluorophore can be readily observed.

Three Tryptophan Analogs with Different Photophysical Properties

As indicated in Fig. 1, there are a number of potentially useful Trp analogs. The three analogs described here, 4-FTrp, 5-OHTrp, and 7-ATrp, have distinct photophysical features that make them useful for the studies of protein–protein and protein–nucleic acid interactions. 4-Fluorotryptophan is nonfluorescent, making a "silent" analog. 5-Hydroxytryptophan fluorescence has a high limiting anisotropy when excited at wavelengths longer than 310 nm, and 7-ATrp has an emission spectrum that is sensitive to the local environment, showing dramatic changes in quantum yield and large wavelength shifts depending on exposure to water.

4-Fluorotryptophan: A Silent Analog

Bronskill and Wong[75] demonstrated that 4-FTrp in water is essentially nonfluorescent at room temperature. On this basis, they suggested that 4-FTrp incorporation could be used as a general technique to eliminate Trp fluorescence, allowing observation of other fluorophores in protein–protein interactions. Hott and Borkman[76] found that 4-FTrp tends to decompose on exposure to ultraviolet (UV) light, and for this reason cautioned about its potential uses for biological investigations. Their main conclusion was that 4-fluoro substitution made the indole ring of Trp more photoreactive at room temperature. However, for protein studies making use of Trp fluorescence, the photoreactivity of 4-FTrp can be minimized, because the absorption of 4-FTrp is blue shifted compared with that of Trp, by using

[74] V. W. Cornish, D. R. Benson, C. A. Altenbach, K. Hideg, W. L. Hubbell, and P. G. Shultz, *Proc. Natl. Acad. Sci. U.S.A.* **91,** 2910 (1994).

[75] P. M. Bronskill and J. T. Wong, *Biochem. J.* **249,** 305 (1988).

[76] J. L. Hott and R. F. Borkman, *Biochem. J.* **264,** 297 (1989).

excitation wavelengths at or longer than 300 nm to excite selectively unsubstituted Trp residues.

Substitution at the indole C-4 position seems more relevant to the observed low fluorescence yields of these Trp analogs, than the chemical nature of the substituent.[77] Absorbance studies on indole derivatives in oriented films showed that the ^1L_a and ^1L_b transition dipoles, which are nearly orthogonal in indole, are almost parallel to each other in 4-methoxyindole with their vectors nearly directed between C-4 and N-1 of the indole ring.[78] Hott and Borkman[76] also examined the low-temperature (77 K) fluorescence and phosphorescence of 4-FTrp. They observed a structured blue-shifted fluorescence with an emission maximum at 297 nm (under the same solvent conditions, the low-temperature fluorescence emission maximum of Trp is 313 nm). The phosphorescence also was blue shifted compared with that of Trp. In addition, the fluorescence yield was lower and the phosphorescence yield was higher compared to Trp, although the total emission yield (fluorescence plus phosphorescence) from 4-FTrp was of the same magnitude as that of Trp. Hott and Borkman[76] argued that photochemical pathways can contribute to the nonfluorescent depopulation of the excited state. Complete photodegradation was not observed after a 20-min irradiation time in their study. Thus, additional possibilities should be considered. For example, a decrease in fluorescence yield accompanied by an increase in phosphorescence yield could suggest that the 4-fluoro substituent promotes intersystem crossing. Finally, it is notable[23] that the crystal structure of 4-FTrp[79] shows variations in bond distances compared to Trp,[80] which could be consistent with altered electron resonance and slightly different ground-state nuclear coordinates.

5-Hydroxytryptophan: Single Transition Excitation and Emission

5-Hydroxytryptophan is a naturally occurring amino acid and is perhaps best known as an intermediate in the biosynthesis of serotonin (5HT, 5-hydroxytryptamine) from Trp. Compared to Trp, the absorbance spectra of 5-OHTrp and 5HT above 300 nm have well-resolved, red-extended shoulders (Fig. 2a and b). The red-extended shoulder in the 5HT absorption has been shown from spectroscopy on stretched, oriented films to be due only to the ^1L_b transition.[81] Thus, while the ^1L_a and ^1L_b transitions of Trp

[77] J. W. Bridges and R. T. Williams, *Biochem. J.* **107,** 225 (1968).

[78] B. Albinsson and B. Nordén, *J. Phys. Chem.* **96,** 6204 (1992).

[79] Z. Xu, M. L. Love, Y. Y. L. Ma, M. Blum, P. M. Bronskill, J. Bernstein, A. A. Grey, T. Hofmann, N. Camermanm, and J. T. Wong, *J. Biol. Chem.* **264,** 4304 (1989).

[80] G. D. Fasman (ed.), "Handbook of Biochemistry and Molecular Biology" 3rd Ed., Vol. II, p. 233. CRC Press, Boca Raton, Florida, 1988.

[81] T. Kishi, M. Tanaka, and J. Tanaka, *Bull. Chem. Soc. Jpn.* **50,** 1267 (1977).

overlap across most of the absorbance spectrum,[63,82] the ^1L_b transition of 5-hydroxyindoles or 5-methoxyindoles occurs at much lower energies.[66,83] As a result, the ^1L_b state can be selectively excited at wavelengths of 315 nm and longer.

The ^1L_b fluorescence of 5-hydroxyindoles is expected to be different from the ^1L_a fluorescence that is observed for Trp in polar solvents. In particular, the effect of solvent interaction on 5-OHTrp fluorescence should be diminished. This is consistent with the observed variations in the fluorescence emission maxima of indole in different solvents (cyclohexane, λ_{em} = 298 nm; water, λ_{em} = 342 nm) compared with smaller variations for the emission maxima of 5-hydroxyindole (cyclohexane, λ_{em} = 322 nm; water, λ_{em} = 331 nm).[63,64,83] The fluorescence spectra of the model compounds NATrpA and *t*-Boc(α-amino)5-OHTrp in solvents with different dielectric constants are shown in Fig. 5a and b, respectively; steady-state emission data for the solvent series are given in Table I. The results for the solvent series indicate that 5-OHTrp in proteins is not expected to exhibit the same magnitude shifts in emission as those observed for Trp (see review by Eftink[6]).

Although 5-OHTrp is not as sensitive to solvent exposure as Trp, it has other distinct advantages as a fluorophore. Because there is essentially no overlap of the ^1L_b and ^1L_a transitions above 310 nm, the fluorescence emission of 5-OHTrp will not be depolarized by internal conversion. Consequently higher fluorescence anisotropy values should be observed from proteins with 5-OHTrp instead of Trp. This prediction is borne out by anisotropy data for the model compound *t*-Boc(α-amino)5-OHTrp[48] and the single 5-OHTrp residue in the mutant Y57W oncomodulin.[47]

5-Hydroxytryptophan is a better Förster energy transfer acceptor for tyrosine fluorescence than Trp because 5-OHTrp absorbance between 300 and 320 nm has greater spectral overlap with the Tyr fluorescence emission,[48] which has its maximum near 303 nm.[71,72] The calculated overlap integral for Förster energy transfer from Tyr to 5-OHTrp in water is 2.2×10^{-12} cm^6 mol^{-1}, whereas that from Tyr to Trp is 6.6×10^{-13} cm^6 mol^{-1}.[48] Thus, for the same distance of separation and mutual orientation of the donor emission and acceptor absorption transition dipoles, the efficiency of resonance energy transfer from Tyr to Trp is expected to be about one-third that from Tyr to 5-OHTrp.

Hudson and co-workers[33] have suggested that an amino acid chromophore with simpler photophysical properties than those of Trp would make the interpretation of intrinsic fluorescence decay kinetics in proteins more

[82] Y. Yamamoto and J. Tanaka, *Bull. Chem. Soc. Jpn.* **45,** 1362 (1972).

[83] H. Lami, *J. Chem. Phys.* **67,** 3274 (1977).

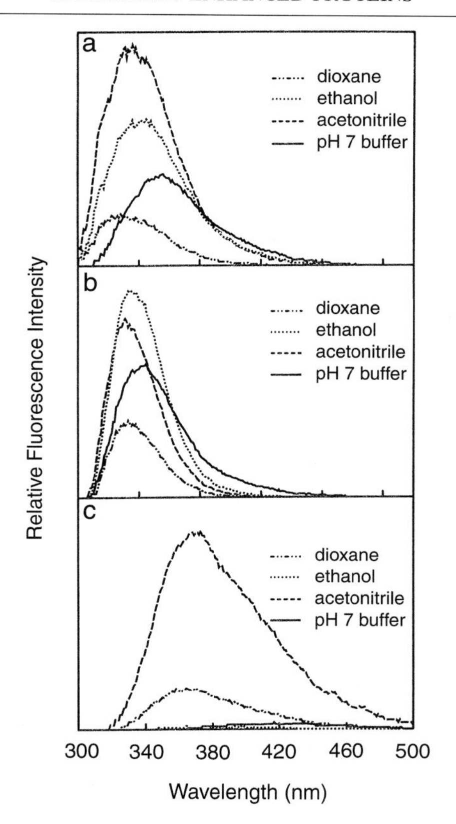

FIG. 5. Fluorescence emission spectra of (a) NATrpA, (b) (*t*-Boc)(α-amino)5-OHTrp, and (c) (*t*-Boc)(α-amino)7-ATrp in solvents of different dielectric constant.

straightforward. 5-Hydroxytryptophan may help meet this need in the following ways: (1) the ^1L_b transition is shifted substantially toward lower energy, becoming separated from the envelope of the ^1L_a transition; (2) the ^1L_b fluorescence emission is less sensitive to changes in polarity than is the ^1L_a emission of Trp; and (3) it has single-exponential fluorescence decay kinetics as the isolated amino acid in aqueous solution at physiological pH.[24,41,47,48] Using 5-OHTrp fluorescence with red-edge excitation wavelengths greater than 315 nm, contributions from dual emission states can be discounted as a possible explanation for any observed multiexponential intrinsic fluorescence decay behavior in proteins. In addition, the ^1L_b state of the 5-hydroxyindole group is much less likely to participate in solvent interactions or exciplex formation, as proposed by Lumry and Hersh-

TABLE I
FLUORESCENCE OF TRYPTOPHAN ANALOGS IN SOLVENTS OF DIFFERENT DIELECTRIC CONSTANT

Solvent	ε^a	Analog	λ_{max} (nm)^b	ϕ^c	$\langle\tau\rangle$ (nsec)^d
Dioxane	2.2	NATrpA	329	0.06	0.7
		t-Boc(α-amino)7-ATrp	360	0.06	1.9
		t-Boc(α-amino)5-OHTrp	332	0.08	0.8
Ethanol	25	NATrpA	340	0.21	3.7
		t-Boc(α-amino)7-ATrp	370	0.004	(0.2)
		t-Boc(α-amino)5-OHTrp	334	0.22	4.3
Acetonitrile	37	NATrpA	334	0.25	4.9
		t-Boc(α-amino)7-ATrp	370	0.34	9.1
		t-Boc(α-amino)5-OHTrp	331	0.18	4.0
pH 7 (0.1 *M* Tris)	~80	NATrpA	352	0.14	3.0
		t-Boc(α-amino)7-ATrp	412	0.01	(0.6)
		t-Boc(α-amino)5-OHTrp	342	0.21	4.2

[a] Values for the dielectric constant ε (at 20°) were obtained from the "Handbook of Chemistry and Physics," 76th Ed., CRC Press, Boca Raton, Florida, 1995.

[b] The maximum emission wavelengths are from steady state spectra obtained on an SLM/Aminco SPF-500C fluorometer, corrected for the wavelength-dependent efficiency of the detection system. Excitation was at 293 nm, bandpass of 2.5 nm. The emission bandpass was 10 nm.

[c] Quantum yields ϕ were obtained by comparing integrated intensities of steady state emission spectra excited at 287 nm with that of NATrpA, using a value of 0.14 for the latter [cf. M. R. Eftink, *Methods Biochem. Anal.* **35,** 127 (1991)].

[d] The mean lifetime $\langle\tau\rangle = \Sigma\alpha_i\tau_i^2/\Sigma\alpha_i\tau_i$, where α_i and τ_i are the amplitude and lifetime of the *i*th component. Values reported in parentheses are known to involve excited-state reactions.

berger[84] for the ^1L_a state of Trp. The 5-hydroxyl, however, does introduce the possibility of a new hydrogen bond interaction with the indole ring as well as excited-state ionization. The singlet excited state of an aromatic alcohol is typically several orders of magnitude more acidic than the singlet ground state,[1] introducing the possibility of ionization of 5-OHTrp in the excited state. In contrast to Tyr, which has a ground-state pK_a between 9 and 10 for the phenolic hydroxyl,[72] the ground-state pK_a of 5-OHTrp is 11.4 for the indolyl hydroxyl.[85] Thus, in the ground state, the indolyl hydroxyl is less acidic than the phenolic hydroxyl. Ionization of the indolyl hydroxyl quenches *t*-Boc(α-amino)5-OHTrp fluorescence, and titration curves monitored by absorbance or fluorescence yield the same pK_a of 11.4, indicating

[84] R. Lumry and M. Hershberger, *Photochem. Photobiol.* **27,** 819 (1978).

[85] S. V. Jovanovic, S. Steenken, and M. G. Simic, *J. Phys. Chem.* **94,** 3583 (1990).

that excited-state ionization is slow compared to the rate of fluorescence decay.[86] As a result, interpretation of a multiexponential fluorescence intensity decay for a single 5-OHTrp in a protein is simplified compared to that of Trp. A specific interaction resulting in a complex between the indolyl hydroxyl in the ground state and a strong proton acceptor in the protein, however, will tend to shift both the emission and absorption to lower energy, and the fluorescence lifetime of the new emission band will be characteristic of the complex.[87]

7-Azatryptophan: A Water-Quenched Fluorescence Probe

It is important to recognize the seminal contribution of Schlesinger on 7-ATrp fluorescence in proteins. Many of the useful properties of 7-ATrp as a fluorescence probe (see Table I) can be identified in the careful observations of Schlesinger on alkaline phosphatase.[26] The results show that 7-ATrp is an extremely sensitive probe of local environment. The emission of 7-ATrp fluorescence in alkaline phosphatase had intense, blue-shifted fluorescence (uncorrected emission maximum near 370 nm). Further, the results for guanidine-HCl denaturation of the 7-ATrp-containing enzyme revealed a dramatic quenching of 7-ATrp fluorescence, beyond that of either the 2-ATrp or Trp forms of the protein.

The intense, long-lived, blue-shifted fluorescence characteristic of 7-ATrp when buried in the protein matrix now has been observed for other proteins.[23,24] The large fluorescence quantum yield changes (10-fold or more) of proteins with incorporated 7-ATrp indicate that this analog can be a useful and specific probe for the examination of protein folding or conformational changes resulting from protein–protein interaction and ligand binding.

To understand the origin of the large fluorescence yield differences for 7-ATrp in different environments, relevant results from simpler model systems need to be examined. The parent chromophore 7-azaindole (7-AI) has been the subject of considerable study. The relative fluorescence intensities of various azaindoles were first reported in 1962 by Adler,[88] who noted a low fluorescence yield of 7-AI in water. The fluorescence of 7-AI in ethanol and water was examined in more detail in 1976 by Avouris *et al.*[89] A number of important conclusions were made about the photophysics, in particular regarding the phenomenon that the water spectrum shows one

[86] E. Rusinova and J. B. A. Ross, unpublished data (1996).

[87] C. A. Hasselbacher, E. Waxman, L. T. Galati, P. B. Contino, J. B. A. Ross, and W. R. Laws, *J. Phys. Chem.* **95,** 2995 (1991).

[88] T. K. Adler, *Anal. Chem.* **34,** 685 (1962).

[89] P. Avouris, L. L. Yang, and A. El-Bayoumi, *Photochem. Photobiol.* **24,** 211 (1976).

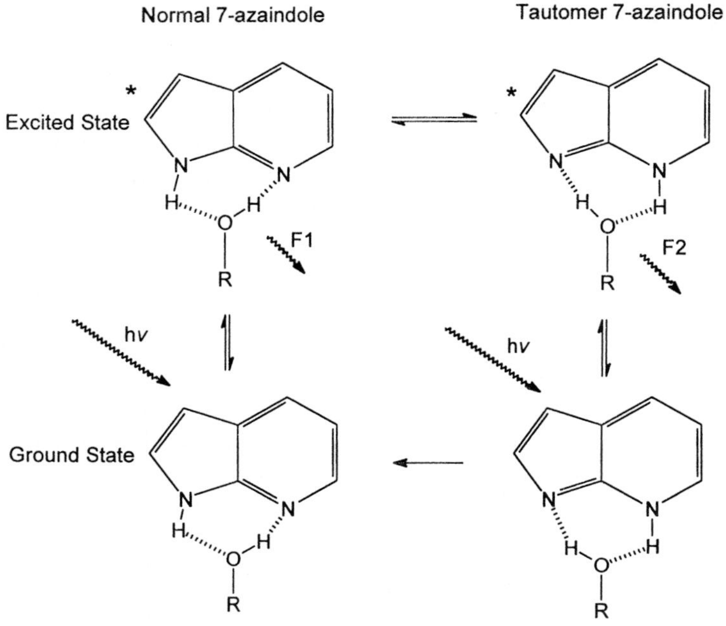

FIG. 6. Photophysical pathways of azaindole.

band (emission maximum near 400 nm) whereas the ethanol spectrum shows two well-resolved bands, termed F1 (emission maximum near 360 nm) and F2 (emission maximum near 530 nm). In both ethanol and water, 7-AI exhibits a weak emission (quantum yields of about 0.01 and 0.03, respectively).[89,90]

The complex fluorescence of 7-AI in alcohol has been the subject of competing photophysical theories. A consensus mechanism involves a cyclic intermediate that involves formation of hydrogen bonds N^1-H–OR and N^7–HOR between the 7-AI ring and the alcohol, as shown in Fig. 6. By comparing emission spectra of 7-AI, N^1-methyl-7-azaindole, and N^7-methyl-7*H*-pyrrolo[2,3-*b*]pyridine (called N^7-methyl tautomer), F1 was assigned by Avouris *et al.*[89] to the N^1-protonated 7-AI fluorescence, and F2 was assigned to fluorescence arising from tautomer 7-AI formed in the excited state. Once the complex forms, tautomerization can produce two different bonds: N^1–HOR and N^7-H–OR. It is important to note that the tautomer formed in the excited state does not accumulate in alcohol solutions. To account for this, it has been suggested that protonated 7-AI

[90] C. F. Chapman and M. Maroncelli, *J. Phys. Chem.* **96,** 8430 (1992).

reforms rapidly from tautomer 7-AI in the ground state via a double-proton transfer mechanism.[89] The photophysical pathways that were considered for 7-AI are summarized in Fig. 6. On the basis of the properties of 7-AI, Petrich and co-workers have suggested that 7-ATrp could be a potent and novel noninvasive fluorescent probe of protein structure and dynamics, and they have further investigated 7-AI and 7-ATrp photophysics.[35–45]

An important feature of the 7-AI ring is that it becomes positively charged at acid pH. The pyrrolopyridine N-7 of 7-AI has a pK_a of about 4.5 at room temperature, and the pK_a of the pyrrole ring N-1 is reduced dramatically from ~17 to ~12.[39] It also appears that there is negligible difference in the excited-state and ground-state ionization of N-7. Because the free amino acid 7-ATrp has four potentially titratable functions (α-amino, α-carboxyl, N-7, and N-1), a complete interpretation of 7-ATrp fluorescence must take into account coupled equilibria among a number of differently charged ground-state species as well as the possible excited-state reactions among these species (e.g., double-proton transfer, single-proton transfer, solvent reorganization).

Like 7-AI, 7-ATrp in water is quenched compared with Trp and the blocked analog 1-methyl-7-AI.[89] The deuterium isotope effect on 7-AI results in equivalent increases in the steady state quantum yield (3.6-fold) and the fluorescence lifetime (3.5-fold).[39] An equivalent increase (or decrease) in quantum yield and lifetime is the characteristic hallmark of a dynamic mechanism.[12] Regardless of the details of mechanism, it is clear that H_2O and alcohols are effective dynamic quenchers of 7-AI and 7-ATrp fluorescence.

In several studies by Petrich and co-workers, the authors suggest that 7-ATrp would be useful as a fluorescence probe in proteins because it exhibits a single-exponential fluorescence decay in experiments in which wide bandwidth filters are used to collect the decay of entire fluorescence emission band.[35,36,39,44] However, if discrete emission wavelengths are monitored (narrow bandpass observation), Petrich and co-workers,[36,38] as well as others,[24,91] find that 7-AI and 7-ATrp both exhibit multiexponential fluorescence decay. Moreover, the short decay component has a negative preexponential term at long emission wavelengths, which is the signature of an excited-state reaction process.[1,2] It is clear that using a wide-pass filter to observe the fluorescence decay masks important information. Because the fluorescence decay kinetics of 7-ATrp are complex, time-resolved fluorescence studies of proteins containing 7-ATrp will require collecting emission wavelength-resolved data.

[91] J. D. Brennan, I. D. Clark, C. W. V. Hogue, A. S. Ito, L. Juliano, A. C. M. Paiva, B. Rajendran, and A. G. Szabo, *Appl. Spectrosc.* **49,** 51 (1995).

The results of a study of 7-ATrp in the tripeptide Lys-7-ATrp-Lys indicate that 7-ATrp is far more sensitive than Trp to changes of pH over the range of 3–4.[91] This was also demonstrated in pH studies of 7-AI[36] and its methyl derivatives.[39] This behavior is expected on the basis of the possible protonation of the N-7 of the pyrrolopyridine ring of azaindole, which is not possible with the pyrrolobenzene ring of indole.

7-Azatryptophan should prove to be useful as a biological probe, not for simpler photophysical behavior such as is evident for 5-OHTrp, but rather for its range of responses to the local environment including spectral shifts, quantum yield changes, and possibilities for excited-state reactions. An important conclusion from the time-resolved data for 7-ATrp is that its fluorescence decay behavior provides no inherent advantage over that of Trp for use in biological systems. In fact, the photophysics of 7-ATrp seem more complex. Nevertheless, the rich photophysical behavior should help fingerprint each environment in a protein uniquely. The extreme sensitivity of 7-ATrp fluorescence to water renders it a highly useful probe for monitoring conformational changes that involve changes in solvation.

Methods for Incorporation of Tryptophan Analogs in Proteins

The following section describes general methods for *in vivo* biosynthetic incorporation of Trp analogs into proteins, methods for incorporation by direct chemical synthesis, and methods for characterization of products to determine the extent of analog incorporation. The methods used for *in vitro* incorporation of amino acid analogs have been reviewed by Steward and Chamberlin.[92]

Biosynthetic Incorporation of Tryptophan Analogs

Two procedures are described here for biosynthetic incorporation of Trp analogs into recombinant proteins by using plasmids in *E. coli* host cells. It should be noted that these procedures are applicable to any other analog amino acids that are capable of being activated by an aminoacyl-tRNA synthetase. A prerequisite is that a plasmid-borne bacterial expression system has been developed for the protein of interest, and that the plasmid can be transferred to and expressed by an organism that is a Trp auxotroph. To minimize synthesis of non-analog-containing protein, it is desirable that the expression system be under the control of a nonleaky promotor. For information about the standard molecular biology techniques

[92] L. E. Steward and A. R. Chamberlin, Protein expression by expansion of the genetic code. *In* "Encyclopedia of Molecular Biology and Molecular Medicine." (R. A. Meyers, ed.), pp. 135–147. VCH Publishers, New York, 1996.

used to create expression systems and carry out transformation of host *E. coli* strains, the reader may consult *Methods in Enzymology* Volumes 154, 155, 185, and 217,[92a] as well as the excellent laboratory manual by Sambrook *et al.*[93]

Escherichia coli Auxotroph Strains and Expression Vectors. The *in vivo* incorporation of amino acid analogs into recombinant proteins is carried out by using auxotrophic organisms. The selection of an *E. coli* auxotroph host strain depends on the promoter system of the expression vector. We and others have used Trp-auxotrophic *E. coli* W3110 TrpA88 (a suppressible *amber* mutation),[94] W3110 TrpA33 (Glu48Met),[94] and CY15077 (W3110 *tnaA2*ΔTrpEA2; mutation in the *tra* gene and deletion of the Trp operon), which are from the laboratory of C. Yanofsky at Stanford University. These strains, which also are available from the authors of this chapter, have been transformed successfully with a variety of expression plasmids. A partial listing is given in Table II. Other hosts that are Trp auxotrophs can be employed, following the principles set forth here and elsewhere.[25,30,31,75]

Strains suitable for analog incorporation have been constructed for a number of different promoter systems, including *tac,* T7, T5, and temperature-sensitive and oxygen-sensitive promoters. As indicated in Table II, the efficiency of incorporation varies with each plasmid construct. Although our experience is still limited, the efficiency of incorporation seems to depend on several factors. First, as indicated previously, a promoter that has low constitutive activity helps prevent accumulation of unlabeled protein. Once the selected auxotroph host has been transformed with the expression vector, a test should be carried out to determine basal and induced protein expression levels. The most desirable circumstance is one in which there is no detectable basal expression prior to induction. Second, any cellular Trp that is available after induction will compete effectively with a Trp analog for charging of the tRNA. For this reason, depletion of cell Trp pools prior to induction is important. The amount of time allowed for expression is also important because Trp can become available from turnover of unlabeled cell proteins after induction. Third, expression under control of a particular promoter may require newly synthesized proteins for translation, for example the T7 promoter, which requires T7 RNA polymerase.[95] If an essential Trp residue of one or more

[92a] *Methods Enzymol.* **154** (1987); **155** (1987); **185** (1990); **217** (1993).

[93] J. Sambrook, E. F. Fritsch, and T. Maniatis, "Molecular Cloning: A Laboratory Manual," 2nd Ed., Vols. I, II, and III. Cold Spring Harbor Laboratory Press, Cold Spring Harbor, New York, 1989.

[94] G. R. Drapeau, W. J. Brammar, and C. Yanofsky, *J. Mol. Biol.* **35,** 357 (1968).

[95] F. W. Studier, A. H. Rosenberg, J. J. Dunn, and J. W. Dubendorff, *Methods Enzymol.* **185,** 60 (1990).

TABLE II
PROTEINS EXPRESSED WITH TRYPTOPHAN ANALOGS[a]

Protein	Trps	Promoter	Percent analog incorporation			Function	Ref.[b]
			5-OHTrp	7-ATrp	4-FTrp		
Rat oncomodulin (Y57W)	1	OXYPRO	<50			Wild type	1
λ *c*I repressor	3	*tac*	95			Wild type	2
cAMP regulatory protein	2	λP^L	50–90			Wild type	3, 4a,b
α subunit RNA polymerase	1	T7	50–90			Wild type	4a
11 mutants of W321F {W260, ..., W270}	1	T5	>95			Wild type	4b
σ subunit RNA polymerase	4	T7	50–60			Wild type	4a,b
Cytidine repressor (M151W)	1	T7	30–50			Wild type	5
Biotin repressor	7	*tac*	85			Inactive	6
Soluble human tissue factor	4	*tac*	<20	<30		?	7
Trp tRNA synthetase	1	*tac*	>95	>95	>95	Altered	8
Herpesvirus protein VP16 W442, W473	1	*tac*	50–95	50–95		Wild type	9
MyoD	1	T5	>95	>95		Wild type	10
Rat parvalbumin F102W	1	T7/pLysE	~50	~50	~50	Wild type	11
TBP	1	T7	30–50			?	12

[a] Proteins were expressed in *E. coli* Trp auxotrophic cells W3310TrpA88 and/or CY15077ΔEA2.

[b] Key to references: (1) C. W. V. Hogue *et al., FEBS Lett.* **310,** 269 (1992); (2) J. B. A. Ross *et al., Proc. Natl. Acad. Sci. U.S.A.* **89,** 12023 (1992); (3) J. C. Lee and T. Heyduk, personal communication (1996); (4a) E. Heyduk and T. Heyduk, *Cell. Mol. Biol. Res.* **39,** 401 (1993); (4b) T. Heyduk, personal communication (1996); (5) D. F. Senear, personal communication (1994); (6) D. Beckett, personal communication (1994); (7) C. A. Hasselbacher, personal communication (1994); (8) C. W. V. Hogue, Ph.D. dissertation, University of Ottawa, Ottawa, Ontario, Canada, 1994; (9) F. Shen *et al., J. Biol. Chem.* **271,** 4827 (1996); (10) S. Khotz, personal communication (1996); (11) C. W. V. Hogue *et al., Biophys. J.* 68, A193 (1995). The 5-OHTrp-containing protein failed to fold; (12) M. Brenowitz, personal communication (1994).

of the new proteins is replaced by an analog, function, and therefore expression, may be compromised. Fourth, the expressed protein itself may not tolerate uniform replacement of its Trp residues. As indicated in Table II, the efficiency of incorporation of analogs into the soluble extracellular domain of human tissue factor (sTF), was quite low although expression was under control of the *tac* promoter, which has proved efficient for analog incorporation into other proteins. A possible explanation for this is evident in the observation that mutation individually of two of its four Trp residues results in low yields of expressed protein.[14] Thus, substitution of these two Trp residues with analogs might result in an unstable protein, thereby accounting for both low yields and inefficient analog incorporation.

Media for Culture and Expression Tests. LB medium and LB plates are produced according to Sambrook *et al.*[93] Alternatives are described by Studier *et al.*[95] Antibiotics are added to final concentrations normally used for expression.

MEDIA USED FOR TWO-STEP ANALOG INCORPORATION. The first step, growth prior to expression, requires a medium containing Trp, for example, "Terrific Broth," prepared according to Sambrook *et al.*,[93] but with glycerol (10 ml/liter) or LB-rich medium. Alternatively, M9 minimal medium, prepared according to Sambrook *et al.*,[93] supplemented with 0.5% (w/v) glucose, 1% (w/v) casein acid hydrolysate (Casamino Acids), 0.1% thiamin (w/v), and 0.25 m*M* Trp can be used. Antibiotics are added to the cooled medium to concentrations normally used for growth.

The second step, protein expression, requires a growth medium that does not contain Trp, such as M9 minimal medium supplemented with 1% (w/v) Casamino Acids. Antibiotics are added to concentrations normally used for growth. In addition, the following sterile solutions are added per liter of medium: 2 ml of 1 *M* $MgSO_4$; 0.1 ml of 1 *M* $CaCl_2$; and 1 ml of thiamin (100 mg/ml; filter sterilized). (This medium is also suitable for making plates to test for Trp auxotropy; Trp auxotrophs will not grow on this medium.)

For analog incorporation, freshly prepared solutions, filter sterilized before use, are made with 50 mg of L-Trp or 100 mg of DL-Trp analogs in 10 ml of deionized water, using 100 μl of 1.0 *M* NaOH to help solubilize the analogs. When added to 1 liter of medium, this gives an effective concentration of 0.25 m*M* for the L-amino acid. Alternatively, just prior to induction, the analogs can be added directly to the growth medium in dry form with minimal risk.The analogs DL-7-ATrp, L-5-OHTrp, and DL-4-FTrp can be obtained from Sigma (St. Louis, MO). Other Trp analogs may be obtained from Aldrich (Milwaukee, WI).

MEDIA FOR ONE-STEP ANALOG INCORPORATION. M9 minimal medium is prepared according to Sambrook *et al.*,[93] but supplemented prior to autoclaving with 2% (w/v) Casamino Acids and twice the recommended amount of carbon source (half glucose and half glycerol). Antibiotics are added to the cooled medium to final concentrations generally used for growth. In addition, the following sterile solutions are added per liter of medium: 2 ml of 1 *M* $MgSO_4$, 0.1 ml of 1 *M* $CaCl_2$, and 1 ml of thiamin (100 mg/ml; filter sterilized). A solution is prepared of 4 mg of L-Trp in 10 ml of deionized water, using 100 μl of 1.0 *M* NaOH to help solubilize the amino acid, filter sterilized. The final L-Trp concentration is 4.0 μg/ml (0.02 m*M*, which is growth limiting for these auxotrophs).[47]

Protein Expression Minipreparation STET Lysis Buffer. A pH 8.0 solution is made of 0.1 *M* NaCl, 10 m*M* Tris-HCl, 1 m*M* EDTA, and 5% (v/v) Triton X-100.[93]

Lysozyme Solution. A pH 8.0 buffer solution is made of 10 m*M* Tris-HCl, 1 m*M* EDTA. For the working solution, 1 mg of lysozyme (Sigma) is added to 1 ml of the buffer.

Protein Expression Minipreparation Test. Cells are grown in 20 ml of LB medium with the appropriate antibiotics in sterile, 50-ml disposable culture tubes with conical ends. Lids are fitted loosely, but secured with tape to allow aeration while they are shaken in an incubator. Duplicate cultures of each strain are grown. Once the cells have reached late-log phase of growth, one of the cultures is induced [e.g., with isopropyl-β-D-thiogalactopyranoside (IPTG)], the other is not. At the same time, an additional 5 ml of LB medium is added. After an additional 3–5 hr of shaking, the cells are pelleted at the bottom of the culture tubes by centrifugation. For the test, the medium is poured off, and the last remaining drop of medium is removed with the point of a sterile wipe. Each cell pellet is resuspended in 300 μl of STET buffer, and then transferred by pipette to a marked, 1.5-ml plastic disposable centrifuge tube. To effect lysis of the cells, 100 μl of the lysozyme solution is added. After 10 min at room temperature, the tubes are placed into a vigorously boiling water bath for 50 sec, then transferred to an ice–water bath and chilled. This step causes the large DNA and cellular debris to aggregate. The tubes are then centrifuged (15,000 *g*) for 10 min, and the supernatants of the induced and noninduced samples are transferred into fresh tubes. Small aliquots of the supernatants are used to determine protein concentration (e.g., by absorbance or protein assay). If necessary, the more concentrated (induced or noninduced) sample can be diluted to match the concentrations. The induced and noninduced samples are compared by sodium dodecyl sulfate-polyacrylamide gel electrophoresis (SDS–PAGE) using purified protein as a standard. With appropriate loading concentrations of bacterial protein, the resulting gel should have a readily identified band corresponding to the expressed protein in the induced sample lane and lack a corresponding band in the noninduced sample lane. In our experience, this test is generally a good indicator of whether an *E. coli* expression system will be suitable for analog incorporation and yield useful amounts of protein. More detailed expression tests are described elsewhere.[96]

Analog Incorporation. The appropriate auxotrophic host strain is selected, and transformed with the desired plasmid containing the expression system. Expression tests are performed using this strain to determine that basal expression is sufficiently low and that an acceptable amount of protein is expressed after induction. Cells are inoculated into medium from freshly grown plates to avoid introducing cells from confluent subcultures that may have lost their plasmids. Such plasmid-free cells can overrun a slow-growing culture after the enzymatic depletion of the antibiotic from the medium.

Several different protocols have been used successfully for expression of analog-containing proteins. One protocol described here is a single-step method that relies on exhausting the available Trp without changing the culture medium prior to induction in the presence of added analog. This protocol is required for certain expression system constructs. However, it has not been as successful as the two-step protocol, which involves growing the auxotrophic cells in medium containing a Trp source and then prior to induction switching them to a minimal medium that contains the analog of interest in place of Trp. Expression and analog incorporation depend on expression strain, type of expression system, and the particular recombinant protein. Suitable protocols for particular expression systems are likely to require further optimization.

TWO-STEP ANALOG INCORPORATION. The following procedure was used to express TrpRS[23] and is similar to that used to express λ *c*I repressor.[48] Culture vessels may vary. We employ 4-liter Erlenmeyer flasks with aeration vanes added by our glassblower. Similarly constructed flasks are commercially available. Growth medium (2 liters) and expression medium (2 liters) are prepared in separate flasks. Cultures are inoculated as previously described, and grown in a shaker at 37°. Cell growth is monitored by optical density at 600 nm. When the optical density reaches 1.5, the cells are collected by centrifugation. (A washing step to reduce Trp availability can be included at this point.) The cells are resuspended into the expression medium, and returned to the shaker for 0.5 hr at 37° to recover and to deplete residual Trp. At this point, a 25-ml aliquot should be removed from the culture for the expression test described above, and stored at 4°. The Trp analog is then added to the culture either as a solution or in dry form. After 10 min, the culture is induced. After an additional 5–6 hr of shaking, another 25-ml aliquot is taken from the culture and the remainder of the cells is collected by centrifugation and weighed. A minipreparation

[96] T. L. Pauls, I. Durussel, J. A. Cox, I. D. Clark, A. G. Szabo, S. M. Gagne, B. D. Sykes, and M. W. Berchtold, *J. Biol. Chem.* **268,** 20897 (1993).

test, as described, should be carried out with the two 25-ml aliquots collected from the culture pre- and postinduction. The final cell pellet can be frozen at $-80°$ until purification.

An example growth curve from TrpRS in the CY15077 pMS421 host strain is presented in Fig. 7. The purification of TrpRS is detailed elsewhere.[23] In the case of TrpRS, the protein yield by the two-step procedure with Trp was superior to that obtained by conventional protein expression: one 2-liter culture using the two-step procedure gave as much TrpRS (122 mg) as three 2-liter cultures using Terrific Broth, and also gave 65 mg of 5-OHTrp containing protein. In the case of λ *c*I repressor,[48] with growth before induction to an OD_{550} of 1.0, the resultant yield of purified analog-containing protein (greater than 95% incorporation) was about 65% that typically obtained from expression of non-analog-containing protein in LB-rich medium.

ONE-STEP ANALOG INCORPORATION. The one-step analog incorporation procedure is necessary when a T7 RNA polymerase promoter[95] is used together with the pLysE plasmid, owing to the presence of T7 lysozyme in these strains. Treatments that damage a small fraction of these cells, such as the centrifugation step in the two-step incorporation procedure, allow T7 lysosyme to escape. This causes rapid lysis of the entire culture. Typically, 2 liters of 1-step medium is prepared in a 4-liter Erlenmeyer flask with aeration vanes. Cultures are inoculated as previously described,

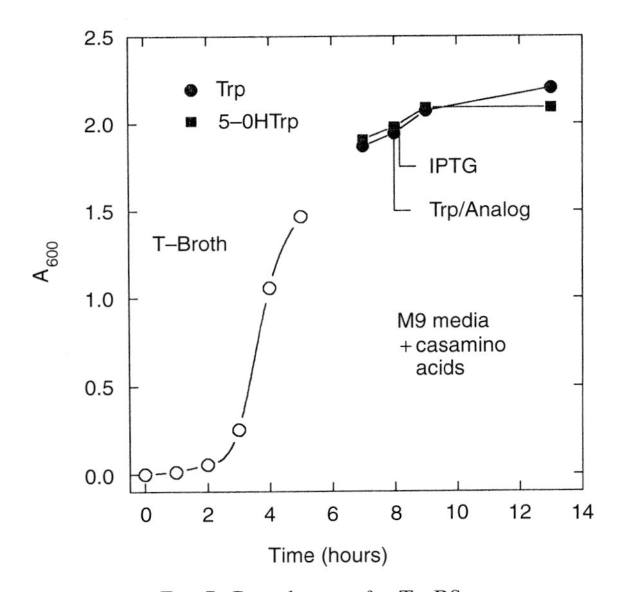

FIG. 7. Growth curve for TrpRS.

usually at the end of the work day, and grown in a shaker at 37° with ample agitation. Cells are allowed to culture overnight, reaching their limiting concentration by consuming the Trp present in the medium. This can be confirmed by measuring the optical density of the culture at 600 nm and then again 1 hr later. A 25-ml aliquot of the culture should be taken and stored at 4° for the minipreparation test, to ascertain the basal level of expression. The Trp analog solution is added to the culture. After 10 min, the culture is induced. Cells are allowed to culture for an additional 5 hr. A 25-ml culture aliquot should be saved at the end for the expression test. An expression test should be carried out with the two aliquots representing pre- and postinduction before attempting to purify the expressed protein. The remainder of the cells is harvested by centrifugation, weighed, and can be frozen at −80° until purification.

Incorporation of Tryptophan Analogs at Specific Sites

Incorporation of Trp analog amino acids into proteins *in vivo* does not control which Trp residues are replaced. Two methods that provide a way to replace a Trp (or other residue) selectively in a multi-Trp-containing protein are *in vitro* transcription/translation using a semisynthetic nonsense supressor tRNA[51–56] or direct peptide synthesis using traditional chemical techniques.[57] The *in vitro* approach allows one to place an amino acid analog at a specific site in a large protein readily because polypeptide synthesis is carried out biochemically. It is a more difficult technical challenge to generate large proteins by direct chemical synthesis. The reaction yields for *in vitro* incorporation, however, are limited and variable. However, in favorable circumstances protein synthesis levels may approach about 20 μg/ml. Typical reaction volumes are 30–50 μl, and in the effort to scale up protein production, continuous-flow methods have been developed.[56] By contrast, it is less of a challenge by chemical synthesis to produce milligram-to-gram quantities of polypeptides with molecular weights up to several kilodaltons. Direct synthetic methods, which are outlined as follows, have been used to prepare small peptides containing 5-OHTrp[48] or 7-ATrp[40,91] and the insulin analog, [A14-5-OHTrp]insulin.[57]

Peptide Synthesis with 5-Hydroxytryptophan. 5-Hydroxytryptophan is commercially available in the L form. After converting the analog to the *t*-Boc(α-amino) derivative, the coupling can be carried out by standard solid-phase methods[97] without using blocking groups to protect the secondary functions of the indole ring.[57] Addition of pentachlorophenol (0.4 m*M*)

[97] G. Barany and R. B. Merrifield, *in* "The Peptides" (E. Gross and J. Meienhofer, eds.), Vol. 2, pp. 3–284. Academic Press, New York, 1980.

during the coupling reaction with the analog, as well as during subsequent coupling reactions, minimizes reaction with the hydroxyl group.[98] Removal of *t*-Boc groups after incorporation of the analog with minimal side reactions of the reactive 5-hydroxyindole ring can be accomplished with a mixture of trifluoroacetic acid, anisole, ethanedithiol, and dichloromethane (20 : 2 : 1 : 17, v/v). In this way, the peptide Gly-L-5-OHTrp-$(Gly)_3$-L-Glu-$(Gly)_3$-L-Tyr-Gly[48] and the A chain of human insulin with 5-OHTrp replacing Tyr at position 14[57] have been prepared.

Peptide Synthesis with 7-Azatryptophan. 7-Azatryptophan is available only as a DL mixture. However, pure tripeptides containing a single L-7-ATrp have been prepared by using high-performance liquid chromatography (HPLC) to separate the different enantiomers.[40,91] After converting the analog to the *t*-Boc(α-amino) derivative, the tripeptides *N*-Ac-L-Pro-DL-7-ATrp-L-Asn-NH_2 and L-Lys-DL-7-ATrp-L-Lys can be prepared by solid-phase synthesis[41] and solution methods,[91] respectively. For the solid-phase synthesis,[41] *t*-Boc chemistry is employed, and deblocking is effected by standard procedures. Identification of the L form is made on the basis of relative elution times for racemic mixtures of other amino acids. For the solution-phase synthesis,[91] *t*-Boc groups are removed from the dipeptide containing the analog with trifluoracetic acid in the presence of anisole. The final step involves coupling with benzyloxycarbonyl-blocked Lys. The resulting tripeptide is saponified in NaOH and hydrogenated in the presence of palladium to remove the protecting groups. After HPLC separation, the L and D enantiomers are identified by an assay involving enzymatic digestion followed by reaction with TrpRS to form the L-7ATrp adenylate. This is done as described by Hogue and Szabo.[24] The peptides are digested with a cocktail of proteases and endopeptidases. The enzymes are removed from the solution by boiling, followed by filtration. The remaining filtrate is assayed by reacting with TrpRS and ATP in the presence of inorganic pyrophosphatase. The reaction of L-7-Trp with ATP results in L-7-ATrp 5′-adenylate and pyrophosphate, which is converted to monophosphate by pyrophosphatase. The adenylate binds in the active site of the enzyme in a stereospecific manner. The complex exhibits a 20-fold increase in fluorescence when using 310-nm excitation and monitoring the 360-nm emission, characteristic of 7-ATrp buried in the interior of a protein.

Characterization of Tryptophan Analog Incorporation in Proteins

A potential limitation for direct synthesis is the commercial availability of L analogs. If DL mixtures of analogs must be used, the enzymatic assay

[98] J. Martinez and M. Bodanszky, *Int. J. Peptide Protein Res.* **12,** 277 (1978).

described immediately above is recommended for assuring identification of the enantiomers. In the case of 7-ATrp, it appears that separation of enantiomers can be effected by HPLC. Otherwise, if it is impossible to achieve a separation, a way to resolve the D and L forms of the amino acid might be first to chemically acylate the DL mixture and then selectively deacylate by using an enzyme.[99] Alternatively, the L form of the amino acid must be synthesized directly.

Incorporation of analogs by using expression with a Trp auxotroph circumvents the issue of availability of L analogs because enzymes in the cell discriminate the D and L forms effectively. However, the enzymes will tend to facilitate incorporation of any available Trp rather than an analog. Thus, for *in vivo* synthesis, it is essential to assess carefully the degree of analog incorporation.

Analysis of Analog Incorporation Using Absorbance Spectra. Determination of amino acid content in proteins and polypeptides is generally done by acid hydrolysis and chromatographic separation of a colored or fluorescent adduct. Tryptophan, however, is destroyed by acid hydrolysis. A classic method developed by Edelhoch[100] to quantitate Trp content is to measure the absorbance spectrum of the protein denatured in guanidinium. On the basis of the extinction coefficients at several wavelengths of the model compounds *N*-acetyltryptophanamide (NATrpA) and *N*-acetyltyrosinamide (NATyrA) in the same solvent, a calculation is made of the ratio of Trp residues to Tyr residues. The Tyr content is determined separately by amino acid analysis. The Trp content can then be calculated from the Trp:Tyr ratio. The accuracy of the calculation not only depends on knowing the extinctions of the model compounds accurately, but also on determining whether they are good representations for the aromatic groups when incorporated in a peptide chain.

While extinction coefficients can be measured with great precision, accuracy is limited by systematic errors that depend on the purity/stability of the compound and the ability to know how many moles of the compound are in the sample. One way this issue has been addressed for determining the Trp:Tyr ratio in proteins is to scale the absorption spectra of the model compounds (in guanidinium chloride) by least-squares fitting to the absorption spectra of synthetic peptides (in guanidinium chloride) that contain both Trp and Tyr in different ratios fixed by composition.[101] In this way, properly scaled basis set spectra can be generated for the "characteris-

[99] H. K. Chenault, J. Dahmer, and G. M. Whitesides, *J. Am. Chem. Soc.* **111,** 6354 (1989).

[100] H. Edelhoch, *Biochemistry* **6,** 1948 (1967).

[101] E. Waxman, E. Rusinova, C. A. Hasselbacher, G. P. Schwartz, W. R. Laws, and J. B. A. Ross, *Anal. Biochem.* **210,** 425 (1993).

tic" Trp and Tyr absorption in a fully denatured protein. The accuracy of the scale in terms of absolute concentration depends ultimately on the "best" estimate of the extinction coefficient for one of the model compounds. Similarly, the absorption spectrum of an analog model compound can be properly scaled to the previously determined basis set spectra, for example NATrpA and NATyrA, by least-squares fitting of the absorption spectra of synthetic peptides that contain in addition to the analog of interest either Trp or Tyr.[48]

Reasonably accurate estimates of the efficiency of analog incorporation can be obtained by fitting the denatured protein spectrum to linear combinations of basis spectra, provided there are no contaminating chromophores (e.g., a large number of disulfide bonds relative to the number of Trp residues), the sample is otherwise "pure," and there are no strong perturbations of the aromatic side chains owing to specific interactions with neighboring groups. Generally, the last condition can be met by denaturation in 6 *M* guanidinium chloride. It is also important that the sample solvent correspond exactly to that used for the basis set spectra. This can be achieved by dialysis of the sample.

Analysis of Analog Incorporation by HPLC. The general expectation is that the frequency of analog incorporation will be approximately equivalent for all Trp sites in a protein, reflecting the overall efficiency of incorporation. However, in principle, if analog incorporation at a particular site results in increased protein degradation, then polypeptide chains mainly with Trp at that particular site will be recovered after purification. In addition, if occupation of the other Trp sites by analog has little or no effect on folding or degradation, then the frequency of incorporation at these sites will be significantly greater, reflecting the overall kinetics of the protein synthetic machinery toward utilization of particular analogs. The net result will be a site-biased frequency of analog incorporation.

The efficiency of overall incorporation and the frequency of substitution at a particular site can be quantified by peptide mapping using HPLC. First, a map of the Trp-containing peptides is established for the wild-type protein on the basis of sequence and selection of optimal protease cleavage sites. A protease that yields peptides containing only one Trp site facilitates interpretation. Because Tyr and Trp both absorb light at 280 nm, Trp-containing peptides are best identified by absorption at 295 nm. The identity of the Trp-containing peptide fractions can be made by amino acid and/or sequence analysis. The relative recovery of different Trp-containing peptides can be established by comparing the ratios of the integrated absorbances at 295 nm of the appropriate fractions corresponding to the known sequence.

Second, the HPLC elution profiles for the protein expressed in the presence of Trp or analog are compared at 295 and 315 nm. The expectation is that Trp absorption at 315 nm should be negligible compared to that of either 5-OHTrp or 7-ATrp. Generally, peptides containing analog will elute at new positions owing to the altered polarity of the side chain. Assuming that analog incorporation does not affect proteolysis, the inefficiency of incorporation can be assessed from the difference in the recovery of Trp-containing peptides from the protein expressed in the presence of Trp versus the recovery of the identical peptides when expressed in the presence of analog. If incorporation of analog is 100% efficient, there should be no Trp-containing peptides. Instead, the chromatogram should show identical sets of peaks when monitored at 295 and 315 nm, and they should be migrating at new positions. [This type of analysis was carried out with Y57(5-OHTrp) oncomodulin.[47]] The amino acid composition of the fractions should reveal analog-containing peptides corresponding to the Trp-containing peptides. 5-Hydroxytryptophan and 7-ATrp both are significantly more stable than Trp toward acid hydrolysis, and they should be readily detectable when the HPLC fractions are assayed by amino acid analysis.

Spectroscopy of Tryptophan Analogs in Proteins

The 1992 papers reporting *in vivo* incorporation of 5-OHTrp into the recombinant proteins Y57W oncomodulin[47] and λ *c*I repressor[48] demonstrated a new, general approach for introducing site-specific, minimally perturbing probes for absorption and fluorescence spectroscopy of protein–protein and protein–nucleic acid interactions. As indicated in the previous sections, 4-FTrp and 7-ATrp are also useful analogs for this purpose. A list of recombinant proteins that have been expressed in *E. coli* Trp auxotrophs in the presence of 4-FTrp, 5-OHTrp, or 7-ATrp is provided in Table II. In the following section, we review the absorption and fluorescence properties of several of these analog-containing proteins to illustrate important features of spectral enhancement.

Examples of Spectra of Analog-Containing Proteins

The number of Tyr residues in a protein must be considered when examining the contribution of each analog to the overall spectrum. Figure 8 depicts the normalized absorbance spectra of *B. subtilis* TrpRS expressed with the analog series 4-FTrp, 5-OHTrp, or 7-ATrp incorporated at 95% efficiency. This protein has a 14:1 Tyr-to-Trp ratio. Thus, the absorbance spectrum of the wild-type protein is dominated by Tyr, giving it a λ_{max} of

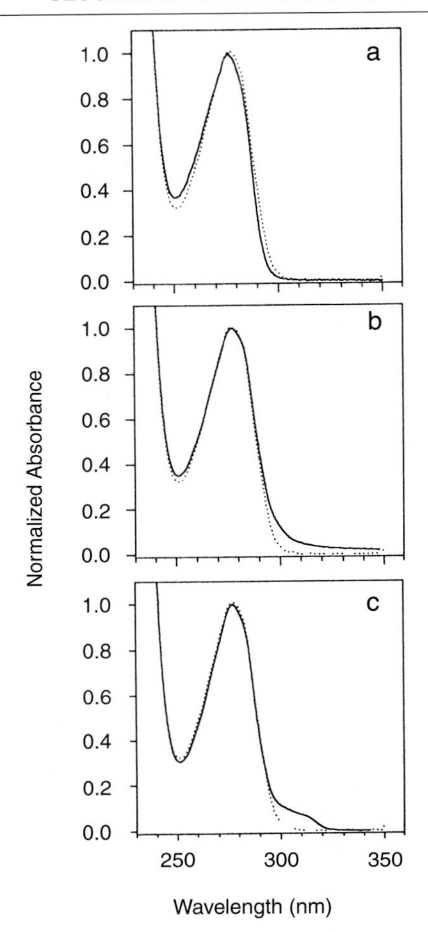

FIG. 8. Normalized absorption spectra of TrpRS containing (a) 4-FTrp, (b) 7-ATrp, or (c) 5-OHTrp.

277 nm. As a result, the absorbance of the spectral shoulder of 5-OHTrp at 310 nm is only about one-tenth the absorbance at 277 nm, which is the expected proportion. For proteins such as Y57W oncomodulin,[47] which has a 1:1 Tyr-to-Trp ratio, or λ *c*I repressor,[48] which has a 7:3 Tyr-to-Trp ratio, the red shoulder of 5-OHTrp is a considerably more dominant feature.

The red-edge absorbance at 310 nm of 7-ATrp in TrpRS is larger than expected on the basis of the absorbance of 7-ATrp zwitterion; the red-edge absorption of 7-ATrp has been noted to increase after incorporation into a polypeptide.[23,91] Depending on the local environment of 7-ATrp when incorporated in a protein, the absorbance spectrum may include contributions from the $^{1}L_{b}$ transition dipole that results in a resolved band in this wavelength region.

Tryptophan at position-92 in TrpRS resides in a buried environment. The effect of replacing Trp-92 with 4-FTrp is to shift the absorbance of the protein by a small amount, from 277 to 276 nm. Because 4-FTrp in TrpRS has significantly reduced absorbance at 295 nm, the inherent sensitized photoreactivity of 4-FTrp[76] can be minimized when excitation wavelengths longer than 295 nm are used.

To evaluate the influence of Tyr on the spectroscopy of TrpRS containing the analog 4-FTrp, 5-OHTrp, or 7-ATrp, the fluorescence emission spectra obtained with excitation at 280 nm can be compared with that obtained when using red-edge excitation wavelengths in the range 300–315 nm. Using 280-nm excitation (Fig. 9), the wild-type protein has an uncorrected emission maximum at 317 nm, which is intermediate between the 305-nm maximum expected for Tyr[72] and the 325-nm maximum obtained when using 295-nm light for selective excitation of Trp-92.[23] The emission obtained for 280-nm excitation of the protein containing 4-FTrp has a maximum at 304 nm, indicating loss of the longer wavelength Trp fluorescence. This spectrum was similar to that of the inactive, Trp-free mutant W92F TrpRS, which had an uncorrected emission maximum at 307 nm.[15] The emission intensity at 300 nm of TrpRS containing 4-FTrp is greater than that of the wild-type enzyme at the same protein concentrations. This is most likely owing to less efficient energy transfer from Tyr to the blue-shifted 4-FTrp-92 compared to transfer to Trp-92. Tryptophanyl-tRNA synthetase containing 4-FTrp had a low level of residual Trp-92 fluorescence, as judged by fluorescence lifetime measurements when excited at 300 nm (see Fig. 10). This probably represents a low level of available Trp competing with the analog in the cells during expression, resulting in formation of some wild-type TrpRS.

The fluorescence emission of 7-ATrp-92 in TrpRS when using 280-nm excitation (Fig. 9) is less than that of Trp in the wild-type protein. It is evident from the 306-nm peak in the emission spectrum of the analog-containing protein that its emission is dominated by Tyr fluorescence. The 7-ATrp-92 emission appears as a shoulder between 340 and 370 nm, with a tail extending past 450 nm. It is important to note, however, that this shoulder indicates a substantial blue shift from that of fully solvated 7-ATrp emission (see Table I and Fig. 5c). Thus, the 7-ATrp emission maximum is highly sensitive to the local environment. The excitation spectrum of 7-ATrp in TrpRS shows an enhanced red edge at 310 nm compared to that of the wild-type protein, and as seen in Fig. 10, the fluorescence intensity resulting from 300-nm selective excitation of 7-ATrp-92 is greater

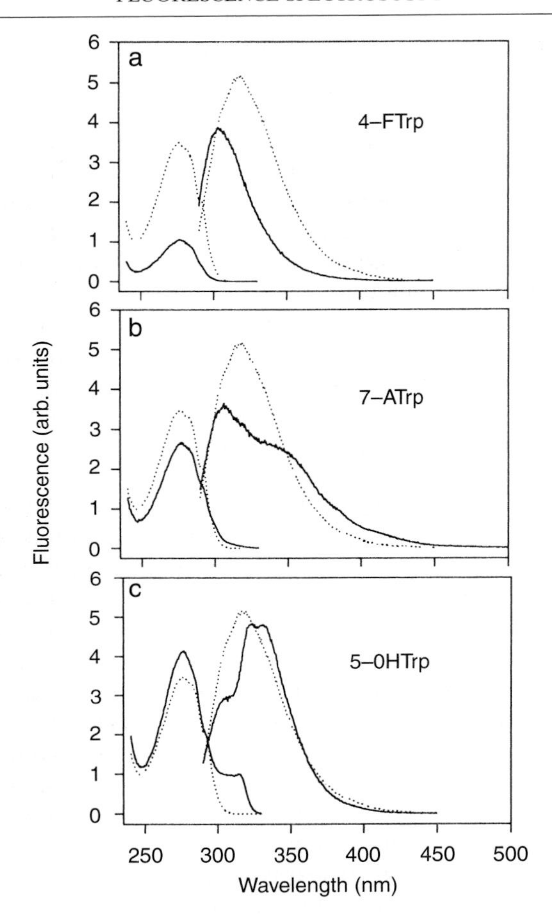

FIG. 9. Fluorescence emission spectra of wild-type (dotted line) and analog-containing TrpRS with excitation at 280 nm. Each solution had an optical density of 0.1 at 280 nm.

than that of Trp-92 at the same protein concentration. This is owing to both the greater extinction of 7-ATrp-92 at the excitation wavelength, and its substantially greater quantum yield (ϕ = 0.15) compared to that of solvated 7-ATrp (Table I). Similar high quantum yields were observed for the fluorescence of 7-azatryptophanyl adenylate bound in the active site of wild-type TrpRS[24] and for the fluorescence of 7-ATrp-102 rat parvalbumin. This phenomenon is also observed in other proteins containing 7-ATrp residues, such as alkaline phosphatase,[26] and in spectra of partially purified protein extracts.[14] Thus, burial of 7-ATrp into a less solvent accessible portion of the protein matrix generally results in a significant enhance-

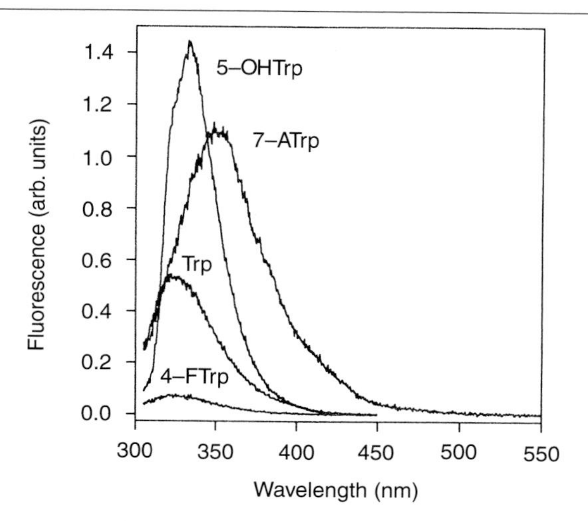

FIG. 10. Fluorescence emission TrpRS with red-edge excitation at 300 nm. Each solution had an absorbance of 0.1 at 280 nm.

ment in quantum yield, concomitant increase in fluorescence lifetime,[23,24] and large (30–50 nm) blue shift in the emission maximum.

The fluorescence spectrum obtained with 280-nm excitation (Fig. 9) of TrpRS containing 5-OHTrp has an unusually structured shape compared to typical fluorescence spectra of Trp-containing proteins. The unusual features can be accounted for in part by the spectroscopic characteristics of 5-hydroxyindole compared with indole, in part by the large Tyr-to-Trp ratio of TrpRS. Compared to Trp or 7-ATrp fluorescence (see Table I), 5-OHTrp fluorescence exhibits smaller shifts in response to differences in the local environment; the emission maximum of Trp-92 is near 325 nm whereas that of 5-OHTrp-92 is near 334 nm. In addition, as seen in Fig. 10, 5-OHTrp-92 emission is relatively negligible at wavelengths shorter than 310 nm, where Tyr emission is maximum. The net result is a less obstructed view of the Tyr fluorescence of the 5-OHTrp-containing protein, which is observed as a shoulder between 300 and 310 nm. It should be noted that the overlap of the emission of Tyr at 305 nm with the absorbance of 5-OHTrp is greater than with that of Trp, increasing the probability of quenching by energy transfer.[48] The effect of more efficient energy transfer is evident in TrpRS containing 5-OHTrp because the Tyr emission at 310 nm is less than that of wild-type TrpRS, and less than the nearly pure Tyr emission of TrpRS when containing 4-FTrp. Thus, the unusual peak shape of the spectrum from TrpRS containing 5-OHTrp, obtained with 280-nm excitation, arises from the resolved Tyr emission component.

Time-Resolved Fluorescence of Analog-Containing Proteins

As discussed previously, the fluorescence decay behavior of 7-ATrp is quite complex, whereas that of 5-OHTrp appears simple, exhibiting single-exponential fluorescence decays in solution. This is important for understanding the well-documented multiexponential decay behavior of most single Trp proteins,[5] because Trp itself exhibits complex fluorescence decay kinetics.[102–105] In principle, interpretation of the fluorescence lifetime behavior of 5-OHTrp in proteins should be more straightforward. As is discussed, multiexponential decay behavior extends to the Trp analogs when they are incorporated as residues in a polypeptide chain.

Time-resolved fluorescence of the single 5-OHTrp residue in Y57W oncomodulin exhibits dual-exponential fluorescence decay.[47] Each of the time-resolved fluorescences of the single Trp-92, 5-OHTrp-92, and 7-ATrp-92 in TrpRS exhibits triple-exponential decay kinetics, but with different decay times and amplitudes.[23] Similarly, the two insulin analogs, [A14-Trp]- and [A14-OHTrp]insulin, also exhibit triple-exponential intensity decay kinetics.[57] The single Trp-102 residue in F102W rat parvalbumin has a single exponential fluorescence decay, as does 7-ATrp-102 in the same protein. Thus, despite 5-OHTrp and 7-ATrp being different fluorophores from Trp, the same hallmarks of multiexponential fluorescence decay are observed when they are incorporated as a single residue in a protein.

Because 5-OHTrp and 7-ATrp have different ground-state and excited-state properties, they may help provide a better understanding of the physical basis for multiexponential fluorescence decay times of Trp in proteins. One critical experiment, suggested by Hudson *et al.*,[33] is to examine in the context of the protein matrix a fluorophore with photophysics simpler than those of Trp. As previously described, this experiment has now been performed by using 5-OHTrp. Interestingly, the results are strikingly similar to those obtained for Trp. We conclude from this that the factors affecting the fluorescence decay kinetics of Trp, 5-OHTrp, and 7-ATrp in proteins may have less to do with the photophysics of the fluorophore and more to do with the protein matrix itself, such as side-chain conformation,[7,10,103–105] secondary structure,[8,9] and local dipole moment.[23,106]

[102] D. M. Rayner and A. G. Szabo, *Can. J. Chem.* **56,** 743 (1978).

[103] A. G. Szabo and D. M. Rayner, *J. Am. Chem. Soc.* **102,** 554 (1980).

[104] J. B. A. Ross, K. W. Rousslang, and L. Brand, *Biochemistry* **20,** 4361 (1981).

[105] M. C. Chang, J. W. Petrich, D. B. McDonald, and G. R. Fleming, *J. Am. Chem. Soc.* **105,** 3819 (1983).

[106] It has been noted that although 4-F indole is relatively nonfluorescent in water, it is fluorescent in hydrocarbon solvents. Therefore, it is possible that 4-FTrp might have a measurable fluorescence when completely buried in the protein matrix (P. Callis, personal communication, 1996).

Masking Tryptophan Fluorescence

In principle, intrinsic Trp fluorescence can be eliminated by replacing Trp residues with 4-FTrp. Masking Trp fluorescence would facilitate the use of intrinsic Trp fluorescence in experimental systems involving multiprotein complexes, for example the proteins involved in transcription assemblies, complement and clotting cascades, or chaperonin-mediated refolding. However, for 4-FTrp to be useful in masking the Trp fluorescence of proteins it is desirable to achieve complete incorporation.[107] Any expression of proteins containing Trp diminishes the utility of this analog, as it is unlikely that a mixture of Trp- and 4-FTrp-containing protein could be easily separated by purification, especially if the residues are buried. Tightly regulated promoters, such as the T7-*tac*,[95] may help diminish the persistant 5% leakage observed. To assure complete incorporation, Bronskill and Wong[75] have suggested the use of a *B. subtilis* strain that is deadapted away from Trp and can grow on 4-FTrp. Although this method has not yet been used, it might be useful to restate it here. A similarly deadapted auxotrophic strain of *E. coli* would be highly desirable, as it would allow the use of plasmids for complete 4-FTrp incorporation.

Expression Systems for Incorporation of Analogs

The best results to date for analog incorporation have been obtained by using *tac* or T5 promotors (see Table II). (T5 is used with the His-Tag protein expression/purification system.[108,109]) By contrast, the T7 polymerase promoter has not proved efficient for analog incorporation. Heyduk and co-workers[110] have compared expression of a series of α-subunit RNA polymerase mutants under control of either the T7 or the T5 promotor. After induction in the presence of analog, the T7 promotor often requires for efficient expression the addition of a small amount of Trp up to a concentration a few percent relative to that of the analog. The T5 promotor, however, which utilizes bacterial RNA polymerase, needs no additional Trp. Thus, it appears that the T7 polymerase does not tolerate 5-OHTrp or 7-ATrp well.

[107] C. W. V. Hogue, S. Cyr, J. D. Brennan, T. L. Pauls, J. A. Cox, M. W. Berchtold, and A. G. Szabo, *Biophys. J.* **68,** A193 (1995).

[108] H. Bujard, R. Gentz, M. Lanzer, D. Stüber, M. Müller, I. Ibrahimi, M. T. Häuptle, and B. Dobberstein, *Methods Enzymol.* **155,** 416 (1987).

[109] R. Janknecht, G. de Martynoff, J. Lou, R. A. Hipskind, A. Nordheim, and H. G. Stunnenberg, *Proc. Natl. Acad. Sci. U.S.A.* **88,** 8972 (1991).

[110] T. Heyduk, personal communication (1996).

Concluding Remarks

In 1992, our respective laboratories demonstrated that 5-OHTrp, a Trp analog with a red-shifted absorption spectrum, could be readily introduced into recombinant proteins *in vivo* by expression using *E. coli* Trp auxotrophs.[47,48] As indicated in Table II, spectrally enhanced proteins containing 5-OHTrp, 7-ATrp, or 4-FTrp now are being used by an increasing number of investigators. In many cases, these analog-containing proteins retain wild-type function. Thus, the methodology promises to be broadly applicable for the study of protein–protein and protein–nucleic acid interactions by either absorption or fluorescence spectroscopy. On the basis of our experience, as well as that of other investigators, we anticipate that spectrally enhanced proteins will provide important new insights and advances in understanding how the thermodynamics and structural dynamics of these multimeric macromolecular systems help regulate their biological functions.

Acknowledgments

J. B. A. Ross thanks his collaborators Drs. D. F. Senear, C. A. Hasselbacher, W. R. Laws, and Ms. E. Rusinova for contributions toward parts of the research reviewed here. C. W. V. Hogue and A. G. Szabo thank their collaborators, Drs. J. T.-F. Wong, J. D. Brennan, and Mr. D. Krajcarski, for contributions toward parts of the research reviewed here. The authors together thank Drs. P. Callis and M. Eftink for detailed discussions regarding the spectroscopy and properties of indole analogs. This work was supported in part by NIH Grants GM-39750 and HL-29019 to J. B. A. Ross, in part by an operating NSERC grant to A. G. Szabo, and in part by NSERC postgraduate and postdoctoral awards to C. W. V. Hogue.

[9] Time-Resolved Fluorescence of Constrained Tryptophan Derivatives: Implications for Protein Fluorescence

By MARK L. MCLAUGHLIN and MARY D. BARKLEY

Introduction

Fluorescence spectroscopy requires appropriate fluorescence probes. Nature has been kind and included fluorophores as intrinsic components of many biological molecules, such as the aromatic amino acids tryptophan, tyrosine, and phenylalanine. Tryptophan is especially useful, being present in low abundance in most proteins. Tryptophan has the potential to be a superb structural probe because its fluorescence is so sensitive to the

0076-6879/97 $25

environment. Unfortunately, its complex photophysics have made structural analysis based on tryptophan fluorescence challenging.[1] There are several reasons for the complex photophysics exhibited by tryptophan. Two overlapping electronic transitions have different polarity in the first absorption band of the indole chromophore, whose relative energy depends on solvent polarity and solvent relaxation.[2] More important, there are multiple competing nonradiative decay pathways that respond to the microenvironment of the indole ring in different ways.[3] Finally, there are multiple microconformational states that produce a heterogeneous environment for the indole ring. Therefore, to glean structural information from the fluorescence decay of tryptophan, it is necessary to measure each of the environmentally sensitive processes independently for each microenvironment. The radiative rate k_r varies little among the indole derivatives that have been studied, whereas the overall nonradiative rate k_{nr} varies widely within the same series of compounds. The task of measuring structurally relevant quenching rates is eased somewhat because some environmentally insensitive nonradiative rates can be lumped together and the temperature-dependent terms can be determined separately. The reciprocal of the fluorescence lifetime τ gives the total rate for deactivation of the excited state.

$$\tau^{-1} = k_r + k_{nr} \tag{1}$$

The radiative and nonradiative rates k_r and k_{nr} are determined by combining fluorescence quantum yield Φ_f and lifetime measurements.

$$\Phi_f = k_r \tau \tag{2}$$

The fluorescence decay rate can also be divided into temperature-independent and -dependent rates. The temperature-independent rate k_0 comprises both radiative k_r and temperature-independent nonradiative k_{nr}^0 rates.

$$k_0 = k_r + k_{nr}^0 \tag{3}$$

The temperature-dependent nonradiative rate can be described by a sum of Arrhenius terms with frequency factors A_i and activation energies E_i^*.

$$\tau^{-1} = k_0 + \sum_i A_i \exp(-E_i^*/RT) \tag{4}$$

[1] J. M. Beechem and L. Brand, *Annu. Rev. Biochem.* **54,** 43 (1985).

[2] H. Lami and N. Glasser, *J. Chem. Phys.* **84,** 597 (1986).

[3] H.-T. Yu, M. A. Vela, F. R. Fronczek, M. L. McLaughlin, and M. D. Barkley, *J. Am. Chem. Soc.* **117,** 348 (1995).

Quenching Mechanisms

Indole fluorescence decay rates are not affected by photoionization, which occurs on the femtosecond time scale.[4] Photoionization reduces the fluorescence quantum yield, but has no effect on the fluorescence lifetime because it occurs from a prefluorescent state. This results in overestimation of the radiative rate k_r and underestimation of the nonradiative rate k_{nr} if the fluorescence quantum yield in Eq. (2) is not corrected for photoionization. Photoionization yields of indole derivatives are low in aqueous solution. Internal conversion is apparently small or absent in indoles.[5] Intersystem crossing is independent of temperature. Intersystem crossing rates can be measured separately, either directly from transient absorbance measurements of triplet yields or indirectly from the temperature-independent nonradiative rate k^0_{nr}.[5] The rate of intersystem crossing is enhanced by inter- or intramolecular collisions of the indole ring with heavy atoms, such as iodide and metals. The report of emission from two excited triplet states in the zwitterion but not the anion of a constrained tryptophan suggests that intersystem crossing rates may depend on environment,[6] but other decay pathways are probably more informative about structure in the absence of heavy atoms. The remaining quenching mechanisms involve excited-state reactions, such as solvent quenching, excited-state proton transfer, exciplex formation, and excited-state electron transfer. All of these processes may be expected to depend on the proximity of appropriate species—water molecules, proton donors and acceptors, aromatic rings, electron acceptors—and on temperature. Rotation about the tryptophan side chain changes the local environment of the indole ring. Side-chain rotations in tryptophan derivatives can happen on time scales ranging from much slower to much faster than the fluorescence time scale. The effect of side-chain rotation on tryptophan fluorescence is greatly simplified and more easily studied using constrained tryptophan derivatives. This approach has stood us in good stead, allowing us to examine how ground-state conformation affects fluorescence decay rates and giving us a handle on various excited-state quenching mechanisms.

Ground-State Conformation

To use fluorescence decay rates as probes of the indole microenvironment, a basis set of tryptophan derivatives with well-characterized structures is required to map fluorescence properties to ground-state structure. A

[4] J. C. Mialocq, E. Amouyal, A. Bernas, and D. Grand, *J. Phys. Chem.* **86,** 3173 (1982).
[5] Y. Chen, B. Liu, H.-T. Yu, and M. D. Barkley, *J. Am. Chem. Soc.* **118,** 9271 (1996).
[6] K. Sudhakar, C. M. Phillips, S. A. Williams, and J. M. Vanderkooi, *Biophys. J.* **64,** 1503 (1993).

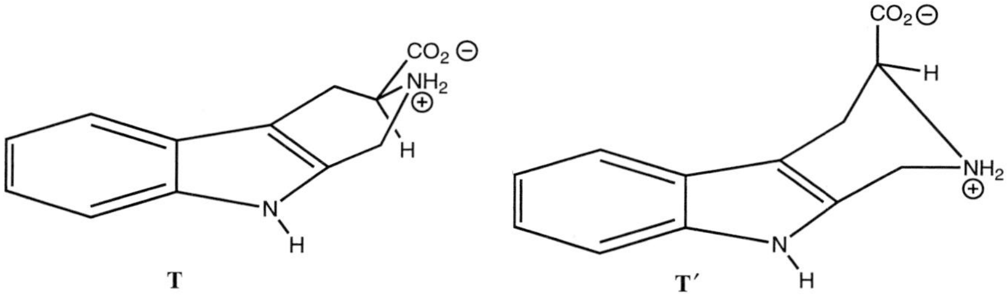

Fig. 1. T and T′ conformers of W(1).

combination of X-ray crystal and solution nuclear magnetic resonance (NMR) structures supplemented with molecular mechanics calculations has been used to determine the ground-state structures. The relative populations of other low-energy conformers can be estimated from NMR data and energy barriers between minima can be calculated using molecular mechanics. The constrained tryptophan derivatives studied to date have only two main conformations in solution, which reduces the uncertainty of estimates based on NMR and molecular mechanics. Figure 1 shows the T and T′ conformers for the tryptophan derivative W(1).

Deuterium Isotope Effects

Water quenching and excited-state proton transfer rates are readily studied and in some cases confidently assigned to individual conformers of constrained tryptophan derivatives. Excited-state proton transfer is a major fluorescence decay pathway in the tryptophan zwitterion.[7] The ammonium group of tryptophan specifically transfers a proton to C-4 of the excited indole ring, quenching the fluorescence via a seven-membered ring transition state as shown in Fig. 2. This intramolecular proton transfer is blocked if the ammonium is deprotonated or incorporated into an amide bond. Any constraint that prevents the ammonium group from adopting a conformation that could form a similar transition state should also block this decay pathway. The tryptophan derivative, W(1), has restricted rotation about the C_α–C_β and C_β–C_γ bonds and the amino group. Intramolecular excited-state proton transfer is precluded in W(1), but another unanticipated change was observed when the deuterium isotope effects on the fluorescence decay of W(1) were analyzed. In tryptophan, the ammonium group clearly enhances the solvent isotope effect on the fluorescence quantum yield and

[7] H. Shizuka, M. Serizawa, T. Shimo, I. Saito, and T. Matsuura, *J. Am. Chem. Soc.* **110,** 1930 (1988).

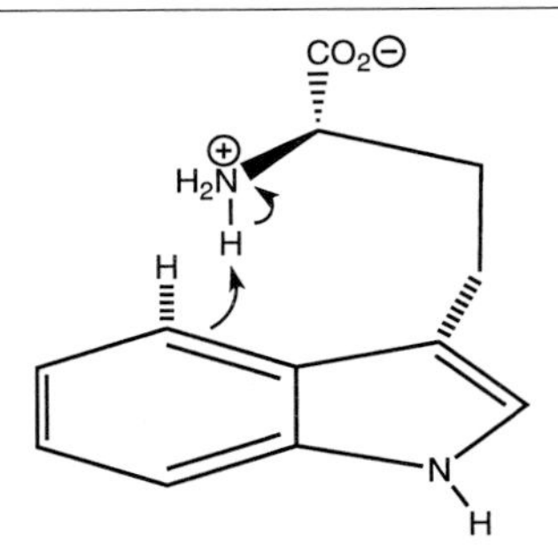

FIG. 2. Saito mechanism of intramolecular excited-state proton transfer for tryptophan.

lifetime. Figure 3 compares the pH profiles of the fluorescence quantum yield for tryptophan and W(1) in H_2O and D_2O. At pH $\leq$ 3, indole fluorescence is quenched via excited-state proton transfer from hydronium to the aromatic ring. At pH $\geq$ 11, the fluorescence is quenched by excited-state deprotonation of the indole nitrogen. *N*-Alkylated indoles are not quenched at high pH.[8] In the range pH 4–8, the ratio of the quantum yields in D_2O and H_2O is about 2 for tryptophan, decreasing to 1.4 for pH 9–10.[9] The actual isotope effect on the intramolecular excited-state proton transfer rate for pH 4–8 is greater than 2, because other isotopically insensitive decay rates are competing with the isotopically sensitive processes. The isotope effect on the fluorescence quantum yield seen at alkaline pH with tryptophan is typical of simple indoles and is pH independent from pH 3–11 for 3-methylindole. In W(1), the deuterium isotope effect is much smaller in the range pH 4–7, but increases as the ammonium group deprotonates at alkaline pH.[10]

The apparent diminished isotope effect on the fluorescence quantum yield of W(1) probably results from the constrained ammonium, which is held two bonds from C-2 of the indole ring. Presumably, the inductive effect of the ammonium reduces the excited-state basicity of the indole ring of W(1), which reduces the rate of isotopically sensitive quenching processes. Similar studies of simple indoles and derivatives of W(1) shown in Fig. 4 support the hypothesis that ammonium group proximity is responsible for the reduced isotope effect in W(1).[10] W(1a) has an ammonium group in the same position relative to the indole ring and has a reduced isotope effect in the range pH 4–8 that increases when the ammonium

[8] E. Vander Donckt, *Bull. Soc. Chim. Belg.* **78,** 69 (1969).

[9] S. S. Lehrer, *J. Am. Chem. Soc.* **92,** 3459 (1970).

[10] L. P. McMahon, W. J. Colucci, M. L. McLaughlin, and M. D. Barkley, *J. Am. Chem. Soc.* **114,** 8442 (1992).

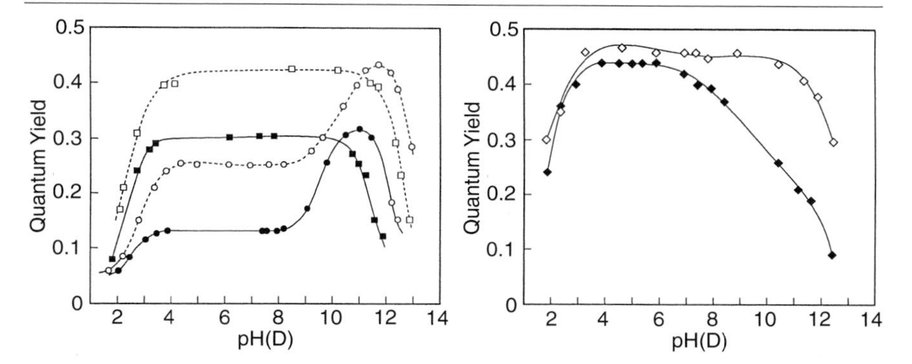

FIG. 3. Isotope effects on the fluorescence quantum yield for tryptophan,[9] 3-methylindole,[9] and W(1)[10]: W in H_2O (●), D_2O (○); 3-methylindole in H_2O (■), D_2O (□); W(1) in H_2O (◆), D_2O (◇).

group deprotonates at alkaline pH. W(1c) has the carboxylate but not the ammonium group of W(1). It behaves like THC, which shows the typical deuterium isotope effect of simple indoles in the range pH 3–11. W(2) constrains rotation about the C_α–C_β and C_β–C_γ bonds as in W(1), but does not restrict the ammonium to be proximate to the indole ring. Nevertheless, the conformational constraints block intramolecular excited-state proton transfer and the larger distance between the ammonium group and the

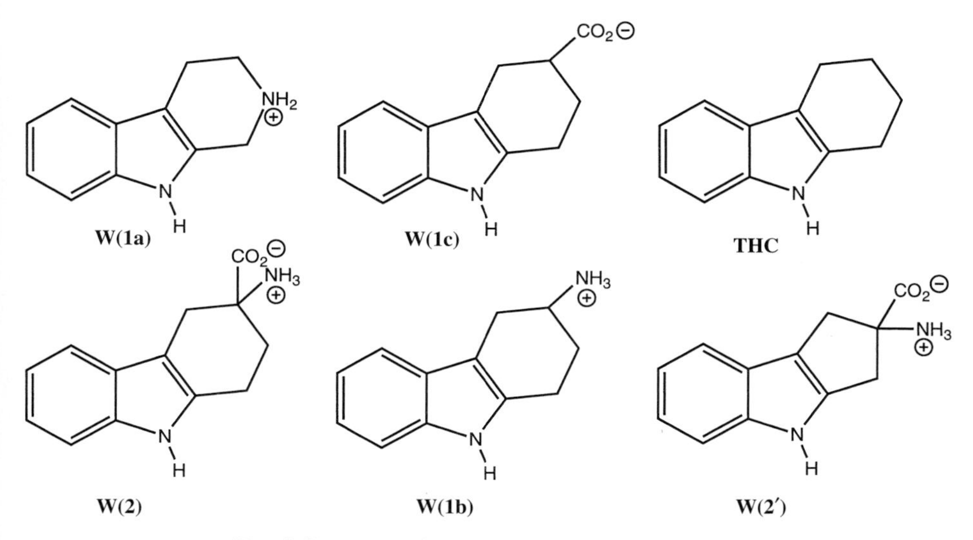

FIG. 4. Structures of constrained tryptophan analogs.

indole ring reduces the solvent isotope effect to 1.5 in the range pH 4–8. W(2) anion has a slightly increased isotope effect of 1.8.

Water Quenching

The dependence on solvent isotope can be masked by isotopically insensitive quenching mechanisms. For instance, iodide quenching reduces the apparent deuterium isotope effect on the fluorescence quantum yield and lifetime. This illustrates how important it is to isolate different quenching rates whenever possible. Fortunately, the isotopically sensitive nonradiative process that occurs in the range pH 3–11 in simple indoles as well as other tryptophan derivatives has a characteristic temperature dependence with frequency factors of 10^{15}–10^{17} sec^{-1} and activation energies of 11–13 kcal/mol. The rate of this temperature-dependent process can be calculated from the temperature dependence of the fluorescence lifetime with a single Arrhenius term in Eq. (4). The slight differences in fluorescence quantum yields in H_2O and D_2O for W(1) zwitterion and *N*-methylindole are a case in point. Indole and most other simple indoles show sizable isotope effects by simply comparing the fluorescence quantum yield in H_2O and D_2O. This is because the isotopically sensitive quenching process makes up most of the overall nonradiative rate k_{nr} in these compounds at room temperature. When the Arrhenius terms for W(1) zwitterion, *N*-methylindole, W(1) anion, and other simple indoles are compared, the isotope effects are found to be in the frequency factors, which are all two to three times greater for H_2O compared to D_2O. This suggests that all indoles undergo an intrinsic quenching process in aqueous solution that depends on solvent isotope and temperature.

Excited-State Proton Transfer

Although the discovery by Saito *et al.*[11] of intramolecular excited-state proton transfer in tryptophan is important, it is inapplicable to peptide or protein fluorescence except for N-terminal tryptophans. However, the intermolecular excited-state proton transfer reactions of indoles examined to date suggest that solvent-mediated or intramolecular excited-state proton transfer could be an important quenching mechanism in peptides and proteins. Pure and aqueous trifluoroethanol[12] and the ammonium of glycine[13]

[11] I. Saito, H. Sugiyama, A. Yamamoto, S. Muramatsu, and T. Matsuura, *J. Am. Chem. Soc.* **106,** 4286 (1984).

[12] Y. Chen, B. Liu, and M. D. Barkley, *J. Am. Chem. Soc.* **117,** 5608 (1995).

[13] H.-T. Yu, W. J. Colucci, M. L. McLaughlin, and M. D. Barkley, *J. Am. Chem. Soc.* **114,** 8449 (1992).

quench the fluorescence of simple indoles by excited-state proton transfer to aromatic carbons on the indole ring. The proton transfer rate can be measured directly from the photochemical isotope exchange monitored by NMR and mass spectrometry or indirectly from the Stern–Volmer constant. Trifluoroethanol is about 3–4 logs more acidic than water. The ammonium of glycine is even more acidic. The temperature dependence is much weaker for glycine quenching than for water quenching. The glycine-induced intermolecular excited-state proton transfer at neutral pH has a frequency factor of 10^{10} sec^{-1} and activation energy of about 4 kcal/mol. The higher frequency factors of 10^{15}–10^{17} sec^{-1} for the water-quenching mechanism are indicative of electronic movement with little nuclear movement in the transition state. The proton–hydroxide bond is lengthened, but no rehybridization of the indole ring is incumbent in the proposed water-quenching transition state. The lower frequency factors found for both intermolecular[13] and intramolecular[14] excited-state proton transfer quenching mechanisms suggest greater nuclear movement in the transition state.

The proton transfer transition state involves greater cleavage of the proton–donor bond than in the water-quenching mechanism and partial rehybridization of the atom in the indole ring that accepts the proton. Possible mechanisms for this isotopically sensitive process were considered and the mechanism that best suits the data is illustrated in Fig. 5. With water as proton donor, the transition state must have greater bonding between the transferring proton and the hydroxide. This exciplex then decomposes by back electron transfer to ground-state indole and water, because there is no detectable photochemical H–D exchange on the indole ring. On the other hand, with better proton donors there is photochemical H–D exchange. The transition state must have greater bond cleavage between the incoming proton and its donor as well as greater proton bonding to the excited indole ring, so that the proton completely transfers from donor to indole. This model unifies the isotopically sensitive quenching mechanisms for indoles and allows qualitative predictions about the likelihood of H–D photochemical exchange on the basis of the excited-state basicity of the indole ring and the acidity of solvent or quencher. This is important, because many protein side chains have pK_a values ≤ 12 that may quench tryptophan fluorescence via excited-state proton transfer at neutral pH: histidine, cysteine, tyrosine, lysine, and arginine in order of decreasing acidity. The rates of these putative excited-state proton transfer quenching processes will depend on whether the proton donor can access the indole ring during the excited-state lifetime. Even

[14] H. Shizuka, M. Serizawa, H. Kobayashi, K. Kameta, H. Sugiyama, T. Matsuura, and I. Saito, *J. Am. Chem. Soc.* **110,** 1726 (1988).

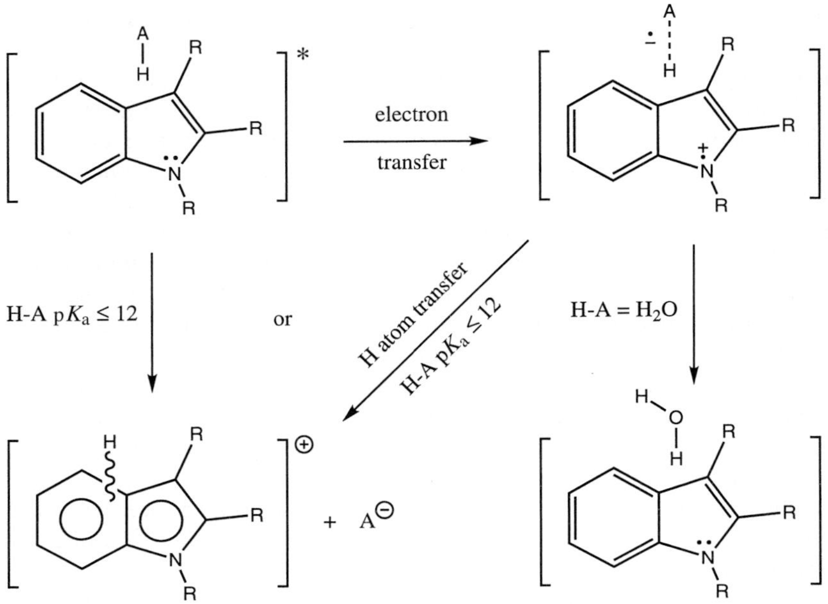

FIG. 5. Unifying mechanism for isotopically sensitive fluorescence quenching of indole derivatives.

if these functional groups have access to excited tryptophan, hydrophobic pockets in proteins may be too poorly hydrated to allow H–D exchange yet still show an isotope effect. Although water quenching and excited-state proton transfer are both temperature-dependent quenching mechanisms, the temperature dependence is sufficiently different to resolve in a few cases.

Conformer Interconversion

The work with constrained tryptophan derivatives eliminated intramolecular excited-state proton transfer and produced other dividends. By reducing the number of conformers to two, we were able to dissect the different nonradiative processes and to devise ways of measuring and studying them individually. W(1)[15,16] and W(2)[3] have two stable ground-state conformations with relative populations that correspond to the amplitudes

[15] L. Tilstra, M. C. Sattler, W. R. Cherry, and M. D. Barkley, *J. Am. Chem. Soc.* **112,** 9176 (1990).

[16] W. J. Colucci, L. Tilstra, M. C. Sattler, F. R. Fronczek, and M. D. Barkley, *J. Am. Chem. Soc.* **112,** 9182 (1990).

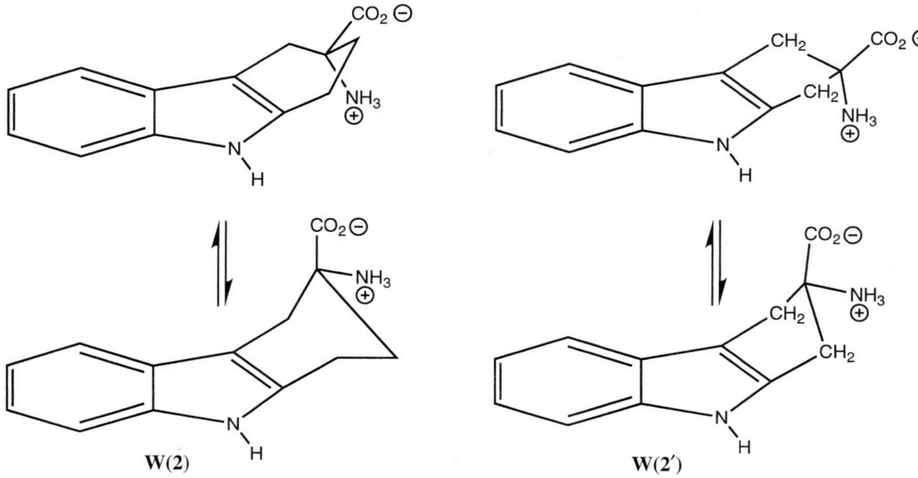

FIG. 6. T and T′ conformers of W(2) and the envelope conformers of W(2′).

of a biexponential fluorescence decay. The assignments of fluorescence lifetimes to conformers on the basis of ground-state populations were consistent with microenvironmental differences in the two conformers. Molecular mechanics calculations showed that the energy barrier should have been sufficient to prevent conformer interconversion during the lifetime of the excited state. These experiments confirm that the complex decays seen for tryptophan can reflect ground-state heterogeneity that leads to excited-state heterogeneity. Time-resolved fluorescence and NMR studies of oxytocin with a single tryptophan also correlated fluorescence lifetimes with specific side-chain rotamers.[17] Jet expansion studies showed that individual tryptophan conformers have single fluorescence lifetimes.[18,19] All nonradiative pathways examined so far except conformer interconversion have nonfluorescent products.

While the two individual conformers of W(1) and W(2) shown in Figs. 1 and 6 apparently decay as separate species, conformer interconversion can be much faster than the fluorescence time scale. The biexponential fluorescence decay of the two conformers should collapse to a single expo-

[17] J. B. A. Ross, H. R. Wyssbrod, R. A. Porter, G. P. Schwartz, C. A. Michaels, and W. R. Laws, *Biochemistry* **31,** 1585 (1992).

[18] J. Sipior, M. Sulkes, R. Auerbach, and M. Boivineau, *J. Phys. Chem.* **91,** 2016 (1987).

[19] L. A. Philips, S. P. Webb, S. J. Martinez III, G. R. Fleming, and D. H. Levy, *J. Am. Chem. Soc.* **110,** 1352 (1988).

nential with a dynamically averaged fluorescence lifetime. This is inherently difficult to demonstrate, but we believe that W(2′) provides an example.[20] The five-membered ring of W(2′) has two envelope conformers shown in Fig. 6 that should interconvert on the picosecond time scale according to molecular mechanics calculations. Unlike W(2), W(2′) exhibits a single exponential decay.

Conformer interconversion can also be competitive with the fluorescence decay of individual conformers. The effect of competing rates for conformer inversion and deactivation of the excited state is more involved than the well-studied dynamic averaging seen in NMR spectroscopy. Many examples from NMR spectroscopy show that, as the temperature is raised, two distinct types of protons coalesce and then develop a dynamically averaged signal. With time-resolved fluorescence, the fluorescence decay is a sum of exponentials with decay times and preexponentials that depend on both conformer lifetimes and interconversion rates.[21] A negative amplitude or a rise time in the fluorescence decay is the hallmark of an excited-state reaction with a rate competitive with the fluorescence decay rate. Although Donzel *et al.*[22] did not report a negative amplitude, they analyzed fluorescence decay data of tryptophanyl diketopiperazines assuming a two-state excited-state reaction for the conformer interconversion. All of the constrained tryptophan derivatives described herein have two ground-state conformers. Figure 7 shows the two-state excited-state scheme used to analyze the fluorescence decay data of constrained tryptophan derivatives W(1b) and W(1c), for which the rates of fluorescence decay and conformer inversion are comparable. For W(1c), the fluorescence decay rates are faster than conformer inversion rates in D_2O. The water-quenching rate becomes even faster in H_2O, making conformer inversion less competitive and more difficult to resolve. For W(1b), conformer inversion rates are somewhat faster than fluorescence decay rates in H_2O and D_2O, so that the excited-state reaction is a competitive nonradiative process in both solvents. The results suggest that relative populations of ground-state conformers will not always correlate with excited-state conformer populations. Moreover, a negative amplitude may not always be apparent in cases such as the constrained tryptophans, in which the two species in an excited-state reaction have highly overlapped emission spectra. Model calculations show that the appearance of a short decay time with a small preexponential factor is

[20] L. P. McMahon, H.-T. Yu, M. A. Vela, G. A. Morales, L. Shui, F. R. Fronczek, M. L. McLaughlin, and M. D. Barkley, *J. Phys. Chem.* 1997 (in press).

[21] N. Mataga and M. Ottolenghi, *in* "Molecular Association" (R. Foster, ed.), Vol. 2, p. 1. Academic Press, New York, 1979.

[22] B. Donzel, P. Gauduchon, and P. Wahl, *J. Am. Chem. Soc.* **96,** 801 (1974).

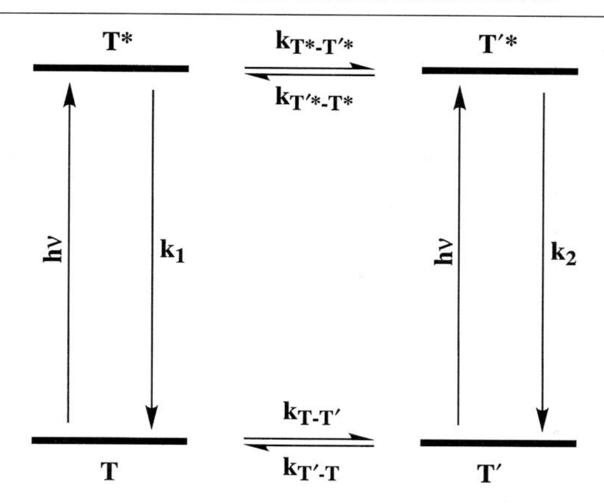

FIG. 7. Two-state excited-state reaction scheme.

also a signal of an excited-state reaction on the fluorescence time scale. Multiexponential fluorescence decays with a minor component of short lifetime have been observed for tryptophans in proteins.[1] In such cases, anisotropy experiments should establish whether the lifetime heterogeneity could be explained by conformer interconversion.

Conclusion

Nonradiative decay pathways that compete with emission for deactivation of the excited chromophore are key to fluorescence as a structural tool. In the case of tryptophan, these include a number of environmentally sensitive processes, such as solvent quenching and excited-state proton and electron transfer reactions. Our strategy has been to design simple tryptophan derivatives that allow us to relate excited-state properties directly to structure. These fundamental studies of tryptophan photophysics have broad implications by providing a framework for interpreting protein fluorescence. For example, knowing the identity and quenching mechanism of protein functional groups will greatly improve our capacity to draw structural conclusions from fluorescence data. These methods establish criteria for determining the important nonradiative processes and in favorable cases for quantifying individual nonradiative rates. For example, water quenching rates can be monitored independently of excited-state proton transfer rates. We continue to synthesize new tryptophan derivatives in our ongoing effort to unravel the multiple pathways of tryptophan photophysics.

Acknowledgments

Research was supported by NIH Grant GM42101. We thank our collaborators Drs. Yu Chen, William R. Cherry, William J. Colucci, Frank R. Fronczek, Guillermo A. Morales, Li Shui, Luanne Tilstra, Marco A. Vela, and Hong-Tao Yu, Mr. Bo Liu, Mr. Lloyd P. McMahon, and Ms. Melissa C. Sattler.

[10] Conformational Heterogeneity in Crystalline Proteins: Time-Resolved Fluorescence Studies

By TANYA E. S. DAHMS and ARTHUR G. SZABO

Introduction

Single tryptophan (Trp)-containing proteins have often been shown to exhibit multiexponential fluorescence decay kinetics in solution.[1,2] In many of these cases, the heterogeneous decay behavior observed for single-Trp proteins was attributed to different local environments of the indole ring as a result of alternate protein conformations.[1] The observation of single-exponential fluorescence decay kinetics for a single-Trp protein in solution is rare. A single decay time has been reported for apoazurin, nuclease B, melittin,[1] ribonuclease T_1[2] (pH 5.0–5.5), and F102W parvalbumin.[3] In each case, the single-Trp indole ring is buried in a motion-restricted environment of the protein. Increasingly, single-Trp mutants of selected proteins have been investigated by time-resolved fluorescence methods,[4–6] whereby the Trp residue has been site-specifically incorporated into different segments of the protein. The interpretation of the fluorescence decay behavior of Trp in terms of local structural elements remains to be clearly defined.

The observation of multiexponential decay from single-Trp proteins has more recently been rationalized in terms of Trp conformational heterogeneity originating from different ground-state, side-chain rotamer conforma-

[1] J. M. Beechem and L. Brand, *Annu. Rev. Biochem.* **54,** 43 (1985).

[2] M. R. Eftink, *in* "Protein Structure Determination" (C. H. Suelter, ed.), p. 127. John Wiley & Sons, New York, 1990.

[3] T. L. Pauls, I. Durussel, J. A. Cox, I. D. Clark, A. G. Szabo, S. M. Gagné, B. D. Sykes, and M. W. Berchtold, *J. Biol. Chem.* **28,** 20897 (1993).

[4] D. L. Harris and B. S. Hudson, *Biochemistry* **29,** 5276 (1990).

[5] C. M. L. Hutnik, J. P. MacManus, D. Banville, and A. G. Szabo, *Biochemistry* **30,** 7652 (1991).

[6] M. C. Kilhofer, M. Kubina, F. Travers, and J. Haiech, *Biochemistry* **31,** 8098 (1992).

0076-6879/97 $25

tions.[7,8] The "rotamer model" for Trp in a protein proposes that the Trp side chain may adopt low-energy conformations in the ground state, owing to rotation about the C_α–C_β and/or the C_β–C_γ bonds,[9] with each conformation displaying a distinct decay time. Each decay time would correspond to a unique protein environment for the indole ring and the decay time relative proportions would reflect the ground-state population of each low-energy conformation. The rotamer model originated from studies on Trp in a diketopiperazine,[10] and from studies utilizing Trp zwitterion,[11,12] and was shown to be applicable to fluorescence decay studies of small peptides, in which the relative proportions of each decay component corresponded to the individual rotamer populations measured by nuclear magnetic resonance (NMR).[7] Potential energy minimization calculations[9] revealed three low-energy χ_1 Trp conformers effected by rotation about the C_α–C_β bond (rotation about the peptide backbone) and two low-energy χ_2 Trp conformers that correspond to a rotation about the C_β–C_γ bond (Trp ring flip[13]).

The fluorescence lifetime and quantum yield of Trp in a protein show significant variation, which corresponds to the many possible quenching contributions from a proteinaceous environment. Certain species have been found to be responsible for quenching of Trp, such as metal ions, heme groups and other cofactors,[14] the charged groups of histidine, aspartate, and glutamate,[15] or nearby cysteine and disulfides.[16] By analyzing crystallographic structures, Burley and Petsko[17] observed that the (partially) positively charged amino groups of lysine, arginine, asparagine, glutamine, and histidine are preferentially located within 6 Å of the phenylalanine, tyrosine, and tryptophan ring centroids, where they are in van der Waals contact with the delta-negative π electrons, thus avoiding the delta-positive ring edge. Despite the efforts to identify potential Trp fluorescence quenching species in proteins, the interpretation of relative quantum yields and fluorescence lifetime values remains unclear.

[7] J. B. A. Ross, H. R. Wyssbrod, R. A. Porter, G. P. Schwartz, C. A. Michaels, and W. R. Laws, *Biochemistry* **31,** 1585 (1992).

[8] K. J. Willis and A. G. Szabo, *Biochemistry* **31,** 8924 (1992).

[9] H. L. Gordon, H. C. Jarrell, A. G. Szabo, K. J. Willis, and R. L. Somorjai, *J. Phys. Chem.* **96,** 1915 (1992).

[10] B. Donzel, P. Gauduchon, and Ph. Wahl, *J. Am. Chem. Soc.* **96,** 801 (1974).

[11] A. G. Szabo and D. M. Rayner, *J. Am. Chem. Soc.* **102,** 554 (1980).

[12] M. C. Chang, J. W. Petrich, D. B. MacDonald, and G. R. Fleming, *J. Am. Chem. Soc.* **105,** 3819 (1983).

[13] K. J. Willis, A. G. Szabo, and D. T. Krajcarski, *Chem. Phys. Lett.* **182,** 614 (1991).

[14] R. W. Cowgill, *Arch. Biochim. Biophys.* **100,** 36 (1963).

[15] R. W. Cowgill, *Biochim. Biophys. Acta* **133,** 6 (1967).

[16] R. W. Cowgill, *Biochim. Biophys. Acta* **207,** 556 (1970).

[17] S. K. Burley and G. A. Petsko, *FEBS Lett.* **203,** 139 (1986).

For protein crystals, the electron density of the tryptophan (Trp) side chain (when compared with other residues) is considered to be well defined by crystallographic criteria and is almost exclusively modeled as a single conformer (Brookhaven Data Bank[18]). Therefore, one might predict single-exponential decay kinetics for an individual tryptophan residue in a crystalline protein. In this case, measuring the fluorescence parameters of single protein crystals could provide a direct comparison between the fluorescence lifetime of the Trp residue and local structural features of its environment, as described by the crystallographic model. Such an approach could potentially provide a better understanding of the factors governing Trp fluorescence in solution and increase the information available from protein fluorescence parameters. The first time-resolved Trp fluorescence measurements, from the crystalline heme protein myoglobin, demonstrated the feasibility of accurately measuring Trp fluorescence decay parameters from protein crystals.[19]

When selecting the proteins for this particular project, an attempt was made to eliminate contributions from nonproteinaceous fluorescent moieties (i.e., heme groups, vitamins, cofactors, metals). Erabutoxin b was the first nonheme, single-Trp protein to be studied in the crystalline state by time-resolved fluorescence spectroscopy.[20] Those experiments indicated alternate Trp rotamers in the crystalline state, which had not been detected by X-ray crystallography, providing strong support for the rotamer model of Trp photophysics as it relates to aqueous proteins. In this chapter we provide the details associated with measuring protein crystals by time-resolved fluorescence, a summary of the fluorescence data collected for erabutoxin b, crotonase, apo- and holoazurin, and a brief summary of the implications arising from those experiments.

Methods

Protein Isolation and Identification

Erabutoxin b is either isolated from snake venom (*Lacticauda semifasciata*) according to published procedures[21] or purchased in a fully purified form (Sigma Chemical Co., St. Louis, MO). The crotonase expression system is a kind gift of P. Tonge (NRC, Ottawa, Canada) and *Pseudomonas*

[18] F. C. Bernstein, T. F. Koetzle, G. J. B. Williams, E. F. Meyer, Jr., M. D. Brice, J. R. Rodgers, O. Kennard, T. Shimanouchi, and M. Tasumi, *J. Mol. Biol.* **112,** 535 (1977).

[19] K. J. Willis, A. G. Szabo, and D. T. Krajcarski, *J. Am. Chem. Soc.* **113,** 2000 (1991).

[20] T. E. S. Dahms, K. J. Willis, and A. G. Szabo, *J. Am. Chem. Soc.* **117,** 2321 (1995).

[21] T. E. S. Dahms and A. G. Szabo, *Biophys. J.* **69,** 569 (1995).

TABLE I
ELECTROSPRAY IONIZATION MASS SPECTRAL ANALYSIS OF HIGH-PERFORMANCE LIQUID CHROMATOGRAPHY-PURIFIED ERABUTOXIN b, CROTONASE, AND APOAZURIN

	Molecular mass		
Protein	Calculated[a] (g/mol)	ESIMS (Da)	Difference (%)
Erabutoxin b	6,860.8	6,860.1	0.01
Crotonase	28,287.0	28,290.3	0.01
Apoazurin	13,625.6	13,627.6	0.00[b]

[a] Isotope average calculated molecular mass.
[b] The percentage difference, between the theoretical and measured molecular mass, was calculated following the addition of 2.016 to the theoretical molecular mass to account for the addition of two hydrogens (refer to text).

fluorescens (*Pfl,* ATCC 13525) azurin is isolated directly from the bacterial source according to published procedures.[22]

Time-resolved fluorescence has proven to be a highly sensitive method for monitoring conformational properties of proteins. *In vivo* concentrations in the nanomolar to micromolar range are often more than sufficient for an adequate fluorescence signal, meaning that even trace amounts of fluorescent contaminants will contribute to the fluorescence data. To establish the highest sample homogeneity, anion- or cation-exchange (DEAE- or CM-5PW, TSK; Supelco, Toronto, Ontario, Canada) high-performance liquid chromatography (HPLC) is the ultimate purification step for all proteins studied. All HPLC fractions are pooled, and a small sample is reanalyzed by HPLC to ensure the homogeneity of the pooled material. The purity and identity of the protein are subsequently evaluated by nebulization-assisted electrospray ionization (ESI) mass spectrometry (API III quadrupole; Sciex, Mississauga, Ontario, Canada) using approximately a 0.5-mg/ml protein solution in 10% acetic acid–water (Table I).

Copper-free (apo)azurin is obtained using the modified KCN dialysis method as described elsewhere.[22] The complete absence of an absorption band, which peaks at approximately 625 nm, is used to verify successful Cu^{2+} removal. The preparation of apoazurin is carried out in a fume hood owing to the production of toxic HCN gas.

[22] C. M. Hutnik and A. G. Szabo, *Biochemistry* **28,** 3923 (1989).

Protein Crystallization

All crystals are grown by vapor diffusion methods from hanging or sitting droplets.

Crystals of erabutoxin b are grown from a 1:1 mixture of aqueous protein (26.3 mg/ml) with 1.4–1.48 *M* ammonium sulfate as the reservoir solution.[20,23]

Crotonase crystals are grown from equal volumes (5 μl) of crotonase (5.9 mg/ml) and 15–22% ethanol (v/v) in 0.1 *M* $NaKHPO_4$ (pH 7.5) as the reservoir solution.

Pfl holoazurin crystals are grown from equal volumes of HPLC peak 2 (see Results) concentrated to 8.2 mg/ml and 2.9–3.0 *M* ammonium sulfate in Tris buffer (200 m*M*, pH 7.5), according to published procedures.[24] A higher protein concentration is required to grow the crystals from HPLC-pure material as compared to that which had been prepared according to Hutnik and Szabo.[22] Apoazurin crystals are grown under identical vapor diffusion conditions as the holoprotein, using a protein concentration of 9.2–10.0 mg/ml.

Fluorescence

The fluorescence measurements of buffered, aqueous protein samples are collected from protein samples directly eluted from the HPLC. An analogous fraction from an HPLC "blank run" is used as the fluorescence blank.

The measurement of protein crystals by fluorescence involves special considerations owing to the small size and fragile nature of the crystalline material. There is no need to measure a buffer blank because the crystals contain such a small amount of buffer solution, and as well, the fluorescence signal from the protein crystal is intense. The novel procedures that have been developed for the acquisition of fluorescence data from crystalline proteins are outlined in the following sections.

Steady-State Fluorescence. The large rectangle-shaped crystals of the myoglobins used in previous studies (Willis *et al.*[13]) allow steady-state fluorescence measurements to be made using the SLM 8000 (SLM Instruments, Urbana, IL) fluorimeter. This is accomplished by aligning several crystals side by side in a rectangular cell (0.1 × 10 × 30 mm) mounted in a triangular cell holder such that the cell face is oriented at 45° to the excitation source, and detection optics. The protein crystals used in the present studies are

[23] D. Tsernoglou and G. A. Petsko, *FEBS Lett.* **68,** 1 (1976).

[24] D. W. Zhu, T. Dahms, K. Willis, A. G. Szabo, and X. Lee, *Arch. Biochem. Biophys.* **308,** 469 (1994).

either too small and fragile for such a manipulation, or alignment is not possible owing to the shape of the crystal.

The steady-state fluorescence of protein crystals is successfully measured using the time-resolved instrument in the following manner. The fluorescence intensity, at a series of emission wavelengths, is determined by integration of the fluorescence decay curve for a fixed time period (120 sec). The intensity is subsequently corrected for the wavelength-dependent sensitivity of the instrument by measuring the fluorescence spectrum of an aqueous solution of *N*-acetyltryptophan amide, pH 7 (as the fluorescence standard), on both the time-resolved instrument and the SLM 8000 (fully corrected steady-state spectrum).

Time-Resolved Fluorescence. Time-resolved fluorescence measurements are performed using time-correlated single-photon counting with laser/microchannel plate-based instrumentation. The procedures and instrumentation used for steady-state, and time-resolved fluorescence measurements of aqueous protein samples are detailed elsewhere.[25]

A crystal mounted in a quartz capillary tube was tested during the first attempt to measure time-resolved fluorescence parameters from single myoglobin crystals, and, would have facilitated the measurement of crystallographic and fluorescence parameters from a single crystal. The regular curvature of the capillary tube resulted in an intense scatter that impeded data analysis.[26] The rectangular cells, described by Willis *et al.*,[19] were used for the preliminary time-resolved fluorescence measurements of erabutoxin b crystals. However, the sides of the cells produced strong scatter at any orientation that significantly deviated from the vertical cell position, and thus, the rectangular cells were inappropriate for the detailed orientation experiments presented herein. The following section describes the cell and apparatus that were specifically designed to measure steady-state and time-resolved fluorescence parameters from single protein crystals at multiple orientations with respect to the vertically polarized excitation source.

Single crystals, washed with protein-free mother liquor, are placed centrally on the inside front face of a long-necked, round circular dichroism (CD) cell (width, 2 mm; diameter, 2 cm) containing 50 μl of protein-free mother liquor to provide solvent atmosphere. Any liquid directly in contact with the crystal is removed using thin absorbant strips of paper. The cell is sealed with Parafilm and mounted on a device consisting of a goniometer attached to an *x, y, z* translation stage (Fig. 1), such that the front face of the cell is perpendicular to the goniometer rotational

[25] K. J. Willis and A. G. Szabo, *Biochemistry* **28,** 4902 (1989).

[26] K. J. Willis, unpublished results (personal communication, 1990).

FIG. 1. Photographic representation of the apparatus utilized for time-resolved fluorescence measurements of single protein crystals. A 1-mm CD cell is mounted on a device consisting of a goniometer head attached to an *x, y, z* translation stage.

axis. The apparatus is designed so that the crystal, when placed in the center of the cell, is in the axis of rotation of the goniometer. The cell clamp with plastic screw ensures that the crystal position can be maintained after cell removal and subsequent replacement. The rod attached to the cell clamp is made of sufficient length to assure that both the goniometer and goniometer rod do not interfere with either the excitation beam or fluorescence signal. In either of the latter cases, the fluorescence data would be complicated by a scatter function.

The crystal is first aligned with the laser beam by eye, using the *x, y, z* translations until a spot of fluorescent marker, placed on the outside of the cell directly above the crystal, is illuminated. The fluorescent marker is carefully washed from the outside of the cell, using spectral-grade methanol (3×). As a control experiment fluorescent marker is also placed onto the outside of an empty cell and subsequently removed (as previously described), to ensure that no residual fluorescent material is contributing to the fluorescence signal (data not shown). The fluorescence signal from the protein crystal is optimized by monitoring the total photon counts per second while adjusting the crystal position in the *x, y,* and *z* directions.

The excitation beam is incident on the cell at an angle of 35° to the normal of the cell face to prevent specularly reflected light from entering the detection optics.[27]

The data are analyzed as a sum of exponentials with discrete lifetimes, according to the integrated rate equation:

$$I(\lambda, t) = \sum_{i=1}^{n} \alpha_i(\lambda)e^{-t/\tau_i} \tag{1}$$

where I is intensity, λ is wavelength, t is time, α_i is the normalized preexponential term, and τ_i is the fluorescence decay time.

Orientation Experiments. Experiments measuring orientation effects of the fluorescence decay behavior are performed by rotating the crystal (at 10° rotation intervals) with respect to the vertical polarization plane of the excitation beam. After several data sets are collected for the same crystal, the sample is returned to the original orientation and the fluorescence decay parameters are again measured to assess whether crystal photodamage has occurred. If the parameters obtained are significantly different from those originally measured, then the data sets collected between the previous reproducible data set and the nonreproducible data set are discarded, and the experiment terminated.

Results

Protein Identity, Purity, and Crystallization

Erabutoxin b. It was necessary to evaluate the purity of the protein crystals, because the protein was sometimes crystallized from toxin (Sigma purified) without further purification. Erabutoxin b produced rod-shaped crystals within 24 hr (Fig. 3A) that were >99% pure by HPLC (Fig. 2A). The molecular weight of the crystalline material was identical to that found for the HPLC-pure material (Table I). The erabutoxin crystals were of diffraction quality; however, determination of all three crystallographic cell parameters was impeded by their small size (0.2 × 0.01 mm) and a limitation of the X-ray instrumentation available to the authors. Two unit cell dimensions were consistent with the original X-ray data.[23]

[27] J. Eisinger and J. Flores, *Anal. Biochem.* **94,** 15 (1979).

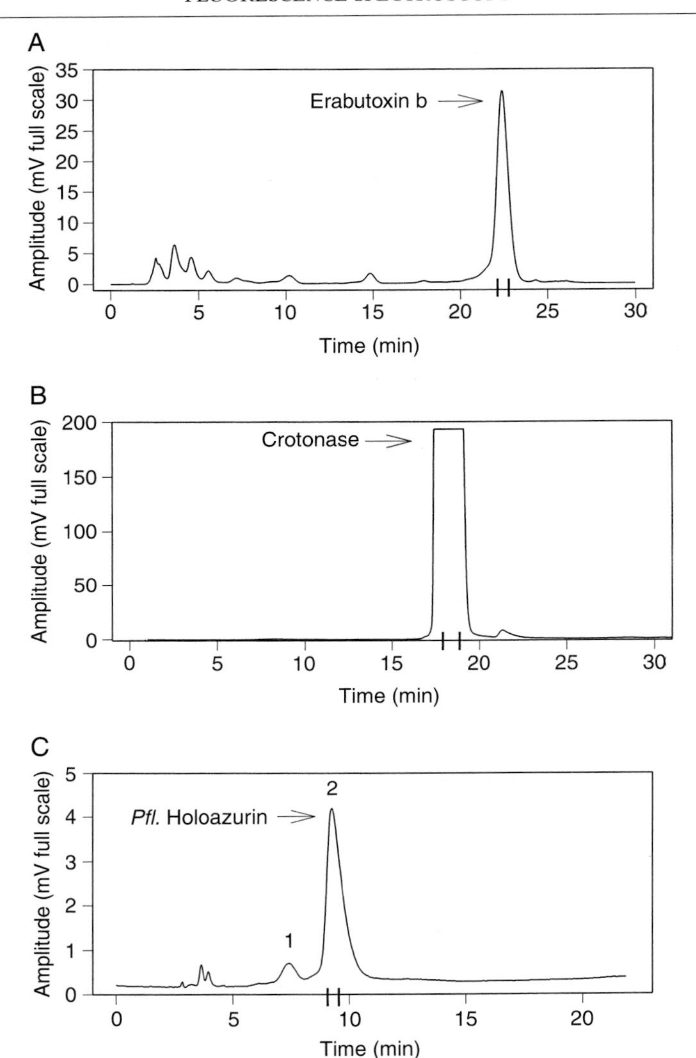

FIG. 2. HPLC profiles: (A) CM batch-treated venom from *Lacticauda semifsciata* (see Dahms and Szabo[21]); erabutoxon b]. Buffer A: 10 m*M* sodium acetate, pH 6.5. Buffer B: Buffer A plus 0.19 *M* NaCl. Gradient: 0–2 min (0% B), 2–20 min (0–15% B), 20–25 min (15–100% B). (B) Affinity-treated, overexpressed crotonase from *Escherichia coli* (see Dahms[28]). Buffer A: 10 m*M* KH_2PO_4, pH 6.0. Buffer B: Buffer A plus 0.5 *M* KCl. Gradient: 0–10 min (0% B), 10–25 min (0–100% B), 25–30 min (100% B). (C) CM-treated extract from *Pseudomonas fluorescens* (see Dahms[28]). Buffer A: 10 m*M* sodium acetate, pH 4.0. Buffer B: Buffer A plus 0.19 *M* NaCl. Gradient: 0–2 min (30% B), 2–20 min (30–45% B), 20–25 min (45% B). The collected fraction is indicated by the vertical boundary lines (λ_{max} = 280 nm).

Crotonase. The details of crotonase purification have been detailed elsewhere.[28] Following the affinity purification (crotonyl-CoA) of overexpressed crotonase and precipitation with ethanol, the enzyme was >99.9% pure by HPLC (Fig. 2B) and the measured molecular weight was within 0.01% of the calculated molecular weight (Table I). Both sequencing of the gene and N-terminal amino acid analysis confirmed the correct sequence of the enzyme.[28] The enzyme was homogeneous by activity assay with crotonyl-CoA.[29]

Crotonase crystals grew within 1 week (Fig. 3B) and diffracted to approximately 2.8 Å. The unit cell parameters of the recombinant crotonase are as follows: $a = b = 68.0$ Å, and $c = 214.2$ Å, with two hexamers in the crystallographic unit cell. The space group was determined to be $P6_222$ or $P6_422$ with one hexamer in the asymmetric unit.

Azurins. A previous report from this laboratory,[22] which employed a similar purification procedure (excluding HPLC), determined a 1:1 copper–protein complex by atomic absorption and amino acid analysis for *Pfl* holoazurin with a spectral ratio of 0.50 or greater (625 nm/280 nm). The spectral ratio of the protein is used as a criterion with which to establish purity of the holoform. The highest spectral ratio previously reported for *Pfl* holoazurin[22] was 0.55, and this was considered homogeneous.

The HPLC-purified holoazurin (Fig. 2C) absorption spectrum displays maxima at approximately 280 and 625 nm, with no evidence of a Soret band (420 nm, absorption from contaminating cytochromes), and the spectral ratio (0.58–0.6) indicated a homogeneous protein preparation that was superior to that previously reported.[22] The holoazurin amino acid sequence has been determined in conjunction with this work[30] and the isoelectric point determined to be 6.7. This is in accordance with the value approximated by isoelectric focusing.[22] The sequence of *Pfl* holoazurin is reported elsewhere.[30] The mass spectral data for this peak are in excellent agreement with the molecular weight calculated from the sequence (Table I). The 2-mass unit difference is expected, because the protein used for mass spectral analysis had been reduced, and the native protein has one disulfide linkage, giving a theoretical difference of 2.016 Da. Adding this difference to the mass measured by electrospray ionization-mass spectrometry (ESIMS) gives an observed mass of 13,625.6 Da, which is in exact agreement with the sequence data (Table I).

[28] T. E. S. Dahms, Probing Protein Secondary Structure and Conformational Heterogeneity (Ph.D. Thesis), University of Ottawa Press, Ottawa, Ontario, Canada, 1996.

[29] H. M. Steinman and R. L. Hill, *Methods Enzymol.* **35,** 136 (1975).

[30] X. Lee, T. E. S. Dahms, D. W. Zhu, P. Lanthier, M. Yaguchi, and A. G. Szabo, *Crystallogr. Acta* 1997 (in press).

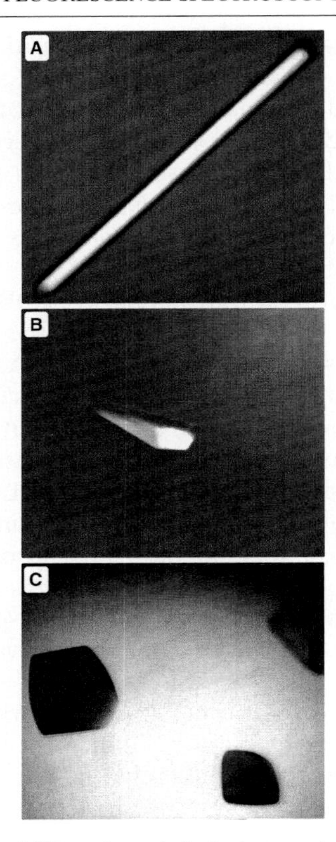

FIG. 3. Microphotograph of (A) erabutoxin b single crystals (longest dimension, 0.2 mm); (B) crotonase crystals (largest dimension, 0.6 mm); (C) *Pfl* holoazurin crystals (largest dimension, 0.6 mm). The apoazurin crystals were identical in shape to the holoazurin crystals, but were smaller (largest dimension, 0.2 mm) and colorless owing to the absence of Cu^{2+}.

Large crystals were obtained for both the holo- ($0.6 \times 0.6 \times 0.2$ mm) and apoprotein ($0.2 \times 0.2 \times 0.07$ mm) after allowing them to sit at room temperature for 10 days (Fig. 3C). The cell parameters have been measured for both the holo- (a = 31.95 Å, b = 43.78 Å, c = 78.81 Å) and apoprotein (a = 31.91 Å, b = 43.85 Å, c = 79.19 Å) and can be considered identical. The larger holoazurin crystals diffracted to 2.05 Å, whereas the smaller apoazurin crystals diffracted to 2.4 Å.

Fluorescence

Erabutoxin b. Only the details pertinent to fluorescence data collection from single protein crystals will be included herein (for other details see Dahms *et al.*[20]).

The excitation of Trp-29 from a single erabutoxin b crystal produced a significant fluorescence signal. The photon count rate was dependent on crystal size and orientation. When there was a decrease in count rate of greater than 15% during the experiment, a scatter element was detected in the data and consequently the decay parameters. This fluorescence behavior was indicative of crystal photodamage and could be correlated with crystal damage as observed by crystal cracking and loss of birefringence (viewed under a microscope equipped with polarizing filters).

Contrary to expectations, triple-exponential decay kinetics were observed for erabutoxin b in the crystalline state (Table II). The fluorescence decay times could be accurately obtained from the global analysis of a combination of 20 data sets from measurements made at 10 different emission wavelengths (305–400 nm) and 10 angular orientations of a single crystal.

Crotonase. The fluorescence of crystalline crotonase is best described by triple exponential decay kinetics (Table II), suggesting that three different Trp rotamers exist within different molecules of crotonase in the crystal.

Azurins. Apoazurin crystals, as in solution, exhibit a spectral maximum at 308 nm and display single exponential decay kinetics (Table II). Preliminary investigation of the crystalline apoazurin revealed double-exponential decay kinetics with a long lifetime (2.82 nsec) and an intermediate decay time value (0.86 nsec). Given the similarity between these values and those measured for the crystalline holoazurin, it was postulated that the apoazurin crystals had leached Cu^{2+} from the siliconized microscope slides used in the vapor diffusion process. This hypothesis was tested by placing a sample of aqueous apoazurin, which had displayed single exponential decay kinetics, into a glass test tube that was stored in the refrigerator for several weeks (crystallization time). The resultant material displayed double-exponential decay kinetics and exhibited a small 625-nm absorbance. When the apoazurin crystals were grown on plastic microscope slides using crystallization buffers that had been pretreated with Chelex, this problem was avoided and single-exponential decay kinetics were observed (Table II). The detection of such a small contaminant demonstrates the high sensitivity of the single photon-counting method using laser-based instrumentation for fluorescence measurements. The lifetime value for crystalline apoazurin (3.786 nsec) is shorter than that observed in solution (4.683 nsec).

The global analysis of crystalline holoazurin at eight wavelengths (300–350 nm) revealed that the best statistical fit was to four decay times (Table II).

Orientation Experiment

The orientation experiments were designed on the basis of the following reasoning: The absorption of the Trp residue is dependent on the angular

TABLE II
TIME-RESOLVED TRYPTOPHAN FLUORESCENCE OF ERABUTOXIN b, CROTONASE, APOAZURIN, AND HOLOAZURIN[a]

Protein	τ_1[b] (nsec)	τ_2 (nsec)	τ_3 (nsec)	τ_4 (nsec)	λ_{max}[c] (nm)	SVR[d]	σ[d] (χ^2)
Erabutoxin b							
Solution	3.64	1.171	0.276		340	1.81	1.09
Crystal	1.78	0.734	0.113		340	1.71	1.12
Crotonase							
Solution	5.57	2.06	0.51		325	1.8	1.09
Crystalline	4.92	2.55	0.63		325	1.9	1.07
Holoazurin							
Solution	4.49	0.46	0.101		308	1.91	1.02
Crystalline	2.769	1.36	0.167	0.0537		1.80	1.08
Apoazurin							
Solution	4.683					1.88	1.06
Crystalline	3.786					1.90	1.07

[a] In solution and in the crystalline state: Fluorescence decay times from the global analysis of multiple emission wavelengths (20°, pH 7).

[b] The fluorescence lifetime values of erabutoxin b (λ_{ex} = 295 nm), crotonase (λ_{ex} = 300 nm), and holoazurin (λ_{ex} = 292 nm) in solution were generated by the global analysis of a multiple wavelength experiment (12 data sets, 11 and 12 data sets, and 8 data sets, respectively). The fluorescence lifetimes of the erabutoxin b crystal were determined by the global analysis of combined data from a multiple (10 data sets) wavelength and a multiple orientation (10 data sets) experiment. Typical errors associated with τ_1, τ_2, τ_3 were ±0.02, 0.002, and 0.004 for erabutoxin b in solution, ±0.002, 0.004, and 0.002 for erabutoxin b in the crystalline state, ±0.02, 0.007, and 0.006 for crotonase in solution and in the crystalline state, ±0.04, 0.04, and 0.005 for holoazurin in solution, ±0.005, 0.01, 0.001, and 0.0005 (τ_4) for crystalline holoazurin, and ±0.005 for crystalline apoazurin.

[c] The emission maximum for crystalline holoazurin was obscured by the Raman scatter and that of apoazurin was not measured in this study.

[d] The serial variance ratio (SVR) provides a measure of the correlation between successive residuals. A good statistical fit corresponds to SVR values between 1.7 and 2.0 [A. E. McKinnon, A. G. Szabo, and D. R. Miller, *J. Phys. Chem.* **81,** 1564 (1977)]. The sigma value (σ) is defined as the square root of the reduced χ^2 and an optimal fit is indicated when σ is equal to unity.

orientation of the excited-state transition moment with respect to the polarization direction of the excitation beam. Maximum absorption will occur when the transition dipole is aligned with the excitation plane of polarization. At 295 nm, the L_a transition of Trp is likely the dominant one, with the L_b transition making only a minor contribution to the absorbance.[31] Therefore, the fluorescence should be primarily from the L_a singlet state.

[31] B. Valeur and G. Weber, *Photochem. Photobiol.* **25,** 441 (1977).

The fluorescence intensity is proportional to the square of the absorption transition probability, and thus the fluorescence intensity of a fluorophore should vary with an angular dependence of its dipole with respect to the polarization direction (vertical) of the excitation source. In this case, there are effectively three fluorophores (one for each rotamer) and the relative intensity of each will be reflected in the normalized preexponential terms derived from the fluorescence decay curve [Eq. (1)]. An orientational dependence of the normalized preexponential terms for the fluorescence decay should be expected if there are Trp side-chain rotamers, because each rotamer (and therefore each dipole) would have different spatial coordinates. If the three fluorescence decay times were the result of three distinct Trp conformers in separate molecules of the crystal, then the relative proportions (normalized preexponential terms) of the decay components would vary with crystal orientation but the decay time values would not.[20]

Erabutoxin b. The fluorescence decay times observed for erabutoxin b single crystals remained constant at all orientation angles but the normalized preexponential terms varied. Full 360° rotation experiments were precluded by eventual crystal photodamage. Significantly, the normalized preexponential terms (which correspond to the relative contributions of each decay component) were dependent on the crystal orientation (Fig. 4A). This provided direct evidence for at least three Trp side-chain rotamer conformations within separate protein molecules of the crystals.

Crotonase. There was no significant variation of the normalized preexponential terms with crystal orientation (data not shown; >180° rotation) for all three crystals measured.

Azurin. When single *Pfl* holoazurin crystals were rotated with respect to the plane of polarization, variation of the preexponential terms was observed. It is interesting to note that the patterns in variation can be divided into two subsets that include the two longer decay times (2.769 and 1.36 nsec) and the two shorter decay times (167 and 53.7 psec). Within each set of values, the preexponential terms varied inversely (data not shown).

Simulations and Calculations

It was possible to simulate the pattern for the orientational dependence of the normalized preexponential terms using erabutoxin b as the structural model. These calculations took into account the exact experimental geometry, the four molecules of erabutoxin b in the crystallographic unit cell, different possible populations of each Trp conformer within the protein crystal, and alternate starting positions of the crystal in the sample cell (with respect to flat crystal surfaces).[20]

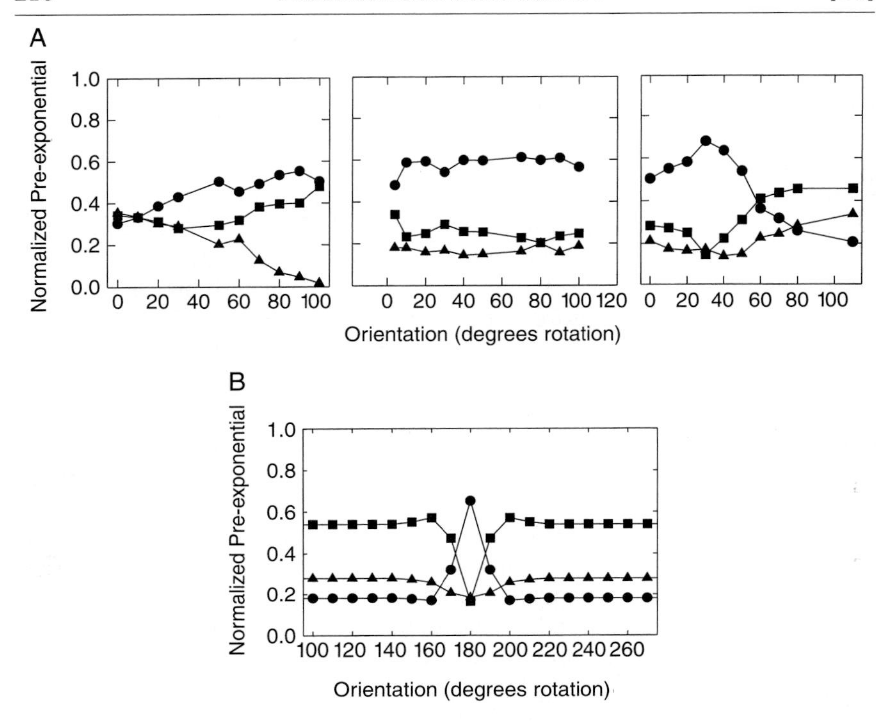

FIG. 4. (A) Dependence of the relative fluorescence decay-time component proportion on crystal orientation with respect to the polarization direction of the excitation beam. *Left, middle,* and *right:* Data acquired from individual erabutoxin b protein crystals. (●) 1.78 nsec; (■) 0.73 nsec; (▲) 0.11 nsec. (B) Theoretical orientational dependence (180°) of the relative fluorescence decay-time proportions for erabutoxin b protein crystals. Data were simulated where Trp $\chi_1 = 72.8°$, $\chi_2 = -88.7°/\chi_1 = -168.94°$, $\chi_2 = -94.7°/\chi_1 = -36.62°$, $\chi_2 = -88.9°$ (g^{+-}, g^{--}, and t^-, respectively) with 0.7, 0.1, and 0.2 as the relative proportions of the three rotamers and excitation polarization perpendicular to the z plane of the crystal unit cell (corresponding to the largest face of the rodlike crystals).

A program was written to calculate the theoretical dependence of the relative decay-time proportions on protein crystal orientation, in which the other two tryptophan side-chain rotamers, not observed by X-ray crystallography, were included, and the initial population of each rotamer could be varied. The rotamer populations could not be absolutely determined from the crystal fluorescence data. The simulation curve (Fig. 4B), which was produced by assuming the relative proportions of the three rotamer populations to be similar to those determined by solution fluorescence,[21] showed remarkable similarities to the experimental curves (Fig. 4A).

Anisotropy

One might expect the anisotropy value to be unity for a protein crystal if the system were rigid, and any Trp vibrations would likely occur on a subpicosecond time scale. From the anisotropy decay, there appears to be minimal steady-state anisotropy in the crystal. However, the anisotropy results were inconclusive, which is not surprising if one considers the system: The principle of measuring fluorescence anisotropy from a protein in solution is based on a system in which there is an isotropic orientation of the absorption transitions and, following excitation, the anisotropy of the excited molecules is monitored. Assuming the rotamer model, there would be three anisotropic rotamers in the crystal, each having a different alignment with the polarized excitation beam and with the vertical and horizontal detection optics. Each rotameric species will contribute to the fluorescence signal and therefore interpretation of the anisotropy becomes highly complicated.

Interpretation

For reasons stated earlier in this chapter, the authors expected to observe a single exponential decay function for any single Trp protein in the crystalline state. Significantly, the Trp-29 fluorescence of erabutoxin b single crystals was best described by multiple exponential decay behavior (Table II), which suggested heterogeneity of the fluorescence originating from Trp side-chain rotamers.[20] The crystal structure of erabutoxin b had been determined by two groups and was reported almost simultaneously.[23,32] The structure was later refined to 1.4-Å resolution[33] with the Trp residue modeled in a single conformation ($\chi_1 = 72.8$, $\chi_2 = -88.7$; 3ebx.pdb[18]). Therefore, the results of these initial experiments were both surprising and exciting to the investigators and inspired the formulation of the orientation experiments and theory-based computer simulations.

The orientation experiment conducted on erabutoxin b crystals provided further evidence in support of Trp side-chain rotamers (Fig. 4A). Each curve, measured experimentally on a different crystal, samples a different area of the 360° rotation, and is in good agreement with portions of the entire curve generated by theoretical simulation (Fig. 4B; for detailed differences see Dahms *et al.*[20]). It was nonetheless a possibility that the orientation phenomena could have arisen from some unknown source. To rule out this

[32] B. W. Low, H. S. Preston, A. Sato, L. S. Rosen, J. E. Searl, D. Rudko, and J. S. Richardson, *Proc. Natl. Acad. Sci. U.S.A.* **73,** 2991 (1976).

[33] J. L. Smith, P. W. R. Corfield, W. A. Hendrickson, and B. W. Low, *Acta Crystallogr.* **44**(A), 357 (1988).

possibility, crotonase crystals were used as a "negative control" for the orientation experiments.

The X-ray crystallographic structure of crotonase has recently been determined.[33a] This protein, made available to the authors, contains a single Trp residue and was easily crystallized, providing a suitable candidate for this project. According to some preliminary data,[20,34] crotonase had been shown to display triple-exponential decay kinetics in solution and in the crystalline state (Table II). Biochemical evidence[29] combined with preliminary crystallographic data,[34] suggested that crotonase grown under the conditions described herein exists as a hexamer in solution and as a dimer of hexamers in our crystallographic unit cell. The observation of triple-exponential decay kinetics for crotonase crystals implied that 3 distinct Trp rotamers existed for each crotonase molecule, and provided a model in which there would be a minimum of 36 possible Trp dipole orientations (3 Trp rotamers $\times$ 12 protein orientations) in the crystallographic unit cell. This type of Trp side-chain arrangement should effectively mimic an isotropic system in which no orientational dependence would be observed for any of the fluorescence parameters. The normalized preexponential terms of crotonase crystal fluorescence did not vary significantly with crystal orientation[28] (data not shown), lending further support to the interpretation of the erabutoxin b orientation experiments in terms of Trp side-chain rotamers.[20] These results from crotonase demonstrate that it is unlikely that the variation found for erabutoxin was the result of unusual light scatter from the crystal faces or any other experimental artifacts characteristic of measuring crystals by time-resolved fluorescence.

It became apparent that the observation of multiple exponential decay kinetics from single Trp crystalline proteins was more common than originally expected. Yellowfin tuna metmyoglobin was the only crystalline single-Trp protein for which a single decay time value (24.0 $\pm$ 0.1 psec) had been measured, and a single decay time could be observed for each of the two Trp residues (14.6 $\pm$ 0.2 and 70.4 $\pm$ 0.3 psec) contained in sperm whale metmyoglobin.[19] In this case, the fast Trp-to-heme energy transfer precluded the observation of effects from the slower deactivation pathways. Therefore, it became important to show that multiple exponential decay kinetics for single Trp crystalline proteins was not simply an artifact and was truly a result of measuring time-resolved fluorescence from crystalline proteins. There are few single-Trp proteins that display single-exponential decay kinetics in solution; one is apoazurin.

[33a] C. K. Engel, M. Mathieu, J. Ph. Zeeland, J. K. Hiltunen, and R. K. Wierenga, *EMBO J.* **15,** 513 (1996).

[34] P. J. Tonge, T. E. S. Dahms, R. To, and A. G. Szabo, *Biophys. J.* **66,** A263 (1994).

Azurin is classified as a type I copper protein based on its characteristic blue absorption band; λ_{max} = 625 nm. The elucidation of the Cu^{2+} ligand site and the mechanism of the redox reaction have been the objective for most spectroscopic and structural studies of the various azurins (for review see Canters *et al.*[35]). *Pfl* holoazurin contains only one Trp residue at position-48. Several studies from the laboratory of the present authors have focused on the metal-binding properties and conformational heterogeneity of *Pfl* holoazurin.[22,36,37] *Pfl* holoazurin in the presence of copper, cobalt, or nickel, displays triple-exponential fluorescence decay kinetics in solution, but once the metal ligand has been removed one long fluorescence decay time is observed.[37] Thus, azurin provided a suitable model with which to test the fidelity of time-resolved fluorescence measurements from crystalline systems.

The observation of single-exponential decay kinetics for crystalline apoazurin provides further support for the rotamer model of Trp photophysics as it relates to proteins in solution. Single-exponential decay kinetics were predicted on the basis of the fluorescence properties of apoazurin in solution and were observed in the crystalline state.

The lifetime values for holoazurin in the crystalline state are different from those in solution and prompted the authors to postulate that in the crystalline state, Trp-48 is altered in proximity to the Cu^{2+} ligand and/or other potential quenchers. The crystal structure revealed that alternate low-energy Trp rotamers could be modeled for holoazurin (For details see Lee *et al.*[30]). Apoazurin displays a shorter fluorescence lifetime value in the crystalline state (as compared to solution values; Table II), indicating that quenching may arise from increased contact of the indole ring with surrounding amino acid side chains (within 7 Å), owing to constraints placed on the protein by the crystal lattice.

As predicted, there was significant orientational dependence of the normalized preexponential terms for the holoazurin crystal. There were two subsets of lifetimes, a long lifetime subset (2.769 and 1.36 nsec) and a short lifetime subset (167 and 53.7 psec), and within each subset the normalized preexponential terms varied inversely. This inverse relationship could be the result of two distinct χ_2 Trp conformers (180° ring flip) in which the L_a dipoles of the two indole conformers would lie orthogonal to one another, producing an inverse variation of the preexponential terms. The two subsets

[35] G. W. Canters, A. Lommen, M. Van de Kamp, and C. W. G. Hoitink, *Biol. Metab.* **3,** 67 (1990).

[36] A. G. Szabo, T. M. Stepanik, D. M. Rayner, and N. M. Young, *Biophys. J.* **41,** 233 (1983).

[37] C. M. Hutnik and A. G. Szabo, *Biochemistry* **28,** 3935 (1989).

(longer and shorter) of lifetimes may arise from two different ligand complexes or protein conformations in which the indole ring is farther from or closer to perspective quenchers, respectively.

Conclusion

The orientation-dependent fluorescence decay experiments for single protein crystals, combined with the simulated orientational dependence of the preexponential terms, provide strong evidence to support the rotamer model for the interpretation of time-resolved fluorescence parameters in proteins. Given the precise relationship between the Trp rotamer coordinates and the coordinates of the crystal faces, it should be possible to calculate the other rotamer angles. This relationship could be obtained from the crystal cell parameters and time-resolved fluorescence measurements at specific crystal orientations (with respect to the polarized excitation source). Identification of the alternate Trp rotamers may lead to the eventual rationalization of nonradiative deactivation pathways for the excited singlet state of Trp. The observation of conformational heterogeneity for Trp in a protein crystal, which was previously reported to be in a single side-chain configuration, has implications for protein X-ray crystallographic studies. It suggests that a careful examination of high-resolution X-ray data for evidence of side-chain rotamers is warranted. There are only two protein crystal structures in the Brookhaven Data Bank[18] that display two distinct conformations for the Trp side chain, and both these structures are highly resolved [0.86 Å for gramicidin[38] (1gma.pdb) and 1.5 Å for relaxin[39] (6rlx.pdb)]. Perhaps as more protein crystal structures are determined to a higher resolution, modeling of alternate Trp side-chain rotamers will become more frequent. Identification of conformational heterogeneity by sensitive time-resolved fluorescence techniques and other methods may provide improved levels of structural detail and lead to new insights into the structure–function relationships of proteins.

From the few protein crystals that have been studied by time-resolved fluorescence, a pattern has begun to emerge: Proteins that display multiexponential fluorescence decay in solution will exhibit a similar phenomenon in the crystalline state, and these different decay-time values can be attributed to different Trp rotamers. The only nonheme protein to be measured in the crystalline state, which displays single-exponential decay kinetics in solution, exhibits a single fluorescence lifetime. Most single-Trp proteins

[38] D. A. Langs, *Science* **241,** 188 (1988).

[39] C. Eigenbrot, M. Randal, C. Quan, J. Burnier, O. Connell, E. Rinderknecht, and A. A. Kossiakoff, *J. Mol. Biol.* **221,** 15 (1991).

display more than one decay time in solution, therefore, it follows that it is common for the Trp side chain to adopt more than one low-energy conformation both in solution and in the crystalline state. On the other hand, crystallographic studies have modeled Trp almost exclusively in a single conformation. This discrepancy can be explained by limitations in the data obtained by X-ray crystallography. It has been suggested that a conformer that is present at a frequency of less than 33% cannot be detected using data with a resolution of 1.4 Å or greater.[40] If conformational heterogeneity is so common, then this phemomenon is likely to be important to the function of most proteins. In the case of holoazurin, the conformational heterogeneity imparted by the copper ligand may be key to the process of electron transfer.

Acknowledgments

The authors thank Dr. Kevin Willis for providing valuable guidance and intellectual discussion during the initial experiments on erabutoxin b. Don Krajcarski provided expert technical support throughout the work on erabutoxin b and crotonase, and for this the authors are grateful. We thank Dr. Makito Yaguchi for performing the N-terminal sequencing and amino acid analysis of crotonase and Dave Watson for the ESI mass spectrometric measurements. Dr. Peter Tonge provided the expression system for crotonase and valuable guidance during the purification of this protein.

For the duration of this study, T. Dahms was a graduate student in the Department of Biochemistry, University of Ottawa, and was a holder of a National Science and Engineering Research Council Scholarship.

[40] J. L. Smith, W. A. Hendrickson, R. B. Honzatko, and S. Sheriff, *Biochemistry* **25,** 5018 (1986).

[11] Fluorescence Methods for Studying Equilibrium Macromolecule–Ligand Interactions

By MAURICE R. EFTINK

Introduction

The function of almost all proteins is related to their ability to bind ligands. A major effort of biochemists is to characterize the strength, stoichiometry, and specificity of ligand-binding reactions, to characterize the existence of linkages in binding reactions, to understand the effect of binding on the structure of the proteins, and, ultimately, to obtain insights about biological function.

0076-6879/97 $25

There are a number of methods that can be used in such investigations. This chapter discusses the use of fluorescence spectroscopy to study binding equilibria in a direct manner (i.e., in a cuvette, without the need for physical separation of bound and free states). There are several advantages to the use of fluorescence methods. The high sensitivity allows the study of micromolar or lower (depending on the fluorophore involved) concentrations of the protein and ligand, a consequence of which is that high binding affinities can be determined. Because it is a solution method, fluorescence is amenable to studies of the effects of varying solution conditions, pH, ionic strength, and temperature. Relatively small sample sizes are required, thus further minimizing the amount of sample needed. Other important characteristics of fluorescence are that the method can be adapted to a variety of sample cell designs, the data are readily digitized, and the data (steady-state fluorescence) can be rapidly acquired.

After discussing the causes of fluorescence changes (a requisite to the use of the method to study binding reactions), some simple thermodynamic models are described for binding reactions. Next, some practical matters and problems of the method are addressed. Finally, selected applications of fluorescence methods for studying ligand-binding reactions are briefly reviewed.

Causes of Fluorescence Changes in Ligand–Macromolecule Interactions

Fluorescence is a multiparametric type of spectroscopy. The fluorescence of a compound can be described by its excitation and emisison spectra (and its wavelength of maximum emission λ_{max}), its quantum yield Φ, anisotropy r, and decay time τ (or decay profile, because decays are frequently not monoexponential). In addition, the fluorescence of a compound can be quenched by various quenchers, including those that act as electron acceptors, proton donors, enhancers of spin–orbital coupling, or energy transfer acceptors. With all these parameters and excited-state reactions, the fluorescence of a fluorophore will usually be sensitive to changes in its environment. It is not the purpose of this chapter to go into detail about the basics of fluorescence spectroscopy. For general background reading readers are referred to Refs. 1–5. Instead, some aspects that are specific to protein–ligand interactions are discussed.

[1] I. Weinryb and R. F. Steiner, *in* "Excited States of Proteins and Nucleic Acids," (R. F. Steiner and I. Weinryb, eds.), pp. 277–318. Plenum, New York, 1971.

[2] J. R. Lakowicz, "Principles of Fluorescence Spectroscopy." Plenum, New York, 1983.

[3] J. M. Beechem and L. Brand, *Annu. Rev. Biochem.* **54,** 43 (1985).

[4] M. R. Eftink, *Methods Biochem. Anal.* **35,** 127 (1991).

[5] T. G. Dewey, "Biophysical and Biochemical Aspects of Fluorescence Spectroscopy." Plenum, New York, 1991.

There are two possible situations that can be exploited to study protein–ligand interactions. Either the fluorescence of the protein can change on binding, or the fluorescence of the ligand can change. These two situations require slightly different data analysis routines, as is discussed. Of course, the fluorescence of both macromolecule and ligand can change on binding; this would lead to additional complications in treating the data and this situation can be avoided by the choice of molecules and by selecting excitation and emission wavelengths at which the contributions of the two species do not overlap.

When changes in the fluorescence of a protein are observed, the fluorophore may be one or more of the intrinsic tryptophan residues, because emission from this amino acid usually dominates the fluorescence of proteins (except for proteins that do not have tryptophans, in which case emission from tyrosines may be observed). There are cases in which the fluorescence of another covalently attached (or tightly bound) fluorophore (intrinsic or extrinsic) is observed as a ligand binds. Examples of such attached intrinsic fluorophores include NADH, flavins, lumazine, and pyridoxal phosphate. There are a variety of highly fluorescent extrinsic probes, including those based on dansyl (diethylaminonaphthalenesulfonic acid), fluorescein, coumarin, and pyrene, that can be attached to reactive sulfhydryl or amino groups in proteins. This chapter focuses on intrinsic tryptophan fluorescence.

Tryptophan is a relatively sensitive fluorophore. Its λ_{max} ranges from 308 nm, for the deeply buried tryptophan in azurin,[6,7] to 350 nm for solvent-exposed residues in unfolded proteins.[4] Also, the fluorescence quantum yield for tryptophan can range from about 0.01 to 0.4. The average decay times usually change in concert with changes in quantum yield, because most quenching reactions are predominantly collisional in nature. The binding of a ligand to a protein may directly affect the fluorescence of a tryptophan residue by acting as a quencher (i.e., by a collisional or energy transfer mechanism) or by physically interacting with the fluorophore and thereby changing the polarity of its environment and/or its accessibility to solvent. Alternatively, a ligand may bind at a site that is remote from the tryptophan residue(s) and may induce a change in conformation of the protein that alters the microenvironment of the tryptophan(s). These changes in microenvironment may result either in enhancement or quenching of fluorescence or in shifts in the spectrum to the red or blue. If there are multiple tryptophan residues in a protein, this will make it difficult to assign any fluorescence changes to a specific residue. However, for the purpose of studying the

[6] A. Finazzi-Agro, G. Rotillo, L. Avigliano, P. Guerrieri, V. Boffi, and B. Mondovi, *Biochemistry* **9,** 2009 (1970).

[7] A. G. Szabo, T. M. Stepanik, D. M. Wagner, and N. M. Young, *Biophys. J.* **41,** 233 (1983).

thermodynamics of ligand binding, the presence of multiple fluorophores is a minor consideration. Although the fluorescence signals of tryptophans are not as intense of some of the previously mentioned extrinsic probes (due to a lower extinction coefficient and modest quantum yield of tryptophan and the lower output of lamps and lasers below 300 nm), an advantage of using tryptophan fluorescence is that no chemical modification of the protein is needed.

The other possible situation is that the ligand fluoresces and that its emission changes on binding. The well-known example of such a ligand is 8-anilino-1-naphthalenesulfonic acid (ANS), for which the fluorescence increases on binding to various proteins.[8,9] The photophysical bases for changes (whether it is an increase or decrease in intensity or it is a change in the spectral position or shape) in the fluorescence of a ligand on binding are similar to those already mentioned for tryptophan residues. The case of ANS is extreme, because the fluorescence of this molecule is largely quenched by polar water molecules and its fluorescence yield is much larger in apolar media. Consequently, there can be a large enhancement in the fluorescence of ANS on binding to proteins. Besides changes in λ_{max} and/or Φ, there may also be an increase in the fluorescence anisotropy of bound ligand, owing to an increase in the rotational correlation time of the ligand when it is bound to the macromolecule. A later section specifically discusses applications involving changes in the fluorescence anisotropy of a ligand on binding.

The changes in either tryptophan or ligand fluorescence may involve resonance energy transfer processes. An example of such a fluorescence change is also presented in a later section.

General Binding Models and Their Relationship to Fluorescence Signal Changes

Several models and the related equations for protein–ligand interactions are presented. The intent here is not to give an exhaustive treatment of the subject, but to provide a framework for relating fluorescence data to thermodynamic parameters and to highlight the limitations and assumptions of the method.

Consider the case in which there is a monomeric macromolecule, M, that binds n ligand molecules, L. The expressions for the mole fraction, X, of macromolecule in the free (M) and complexed forms (ML_1, ML_2, etc.) are given as follows.

[8] E. Daniel and G. Weber, *Biochemistry* **5,** 1893 (1966).

[9] L. Brand and J. R. Gohlke, *Annu. Rev. Biochem.* **41,** 843 (1972).

$$\mathrm{M} + \mathrm{L} \underset{}{\overset{\overline{K}_1}{\rightleftharpoons}} \mathrm{ML}_1 \underset{}{\overset{\mathrm{L}\ \overline{K}_2}{\rightleftharpoons}} \mathrm{ML}_2 \cdots \underset{}{\overset{\mathrm{L}\ \overline{K}_n}{\rightleftharpoons}} \mathrm{ML}_n \tag{1}$$

$$\overline{K}_i = [\mathrm{ML}_i]/[\mathrm{ML}_{i-1}][\mathrm{L}] \tag{2}$$

$$X_\mathrm{M} = 1/Q; \qquad X_{\mathrm{ML}_i} = \beta_i[\mathrm{L}]^i/Q \tag{3}$$

$$Q = 1 + \sum \beta_i[\mathrm{L}]^i \tag{4}$$

where $\overline{K}_i$ is the macroscopic association constant for binding of the ith L molecule to the macromolecule, $\beta_i = \overline{K}_1\overline{K}_2\overline{K}_3 \cdots \overline{K}_i$, and Q is the partition function for the macromolecule. These general expressions can lead to the following specific binding models, depending on the number of sites, on whether they are identical, and on whether there is cooperativity in binding: (1) single site ($n = 1$); (2) identical, noninteracting sites ($n > 1$); (3) identical, interacting sites ($n > 1$); (4) nonidentical, noninteracting sites ($n > 1$); and (5) nonidentical, interacting sites ($n > 1$).

The intent, of course, is to relate a model, and the associated equations, to experimental data. Whereas macroscopic association constants are useful for certain types of binding data, such as that obtained from equilibrium dialysis, it is usually desirable to use models that include microscopic association constants when dealing with spectroscopic binding data. To illustrate this point, consider a case of nonidentical, noninteracting sites with $n = 2$. As shown in Eq. (5), there can exist two singly ligated species, M_1L and M_2L, with the ligand bound to the two different sites[10] with corresponding microscopic (or site) association constants K_1 and K_2.[10] It is desired to consider M_1L and M_2L as distinct species, rather than lumping them together as singly ligated species, as is done when using macroscopic models, because binding at the two sites may cause different fluorescence changes. That is, with a microscopic treatment, the fluorescence intensities of M_1L and M_2L, as well as the individual association constants, are allowed to be different.

[10] The following labeling scheme is used for site (microscopic) association constants. For a multisite system, subscript j (ranging from 0 to n) between the M and L identifies a specific binding site, i.e., M_1L versus M_2L. The number of bound ligands is indicated by subscript i (also ranging from 0 to n) following the "L," i.e., ML_2 for $n = 2$ with two ligands bound or $M_{1,2,4}L_3$ for $n = 4$ and three ligands bound to j sites 1, 2, and 4. Either j or i will be dropped when their value is obvious, as in ML_2 (instead of $M_{1,2}L_2$) for $n = 2$ and both sites are filled. Likewise, the subscripts following a microscopic association constant refer to a particular binding site, i.e., K_2 is for site $j = 2$ in a multisite system. Again, subscripts are not given when they are obvious for the model, i.e., K is used for a single-site model or for a model in which all the binding sites are identical.

$$\begin{array}{ccccc} & & M_1L & & \\ & \overset{K_1}{L \nearrow\!\!\!\swarrow} & & \overset{K_2}{L \searrow\!\!\!\nwarrow} & \\ M & & & & ML_2 \\ & \underset{K_2}{L \searrow\!\!\!\nwarrow} & & \underset{K_1}{L \nearrow\!\!\!\swarrow} & \\ & & M_2L & & \end{array} \tag{5}$$

In Table I equations are given for the previously described types of binding models, with microscopic association constants being used in the models. Included in Table I are equations that relate the free ligand concentration to the total ligand concentration, $[L]_0$, and total macromolecule concentration, $[M]_0$. Most fluorescence binding studies are performed by measuring a fluorescence signal as a function of total ligand (or total macromolecule) concentration and the calculation of free ligand concentration, from the total ligand concentration, is the most difficult step in the analysis. When the macromolecule concentration is maintained at a level that is much lower than $1/K$, then it is safe to assume that $[L]_0 \approx [L]$, and the fitting of equations to data is much easier.

Other binding models are also important. For example, the macromolecule may undergo a change in its state of self-association in response to ligand binding. In Table I equations are presented for a model in which ligand binding is coupled to a monomer $\rightleftharpoons$ dimer equilibrium of the macromolecule. Also, there may be two different types of ligands that bind to a macromolecule.

In each of these cases equations are given for the partition function of the macromolecule, the mole fraction of the various states of the macromolecule, including that for the ligated species, equations for the total ligand and macromolecule concentrations, and an equation needed to calculate the free ligand concentration. Finally, equations are given that relate the binding model to the actual data, the steady-state fluorescence intensity, for the case in which a fixed concentration of the fluorescing species (e.g., the macromolecule) is titrated with the nonfluorescing species (e.g., the ligand). The latter type of equation, together with that needed to calculate free ligand concentration, is used in the fitting of a model to experimental data to obtain the pertinent fitting parameters for the model.

The following pages present a discussion of specific binding models and the way that they can be related to changes in fluorescence signals. Actually, there can be changes in either the fluorescence of the macromolecule or the ligand on binding. The strategies for collecting and analyzing data are slightly different in each case, and the discussion is divided into two sections. Although the author's preference in fitting data is to use the various equations/models in Table I, together with a nonlinear least-squares procedure, some traditional equations and procedures that have been used to

TABLE I
BINDING MODELS AND THEIR RELATIONSHIP TO STEADY STATE FLUORESCENCE OF MACROMOLECULE[a]

Model and parameters	Reactions and equations
Single site	$M + L \overset{K}{\rightleftharpoons} ML$
Partition function and mole fractions	$Q = 1 + K[L]$ $X_M = 1/Q; X_{ML} = K[L]/Q$
Conservation of mass	$[M]_0 = [M] + [ML]; [M] = [M]_0X_M$ $[L]_0 = [L] + [ML]; [L] = [L]_0 - [M]_0X_{ML}$
Free ligand	$[L]^2 + [L]([M]_0 - [L]_0 + K^{-1}) - [L]_0/K = 0$ Solve quadratic equation for [L]
Relation to fluorescence signal[a]	$F_{rel} = (1 + K[L]\,F_{ML})/Q$ (where $F_{rel} = 1$ at $[L] = 0$, by definition) Variables: $F_{(exp)}$, $[L]_0$, $[M]_0$ Parameters; K, F_{ML}
Identical, noninteracting sites ($n > 1$ sites)	
	$M + L \overset{K}{\longleftrightarrow} M_1L, M_2L, M_3L, \ldots, M_nL \overset{K[L]}{\longleftrightarrow} M_{1,2}L_2, M_{1,3}L_2, \ldots, M_{1,n}L_2 \overset{K[L]}{\longleftrightarrow} \cdots \longleftrightarrow ML_n$
Partition function and mole fractions	$Q = 1 + \Sigma\omega_i K^i[L]^i$, where $\omega_i = n!/[i!(n-i)!]$ $X_M = 1/Q; X_{ML_i} = \omega_i K^i[L]^i/Q$
Conservation of mass	$[M]_0 = [M] + \Sigma[ML_i]; [M] = [M]_0X_M$ $[L]_0 = [L] + \Sigma i\omega_i K^i[L]^i[M]_0/Q; [L] = [L]_0 - [M]_0\Sigma iX_{ML_i}$
Free ligand concentration	$[L]^{n+1} + \Sigma[L]^i[\omega_{i-1}K^{i-1-n} + \omega_i K^{i-n}(i[M]_0 - [L]_0)] - [L]_0/K^n = 0$ (solve $n + 1$ order equation by successive approximation)
Relationship to fluorescence signal[b]	$F_{rel} = (1 + K[L](nF_{ML_i} - n + 1))/(1 + K[L])$ $= 1 + nK[L]\Delta F_{ML_i}/(1 + K[L])$ Variables: $F_{(exp)}$, $[L]_0$, $[M]_0$ Parameters: n, K, F_{ML_i}
Nonidentical, noninteracting sites ($n = 2$)[c]	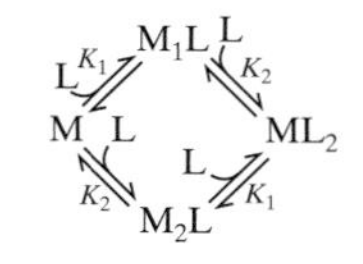
Partition function and mole fractions	$Q = 1 + K_1[L] + K_2[L] + K_1K_2[L]^2$ $X_M = 1/Q; X_{M_1L} = K_1[L]/Q; X_{M_2L} = K_2[L]/Q$ $X_{ML_2} = K_1K_2[L]^2/Q$

(*continued*)

TABLE I (*continued*)

Model and parameters	Reactions and equations
Conservation of mass	$[M]_0 = [M] + [M_1L] + [M_2L] + [ML_2]$ $[M] = X_M[M]_0$ $[L]_0 = [L] + [M_1L] + [M_2L] + 2[ML_2]$ $[L] = [L]_0 - [M]_0X_{M_1L} - [M]_0X_{M_2L} - 2[M]_0X_{ML_2}$
Free ligand	$[L]^3K_1K_2 + [L]^2(2K_1K_2[M]_0 - K_1K_2[L]_0 + K_1 + K_2)$ $+ [L]\{1 + (K_1 + K_2)([M]_0 - [L]_0)\} - [L]_0 = 0$ Solve cubic equation for [L] by successive approximation
Relationship to fluorescence signal	$F_{rel} = (1 + (K_1F_{M_1L} + K_2F_{M_2L})[L] + K_1K_2[L]^2F_{ML_2})/Q$ Variables: F_{exp}, $[L]_0$, $[M]_0$ Parameters: K_1, K_2, F_{M_1L}, F_{M_2L}, F_{ML_2}
Identical, interacting sites (Cooperative binding)[d]	$M + nL \overset{K}{\rightleftharpoons} ML_n$
Partition function and mole fractions	$Q = 1 + K[L]^n$ $X_M = 1/Q; X_{ML_n} = K[L]^n/Q$
Conservation of mass	$[M]_0 = [M] + [ML_n]; [M] = [M]_0X_M$ $[L]_0 = [L] + n[ML_n]; [L] = [L]_0 - n[M]_0X_{ML_n}$
Free ligand	$[L]^{n+1} + [L]^n(n[M]_0 - [L]_0) + [L]/K - [L]_0/K = 0$ Solve for [L] by successive approximation process
Relationship to fluorescence signal	$F_{rel} = (1 + F_{ML_n}K[L]^n)/Q$ Variables: F_{exp}, $[L]_0$, $[M]_0$ Parameters: K, n, F_{ML_n}
Identical, noninteracting sites, coupled to dimerization of macromolecule	$D \underset{}{\overset{L\ {}^DK}{\rightleftharpoons}} D_1L \overset{L\ {}^DK}{\rightleftharpoons} DL_2$ (degenerate) $K_D \Updownarrow$ $M + L \overset{{}^MK}{\rightleftharpoons} ML$ + $M + L \overset{{}^MK}{\rightleftharpoons} ML$

(*continued*)

fit such data are also presented. A third section discusses the matter of how to obtain information about the stoichiometry of complexes from fluorescence binding data. A fourth section discusses the use of fluorescence signals other than the fluorescence intensity or quantum yield. Several simulations are shown to illustrate the appearance of typical data. Hyperbolic binding profiles are frequently encountered, but the simulations show that such profiles can be distorted and can even appear to be sigmoidal (in the absence of cooperativity), depending on the binding model, the

TABLE I (*continued*)

Model and parameters	Reactions and equations
Partition function and monomer mole fractions	$Q = 1 + {}^{M}K[L] + 2K_D[M] + 2{}^{D}KK_D[M][L](1 + {}^{D}K[L])$ $X_M = 1/Q; X_D = 2K_D[M]/Q; X_{ML} = {}^{M}K[L]/Q;$ $X_{DL} = 2K_D{}^{D}K[M][L]/Q; X_{DL_2} = 2K_D{}^{D}K^2[M][L]^2/Q$
Conservation of mass	$[M]_0 = [M] + [ML] + 2[D] + 2[DL] + 2[DL_2]$ $[M] = X_M[M]_0 = \{-1 - {}^{M}K[L] + \mathrm{sqrt}[(1 + {}^{M}K[L])^2 + 8[M]_0K_D(1 + {}^{D}K[L] + {}^{D}K^2[L]^2)]\}/\{4K_D(1 + {}^{D}K[L] + {}^{D}K^2[L]^2)\}$ $[L]_0 = [L] + [ML] + [DL] + 2[DL_2]$ $[L] = [L]_0 - [M]_0X_{ML} - [M]_0X_{DL} - [M]_0X_{DL_2}$
Free ligand	Substitute expression for M into that for Q and [L]; solve high-order equation by successive approximation
Relationship to fluorescence signal	$F_{rel} = \{1 + {}^{M}K[L]F_{ML} + [M]_0(K_DF_D + K_D{}^{D}K[L]F_{DL} + K_D{}^{D}K^2[L]^2F_{DL_2})\}/Q$ Variables: F_{exp}, $[L]_0$, $[M]_0$ Parameters: ${}^{M}K$, ${}^{D}K$, K_D, F_{ML}, F_D, F_{DL}, and F_{DL_2}

[a] Where all the fluorescence signals are relative to that of the unliganded monomeric species, M.

[b] This last equation assumes that the change in the fluorescence signal is linearly related to the number of bound ligand molecules (i.e., that there is no quenching between sites). For the simple case of two identical, noninteracting sites, the model simplifies to $M + L \rightleftharpoons ML \rightleftharpoons ML_2$ (with $M_1L = M_2L = ML$, i.e., the two possible M_1L and M_2L species are identical) and $F_{rel} = 1 + 2K[L]\Delta F_{ML}/(1 + K[L])$.

[c] If $K_1 \gg K_2$, these equations collapse somewhat, to a case of sequential binding, with parameters K_1, K_2, F_{ML}, and F_{ML_2}. If the two binding sites are identical and noninteracting, then $K_1 = K_2$ and $F_{M_1L} = F_{M_2L} = F_{ML}$, reducing the model to three parameters: K, F_{ML}, and F_{ML_2}. If there is cooperativity between the two sites, this model and relationships are changed in the following way. The two association constants on the right side of the reaction scheme become $K_{2,1}$ and $K_{1,2}$ and may be larger or smaller than K_2 and K_1, respectively, by the cooperativity factor, which is defined as $K_{1,2}/K_1 = K_{2,1}/K_2$ (using the convention that $K_{1,2}$ refers to the binding to site 1 when ligand is already present in the second site, etc). The last equation, describing the dependence of the fluorescence signal on ligand concentration, is modified to contain $\alpha K_1K_2[L]^2F_{ML_2}$ in place of $K_1K_2[L]^2F_{ML_2}$. Such a cooperative model thus contains an additional fitting parameter α. If K_2 is much less than K_1, making the M_2L state insignificantly populated, then the relationship can be simplified to have only K_1, αK_2 ($=K_{2,1}$), F_{M_1L}, and F_{ML_2} as fitting parameters.

[d] This model assumes perfect cooperativity for ligand binding.

magnitudes of the association constants, the fluorescence changes with each binding step, and the conditions under which the data are collected.

Changes in Intrinsic Protein Fluorescence on Ligand Binding

First are considered cases in which the fluorescence of the protein changes in some way on ligand binding. The fluorescence observable that is most straightforwardly related to the previously described binding functions is the steady-state fluorescence intensity, F, of the macromolecule, which is measured at some pair of excitation and emission wavelengths,

λ_{ex} and λ_{em}, at a constant concentration of total macromolecule,[11] $[M]_0$, and as a function of the total concentration of ligand, $[L]_0$. The observed intensity[12] can be expressed by either of the following relationships, where F_M and F_{ML_i} are the relative intensities of the free and complexed forms of the macromolecule and $\mathbf{F}_M$ and $\mathbf{F}_{ML_i}$ are the molar fluorescence values of the respective forms.[13]

$$F = X_M F_M + \Sigma\, X_{ML_i} F_{ML_i} \tag{6a}$$

$$F = [M]\mathbf{F}_M + \Sigma\, [ML_i]\mathbf{F}_{ML_i} \tag{6b}$$

The most direct way to relate fluorescence intensity changes to a binding model is to combine either Eq. (6a) or (6b) with one of the binding models

[11] In typical titrations of a concentrated ligand solution into a more dilute solution of macromolecule, there will be some dilution of the macromolecule as ligand is added. A reasonable experimental design is one in which the total volume of added ligand aliquots is about 10% of the initial volume of macromolecule. For example, if an initial volume of macromolecule, $V_{i,M}$, of 2000 μl is used, then a sum of ligand aliquots, $\Sigma\, V_L$, equal to 200 μl (e.g., in 10- to 20-μl steps) would result in a progressive dilution of the macromolecule during the course of the titration. The total macromolecule concentration at any point during the titration can be calculated as $[M]_0' = [M]_0 V_{i,M}/(V_{i,M} + \Sigma\, V_L)$. The value of 10% is only a guide; in principle, greater dilutions can be handled, provided that the fluorescence signal is truly linear in the concentration of macromolecule under the experimental conditions. Likewise, when the macromolecule is diluted during the course of the titration, it is necessary to correct the relative fluorescence signal [i.e., F' in Eq. (9)] for this dilution by multiplying the uncorrected relative fluorescence by the factor $(V_{i,M} + \Sigma V_L)/V_{i,M}$.

[12] Any background signal (fluorescence or scattered light) from the buffer should be subtracted, if it is significant. Likewise, if the ligand titrant contributes to the signal (when the fluorescence from the macromolecule is being observed), then these contributions should be subtracted. The observed fluorescence signal of the sample either can be measured at a single emission wavelength or can be integrated over part or all of the emission envelope. If measurements are made at one emission wavelength, then it should be realized that the enhancement factors may depend on the selected wavelength, if there is a red or blue shift in fluorescence on binding. In either case, it is usually not necessary to correct the signals for any wavelength dependence of the photomultiplier response, since only relative signals are needed for the purpose of studying ligand binding.

[13] These F, F_M, and F_{ML_i} values are relative intensity values, for the experimental conditions and the instrument employed, that would be obtained at a constant concentration of macromolecule (the fluorescing species; see note 10 for corrections for dilution). The boldface values, $\mathbf{F}_M$ and $\mathbf{F}_{ML_i}$ (and the $\mathbf{F}_L$ to be used later) are simply defined as the corresponding molar fluorescence (at constant experimental conditions) of each species. For example, F_M and $\mathbf{F}_M$ can be related as $F_M = \mathbf{F}_M[M]$, that is, by multipling the molar fluorescence by the molar concentration of a species. One could use the flourescence quantum yield of each species, but this would be more difficult to measure and provides no advantage over the use of an arbitrarily defined relative or molar fluorescence. Likewise, when the fluorescence of the ligand is monitored, the boldface $\mathbf{F}_{ML_i}$ values refer to the molar fluorescence of each bound ligand.

and then to perform nonlinear least-squares regression analysis. For example, for the case in which there is only a single binding site, Eqs. (6a) and (6b) become

$$F = (F_M + F_{ML}K[L])/(1 + K[L]) \tag{7a}$$

$$F = [M]_0(\mathbf{F}_M + \mathbf{F}_{ML}K[L])/(1 + K[L]) \tag{7b}$$

Depending on whether the fluorescence intensity decreases or increases, F_{ML} (or $\mathbf{F}_{ML}$) will be smaller or larger than F_M (or $\mathbf{F}_M$). F values are generally measured in relative numbers and it is usually convenient to normalize the initial intensity of the macromolecule in the absence of ligand, F_0 (which is also equal to $F_M = \mathbf{F}_M[M]_0$), to a value of 1.0 (or 100). If all fluorescence intensities are normalized by dividing by F_0 (= F_M), then Eq. (7b) becomes (using only molar fluorescence quantities from this point on)

$$F' = (1 + \mathbf{F}'_{ML}K[L])/(1 + K[L]) \tag{8}$$

where $F' = F/F_0 = F/F_M = F/(\mathbf{F}_M[M]_0)$ and $\mathbf{F}'_{ML} = \mathbf{F}_{ML}/\mathbf{F}_M = \mathbf{F}_{ML}[M]_0/F_0$. The parameter $\mathbf{F}'_{ML}$ is a fluorescence enhancement factor, being greater than unity if fluorescence increases and less than unity if fluorescence decreases on ligand binding. If $[L]_0$ is not much greater than $[M]_0$, which will be a necessary condition in titrations for tight binding systems, then free ligand concentration, [L], must be calculated by solving the quadratic equation, $[L] = 0.5 \times \{-b + (b^2 - 4[L]_0/K)^{1/2}\}$, where $b = ([M]_0 - [L]_0 + 1/K)$. Figure 1A shows a plot of simulated F versus $[L]_0$ data described by Eq. (8) for various values of the total macromolecule concentration, $[M]_0$; this model can be fitted to such hyperbolic data with two fitting parameters, K and $\mathbf{F}'_{ML}$.

Alternatively, instead of analyzing direct F measurements (or their normalized values, F') versus $[L]_0$, one can choose to measure and analyze the fluorescence difference values, $\Delta F = F - F_0 = F - \mathbf{F}_M[M]_0$ (or $\Delta F' = F'_{ML} - 1$). Equation (7) [or Eq. (8)] can be rearranged to the following. A plot of simulated data described by Eq. (9) is shown in Fig. 1B.

$$\Delta F = [M]_0(\mathbf{F}_{ML} - \mathbf{F}_M)K[L]/(1 + K[L]) = [M]_0\,\Delta\mathbf{F}_{ML\text{-}M}K[L]/(1 + K[L]) \tag{9a}$$

$$\Delta F' = (F'_{ML} - 1)K[L]/(1 + K[L]) \tag{9b}$$

where $\Delta\mathbf{F}_{ML\text{-}M} = \mathbf{F}_{ML} - \mathbf{F}_M$.

When there is a single binding site, the fitting of Eq. (7), (8), or (9) to data is straightforward. However, if there are two or more sites, the analysis

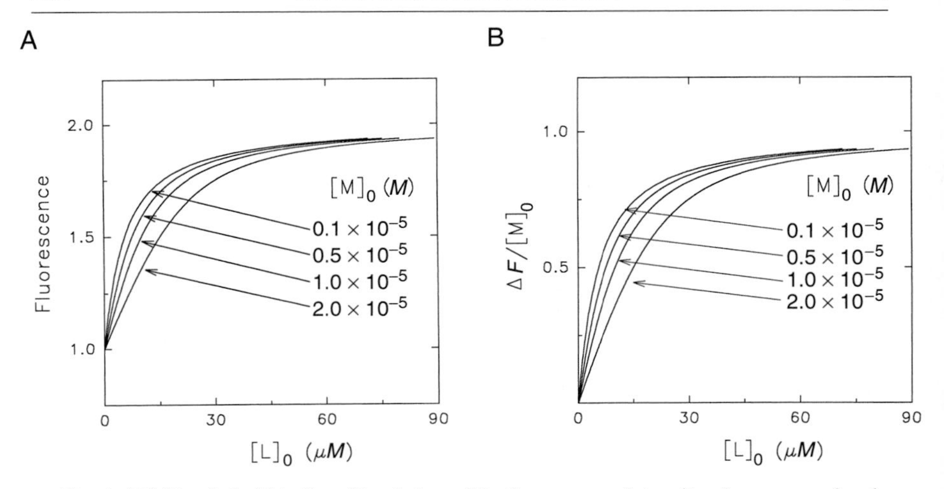

FIG. 1. (A) Simple 1:1 binding. Simulation of the fluorescence intensity of a macromolecule as a function of the concentration of added total ligand, $[L]_0$, for various total concentrations of macromolecule, $[M]_0$ (typical range for fluorescence measurements). Single-site binding model with $K = 2 \times 10^5\ M^{-1}$. For the case $F_M = 1.0$ and $F_{ML} = 2.0$. (B) Plot of the above simulations in terms of ΔF versus $[L]_0$ [Eq. (9)].

of data becomes much more difficult. This is due not only to the need to use a more complicated binding function, which may include different K_j for the different sites or cooperativity between sites (either phenomenon requiring additional thermodynamic fitting parameters), but to the additional complication that the fluorescence signal will not necessarily be linearly dependent on the number of ligands that are bound. That is, the change in the fluorescence intensity of an ML_2 complex may not be one-half (or twice) the intensity of an ML_1 complex (and, of course, M_1L_1 and M_2L_1 species may have different signals when binding sites $j = 1$ and 2 are different). This is particularly true when the binding of ligand causes resonance energy transfer quenching of tryptophan fluorescence, as is illustrated by an example as follows.

To show some of the complexity and assumptions that are involved in a multisite system, consider a two-site model in which there are nonidentical, noninteracting sites (Table I). If the association constant for the first site, K_1, is much greater than the association constant for the second site, K_2, the model simplifies to the following sequential model:

$$M + L \underset{}{\overset{K_1}{\rightleftharpoons}} ML \overset{L}{\underset{}{\overset{K_2}{\rightleftharpoons}}} ML_2 \tag{10}$$

and the observed relative fluorescence of the macromolecule is given by

$$F = (1 + \mathbf{F}_{ML}K_1[L] + \mathbf{F}_{ML_2}K_1K_2[L]^2)/(1 + K_1[L] + K_1K_2[L]^2) \quad (11)$$

where F, $\mathbf{F}_{ML}$, and $\mathbf{F}_{ML_2}$ are all normalized by dividing by $F_M = \mathbf{F}_M[M]_0$ [as in Eq. (8), with the primes dropped from this point on] and where [L] is related to $[L]_0$, $[M]_0$, and the K values by a cubic equation:

$$[L]^3K_1K_2 + [L]^2(K_1 + 2K_1K_2[M]_0 - K_1K_2[L]_0) + [L](1 + K_1[M]_0 - K_1[L]_0) - [L]_0 = 0 \quad (12)$$

Equations (11) and (12) [or Eq. (11) alone, if $[M]_0$ is much lower than $[L]_0$] can be fitted to data with four fitting parameters, K_1, K_2, $\mathbf{F}_{ML}$, and $\mathbf{F}_{ML_2}$; solving Eq. (11) for free ligand concentration can be achieved by a Newton–Raphson iterative procedure.[14] Figure 2A shows simulated data for this $n = 2$ nonidentical, noninteracting (sequential binding, $K_1 > K_2$) model for a few choices of $\mathbf{F}_{ML}$ and $\mathbf{F}_{ML_2}$. If $\mathbf{F}_{ML}$ is between 1 and $\mathbf{F}_{ML_2}$ (curve a, Fig. 2A) the profile looks much like a simple hyperbolic plot in Fig. 1A; if $\mathbf{F}_{ML}$ is approximately 1 (i.e., no change from F_M, but $\mathbf{F}_{ML_2}$ is >1), the profile can appear to be sigmoidal (curves b and c, Fig. 2A); if $\mathbf{F}_{ML}$ is > $\mathbf{F}_{ML_2}$, one will see a maximum in the F versus $[L]_0$ profile (curve d, Fig. 2A); and, if $\mathbf{F}_{ML} < 1$, a dip can appear in the profile (curve e, Fig. 2A). (Here all $\mathbf{F}_i$ are normalized by dividing by F_M. Most of the simulations are for the case that $\mathbf{F}_{ML_2}$ is greater than 1, i.e., enhancement of fluorescence on ligand binding; a similar behavior occurs if there is quenching on ligand binding, with $\mathbf{F}_{ML_2}$ being a number less than 1.) Shown in Fig. 2B is a simulation for the case in which the binding of ligand (for a multisite, n = 2 identical, noninteracting sites model) causes a signal change that is not linear in the coverage of sites. That is, in this simulation, the binding of the first ligand may quench more than half of the fluorescence of the protein. This situation can be encountered with multimeric proteins, where the ligand quenches the fluorescence of the protein by long-range resonance energy transfer and if binding of ligand to one subunit can quench the fluorescence of other subunits. The simulation in Fig. 2B is for a dimeric protein with identical, noninteracting sites. As can be seen, the drop in signal from the macromolecule can have a larger initial slope than the stoichiometric limit slope (dashed line in Fig. 2B, which is the degree of quenching expected when the signal change is linear in the coverage of sites). The existence of such nonlinear quenching must be recognized for proper analysis of data. An example of this type of system is given in the section Example Studies.

[14] A. Constantinides, "Applied Numerical Methods with Personal Computers," Chap. 2. McGraw-Hill, New York, 1987.

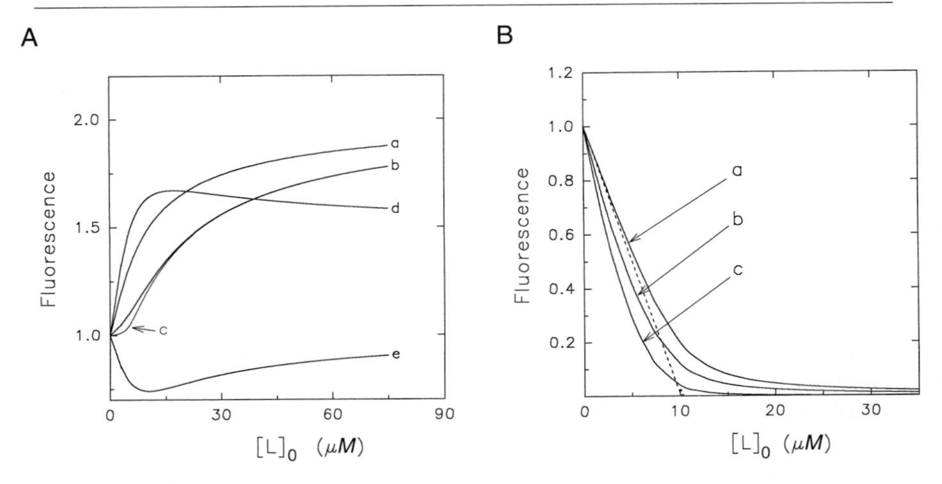

FIG. 2. (A) Nonidentical, noninteracting sites ($n = 2$). Simulation of the fluorescence intensity of a macromolecule vs total ligand concentration for a two-site nonidentical, noninteracting model, with $K_1 = 5 \times 10^5\ M^{-1}$, $K_2 = 5 \times 10^4\ M^{-1}$, $[\mathrm{M}]_0 = 5 \times 10^{-6}\ M$, and for the following values of $F_{\mathrm{M_1L}}$, $F_{\mathrm{M_2L}}$, and $F_{\mathrm{ML_2}}$: (a) 1.5, 1.5, 2.0; (b) 1.0, 2.0, 2.0; (c) 0.5, 2.5, 2.0; (d) 2.0, 0.5, 1.5; and (e) 0.5, 1.5, 1.0. Curve a has a linear increase (enhancement) in fluorescence for binding to the two sites. Curve b has no change for binding to the first site, but an enhancement for binding to the second. Curve c has a decrease (quenching) in fluorescence for binding to the first site and an increase on binding to the second. Curve d has an increase in binding to the first site and a decrease on binding to the second. Curve e has a large decrease on binding to the first site and an increase on binding to the second. All quenching and enhancement contributions are considered additive (i.e., the signal of $F_{\mathrm{ML_2}}$ is an average). (B) Nonlinear quenching (i.e., by energy transfer). Simulation of a two-site model with identical, noninteracting sites, $K_1 = 2 \times 10^6\ M^{-1}$, $[\mathrm{M}]_0 = 5 \times 10^{-6}\ M$ (i.e., as dimer), where the binding of ligand causes a quenching of the fluorescence of the macromolecule. Curve a, linear quenching with $F_{\mathrm{M_1L}} = F_{\mathrm{M_2L}} = 0.5$ and $F_{\mathrm{ML_2}} = 0$. Curve b, nonlinear quenching (quenching between subunits) with $F_{\mathrm{M_1L}} = F_{\mathrm{M_2L}} = 0.25$ and $F_{\mathrm{ML_2}} = 0$. Curve c, nonlinear quenching (total quenching on binding of the first ligand) with $F_{\mathrm{M_1L}} = F_{\mathrm{M_2L}} = 0$ and $F_{\mathrm{ML_2}} = 0$. The dashed line is the "stoichiometric quenching" that would be expected if the two subunits of the macromolecule were completely independent (i.e., if the protein were a monomer). Notice the enhanced quenching observed in simulations b and c when quenching between subunits occurs. Failure to use the correct model ($n = 2$ identical, noninteracting sites with a nonlinear signal change) in the latter cases would lead to an overestimate of the actual association constant. (C) Simulation of an $n = 2$ identical, cooperative binding model [Eq. (14)], with $K = 2 \times 10^9$ and the indicated values of $[\mathrm{M}]_0$.

If the multiple binding sites have the same K (n identical, noninteracting sites) and have the same incremental signal change, $\Delta \mathbf{F}_{\mathrm{ML}_i}$, on binding at each site, the binding function for the normalized fluorescence intensity simplifies to[15]

[15] J. J. Holbrook, *Biochem. J.* **128,** 921 (1972).

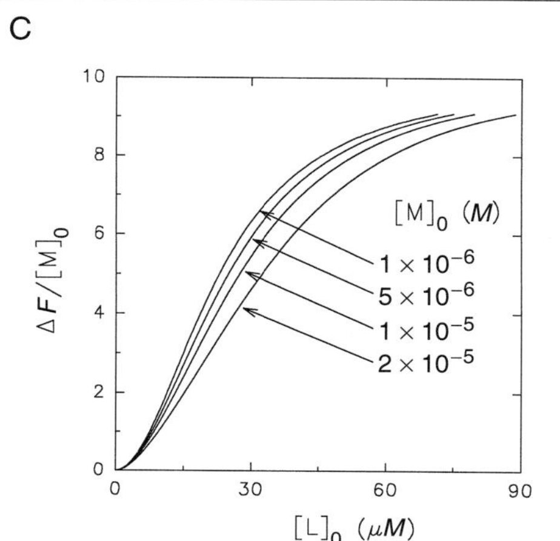

FIG. 2. (*continued*)

$$F = [1 + K[\mathrm{L}](n\mathbf{F}_{\mathrm{ML}_i} - n + 1)]/(1 + K[\mathrm{L}])$$
$$= 1 + nK[\mathrm{L}]\,\Delta\mathbf{F}_{\mathrm{ML}_i}/(1 + K[\mathrm{L}]) \quad (13a)$$
$$\Delta F = nK[\mathrm{L}]\,\Delta\mathbf{F}_{\mathrm{ML}_i}/(1 + K[\mathrm{L}]) \quad (13b)$$

The expression relating [L] to $[\mathrm{L}]_0$ and $[\mathrm{M}]_0$ is complicated, being a polynomial of order $n + 1$ (see Table I). For such a system, a plot of F vs $[\mathrm{L}]_0$ or ΔF vs $[\mathrm{L}]_0$ should show a simple hyperbolic shape. Whether solving Eqs. (13a) and (13b) for [L] or working under conditions where $[\mathrm{L}] \approx [\mathrm{L}]_0$, Eqs. (13a) and (13b) can be fitted to data to obtain K. The values of n and $\Delta\mathbf{F}_{\mathrm{ML}_i}$ are highly correlated, however, and to determine n it is necessary to collect data at different $[\mathrm{M}]_0$. A following section focuses on the problem of determining the stoichiometry of complex formation for such cases of identical, noninteracting sites.

Another frequently encountered binding system is one in which there is cooperative binding of ligands to n identical sites on a protein. Equations are given for a perfectly cooperative binding model in Table I. A more realistic system is one in which the cooperativity is imperfect, that is, the Hill model, in which the concentration of ligand is raised to some exponent, c, which has a value between 1 and n. In this case the equation to describe the relative fluorescence becomes

$$F = 1 + K[\mathrm{L}]^c\,\Delta\mathbf{F}_{\mathrm{ML}_n}/(1 + K[\mathrm{L}]^c) \quad (14a)$$
$$\Delta F = \Delta\mathbf{F}_{\mathrm{ML}_n}K[\mathrm{L}]^c/(1 + K[\mathrm{L}]^c) \quad (14b)$$

Such a relationship corresponds to an F vs $[L]_0$ plot that is sigmoidal, as shown in Fig. 2C, and a three-parameter (c, K, and $\Delta \mathbf{F}_{ML_n}$) fit should be possible. Whereas sigmoidal data would seem to be diagnostic of such a cooperative binding model, Fig. 2A shows that it is also possible that a sigmoidal F vs $[L]_0$ plot can occur for the model of nonidentical, noninteracting sites. For the latter type of model [i.e., Eq. (11) with $K_1 > K_2$, sequential binding], a sigmoidal plot can occur if the fluorescence does not change much on binding of the first ligand, but changes greatly on binding the second ligand.

Changes in Fluorescence of Ligand on Binding

In this type of experiment the fluorescence of the macromolecule is either avoided (using λ_{ex} longer than that which excites chromophores in the macromolecule) or is unchanged, but a change in the fluorescence of the ligand is observed on its binding. Just as above, the observed fluorescence intensity, F, can be related to the concentration of free and bound ligand. Returning to the general model in which there may be multiple binding sites for ligand on the macromolecule

$$M + L \underset{}{\overset{\overline{K}_1}{\rightleftharpoons}} ML_1 \overset{L}{\underset{}{\overset{\overline{K}_2}{\rightleftharpoons}}} ML_2 \cdots \overset{L}{\underset{}{\overset{\overline{K}_n}{\rightleftharpoons}}} ML_n \tag{15}$$

$$\overline{K}_i = [ML_i]/[ML_{i-1}][L] \tag{16}$$

$$X_L = [L]/[L]_0; \qquad X_{ML_i} = i[M]\beta_i[L]^i/[L]_0 \tag{17}$$

$$[L]_0 = [L] + \sum_i i[M]\beta_i[L]^i. \tag{18}$$

where, as in Eqs. (1)–(4), the ML_i are macroscopic species, $\overline{K}_i$ are ith macroscopic association constants, and $\beta_i = \overline{K}_1\overline{K}_2\overline{K}_3 \cdots \overline{K}_i$. In the treatment of changes in the signal of the macromolecule, we quickly abandoned the macroscopic approach when dealing with multisite systems, because it is likely that the binding of successive ligands will lead to different incremental change in fluorescence signal of the macromolecule, particularly if the sites are not identical or if energy transfer quenching occurs. When the fluorescence of the ligand is monitored, the latter effect will not exist, but a microscopic site treatment is still the most rigorous and is presented here. The problem with relating any of the binding models in Table I to fluorescence data is again one of solving a polynomial equation for the free ligand concentration corresponding to any $[L]_0$ and $[M]_0$. The derivation of the appropriate polynomial equations (such as those in Table I for the case of monitoring changes in the fluorescence of M) is tedious but straightforward and will not be given. Two frequently encountered models are described as follows.

When there is a single binding site on the macromolecule, the fluorescence of the ligand is described by

$$F = \mathbf{F}_L[L] + \mathbf{F}_{ML}K[L][M]_0/(1 + K[L]) \tag{19}$$

where, again $\mathbf{F}_L$ and $\mathbf{F}_{ML}$ are the molar fluorescence of each species.[13] The free ligand concentration is determined from the solution to the quadratic equation $[L]^2 + [L]([M]_0 - [L]_0 + 1/K) - [L]_0/K = 0$ (unless, of course, $[M]_0 \ll [L]_0$, so the assumption can be made that $[L] \approx [L]_0$). Equation (19) has three fitting parameters, $\mathbf{F}_L$, $\mathbf{M}_{ML}$, and K, but the value of $\mathbf{F}_L$ can usually be determined from separate measurements in the absence of macromolecule. A commonly seen version of Eq. (19) is that obtained by defining $\Delta F = F - \mathbf{F}_L[L]_0$.

$$\Delta F = F - \mathbf{F}_L[L]_0 = \mathbf{F}_L([L] - [L]_0) + \mathbf{F}_{ML}[ML] \tag{20a}$$
$$\Delta F = [ML](\mathbf{F}_{ML} - \mathbf{F}_L) \tag{20b}$$

At saturating ligand concentration, $[ML] = [M]_0$ and the maximum signal change can be defined as $\Delta F_{max} = [M]_0(\mathbf{F}_{ML} - \mathbf{F}_L)$; substituting the latter relationship, as well as $K[M]_0[L]/(1 + K[L])$ for [ML], we obtain

$$\Delta F = K[L]\,\Delta F_{max}/(1 + K[L]) \tag{21}$$

This familiar equation has K and ΔF_{max} as fitting parameters. In using Eq. (21) one measures the difference in the fluorescence of the ligand as it is titrated into two solutions, one with and one without macromolecule [in lieu of describing the fluorescence of free ligand as $\mathbf{F}_L[L]$, as done in Eq. (19)]. In some experimental designs, ΔF_{max} is determined independently by performing a reverse titration, that is, by titrating a fixed ligand concentration with macromolecule and extrapolating to the maximum signal change.

Figure 3A illustrates simulated data for the binding ($n = 1$, $K = 2 \times 10^5\ M^{-1}$) of a fluorescent ligand to a macromolecule, for different total concentrations of the latter, for the case in which the fluorescence of the ligand is enhanced on binding. The dashed line represents the fluorescence of the free ligand and the difference, ΔF, would be obtained from the titrations in the absence and presence of M. Plots of ΔF versus $[L]_0$ are shown in Fig. 3B for different $[M]_0$ values. The effect of variation in K on the appearance of the isotherm is shown in Fig. 3C (for $[M]_0 = 1 \times 10^{-5}$ M). In Fig. 3B and C for the dashed lines are the "stoichiometric limit" titration curves that would result at high $[M]_0$ or K.

When there are two or more binding sites on a macromolecule, there are questions as to whether the fluorescence is the same for ligand bound to the different sites, as well as questions as to whether the sites have

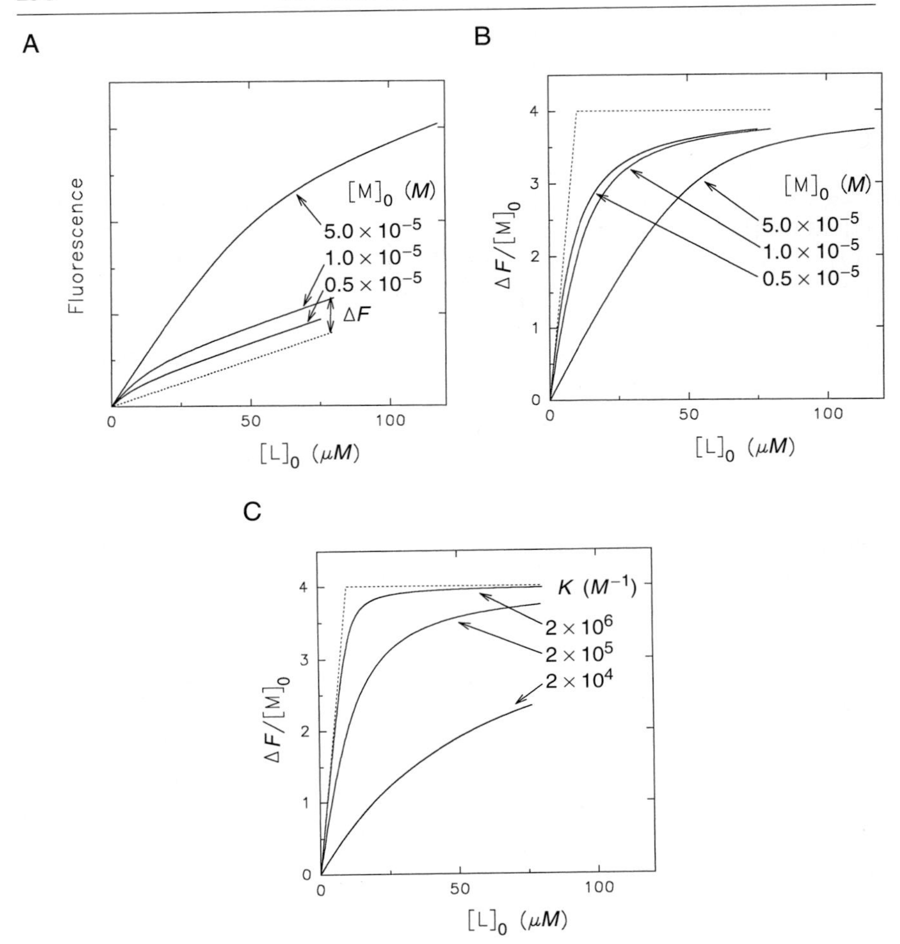

FIG. 3. Simulated plots of fluorescence intensity versus $[L]_0$ when the fluorescence of the ligand is changed (is enhanced) on binding to the macromolecule. (A) Single site-binding model, with $K = 2 \times 10^5\ M^{-1}$, $F_{ML} = 5$, $F_L = 1$ ($\Delta F = 4$), and for various concentrations of total macromolecule, $[M]_0$. The dashed line is the signal for free ligand. Values of ΔF versus $[L]_0$ are determined by subtracting one of the solid lines from the dashed line. (B) Plot of ΔF versus total ligand concentration, $[L]_0$, for the simulations in (A). (C) Plot of ΔF versus total ligand concentration, $[L]_0$, for various values of K. The dashed curves in (B) and (C) represent the stoichiometric limit when K is infinitely large [or $[M]_0$ is large in (B)].

identical K and/or interact. For the next simplest case of two nonidentical, noninteracting sites [reaction scheme in Eq. (5)], the appropriate equations become

$$F = \mathbf{F}_L[L] + \mathbf{F}_{M_1L}K_1[M][L] + \mathbf{F}_{M_2L}K_2[M][L] + 2\mathbf{F}_{ML_2}K_1K_2[M][L]^2 \quad (22)$$

where the $\mathbf{F}_{ML_i}$ are the molar fluorescence values for each bound ligand and $\mathbf{F}_{ML_2}$ is an average value for the two bound ligands, and where $[M] = [M]_0/(1 + K_1[L] + K_2[L] + K_1K_2[L]^2)$ and [L] is determined by solving a cubic equation. If $\mathbf{F}_L$ is determined independently, there are five fitting parameters for this model [four, if one assumes that $\mathbf{F}_{ML_2} = (\mathbf{F}_{M_1L} + \mathbf{F}_{M_2L})/2$]. Usually this would be too many floating parameters. If, however, the assumption is made that $K_1 > K_2$ (i.e., sequential binding with the M_2L species, which has one ligand bound to the second site, not being populated and with $\mathbf{F}_{M_2L}$ being dropped), there would be a more manageable four-parameter fit.

Using the ΔF approach, the corresponding equation for a two nonidentical, noninteracting site model, with $K_1 > K_2$, would be

$$\Delta F = (K_1[L]\,\Delta\mathbf{F}_{M_1L} + 2K_1K_2[L]^2\,\Delta\mathbf{F}_{max})/(1 + K_1[L] + K_1K_2[L]^2) \quad (23)$$

where $\Delta\mathbf{F}_{max} = \mathbf{F}_{ML_2} - \mathbf{F}_L$ and $\Delta\mathbf{F}_{M_1L} = \mathbf{F}_{ML_1} - \mathbf{F}_L$, the latter being the molar fluorescence change on binding ligand to the first site. If $\Delta\mathbf{F}_{max}$ can be separately determined from a reverse titration, then Eq. (23) contains three fitting parameters.

For such a two-site (nonidentical, noninteracting) system, simulated fluorescence titration curves are shown in Fig. 4A for different $[M]_0$. In the simulations, only a factor of two difference was assumed between the two K values (i.e., $K_1 = 2 \times 10^5$ and $K_2 = 1 \times 10^5$). As can be seen, the curves have a hyperbolic shape and it would be difficult to determine the two K values. In Fig. 4B the difference between the two K values is a factor of 10 and simulations are shown for various values of $\Delta\mathbf{F}_{M_1L}$ and $\Delta\mathbf{F}_{M_2L}$. A noteworthy result is that the binding curve can appear to be sigmoidal if the signal change is small for the higher affinity site.

The analysis of more complicated models is more difficult, of course. In general, the stoichiometry of the binding process will not be known. Equation (23) can be generalized for an indefinite value of n, but it is impractical to fit such a model to data unless it is assumed that all the K_j are the same and that all the incremental signal changes, $\Delta\mathbf{F}_{ML_i}$, are the same ($= \Delta\mathbf{F}_{max}$). With the latter assumptions we arrive at the familiar expression for a multisite binding process

$$\Delta F = nK[L]\,\Delta\mathbf{F}_{max}/(1 + K[L]) \quad (24)$$

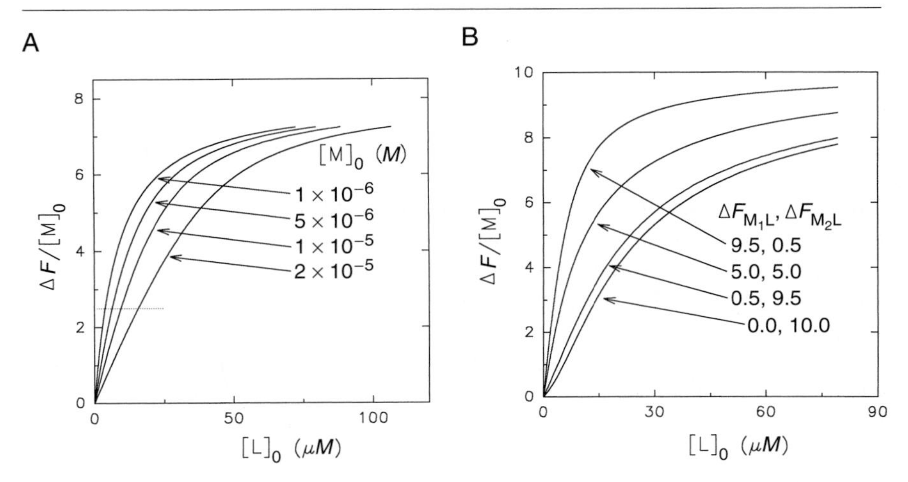

FIG. 4. Simulated ΔF versus $[L]_0$ data for an $n = 2$ nonidentical, noninteracting site model, where changes in the fluorescence of the ligand are measured [Eq. (22)]. (A) Equal and linearly additive signal changes for the two sites. $K_1 = 2 \times 10^5\ M^{-1}$, $K_2 = 1 \times 10^5\ M^{-1}$, and $\Delta\mathbf{F}_{M_1L} = \Delta\mathbf{F}_{M_2L} = \Delta\mathbf{F}_{ML_2}$. (B) Different signal changes for individual sites. $K_1 = 5 \times 10^5\ M^{-1}$, $K_2 = 5 \times 10^4\ M^{-1}$, and $\Delta\mathbf{F}_{M_1L}$ and $\Delta\mathbf{F}_{M_2L}$ values as given here.

where the relationship between free ligand, [L], and total ligand, $[L]_0$, is a polynomial of order $n + 1$. Consequently, Eq. (24) is only tractable under the condition that $[L] \approx [L]_0$. Note that n and $\Delta\mathbf{F}_{max}$ are correlated in this relationship. To determine n, one can separately determine $\Delta\mathbf{F}_{max}$ by a reverse titration of ligand with saturating macromolecule. If this can be done, then it is possible to determine the stoichiometry of the complex, in the context of the assumptions implicit in Eq. (24) (also, see the following section).

A version of Eq. (24) has been widely used in the study of the interactions of fluorescent drugs with DNA. The form of Eq. (24) usually used in DNA-binding studies is [ML] = [bound ligand] = $(F/F_0 - 1)[L]_0/(\mathbf{Q} - 1)$, where F and F_0 are the fluorescence intensities of the drug (e.g., ethidium bromide) in the presence and absence of DNA and $\mathbf{Q}$ is a fluorescence enhancement ratio, which, in the notation of this author, is equal to $\mathbf{F}_{ML}/\mathbf{F}_L$. Values of [ML] vs $[L]_0$ are usually then analyzed via a Scatchard or neighbor exclusion model.[16,17] A similar treatment has also been used in studies of the interaction of NADH with dehydrogenases.[15,18]

[16] J. B. LePecq and C. Paoletti, *J. Mol. Biol.* **27,** 87 (1967).
[17] A. Blake and A. R. Peacocke, *Biopolymers* **6,** 1225 (1968).
[18] H. Theorell and K. Tatemoto, *Arch. Biochem. Biophys.* **142,** 69 (1971).

Determining Binding Stoichiometries

The previous discussion has focused on fitting specific models to data with the presumption that the researcher has some knowledge about the stoichiometry of the complex. In many studies, of course, the binding model and stoichiometry of the complex will be unknown. When monitoring the fluorescence of the macromolecule as ligand is varied, it will usually not be possible to determine the binding stoichiometry, unless the $K[\mathrm{M}]_0$ product is high and the stoichiometric limit is reached. If the binding profile of F vs $[\mathrm{L}]_0$ has a hyperbolic shape, this can mean that $n = 1$ or that $n > 1$ with either the $\mathbf{F}_{\mathrm{ML}_i}$ being linearly related to n [e.g., Eq. (13)] or with the entire fluorescence change occurring on binding of the first ligand. If there is a distortion from a hyperbolic shape, this gives an indication that $n > 1$ and that either the sites are intrinsically nonidentical or there is some interaction between the sites. A researcher may then attempt to fit a model in Table I [e.g., Eqs. (11) or (14)] to the data for assumed values of n.

If a change in the fluorescence of the ligand is measured as macromolecule is titrated with ligand, there is usually a better chance of determining the stoichiometry of an unknown complex. Again, if ΔF vs $[\mathrm{L}]_0$ data are not hyperbolic, then this indicates heterogeneity of sites and a model with $n > 1$ [e.g., Eq. (22)] can be fitted to the data. If a hyperbolic binding curve is obtained, this strongly indicates that the binding sites are identical in both their value of K and $\Delta\mathbf{F}_{\mathrm{ML}_i}$ and the value of n can then be determined. As previously explained, this can be done by separately titrating ligand with excess macromolecule to determine $\Delta\mathbf{F}_{\mathrm{ML}_i}$ and then fitting Eq. (24) to F vs $[\mathrm{L}]_0$ data to determine n and K. A version of this strategy is to obtain a ΔF vs $[\mathrm{M}]_0$ at constant $[\mathrm{L}]_0$ data set and a ΔF vs $[\mathrm{L}]_0$ at constant $[\mathrm{M}]_0$ data set and to simultaneously fit Eqs. (9a) and (24), respectively, to these two data sets with the same value of K, $\Delta\mathbf{F}_{\mathrm{ML}_i}$, and n.[19] Also, we have developed a computer algorithm for simultaneously fitting ΔF vs $[\mathrm{L}]_0$ data sets obtained at different $[\mathrm{M}]_0$, for a general model of n equivalent, noninteracting sites, to obtain K, $\Delta\mathbf{F}_{\mathrm{ML}_i}$, and n (compiled copy of program available on request). Essentially, data obtained at higher $[\mathrm{M}]_0$ help to define the value of n, and data at lower $[\mathrm{M}]_0$ help to define K in the fit. The latter procedures can work either when monitoring changes in the fluorescence of the macromolecule or the ligand.

A method for graphically estimating the stoichiometry of an unknown complex is the Job method of continuous variation. In an example of this approach, one would measure the fluorescence of the ligand as a function of the $[\mathrm{L}]_0/[\mathrm{M}]_0$ ratio. This is typically done by mixing various portions of

[19] P. M. Horowitz and N. L. Criscimagna, *Biochemistry* **24,** 2587 (1985).

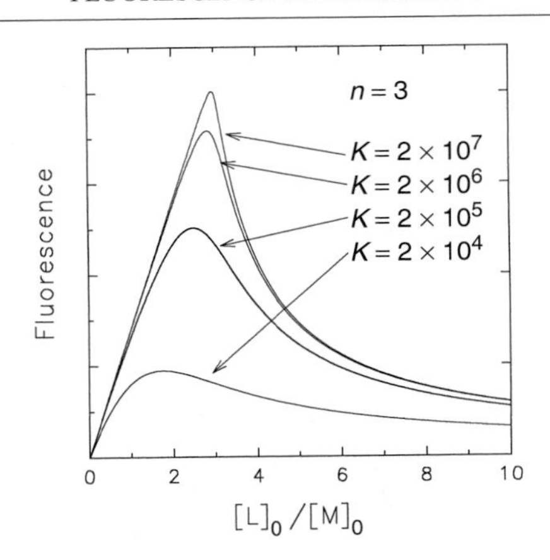

FIG. 5. Simulated Job plot of fluorescence intensity of a ligand as a function of the $[L]_0/[M]_0$ ratio for an $n = 3$ identical, noninteracting site model with the K values given. Notice that when K is large, the peak in the plot is approximately equal to the expected [L]/[M] ratio of 3. However, when K is small, the apparent position of the peak can lead to mistaken estimates of n. The simulation assumes that the free ligand fluoresces and its emission is enhanced on binding.

approximately equimolar solutions of $[L]_0$ and $[M]_0$. From the maximum of such a plot (see Fig. 5 for a simulation of an $n = 3$ identical, noninteracting site model), the stoichiometry of the complex is estimated. This method works when the system approaches the stoichiometric limit (i.e., when K and the sum of $[M]_0 + [L]_0$ are large) and when the $\mathbf{F}_{ML_i}/\mathbf{F}_L$ ratio is large. As shown in Fig. 5, the stoichiometry is poorly determined by this graphic procedure when K is small.

An alternate, general procedure for analyzing raw ΔF vs $[L]_0$ data to evaluate the value of n and the most appropriate binding model has been presented by Halfman and Nishida[20] and Bujalowski and Lohman.[21] This procedure involves collecting ΔF vs $[L]_0$ data at two or more different $[M]_0$ concentrations and then interpolating the curves to obtain $\sum \upsilon_i$ vs [L] as follows, where $\sum \upsilon_i = \sum iX_{ML_i}$ is the sum of the degree of saturation of each individual site. The procedure is based on the fact that the concentration of free ligand, [L], will be the same for any points along the binding profiles that have the same $\Delta F/[M]_0$. That is, for a pair of curves such as those in Fig. 4A, a particular value of $\Delta F/[M]_0$ (see horizontal line) will occur at pairs of $[L]_0$ and $[M]_0$, in the two separate titrations, such that

$$[L] = [L]_0 - [M]_0 \sum \upsilon_i \tag{25}$$

[20] C. J. Halfman and T. Nishida, *Biochemistry* **11,** 3493 (1972).
[21] W. Bujalowski and T. M. Lohman, *Biochemistry* **26,** 3099 (1987).

will be the same. Equating this unknown value of [L] for two interpolated pairs of points on the graph (corresponding to $[L]_{0,1}$, $[M]_{0,1}$ and $[L]_{0,2}$, $[M]_{02}$), we obtain

$$\sum \upsilon_i = ([L]_{0,1} - [L]_{0,2})/([M]_{0,2} - [M]_{0,1}) \tag{26}$$

Using this value of $\sum \upsilon_i$, [L] can then be calculated from Eq. (25). A series of $\sum \upsilon_i$ vs [L] points can then be analyzed to characterize the binding process.

Other Fluorescence Signals

Other fluorescence parameters that can be measured as a function of degree of binding are the emission, λ_{max}, the emission anisotropy, r, and the mean fluorescence decay time, τ. As discussed elsewhere for the case of protein-unfolding reactions,[22] λ_{max} and r data are complicated by the fact that their value is determined both by the mole fraction of states and by the fluorescence quantum yield of the states. This problem can be easily appreciated for the λ_{max} of spectra; if the emission of a free macromolecule occurs at 350 nm and that of a ligand-bound state occurs at 330 nm, then the apparent λ_{max} at intermediate states of ligation (i.e., when there is a mixture of free and bound states) will reflect the mole fraction of these states only if the fluorescence quantum yield of the two states is the same. If, for sake of argument, the bound state fluoresces only 10% as much as does the free state, then the latter will dominate the apparent λ_{max}. A rigorous analysis of the shift in emission maximum on binding would require a fitting of the complete spectrum as the sum of basis spectra for the free and complexed species (with the contribution of each component spectrum being related to the mole fraction of species).

The problem with measurements of emission λ_{max} is probably obvious to most researchers, but there is a similar problem with emission anisotropy (and this problem seems not to be widely appreciated). The steady-state anisotropy of a mixture of species is given by

$$r \approx \sum X_i \Phi_i r_i / \sum X_i \Phi_i \tag{27a}$$

$$r \approx \sum f_i r_i \tag{27b}$$

where Φ_i is the fluorescence quantum yield of species i [or one could use the molar fluorescence of species i, as we have done in Eqs. (18)–(24)], and $f_i \approx X_i \Phi_i / \sum X_i \Phi_i$ is the fractional contribution to the steady-state fluorescence of component i at λ_{ex} and λ_{em}. Thus, the total contribution of a given species to the observed steady-state fluorescence anisotropy de-

[22] M. R. Eftink, *Biophys. J.* **66,** 482 (1994).

pends on both its mole fraction and its relative fluorescence intensity. Species that fluoresce more intensely contribute disproportionally to the observed r. Consequently, to use anisotropy measurements to monitor a ligand–macromolecule interaction, a researcher should either show that the fluorescence intensities of the free and bound state(s) are nearly the same or include the relative fluorescence intensity f_i as a parameter in fitting r versus $[L]_0$ data.

Even if the binding process is 1:1, the above distortions in λ_{max} and r data (due to disproportional contributions from one species to the signal) can significantly affect the recovered association constant, if the free and bound states do not have nearly equal fluorescence intensity. This is illustrated by the simulations in Fig. 6. Here we have assumed that the binding of a ligand to a fluorescent protein is monitored via changes in the anisotropy of the latter as a function of ligand concentration. We assume that the binding is 1:1, with true association constants of $2 \times 10^5\ M^{-1}$ and $2 \times 10^7\ M^{-1}$ in Fig. 6A and B. If the relative fluorescence intensities of free and bound macromolecule are the same, then the correct K is recovered in fitting r vs $[L]_0$ data. However, if the liganded macromolecule is more

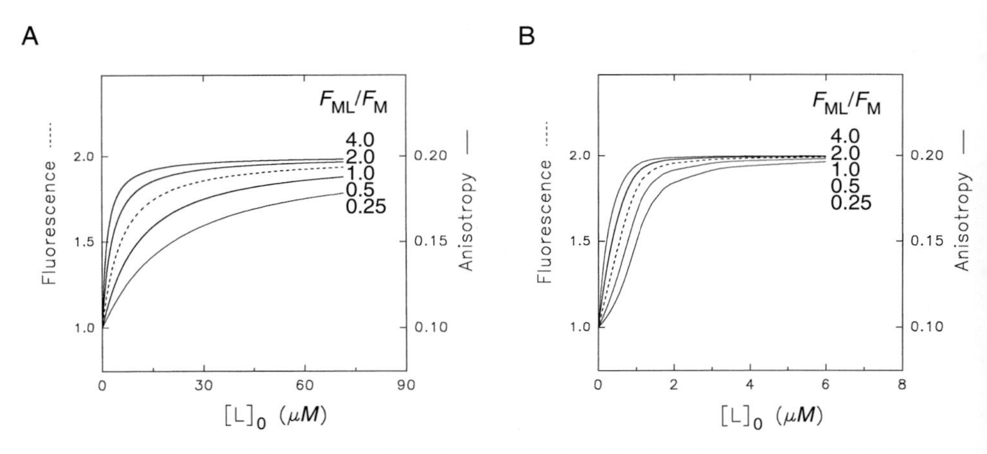

FIG. 6. Simulated fluorescence intensity and anisotropy binding data for a single binding site model, illustrating how differences in intensities can skew the anisotropy profiles. (A) $K = 2 \times 10^5\ M^{-1}$ (modest affinity) and (B) $K = 2 \times 10^7\ M^{-1}$. In each case $r_M = 0.1$ and $r_{ML} = 0.2$. Ratios of the fluorescence intensity (quantum yield) of the ML complex to that of M are (curves from top to bottom) 4, 2, 1, 0.5, and 0.25. The dashed curve for F_{ML}/F_M is also the curve that would be obtained from intensity data and gives a true tracking of the population of the complex. If Eq. (27) is not used, an overestimate of K will be obtained when $F_{ML}/F_M > 1$ and an underestimate of K will be obtained when $F_{ML}/F_M < 1$. Note that for high-affinity binding (i.e., when $K > [M]_0$) the apparent binding isotherms from anisotropy data can be distorted and can appear to be sigmoidal when $F_{ML}/F_M < 1$.

fluorescent (or less fluorescent), then the recovered K values will be larger (or smaller) than the true value. For tight binding systems, this problem is more serious, because the apparent isotherms can be distorted and can even appear to be sigmoidal (when $\mathbf{F}_{ML}/\mathbf{F}_M$ is less than 1). Because it has become popular to use fluorescence anisotropy in binding assays, this consideration is important and researchers should be aware that apparent K values are valid only if the bound and free forms of the species have similar fluorescence intensities.

Practical Considerations

When using fluorescence to study a binding reaction, a requirement is that the signal be linearly related to the concentration of species. A nonlinear dependence of the fluorescence intensity on concentration will most likely be due to self-aggregation of the species. Nonlinearity can also be caused by absorptive screening (inner filter) effects. Corrections for the latter have been widely discussed.[23–25] The simplest correction factor, which assumes that fluorescence emanates from a central point in a spectroscopic cell, is

$$C = 10^{(\Delta A_{ex}/2 + \Delta A_{em}/2)} \tag{28}$$

where ΔA_{ex} and ΔA_{em} are the additional absorbance at the excitation and emission wavelengths owing to the added titrant. A more rigorous correction is provided by

$$C = 2.303A(x_2 - x_1)/(10^{-Ax_1} - 10^{-Ax_2}) \tag{29}$$

where A is the total absorbance and x_1 and x_2 are the distances from the front of the cell to the points at which the collection of the emitted light begins and ends (i.e., the unmasked region of the cell). Other correction factors have also been presented. In each case, the observed fluorescence signal must be multiplied by the correction factor, C, to obtain the corrected signal. For Eqs. (28) and (29) for C, the simpler relationship agrees with the second, within 5%, up to an absorbance of 0.5. Of course, one can minimize the need for such corrections by using smaller path cells and/or by selecting an excitation wavelength (e.g., at the red edge of the absorbance band) where the total absorbance is relatively small.

[23] C. A. Parker, "Photoluminescence of Solutions," Chap. 3. Elsevier, Amsterdam, 1968.
[24] B. Birdsall, R. W. King, M. R. Wheeler, C. A. Lewis, S. R. Goode, R. B. Dunlap, and G. C. K. Roberts, *Anal. Biochem.* **132,** 353 (1983); J. F. Holland, R. E. Teets, P. M. Kelly, and A. Timnick, *Anal. Chem.* **49,** 706 (1977).
[25] L. D. Ward, *Methods Enzymol.* **17,** 400 (1985).

When monitoring apparent changes in the fluorescence of a macromolecule (by titrating with nonabsorbing ligand), it is possible that binding will induce changes in the absorption spectrum of chromophores in the macromolecule, as well as changes in fluorescence efficiency. Such changes in absorbance of the fluorescing chromophore will be most dramatic in ascending/descending regions of the spectrum. Other than the fact that they may enhance or attenuate a fluorescence change, such changes in absorbance are usually not a problem, because absorbance changes are also linearly related to the population of species.

Control of temperature and solution conditions is important in binding studies. The fluorescence of most fluorophores, as well as association constants, is sensitive to temperature. Also, care should be taken to control pH during a titration, because some fluorescence signals can be intrinsically pH dependent. Most commonly used buffer components can be obtained in forms of adequate optical purity (i.e., negligible absorbance above 230 nm) for fluorescence studies. Imidazole and histidine should be avoided as buffers, because they can act as fluorescence quenchers. Likewise, phosphate can be a quencher of tyrosinate fluorescence. Certain proteins and ligands require particular precautions. For example, NADH tends to oxidize and fresh samples are required daily. In studies of the binding of NAD^+ to alcohol dehydrogenase, it is essential that ethanol be avoided or exhaustively removed by dialysis (so that NADH is not formed). In studies with membrane proteins, certain detergents have appreciable fluorescence. The experimenter must also be aware and take action to minimize the possibility that a sample becomes photolytically or chemically degraded during the course of a titration study. An example of such a photoinduced reaction is given in connection with Fig. 7.

One of the advantages of fluorescence binding studies is that they enable large association constants to be determined, but this will be possible only if the experiment is done under conditions in which the concentration of free ligand is approximately equal to $1/K$. This requires that the total concentration of macromolecule sites also be in the range of or less than $1/K$. For large values of K, this can lead to experimental difficulties even for fluorescence. To make fluorescence measurements at concentrations below 10 nM, it is best to use fluorophores such as fluorescein, rhodamine, or pyrene, which have higher molar extinction coefficients and quantum yields (usually) than have intrinsic tryptophan residues. There are additional demands on the purity of reagents and the need to recognize and minimize scattered light, photolysis, and adsorption to the cells, tubing, syringes, etc., and the need to achieve equilibrium (i.e., due to slow off rate constants; see below). Syringes and pipetting devices should be "poisoned" with the reactants or treated with a siliconizing agent to control adsorption. Dust particles should be removed from both ligand and proteins solutions by Millipore filtration or by centrifugation with a microcentrifuge. The latter method is preferred when attempting to measure weak fluorescence signals, because fluorescent material can be washed from the filters. Also, it is necessary to have accurate values for $[L]_0$ and $[M]_0$, which may be the greatest limitation for the determination of accurate values of K. In cases of extremely large association constants, it is often possible to determine the dissociation rate constant (e.g., by displacing a bound fluorescent with an excess of a nonfluorescent competing ligand). If the association rate constant can also be measured (or estimated), then an apparent K can be calculated as the ratio of the rate constants.

A minor problem in performing fluorescence measurements is polarization bias related to the photoselection that occurs owing to the natural polarization of the exciting light; the effect that this photoselection has on intensity measurements in standard 90° detection geometry; and the consequences that changes in the anisotropy of the emitting fluorophore, on binding, have on the measured fluorescence intensity. This problem has been discussed by Shinitzky,[26] Mielenz and co-workers,[27] and Badea and Brand.[28] In ligand-binding studies it becomes most significant when monitoring the fluorescence of a small ligand as it binds to a large macromolecule (and for which the magnitude of the true fluorescence change is not large). The free ligand will be able to rotate rapidly, depolarizing its fluorescence so that its intensity will show a negligible effect owing to photoselection. When bound to the macromolecule the ligand will be more immobilized and will have its fluorescence attenuated by photoselection. This photoselection occurs even when using natural light (because excitation occurs preferentially in a plane that is perpendicular to the excitation axis). In the 90° detection geometry and with natural light for excitation, the effects of photoselection can be eliminated by using magic angle detection with an emission polarizer oriented at 35° to the vertical axis. This polarization bias is almost always ignored in steady-state measurements and probably causes only a small additional contribution to a change (maybe a few percent) in fluorescence intensity when a fluorescent ligand binds to a macromolecule.

When analyzing fluorescence binding data, regardless of the strategy employed (i.e., change in macromolecule or ligand fluorescence), then F versus $[L]_0$ and $[M]_0$ data should be fitted by nonlinear least squares (NLLSQ) with an equation related to a relevant model. We routinely use programs that employ Johnson and Fraiser's NONLIN.[29] In each case, the

[26] M. Shinitzky, *J. Chem. Phys.* **56,** 5979 (1972).

[27] K. D. Mielenz, E. D. Cehelnik, and R. L. McKenzie, *J. Chem. Phys.* **64,** 370 (1976).

[28] M. G. Badea and L. Brand, *Methods Enzymol.* **61,** 378 (1979).

[29] M. L. Johnson and S. G. Fraiser, *Methods Enzymol.* **117,** 301 (1985).

analysis includes the calculation of free ligand concentration, [L], from $[L]_0$ and $[M]_0$ by solution of the corresponding polynomial equation (via the Newton–Raphson procedure, when necessary). NONLIN is well designed for performing these analyses because it readily allows for two independent variables (e.g., $[L]_0$ and $[M]_0$) in its input files. This fitting routine also provides a rigorous calculation of the confidence intervals for the fitting parameters and it can perform global analyses over multiple data sets (e.g., F versus $[L]_0$ data obtained with different $[M]_0$). We have compiled programs for most of the models in Table I and we will provide them on request. For routine and rapid analyses we also use other commercial NLLSQ programs. Another noteworthy NLLSQ program for analyzing fluorescence binding data is the BIOEQS program of Royer and co-workers,[30,31] which uses a numerical solver-based approach and is designed for the simultaneous analysis of multiple, linked data sets. It goes without saying that direct NLLSQ fits are preferred over linearly transformed data, although the use of linear plots (e.g., Scatchard, Hill) can aid in the visual display of data. Also, when presenting data over a broad range of concentration, a semilogarithmic plot is useful.

A standard procedure for performing a fluorescence binding experiment is to place one of the reactant solutions in a cuvette, add (with mixing) aliquots of the second reactant, and measure the fluorescence signal after a brief period of time (correcting the signal for dilution of the fluorescent species and for the inner filter effect). Measurements can also be made with a series of prepared solutions, having various ratios of ligand and macromolecule. It is possible to automate the titration procedure by the use of computer-controlled syringe pumps and A/D acquisition of the fluorescence signals. In our laboratory we often use a computer-controlled syringe pump in combination with different spectrophotometers. With such automation strategies, one must be concerned that equilibrium is reached after each addition of reactant. The association rate constant of a ligand with a macromolecule is often near the diffusion limit, but dissociation rate constants are much smaller. For a tight binding system, there should be some concern about reaching equilibrium, and the exact procedure (extent of stirring, order of mixing, for ternary systems) should be examined if there is doubt about whether the signals are equilibrated.

An example of a fluorescence binding study is presented as follows (see caption to Fig. 7).

[30] C. A. Royer, W. R. Smith, and J. M. Beechem, *Anal. Biochem.* **191,** 287 (1990).
[31] C. A. Royer, *Anal. Biochem.* **210,** 91 (1993).

Example Studies

This section briefly discusses several types of studies in which fluorescence has been used to monitor binding to a protein. Some of these examples involve a change in the intrinsic fluorescence of the protein; other examples involve changes in the fluorescence of the ligand.

Calcium Binding to Calcium-Binding Proteins: Example of Change in Intrinsic Fluorescence on Binding, Complicated by Existence of Multiple Binding Sites

Parvalbumin, troponin C, calmodulin, calbindin, oncomodulin, and α-lactalbumin are among a family of calcium-binding proteins.[32–34] Forms of these proteins from certain species contain tryptophan residues and the fluorescence intensity of these tryptophans has been found to change dramatically on association/dissociation of Ca^{2+}. For example, the fluorescence of the single tryptophan in cod parvalbumin shifts from a λ_{max} of 316 to 340 nm and its quantum yield decreases on removal of the two Ca^{2+} that bind to this protein. This change provides a convenient way to study the thermodynamics and kinetics of Ca^{2+} binding. A complication is that these proteins contain multiple Ca^{2+} sites, two for parvalbumin and oncomodulin, four for troponin C and calmodulin; this raises the question of whether the binding sites are identical and interacting.

Most common forms of calmodulin, troponin C, and parvalbumin do not have tryptophan residues. Several research groups have taken advantage of this feature to engineer single tryptophan residues into positions near the known binding sites for the Ca^{2+} ions.[32,34,35] The binding of Ca^{2+} causes a change in fluorescence of these single tryptophan mutants. Arguing that the fluorescence changes should primarily reflect binding to the nearest Ca^{2+} site, these workers have determined the corresponding "site" association constants. For example, Trigo-Gonzalez and co-workers[34] have prepared mutants of troponin C having Trp residues in each of two different Ca^{2+}-binding domains. A reporter group was placed before the Ca^{2+}-binding loops I (F29W) and III (F105W) in this four-site protein. Titration of Ca^{2+} into these mutants causes an increase or a shift in fluorescence of these tryptophan probes, enabling binding isotherm data to be collected and

[32] M.-C. Kilfoffer, M. Kubina, F. Travers, and J. Haiech, *Biochemistry* **31,** 8098 (1992).

[33] K. P. Koshe and L. M. G. Heilmeyer, Jr., *Eur. J. Biochem.* **117,** 507 (1981).

[34] G. Trigo-Gonzalez, K. Racher, L. Burtnick, and T. Borgford, *Biochemistry* **31,** 7009 (1992).

[35] T. L. Pauls, I. Durussel, J. A. Cox, I. D. Clark, A. B. Szabo, S. M. Gagné, B. D. Sykes, and M. W. Berchtold, *J. Biol. Chem.* **268,** 20897 (1993).

analyzed. The tryptophan in F29W senses a weaker binding domain and the tryptophan in F105 senses a stronger binding domain for Ca^{2+} ions. When both tryptophans were incorporated into the protein (F29W/F105W), the fluorescence titrations became biphasic, enabling the association constant for both high- and low-affinity sites to be determined. In a similar study, Kilhoffer and co-workers[32] have placed tryptophans at five different places in calmodulin and have used these probes to site-specifically monitor Ca^{2+} binding (equilibrium and kinetics).

Binding of Indoleacrylic Acid to Tryptophan Aporepressor: Quenching by Energy Transfer, Complicated by Existence of Two Sites

Tryptophan aporepressor from *Escherichia coli* is a homodimeric protein. It binds tryptophan, and its small analogs, and also forms a ternary complex with specific operator DNA. Indoleacrylic acid (IAA) is a tryptophan analog that happens to have a fairly high affinity for the protein. On binding, IAA quenches almost all of the intrinsic (tryptophanyl) fluorescence of Trp repressor.[36] Owing to the favorable overlap between the emission of the protein and the absorbance of IAA in the 300- to 340-nm range, this quenching is believed to be due to resonance energy transfer. In fact, the four intrinsic tryptophan residues (two in each subunit) are each well within the $R_0^{2/3}$ distance for 50% efficiency of energy transfer. Consequently, the binding of IAA to one binding site quenches the fluorescence of both subunits. This type of situation (i.e., binding to one site causing a quenching over multiple subunits) has been shown to occur with other protein–ligand systems, including NADH binding to deydrogenases (see below). As a result of this phenomenon, the analysis of fluorescence vs $[L]_0$ data can easily be misinterpreted, as is illustrated later in this chapter and as was discussed in connection with Fig. 2B.

Shown in Fig. 7 are raw binding data (drop in fluorescence vs $[IAA)_0$) for the addition of IAA to Trp aporepressor. Figure 7 gives a typical experimental procedure for collecting such data. Note that this protein–ligand system presents the additional problem that the ligand undergoes a photoinduced isomerization; consequently we had to minimize the time of exposure. Shown in Fig. 8 are relative fluorescence data (corrected for screening and dilution) versus total IAA concentration for two concentrations of the protein. The dashed lines are fits of the data to a simple 1:1 binding model [i.e., Eq. (8), with a single K and $\mathbf{F}_{ML}$, neglecting the fact that the protein is a homodimer]. At low protein concentration, the fit is good, with $K = 2.7 \times 10^5\ M^{-1}$. At high protein concentration the fit is poor

[36] D. Hu and M. R. Eftink, *Arch. Biochem. Biophys.* **305,** 588 (1993).

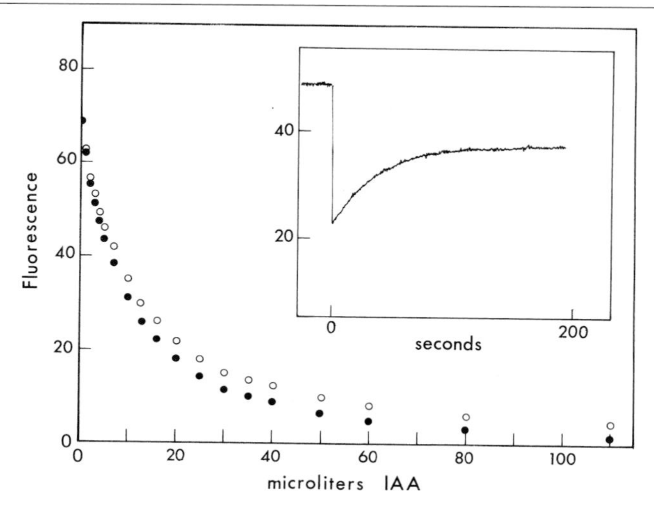

FIG. 7. Fluorescence titration of a 2.0-ml solution of *E. coli* Trp aporepressor with IAA at 25° in 0.01 M potassium phosphate, pH 7.5 buffer (solid circles, raw data). The protein solution, which was 2.5×10^{-6} M (expressed as monomer), was placed in a standard 1×1 cm cell within a Perkin-Elmer MPF44A spectrofluorometer, equipped with a thermostatted cell holder and magnetic stirrer. Fluorescence was measured with excitation at 295 nm and emission at 340 nm (5-nm slits). Each signal was measured for about 10–15 sec. Following measurement of the initial signal (F_M) of the apoprotein, small aliquots (5–25 μl) of a concentrated (2×10^{-3} or 1×10^{-4} M) IAA solution were added, using a positive displacement micropipette, and the signals were measured. A total of 200 μl of titrant was added. The addition of IAA causes almost a total quenching of the fluorescence of the protein. The signals, protein, and ligand concentrations were corrected for dilution effects, using the equations in note 11. Because IAA absorbs light at both the excitation and emission wavelengths, the absorptive screening correction factor from Eq. (28) was applied. The open circles are data corrected for both dilution and screening. The relative F versus $[\mathrm{IAA}]_0$ data were then fitted as described in the caption to Fig. 8.

The inset shows the time-dependent increase in fluorescence that results from exposing an unstirred solution to UV light. This light-sensitive recovery of signal was found to be greatly diminished by stirring. We have shown that IAA undergoes a photoinduced *trans* → *cis* isomerization,[36] which apparently converts the ligand to a form that does not bind as well to the protein. To minimize the effect of this reaction, we have stirred the solutions, used narrow slits, closed the shutter during additions, and measured the fluorescence signal for only 10–15 sec. This example illustrates a possible complication in fluorescence titration studies. (Reprinted from Ref. 36, with permission.)

with this model; a forced fit (dashed line) yields $K = 29.4 \times 10^5\ M^{-1}$, much higher than that at low concentration. This apparent dependence of K on protein concentration might be misinterpreted as indicating some type of change in the protein (i.e., self-association) at higher concentration. Actually, the problem is that the 1:1 model first used to analyze the data is

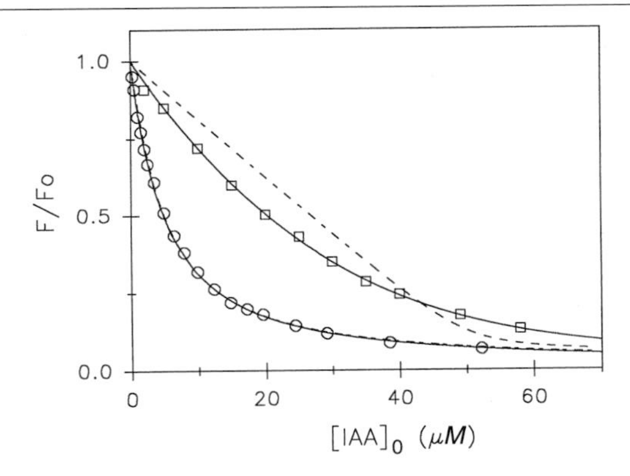

FIG. 8. Fluorescence titration data for the binding of IAA to Trp aporepressor at two total protein (monomer) concentrations, $[M]_0 = 2.5 \times 10^{-6}$ *M* (○) and 50.0×10^{-6} *M* (□). The dashed curves (which are drawn over by the fit below for the lower concentration data) are fits of the 1:1 binding model. An excellent fit is obtained for the lower $[M]_0$ data with $K = 2.72 \pm 0.07 \times 10^5\ M^{-1}$. A very poor fit (dashed curve) is obtained for the higher $[M]_0$ data with $K = 29 \pm 60 \times 10^5\ M^{-1}$. When we take into account the fact that the protein is a dimer and fit the data to an $n = 2$ identical, noninteracting site model, with the further allowance that changes in fluorescence are not necessarily additive for the two binding steps (i.e., there can be a greater degree of quenching for binding the first ligand), then we obtain the solid curves at both low and high protein concentrations. This fit is in terms of the following equation, $F = (1 + 2\mathbf{F}_{ML}K[L] + \mathbf{F}_{ML_2}K^2[L]^2)/(1 + 2K[L] + K^2[L]^2)$, where [L] is related to $[L]_0$ and $[M]^0$ by a third-order polynomial given in Table I (under identical, noninteracting sites, $n > 1$). For both the low and high concentration data we obtained a good fit with $K = 1.8 \pm 0.4 \times 10^5$ and $1.20 \pm 0.12 \times 10^5\ M^{-1}$, respectively, and with $\mathbf{F}_{ML} \approx 0.3$ and $\mathbf{F}_{ML_2} \approx 0.02$ (i.e., binding of the first IAA molecule quenches approximately 70% of the fluorescence, subsequent binding of the second IAA quenches 98% of the fluorescence). (Reprinted from Ref. 36, with permission.)

incorrect. Instead, a good fit is obtained with a 1:2 model (two identical, noninteracting sites; equations in Table I with $K_1 = K_2$ and $\mathbf{F}_{M_1L} = \mathbf{F}_{M_2L}$), which includes parameters for the fluorescence of the singly and doubly ligated states, $\mathbf{F}_{ML}$ and $\mathbf{F}_{ML_2}$. This model fits the data at low and high protein concentration (see solid lines in Fig. 7) with nearly the same *K* values of $\sim 1.0 \times 10^5\ M^{-1}$ for both data sets and with $\mathbf{F}_{ML} \approx 0.3$ an $\mathbf{F}_{ML_2} \approx 0.02$. The reason that the simple 1:1 model fails to describe the data at high protein concentration is that one is essentially performing a stoichiometric titration of sites with IAA, and, as the first site is filled with IAA, it quenches most of the fluorescence of the entire protein. An initial tangent drawn to the *F* vs $[IAA]_0$ data at high protein concentration intercepts the $[IAA]_0$ axis at a point that is approximately two-thirds the total protein concentration (as monomer), which is possible only if the ligand causes quenching of both subunits (i.e., the signal change is not linear in the coverage of sites). We have measured IAA binding to Trp aporepressor over a 200-fold range of protein concentration and have found an excellent, global fit of the 1:2 identical, noninteracting sites model to a single K value. This phenomenon of a signal change being nonlinear in the coverage of sites has previously been addressed in studies of the binding of NADH to dehydrogenases, as is discussed in the next section.

Binding of Coenzymes and Nucleotides to Proteins, Including Enhancement of Fluorescence of NADH on Binding

The coenzymes NADH and NADPH, and their oxidized forms, NAD^+ and $NADP^+$, are important ligands for several dehydrogenases and other oxidoreductases. NAD^+ (and $NADP^+$) is nonfluorescent, but it usually causes a modest quenching of tryptophanyl fluorescence of proteins on binding. NADH (and NADPH), however, has an absorbance band centered at 339 nm and fluoresces with a maximum at 470 nm in water. The emission maximum of NADH blue shifts and its quantum yield usually increases dramatically on binding to proteins.[37,38] The extent of the blue shift and the amount of fluorescence enhancement also are sensitive to the presence of a second ligand and have been used to describe the microenvironment of the bound NADH. This fluorescence enhancement, as well as resonance energy transfer quenching of tryptophanyl residues, provide a convenient means of studying the binding of NADH to dehydrogenases. Theorell and Tatemoto[18] and Stinson and Holbrook[39] have given extensive treatment and examples of the use of such fluorescence methods to study NADH binding.

Studies using the enhancement of NADH emission on binding are relatively straightforward.[40] Dehydrogenases are multisubunit enzymes, but for many of them there does not appear to be any interaction between the subunits. Consequently, a simple 1:1 model often can be applied when monitoring the fluorescence enhancement of NADH (because there is not a problem regarding the signal change being nonlinear with coverage of sites, as is the case for energy transfer quenching), with a single K and with $\mathbf{F}_{ML}/\mathbf{F}_L$ being the fluorescence enhancement ratio. For example, Theorell and McKinley-McKee[40] have used the 13-fold enhancement of NADH fluorescence on binding of this coenzyme to horse liver alcohol dehydrogenases, LADH, to determine the association constant of $\sim 3 \times 10^6\ M^{-1}$ to this dimeric protein. When the inhibitor isobutyramide is also present, the

[37] T. G. Scott, R. D. Spencer, N. J. Leonard, and G. Weber, *J. Am. Chem. Soc.* **92,** 687 (1970).
[38] B. Baumgarten and J. Hönes, *Photochem. Photobiol.* **47,** 201 (1988).
[39] R. A. Stinson and J. J. Holbrook, *Biochem. J.* **131,** 719 (1973).
[40] H. Theorell and J. S. McKinley-McKee, *Acta Chem. Scand.* **15,** 1811 (1961).

degree of fluorescence enhancement is 40-fold and the association constant for NADH increases by approximately 50-fold.

Alternatively, when NADH binding is monitored via quenching (resonance energy transfer) of intrinsic protein fluorescence, the phenomenon of the signal not being linear in the coverage of sites must be considered. This can be done, as discussed above for IAA binding to Trp aporepressor, by fitting the equations in Table I (two identical, noninteracting sites) to data obtained at different protein concentrations. Alternative approaches, in which the binding of NADH to an oligomeric protein is monitored by both its quenching of protein fluorescence and by the enhancement of NADH fluorescence (at high-affinity conditions; the latter signal changes are linear in the coverage of sites), have been used by Theorell and Tatemoto[18] and Stinson and Holbrook.[39] For example, for dimeric horse liver alcohol dehydrogenase, it has been shown that the binding of NADH to the first subunit quenches approximately twice as much as the protein fluorescence as does binding of the second NADH. That is, 65–70% of the total quenching of the protein's fluorecence occurs on binding to the first subunit, as opposed to 50% of the total quenching that would be expected if the quenching effect of NADH were localized to the subunit to which it is bound. This indicates that the quenching effect occurs across the subunit interface. This implicates the quenching process to involve resonance energy transfer, which has been shown by other studies.

Etheno analogs of NAD^+ (i.e., 1,N^6-ethenoadenine dinucleotide) and the other coenzymes and nucleotides (i.e., 1,N^6-ethenoadenosine triphosphate) are fluorescent (λ_{ex} ~325–350 nm; λ_{em} ~400 nm).[41,42] Their fluorescence is modestly sensitive to binding to proteins. Changes in ethenoadenosine fluorescence and quenching of tryptophanyl fluorescence (via energy transfer) provide complementary means of studying the binding of this type of ligand to proteins.[43] The binding of the nonfluorescent (e.g., NAD^+ and ATP) forms of the ligands can be determined by competition studies.[44] Examples of other fluorescent nucleotide analogs that have been used in binding studies are 3′(2′)-*O*-methylanthraniloyl-ATP, 3′-*O*-(dimethylamino)naphthoyl-ATP, 2′(3′)-*O*-(2,3,6-trinitrophenyl)-ATP, and dansyl-ATP.[45,46] The trinitrophenyl (TNP) group has little fluorescence when free in solution and, along with the so-called MANT and DAN nucleotide analogs, can show a large fluorescence increase on binding to proteins.

[41] J. A. Secrist, J. R. Barrio, N. J. Leonard, and G. Weber, *Biochemistry* **11,** 3499 (1972).
[42] N. J. Leonard, *Crit. Rev. Biochem.* **15,** 125 (1984).
[43] V. G. Neff and F. M. Huennekens, *Biochemistry* **15,** 4042 (1976).
[44] B. Birdsall, A. S. V. Burgen, and G. C. K. Roberts, *Biochemistry* **19,** 3723 (1980).
[45] L. Mayer, A. S. Dahms, W. Riezler, and M. Klingenberg, *Biochemistry* **23,** 2436 (1984).
[46] W. Bujalowski and M. M. Konowska, *Biochemistry* **32,** 5888 (1993).

ANS Binding to Proteins: Fluorescence Enhancement on Binding

8-Anilino-1-naphthalenesulfonic acid (ANS) and its dimeric form, bis-ANS, are widely used fluorescence probes for studies with proteins and membranes.[8,9,19] ANS has a low fluorescence quantum yield in water ($\Phi = 0.003$), but its yield increases dramatically ($\Phi = 0.4$ in ethanol) and blue shifts when it is in an apolar microenvironment. ANS binds to a variety of proteins. It is particularly prone to bind to nucleotide-binding sites[47,48] and to apolar pockets or surfaces of proteins, including the apolar surfaces found in partially unfolded proteins.[49–53] For example, ANS and similar fluorophores have been found to bind a Ca^{2+}-induced state of calmodulin (having an exposed hydrophobic surface), with there being a 27-fold enhancement in the fluorescence of ANS and a blue shift from 540 to 488 nm.[49] There may be multiple binding sites of ANS to a protein, which complicates matters. Shi and co-workers have studied the binding of ANS to the native and molten globule state of DnaK.[52] Whereas a single binding site for ANS exists in the native state, three binding sites were found for the molten globule state (at 0.8 *M* guanidine hydrochloride, assuming that the fluorescence enhancement of ANS is the same for each site).

Haptens Binding to Antibodies

The binding of fluorescein to monoclonal fluorescyl antibody (IgG) results in the quenching of 90% of the fluorescence of the ligand; this quenching can be used to monitor the thermodyamics and kinetics of binding.[54,55] 2,4-Dintrophenyl-labeled haptens quench the intrinsic tryptophanyl fluorescence of anti-DNP antibodies[56] or the fluorescence of fluorescein isothiocynate-labeled antibodies.[57] Quenching by DNP is believed to occur by resonance energy transfer. For either the fluorescein or 2,4-DNP-based hapten, fluorescence provides a convenient method for determining association constants in the range of 10^6–10^{10} M^{-1}.

[47] S. H. Yoo, J. P. Albanesi, and D. M. Jameson, *Biochim. Biophys. Acta* **1040,** 66 (1990).
[48] G. A. Griess, S. A. Khan, and P. Serwer, *Biopolymers* **31,** 11 (1991).
[49] D. C. LaPorte, B. M. Wierman, and D. R. Storm, *Biochemistry* **19,** 3814 (1980).
[50] Y. Goto and A. L. Fink, *Biochemistry* **28,** 945 (1989).
[51] G. V. Semisotnov, N. A. Rodionova, O. L. Razgulyaev, V. N. Uversky, A. F. Gripas, and R. I. Gilmanshin, *Biopolymers* **31,** 119 (1991).
[52] L. Shi, D. R. Palleros, and A. L. Fink, *Biochemistry* **33,** 7536 (1994).
[53] W. R. Kirk, E. Kurian, and F. G. Prendergast, *Biophys. J.* **70,** 69 (1995).
[54] E. V. Voss, Jr., R. M. Watt, and G. Weber, *Mol. Immunol.* **17,** 505 (1980).
[55] R. M. Watt and E. W. Voss, Jr., *Immunochemistry* **14,** 533 (1977).
[56] H. Eisen and J. McGuigan, *in* "Methods in Immunology and Immunochemistry," (W. C. Williams and M. Chase, eds.). Academic Press, New York, 1968.
[57] J. Erickson, P. Kane, D. Goldstein, and B. Baird, *Mol. Immunol.* **23,** 769 (1986).

Binding of Proteins to DNA

There is much interest in being able to determine the association constant for complexes between specific DNA sequences and proteins that bind DNA, including repressors, transcription factors, helicases, topoisomerases, and other proteins involved in DNA replication. Some of these complexes involve specific, high-affinity binding, whereas complexes with other proteins are nonspecific and weaker. Fluorescence methods have been developed to characterize such interactions in solution, as an alternative to filter binding and electrophoresis-based methods.

In several cases, the binding of proteins to DNA quenches the intrinsic tryptophanyl fluorescence of the protein. Examples of such quenching occur in the complexes formed between single-stranded binding protein and single-stranded nucleic acids.[21,58]

Because DNA has such weak intrinsic fluorescence, it is necessary to covalently attach (or insert) a fluorescent probe to DNA. One means of covalent attachment involves incorporation of a 5-(3-amino-l-propyl)-2′-deoxyuridine into an oligonucleotide, in place of thymidine, by standard phosphoramidite chemistry, followed by the derivatization of the propylamine group by a fluorescence probe (e.g., dansyl or fluorescein).[59,60] Fluorescent probes have also been attached by covalent linkage to a 5′-amino nucleotide analog placed at the end of one or both strands of an oligonucleotide.[61,62] Alternatively, the nucleic acid bases can be modified to a fluorescent analog or an analog can be synthetically substituted in the sequence. Treatment with chloroacetaldehyde can convert adenine in single-stranded poly(dA) to fluorescent ethenoadenosine bases.[63] 2-Aminopurine is a fluorescent base that can be incorporated into synthetic oligonucleotides.[64]

Interaction between such labeled DNA and a protein may cause a change in the fluorescence intensity or anisotropy of the probe. For example, LeTilly and Royer[65] have studied the interaction between Trp repressor and a specific operator DNA 25-mer oligonucleotide that was end labeled with fluorescein. On titrating the fluorescent 25-mer with Trp repressor, binding was observed as an increase in the anisotropy of the fluorescein.

[58] U. Curth, J. Greipel, C. Urbanke, and G. Maass, *Biochemistry* **32,** 2585 (1993).

[59] D. J. Allen, P. L. Darke, and S. J. Benkovic, *Biochemistry* **28,** 4601 (1989).

[60] C. R. Guest, R. A. Hochstrasser, C. G. Dupuy, D. J. Allen, S. J. Benkovic, and D. P. Millar, *Biochemistry* **30,** 8759 (1991).

[61] L. E. Morrison, T. C. Halder, and L. M. Stols, *Anal. Biochem.* **183,** 231 (1989).

[62] T. Heyduk and J. C. Lee, *Proc. Natl. Acad. Sci. U.S.A.* **87,** 1744 (1990).

[63] W. Bujalowski and M. J. Jezewska, *Biochemistry* **34,** 8513 (1995).

[64] T. M. Norlund, S. Andersson, L. Nilsson, R. Rigler, A. Gräslund, and L. W. McLaughlin, *Biochemistry* **28,** 9095 (1989).

[65] V. LeTilly and C. A. Royer, *Biochemistry* **32,** 7753 (1993).

With the high sensitivity of fluorescein emission (fluorescence anisotropy measurements made down to the 100 pM range), a protein–DNA association constant on the order of 10^{10} M^{-1} was determined. In another example, the binding of TATA-binding protein to an oligonucleotide 5′ end labeled with rhodamine X causes a change in the anisotropy of the label, thus enabling equilibrium and kinetic binding studies.[66] In these two cases, the fluorescence intensity of the probe was found not to change much on interacting with the protein, thus enabling the anisotropy data to be directly analyzed (see Fig. 6 and the related discussion for the problem that arises in anisotropy studies when the fluorescence intensity is much different for free and bound states). Bailey and co-workers[67] have performed a thermodynamic fingerprinting study of the interaction of the regulatory protein TyrR with a 42-bp oligonucleotide containing the TyrR strong box recognition sequence. They systematically moved the position of a fluorescein label (attached via a propyl amine linker) along the recognition sequence and performed a "fluorescence footprinting" analysis of the positional dependence of the binding constant and the effect of binding on the fluorescence parameters (decay times and rotational correlation times) of the probe.

Final Comment

Fluorescence spectroscopy is one of the most versatile, sensitive, and convenient methods for the characterization of the binding of ligands to macromolecules. This chapter has discussed experimental strategies for the use of fluorescence for this purpose, including a description of general binding schemes, a discussion of a number of experimental considerations, and a brief description of example studies that demonstrate various possibilities and limitations in the use of fluorescence.

Acknowledgments

This work was supported by NSF Grant MCB 94-07167. I thank Roxana Ionescu for useful advice in preparing this chapter.

[66] G. M. Perez-Howard, P. A. Weil, and J. M. Beechem, *Biochemistry* **35,** 8005 (1995).

[67] M. Bailey, P. Hagmar, D. P. Millar, B. E. Davidson, G. Tong, J. Haralambidis, and W. H. Sawyer, *Biochemistry* **34,** 15802 (1995).

[12] Fluorescence Methods for Studying Kinetics of Protein-Folding Reactions

By MAURICE R. EFTINK and M. C. R. SHASTRY

Introduction

Proteins adopt a unique three-dimensional structure as the result of the folding of a polypeptide that comprises a linear sequence of amino acids made under genetic direction. The "protein-folding problem" deals with how such a linear polymer can spontaneously acquire its tertiary structure. Although the study of this problem originally was motivated by fundamental interests, the advent of protein engineering has made this problem one of great practical importance. Equilibrium studies primarily characterize differences in the structure and stability of the native and unfolded states of a protein. Such equilibrium studies are essential in the characterization of a given protein, but it is necessary to study the kinetics of the folding process to understand further the mechanism or pathway by which the randomly coiled polypeptide is transformed to the unique native state.

There are a number of biophysical methods that can be used to monitor the kinetics of the folding and unfolding of a protein. Optical spectroscopic methods enable real-time tracking of the process, and, among spectroscopic methods, fluorescence has many advantages and applications.[1–3]

In this chapter we discuss some advances in the use of stopped-flow fluorescence spectroscopy (SFFS) to study the protein-folding problem. Although the purpose of this chapter is not to review protein folding per se, first we briefly give an overview of the topic and some of the important questions in this field. (We refer the reader to Schmid,[4,5] Hagerman and Baldwin,[6] and Utiyama and Baldwin[7] for detailed treatments on how to determine folding mechanisms from SF data. Also, we recommend the

[1] T. Jovin, *in* "Biochemical Fluorescence: Concepts" (R. Chen and H. Edelhoch, eds.), pp. 305–374. Marcel Dekker, New York, 1975.

[2] S. A. Levinson, *in* "Biochemical Fluorescence: Concepts" (R. Chen and H. Edelhoch, eds.), pp. 375–405. Marcel Dekker, New York, 1975.

[3] W. B. Dandiliker, J. Dandiliker, S. A. Levinson, R. J. Kelly, A. N. Hicks, and J. U. White, *Methods Enzymol.* **48,** 380 (1978).

[4] F. X. Schmid, *Methods Enzymol.* **131,** 70 (1986).

[5] F. X. Schmid, *in* "Protein Folding" (T. E. Creighton, ed.), pp. 197–241. W. H. Freeman and Co., New York, 1992.

[6] P. J. Hagerman and R. L. Baldwin, *Biochemistry* **15,** 1462 (1976).

[7] H. Utiyama and R. L. Baldwin, *Methods Enzymol.* **131,** 51 (1986).

METHODS IN ENZYMOLOGY, VOL. 278
0076-6879/97 $25

books edited by Creighton,[8] Pain,[9] and Vol. 131 of *Methods in Enzymology*[10] for several excellent chapters on various aspects of the protein-folding problem.) Second, we discuss advantages/disadvantages of fluorescence, as compared to other SF methods. We then present a series of sections dealing with new developments in SFFS. These include double jump strategies, studies involving the dye 8-anilino-1-naphthalenesulfonic acid (ANS), so-called "double kinetics" experiments that measure intensity and anisotropy decays following stopped-flow mixing, the addition of solute quenchers and specific ligands to perturb the fluorescence signals, kinetic measurements of the efficiency of energy transfer between sites on a protein, the use of diode array and rapid scanning detectors, the global analysis of SF fluorescence intensity and anisotropy data, and the combined acquisition and analysis of SF fluorescence and circular dichroism data. We present worked examples of the last three types of experiments.

Overview of Kinetics of Protein Folding

Proteins can be unfolded by addition of chemical denaturants, addition of acid or base, removal of a stabilizing ligand, or elevation of temperature or pressure. In studies of the kinetics of protein folding/unfolding, the most frequently employed experimental design involves rapid mixing of the protein with (or dilution from) urea or guanidine hydrochloride solutions. The examples that we discuss primarily involve such denaturant-induced reactions, in combination with rapid mixing devices. Temperature-jump[1] and pressure-jump[11] studies, ultrafast mixing,[12] and even laser-induced reactions (e.g., causing the dissociation of heme–CO bonds or causing a transient increase in temperature[13–17] are important types of measurements. Laser-induced reactions can provide the most rapid trig-

[8] T. E. Creighton (ed.), "Protein Folding." W. H. Freeman and Co., New York, 1992.

[9] R. H. Pain (ed.), "Mechanism of Protein Folding." IRL–Oxford University Press, New York, 1994.

[10] *Methods Enzymol.* **131** (1986).

[11] G. J. A. Vidugiris, J. L. Markley, and C. A. Royer, *Biochemistry* **34,** 4909 (1995).

[12] C.-K. Chan, Y. Hu, S. Takahashi, D. L. Rousseau, W. A. Eaton, and J. Hofrichter, *Biophys. J.* **70,** A177 (1996).

[13] C. M. Jones, E. R. Henry, Y. Hu, C. Chan, S. D. Lusk, A. Bhuyan, H. Roder, J. Hofrichter, and W. A. Eaton, *Proc. Natl. Acad. Sci. U.S.A.* **90,** 11860 (1993).

[14] R. B. Dyer, S. Williams, W. H. Woodruff, R. Gilmanshin, and R. H. Callender, *Biophys. J.* **70,** A177 (1996).

[15] A. Gershenson, C. J. Fisher, J. A. Schuaerte, R. M. Wolanin, A. Gafni, and D. G. Steel, *Biophys. J.* **70,** A175 (1996).

[16] C. M. Phillips and R. M. Hochstrasser, *Biophys. J.* **70,** A177 (1996).

[17] P. A. Thompson, W. A. Eaton, and J. Hofrichter, *Biophys. J.* **70,** A177 (1996).

gering of the protein transitions. However, we limit this chapter to stopped-flow, chemically induced folding/unfolding transitions.

For small, globular proteins, the kinetics of denaturant-induced unfolding reactions is often found to be a monoexponential process, indicating the existence of a single, dominant activation free energy barrier. More complicated unfolding kinetics responses (i.e., bi- or triexponential kinetic phases) can be observed for some proteins, particularly those that have multiple domains and/or native conformations. The kinetics of folding transitions usually are multiexponential, either because of the existence of multiple, slowly interconverting unfolded states or the existence of multiple energy barriers of similar height. The identification of the time constants with individual steps in a folding/unfolding mechanism and with particular molecular events is the challenge in such studies. Addressing the latter question involves characterizing the structure of folding intermediates. As we discuss in the following sections, certain fluorescence techniques provide ways to obtain low-resolution structural information about folding intermediates.

In typical denaturant-induced studies, a researcher should first obtain equilibrium data for the unfolding transition, using the particular spectroscopic method to be used in subsequent kinetics studies. We direct the reader to Refs. 18–23 for information regarding the practical aspects of performing such equilibrium studies and the possible interpretations of the resulting thermodynamic parameters. One of us has presented a discussion of advantages and limitations of the use of fluorescence methods in such equilibrium unfolding studies.[24]

The following equations are generally accepted to describe the thermodynamics of a denaturant-induced unfolding of a protein, in the simple case in which the transition is two-state in an equilibrium sense (i.e., if there are intermediates, they have an insignificantly low population at equilibrium).

$$\mathrm{N} \underset{k_{\mathrm{fold}}}{\overset{k_{\mathrm{unf}}}{\rightleftharpoons}} \mathrm{U} \tag{1}$$

[18] C. Tanford, *Adv. Protein Chem.* **21,** 1 (1970).

[19] J. A. Schellman, *Biopolymers* **17,** 1305 (1978).

[20] C. N. Pace, *Methods Enzymol.* **131,** 266 (1986).

[21] C. N. Pace, B. A. Shirley, and J. A. Thomson, *in* "Protein Structure and Function: A Practical Approach" (T. E. Creighton, ed.), pp. 311–330. IRL Press, Oxford, 1989.

[22] M. M. Santoro and D. W. Bolen, *Biochemistry* **27,** 8063 (1988).

[23] F. X. Schmid, *in* "Protein Structure and Function: A Practical Approach" (T. E. Creighton, ed.), pp. 251–285. IRL Press, Oxford, 1989.

[24] M. R. Eftink, *Biophys. J.* **66,** 482 (1994).

$$K_{un} = [U]/[N] \tag{2}$$

$$X_N = 1/Q, \qquad X_U = K_{un}/Q \tag{3}$$

where Q equals $1 + K_{un}$.

$$\Delta G^\circ(d) = \Delta G_0^\circ - m[\mathbf{d}] = -RT \ln K_{un} \tag{4}$$

K_{un} is the unfolding equilibrium constant (which will depend on the concentration of denaturant and other conditions), k_{unf} and k_{fold} are the apparent rate constants for the unfolding and folding reactions, X_i are mole fractions of the protein in the native and unfolded state, and Q is the partition function. In Eq. (4), $\Delta G^\circ(d)$ and ΔG_0° are the standard free energy changes for the unfolding transition in the presence and absence of denaturant, **d**, and $m(=-\delta\Delta G^\circ(d)/\delta[\mathbf{d}])$ is a parameter that describes the dependence of $\Delta G^\circ(d)$ on denaturant concentration. If fluorescence is used to monitor the transition and if F_N and F_U are the relative fluorescence intensities of the N and U states, respectively, then the observed fluorescence intensity at any denaturant concentration, $F(d)$, will be given by (assuming an identical absorbance of the N and U states)

$$F(d) = \sum X_i F_i = X_N F_N + X_U F_U \tag{5}$$

Shown in Fig. 1A are simulated data of $F(d)$ as a function of denaturant concentration as described by Eqs. (2)–(5). A nonlinear least-squares fit of such data can retrieve the thermodynamic parameters ΔG_0° and m. (Note that it is usually necessary also to include additional terms for the "baseline" slopes, i.e., the dependence of the fluorescence of each pure state on the concentration of denaturant, $F_N = F_{0,N} + s_N[\mathbf{d}]$ and $F_U = F_{0,U} + s_U[\mathbf{d}]$, where $F_{0,N}$ and $F_{0,U}$ are the fluorescence of N and U states in the absence of denaturant and s_N and s_U are slopes that describe the dependence of the fluorescence signal of N and U on denaturant concentration.[23,24]) As is discussed, it is necessary first to obtain equilibrium data to determine whether the expected amplitudes are recovered in transient unfolding data and to provide a value for $K_{un}(d)$ for testing different kinetic mechanisms. The existence of a significantly populated folding intermediate (in which case the transition is not two-state) can be discerned from a careful analysis of equilibrium data. Also, such data are needed to design kinetics studies so that the mixing experiments will result in jumps in conditions that cause a significant change in the population of conformational states.

An unfolding reaction can be initiated by rapid mixing of protein in a "native" solution with an "unfolding" solution, so that the final denaturant concentration is at or above the midpoint in Fig. 1; likewise, a folding

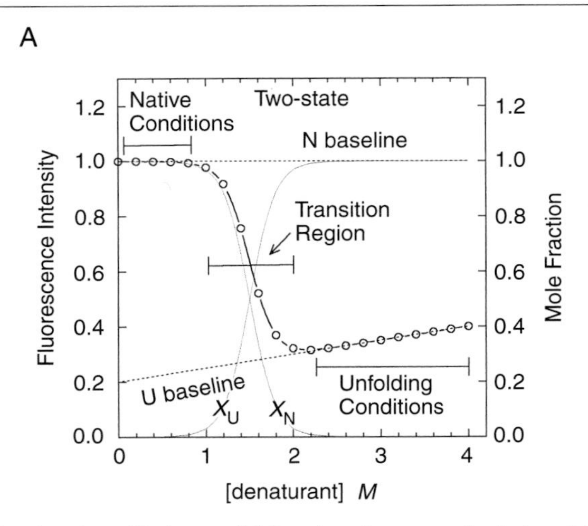

FIG. 1. (A) Simulated equilibrium unfolding data for a protein, induced by a denaturant, **d**, in a two-state manner and monitored by steady-state fluorescence intensity, F. Note the pre- and posttransition baseline slopes, with the latter usually being larger.[24] The simulation is for a case in which the fluorescence of the protein (i.e., intrinsic tryptophan residues) is quenched on unfolding. It is also possible for there to be an enhancement of fluorescence on unfolding. The simulation is calculated via Eqs. (3)–(5), with an assumed $\Delta G_0^\circ = 6$ kcal/mol, $m = 4$ kcal/mol/M, $F_{0,\mathrm{N}} = 1.0$, $s_\mathrm{N} = 0$, $F_{0,\mathrm{U}} = 0.2$, and $s_\mathrm{u} = 0.05$. The graph also shows a plot of the mole fraction of the N and U species, X_N and X_U, as a function of denaturant concentration (faint dotted curves and right axis) and indicates the concentration regions that are considered native conditions, unfolding conditions, and the transition region. (B) Simulated time constant, τ, for the refolding and unfolding of a two-state (kinetic sense) protein as a function of denaturant concentration. The simulation was in terms of Eq. (7) with $k_{0,\mathrm{fold}} = 10\ \mathrm{sec}^{-1}$, $m_\mathrm{fold} = -3$ kcal/mol/M, $m_\mathrm{unf} = 1$ kcal/mol/M, and with $k_{0,\mathrm{unf}}$ related to the above unfolding thermodynamics as $K_\mathrm{un} k_\mathrm{fold}$. Note that the intercepts at $[\mathbf{d}] = 0$ will be the values of log $k_{0,\mathrm{unf}}$ and log $k_{0,\mathrm{fold}}$, the slopes of the plots are $m_\mathrm{unf}/(2.303RT)$ and $m_\mathrm{fold}/(2.303RT)$, and the two lines will cross at [**d**] where the unfolding equilibrium constant is 1. For a simple kinetic two-state model, there will be a single time constant for both refolding and unfolding reactions.

reaction can be initiated by rapid mixing of protein in an "unfolding" solution with a "native" solution, so that the final denaturant concentration is at or below the midpoint in Fig. 1. In general, both types of reactions should result in a reaction that, when monitored by fluorescence or another method, will result in data that are described by the following multiexponential decay law

$$F(t)_{[\mathbf{d}]} = F(\infty)_{[\mathbf{d}]} + \sum \Delta F_{i,[\mathbf{d}]} \exp(-t/\tau_i) \tag{6}$$

where $F(\infty)_{[\mathbf{d}]}$ is the fluorescence intensity at infinite time (i.e., at equilibrium) at the particular denaturant concentration, $\Delta F_{i,[\mathbf{d}]}$ is the fluorescence

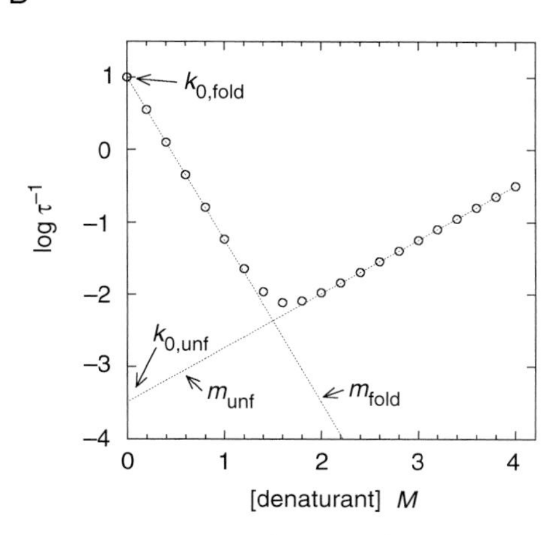

FIG. 1. (*continued*)

amplitude associated with time constant τ_i. In an unfolding experiment, $\Sigma\ \Delta F_{i,[\mathbf{d}]} = F_{\mathrm{N},[\mathbf{d}]} - F(\infty)_{[\mathbf{d}]}$, and, by definition, $\Sigma\ \Delta F_{i,[\mathbf{d}]}$ will be positive if the fluorescence intensity drops on unfolding (as in the simulation in Fig. 1) and will be negative if the intensity increases. (Note that for a multiphasic process the individual $\Delta F_{i,[\mathbf{d}]}$ may be either positive or negative.) Likewise, for a folding transition, in which the protein is jumped from a high [**d**] to a lower [**d**] (below the transition midpoint, $[\mathbf{d}]_{1/2}$), the amplitudes are given by $\Sigma\ \Delta F_{i,[\mathbf{d}]} = F_{\mathrm{U},[\mathbf{d}} - F(\infty)_{[\mathbf{d}]}$. The values of $F_{\mathrm{N},[\mathbf{d}]}$, $F_{\mathrm{U},[\mathbf{d}]}$, and $F(\infty)_{[\mathbf{d}]}$ should be consistent with the equilibrium titration intensities. If there is missing amplitude in the kinetic measurements [i.e., if $F(\infty)_{[\mathbf{d}]} + \Sigma\ \Delta F_{i,[\mathbf{d}]}$ does not equal the expected value of $F_{\mathrm{N},[\mathbf{d}]}$ in an unfolding experiment, or if $F(\infty)_{[\mathbf{d}]} + \Sigma\ \Delta F_{i,[\mathbf{d}]}$ does not equal the expected value of $F_{\mathrm{U},[\mathbf{d}]}$ in a folding experiment], this indicates that there is an additional rapid kinetic phase that is not detected by the instrument.

In the simplest two-state case, both the unfolding and folding reactions will be monoexponential, with time constant τ. The inverse of this time constant will be the sum of the forward and reverse rate constants in Eq. (1). That is, $\tau^{-1} = k_{\mathrm{unf}} + k_{\mathrm{fold}}$ at any particular final concentration of denaturant. Shown in Fig. 1B is a simulation for such a simple case, which corresponds to the equilibrium simulation in Fig. 1A. Reactions that jump to a final concentration above $[\mathbf{d}]_{1/2}$ are in the unfolding regime in which $\tau^{-1} \approx k_{\mathrm{unf}}$. Reactions that jump to a final concentration below $[\mathbf{d}]_{1/2}$ are in the folding regime in which $\tau^{-1} \approx k_{\mathrm{fold}}$. In denaturant-induced unfolding reactions, the values of k_{unf} and k_{fold} can be related to the denaturant concentration by[18]

$$\log \tau^{-1}_{\text{at high }[\mathbf{d}]} = \log k_{unf} = \log k_{0,unf} + m_{unf}[\mathbf{d}]/(2.303RT) \quad (7a)$$
$$\log \tau^{-1}_{\text{at low }[\mathbf{d}]} = \log k_{fold} = \log k_{0,fold} + m_{fold}[\mathbf{d}]/(2.303RT) \quad (7b)$$

where $k_{0,unf}$ and $k_{0,fold}$ are the values of the unfolding and folding rate constants in the absence of denaturant and m_{unf} and m_{fold} are analogous to the parameters in Eq. (4), except that they now refer to dependence of the respective activation free energy changes on denaturant concentration. In a "chevron" plot of log τ^{-1} versus [**d**] in Fig. 1B, the slope at high [**d**] is $m_{unf}/(2.303RT)$ and the slope at low [**d**] is $m_{fold}/(2.303RT)$. The intercepts at [**d**] = 0 for the two limbs are $\log(k_{0,unf})$ and $\log(k_{0,fold})$, from which the value of the equilibrium constant $K_{0,un}$ in the absence of denaturant can be calculated as $k_{0,unf}/k_{0,fold}$. For such a simple two-state case, the values of m_{unf} and m_{fold} are related to the equilibrium m value by $m = m_{unf} - m_{fold}$.

Unfolding and folding reactions are usually more complicated than the previous simulation. Consider the possibility that there is an unfolding intermediate, I, and that the reaction occurs by the following sequential unfolding scheme:

$$\mathrm{N} \underset{k_{1,fold}}{\overset{k_{1,unf}}{\rightleftharpoons}} \mathrm{I} \underset{k_{2,fold}}{\overset{k_{2,unf}}{\rightleftharpoons}} \mathrm{U}$$

where $K_{I/N} = [I]/[N]$, $K_{U/I} = [U]/[I]$, and the forward and reverse rate constants are indicated in the scheme. If the population of I is sufficiently large (and its signal is sufficiently distinct from the other states), then the existence of the intermediate may be observed in equilibrium studies by the appearance of a biphasic transition (i.e., the presence of a plateau or hump in a plot of F versus [**d**]). However, the population of the intermediate may be small and it may then be difficult to determine its presence. For example, consider the simulated equilibrium unfolding data in Fig. 2A, where the mole fraction of intermediate X_I reaches a level of only 5–10%. The overall transition could easily be mistakenly considered two-state. However, denaturant-induced transient folding/unfolding studies should clearly identify a three-state process by the observation of two time constants, τ_1 and τ_2, which can be potentially detected over the entire range of denaturant concentration. Shown in Fig. 2B is a simulated example of a three-state process in which the I $\rightleftharpoons$ U process occurs more rapidly than does the N $\rightleftharpoons$ I process.

Kinetic experiments frequently show multiple kinetic phases that can be categorized as slow ($\tau \sim$ 1–500 sec), fast ($\tau \sim$ 10–1000 msec), and very fast ($\tau <$ 1 msec). For a number of proteins, slow phases observed in folding reactions have been attributed to the interconversion of *cis* and *trans* isomeric states of X–proline residues. This slow reaction was first demonstrated in the case of ribonuclease A.[5,25–28] The peptide bonds formed by all other amino acids exist almost exclusively in the *trans* configuration in both native and unfolded states of protein. It is believed that peptide bonds for proline can exist about 10–30% of the time in the *cis* form in unfolded proteins. Analysis of the crystal structures of native proteins shows several examples of *cis* X–prolines (e.g., in ribonuclease A, ribonuclease T_1, barstar, and thioredoxin), although *trans* X–prolines are the norm (e.g., all *trans* in hen and T4 lysozyme, and cytochrome c). Because the time constant for the *cis* $\rightleftharpoons$ *trans* interconversion in the unfolded state is on the order of 100 sec, this reaction is a likely cause of slow refolding phases in proteins (in which case, the above three-state model can be written as N $\rightleftharpoons U_f \rightleftharpoons U_s$, where U_f and U_s are fast and slow refolding forms of the unfolded state; see below). Such a slow interconversion is often not seen in unfolding reactions, because the fluorescence [and other signals, except possibly for high-resolution nuclear magnetic resonance (NMR)] of an unfolded protein will usually not be sensitive to whether a particular X–proline bond is *cis* or *trans*. One system for which fluorescence is sensitive to the *cis* $\rightleftharpoons$ *trans* state of a proline is the protein ribonuclease A, where there is a Tyr^{92}–Pro^{93} sequence and the fluorescence of the tyrosine senses the isomeric state.[27,29] Double jump experiments (see below), mutational studies, and/or the use of prolyl isomerase[28,30] or certain proteolytic enzymes[26] are means of identifying a slow phase with a proline isomerization.

Fast reactions, which occur within the dead time of most stopped-flow instruments, have received much attention. The presence of these phases can be detected by the existence of "missing amplitude" in a kinetic experiment. Expanding on the previous description, if transient folding data are fitted with Eq. (7) and the sum of $F(\infty)_{[\mathbf{d}]} + \sum F_{i,[\mathbf{d}]}$ at a particular final [**d**] is more or less than the $F_{N,[\mathbf{d}]}$ expected at this [**d**] [from Eq. (5) or from actual equilibrium data], then there is a missing amplitude that must be associated with a kinetic step that occurs much faster than the dead time of the instrument. The existence of such fast phases can be detected in this manner, but limited information can be obtained about the structural

[25] J. F. Brandts, H. R. Halverson, and M. Brennan, *Biochemistry* **14,** 4953 (1975).
[26] J. F. Brandts and L.-N. Lin, *Methods Enzymol.* **131,** 107 (1986).
[27] L.-N. Lin and J. F. Brandts, *Biochemistry* **22,** 564 (1983).
[28] B. T. Nall, *in* "Mechanisms of Protein Folding" (R. H. Pain, ed.), Chap. 4. IRL Press–Oxford University Press, New York, 1994.
[29] F. X. Schmid and H. Blaschek, *Eur. J. Biochem.* **114,** 111 (1981).
[30] K. Lang, F. X. Schmid, and G. Fisher, *Nature (London)* **329,** 268 (1987).

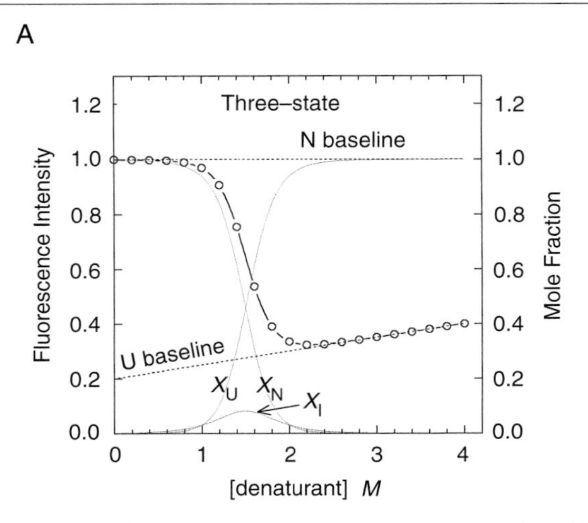

FIG. 2. (A) Simulated equilibrium unfolding data for a protein, induced by a denaturant, **d**, in a three-state manner (N $\rightleftharpoons$ I $\rightleftharpoons$ U) and monitored by steady-state fluorescence intensity, F. Note that the population of intermediate, I, reaches a maximum mole fraction of only 5–10% in this simulation, which will probably be too little for detection in equilibrium unfolding experiments. The simulation is with $\Delta G^{\circ}_{0,\text{I/N}} = 4$ kcal/mol, $m_{\text{I/N}} = 2$ kcal/mol/M, $\Delta G^{\circ}_{0,\text{U/I}} = 2$ kcal/mol, $m_{\text{U/I}} = 2$ kcal/mol/M, $F_{0,\text{N}} = 1.0$, $s_{\text{N}} = 0$, $F_{0,\text{I}} = 0.70$, $s_{\text{I}} = 0.025$, $F_{0,\text{U}} = 0.2$, and $s_{\text{U}} = 0.05$. (B) Simulated time constants for the refolding and unfolding of a three-state (N $\rightleftharpoons$ I $\rightleftharpoons$ U) protein as a function of denaturant concentration [simulations performed using Eqs. (7) and (8) from Hagerman and Baldwin[6]]. There will be two time constants, τ_1 for the slower step and τ_2 for the faster step, over the entire range of [**d**], although the amplitudes of these time constants will depend on the initial and final conditions, the magnitude of the rate constants, and the relative difference in the fluorescence signals for the different species. The asymptotic slopes of the ascending and descending arms are related to the m values for the unfolding and folding reactions for each step as indicated. This is only one example of the appearance of such a plot for a three-state reaction and is obtained with the following parameters: $k_{0,1,\text{fold}} = 10$ sec^{-1}, $m_{1,\text{fold}} = -1$ kcal/mol/M, $m_{1,\text{unf}} = 1$ kcal/mol/M, $k_{0,2,\text{fold}} = 100$ sec^{-1}, $m_{2,\text{fold}} = -1$ kcal/mol/M, and $m_{2,\text{unf}} = 1$ kcal/mol/M (i.e., with the I $\rightleftharpoons$ U step being more rapid than the N $\rightleftharpoons$ I step; again, $k_{1,\text{unf}}$ and $k_{2,\text{unf}}$ are taken to be consistent with the above thermodynamic simulation). This simulation is for a case in which there is some population of the intermediate state at equilibrium (X_{I} reaches a maximum of 5–10%). It is possible that X_{I} may never reach an equilibrium level as high as 1%, in which case the equilibrium transition would appropriately be described as two-state, whereas the transient folding/unfolding reaction could be three-state.

characteristics of the rapidly formed species by fluorescence or other methods. Hydrogen–deuterium (H–D) pulse labeling and stopped-flow circular dichroism (CD) studies have shown that these rapidly formed species can contain a significant amount of secondary structure. Results of this type have led to proposed folding mechanisms that emphasize a hydrophobic

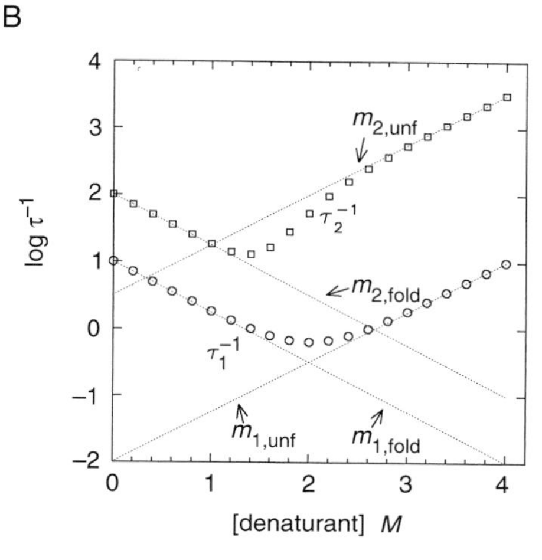

FIG. 2. (*continued*)

collapse and/or the rapid formation of local secondary structure. References 5 and 31–35 discuss recent models and interpretations for protein-folding reactions.

Advantages of Stopped-Flow Fluorescence

Among the spectroscopic methods used in combination with stopped-flow mixing, fluorescence is probably the most simple, yet most sensitive and widely employed. Fluorescence measurements with micromolar protein solutions can yield data with a good signal-to-noise (S–N) ratio and with a temporal resolution limited only by the mixing dead time of the SF instrument. Circular dichroism (CD) spectroscopy is also widely employed and can be related to changes in α-helix and β-sheet content. However, CD has a much lower S/N, is not specific regarding the location of the secondary structures, is less useful for proteins with high β-sheet content, and can suffer from optical artifacts resulting from strain on windows caused by the rapid mixing. Also, urea and guanidine hydrochloride absorb ultraviolet (UV) light below 210 nm, which limits the use of low wavelengths when a high concentration of these denaturants is required. Detection

[31] K. Kuwajima, *Proteins Struct. Funct. Genet.* **6,** 87 (1989).

[32] P. S. Kim and R. L. Baldwin, *Annu. Rev. Biochem.* **59,** 631 (1990).

[33] O. B. Ptitsyn, *Protein Eng.* **7,** 593 (1994).

[34] C. R. Matthews, *Annu. Rev. Biochem.* **62,** 653 (1993).

[35] A. L. Fink, *Annu. Rev. Biophys. Biomol. Struct.* **24,** 495 (1995).

by absorption spectroscopy is intrinsically less sensitive than fluorescence spectroscopy and the absorbance changes between N and U states are often small. Pulse H–D exchange, followed by NMR analysis, is a powerful and structurally informative technique (i.e., it provides information about local secondary structure, in terms of the protection of amide NH from exchange as a result of hydrogen bonding).[36–38] However, it is a difficult method, requires a large quantity of proteins, and is not capable of the higher data density and high *S/N* of fluorescence detection. Such pulsed H–D exchange studies are usually performed in combination with or after characterization of the relaxation profile obtained with SF fluorescence.

The intrinsic fluorescence of a protein will be dominated by the contribution from tryptophan residues; proteins lacking tryptophans will show tyrosine or phenylalanine fluorescence. Tryptophan is the most important of these intrinsic probes, because it has a larger molar extinction coefficient, serves as an energy transfer acceptor for the other aromatic amino acids, can be selectively excited at longer wavelengths (e.g., >295 nm), and because its fluorescence quantum yield and emission maximum (λ_{max}) are sensitive to the microenvironment of the indole side chain.[39–42] Solvent-exposed tryptophan residues usually fluoresce with a λ_{max} of 340–350 nm. Tryptophan residues that are buried in the interior of native proteins can have a λ_{max} ranging from 308 to 340 nm and can have a quantum yield ranging from 0.01 to 0.4. Bluer emission seems to arise when the indole ring is surrounded by nonpolar aliphatic side chains. The wide range of quantum yields in the native state is due to the variable location of the indole ring with respect to different quenching side chains (e.g., disulfide, protonated histidine, peptide bonds) and other groups (e.g., certain metal ions, heme groups, coenzymes). Consequently, when a protein unfolds, its intrinsic tryptophanyl fluorescence will usually shift to the red and may show either an increase or a decrease in quantum yield, thus providing a valuable structural probe.

Tryptophan is also one of the most rare of the amino acids and many

[36] P. S. Kim and R. L. Baldwin, *Biochemistry* **19,** 6124 (1980).

[37] J. B. Udgaonkar and R. L. Baldwin, *Nature (London)* **335,** 694 (1988).

[38] H. Roder, G. A. Elove, and S. W. Englander, *Nature (London)* **335,** 700 (1988); S. W. Englander and L. Mayne, *Annu. Rev. Biophys. Biomol. Struct.* **21,** 243 (1992).

[39] J. M. Beechem and L. Brand, *Annu. Rev. Biochem.* **54,** 433 (1985).

[40] I. Weinryb and R. F. Steiner, *in* "Excited States of Proteins and Nucleic Acids" (R. F. Steiner and I. Weinryb, eds.), pp. 277–318. Plenum, New York, 1971.

[41] J. R. Lakowicz, "Principles of Fluorescence Spectroscopy." Plenum, New York, 1983.

[42] M. R. Eftink, *Methods Biochem. Anal.* **35,** 127 (1991).

proteins have only one or two tryptophan residues. This simplifies the fluorescence signal. With mutagenesis techniques, it is possible to move the "intrinsic" tryptophan probe to different places within a protein and/or to place a single tryptophan residue in a protein previously lacking such a residue.

In addition to fluorescence intensity measurements, other fluorescence signals that can be measured are emission anisotropy and the intensity decay profile (i.e., the fluorescence lifetime). In the following sections we discuss the use of these fluorescence signals in combination with SF mixing devices. Anisotropy studies (in combination with lifetime data) can provide information about the rotational freedom of a fluorophore in various states of a protein.

Fluorescence from other reporter groups, intrinsic or extrinsic, can also be monitored. Extrinsic probes, such as the dansyl group, fluorescein, coumarin, or pyrene, can be covalently attached to reactive sulfhydryl or amino groups. Although these probes involve covalent modification of the protein, which may alter the stability and folding kinetics of the protein, these probes have the advantage of having strong fluorescence signals.

Extrinsic fluorophores can also be noncovalently associated with the protein. The prime example of this is ANS (and bis-ANS), which is known to bind to hydrophobic surfaces and that has been used in protein-folding studies to characterize structural features of folding intermediates. Examples of this use are presented as follows. The advantage of using ANS in such studies is that this fluorophore has a low fluorescence signal when free in solution and its signal increases dramatically on binding. Specific, fluorescently labeled ligands can also be used to sense the refolding of proteins and to characterize the presence of binding sites in folding intermediates.

Other types of information available in fluorescence studies are the accessibility of fluorophores to solute quenchers and the distance between energy transfer donor–acceptor pairs positioned at different sites on the protein. Application of solute-quenching and energy transfer measurements to protein-folding studies is also presented.

To summarize, SF fluorescence is a transient method that has high *S/N*, has a time response limited only by the mixing process, requires relatively small amounts of protein, can be used with both intrinsic and extrinsic probes, is site selective for the microenvironment of these probes, and can provide low-resolution structural information (e.g., exposure of fluorophores to solvent and quenchers, global and local rotational motion, existence of hydrophobic surfaces, distances between sites) about folding intermediates.

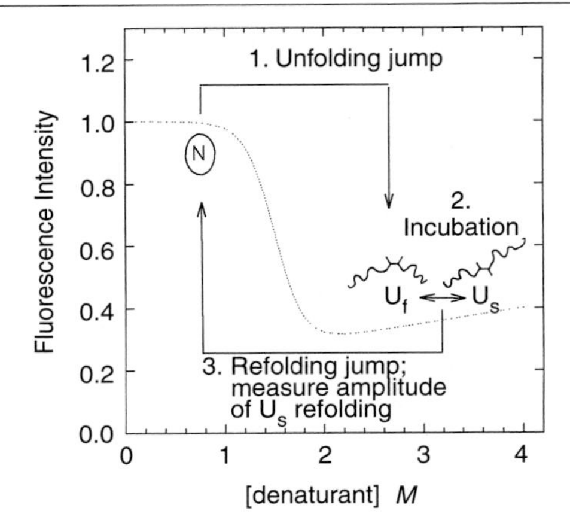

FIG. 3. Illustration of the steps in the first type of "double jump" experiment, which is used to determine the existence (and kinetics) of the slow interconversion of U_f and U_s unfolded states. In many instances this slow interconversion is due to the existence of a *cis* *X*–Pro bond in the N state, with the U_f species having the corresponding *cis* bond and U_s having a *trans* bond.

Developments in Stopped-Flow Fluorescence

Double Jump Studies

Figures 3 and 4 illustrate two different types of double jump (DJ) experiments that can be performed on a stopped-flow instrument. As noted

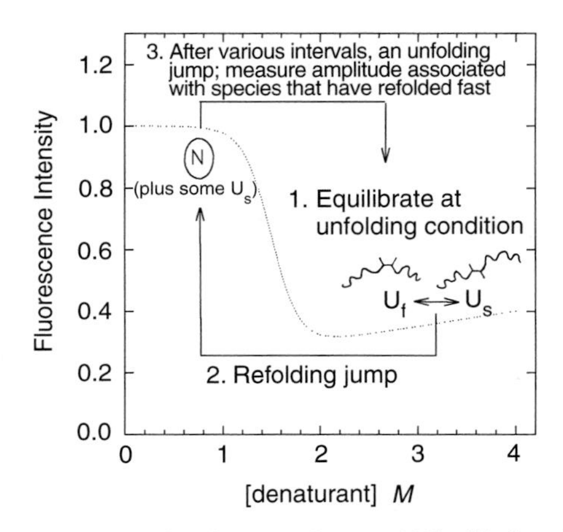

FIG. 4. Illustration of the steps for the second type of "double jump" experiment, which also is used to determine the existence of U_f and U_s species and can provide information about the kinetics of the $U_f \rightarrow N$ reaction.

earlier, unfolded species that refold at different rates have been attributed primarily to the existence of a *cis X*–Pro bond in the folded state of a protein, with the *X*–Pro bond slowly adopting a predominantly *trans* conformation in the unfolded state.[25,43]

The rate of interconversion of different unfolded species can be monitored using the DJ strategy depicted in Fig. 3.[4,23,28] The protein in the folded form is first mixed with denaturant, transferring the protein to an unfolding region of the equilibrium curve. After incubating the protein for various lengths of time in the unfolding condition (to allow interconversion of different unfolded states), a second jump is made to dilute the denaturant and to return to a folding region of the equilibrium curve. If fast-refolding (U_f) and slow-refolding (U_s) species exist at the end of the incubation period in the denaturing condition, then there will be a biphasic refolding process after the second jump. As the names imply, the fast-refolding and slow-refolding species will refold rapidly and slowly, respectively, and the amplitudes associated with each phase can usually be easily separated. Fluorescence, or almost any other spectroscopic signal, can then be used to track the refolding process, even though the fluorescence signal is not intrinsically sensitive to the structural differences between the U_f and U_s species. The amplitudes associated with the fast- and slow-refolding species are proportional to the concentrations of the U_f and U_s species, respectively, that were present at the end of the incubation time. The dependence of these fast- and slow-folding amplitudes on the length of the incubation time yields the apparent time constant for the interconversion (in the denaturing condition) of the U_f and U_s species.

As an example of a DJ experiment, consider the case of barstar, a small (89 amino acids) globular protein. Double jump slow-refolding traces were used to establish the presence of U_f and U_s forms of the unfolded protein.[44,45] Although the optical properties of the U_s and U_f molecules are identical, the fact that the U_f species refolds at least 500 times faster than does U_s makes the two species kinetically distinguishable. The interpretation given for these results was that U_s possesses the Tyr[47]–Pro[48] bond in the nonnative *trans* conformation, whereas U_f has this bond in the native *cis* conformation.[44,46] The apparent rate constant of the $U_s \rightarrow U_f$ reaction was determined by DJ experiments to be 16.5×10^{-3} sec^{-1} over a broad range of denaturant concentration.

A second type of DJ experiment, which can determine the relative

[43] L.-N. Lin and J. F. Brandts, *Biochemistry* **17,** 4102 (1978).
[44] G. Schreiber and A. R. Fersht, *Biochemistry* **32,** 11195 (1993).
[45] M. C. R. Shastry and J. B. Udgaonkar, *J. Mol. Biol.* **247,** 1013 (1995).
[46] M. J. Lubienski, M. Bycroft, S. M. V. Freund, and A. R. Fersht, *Biochemistry* **33,** 8866 (1994).

amounts of fast- and slow-refolding species at equilibrium or in the time course of unfolding, is represented in Fig. 4. As above, success of this procedure depends on a large difference between the rates of refolding of the individual unfolded species. In this procedure, the protein is first allowed to unfold completely at a desired unfolding condition (normally the protein is incubated in a high concentration of denaturant for 8–12 hr) and to achieve its equilibrium ratio of U_f and U_s forms. The unfolded protein sample is then diluted by mixing into a refolding buffer (refolding jump), and, after a refolding interval, a second denaturing jump is made into the denaturing buffer. The relaxation that is observed is primarily due to the $U_f \rightarrow N \rightarrow U_f$ reactions that occur during the first and second jumps and the resulting amplitude (associated with the $N \rightarrow U_f$ process for the second jump) is a measure of the concentration of N formed (from U_f) during the first jump. Measurement of the amplitude of the $U_f \rightarrow N$ reaction for various refolding intervals provides selective information about the kinetics of this refolding step.

Double jump strategies of the type shown in Fig. 4 have been extensively used to monitor the relative amount of U_f and U_s species in proteins.[4,23,44,45,47] For example, the refolding of barstar has been systematically studied using DJ measurements over a range of refolding conditions. It was concluded that refolding of the protein at high concentrations of guanidine hydrochloride ($>2\ M$) follows a $U_s \rightleftharpoons U_f \rightleftharpoons N$ pathway and that refolding at low concentrations of guanidine hydrochloride ($<1\ M$) is more complicated, involving two or three additional intermediates.[44,45,47]

ANS Binding and Other Ligand-Binding Studies

A strategy that is frequently employed with SF fluorescence measurements is to study the unfolding/refolding reaction in the presence of an added ligand. In most situations a ligand will bind only to the native state and any intermediates that have sufficient globular structure to form a binding site. The effect of the added ligand will be one or more of the following: (1) to provide a spectral signal of bound ligand to be used as a probe of the folding process (i.e., the fluorescence of protein-bound ANS is much greater than that of free ANS), (2) to alter the apparent rate constant for the refolding and/or unfolding processes, and (3) to alter the overall stability of the folded protein (and/or other binding-competent forms of the protein), with concomitant shift in the position of the minimum in chevron plots.

Consider the general scheme that follows for the three-state transition of a protein in the absence (top row) and presence (bottom row) of a

[47] M. C. R. Shastry, V. R. Agashe, and J. B. Udgaonkar, *Protein Sci.* **3,** 1409 (1994).

specific ligand, L, that binds to the native (N), intermediate (I), and unfolded (U) states of the protein with association constants K_N, K_I, and K_U, respectively.

$$\begin{array}{ccccc} N & \underset{k_{-1}}{\overset{k_1}{\rightleftharpoons}} & I & \underset{k_{-2}}{\overset{k_2}{\rightleftharpoons}} & U \\ K_N \updownarrow L & & K_I \updownarrow L & & K_U \updownarrow L \\ N\cdot L & \underset{k'_{-1}}{\overset{k'_1}{\rightleftharpoons}} & I\cdot L & \underset{k'_{-2}}{\overset{k'_2}{\rightleftharpoons}} & U\cdot L \end{array}$$

In most cases the binding of L to the U state can be ignored. For transfer of a protein from native to a strongly unfolding condition, the apparent unfolding rate constant will be (assuming that the I $\rightleftharpoons$ U step occurs rapidly, that there is no binding of L to U, and that all ligand-binding reactions occur rapidly compared to conformational transitions)

$$k_{\text{unfold,app}} = k_1(1 + K_N[L]k'_1/k_1)/(1 + K_N[L]) \tag{8}$$

That is, the value of $k_{\text{unfold,app}}$ will depend on the concentration of L and will change from a value of k_1 at [L] = 0 to k'_1 at high [L]. Normally, one would expect k'_1 to be smaller than k_1, that is, the binding of ligand to the N state slows the unfolding reaction by lowering the free energy level of the N · L complex with respect to the transition state for the unfolding reaction. In the extreme case that $k'_1 = 0$, the binding of L to the protein traps it in the native N · L complex. If, however, k'_1 is nonzero, this actually indicates that L interacts to a certain extent with the transition state. If k'_1 is larger than k_1, this indicates that L interacts more with the transition state than it does with N. From transition state theory, the $K_N k'_1/k_1$ factor in Eq. (8) is equal to K_{TS}, the association constant for L to the transition state that exists between N and I.

For transfer of completely unfolded protein to strong refolding conditions, the apparent rate constant for refolding will be

$$k_{\text{fold,app}} = k_{-1}(1 + K_I[L]k'_{-1}/k_{-1})/(1 + K_I[L]) \tag{9}$$

The value of $k_{\text{fold,app}}$ will vary from k_{-1} at [L] = 0 to k'_{-1} at high [L] if there is an interaction between L and the I state. If there is no interaction between L and the I state (or U state), then the addition of L will not alter the time constants for refolding under strong refolding conditions. Again, if there is an interaction between L and I and if the value of k'_{-1} is zero, the presence of saturating L will quench the refolding reaction (trapping the I · L species). If k'_{-1} is nonzero, this indicates an interaction between L and the I → N transition state; if $k'_{-1}/k_{-1} < 1$, the ligand interacts more strongly with I than it does with the transition state; if $k'_{-1}/k_{-1} > 1$, this indicates that the ligand interacts more strongly with the transition state than it does with I (regardless of the strength of the interaction of L with N). The $K_I k'_{-1}/k_{-1}$ factor in Eq. (9) is also equal to the same K_{TS} value (assuming the above mechanism and with the caveats that the K values pertain to the final denaturant conditions, and that, for strong associations, the rate constants for ligand dissociations may become rate limiting and thus alter the previous equations).

Shown in Fig. 5 is a simulation of the effect of interaction of ligand with the N, I, and transition state species on the apparent rate constant of unfolding and folding for an N $\rightleftharpoons$ I $\rightleftharpoons$ U system (where L does not bind to U and for which the I $\rightleftharpoons$ U transition is rapid). The simulation is for the case of 1:1 binding with association constants of $K_N = 1 \times 10^7\ M^{-1}$, $K_{TS} = 1 \times 10^5\ M^{-1}$, and $K_I = 1 \times 10^3\ M^{-1}$, a reasonable case where the affinity for the ligand is greatest for the N state, least for the I state, and intermediate for the transition state, which might be expected if the transition state has a structure that is intermediate between that of N and I. By studying the dependence of $k_{unfold,app}$ and $k_{fold,app}$ on the concentration of L, it should be possible to determine K_N and K_I and to estimate the

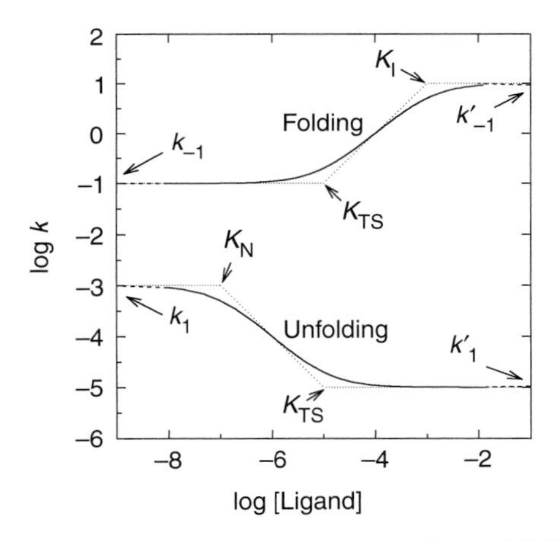

FIG. 5. Simulation of the effect of ligand on the unfolding and folding of a protein (for which the folding is simple, two-state in the absence of ligand). This simulation shows that the addition of L can increase the apparent folding rate constant and can decrease the apparent unfolding rate constant. The simulation is for a simple 1:1 binding system with the following association constants between L and the various protein forms: $K_N = 1 \times 10^7\ M^{-1}$, $K_I = 1 \times 10^3\ M^{-1}$, and $K_{TS} = 1 \times 10^5\ M^{-1}$ (i.e., binding to each state, except U, with greatest affinity for the N state). The simulation is with $k_{fold} = 0.1\ sec^{-1}$ and $k_{unf} = 0.001\ sec^{-1}$. Note that the bend positions are related to the values of the association constants.

affinity of the ligand for the transition state. Of course, these association constants will pertain to the final refolding and unfolding conditions (the latter usually being a high concentration of urea or guanidine hydrochloride). An example of this type of approach is the study of Kuwajima and co-workers[48] of the effect of Ca^{2+} ions on the kinetics of folding/unfolding of carp parvalbumin (albeit using SF circular dichroism).

An important type of ligand for SF fluorescence experiments is the dye ANS (or bis-ANS), the fluorescence of which can be used to monitor protein transitions. ANS is weakly fluorescent in buffer and shows a large increase in quantum yield on binding to many proteins.[49,50] Depending on the protein, the binding of ANS can either be a specific or a fairly nonspecific type of interaction and the stoichiometry of the resulting protein complexes can be quite variable.[51] It seems that there is a tendency for ANS to bind to apolar surfaces or pockets in proteins; consequently, ANS has a tendency to interact with compact intermediate state(s) in a protein-folding pathway.[52,53] In a thorough study of both the thermodynamics and kinetics of binding of bis-ANS to heat shock protein, DnaK, Shi and co-workers[51] have shown that dye binding can shift the conformational state of this protein. This protein can be considered to exist in native, intermediate, and unfolded forms (i.e., $N \rightleftharpoons I \rightleftharpoons U$); data are consistent with the preferential binding of bis-ANS to the I compact intermediate state. Consequently, mixing bis-ANS with the native protein results in a rapid enhancement of fluorescence, owing to diffusion-limited binding of the ligand to N and any existing I species. This is followed by a slower process in which ligand binding shifts the protein to (or traps it in) the I state, with the formation of a highly fluorescing $I \cdot L$ species. Such a preferential binding of bis-ANS to the I state of this system emphasizes that the addition of ligands can alter the thermodynamics and kinetics of transitions between states of a protein. In several cases, however, the addition of a ligand, such as ANS, has been reported not to cause a change in the kinetics profile for protein folding/unfolding, with the ligand simply providing a useful fluorescence signal for tracking the population of the species to which it can bind. From Eqs. (8) and (9), it can be seen that the binding of ligand would not alter the apparent unfolding and folding rate constants if either the ligand

[48] K. Kuwajima, A. Sakuraoka, S. Fueki, M. Yoneyama, and S. Sugai, *Biochemistry* **27,** 7419 (1988).

[49] L. Stryer, *J. Mol. Biol.* **13,** 482 (1965).

[50] L. Brand and J. R. Gohlke, *Annu. Rev. Biochem.* **41,** 843 (1972).

[51] L. Shi, D. R. Palleros, and A. L. Fink, *Biochemistry* **33,** 7536 (1994).

[52] G. V. Semisotnov, N. A. Rodionova, O. I. Razgulyaev, V. N. Uversky, A. F. Gripas, and R. I. Gilmanshin, *Biopolymers* **31,** 119 (1991).

[53] B. E. Jones, P. A. Jennings, R. A. Pierre, and C. R. Matthews, *Biochemistry* **33,** 15250 (1994).

concentration is lower than the reciprocal of the largest pertinent K, or if $K_N = K_{TS}$ or $K_I = K_{TS}$, respectively (i.e., if there is equal binding of the ligand to the transition state and the protein state that precedes the transition state, or if there is no binding to either the I and TS states, in the case of refolding). For example, Jones and co-workers[53] have shown that the presence of ANS does not alter the folding kinetics profile of dihydrofolate reductase. Folding studies of the protein using ANS fluorescence as a reporter showed a rapid collapse (within the mixing dead time) of the unfolded protein to an intermediate state having sufficient apolar surface to bind ANS. Over the next 300 msec, there was a further increase in the fluorescence of ANS, indicating the sequential formation of additional compact intermediate states.

Another example of the use of a fluorescent ligand in SF refolding studies is work by Itzhaki and co-workers[54] on the effect of an umbellifery-labeled disaccharide on the refolding of lysozyme. This ligand binds only to the native state and its fluorescence is enhanced on binding. Thus the fluorescence of this ligand provides a specific "binding capacity" probe for the formation of native lysozyme (note that it was observed that the ligand does not alter the kinetics profile from that observed by other methods, indicating a lack of preferential binding to the intermediate or transition state that directly precedes formation of the native state).

Solute-Quenching Studies

Solute quenchers such as acrylamide and iodide can potentially be used in SF fluorescence studies to provide information about the accessibility of fluorescing species along the folding/unfolding path. As is the case with any added reagent, it first must be considered whether the quenchers alter the kinetics profile by interacting with any form of the protein (as in Fig. 5), rather than simply by altering the F_i values. For any fluorescing site, the intensity in the presence of a solute quencher can be related to the concentration of quencher, [Q], by the Stern–Volmer relationship

$$F = F_0 \exp(-V[\mathrm{Q}])/(1 + K_{SV}[\mathrm{Q}]) \tag{10}$$

where F_0 is the fluorescence intensity of the site in the absence of quencher and V and K_{SV} are referred to as static and dynamic quenching constants, respectively.[55] The most commonly employed solute quenchers, acrylamide and iodide, act primarily as dynamic quenchers, but the static quenching factor, $\exp(V[\mathrm{Q}])$, may be significant. The dynamic quenching constant, K_{SV}, is the product of a bimolecular quenching rate constant, k_q, and a

[54] L. S. Itzhaki, P. A. Evans, C. M. Dobson, and S. E. Radford, *Biochemistry* **33,** 5212 (1994).
[55] M. R. Eftink and C. A. Ghiron, *Anal. Biochem.* **114,** 199 (1981).

fluorescent lifetime, τ_0, in the absence of quencher. The magnitude of K_{SV} can be taken as a measure of the accessibility (not exposure) of a fluorophore to the quencher, with accessibility being a kinetic term that depends on the magnitude of both the k_q and τ_0. Obtaining k_q requires knowledge of τ_0 for a fluorescing site, and the collision frequency reported by k_q will depend on the degree of steric burial of a fluorophore, the size of the quencher, and, in the case of charged quenchers, the presence of charge–charge interactions. The magnitude of the static constant, V, can also be simplistically related to the accessibility of fluorophore to quencher, although other factors, such as binding of the quencher near the fluorophore and excited-state transient terms, can also contribute to what appears to be static quenching. Equation (10) implies that the fluorescence of an individual site has a monoexponential decay, which usually is not the case. With all its imperfections and uncertainties, Eq. (10) still is useful in that it functionally relates a drop in fluorescence to the concentration of quencher. Finally, if there are multiple fluorescing sites on a particular molecular species, one can, in principle, sum Eq. (10) over the number of sites.

In applications of solute quenching with SF fluorescence, the strategy is to measure $F(t, [Q])$ kinetics profiles in the presence of varying concentrations of quencher in order to construct Stern–Volmer plots for each species in a multiexponential relaxation. So long as the quencher does not interact with the protein at any stage, other than by quenching, the time constants should be unaltered (and the recovery of constant τ_i is good validation of the rate equation). An example of this type of study was presented by Itzhari and co-workers,[54] who used iodide as quencher of intrinsic tryptophan fluorescence during the transient folding of hen lysozyme. It was found that the rapidly formed folding intermediate has its tryptophans more accessible to iodide than does the species (including the native state) formed at later times.

In performing these studies, Itzhari and co-workers found that the folding kinetics of lysozyme were dependent on ionic strength, which complicated the use of the ionic quencher iodide. To overcome this problem, these workers developed a double mixing experimental strategy in which the protein was first mixed with refolding buffer, for certain intervals of time (refolding intervals), and then mixed with a solution containing the solute quencher. The initially measured fluorescence signal after the second mix was then used to construct Stern–Volmer plots for each refolding time interval.

Double Kinetics Studies

Additional information can be obtained from fluorescence if time-resolved (or phase-resolved) data can be obtained during the nanosecond decay of the excited state.[56–58] The measurement of time-resolved fluorescence data requires moderately elaborate instrumentation, including pulsed or modulated lasers and time-correlated or phase-sensitive detectors. Developments in the speed of data acquisition have enabled the construction of an impressive SF time-resolved fluorometer.[59] This instrument is capable of monitoring nanosecond decays during subsecond relaxations (e.g., protein foldings), with the latter initiated by an SF mixer, during a time slice as small as 20 msec. As many as a thousand mixing shots are needed, but decay profiles having up to 10^4 counts in the peak channel are possible. Anisotropy decay data can also be obtained for reaction time slices this small.

In an application of this double kinetics instrument, intensity decay and anisotropy decay data have been obtained for the intrinsic tryptophanyl fluorescence and bound ANS fluorescence during the millisecond-to-second refolding of dihydrofolate reductase.[57] The results showed a rapid (<20 msec) formation of a compact species, as sensed by the rapid appearance of a relatively long (~10 nsec) rotational correlation time for bound ANS. In contrast, the intrinsic tryptophanyl fluorescence remains more mobile during this initial refolding phase. During the next few hundred milliseconds, the rotational freedom of bound ANS was found to be further restricted (indicated by a further increase in the rotational correlation time to 20 nsec) and the average fluorescence decay time for the tryptophanyl residues was found to increase (suggesting their exclusion from solvent). During the slowest events that occur in the seconds time range, the rotational correlation times of bound ANS and the tryptophanyl residues were found to approach a common value of 10 nsec, which is related to overall rotation of the protein. This example illustrates the power of SF time-resolved fluorescence experiments to provide molecular details about the folding process.

Energy Transfer Studies

A unique ability of fluorescence spectroscopy is to measure intramolecular distances in the 10- to 60-Å range by resonance energy transfer.[60,61] This requires the existence of a suitable donor (D) and acceptor (A) pair, with sufficient overlap between the fluorescence of the donor and the

[56] D. G. Walbridge, J. R. Knutson, and L. Brand, *Anal. Biochem.* **161,** 467 (1987).
[57] B. E. Jones, J. M. Beechem, and C. R. Matthews, *Biochemistry* **34,** 1867 (1995).
[58] J. M. Beechem, M. A. Sherman, and M. T. Mas, *Biochemistry* **34,** 13943 (1995).
[59] J. M. Beechem, *Proc. S.P.I.E.* **1640,** 676 (1992); see also Chapter 3 in this volume.
[60] R. H. Fairclough and C. R. Cantor, *Methods Enzymol.* **48,** 347 (1978).
[61] L. Stryer, *Annu. Rev. Biochem.* **47,** 819 (1978).

absorbance of the acceptor. Examples of such donor and acceptor pairs are tryptophan → dansyl [or iodoacetyl-5-sulfo-1-naphthylethylenediamine (IAEDANS)] or dansyl → fluorescein, which have effective $R_0^{2/3}$ distances for 50% efficiency of energy transfer of approximately 22 and 33–41 Å, respectively.[60] The observed efficiency of energy transfer, E, can be related to this $R_0^{2/3}$ and the actual separation distance, R, by the equation

$$E = 1 - F_{DA}/F_D = (R_0^{2/3}/R)^6/[1 + (R_0^{2/3}/R)^6] \tag{11}$$

where F_{DA} and F_D are the fluorescence intensities of the donor when the acceptor is attached and not attached to the protein, which is how the efficiency can be measured. (The efficiency of energy transfer can also be measured by changes in fluorescence lifetime of the donor.)

Because these $R_0^{2/3}$ are approximately the dimensions of a globular protein, the measurement of E, and hence R, can be used to compare the D–A distance in a native and unfolded protein.[62,63] In addition to equilibrium studies, the efficiency of energy transfer can be measured during the course of a folding/unfolding transition initiated by SF mixing. As an example of this type of study, Kawata and Hamaguchi[64] have monitored the kinetics of energy transfer between two sites on an immunoglobulin light chain as this protein is refolded. This protein has a single intrinsic tryptophan residue (Trp-148), which served as donor, and an IAEDANS extrinsic acceptor probe was covalently attached to the C-terminal cysteine of the protein. Refolding of this labeled protein showed a rapid recovery of the energy transfer efficiency to the value of the native protein. This rapid decrease in the D–A distance occurred much faster than other phases detected by SF CD and tryptophanyl fluorescence intensity (in the absence of the acceptor) measurements. This indicated that this protein undergoes a rapid collapse to a compact state, which is followed by slower rearrangements to reach the final native structure.

Diode Array and Rapid-Scanning Detection

Whereas most SF fluorescence studies are performed at a single emission wavelength (or a range of wavelengths collected through an interference or bandpass filter), it is possible to use intensified diode array detectors or rapid scanning monochromators to collect multiple wavelength data and thus to characterize the emission spectra of folding intermediates.

In our laboratory we have combined an SF fluorescence system with a

[62] D. S. Gottfried and E. Haas, *Biochemistry* **31,** 12353 (1992).
[63] E. James, P. G. Wu, W. Stites, and L. Brand, *Biochemistry* **31,** 10217 (1992).
[64] Y. Kawata and K. Hamaguchi, *Biochemistry* **30,** 4367 (1991).

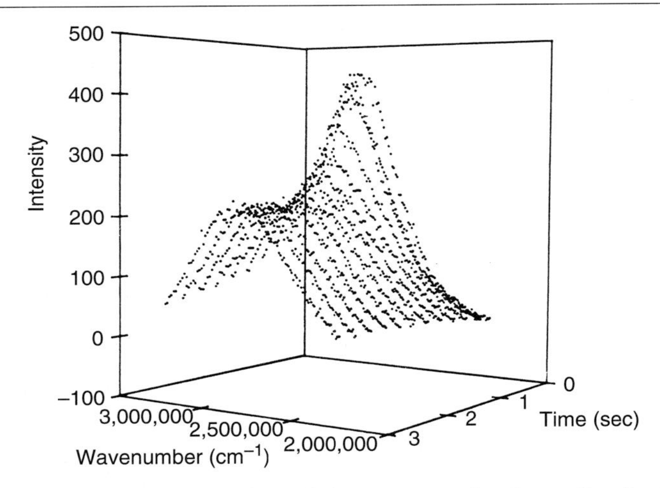

FIG. 6. Diode array detection of the emission spectrum of cod parvalbumin after mixing with a chelating agent, EGTA. Spectra were taken every 50 msec for a total period of 3 sec. Only every fifth spectrum is shown for clarity. Conditions: 20°, 0.1 *M* KCl, 0.001 *M* dithiothreitol, 0.02 *M* sodium borate buffer, pH 9.0, excitation at 295 nm, ~5 μM protein mixed (50:50) with 0.01 *M* EDTA.

spectrograph and Princeton Instruments (Princeton, NJ) intensified (UV) diode array. Shown in Fig. 6 are typical SF spectral data obtained for the reaction of the chelator EGTA with cod parvalbumin, where the fluorescence of the single tryptophan was monitored. The spontaneous dissociation of Ca^{2+} ions (trapped by EGTA) from this protein produces an apoprotein, which appears to have lost most of its tertiary structure. Because there are two Ca^{2+}-binding sites in this protein, the minimum (sequential) mechanism for Ca^{2+} release is

$$\mathrm{Holo}_{2\mathrm{Ca}^{2+}} \xrightarrow{k_1} \mathrm{Inter}_{1\mathrm{Ca}^{2+}} \xrightarrow{k_2} \mathrm{Apo}_{\mathrm{no\ Ca}^{2+}} \quad (12)$$

where an intermediate with one bound Ca^{2+} is formed. The kinetics of the dissociation of Ca^{2+} should be a biexponential process, so long as k_2 is not much larger than k_1.

The family of spectra in Fig. 6 were obtained as 50-msec exposures from over the period of 50–3000 msec following mixing with EGTA. (Actually, only every fifth spectrum is shown in Fig. 6 for clarity.) We have performed a global analysis of these data, using the following approach, to calculate the fluorescence spectrum of the intermediate. The spectrum of the equilibrium holo and apo forms were collected and fitted with Eq. (13) for a Gaussian distribution (in wavenumber space, i.e., after converting diode number to wavelength and then to wavenumber, υ, because, in principle, such spectra should have a symmetrical shape when plotted vs wavenumber), using three fitting parameters, a maximal intensity, F_i, the mean wavenumber, $\bar{\upsilon}_i$, and the width of the distribution, σ_i.

$$F(\upsilon)_i = F_i \exp[-(\upsilon - \bar{\upsilon}_i)^2/\sigma_i^2]/[\sigma_i(2\pi)^{1/2}] \tag{13}$$

After obtaining these three parameters to describe the emission of the holo and apo forms, the time-dependent family of spectral data, $F(\upsilon, t)$, were fitted to kinetic Eq. (14), where the $F(\upsilon)_i$ for each species is given by Eq. (13).

$$F(\upsilon, t) = F(\upsilon)_{holo}X_{holo}(t) + F(\upsilon)_{inter}X_{inter}(t) + F(\upsilon)_{apo}X_{apo}(t) \tag{14}$$

where $X_{holo}(t) = \exp(-k_1t)$; $X_{inter}(t) = [k_1/(k_2 - k_1)][\exp(-k_1t) - \exp(-k_2t)]$; and $X_{apo}(t) = 1 - X_{holo}(t) - X_{inter}(t)$. The fitting parameters for this analysis are the rate constants k_1 and k_2, for the assumed sequential mechanism in Eq. (12), and the F_{inter}, $\bar{\upsilon}_{inter}$, and σ_{inter} that describe, respectively, the amplitude, position, and width of the fluorescence spectrum of the intermediate. Shown in Fig. 7 is a result of this analysis. As can be seen, the intermediate has a spectral position that is between that of the holo and apo forms. (Actually, there are two nearly equivalent fits with widely different intensities for the intermediate, owing to the high correlation between the parameters F_{inter} and the rate constants.)

Commercial SF instruments (Applied Photophysics Limited, Leatherhead, UK, and Hi-Tech Scientific, Salisbury, UK) are available with optional diode array detectors. Alternatively, Olis Instruments (Bogart, GA) has developed a rapid-scanning monochromator for use with SF fluorescence studies.

Global Analysis of Intensity and Anisotropy Data

The simultaneous collection and global analysis of SF fluorescence intensity and steady state anisotropy, r, was first performed by Otto and coworkers,[65] to our knowledge. Such global analyses have the ability to distinguish between kinetic models because the different types of data track the population of emitting species according to different relationships. That is, SF fluorescence intensity data track the population of states as given by Eq. (15), whereas SF fluorescence anisotropy data track the species population by the more complicated Eq. (16).

$$F(t) = \Sigma F_iX_i(t) \tag{15}$$

$$r(t) = \Sigma F_ir_iX_i(t)/\Sigma F_iX_i(t) \tag{16}$$

[65] M. R. Otto, M. P. Lillo, and J. M. Beechem, *Biophys. J.* **67,** 2511 (1994).

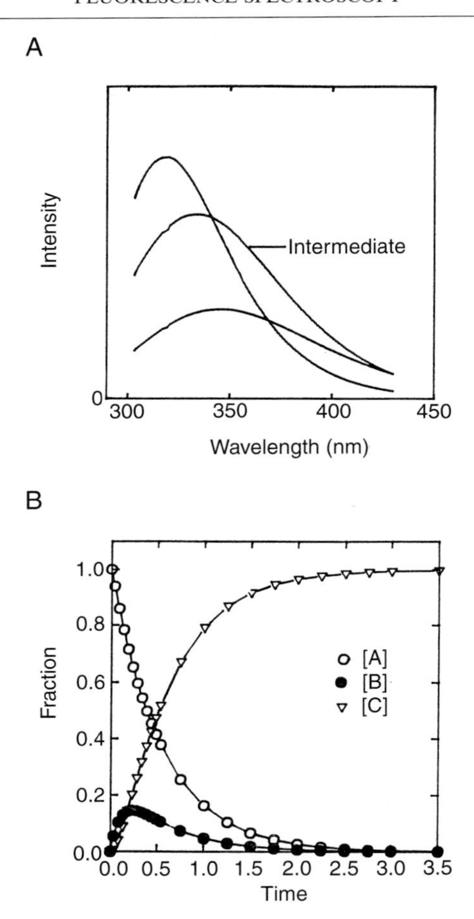

FIG. 7. (A) Recovered spectra for the native (top spectrum), intermediate (middle spectrum), and final state (lower spectrum) from the data in Fig. 6, following analysis of the data in terms of a two-step reaction as described in text. (B) The time dependence of the mole fraction of the native (○), intermediate (●), and final calcium depleted (▽) states of cod parvalbumin.

That is, the intensity data are linearly related to the mole fraction of species, $X(t)$, whereas the anisotropy data depend on both the mole fractions and the relative fluorescence intensity, $F_i/\sum F_i$ of each species. In other words, more strongly fluorescent species contribute disproportionately to the time-dependent anisotropy data. It is important to note that SF anisotropy data will not be a simple sum of exponentials unless the fluorescence intensity of each species is the same. Otto and co-workers[65] have provided simulations for a simple two-state transition to show that $r(t)$ data can appear either to lead or lag $F(t)$ data, depending on whether the final state is more fluorescent or less fluorescent, respectively, than the initial state. Intensity and anisotropy data can be acquired simultaneously (essentially) and the advantage of the complementary trackings in Eqs. (15) and (16) is that global nonlinear least-squares fittings of combined data sets will have steeper χ^2 surfaces, thus enabling a more confident fit of the data and the ability to discrimination between alternate models. In addition, the global analysis will yield the anisotropy of each species, which can provide information about the rotational dynamics of the species.

Shown in Fig. 8 are SF fluorescence intensity and anisotropy data obtained by Otto and co-workers for the pH-induced refolding (Fig. 8A) and unfolding (Fig. 8B) of *Staphylococcal* nuclease. Notice that the anisotropy (r) profile seems to lead the intensity (labeled as S for sum) profile for the refolding reaction, and vice versa for the unfolding reaction. As explained previously, this is due to the difference in fluorescence quantum yield of Trp-140 of the native and final unfolded states of this protein. The solid line drawn through both types of data is a global fit of the anisotropy and intensity data to a three-state (biexponential) folding/unfolding mechanism.

Combination of Circular Dichroism and Fluorescence Data

Although having lower *S/N* and therefore not being quite as good for establishing rate equations, SF CD data can provide useful information about the recovery/loss of secondary structure during folding/unfolding transitions. In addition, CD data usually report information over the entire protein, whereas fluorescence data provide information that is specific to the fluorescent sites. For these reasons the two methods are complementary and we have obtained from Aviv Instruments (Lakewood, NJ) a new SF instrument that is capable of monitoring both fluorescence and CD signals on the same sample.[66]

Shown in Fig. 9 is an example of the data acquired with this instrument for the guanidine hydrochloride-induced unfolding of nuclease A. The fluorescence intensity drops as the protein unfolds and the CD signal approaches zero. It is clear that the fluorescence data have a much higher *S/N* (for an average of five mixing shots in each case), but, in principle, both kinetic traces should be fitted by the same rate equation. The solid line through the data represents a global biexponential fit of both data sets (weighted by the standard deviation of each type of data; see the next

[66] See also our combined CD/fluorometer for monitoring equilibrium unfolding reactions; G. D. Ramsay and M. R. Eftink, *Biophys. J.* **66,** 516 (1994) and G. D. Ramsay, R. Ionescu, and M. R. Eftink, *Biophys. J.* **69,** 701 (1995).

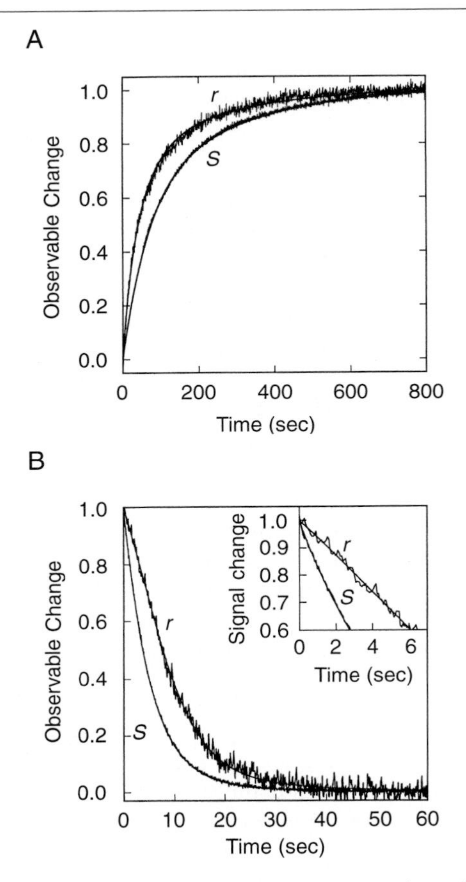

FIG. 8. Simultaneous monitoring and analysis of fluorescence intensity (S) and anisotropy (r) data for the pH-induced refolding (A; pH jump of 3.2 to 7) and unfolding (B; pH jump of 7 to 3.6) of *Staphylococcal* nuclease mutant Δ114–119 (monitoring the fluorescence of Trp-140). The solid lines are a global fit of a biexponential decay law to both intensity and anisotropy data to Eqs. (15) and (16). (Reprinted from Ref. 65, with permission from the *Biophysical Journal.*)

paragraph). For the CD data, the individual data set can be adequately fitted as a monoexponential. However, the fluorescence intensity data require a biexponential fit, when such data are analyzed individually. Shown in the bottom two panels of Fig. 9 are residual patterns, which illustrate that the biexponential fit is required for the fluorescence data. Thus, in the global analysis the higher *S/N* fluorescence data can "pull out" the biexponential decay from the noisier CD data, enabling a recovery of the CD signal for the intermediate species.

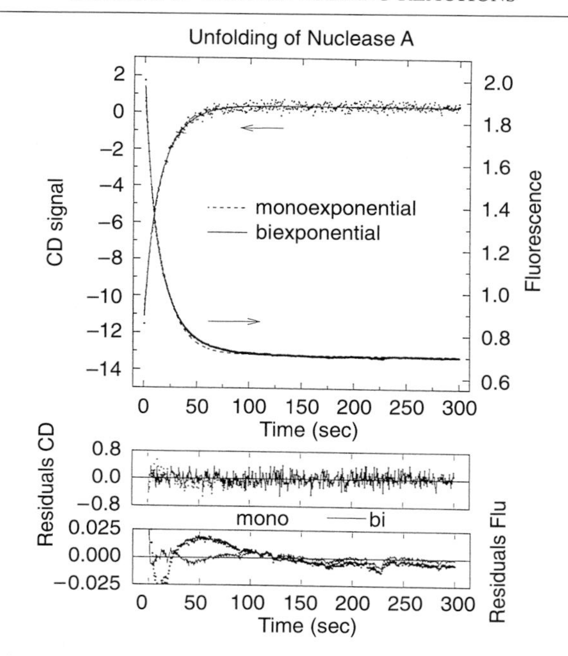

FIG. 9. Simultaneous monitoring and analysis of stopped-flow CD and fluorescence data for the guanidine hydrochloride-induced unfolding of *Staphylococcal* nuclease A using an Aviv Model 63 instrument. The CD (ascending, noisier signal, left axis) and fluorescence (descending signal, right axis) signals are an average of five mixing shots each. The data were first individually fitted with a mono- and biexponential decay law. The CD data were equally well fitted by both models, whereas the fluorescence data support a biexponential fit. The two types of data were then subjected to a global analysis (using standard deviations for each type of data obtained from averaging the final plateau data). The bottom panels show the residual patterns for the global monoexponential (dots, not connected) and a global biexponential (jagged line). The protein was unfolded at 20° by mixing into 1.33 *M* guanidine hydrochloride at neutral pH. The fluorescence was excited at 290 nm, with emission observed through a 340-nm interference filter (emission from Trp-140). The CD data were acquired at 222 nm.

In performing the global analyses of combined CD and fluorescence data (or combined fluorescence intensity and anisotropy data), it is important to weight the different data sets properly.[67] This can be done by determining the absolute standard deviation for a particular signal for a sample that has reached equilibrium. With our instrument, we take the last 10–50 data points at the end of final CD and fluorescence traces (after the reaction has fully relaxed) and calculate the standard deviation for each type of signal. We use the nonlinear least-squares program, NONLIN of Johnson

[67] G. D. Ramsay and M. R. Eftink, *Methods Enzymol.* **240,** 615 (1994).

and Frasier,[68] for the global analysis and the previous standard deviations are included for each type of data.

Final Remarks

Stopped-flow fluorescence techniques provide a number of possible strategies for studying the kinetics of protein folding. The signal-to-noise ratios of such experiments can be good and there are a variety of intrinsic and extrinsic reporter groups that can be employed. Probably the most attractive feature of SF fluorescence is that there are several variations on the basic experimental design and these can provide different types of useful structural information regarding the properties of transient folding intermediates.

Acknowledgments

This work was supported by NSF Grant MCB 94-07167. We thank Roxana Ionescu, R. Kent Gartin, and Evan Tillman for assistance in collecting data reported in this chapter.

[68] M. L. Johnson and S. G. Fraiser, *Methods Enzymol.* **117,** 301 (1985).

[13] Intramolecular Pyrene Excimer Fluorescence: A Probe of Proximity and Protein Conformational Change

By SHERWIN S. LEHRER

Introduction

In addition to emitting fluorescence from the excited monomer state, some fluorophors can form an excited-state dimer or excimer, by a specific interaction between the excited monomer and a ground-state monomer. When attached to proteins at specific sites, fluorophors that can form excimers can be used to provide information about changes in proximity between attachment sites. The most useful fluorophors for this purpose are pyrene and its derivatives. The excimer fluorescence of these compounds was first reported by Kaspar and Förster in 1954 (reviewed by Förster[1]) and their properties have been extensively studied by Birks.[2] Excimer

[1] T. Förster, *Angew. Chem. Int. Ed.* **8,** 333 (1969).
[2] J. B. Birks, "Photophysics of Aromatic Molecules." Wiley-Interscience, London, 1970.

0076-6879/97 $25

fluorescence has also been observed for benzene and toluene[2] and phenol and para-substituted phenols,[3] suggesting that one might observe intrinsic excimer formation in proteins from proximal phenylalanine and tyrosine.

The efficient excimer fluorescence of pyrene derivatives attached to neighboring protein sites can be used to probe conformational changes associated with a variety of protein interactions. With the use of site-directed mutagenesis, amino acid residues that selectively react with pyrene derivatives can be substituted at regions of the sequence that are suspected or known to be in close proximity. Although other spectroscopic properties of pyrene derivatives have been used in protein studies, only a limited number of excimer studies have been reported since the first studies of pyrene excimer fluorescence from probes attached to proteins at specific sites[4,5] were published.

A review has summarized previous studies of excimer fluorescence.[6] This chapter emphasizes the practical problems of measuring and interpreting intramolecular pyrene excimer fluorescence in aqueous solution relevant to studies of protein conformational changes.

Intramolecular Pyrene Excimer Formation

In Organic Solvents

An excimer is formed if an excited pyrene monomer, during its fluorescence lifetime, interacts in a specific manner with a neighboring ground-state (unexcited) pyrene. Early studies in "good" solvents, in which the pyrenes do not interact in the ground state, showed that as the pyrene concentration was increased, instead of the expected decrease in the monomer fluorescence due to concentration quenching, excimer fluorescence replaced the monomer fluorescence, a special type of concentration quenching.[1,2] Intramolecular excimer fluorescence in organic solvents has also been observed: for phenyl groups attached between propane,[7] for pyrenes

[3] S. S. Lehrer and G. D. Fasman, *J. Am. Chem. Soc.* **87,** 4678 (1965).

[4] S. Betcher-Lange and S. S. Lehrer, *J. Biol. Chem.* **253,** 3757 (1978).

[5] M. Zama, P. N. Bryan, R. E. Harrington, A. L. Olins, and D. E. Olins, *Cold Spring Harbor Symp. Quant. Biol.* **42,** 31 (1978).

[6] S. S. Lehrer, Pyrene excimer change as a probe of protein conformational change. *In* "Subcellular Biochemistry" (B. B. Biswas and S. Roy, eds.), Vol. 24, pp. 115–139. Plenum, New York, 1995.

[7] F. Hirayama, *J. Chem. Phys.* **42,** 3163 (1965).

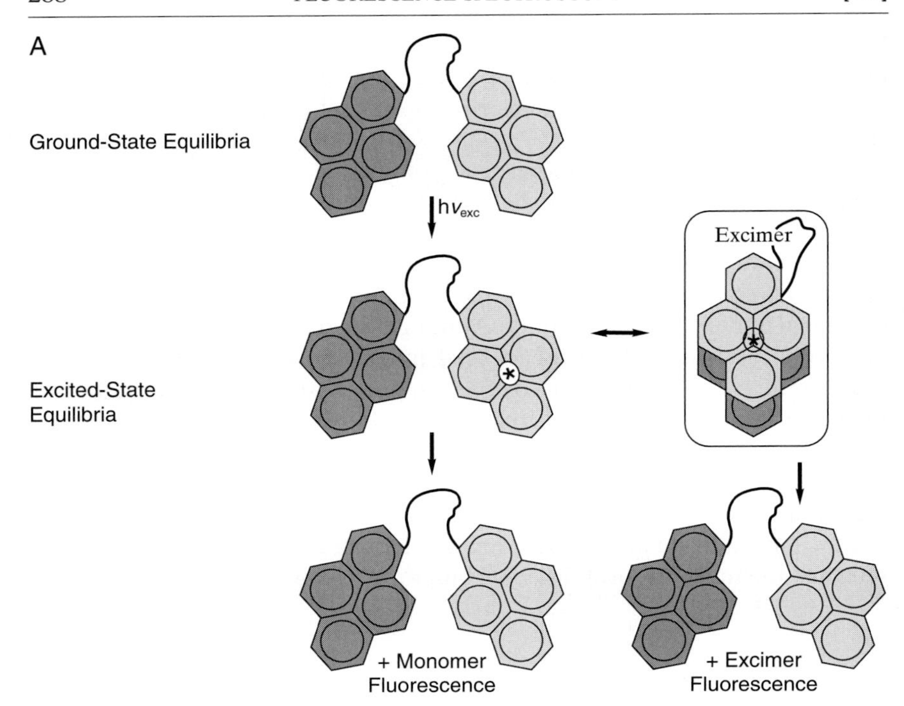

FIG. 1. Schematic diagrams of the production of intramolecular excited dimer (excimer) fluorescence from excited monomer pyrenes in organic solvents (A) and in aqueous solution (B). (A) In an organic solvent, the pyrenes do not interact in the ground state. On excitation (*), if unhindered, the pyrenes can reorient to form the excimer, in competition with monomer fluorescence. The excimer emits fluorescence at a longer wavelength and dissociates in parallel with emission. (B) In aqueous solution, the pyrenes will tend to stack owing to hydrophobic interactions producing an equilibrium between unstacked and stacked configurations determined by the flexibility of the "loop" between them and the nature of the hindrance. For the unstacked configuration, only monomer fluorescence will be produced. For the stacked configuration, on excitation, either excimer fluorescence or quenching (no fluorescence) will result, depending of the ability of the stacked pyrenes to reorient to the excimeric configuration.

attached to propane derivatives,[8,9] and for pyrenes attached to the two SH groups of dithiothreitol.[10]

Intramolecular excimer formation is schematically depicted for two pyrene moieties attached to a hydrocarbon chain in a organic solvent environ-

[8] H. Dangreau, M. Joniau, M. De Cuyper, and I. Hanssens, *Biochemistry* **21,** 3594 (1982).

[9] K. A. Zachariasse, G. Duveneck, and R. Busse, *J. Am. Chem. Soc.* **106,** 1045 (1984).

[10] Y. Ishii and S. S. Lehrer, Intramolecular excimer fluorescence of pyrene maleimide-labeled dithiothreitol. *In* "Fluorescent Biomolecules" (D. M. Jameson and G. D. Reinhart, eds.), Vol. 51, pp. 423–425. Plenum, New York, 1989.

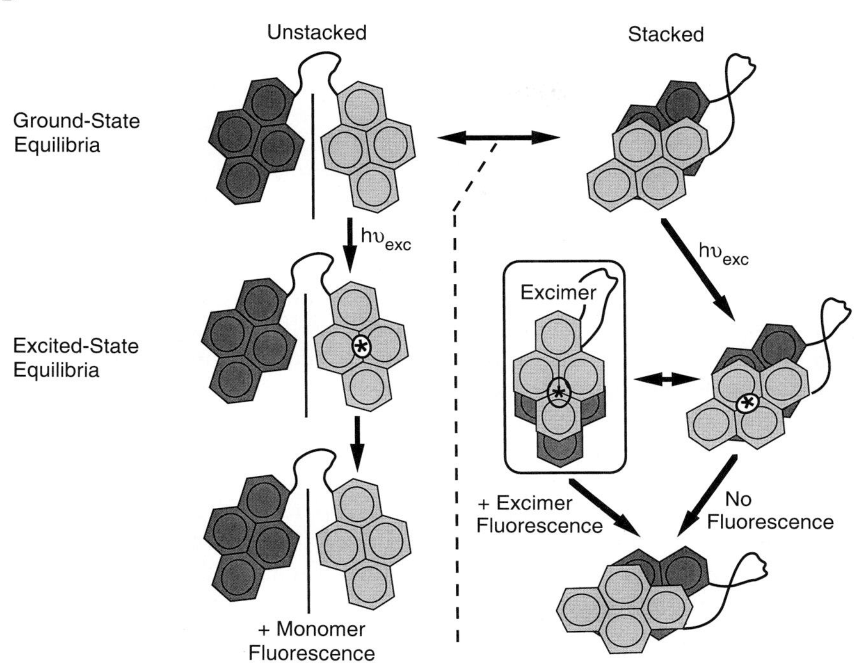

FIG. 1. (*continued*)

ment where the pyrenes do not interact in the ground state (Fig. 1A). On excitation into the absorption band of pyrene (λ_{max} = 340–343 nm) an excited monomer is produced. During its long lifetime, >90 nsec in the absence of oxygen for pyrene maleimide dithiothreitol,[10] the excited monomer can rotationally diffuse near its neighbor ground-state pyrene, and if unhindered, an excited-state dimer with a precise symmetrical configuration can be formed. The excimer emits fluorescence with a red-shifted unstructured spectrum because it originates from a lower energy excited state than the excited monomer and dissociates while remaining tethered on returning to the ground state as indicated (Fig. 1A). Two main properties of this "classic" scheme of excimer fluorescence follow from the photophysics: (1) The excitation spectrum of the excimer is identical to that of the monomer. This follows because both excimer and monomer fluorescence originate from the same source—ultraviolet (UV) absorption by a ground-state uncomplexed pyrene; (2) the time dependence of excimer fluorescence, in the nanosecond time domain, shows both a buildup and decrease; i.e., the excimer is formed after a monomer is excited so that a delay in excimer fluorescence compared to monomer fluorescence is observed.[1,2] However, in aqueous solution, the situation with respect to these two properties is altered as discussed in the next section.

In Aqueous Solution

Studies with pyrene-labeled tropomyosin[4] showed that in aqueous solution the excimer excitation spectrum is not identical to that of the monomer. The excimer excitation spectrum was somewhat broadened compared to that of the monomer, indicating ground-state hydrophobic interactions between pyrenes. Lifetime studies showed little time delay between monomer and excimer emission for pyrene–tropomyosin.[11] Thus, in aqueous solution, attaining the excimer configuration appears to require only a slight reorientation of hydrophobically stacked pyrenes. Data obtained with the model compound, pyrene–dithiothreitol, also indicated that excimer fluorescence originates from stacked pyrenes. The model compound studies also showed that monomer fluorescence can be quenched by pyrene–pyrene interaction without formation of excimer[10] and data with pyrene-labeled proteins indicate the generality of such quenching. It appears that pyrene–pyrene interaction can result in quenching of monomer without production of excimer fluorescence if the stacked pyrenes are inhibited from reorienting because of conformational restraint. Thus, ground-state pyrene–pyrene interaction can either facilitate or inhibit excimer formation depending on the ability of the pyrenes to reorient. For aqueous systems the "classic" schematic diagram of intramolecular excimer formation illustrated in Fig. 1 must be modified to take into account ground-state pyrene–pyrene interactions.

In aqueous solution, depending on the conformation of the protein and the local environment, the pyrenes will equilibrate between unstacked and stacked configurations in the ground state (Fig. 1B). For proteins in which the "loop" separating the pyrenes is quite flexible, the stacked configuration should dominate as it does for pyrene–dithiothreitol.[10] However, for a protein in which the loop represents secondary and tertiary structures that inhibit pyrene interaction, the equilibrium will favor the unstacked pyrene configuration. Changes in conformation of the protein near the pyrenes will influence the equilibrium and change the ability to exhibit excimer fluorescence. Although in Fig. 1B one configuration is schematically indicated for the stacked ground state, it represents a distribution of configurations in view of the nonspecific hydrophobic interactions. It is assumed that excimer formation can originate only from stacked pyrenes and monomer emission can originate only from unstacked pyrenes, because of the different

[11] P. Graceffa and S. S. Lehrer, *J. Biol. Chem.* **255,** 11296 (1980).

excitation spectra obtained in model systems. However, every stacked configuration may not lead to excimer fluorescence. For some stacked configurations the excited monomer state is quenched before reorientation to the excimeric configuration can take place. Another group of stacked states is therefore indicated in Fig. 1B that are nonfluorescent, i.e., in a "dark complex," when the rings are prevented from attaining the precise excimeric configuration.

Protein-Labeling Procedures

N-(1-pyrene) maleimide was one of the first protein pyrene labeling reagents developed,[12] which owing to its maleimide moiety was quite specific for the SH groups of exposed cysteine side chains. It has two useful properties as a general reagent: (1) it provides a low quantum yield in aqueous solutions, which markedly increases as a result of the labeling reaction,[12] and (2) at high pH values (>pH 8) the resulting succinimide ring undergoes hydrolysis or aminolysis (with a neighboring lysine group).[4,13,14] Conversion of type I (intact succinimide ring) to type II (cleaved succinimide ring) by incubation at high pH values,[13] red-shifts the monomer fluorescence and increases the excimer/monomer ratio for labeled tropomyosin and dithiothreitol. *N*-(1-pyrene) iodoacetamide is now also available and both reagents are commonly used to label cysteine groups with pyrene. The iodoacetamide derivative is fluorescent in the unreacted state, in contrast to the maleimide derivative, but has spectra and properties similar to those of the maleimide type II label, with monomer fluorescence peaks at 383 and 402 nm and excimer fluorescence at 480–490 nm (Fig. 2). The excitation spectra for the pyrene iodoacetamide-labeled tropomyosin also show similar broadened peaks of excitation of excimer compared to monomer (Fig. 2) previously seen for the maleimide reagent.[4] Owing to the broadening, a greater excimer/monomer ratio can be realized by excitation near 360 nm instead of at the peak value near 340 nm.

Cysteine groups are the prime choice for specific labeling of proteins for several reasons: (1) pyrene maleimide and iodoacetamide derivatives that preferentially label cysteine are commercially available (e.g., Molecular Probes, Eugene, OR; Sigma, St. Louis, MO; Aldrich, Milwaukee, WI); (2) only a limited number of cysteines are usually present in proteins; (3) cysteines that are present as the disulfide or that can be oxidized to the

[12] J. K. Weltman, R. P. Szaro, A. R. J. Frackelton, R. M. Dowben, J. R. Bunting, and R. E. Cathou, *J. Biol. Chem.* **248,** 3173 (1973).

[13] C. Wu, L. R. Yarbrough, and F. Y. Wu, *Biochemistry* **15,** 2863 (1976).

[14] Y. Ishii and S. S. Lehrer, *Biophys. J.* **50,** 75 (1986).

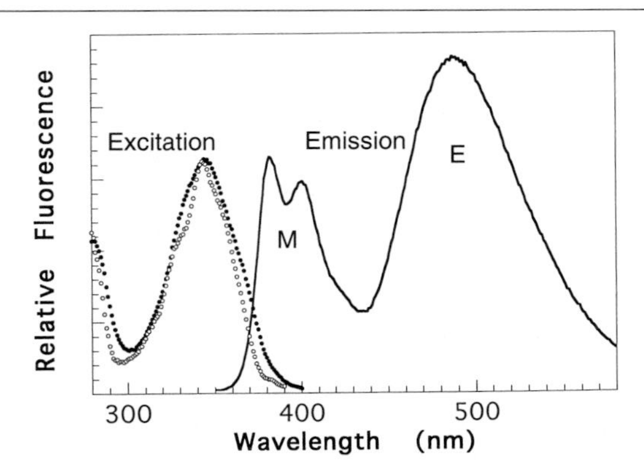

FIG. 2. Fluorescence spectra of pyrene iodoacetamide-labeled β,β-tropomyosin from gizzard muscle in aqueous solution. M, Monomer fluorescence band; E, excimer fluorescence band. (—) Emission spectrum, λ_{exc} = 340 nm. (○) Excitation spectra, λ_{em} = 385 nm; (●), λ_{em} = 480 nm. Note the broadened excimer excitation spectrum. [N. Golitsina, X. Zhou, and S. S. Lehrer, unpublished data, 1996.)

disulfide will offer a high probability of forming excimer when labeled; and (4) cysteine groups can be introduced into the amino acid sequence by site-directed mutation techniques. In the case of cysteine present as disulfide, the disulfide cross-link must first be reduced. This can be accomplished with dithiothreitol, most readily in the unfolded state in the presence of denaturant such as guanidinium hydrochloride (GdmCl) or urea. The dithiothreitol is then removed by dialysis or gel filtration and the labeling reaction is carried out in the presence of denaturant; last, the denaturant is removed to refold the protein. Selective labeling of pairs of cysteines is possible by varying the reaction conditions, e.g., at increased denaturant concentration to selectively reduce a pair that becomes exposed in a partially unfolded intermediate. If cysteine groups are present in the reduced form, labeling can be carried out in the native state if the cysteine groups are accessible or in the unfolded state if they are not. The labeling times are controlled by quenching the reaction with excess dithiothreitol, or by dropping the pH to low values (<pH 4) where the reaction is inefficient. If the labeling reaction is carried out in the native state, the course of the reaction can be monitored fluorometrically, noting that excimer forms only when each molecule is doubly labeled, and the reaction stops when saturation begins to take place. For pyrene maleimide, kinetics can be followed as an increase in monomer and excimer fluorescence.[12] The great wavelength separation between the excitation (340 nm) and the excimer emission (490 nm) will minimize effects of light scattering owing to the insoluble pyrene maleimide or iodoacetamide, as will the use of low concentrations. Mixed

organic–water solutions can be explored to keep the reagent soluble and thereby facilitate reaction. Reaction with either pyrene maleimide or pyrene iodoacetamide is usually performed at pH 6.5–8.0 using a 2–10× excess of reagent over the sulfhydryl concentration, diluted ~20× from dimethylformamide, acetone, methanol, or dimethyl sulfoxide. Reaction times will vary, but 1–5 hr at room temperature or below is often sufficient. Both reagents are not soluble in aqueous solution, so a precipitate usually appears and therefore gentle shaking facilitates the reaction. After quenching the reaction, the solution is filtered or spun to remove excess undissolved reagent and dialyzed or gel filtered to remove unreacted dissolved reagent. Owing to the hydrophobicity of pyrene some noncovalent binding of the reagent may take place, so exhaustive dialysis in the presence of dithiothreitol, which increases the solubility of the reagent on reaction, may be necessary. The presence of noncovalently bound label in the protein solution may lead to extrinsic excimer fluorescence, complicating interpretation. It can be detected on polyacrylamide gel electrophoresis as blue fluorescence migrating with the front. If maleimides are used, the different fluorescent properties of type I- and type II-labeled proteins can be explored (see above). Because the labeled protein may have altered properties, activity or other studies to determine if the labels have affected its properties should be performed.

Although both maleimides and iodoacetamides react preferentially with cysteine, there is the possibility of reaction with other protein side chains such as lysine. Maleimides and iodoacetamides react with the basic form of both cysteine and lysine but cysteine is preferentially labeled if present. The degree of labeling can be determined with the use of the extinction coefficient for pyrene, $\varepsilon_{343\text{ nm}} = 2.2 \times 10^4\ M^{-1}\text{cm}^{-1}$ and the protein concentration determined with a protein colorimetric assay. The protein should be unfolded with denaturant so that effects of putative pyrene stacking on absorption can be eliminated. To verify specific labeling at cysteine, a control labeling reaction with the protein that has had its cysteine group reversibly blocked with 5,5′-dithiobis-2-nitrobenzoate can be performed.[15] Proteins with blocked cysteine groups should have no fluorescence unless other groups reacted. Standard limited protein cleavage methods and peptide analysis can also be used to determine specificity and location of label.

Interpretation of Fluorescence Changes

Regardless of whether the pyrenes are stacked in the ground state, if excimer fluorescence is observed, the pyrenes must be in close proximity. For pyrene iodoacetamide-labeled proteins, the extended distance from

[15] S. S. Lehrer and Y. Ishii, *Biochemistry* **27,** 5899 (1988).

the sulfur of the cysteine to the middle of the pyrene group is about 9 Å. It is therefore possible for excimer fluorescence to be observed for labeled SH groups that are about 20 Å apart. This has been observed in at least one system.[16,17] The ability to observe excimer fluorescence not only depends on the distance between labeling sites, but also on the flexibility of the pyrenes at attachment sites and on the absence of intervening groups. For pyrene-labeled proteins in aqueous solution, excimer fluorescence appears to originate from surface pyrenes that stack hydrophobically in the ground state. However, if the stacked pyrenes cannot reorient on excitation, the fluorescence will be quenched. Thus, in general, the observation of excimer fluorescence provides proximity information but the lack of excimer fluorescence does not preclude pyrene proximity, particularly if absorption spectra are broadened, relative to the singly labeled system, thereby showing evidence of stacked pyrenes.

It is possible that the attached pyrenes could locate in the hydrophobic interior of a protein or in the transmembrane region of reconstituted protein–membrane systems and thereby may not tend to stack. In these situations, somewhat analogous to excimer-forming fluorophors in the lipid environment of a membrane,[18] the probability of excimer formation would be determined by the local viscosity of the medium in addition to proximity. In this case, in which the schematic of Fig. 1A applies, opposite changes in monomer and excimer would occur, resulting in an isoemissive point in the spectrum. Such changes were observed early for thermal perturbation of tropomyosin conformation,[11] which led to an initial model that did not take into account nonfluorescent stacked species. However, a constant amount of stacked nonfluorescent species would also lead to the same result.

In aqueous solution, where conformational changes could change the ability of stacked pyrenes to attain the excimer configuration, both monomer and excimer intensity could, e.g., increase or decrease together depending on changes in the contribution of the nonfluorescent stacked state. An increase in both excimer and monomer fluorescence can be explained, e.g., by a conformational change that "loosens" the pyrene–pyrene interaction, facilitating both the formation of the excimeric configuration and reequilibration toward unstacked configurations.

The following information is useful to help determine the origin of the fluorescence spectrum of a given labeled system.

[16] A. D. Verin and N. B. Gusev, *Biochim. Biophys. Acta* **956,** 197 (1988).

[17] Y.-M. Liou and F. Fuchs, *Biophys. J.* **61,** 892 (1992).

[18] B. Wieb Van der Meer, Biomembrane structure and dynamics viewed by fluorescence. *In* "Fluorescence Studies on Biological Membranes" (H. J. Hilderson, ed.), Vol. 13, pp. 38–41. Plenum, New York, 1988.

1. The degree of labeling: If it is greater than the number of accessible reactive groups, e.g., cysteine, there could be extra extrinsically bound pyrene, either reacted at other groups or noncovalently bound, contributing to the excimer fluorescence. There will also be monomer fluorescence contributions from incompletely labeled proximal cysteine pairs.
2. Excitation spectra of monomer and excimer fluorescence: If the excimer band is broadened, the excimer fluorescence probably originates from stacked pyrenes.
3. Lifetime studies: If there is a lack of appreciable delay between monomer and excimer decay, stacked pyrenes are present. If fluorescence changes do not result in changes in fluorescence decay parameters, changes in ground-state interactions are involved.
4. Effects of the label on protein conformation or activity: The conformation could be distorted by the pyrenes interacting with each other, resulting in altered proximity of the cysteines to which the pyrenes are attached.

Whether excimer fluorescence from labeled proteins arises from stacked or unstacked ground-state configurations, its fluorescence can be used to study conformational changes associated with the binding of substrates and cofactors, complex formation, and unfolding/refolding reactions to obtain binding and equilibrium constants and kinetic parameters.[6]

Acknowledgment

Supported by NIH Grants HL 22416 and AR 41637.

[14] Long-Lifetime Metal–Ligand Complexes as Probes in Biophysics and Clinical Chemistry

By EWALD TERPETSCHNIG, HENRYK SZMACINSKI, and JOSEPH R. LAKOWICZ

Introduction

In the design of a fluorescence experiment one can choose from hundreds of fluorophores that cover a wide range of absorption and emission wavelengths from 300 to 700 nm. However, the diversity of fluorescence decay times is much more limited, with most fluorophores displaying decay times between 1 and 10 nsec. Although this is a useful time scale for many

0076-6879/97 $25

biochemical experiments, it also would be useful to have longer decay times to allow measurement of slower domain-to-domain motions in proteins or rotational motions of membrane-bound proteins. In addition, the sensitivity of most fluorescence assays is limited not by the ability to detect the emission, but rather by the presence of interfering autofluorescence, which also occurs on the 1- to 10-nsec time scale. Once again, the availability of probes with longer decay times would allow increased sensitivity by use of gated detection following decay of the unwanted autofluorescence. The long decay times of lanthanide chelates allow gated detection and thus increased sensitivity for immunoassays.[1] However, to be useful for measurement of protein hydrodynamics the probes must display polarized emission, which to the best of our knowledge does not occur for the lanthanides.

This limited range of decay times (1–10 nsec) and absence of polarization or anisotropy can be circumvented by using transition metal complexes. Such metal–ligand complexes (MLCs) are becoming an increasingly important class of fluorophores for biological applications owing to their favorable properties of long lifetimes (from hundreds of nanoseconds to microseconds), adequate quantum yields, and high thermal, chemical, and photochemical stability. The spectral as well as chemical properties of an MLC can be altered by exchanging the ligands or the metal ion.[2–4] Furthermore, functional groups for covalent linkage to a biomolecule can easily be introduced. So far, MLCs have been successfully used to measure pH,[5] temperature,[6] and oxygen concentrations.[7,8]

We believe there are numerous additional applications for the MLCs in biochemistry, biophysics, and clinical and analytical chemistry. We have shown that certain MLCs display high anisotropy when excited with polarized light.[9] The long decay times of these complexes thus allow the use of steady-state or time-resolved fluorescence methods to measure the rotation of biopolymers in the microsecond time scale. In addition, a variety of clinical assays can be based on MLCs by using simple instrumentation, such as light-emitting diode (LED) light sources and solid-state detectors.

[1] E. P. Diamandis, *Clin. Biochem.* **21,** 139 (1988).

[2] L. A. Sacksteder, M. Lee, J. N. Demas, and B. A. DeGraf, *J. Am. Chem. Soc.* **115,** 8230 (1993).

[3] R. H. Fabian, D. M. Klassen, and R. W. Sonntag, *Inorg. Chem.* **19,** 1977 (1980).

[4] E. M. Kober, J. L. Marshall, W. J. Dressick, B. P. Sullivan, J. V. Caspar, and T. J. Meyer, *Inorg. Chem.* **24,** 2755 (1984).

[5] J. N. Demas and B. A. DeGraff, *Macromol. Chem. Macromol. Symp.* **59,** 35 (1992).

[6] R. W. Harrigan, G. D. Hager, and G. A. Crosby, *Chem. Phys. Lett.* **21,** 487 (1973).

[7] M. E. Lippitsch, J. Pusterhofer, M. J. P. Leiner, and O. S. Wolfbeis, *Anal. Chim. Acta* **205,** 1 (1988).

[8] O. S. Wolfbeis, L. J. Weis, M. J. P. Leiner, and W. E. Ziegler, *Anal. Chem.* **60,** 2028 (1988).

[9] E. Terpetschnig, H. Szmacinski, H. Malak, and J. R. Lakowicz, *Biophys. J.* **68,** 342 (1995).

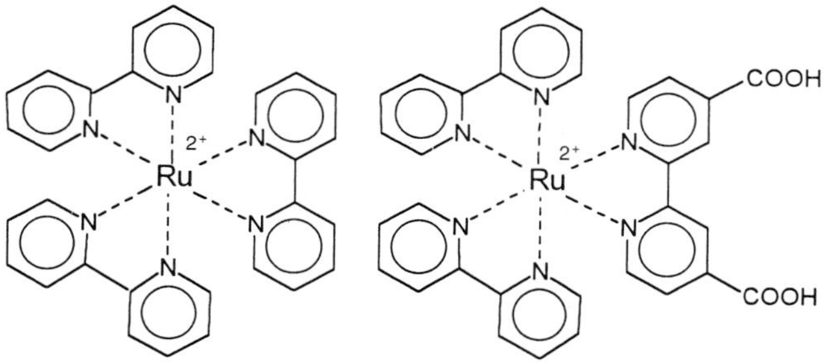

SCHEME I. Chemical structure of $[Ru(bpy)_3]^{2+}$ and of $[Ru(bpy)_2(dcbpy)]$.

Finally, further development of these compounds will provide specific probes for various cations or other analytes. Hence we predict that during the next several years many laboratories will continue our initial efforts to develop and make use of this important class of fluorophores.

The metal–ligand complexes are typified by tris(2,2′-bipyridine)ruthenium(II) ($[Ru(bpy)_3]^{2+}$) (Scheme I). Such complexes have been extensively studied for use in solar energy conversion and photochemical catalysis.[10–13] The previous research on solar energy conversion emphasized excited-state charge separation, rather than optimization of the MLCs for use as luminescence probes. However, it is now known that MLCs can display emission from charge-transfer states with decay times ranging from 100 to 4000 nsec in fluid solutions with reasonable quantum yields.[14] In addition, the range of absorption and emission wavelengths as well as the range of decay times can be selected by choosing complexes of ruthenium, osmium, or rhenium.[2–4] In biochemical applications their use has been limited to MLCs that display enhanced quantum yields on binding to DNA[15] and as simple fluorescent labels in time-resolved immunoassays.[16] However, these complexes have not been used as anisotropy probes, apparently owing to the opinion that symmetrical species, such as $[Ru(bpy)_3]^{2+}$, would display low or zero anisotropies. We have reported that a less symmetrical Ru complex, $[Ru(bpy)_2(dcbpy)]$ (dcbpy, 4,4′-dicarboxyl-2,2′-bipyridine)

[10] A. Juris and V. Balzani, *Coordination Chem. Rev.* **84,** 85 (1988).

[11] V. Balzani, N. Sabbatini, and F. Scandola, *Chem. Rev.* **86,** 319 (1986).

[12] M. Seiler, H. Durr, I. Willner, E. Joselevich, A. Doron, and J. F. Stoddart, *J. Am. Chem. Soc.* **116,** 3399 (1994).

[13] J. N. Demas, E. W. Harris, and R. P. McBride, *J. Am. Chem. Soc.* **99,** 3547 (1977).

[14] J. N. Demas and B. A. DeGraff, *Macromol. Chem. Macromol. Symp.* **59,** 35 (1992).

[15] A. E. Friedman, J.-C. Chambron, J.-P. Sauvage, N. J. Turro, and J. K. Barton, *J. Am. Chem. Soc.* **112,** 4960 (1990).

[16] R. B. Thompson and L. M. Vallarino, *S.P.I.E. Proc.* **909,** 426 (1988).

(Scheme I), displays high anisotropies in the absence of rotational motions,[9] thus allowing the use of MLCs to measure microsecond time-scale rotational motions in macromolecular systems.

Long-lived luminescent probes can be useful for studies of diffusive processes on a time scale presently not accessible by the usual fluorescence probes. There is considerable interest in the rates and amplitudes of domain-to-domain motions in proteins, and there have been repeated attempts to study such motions by time-resolved fluorescence resonance energy transfer (FRET).[17–19] These measurements have been mostly unsuccessful owing to the 5- to 10-nsec decay times and the limited extent of interdomain motions on this time scale. The use of longer lived MLC emission can allow measurement of these motions.[19] Such measurements should not be confused with diffusion-enhanced energy transfer using the lanthanide donors, in which the rate of diffusion is not determined, and the data reveal only the distance of closest approach of the donor and acceptor.[20] It is now known that FRET can be used to reveal both the distribution of distances between donor and acceptor and the rate of donor-to-acceptor diffusion.[18,19]

Applications of Luminescent Metal–Ligand Complexes

Uses of Long-Lived Probes in Biochemistry and Immunoassays

Fluorescence polarization methods are applied to study the interactions of proteins with other macromolecules. The basis for such measurements is the anisotropy (r), which is directly related to the rotational correlation time (θ) and to the molecular weight of a system. The anisotropy of a labeled macromolecule is given by

$$r = \frac{r_0}{1 + \tau/\theta} \tag{1}$$

where r_0 is the value observed in the absence of rotational diffusion. Although the theoretical upper limit for the anisotropy is 0.4, many fluorophores display maximal anisotropies near 0.3, which is the value used in

[17] G. Haran, E. Haas, B. K. Szpikowska, and M. T. Mas, *Proc. Natl. Acad. Sci. U.S.A.* **89,** 11764 (1992).

[18] P. S. Eis, J. Kuśba, M. L. Johnson, and J. R. Lakowicz, *J. Fluoresc.* **3,** 23 (1993).

[19] J. R. Lakowicz, I. Gryczyński, J. Kuśba, W. Wiczk, H. Szmacinski, and M. L. Johnson, *Photochem. Photobiol.* **59,** 16 (1994).

[20] D. D. Thomas, W. F. Carlsen, and L. Stryer, *Proc. Natl. Acad. Sci. U.S.A.* **75,** 5746 (1978).

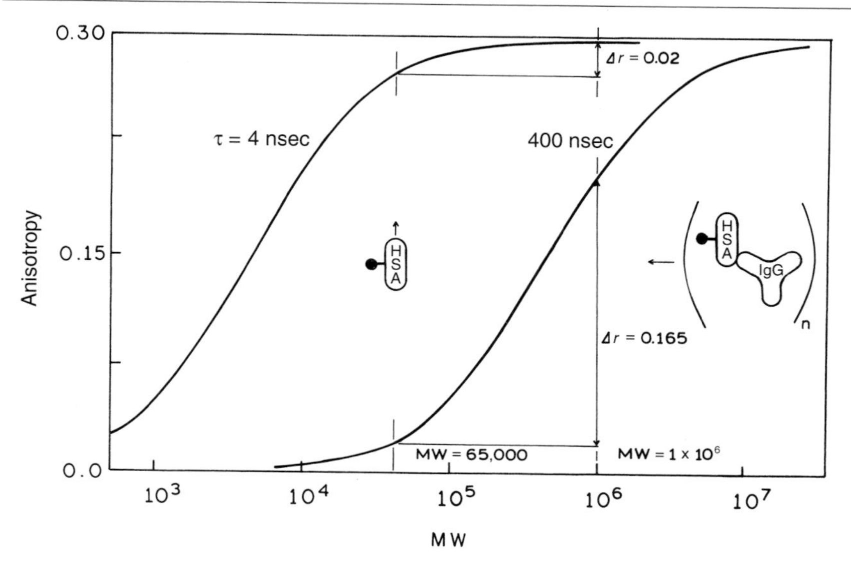

FIG. 1. Molecular weight-dependent anisotropy for a 4- and 400-nsec protein-bound fluorophore. The curves are based on Eqs. (1) and (2) assuming $\overline{v} + h = 1.9$ for the proteins, in aqueous solution at 20° with a viscosity of 1 cP. The arrows and bars indicate molecular weights of the antigen (Ag) and antibodies (Abs) and complexes discussed in text. HSA, Human serum albumin; lgG, human immunoglobulin G.

our simulations (Fig. 1). The molecular volume of the protein is related to the molecular weight (MW) and the rotational correlation time by

$$\theta = \frac{\eta V}{kT} = \frac{\eta \text{MW}}{RT}(\overline{v} + h) \tag{2}$$

where R is the ideal gas constant, $\overline{v}$ is the specific volume of the protein, and h is the hydration, typically 0.2 g of H_2O per gram of protein. In aqueous solution at 20° ($\eta = 1$ cP) one can expect a protein such as human serum albumin (HSA; molecular weight $\simeq$ 65,000, with $\overline{v} + h = 1.9$) to display a rotational correlation time near 50 nsec.

Changes in anisotropy are often caused by changes in the rotational correlation time, for example, on binding or release of a tracer antigen (labeled HSA) from an antibody (Scheme II). Information on the rotational motion is available over a time scale not exceeding three times the lifetime of the fluorophore, after which there is too little signal for rotational motions to affect the steady-state anisotropy. Because the lifetimes of typical fluorophores range from 1 to 10 nsec, it is difficult to measure rotational correlation times greater than 30 nsec. Consequently, it is difficult to deter-

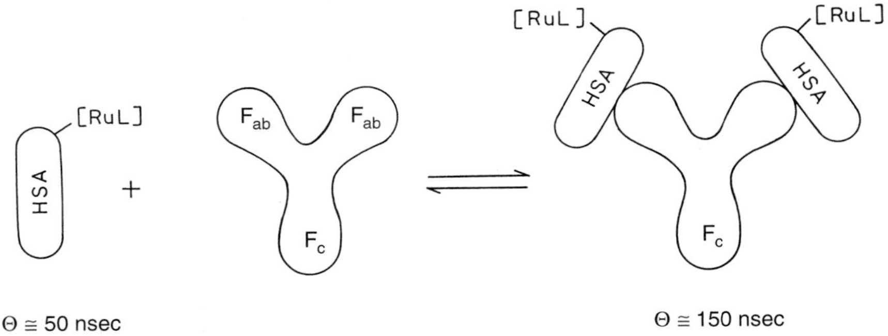

$\Theta \cong 50$ nsec $\Theta \cong 150$ nsec

SCHEME II. Intuitive description of a fluorescence polarization immunoassay. RuL is the Ru–ligand complex, and θ is the rotational correlation time. (Reprinted from Ref. 24, with permission.)

mine the rotational hydrodynamics of larger biomolecules or membrane-bound proteins.

In an attempt to circumvent the limitations imposed by the short fluorescence lifetime, pyrene and its derivatives have been used as extrinsic labels. While pyrene derivatives display lifetimes near 100 nsec, the initial (time = 0) anisotropy of pyrene is low (typically less than 0.1). In addition, pyrene is prone to formation of photoproducts, and requires ultraviolet (UV) excitation with the resulting high levels of autofluorescence from biochemical samples. Hence, pyrene is rather inconvenient as an anisotropy probe. Phosphorescence anisotropy decays have also been used to study the rotational dynamics of membrane-bound proteins.[21–23] Such measurements are based almost exclusively on the triplet probe eosin, which displays a millisecond phosphorescence decay time in the absence of oxygen. However, there are relatively few useful triplet probes. The MLCs are sensitive to slow rotational motions of proteins without the need to rigorously exclude oxygen. The use of phosphorescence is also inconvenient because of the need to rigorously exclude molecular oxygen.

The dependence of the anisotropy on the lifetime and rotational correlation time of a fluorophore is shown in Fig. 1. The sensitivity of the protein-bound fluorophore toward resolving long correlation times is determined by its lifetime. For typical probes with lifetimes near 4 nsec (fluorescein or rhodamine) the anisotropy of low molecular weight antigens (molecular weight < 1000) can be estimated from Fig. 1 to be in the range of 0.05, and that of the antibody-bound antigen (molecular weight 160,000) to be

[21] M. Müller, J. J. R. Krebs, R. J. Cherry, and S. Kawato, *J. Biol. Chem.* **259,** 3037 (1984).
[22] T. Mühlebach and R. J. Cherry, *Biochemistry* **24,** 975 (1985).
[23] M. Bartholdi, F. J. Barrantes, and T. M. Jovin, *Eur. J. Biochem.* **120,** 389 (1981).

about 0.29. Hence, a large change in polarization is found on binding of low molecular weight species to larger proteins or antibodies (Fig. 1).

However, if the molecular weight of the labeled molecule is larger, above 20,000, then the anisotropy changes only slightly on binding to a larger protein. For example, consider an association reaction that changes the molecular weight from 65,000 to 1 million. Such a change could occur for an immunoassay of HSA using polyclonal antibodies, for which the effective molecular weight of the immune complexes could be 1 million or higher. In this case the anisotropy of a 4-nsec probe would change from 0.278 to 0.298, which is too small of a change for quantitative purposes. The small change in anisotropy can be related to the large difference between the lifetime of the fluorophore and the correlation time of the molecules. Owing to the difficulties of using a 4-nsec fluorophore with high molecular molecules, fluorescence polarization measurements are typically used on proteins with molecular weights less than 65,000. In contrast, by use of a 400-nsec probe, which is near the value found for our metal–ligand complex, the anisotropy value of the labeled protein with a molecular weight of 65,000 is expected to increase from 0.033 to 0.198 when the molecular weight is increased from 65,000 to 1 million (Fig. 1). Theoretically, a fluorophore with a lifetime of 400 nsec could allow the analysis of biological systems with molecular weights up to 10 million and correlation times up to 8 μsec.

Development of Ruthenium Metal–Ligand Complex Anisotropy Probe for Microsecond Protein Motions

The first example of an MLC anisotropy probe we synthetized is the *N*-hydroxysuccinimide (NHS)-ester of an asymmetric complex Ru(bpy)(dcbpy)(PF_6) for covalent attachment to proteins (Scheme III). We determined[9] the absorption, emission, and excitation polarization spectra, and measured intensity and anisotropy decays of the Ru complex when covalently bound to proteins. Examples of the absorption spectra of $[Ru(bpy)_2(dcbpy)]$ and $[Ru(bpy)_3]^{2+}$ are shown in Fig. 2 (top). The absorption of $[Ru(bpy)_2(dcbpy)]$ is only moderately dependent on pH. At pH 7, the net charge on the complex is expected to be zero, with two positive charges on the Ru and two negative charges from the two dcbpy ligands. The emission spectrum of Ru-labeled HSA is red shifted about 10 nm relative to that of the Ru complex in aqueous solutions at pH 7 (Fig. 3).

It is important to know whether the intensities and decay times of the MLCs are sensitive to dissolved oxygen. An investigation of the effect of oxygen on the quantum yields of the free and protein bound Ru complexes showed that the sensitivity of the protein-bound form is modest and will not require elimination of oxygen for most applications. As compared to their deoxygenized solutions, the relative fluorescence intensities for

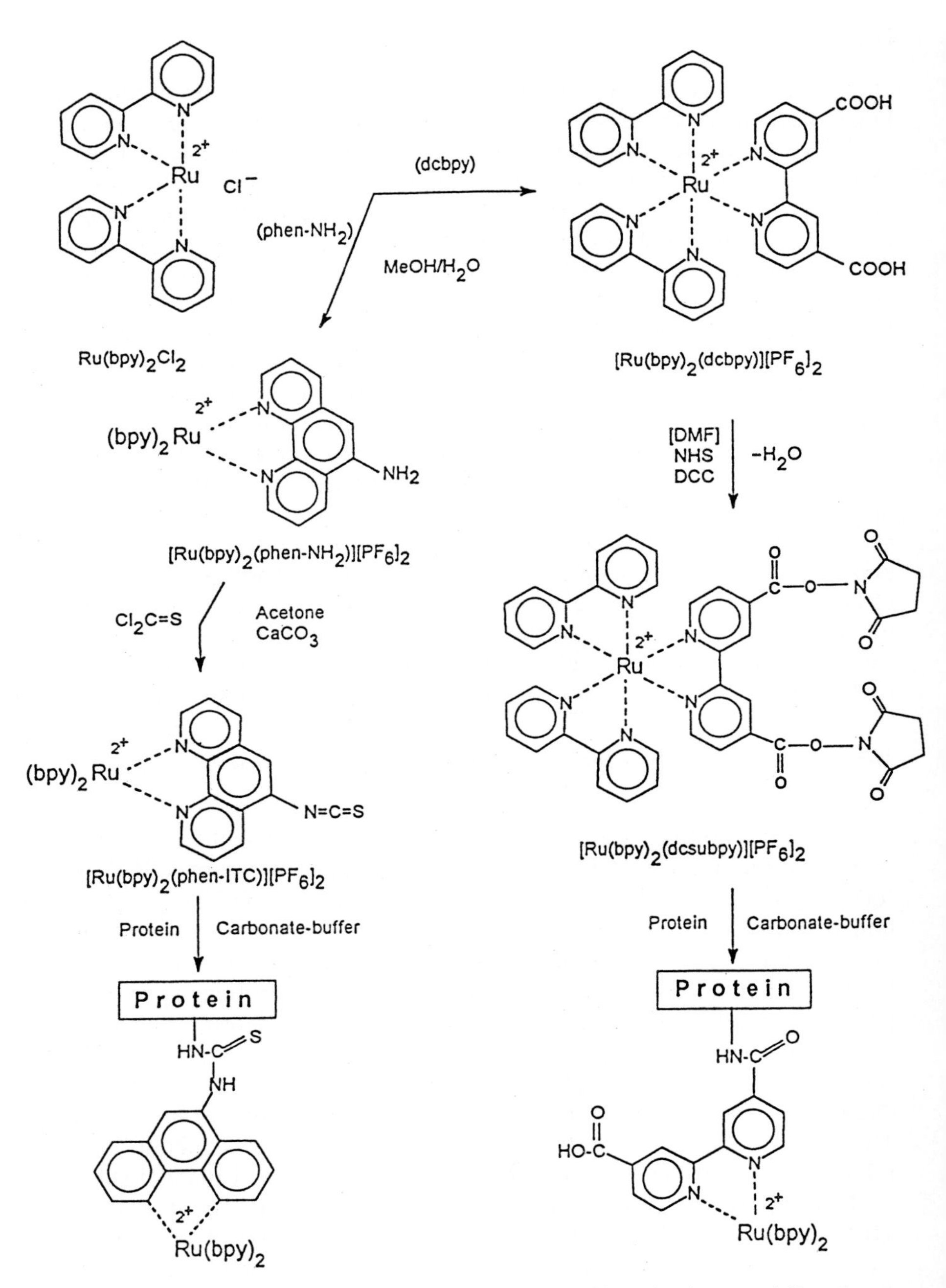

SCHEME III. Synthesis of two representative conjugatable ruthenium metal–ligand complexes.

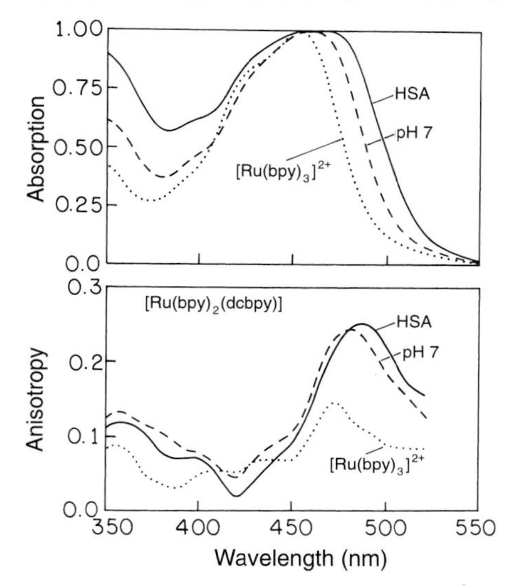

FIG. 2. Absorption (*top*) and excitation anisotropy spectra (*bottom*) of [Ru(bpy)$_2$(dcbpy)], free (---) and when conjugated to HSA (———). The dotted lines (····) represent the symmetrical [Ru(bpy)$_3$]Cl$_2$. The emission wavelength was 650 nm (except for [Ru(bpy)$_3$]Cl$_2$, for which we used 600 nm), with bandpass 8 nm. (Reprinted from Ref. 24, with permission.)

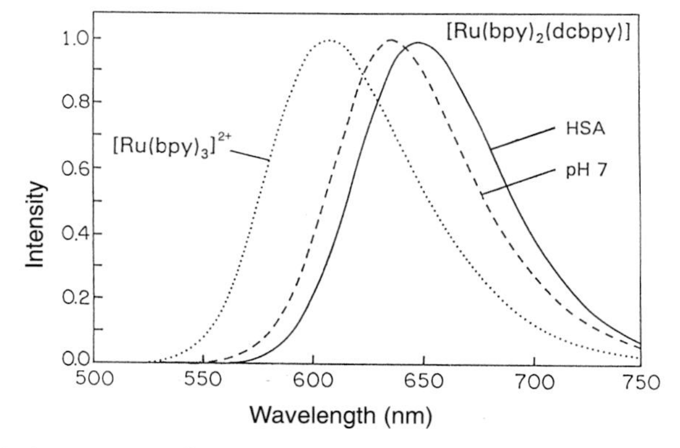

FIG. 3. Emission spectra of [Ru(bpy)$_3$]Cl$_2$ and [Ru(bpy)$_2$(dcbpy)], free (pH 7) and when conjugated to HSA at excitation wavelength 460 nm, 20° in phosphate-buffered saline (PBS). (Reprinted from Ref. 24, with permission.)

$[Ru(bpy)_2(dcbpy)]$ and Ru–HSA in air-equilibrated buffer solutions were 0.77 and 0.89, respectively. The intensity of the free Ru complex is more sensitive to dissolved oxygen than the protein-bound form.

An important aspect of these spectral properties is the convenient excitation and emision wavelengths and the large Stokes shift. These complexes can be excited with visible wavelengths ranging from 400 to 500 nm, which are available from a variety of laser sources and even from simple and inexpensive light-emitting diodes. The emission is shifted by nearly 200 nm, which results in easy separation of the scattered excitation and emission of the sample. The long-wavelength emission centered at 650 nm can potentially be observed in biological samples, and even in whole blood. As is shown below, longer absorption wavelengths are possible with the osmium complexes. In this case the metal–ligand complexes can be excited with simple HeNe lasers or laser diodes.

The steady-state excitation anisotropy spectra for $[Ru(bpy)_3]^{2+}$ and $[Ru(bpy)_2(dcbpy)]$-labeled HSA, in vitrified solution where rotational diffusion does not occur during the excited-state lifetime, are shown in Fig. 2 (bottom). Importantly, the HSA conjugate of $[Ru(bpy)_2(dcbpy)]$ displays a steady-state anisotropy value of 0.26 for excitation near 480–490 nm. In contrast, the more symmetrical $[Ru(bpy)_3]^{2+}$ displays considerably smaller values at excitation wavelengths above 450 nm. Evidently, the presence of a nonidentical ligand is important for obtaining a useful anisotropy probe.

The temperature-dependent anisotropies in a glycerol–water mixture (Fig. 4) indicate that the anisotropies of the Ru complex and the Ru-labeled proteins are sensitive to rotational motions. The steady-state anisotropy of the free Ru complex decreases rapidly above $-50°$, whereas the anisotropies of the Ru-labeled protein decreases more slowly with temperature, and remain relatively high even at 20°. Importantly, the anisotropies of the labeled proteins are always larger than that of the Ru–ligand complex (Fig. 4), which indicates that protein hydrodynamics contributes to the measured anisotropy. This is an important result because there has been considerable controversy concerning the mode of depolarization in these complexes. It was not known whether the emission of MLC solutions become depolarized owing to rotational diffusion of the probe or owing to exchange of the excited-state energy between the organic ligands. The data in Fig. 4 demonstrate that rotational motions of the ligand are dominantly responsible for the anisotropy of the metal–ligand complexes.

The range of measurable correlation times is determined by the lifetime of the excited state. We used time-correlated single-photon counting (TCSPC) to determine the luminescence lifetimes of the Ru complex and the Ru-labeled proteins. The intensity decays were closely approximated by a single decay time (Fig. 5). The decay times of the labeled proteins are

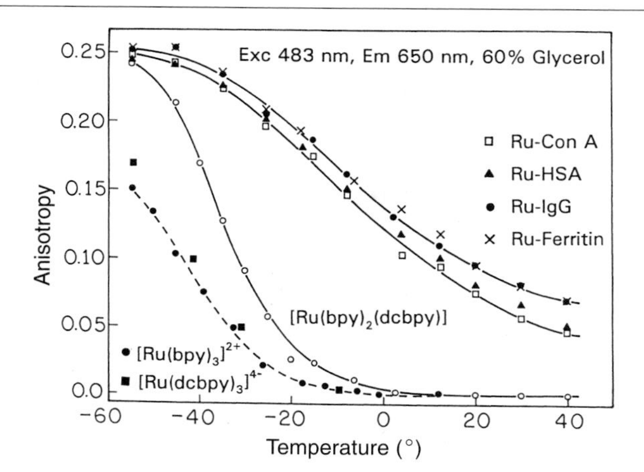

FIG. 4. Temperature-dependent emission anisotropy of the metal–ligand complexes and protein conjugates in glycerol–water (6:4, v/v). Excitation wavelength was 483 ± 4 nm; emission wavelength was 650 nm (and for $[Ru(bpy)_3]^{2+}$ 600 nm), with bandpass 8 nm. (Reprinted with permission from Ref. 9.)

similar to that of the Ru complex alone under comparable experimental condition. The overall range was from 250 to 500 nsec. In addition, the lifetime of the Ru complex appears to be rather independent of the extent of labeling, which indicates minimal interactions between the labels on a given protein molecule. The long lifetime of these labels suggest that the Ru complex can be used to measure rotational correlation times as long as 1.5 μsec, about three times the luminescence lifetime.

Time-dependent anisotropy decays of the free Ru complex and the Ru-labeled proteins are shown in Fig. 6. For the Ru complex alone in buffer at 20° (i.e., not coupled to proteins), the anisotropy decays within the 5-nsec excitation pulse. In contrast, the anisotropy decay is much slower for the Ru-labeled proteins. Importantly, the time-dependent decrease in anisotropy becomes slower as the molecular weight of the labeled protein increases. Specifically, Ru-labeled ferritin displays the slowest anisotropy decay, concanavalin A (ConA) the most rapid anisotropy decay, and human immunoglobulin G (IgG) displays an intermediate decay. The data in Fig. 6 reveal that the anisotropy decay of the Ru-labeled proteins is sensitive to the size and/or shape of the proteins.

Use of Metal–Ligand Complexes in Fluorescence Polarization Immunoassays

The preceding data suggested that the MLC probe is useful in measuring the rotational motions of high molecular weight proteins or complexes.

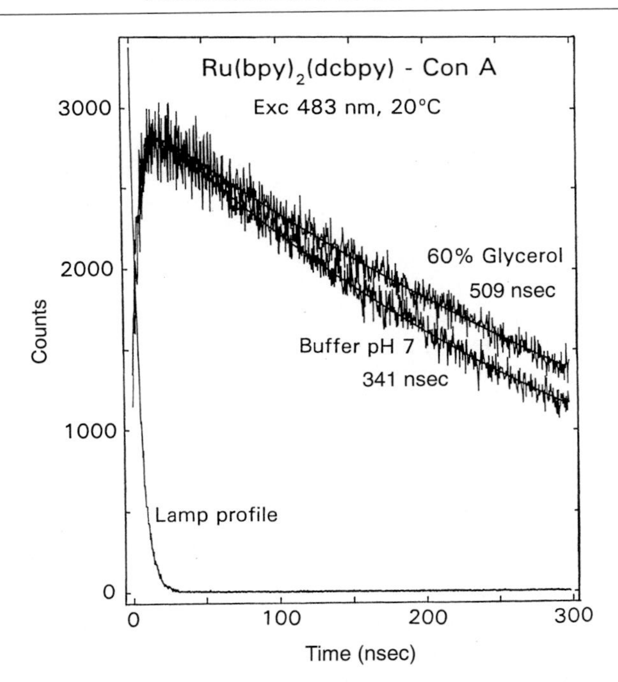

FIG. 5. Intensity decays of $[Ru(bpy)_2(dcbpy)]$ conjugated to concanavalin A (ConA). Similar intensity decays were obtained for $[Ru(bpy)_2(dcbpy)]$ alone and when conjugated to other proteins. (Reprinted with permission from Ref. 9.)

Hence, we chose to use $[Ru(bpy)_2(dcbpy)]^{2+}$ for a fluorescence polarization immunoassay (FPI; Fig. 1) of a high molecular weight analyte, HSA.[24] Such an FPI would not be practical using fluorescein because the rotational correlation time of HSA is near 50 nsec and the decay time of fluorescence is near 4 nsec. Hence we expect its anisotropy to be insensitive to interactions between HSA and IgG because the anisotropy of fluorescein–HSA will be highly polarized even without IgG (Fig. 1). For our FPI the antigen was HSA labeled with $[Ru(bpy)_2(dcbpy)]^{2+}$. We examined the changes in polarization in the presence of increasing amounts of anti-HSA (Fig. 7). The polarization increased about threefold from 0.029 to a plateau at 0.09. Two types of nonspecific IgGs were used as controls. The first was purified human IgG, the second was a mouse IgG. Importantly, no detectable changes in polarization of Ru–HSA were observed in both control experiments (Fig. 7).

We also developed a competitive assay for HSA, wherein labeled and

[24] E. Terpetschnig, H. Szmacinski, and J. R. Lakowicz, *Anal. Biochem.* **227,** 140 (1995).

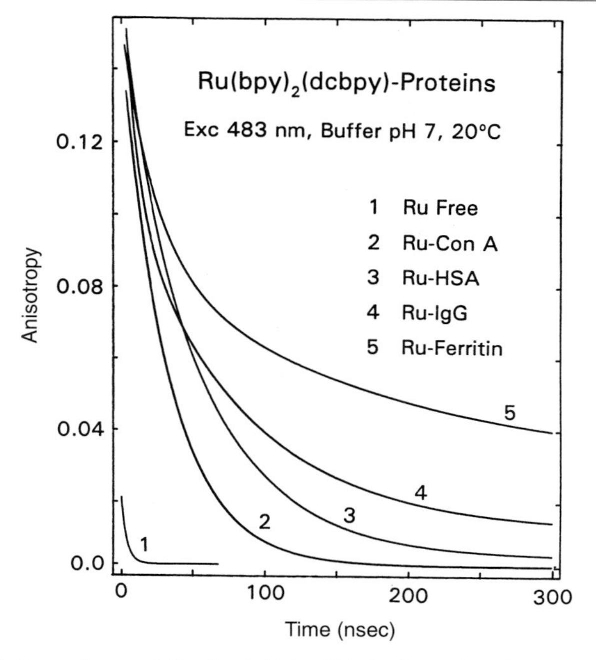

FIG. 6. Anisotropy decays of [Ru(bpy)$_2$(dcbpy)] in buffer and conjugated to proteins. The molecular weights of HSA, concanavalin A (ConA), immunoglobulin G (IgG), and ferritin are 65,000, 102,000, 160,000, and 500,000, respectively. (Reprinted with permission from Ref. 9.)

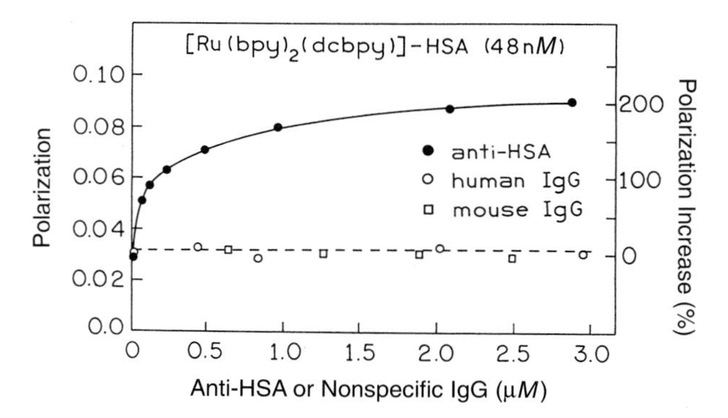

FIG. 7. Steady-state fluorescence polarization of Ru–HSA at various concentrations of IgG specific for HSA (anti-HSA) or nonspecific IgG. The excitation wavelength was 485 nm, with observation at 660 nm, and with a bandpass of 10 nm at 20°. (Reprinted from Ref. 24, with permission.)

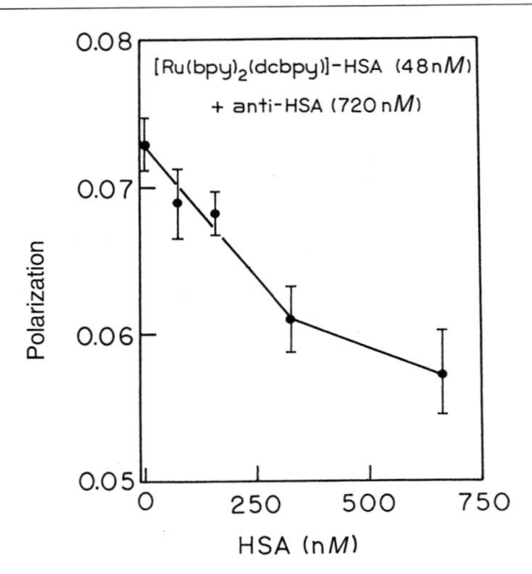

FIG. 8. Competitive immunoassay for HSA, shows the steady-state fluorescence polarization of Ru–HSA. The Ru–HSA was added to preincubated mixtures of anti-HSA with increasing amounts of free HSA. Error bars represent the standard deviations of three polarization readings. See Fig. 7 for experimental conditions. (Reprinted from Ref. 24, with permission.)

unlabeled antigens are competing for the binding sites on the antibody (Ab). In this sequential immunoassay the concentrations for Ru-labeled HSA and anti-HSA were 48 and 720 n*M*, respectively (Fig. 8). The polarization was found to decrease with increasing amounts of unlabeled HSA in the range of 0 to 700 n*M*. Hence, the metal–ligand complexes can be used in competitive immunoassays.

The changes in the steady state polarization (Figs. 7 and 8) were supported by time-resolved anisotropy measurements in Fig. 9. In the absence of anti-HSA (condition 1, Fig. 9) the anisotropy of Ru–HSA has decayed nearly to zero at 200 nsec. In the presence of anti-HSA (condition 3, Fig. 9) the anisotropy has decayed only by 50% at 200 nsec, showing that formation of Ru–HSA : IgG complexes resulted in a slower anisotropy decay. In the presence of a 15-fold excess of unlabeled HSA (condition 2, Fig. 9) the Ru–HSA is partially replaced by unlabeled HSA and the correlation time is reduced from 125 to 109 nsec. These data demonstrate the sensitivity of the Ru label to changes in rotational motions that occur with formation of larger macromolecular associations.

One disadvantage of the present Ru complex is that its rotational motions are partially independent of protein rotational diffusion. The anisotropy decays shown in Fig. 6 display an initially rapid component in the anisotropy decay, which suggests mobility of the present Ru complex that is independent of overall rotational diffusion. We are presently investigating the use of other organic ligands that may improve the anisotropy behavior

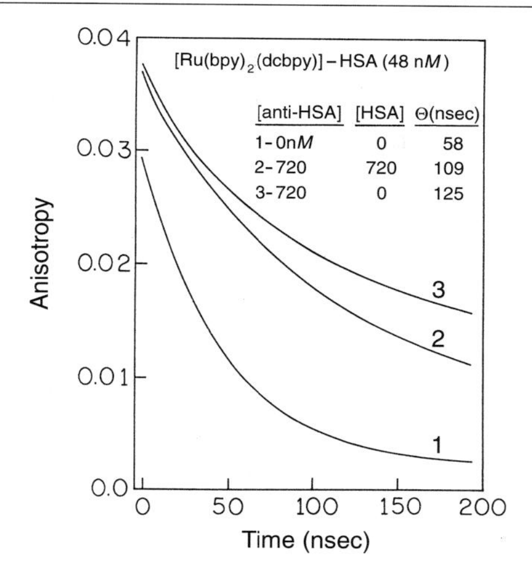

FIG. 9. Time-dependent anisotropy decays of the Ru-labeled HSA (unbound) (1), after addition of anti-HSA (3), and after addition of anti-HSA and unlabeled HSA (2). The excitation wavelength was 360 nm, with observation at 660 nm and with a bandpass of 10 nm at 20°. (Reprinted from Ref. 24, with permission.)

of the metal–ligand complexes. For instance, more hydrophobic organic ligands may enhance noncovalent interactions with proteins and thus less independent probe motion. Alternatively, the use of other organic ligands may improve localization of the excited state.

Metal–Ligand Probes of DNA Dynamics

Another promising application of the MLC probes is for studies of the torsional dynamics of DNA. This topic has been widely investigated as summarized by Schurr *et al.*[25] The majority of these experimental studies were performed using ethidium bromide, which displays a decay time for the DNA-bound state near 30 nsec, or with acridine derivatives, which display shorter decay times.[26–28] The short decay times of most DNA-bound dyes is a serious limitation because DNA is expected to display a wide range of relaxation times.[29,30]

We have demonstrated the possibility of using the MLCs to study DNA

[25] J. M. Schurr, B. S. Fujimoto, P. Wu, and L. Song, Fluorescence studies of nucleic acids: Dynamics, rigidities and structures. *In* "Topics in Fluorescence Spectroscopy: Biochemical Applications" (J. R. Lakowicz, ed.), Vol. 3, pp. 137–229. Plenum, New York, 1992.

[26] D. Genest, P. Wahl, M. Erard, M. Champagne, and M. Daune, *Biochimie* **64,** 419 (1982).

[27] D. P. Millar, R. J. Robbins, and A. H. Zewail, *Proc. Natl. Acad. Sci. U.S.A.* **77,** 5593 (1980).

[28] D. P. Millar, R. J. Robbins, and A. H. Zewail, *J. Chem. Phys.* **76,** 2080 (1982).

[29] S. A. Allison and J. M. Schurr, *Chem. Phys.* **41,** 35 (1979).

[30] M. D. Barkley and B. H. Zimm, *J. Chem. Phys.* **70,** 2991 (1979).

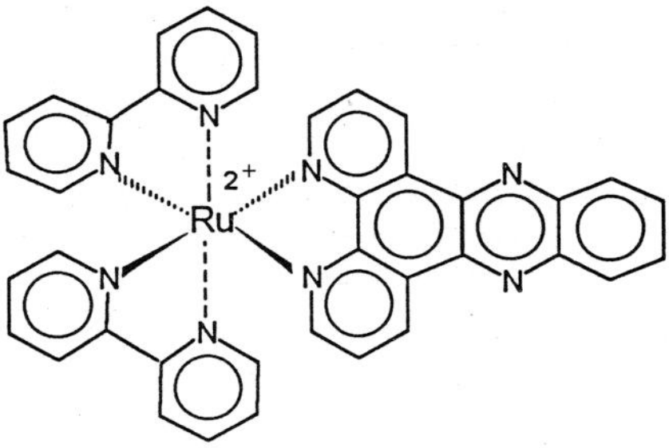

SCHEME IV. Chemical structure of a DNA anisotropy probe. [From J. R. Lakowicz, H. Malak, I. Gryczynski, F. N. Castellano, and G. J. Meyer, *Biospectroscopy* **1,** 163 (1995). Copyright © 1995 Biospectroscopy. Reprinted by permission of John Wiley & Sons, Inc.]

dynamics.[31] To extend the time scale of the measurements to longer times we used $[Ru(bpy)_2(dppz)]^{2+}$ (dppz, dipyrido[3,2-*a*:2′,3′-*c*]phenazine) (Scheme IV). Jenkins and co-workers have pioneered the use of transition metal compounds to probe DNA structure[32] and to study long-range electron transfer.[33] However, to the best of our knowledge these MLCs have not been used to study the hydrodynamics of DNA.

The absorption, emission, and anisotropy spectra of $[Ru(bpy)_2(dppz)]^{2+}$ are shown in Fig. 10. The excitation anisotropy spectra in vitrified solution (glycerol, −60°) display maxima at 365 and 490 nm. The high value of the anisotropy indicates that the excitation is localized on one of the organic ligands, and not randomized among these ligands. This conclusion is supported by observations of Friedman and co-workers, who reported an enhancement of the probe luminescence when bound to DNA, which shields the dppz nitrogens from contact with water.[34] It seems reasonable to conclude that the excitation is localized on the dppz ligand because shielding of the dppz ligand results in an increased quantum yield.

The emission spectrum of $[Ru(bpy)_2(dppz)]^{2+}$ bound to calf thymus DNA is shown in Fig. 10. In aqueous solution the probe luminescence is nearly undetectable. In the presence of DNA the luminescence of $[Ru(bpy)_2dppz]^{2+}$ is remarkably enhanced, an effect attributed to intercalation of the dppz ligand into double-helical DNA. This enhancement of emission on binding to DNA means that the probe emission is observed only from the DNA-bound forms, without contributions from free probe in solution.

[31] J. R. Lakowicz, H. Malak, I. Gryczynski, F. N. Castellano, and G. J. Meyer, *Biospectroscopy* **1,** 163 (1995).

[32] Y. Jenkins, A. E. Friedman, N. J. Turro, and J. K. Barton, *Biochemistry* **31,** 10809 (1992).

[33] C. J. Murphy, M. R. Arkin, Y. Jenkins, N. D. Ghathia, S. H. Bossmehn, N. J. Turro, and J. K. Bemton, *Science* **262,** 1025 (1993).

[34] A. E. Friedman, J.-C. Chambron, J.-P. Sauvage, N. J. Turro, and J. K. Barton, *J. Am. Chem. Soc.* **112,** 4960 (1990).

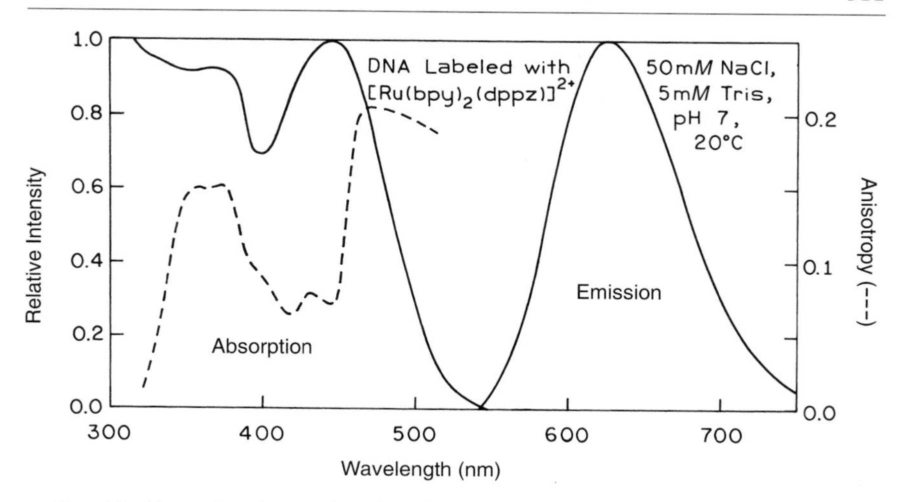

FIG. 10. Absorption (———) and excitation anisotropy spectra of $[Ru(bpy)_2(dppz)]^{2+}$ bound to calf thymus DNA. The dashed line shows the excitation anisotropy spectrum in 100% glycerol at −60°. [From J. R. Lakowicz, H. Malak, I. Gryczynski, F. N. Castellano, and G. J. Meyer, *Biospectroscopy* **1,** 163 (1995). Copyright © 1995 Biospectroscopy. Reprinted by permission of John Wiley & Sons, Inc.]

In this respect $[Ru(bpy)_2(dppz)]^{2+}$ is analogous to ethidium bromide, which also displays significant emission only from the DNA-bound form.[26,27]

The time-resolved intensity decay of $[Ru(bpy)_2(dppz)]^{2+}$ bound to calf thymus DNA is shown in Fig. 11. The intensity decay is best fit by a triple exponential decay with decay times of 12.4, 46.6, and 125.9 nsec, with a mean decay time near 110 nsec. The time-resolved anisotropy decay of DNA-bound $[Ru(bpy)_2(dppz)]^{2+}$ is shown in Fig. 12. The anisotropy decay could be observed to 250 nsec, severalfold longer than possible with ethidium bromide. The anisotropy decay appears to be a triple exponential, with apparent correlation times of 3.1, 22.2, and 189.9 nsec. With the longer lived Ru complexes, and with further experimentation, it should be possible to extend the time scale to several times the intensity decay time, or to more than 2 μsec. In addition, future MLC probes may display longer decay times. Studies of DNA-bound MLC probes offer the opportunity to increase the information content of the time-resolved measurements of nucleic acids.

Second-Generation Metal–Ligand Complexes as Probes

Osmium-Based Complexes with Long-Wavelength Absorption and Emission

The sensitivity of fluorescent measurements in biological systems is determined by the amount of background fluorescence from the biological

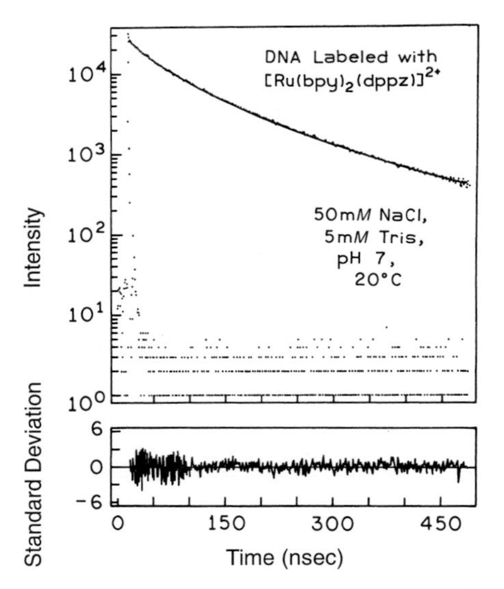

FIG. 11. Time-dependent intensity decays of DNA labeled with $[Ru(bpy)_2(dppz)]^{2+}$. The data are shown as dots. The solid line and deviations (*lower*) are for the best three decay time fits. [From J. R. Lakowicz, H. Malak, I. Gryczynski, F. N. Castellano, and G. J. Meyer, *Biospectroscopy* **1,** 163 (1995). Copyright © 1995 Biospectroscopy. Reprinted by permission of John Wiley & Sons, Inc.]

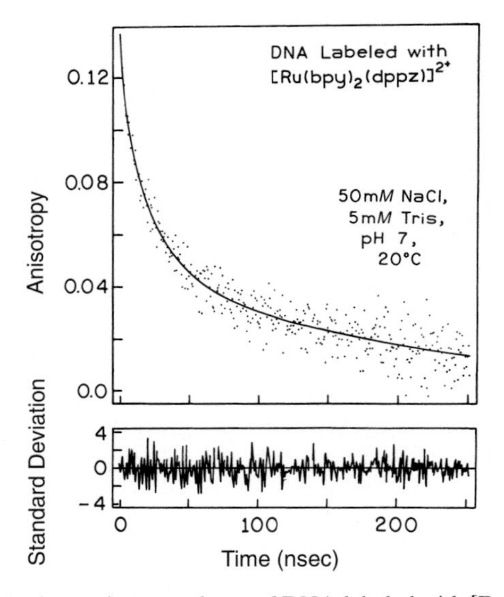

FIG. 12. Time-dependent anisotropy decay of DNA labeled with $[Ru(bpy)_2(dppz)]^{2+}$. The data are shown as dots. The solid line and deviations (*lower*) are for the best three correlation time fits. [From J. R. Lakowicz, H. Malak, I. Gryczynski, F. N. Castellano, and G. J. Meyer, *Biospectroscopy* **1,** 163 (1995). Copyright © 1995 Biospectroscopy. Reprinted by permission of John Wiley & Sons, Inc.]

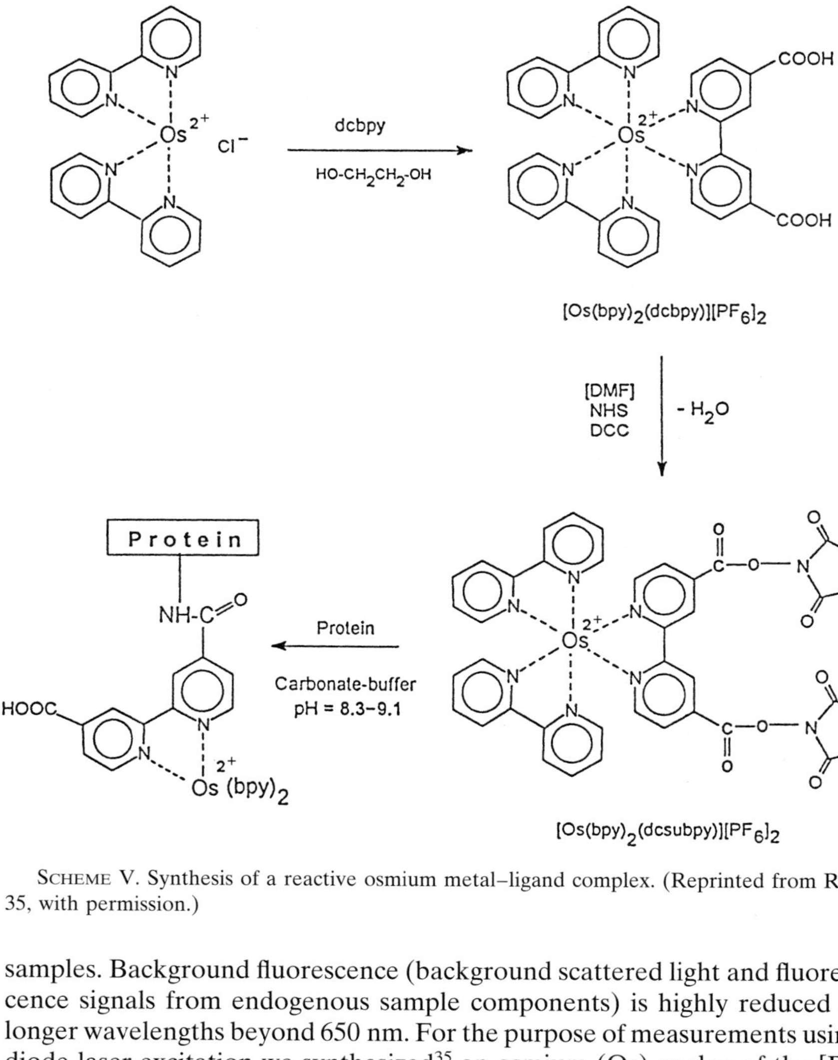

SCHEME V. Synthesis of a reactive osmium metal–ligand complex. (Reprinted from Ref. 35, with permission.)

samples. Background fluorescence (background scattered light and fluorescence signals from endogenous sample components) is highly reduced at longer wavelengths beyond 650 nm. For the purpose of measurements using diode laser excitation we synthesized[35] an osmium (Os) analog of the [Ru $(bpy)_2(dcbpy)$] complex [$Os(bpy)_2(dcbpy)$] (Scheme V), which exhibits long-wavelength absorption up to 750 nm (Fig. 14) and an emission maximum at 780 nm, 130 nm longer than the analogous Ru complex (Fig. 13). The complex was activated and coupled to HSA. The excitation anisotropy spectrum of $Os(bpy)_2(dcbpy)$–HSA is shown in Fig. 14, measured in glyc-

[35] E. Terpetschnig, H. Szmacinski, and J. R. Lakowicz, *Anal. Biochem.* **240,** 54 (1996).

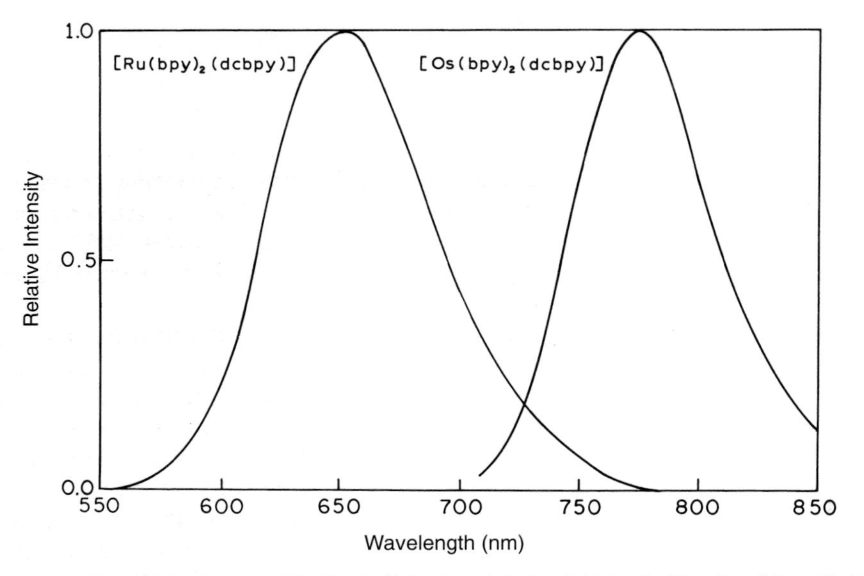

FIG. 13. Emission spectra of $Ru(bpy)_2(dcbpy)$ and $Os(bpy)_2(dcbpy)$. (Reprinted from Ref. 35, with permission.)

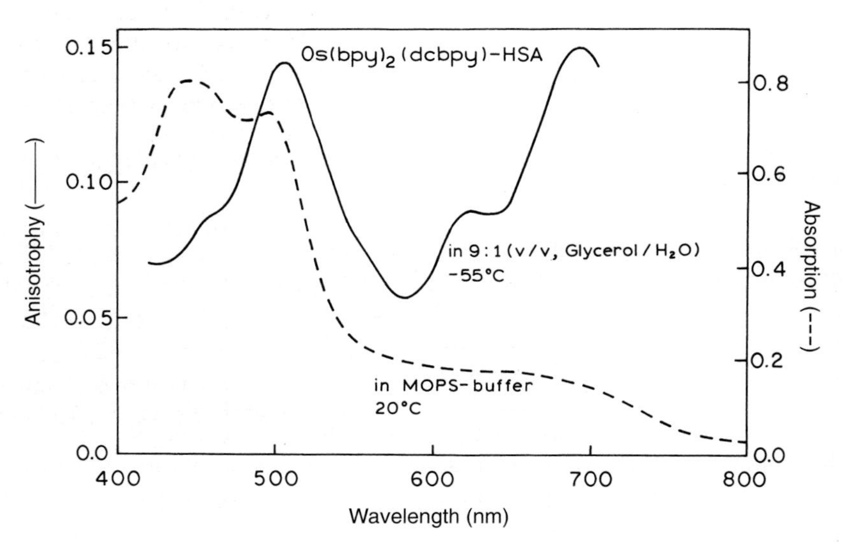

FIG. 14. Absorption (———) and excitation anisotropy spectra (---) of $Os(bpy)_2(dcbpy)$. (Reprinted from Ref. 35, with permission.)

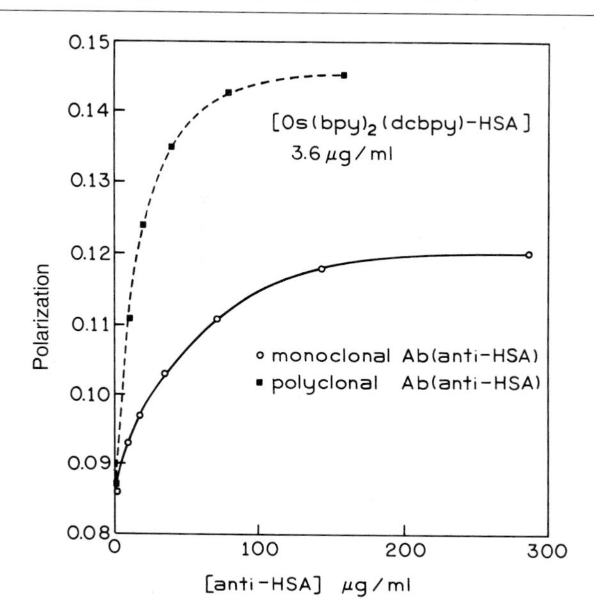

FIG. 15. Steady-state fluorescence polarization of Os–HSA at various concentrations of IgG specific for HSA (anti-HSA). The excitation wavelength was 505 nm.

erol–water (9:1, v/v) at −55°. Importantly, there are two maxima in the anisotropy spectrum near 0.15, one at 505 and one at 685 nm. The initial anisotropy value at 685 nm of 0.14 is ideal for tracing protein interactions, using a 690-nm diode laser. We used an immunoassay of HSA and anti-HSA to demonstrate the ability of the Os system to resolve different correlation times (Fig. 15). Although the lifetime of the Os complex is short (19 nsec), we observed a 100% increase in steady-state anisotropy on binding of the antibody. Longer lifetimes in the range of a few hundred nanoseconds can be obtained by replacing the bpy ligands with tppz [tetrakis(pyridyl)pyrazine], phosphine, or osmium ligands,[36–38] which should allow the measurement of longer correlation times when using diode laser excitation.

Rhenium Metal–Ligand Complexes with Millisecond Lifetimes and High Quantum Yields

Extremely long lifetimes and high quantum yields are the characteristics of some Re(I)-based metal complexes.[2] The isonitrile derivatives (L)Re

[36] R. G. Brewer, G. E. Jensen, and K. J. Brewer, *Inorg. Chem.* **33,** 124 (1994).

[37] C. R. Arana and H. D. Abruna, *Inorg. Chem.* **32,** 194 (1993).

[38] E. M. Kober, J. L. Marshall, W. J. Dressick, B. P. Sullivan, J. V. Caspar, and T. J. Meyer, *Inorg. Chem.* **24,** 2755 (1985).

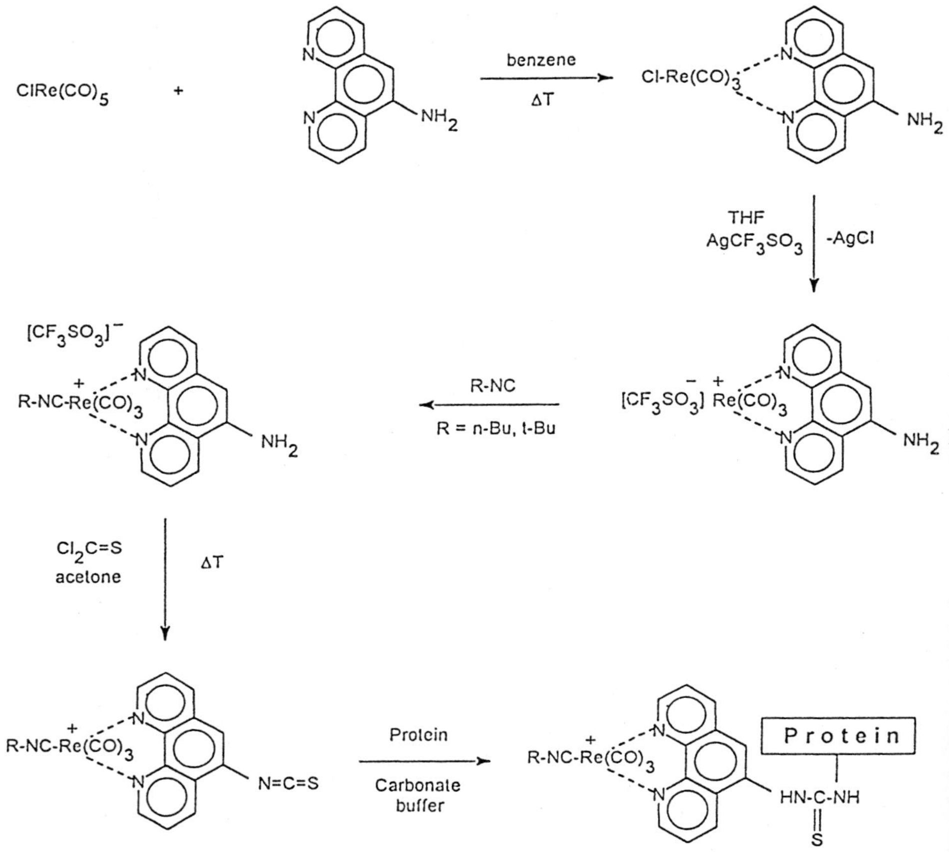

SCHEME VI. Synthesis of a reactive rhenium metal–ligand complex.

$(CO)_3C{=}NR$ [L = 1,10-phenanthroline (phen), bpy; R = n − Bu, t − Bu] of such Re–MLCs have excited-state lifetimes in the range of hundreds of microseconds and quantum yields of >0.7 at room temperature in solution.[2] These properties make them attractive for applications as sensors or molecular probes. Demas and co-workers have described the synthesis and spectral properties of rhenium (Re) MLCs. When such complexes contain an isonitrile ligand the quantum yields are near 0.8 and the decay times as long as 100 μsec. Surprisingly, the long-lived isonitrile derivatives are reported to be stable in alcohol and aqueous solutions.[2] Other reports have also described highly fluorescent and long-lived Re complexes.[39–41] One obvious

[39] T. G. Kotch, A. J. Lees, S. J. Fuerniss, K. I. Papathomas, and R. W. Snyder, *Inorg. Chem.* **32,** 2570 (1993).

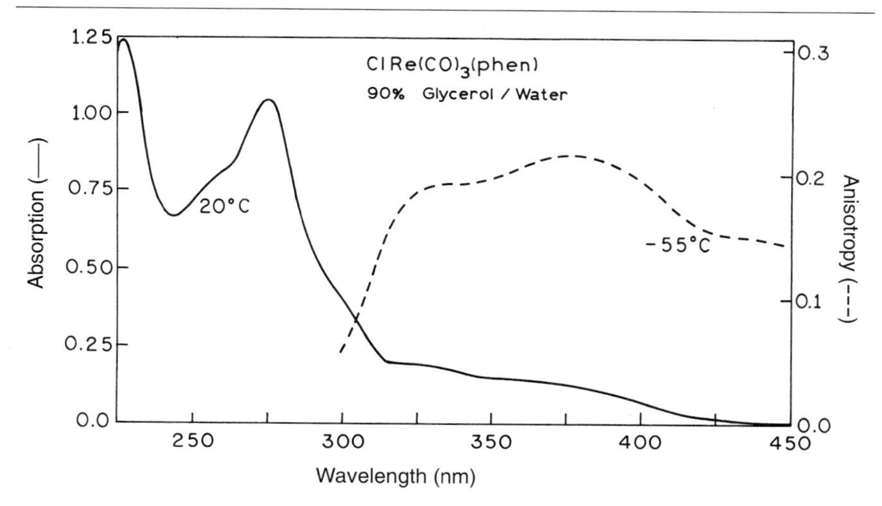

FIG. 16. Excitation anisotropy spectra (---) and absorption spectra with [ClRe(phen)$(CO)_3$] (———) in glycerol–water (9:1, v/v).

application for long-lived Re complexes is in gated detection following decay of the prompt autofluorescence from biological samples. Gated detection is widely used with the long-lived lanthanides, especially in immunoassays.[1] One possible synthesis of reactive Re(I)–isonitrile complexes is shown in Scheme VI. The excitation polarization spectrum of the precursor Cl-Re$(CO)_3$(phen)$^+$, in vitrified solution (Fig. 16), reveals also high polarization values for this compound. The use of Re complexes with lifetimes up to 100 μsec could allow the direct measurement of analytes with molecular weights up to 10 million in a polarization immunoassay.

Metal–Ligand Complexes for Immunoassays Based on Fluorescence Resonance Energy Transfer

Fluorescence immunoassays have been based on a number of spectral properties, including the use of fluorogenic substrates in the enzyme-linked immunosorbent assays (ELISAS),[42] the long-lived emission from lanthanides,[1] chemiluminescence,[43] and the use of fluorescence resonance energy

[40] L. Wallace and D. P. Rillema, *Inorg. Chem.* **32,** 3836 (1993).

[41] A. P. Zipp, L. Sacksteder, J. Streich, A. Cook, J. N. Demas, and B. A. DeGraff, *Inorg. Chem.* **32,** 5629 (1993).

[42] H. T. Karnes, J. S. O'Neal, and S. G. Schulman. *In* "Molecular Luminescence Spectroscopy: Methods and Applications" (S. G. Schulman, ed.), Part 1, pp. 717–779. John Wiley & Sons, New York, 1985.

[43] L. J. Kricka, *Clin. Chem.* **37**(9), 1472 (1991).

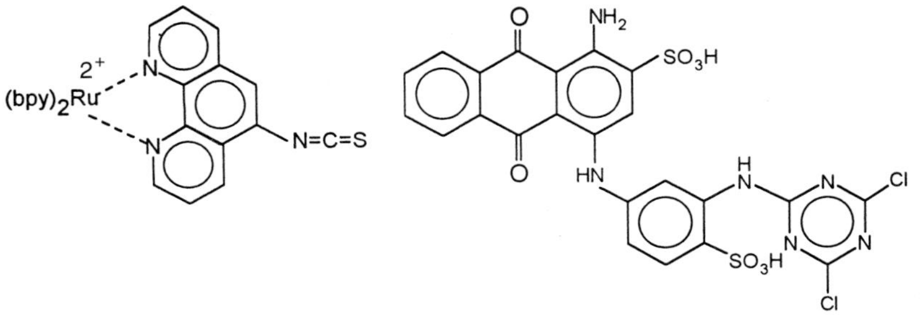

SCHEME VII. Ruthenium metal–ligand donor (left) and Reactive Blue 4 acceptor (right) used in the energy transfer immunoassay. (Reprinted from Ref. 49, with permission.)

transfer (FRET) to detect antigen–antibody association.[44–46] Among the mechanisms previously described, FRET is perhaps the most versatile because it is a through-space interaction that occurs over distances of 20–70 Å,[47,48] which are comparable to the size of typical antigens and antibodies.

We developed[49] a FRET immunoassay based on a ruthenium (Ru) metal–ligand complex as the donor and a nonfluorescent absorber Reactive Blue 4 (Scheme VII). The antigen was human serum albumin (HSA), which was labeled with a ruthenium–ligand complex, $[Ru(bpy)_2(phen\text{-}ITC)]^{2+}$ (phen-ITC, 1,10-phenanthroline 9-isothiocyanate). The antibody (IgG) specific to HSA was labeled with a nonfluorescent absorber, Reactive Blue 4. Absorption and emission spectra of the Ru-labeled HSA and the absorption spectrum of the RB4 acceptor-labeled antibody (RB4–AHA) are shown in Fig. 17. There is sufficient spectral overlap between the donor emission and long-wavelength absorption of the acceptor. The Förster distance (R_0) for energy transfer[47] from the Ru complex to RB4 was calculated to be 30.1 Å.

Association of the Ru-labeled HSA with the antibody was detected by three spectral parameters: a decreased quantum yield of Ru–HSA, a decrease in its fluorescence lifetime, and an increase in its fluorescence anisotropy. Addition of RB4–AHA to donor-labeled HSA led to a progressive

[44] E. F. Ullman, M. Schwarzberg, and K. E. Rubenstein, *J. Biol. Chem.* **251**(14), 4172 (1976).

[45] M. N. Kronick and P. D. Grossman, *Clin. Chem.* **29**(9), 1582 (1983).

[46] W. G. Miller and F. P. Anderson, *Anal. Chim. Acta* **227,** 135 (1989).

[47] Th. Forster, *Ann. Phys.* (*Leipzig*) **2,** 55 (1948). [Translated by R. S. Knox].

[48] H. C. Cheung, Resonance energy transfer. *In* "Topics in Fluorescence Spectroscopy: Principles" (J. R. Lakowicz, ed.), Vol. 2, pp. 127–176. Plenum, New York, 1991.

[49] H. J. Youn, E. Terpetschnig, H. Szmacinski, and J. R. Lakowicz, *Anal. Biochem.* 232, 24 (1995).

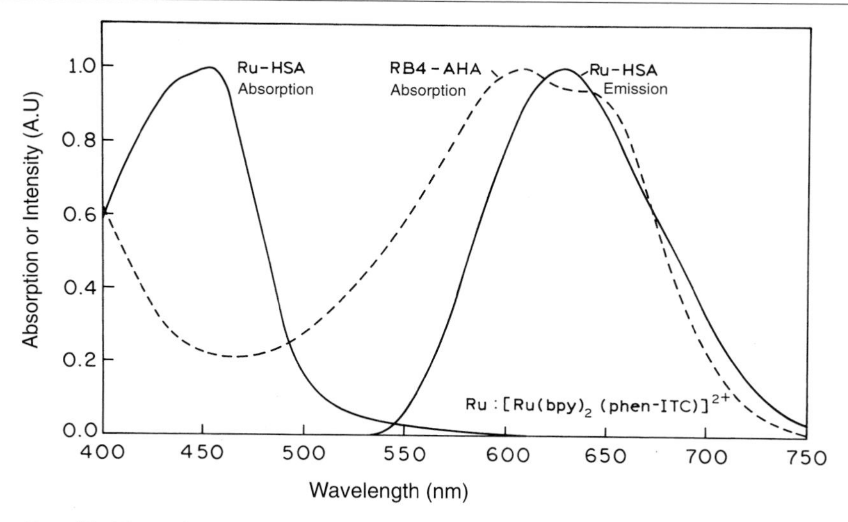

FIG. 17. Absorption spectra of Ru–HSA (———) and RB4–AHA (---) and emission spectrum of Ru–HSA. (Reprinted from Ref. 49, with permission.)

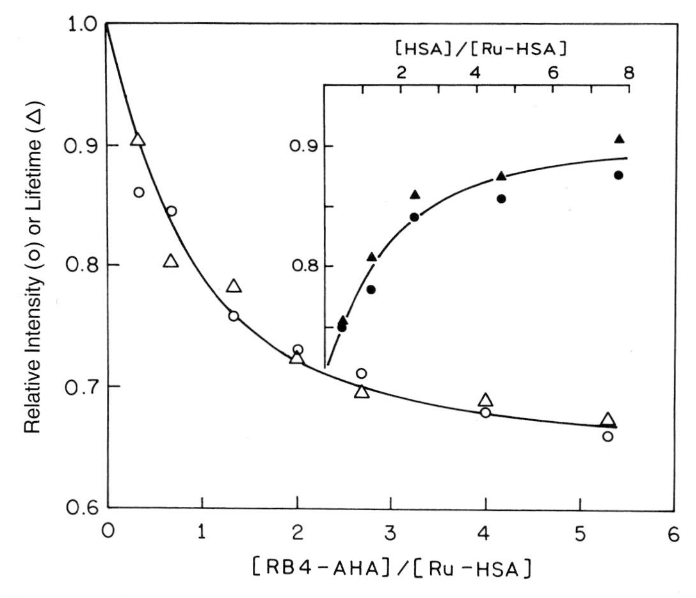

FIG. 18. Effect of RB4–AHA binding on the relative intensities (○) and lifetimes (△) of Ru–HSA. *Inset*: Sequential competitive immunoassay of HSA showing the increase in fluorescence intensity (●) or lifetime (▲) in the presence of unlabeled HSA. (Reprinted from Ref. 49, with permission.)

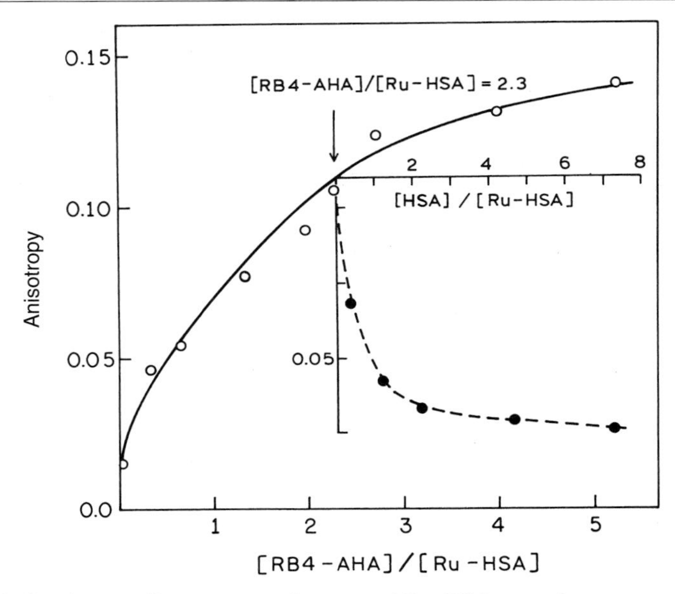

FIG. 19. Steady-state fluorescence anisotropy of Ru–HSA at various concentrations of RB4–AHA. The excitation wavelength was 485 nm, with observation at 600 nm, and with a bandpass of 10 nm at 20°. *Inset*: Effect of unlabeled HSA on the steady-state fluorescence anisotropy of Ru–HSA. The Ru–HSA was added to preincubated mixtures of RB4–AHA with increasing amounts of free HSA. (Reprinted from Ref. 49, with permission.)

decrease in the steady state intensity. This quenching effect shows evidence of saturation (Fig. 18, ○), with a maximum of about 35% quenching at a RB4–AHA to Ru–HSA molar ratio of 5.3:1. The effect of RB4–AHA on the intensity of Ru–HSA can be reversed by addition of unlabeled HSA [Fig. 18, inset (●)], which suggests that the quenching effect is specific for antigenic sites on HSA.

The steady-state anisotropy of Ru–HSA increased approximately eightfold on binding to the antibody (Fig. 19). These spectral effects were observed both in the direct association of the Ru–HSA with Reactive Blue 4-labeled antibody, and in a competitive assay format wherein unlabeled HSA competes with Ru–HSA for the binding sites on the antibody. Some nonspecific interactions of HSA may have occurred with Reactive Blue 4-labeled AHA, a difficulty that can be avoided with a different acceptor. The use of FRET provides a reliable means to alter the spectral properties on antigen–antibody binding. The advantages of a ruthenium–ligand fluorophore include its long-wavelength absorption and emission, long fluorescence lifetime, and high photostability. Long wavelengths minimize problems of autofluorescence from biological samples, and long lifetimes allow off-gating of the prompt autofluorescence.

Conclusion

The polarized emission from metal–ligand complexes offers numerous experimental opportunities in biophysics and clinical chemistry. A wide range of lifetimes and absorption and emission maxima can be obtained by careful selection of the metal and the ligand. For instance, long wavelengths are desirable for clinical applications, such as fluorescence polarization immunoassays. Absorption wavelengths as long as 700 nm can be obtained using osmium, and lifetimes as long as 100 μsec can be obtained using rhenium as the metal in such complexes.[2] The rhenium complexes also display good quantum yields and high initial anisotropies in aqueous solution. Further research is needed to identify which of these metal–ligand complexes display the most favorable spectral properties for a particular application, and to synthesize water-soluble and conjugatable forms of the desired probes.

[15] N-Terminal Modification of Proteins for Fluorescence Measurements

By PENGGUANG WU and LUDWIG BRAND

Introduction

Chemical modifications are often used in protein characterization and in studies of protein interactions with other molecules. For fluorescence measurements in proteins, introduction of probe(s) is sometimes a necessary step to study a particular interaction that is difficult to assess by using intrinsic chromophores such as tryptophan. Many methods are available for labeling different side-chain group proteins.[1–3] Thio, amine, or carboxyl-modifying reagents are the most frequently used and, in general, thiol modification provides more selectivity. The selectivity of thiol groups is due to the combination of several factors, such as less frequent occurrence

[1] A. N. Glaser, R. J. DeLange, and D. S. Sigman, "Chemical Modification of Proteins." Elsevier Biochemical Press, Amsterdam, 1975.

[2] R. L. Lundblad, "Chemical Reagents for Protein Modification," 2nd Ed. CRC Press, Boca Raton, Florida, 1991.

[3] G. E. Means and R. E. Feeney, *Bioconjugate Chem.* **1,** 2 (1990).

0076-6879/97 $25

and more often in the form of disulfide bonds in proteins, as well as differential reactivities of thiol groups in different parts of a protein. In addition, a thiol group can be added to or deleted from a particular site by site-directed mutagenesis for a desired modification.

Another way to modify a protein selectively is through the N-terminal amine. This can be achieved either by directly reacting with the α-amine or by first performing a transamination reaction[4,5] and then attaching a desired probe. In the first approach, the differential reactivity due to pK differences is utilized. In the second aproach, the N-terminal α-amino group is converted into a reactive carbonyl group, which can then react with other molecules. Because the intermediate of transamination involves the participation of peptide backbone, only the terminal amino group is converted, while internal amino groups on lysine residues are not affected.[5] This provides a useful site for attaching chromophores and the method thus is more selective than that relying on pK differences between N-terminal and internal amino groups. The transamination reaction has been used in nuclear magnetic resonance (NMR) studies of bovine pancreatic trypsin inhibitor (BPTI)[6] and fluorescence studies of peptides.[7]

The procedure consists of a transamination reaction to convert the N-terminal α-amino group of a protein into a reactive carbonyl group, a coupling reaction with a chromophore, and a reduction reaction to stabilize the coupling. The scheme is shown in Fig. 1. As an example, we used dinitrophenylhydrazine (DNPH) to react with the carbonyl group of the transaminated staphylococcal nuclease protein and reduced the hydrazone formed between DNPH and the protein by a reducing reagent, borane–pyridine. We then used the attached probe to study the conformational flexibility at the N-terminal region of the nuclease protein by time-resolved resonance energy transfer methods.

Transamination Reaction

The transamination step proceeds under mild conditions that do not denature proteins.[4] The reaction can be completed within 30 to 60 min at room temperature at pH 5 to 7. Three components are needed: an amino group acceptor, a metal ion, and a base. The amine acceptor can be either glyoxylate or pyridoxal phosphate. The metal ion can be either copper or

[4] R. Fields and H. B. F. Dixon, *Biochem. J.* **121,** 587 (1971).

[5] H. B. F. Dixon, *J. Protein Chem.* **3,** 99 (1984).

[6] L. R. Brown, A. De Marco, R. Richarz, G. Wagner, and K. Wuthrich, *Eur. J. Biochem.* **88,** 87 (1978).

[7] R. He and C. L. Tsou, *Biochem. J.* **287,** 1001 (1991).

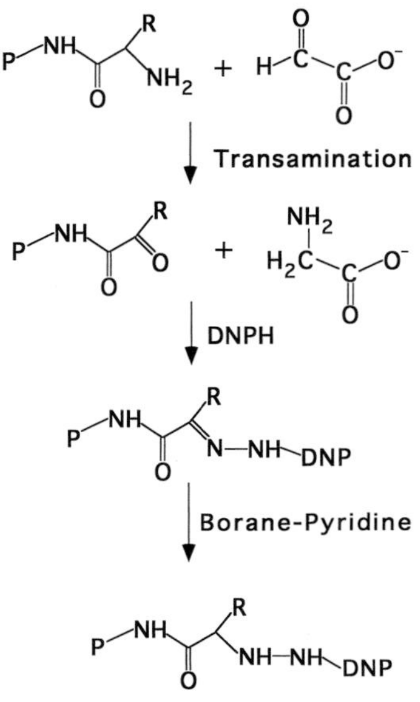

FIG. 1. Scheme of chemical modification of the N-terminal amine. P, Protein; R, amino acid side chain.

nickel. The base can be either pyridine or acetate. A free amino group at the N terminal is necessary for the reaction. If the N terminal of a protein is acetylated or blocked by other groups, this reaction does not occur to a detectable extent and thus the protein can serve as a control. Once the transamination occurs, one positive charge is reduced. The transaminated protein will elute at a lower salt concentration on a cation-exchange column for a positively charged protein and will elute at a higher salt concentration on an anion-exchange column for a negatively charged protein. The transamination reaction thus can be monitored by ion-exchange chromatography, preferentially, high-performance liquid chromatography (HPLC), which can also be used in the subsequent separations.

Procedure

Dissolve the protein to be transaminated in 1 *M* sodium acetate at pH 5.5, 0.1 *M* glyoxylate, and 5 m*M* $CuSO_4$ and stir the reaction mixture for 30 to 60 min at room temperature. Monitor the extent of reaction by ion-exchange HPLC until completion. Stop the reaction either by adding EDTA to a final concentration of 20 m*M* and dialyzing against a buffer of 0.1 *M* HEPES, 50 m*M* NaCl at pH 7.5 (or other buffer), or by a quick gel filtration using a desalting column such as Pharmacia (Piscataway, NJ) PD-10. Perform a control experiment with the same protein, following the same steps in the absence of glyoxylate. This is to make sure that the metal ion in the reaction buffer does not harm the protein (or an alternative metal ion nickel may be used).

Attachment of Chromophore

Reaction with Hydrazine and Reduction

Once a reactive carbonyl is produced, it can then react with amine-containing fluorescent or nonfluorescent chromophores. Many commercially available hydrazines can be used. 2,4-Dinitrophenylhydrazine (DNPH) is often used as a nonfluorescent chromophore. The choice of fluorescent hydrazines depends on a particular application. The rate of hydrazone formation from a hydrazine and a carbonyl group varies with the solution pH. In a strongly acidic pH solution (with an HCl concentration above 0.5 *M*), the reaction can be completed within about 1 hr. Under these conditions, a protein is likely to be denatured. Thus it is necessary that the denaturation be reversible so that the protein can be used later when characterizing its interaction with other molecules. The hydrazone formation can also be achieved at about pH 5, at which pH a protein is likely still to be in its native state. In this pH range, the reaction rate is much slower, generally requiring an incubation of many hours.[8,9] The hydrazone formed is generally not stable, especially at acidic pH, and is not suitable for reliable fluorescence measurements. It can be reduced to a stable hydrazide. Two reducing reagents are frequently used in the reductive amination reaction: sodium cyanoborohydride[10] and borane–pyridine.[11] Each has its advantages and disadvantages. Reductive amination is achieved over the reduction of carbonyl in the range of pH 6–7 for sodium cyanoborohydride[10] and the pH can be extended to pH 4–5 when the hydrazone is formed and separated.[8,9] The release of cyanide during the reaction and the possible side reactions it can cause are the disadvantages of sodium cyanoborohydride. Borane–pyridine has a larger pH range (pH 5–9) and is reported to give a higher yield in reductive amination (methylation). Conversely, it has a lower solubility.

[8] T. P. King, S. W. Zhao, and T. Lam, *Biochemistry* **25,** 5774 (1986).

[9] H. F. Gaertner, K. Rose, R. Cotton, D. Timms, R. Camble, and R. E. Offord, *Bioconjugate Chem.* **3,** 262 (1992).

[10] R. F. Borch, M. D. Bernstein, and H. D. Durst, *J. Am. Chem. Soc.* **93,** 2897 (1971).

[11] W. S. D. Wong, D. T. Osuga, and R. E. Feeney, *Anal. Biochem.* **139,** 58 (1984).

Procedure

Reaction under Native Conditions. Dissolve the transaminated protein in 0.1 *M* sodium acetate at pH 5. Add hydrazine [in dimethyl sulfoxide (DMSO)] to a final concentration 0.1 to 1 m*M*, depending on its solubility. Monitor the extent of reaction by HPLC or by other means. When the formation of hydrazone reaches a satisfactory level, add borane–pyridine (20 to 40 m*M* final concentration). Let the reduction go overnight. Purify the desired product by HPLC.

Reaction under Denatured Conditions. Dissolve the transaminated protein in 0.1 *M* HEPES, 50 m*M* NaCl. Add hydrazine (0.1 to 1 m*M* final concentration). Add concentrated HCl to 0.5 *M*. Stir the reaction mixture at room temperature in the dark for 1 hr. Use a quick desalting column to change the buffer to 0.1 *M* HEPES, 50 m*M* NaCl at pH 7.5. Add borane–pyridine to 20 m*M* and seal the mixture under argon. Allow the reduction to go for 3 to 5 hr (or overnight). Shield the reaction mixture from light by covering with aluminum foil. Once the reaction is finished, remove the excess reducing reagent by means of a desalting column. Perform the same steps with a control protein solution that has been treated with the transamination buffer in the absence of glyoxylate.

Quinoxaline Formation and Fluorescence Amines

Quinoxaline Derivatives

The reactive carbonyl produced by the transamination reaction can react with *o*-phenylenediamine to form fluorescent quinoxaline derivative[7] with a quantum yield similar to that of tryptophan. Its absorption varies with peptide lengths and can shift significantly away from that of tryptophan. Thus it may be selectively excited in fluorescence measurements and may be used as a probe. Because the product can be cleaved at acidic pH, one should exercise caution in using this reaction.

Procedure. Dissolve the transaminated protein in 0.05 *M* NaCl, 0.1 *M* Tris, at pH 9. Add *o*-phenylenediamine to 20 to 50 m*M*. Allow the reaction to proceed at room temperature overnight. Check the extent of reaction by HPLC.

Purify the product on an HPLC or other appropriate column.

Reaction with Fluorescent Amines and Reduction

The reactive carbonyl group produced by the transamination reaction can also react with a fluorescent amine to form a Schiff base, which can then be reduced by reductive amination, either by sodium cyanoborohydride or by borane–pyridine. This type of reaction has been used in the detection of oxidized proteins.[12] The coupling of a fluorescent amine to a reactive carbonyl group in a protein is different from reductive methylation of lysines used in many other studies, in that the carbonyl is located in the protein and can react with lysines within the same protein molecule. Thus it may be necessary to denature the protein in some cases to couple the reaction effectively.

Example of N-Terminal Modification

Staphylococcal nuclease is small protein (17 kDa) with a single tryptophan residue at the 140-position and with no thiol groups. The N-terminal region is disordered and cannot be resolved in X-ray crystal structure. To study this region, we employed the N-terminal modification method to attach a chromophore to be used as an energy acceptor for the donor tryptophan fluorescence.[12a] The protein is positively charged at neutral pH and can be purified on a cation-exchange column. The transaminated protein itself has one positive charge reduced, making it possible to use ion-exchange chromatography for the purification and characterization. We used a weak cation-exchange column: a Brownlee Aquapore CX-300 from Rainin (Ridgefield, NJ). The column is packed with 7-μm silica, 300 Å in pore size. The size of the column is 4.6 mm in inner diameter and 22 cm in length. Buffer A was 10 m*M* sodium acetate, pH 6.5, and buffer B was 1 *M* sodium acetate at pH 6.5. The gradient used was as follows: time, 0–2 min, 95% A; 5 min, 55% A; 20 min, 15% A; and 22 min, 95% A. The flow rate was 1 ml/min at room temperature. The absorbance of the protein was monitored at 280 nm and that of DNPH at 370 nm.

Figure 2A shows the HPLC chromatogram of the intact nuclease protein. We found that the separation on this column is not purely based on ionic interaction for the staphylococcal nuclease. The p*I* of the nuclease protein is between that of cytochrome *c* and that of lysozyme and yet it elutes at a position later (higher salt) than that of lysozyme, while with a regular fast-flow ionic exchange column, it is eluted at the expected position. Thus additional separation mechanisms play a role and this enables a better separation of the transaminated product than that from a fast-flow column with comparable theoretical plates. The transamination reaction reduces the positive charge of the protein by one and thus the modified protein elutes at a lower salt concentration, as shown in Fig. 2B. Because there are 23 lysine residues in the protein, it is important that the amino groups of these lysines not be chemically modified. We did the following controls to

[12] I. Climent, L. Tsai, and R. L. Levine, *Anal. Biochem.* **182,** 226 (1989).
[12a] P. G. Wu and L. Brand, Submitted (1997).

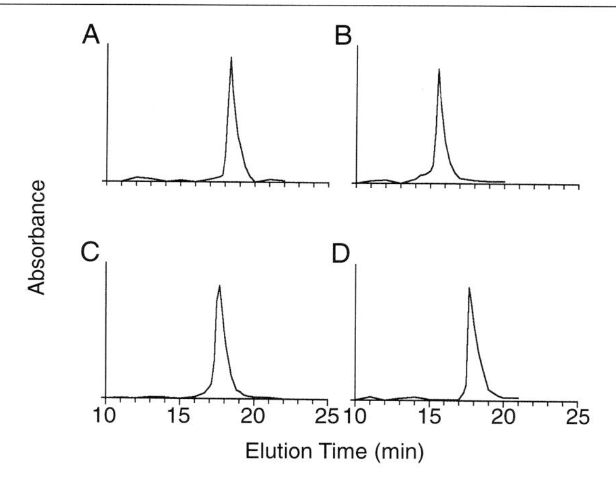

FIG. 2. HPLC elution profiles of the intact and modified nuclease proteins. (A) Native protein monitored at 280 nm; (B) transaminated protein, 280 nm; (C) modified by DNPH and reduced, 280 nm; (D) the same as in (C) but monitored at 370 nm.

make sure that the transamination occurs only at the N-terminal of the protein: (1) Because a free amino group at the N terminal of a protein is required for the transamination reaction, no reaction should occur if it is blocked. The N terminal of horse liver cytochrome *c* is acetylated. Under the same reaction conditions we did not detect any transamination of this protein on HPLC, even with a much longer reaction time; (2) we did the same reaction with the nuclease in the same buffer (1 *M* sodium acetate, 5 m*M* $CuSO_4$ at pH 5.5) in the absence of glyoxylate and stirred the solution for 2 hr at room temperature. We did not observe any transamination of the protein.

In the second step, we used DNPH as a coupling reagent to react with the carbonyl group in the transaminated nuclease protein. For some unknown reason, it was extremely difficult to detect the hydrazone formation beween the transaminated nuclease and the hydrazine at pH 5 or above even with an incubation of 24–48 hr. Because the nuclease is known to be reversible in denaturation and the renaturation occurs within seconds, we used an acidic pH to obtain the hydrazone. The reaction is completed within 1 hr as monitored by HPLC. Once the reaction is finished, we did a quick solution change by a desalting column to a neutral buffer, under which the nuclease becomes native. The hydrazone formed between proteins and DNPH is not stable and can be hydrolyzed. It can be reduced to a more stable hydrazide by reducing reagents such as borane–pyridine. We used this reagent (20 m*M*) to reduce the Schiff base formed between the nuclease protein and DNPH. The product protein–DNP hydrazide can then be used in our subsequent fluorescence measurements. The reduction of the hydrazone by borane–pyridine resulted in a mixture of two components in about equal amounts. These two were then purified twice by HPLC to give pure individual species (>99%). The elution profile of one component is shown in Fig. 2C with 280-nm detection and in Fig. 2D. with 370-nm detection. This is the nuclease modified by DNPH. The second component does not have any absorption at 370 nm. It may be that the transaminated protein with the DNPH was hydrolyzed during the reductive reaction, or the reactive carbonyl itself may be reduced to either to a hydroxyl group or to a nearby lysine group. If the nuclease protein treated with the same reaction buffer in the absence of glyoxylate is reacted with DNPH and borane–pyridine, no modification by DNPH can be detected in the control and the sample elutes at the same position as that of the intact nuclease protein on the HPLC column.

We next characterized the modified protein by absorbance and circular dichroism spectroscopy. The nuclease protein modified by DNPH shows an absorption spectrum characteristic of that of the protein (at 280 nm) and that of the DNPH (at 370 nm). The circular dichroism spectrum of the modified protein is the same as that of the intact protein. Thus the attachment of DNPH has no noticeable effects on the protein secondary structure. The molar absorption coefficient of DNPH covalently attached to the nuclease protein is estimated to be $\varepsilon_{370} = 2.0 \times 10^4 \ M^{-1} \ \text{cm}^{-1}$,

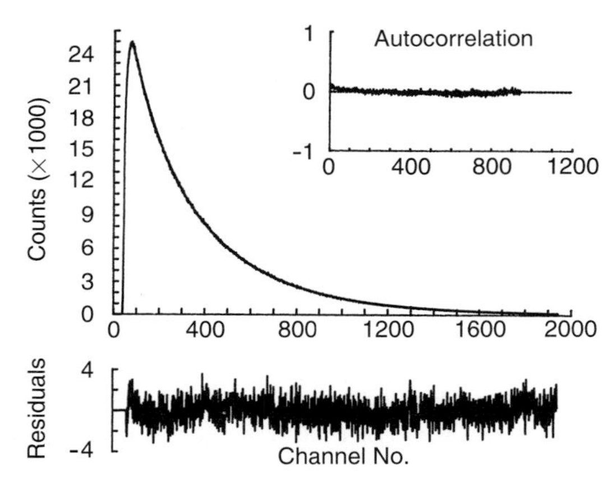

FIG. 3. Fit to the Trp-140 decay in the modified protein with DNPH at 20°: counts as intensity vs channel number as time, with 11 psec/channel. One Lorentzian form was used. Both the experimental data (noisy curve) and fitted data (smooth curve, buried inside the measured data) are shown. The weighted residuals are shown at the bottom and the autocorrelation of the residuals is shown in the inset.

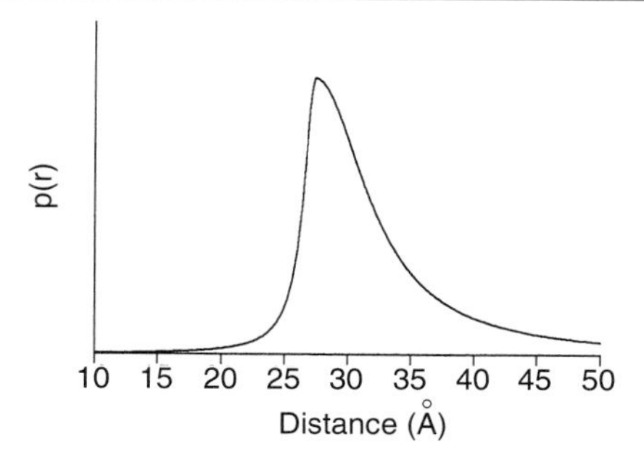

FIG. 4. Results of the fit to the decay shown in Fig. 3. One distance distribution between Trp-140 and the N-terminal DNPH was observed.

assuming that the ratio of the absorbances between 280 and 370 nm remains the same as that of the free DNPH treated with glyoxylate and borane–pyridine.

We utilized the tryptophan fluorescence to study the modified proteins. The transaminated nuclease protein shows a fluorescence emission spectrum identical to that of the native enzyme and its decay characteristics are also the same. The fluorescence decay of Trp-140 in the nuclease can be described by a sum of two exponentials: $\alpha_1 = 0.14$, $\tau_1 = 2.61$ nsec, $\alpha_2 = 0.86$, $\tau_2 = 5.77$ nsec at 20° with an average of 5.35 nsec. The transaminated form also exhibits a decay with two components with an average lifetime of 5.20 nsec. The control nuclease protein that went every step in the absence of glyoxylate, including an incubation in 0.5 *M* HCl for 1 hr, showed the same decay times (average, 5.30 nsec) and circular dichroism as those of the native protein. Thus the experimental conditions for the protein modification had no detectable impact on the protein and the N-terminally modified protein showed the same properties as the native protein, which is expected because the region is disordered even in the crystal structure.

Because the absorption spectrum of DNPH overlaps the emission spectrum of Trp-140 in the protein, we expect a resonance energy transfer between the two chromophores, with a Förster distance[13,14] calculated to be 28.1 Å at 20°. The average distance between Trp-140 and N-terminal DNPH can be determined from the average lifetimes of the donor in the absence and presence of the acceptor. It is about 30 Å and changes little from 1 to 40°. From X-ray structure, the distance between Trp-140 and

[13] T. Förster, *In* "Modern Quantum Chemistry" (O. Sinanoglu, ed.), Vol. III, pp. 9–137. Academic Press, New York, 1965.

[14] P. G. Wu and L. Brand, *Anal. Biochem.* **218,** 1 (1994).

Leu-7 is about 28 Å.[15] The extra six amino acids at the N-terminal are not resolved and appear to contribute little to the average distance.

We used time-resolved resonance energy transfer methods to study the conformational flexibility in the nuclease protein. The disorder in the N-terminal region is reflected in the heterogeneous decays of Trp-140 fluorescence in the presence of DNPH. The decay was fit by a distance distribution[16] as shown in Fig. 3. A Lorentzian distance distribution is used and the curve is shown in Fig. 4. The distance distribution is quite wide, with a full width at half-maximum of the distribution of about 8 Å. The population of distances shorter than 20 Å is unreliable owing to the high degree of quenching of tryptophan fluorescence; a distance longer than 40 Å reflects primarily orientations unfavorable for energy transfer from Trp-140 to DNPH.[17] Because the Trp-140 region is quite rigid, the heterogeneity is primarily from the N-terminal region. Thus our results provide an example of structural studies at the disordered N-terminal region of a protein in solution. This type of disordered structure is difficult to study by other methods such as NMR or X-ray crystallography.

[15] P. J. Loll and E. E. Lattman, *Proteins Struct. Funct. Genet.* **5,** 183 (1989).
[16] P. G. Wu, K. G. Rice, L. Brand, and Y. C. Lee, *Proc. Natl. Acad. Sci. U.S.A.* **88,** 9355 (1991).
[17] P. G. Wu and L. Brand, *Biochemistry* **31,** 7939 (1992).

[16] Fluorescence Studies of Zinc Finger Peptides and Proteins

By PEGGY S. EIS

Introduction

Zinc fingers are a common structural motif found in nucleic acid-binding proteins, and several reviews have appeared in the literature on this subject.[1–5] Zinc finger proteins utilize cysteine (Cys) and histidine (His) residues to coordinate zinc ions tetrahedrally and subsequently form conformationally unique protein domains. Three main classes of zinc fingers have been

[1] A. Klug and D. Rhodes, *Trends Biochem. Sci.* **12,** 464 (1987).
[2] K. Struhl, *Trends Biochem. Sci.* **14,** 137 (1989).
[3] J. M. Berg, *Prog. Inorg. Chem.* **37,** 143 (1989).
[4] J. M. Berg, *Annu. Rev. Biophys. Chem.* **19,** 405 (1990).
[5] D. Rhodes and A. Klug, *Sci. Am.* **268,** 56 (1993).

0076-6879/97 $25

CCHH: **(Y,F)-X-C-$X_{2,4}$-C-X_3-F-X_5-L-X_2-H-X_3-H-X_{2-6}**

CCHC: **C-X_2-C-X_4-H-X_4-C**

CCCC: **C-X_2-C-X_{13}-C-X_2-C-X_{15-17}-C-X_5-C-X_9-C-X_2-C-X_4-C**

SCHEME I. Consensus sequences for the three classes of zinc fingers. Cysteine (C) and histidine (H) side chains tetrahedrally coordinate zinc ion; intervening, nonconserved residues are designated X, and conserved residues are designated by their one-letter code.

identified since the motif was first discovered[6,7]; they are distinguished by their variance in the zinc-coordinating residues (CCHH, CCHC, and CCCC). The CCHH class was identified first; it is found primarily in eukaryotic transcription factors. Retroviral nucleocapsid proteins contain the CCHC motif, and the CCCC motif is primarily found in steroid receptors. Consensus sequences for these three classes of zinc fingers are given in Scheme I.

Numerous biophysical methods have been employed to investigate the structure of zinc finger peptides and proteins, including fluorescence spectroscopy, nuclear magnetic resonance (NMR) spectroscopy, cobalt absorption spectroscopy, circular dichroism, extended X-ray absorption fine structure (EXAFS), and X-ray crystallography. Although each of these methods is useful in characterizing the structural or metal-binding properties of zinc fingers, fluorescence spectroscopy is particularly useful for characterizing both structural and dynamic properties. This chapter provides an overview of the various fluorescence techniques that can be used to investigate zinc finger peptides and proteins.

Zinc-Binding Properties

A common use of fluorescence methodology in zinc finger peptides and proteins is the investigation of their zinc-binding properties. Zinc titrations can be readily performed by monitoring the fluorescence intensity of tyrosine- or tryptophan-containing zinc fingers.[8–12] Zinc finger proteins can also

[6] J. Miller, A. D. McLachlan, and A. Klug, *EMBO J.* **4,** 1609 (1985).

[7] R. S. Brown, C. Sander, and P. Argos, *FEBS Lett.* **186,** 271 (1985).

[8] Y. Mély, F. Cornille, M.-C. Fournié-Zaluski, J.-L. Darlix, B. P. Roques, and D. Gérard, *Biopolymers* **31,** 899 (1991).

[9] M. F. Summers, L. E. Henderson, M. R. Chance, J. W. Bess, Jr., T. L. South, P. R. Blake, I. Sagi, G. Perez-Alvarado, R. C. Sowder III, D. R. Hare, and L. O. Arthur, *Protein Sci.* **1,** 563 (1992).

[10] Y. Mély, H. de Rocquigny, E. Piémont, H. Déméné, N. Jullian, M.-C. Fournié-Zaluski, B. Roques, and D. Gérard, *Biochim. Biophys. Acta* **1161,** 6 (1993).

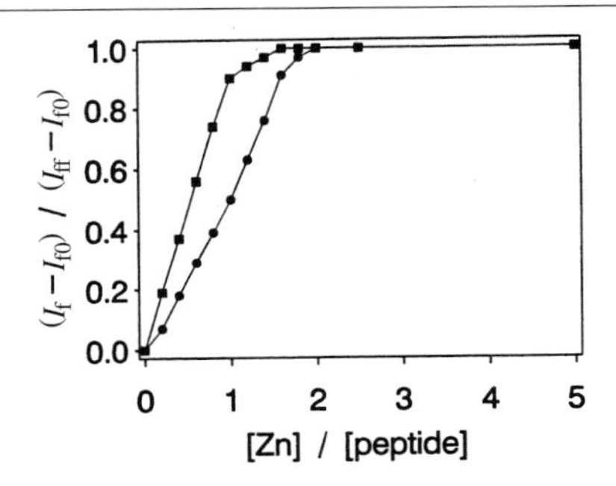

FIG. 1. Zinc-binding curves of (12–53)NCp7 (●) and Ser[28](12–53)NCp7 (■). I_{f0}, I_f, and I_{ff} correspond to the fluorescence of the peptide in the absence and presence of a given and a saturating zinc concentration, respectively. [Reproduced with permission from Ref. 12. Copyright 1994 American Chemical Society.]

be labeled with alternative fluorescence probes such as *N*-iodoacetyl-*N*′-(5-sulfo-1-naphthyl)ethylenediamine (IAEDANS).[13,14] The zinc-induced fluorescence changes can be used to measure zinc-binding constants or to monitor conformational changes.

Zinc Titrations and Binding Constants

The intrinsic fluorescence of zinc finger proteins and peptides can be monitored by steady-state fluorescence methods. Plots of fluorescence intensity vs $[Zn^{2+}]$/[protein] have been used to determine the maximal number of equivalents of zinc bound by the protein, i.e., saturating levels.[9,12] Linearity in the titration curve of proteins with more than one zinc finger is indicative of comparable binding affinities for the individual zinc fingers. Zinc titrations are also useful for characterizing zinc finger mutants. For example, Mély *et al.*[12] demonstrated the impaired zinc binding in the retroviral Ser[28](12–53)NCp7 mutant, in which Ser-28 replaces a zinc-coordinating Cys. The native NCp7 protein contains two CCHC fingers that bind zinc; however, the Ser[28](12–53)NCp7 mutant clearly bound only one equivalent of zinc (Fig. 1). Determination of the zinc-binding constants can be used

[11] Y. Mély, E. Piémont, M. Sorinas-Jimeno, H. de Rocquigny, N. Jullian, N. Morellet, B. P. Roques, and D. Gérard, *Biophys. J.* **65,** 1513 (1993).

[12] Y. Mély, N. Jullian, N. Morellet, H. de Rocquigny, C. Z. Dong, E. Piémont, B. P. Roques, and D. Gérard, *Biochemistry* **33,** 12085 (1994).

[13] J. S. Hanas, A. L. Duke, and C. J. Gaskins, *Biochemistry* **28,** 4083 (1989).

[14] M. K. Han, F. P. Cyran, M. T. Fisher, S. H. Kim, and A. Ginsberg, *J. Biol. Chem.* **265,** 13792 (1990).

as a means for quantitating the impaired zinc-binding affinity in mutants or for investigating the influence of pH.[8,10]

To measure zinc-binding constants using fluorescence intensity, one needs to measure the intensity of the apoprotein (I_{F0}), the zinc-saturated protein (I_{FT}), and intermediate saturation levels of the protein (I_F).[8] The average number of moles of zinc bound, v, per mole of protein can then be calculated using

$$v = (I_F - I_{F0})/(I_{FT} - I_{F0}) \tag{1}$$

Mély *et al.*[8,10] measured the zinc-binding constants in the presence of 1 m*M* EDTA, which buffers the small concentrations of free zinc in the solution. Thus, the concentration of free zinc, [Zn], is then calculated using

$$K_E[Zn^{2+}]^2 + \{K_E([E_t] - [Zn_t^{2+}] + v[P_t]) + 1\}[Zn^{2+}] + v[P_t] - [Zn_t^{2+}] = 0 \tag{2}$$

where $[E_t]$, $[Zn_t^{2+}]$, and $[P_t]$ correspond to the total concentrations of complexant (EDTA), zinc, and protein, respectively; and K_E is the affinity constant of the complexant for zinc. One can now calculate the zinc-binding constant K_{exp} from Eq. (3):

$$v = K_{exp}[Zn^{2+}]/(1 + K_{exp}[Zn^{2+}]) \tag{3}$$

Zinc-Induced Conformational Changes

Fluorescence is an excellent technique for investigating zinc-induced conformational changes in zinc fingers, because the fluorescence intensity or lifetime of many fluorescent probes is sensitive to the environment.[15] Both intrinsic (tyrosine or tryptophan)[8,10,12,16,17] and extrinsic fluorescent probes[13] can be employed for monitoring zinc-induced conformational changes. Hanas *et al.*[13] used the extrinsic probe IAEDANS, which reacts with free sulfhydryl groups, to investigate the conformational effects of zinc on transcription factor IIIA (TFIIIA). Labeling of TFIIIA while it is bound to 5S RNA in the 7S particle yields a singly labeled site in the DNA-binding region of the protein. On removal of zinc and the associated 5S RNA from TFIIIA, the AEDANS fluorescence intensity increased and its maximum blue-shifted, thus indicating that the probe was not in a more hydrophobic environment and/or is more restricted in its motion.

Mély *et al.*[8,10] also reported an enhanced fluorescence intensity on the binding of zinc by the Moloney murine leukemia virus (MoMuLV) NCp10

[15] J. R. Lakowicz, "Principles of Fluorescence Spectroscopy." Plenum, New York, 1983.
[16] P. S. Eis, J. Kuśba, M. L. Johnson, and J. R. Lakowicz, *J. Fluoresc.* **3,** 23 (1993).
[17] P. S. Eis and J. R. Lakowicz, *Biochemistry* **32,** 7981 (1993).

protein and its peptide derivatives. For these studies, a single native sequence tryptophan located within the CCHC finger was used to monitor the zinc-induced conformational changes. Tryptophan was also used to monitor conformational states of the two zinc fingers present in human immunodeficiency virus type 1 (HIV-1) NCp7.[12] The NCp7 derivatives, which retained their zinc-binding ability and their ability for the two fingers to interact, exhibited lower tryptophan quantum yields than did the mutants, which exhibited neither of these actions. For both cases the tryptophan fluorescence maximum (353 nm) indicated that it was solvent exposed; however, the difference in the quantum yields suggests that there are two distinct conformational environments for the tryptophan.

Anisotropy Measurements

Eis and Lakowicz[17] performed time-resolved frequency-domain anisotropy measurements on CCHH zinc finger peptides, which contain intrinsic tryptophan fluorophores, in the absence and presence of zinc. Anisotropy measurements provide information about the global and local motions of a fluorophore.[18] Two rotational correlation times (θ), the longer component associated with global motion of the peptide and the shorter one associated with local motion of the tryptophan fluorophore, were measured for the CCHH peptide. In the zinc-bound peptides, the amplitude of the longer correlation time ($r_0 g_i$) was four to five times greater than the amplitude of the shorter component, thus indicating that the structure of the metal-bound CCHH peptide is compact. The amplitudes for both correlation time components were nearly equivalent in the metal-free peptides, which is consistent with a less compact, unfolded structure. Anisotropy measurements on CCHH peptides labeled separately with two different extrinsic fluorophores were also consistent with an inflexible, compact zinc-bound peptide and a flexible, unfolded metal-free peptide.[17]

Mély *et al.*[10] measured time-resolved and steady-state anisotropies of tryptophan in the CCHC MoMuLV NCp10 protein. Although their time-resolved instrument was unable to detect a subnanosecond correlation time associated with local motion of tryptophan, they did detect this component when sucrose was added to the solution. They were not able to detect a correlation time consistent with global motion of the protein; however, they concluded that the zinc finger domain and surrounding amino acids move in a global manner, whereas the amino- and carboxy-terminal residues move independently of the zinc finger domain. Mély *et al.*[12] also measured

[18] R. F. Steiner, "Topics in Fluorescence Spectroscopy: Principles" (J. R. Lakowicz, ed.), Vol. 2, Chap. 2. Plenum, New York, 1991.

time-resolved anisotropy of tryptophan residues to reveal evidence of an interaction between the two CCHC fingers present in the HIV-1 (12–53)NCp7 derivative. In the case of derivatives that had interacting fingers, the amplitude associated with the longer correlation time component was significantly higher than that measured in derivatives without interacting fingers. It was concluded that the tryptophan residues (in positions 16 and 37) were more constrained in the peptide derivatives with interacting fingers.

Time-resolved anisotropy measurements have also been performed on a 25-residue CCHH peptide that contains a single tyrosine (in position 1).[19] These measurements were then compared to dynamic measurements performed using ^{13}C NMR spectroscopy. Under zinc-free conditions, two rotational correlation times were observed (590 and 47 psec) and the amplitude of the shorter component dominated. The apparent overall rotational correlation time for the zinc-bound peptide was greater than that observed for metal-free peptide, and the amplitude of the longer anisotropy decay time was significantly greater than that for the shorter component; thus, global motions dominate under the zinc-bound conditions. These results are consistent with those reported for another CCHH zinc finger peptide[17]; i.e., the metal-free peptide is flexible and unfolded and the zinc-bound peptide is compact and globular in nature.

Quenching Measurements

Stern–Volmer quenching[15] can be used to characterize the solvent accessibility of fluorophores. Han *et al.*[14] performed KI quenching experiments on zinc-bound AEDANS-labeled TFIIIA protein in the absence and presence of 5S RNA. The Stern–Volmer plots were virtually equivalent for both conditions and indicated that the AEDANS fluorophore was solvent exposed. It was concluded that no significant structural changes occurred in the region of the AEDANS-labeled Cys-287 residue in the absence or presence of 5S RNA. Mély *et al.*[11] used acrylamide quenching to characterize the solvent accessibility of tryptophan in the two CCHC zinc finger peptides of HIV-1 NCp7. The static and bimolecular quenching constants indicated that both components of the tryptophan decays in the two peptides were fully solvent exposed.

Distance Measurements

A more precise measure of conformational changes in zinc fingers can be obtained using fluorescence resonance energy transfer (FRET) meth-

[19] A. G. Palmer III, R. A. Hochstrasser, D. P. Millar, M. Rance, and P. E. Wright, *J. Am. Chem. Soc.* **115,** 6333 (1993).

ods.[15,20] Both time-resolved[16,17] and steady-state[8,12] energy transfer methods have been employed to measure distances in zinc finger peptides. Energy transfer can be measured between the amino acid residues tyrosine (donor) and tryptophan (acceptor); or, frequently, tryptophan can be used as an intrinsic fluorescent donor and the peptide can be labeled with extrinsic acceptor molecules. While steady-state FRET methods enable the measurement of discrete donor–acceptor (D–A) distances, time-resolved FRET can be used to measure distributions of D–A distances.[16,17]

To evaluate the distance (r) between a donor and acceptor, one needs to measure the Förster distance (R_0) for a given D–A pair and the energy transfer efficiency (E):

$$r = R_0[(1/E) - 1]^{1/6} \tag{4}$$

The Förster distance is calculated from the donor and acceptor spectral properties as follows:

$$R_0^6 = \frac{9000(\ln 10)\kappa^2 \phi_{D0}}{128\pi^5 N n^4} \int_0^\infty F_D(\lambda)\varepsilon_A(\lambda)\lambda^4 \, d\lambda \tag{5}$$

where κ^2 is the orientation factor between the dipole moments of the donor and acceptor, ϕ_{D0} is the donor quantum yield in the absence of acceptor, N is Avogadro's number, n is the refractive index of the solution, $F_D(\lambda)$ are the emission spectra of the donor with the area normalized to one, $\varepsilon_A(\lambda)$ are the acceptor absorption spectra, and λ is the wavelength. It is generally assumed that κ^2 equals 2/3, which corresponds to a dynamically averaged orientation between the donor and acceptor dipole moments. Such an assumption is generally valid unless it is known that either the donor or acceptor chromophores are motionally restricted, in which case limiting values of κ^2 can be determined by measuring the donor and acceptor anisotropies.[17,20,21] The transfer efficiency can be calculated by measuring the donor steady-state intensity (I) or time-resolved lifetime (τ) in the absence (I_D or τ_D) and presence (I_{DA} or τ_{DA}) of acceptor as follows:

$$E = 1 - I_{DA}/I_D \tag{6}$$
$$E = 1 - \tau_{DA}/\tau_D \tag{7}$$

The donor lifetimes (τ_D and τ_{DA}) can also be used to determine the distribution of distances between the donor and acceptor. Such an analysis is much more complex; a detailed description is given in Eis and Lakowicz[17] and references therein. In cases where the donor fluorescence is highly quenched

[20] H. Cheung, "Topics in Fluorescence Spectroscopy: Principles" (J. R. Lakowicz, ed.), Vol. 2, Chap. 3. Plenum, New York, 1991.

[21] R. E. Dale, J. Eisinger, and W. E. Blumberg, *Biophys. J.* **26,** 161 (1979).

by the acceptor, one can measure the transfer efficiency using the enhanced fluorescence of the acceptor. For example, Mély *et al.*[12] used the acceptor fluorescence to evaluate the transfer efficiency between tyrosine and tryptophan as follows:

$$E = [\phi_{\text{Trp-280}}/(\phi_{\text{Trp-295}} - f_{\text{Trp-280}})]/f_{\text{Tyr-280}} \tag{8}$$

where $f_{\text{Trp-280}}$ and $f_{\text{Tyr-280}}$ are the fractional absorptions of tyrosine and tryptophan, and $\phi_{\text{Trp-280}}$ and $\phi_{\text{Trp-295}}$ are the tryptophan quantum yields at a given excitation wavelength.

Mély *et al.*[8,12] measured the distance between tyrosine and tryptophan residues in MoMuLV NCp10 and HIV-1 NCp7 using steady-state energy transfer methods. Because the fluorescence of the tyrosine donor is small in the presence of the tryptophan acceptor, the transfer efficiency in these systems was evaluated using the enhanced fluorescence of the acceptor [see Eq. (8)]. The tyrosine donor and the tryptophan acceptor are located seven residues apart within the CCHC-binding domain of the NCp10 protein. An NCp10 peptide (residues 24–42) with the native sequence tryptophan replaced by phenylalanine was used to measure the donor quantum yield for the R_0 calculation. In the absence of zinc, the intrafinger Tyr–Trp distance was 10.8 Å; whereas in the presence of zinc the distance was 12.8 Å. These distances were reproducibly measured (±0.2 Å) in the pH range 5.5–8.

In the HIV-1 NCp7 study, Mély *et al.* measured the interfinger Tyr–Trp distance to demonstrate an interaction between the two CCHC fingers (Fig. 2).[12] The two native finger sequences in NCp7 each contain a single aromatic

FIG. 2. Stereoview of the wild-type (13–51)NCp7 structure: interacting finger model. The α-carbon trace is shown by a thick ribbon, and the zinc atoms are depicted by their van der Waals spheres. The Phe-16 and Trp-37 aromatic residues are shown by thin lines following their respective side chains. [Reproduced with permission from Ref. 12. Copyright 1994 American Chemical Society.]

residue, Phe-16 in finger 1 and Trp-37 in finger 2. Morellet *et al.*[22] previously proposed an interaction between fingers 1 and 2 involving the aromatic rings of Phe-16 and Trp-37. In addition, they found that Pro-31 was responsible for introducing a kink in the peptide chain that enables the two fingers to interact. These conclusions were confirmed by energy transfer measurements on the NCp7 mutants Tyr^{16}(12–53) and Tyr^{16}D-Pro^{31}(12–53). A range of distances was measured for each mutant on the basis of the lower and upper bound limits set on κ^2.[20,21] The interfinger Tyr–Trp distance range in Tyr^{16}(12–53)NCp7 was 7–12 Å, whereas the range measured in Tyr^{16}D-Pro^{31}(12–53)NCp7 was 12.5–18 Å.

Time-resolved FRET distance distribution measurements have been performed on CCHH zinc fingers.[16,17] In these measurements, the distribution of D–A distances is determined as opposed to a single, average D–A distance. Distance probability distributions, $P(r)$, are assumed to be Gaussian; they are characterized by the R_{av} value (most probable distance) and the *hw* parameter (full width of the distribution at half-maximum probability), which indicates the degree of conformational heterogeneity present in the system. A more correct analysis of distance distributions includes determination of the mutual donor-to-acceptor diffusion coefficient (D).[23–29]

Eis *et al.*[16] demonstrated the usefulness of measuring distance distributions as well as site-to-site diffusion coefficients in zinc fingers. In this study, a 28-residue consensus peptide[30] $2F_{28}$ was synthesized with a single tryptophan residue at position 14 and a 5-(dimethylamino)-1-naphthalenesulfonyl (DNS) acceptor was attached to the amino terminus. Distance distributions were determined for the CCHH peptide in the absence and presence of zinc and with and without the inclusion of the donor-to-acceptor diffusion coefficient. When zinc was bound to the finger, the R_{av} distance

[22] N. Morellet, H. de Rocquigny, Y. Mély, N. Jullian, H. Déméné, M. Ottmann, D. Gérard, J. L. Darlix, M. C. Fournie-Zaluski, and B. P. Roques, *J. Mol. Biol.* **235,** 287 (1994).

[23] E. Haas, E. Katchalski-Katzir, and I. Z. Steinberg, *Biopolymers* **17,** 11 (1978).

[24] E. Haas and I. Z. Steinberg, *Biophys. J.* **46,** 429 (1984).

[25] E. Haas, C. A. McWherter, and H. A. Scheraga, *Biopolymers* **27,** 1 (1988).

[26] J. R. Lakowicz, J. Kuśba, W. Wiczk, I. Gryczynski, and M. L. Johnson, *Chem. Phys. Lett.* **173,** 319 (1990).

[27] J. R. Lakowicz, J. Kuśba, I. Gryczynski, W. Wiczk, H. Szmacinski, and M. L. Johnson, *J. Phys. Chem.* **95,** 9654 (1991).

[28] J. R. Lakowicz, J. Kuśba, W. Wiczk, I. Gryczynski, H. Szmacinski, and M. L. Johnson, *Biophys. Chem.* **39,** 79 (1991).

[29] J. R. Lakowicz, J. Kuśba, H. Szmacinski, I. Gryczynski, P. S. Eis, W. Wiczk, and M. L. Johnson, *Biopolymers* **31,** 1363 (1991).

[30] B. A. Krizek, B. T. Amann, V. J. Kilfoil, D. L. Merkle, and J. M. Berg, *J. Am. Chem. Soc.* **113,** 4518 (1991).

between the tryptophan donor and the DNS acceptor was 11.2 Å, whereas under metal-free conditions the R_{av} distance was 19.4 Å (Fig. 3A). The *hw* values for zinc-bound and metal-free peptide were 3.0 and 7.7 Å, respectively, thus indicating the expected increased flexibility in the peptide in the absence of zinc. However, the distributions reported in Fig. 3A did not include the effects of donor-to-acceptor diffusion. In fact, the difference between the zinc-bound and metal-free peptide distance distributions was even more pronounced when the diffusion parameter (D) was included in the analysis (Fig. 3B). Interestingly, the distance distribution of the zinc-bound peptide was virtually unchanged (*hw* decreased 0.2 Å, R_{av} remained the same) when diffusion was included in the analysis. Conversely, the distribution of the metal-free peptide was significantly broader (*hw* = 14.5 Å) and there was a slight increase in its R_{av} value (0.7 Å). In this analysis there was a negligible amount of donor-to-acceptor diffusion detected for the zinc-bound peptide ($D \leq 0.2$ Å^2/nsec), whereas the diffusion rate for the metal-free peptide was 60-fold greater ($D = 12$ Å^2/nsec).

Eis and Lakowicz[17] carried out more extensive studies on the CCHH peptide, including measurement of D–A distributions between the Trp-14 residue and a carboxy-terminal lysine side chain labeled with 7-amino-4-methylcoumarin-3-acetyl (AMCA), as well as measurement of distributions

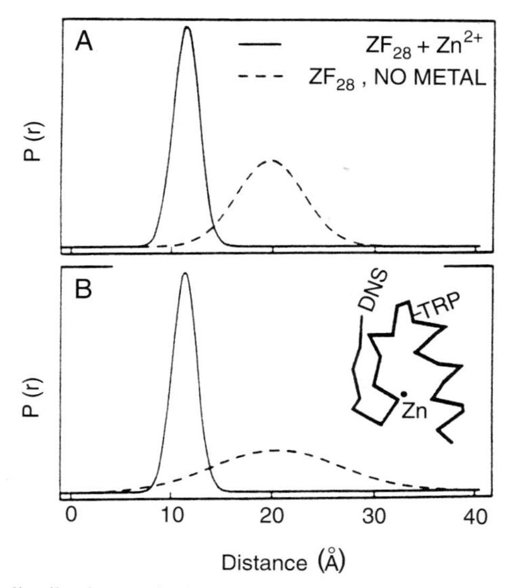

FIG. 3. Distance distributions calculated from the frequency-domain data for metal-free ZF_{28} and zinc-bound ZF_{28}. The distributions were analyzed with (B) and without (A) the apparent donor-to-acceptor diffusion coefficient. The inset in (B) shows a schematic representation of the zinc finger structure and the location of the tryptophan (Trp) donor and dansyl (DNS) acceptor. [Reproduced with permission from Ref. 16.]

between both D–A pairs (Trp^{14}–DNS and Trp^{14}–AMCA) in the presence of the protein denaturant guanidine hydrochloride. Selective labeling of the side chain of the single Lys-29 residue was achieved by first acetylating the amino terminus. Both D–A pair distributions for the zinc-bound peptide were conformationally restricted (Trp^{14}–DNS, hw = 3 Å; Trp^{14}–AMCA, hw = 10 Å; this 7 Å increase in hw was attributed to the flexibility of the lysine side chain), and their R_{av} distances were consistent with the distances present in the NMR structure of a CCHH zinc finger.[31] Donor–acceptor diffusion was limited for both D–A pairs to less than 3 $Å^2$/nsec. Distributions for the metal-free peptides in the presence of 5 *M* guanidine hydrochloride were nearly equivalent, as was expected because the Trp-14 donor is located midway between the two terminal acceptors. However, significant differences were observed between the D–A pair distributions under metal-free conditions. The distribution parameters for the Trp^{14}–DNS pair were as follows: R_{av} = 20 Å, hw = 15 Å, and D = 12 $Å^2$/nsec. The Trp^{14}–AMCA pair parameters were as follows: R_{av} = 15 Å, hw = 28 Å, and D = 20 $Å^2$/nsec. On the basis of these results, it was concluded that the carboxy-terminal half of the CCHH zinc finger was less structured than the amino-terminal half; see Fig. 4 for the proposed structural model. In conclusion, energy transfer distance distribution analysis is useful in characterizing the various conformational states of zinc fingers.

Interactions with Nucleic Acids

Fluorescence methods have also been used to characterize interactions between zinc fingers and nucleic acids.[9,11,13,14,32] Hanas *et al.*[13] reported that AEDANS fluorescence in the singly labeled TFIIIA protein was quenched when the protein was bound to plasmid DNA; however, there was little change in fluorescence when TFIIIA was bound to 5S RNA. It was concluded that the protein adopted an alternative conformation when bound to DNA as opposed to RNA, although the DNA binding in this case was not sequence specific. Han *et al.*[14] also examined TFIIIA structural changes by monitoring the AEDANS fluorescence. They determined that TFIIIA was singly labeled at Cys-287 with AEDANS while the protein was bound to 5S RNA in the 7S particle. They also found that there were virtually no structural changes in the vicinity of Cys-287 when TFIIIA was bound to 5S RNA because Stern–Volmer quenching constants of the AEDANS probe in the absence and presence of the RNA were equivalent.

[31] M. S. Lee, G. P. Gippert, K. V. Soman, D. A. Case, and P. E. Wright, *Science* **245,** 635 (1989).
[32] M. D. Delahunty, T. L. South, M. F. Summers, and R. L. Karpel, *Biochemistry* **31,** 6461 (1992).

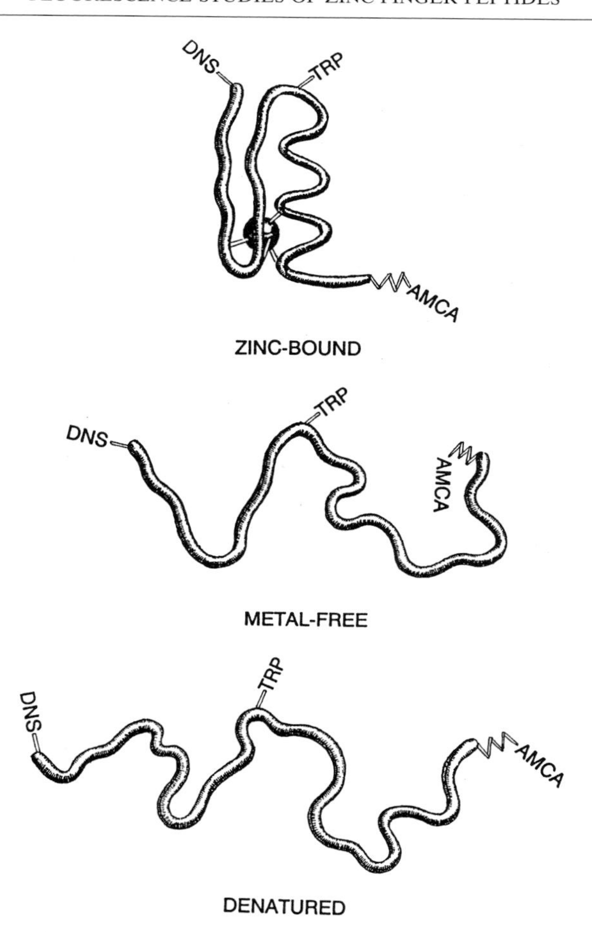

FIG. 4. Proposed structural model for the zinc finger peptide under zinc-bound, metal-free, and denatured conditions. Energy-transfer chromophores are indicated, TRP is the donor, and DNS and AMCA are the two acceptors. [Reproduced with permission from Ref. 17. Copyright 1993 American Chemical Society.]

Delahunty *et al.*[32] used the fluorescent polynucleotide poly(ethenoadenylic acid), poly(εA), to examine the nucleic acid-binding properties of a CCHC peptide corresponding to the first finger of the HIV-1 NCp7 protein (NC-F1). They found that the apparent poly(εA) binding site size in low ionic strength solutions was approximately one-third smaller for the zinc-bound NC-F1 as compared to the metal-free peptide, thus indicating a metal-induced compaction of the structure of the peptide. The binding affinity of NC-F1 for poly(εA) increased one to two orders of magnitude when zinc was bound. Binding of zinc-bound and metal-free NC-F1 to nonfluorescent nucleic acids was investigated by performing competition experiments, i.e., addition of single-stranded (ss) and double-stranded (ds) DNA to poly(εA)-bound NC-F1. The zinc-bound NC-F1-binding affinity for both ssDNA and dsDNA was comparable to its affinity for poly(εA), whereas the metal-free peptide did exhibit some differential binding properties (affinity for ssDNA $<$ poly(εA) $<$ dsDNA).

Summers *et al.*[9] investigated the nucleic acid-binding properties of the full-length HIV-1 NCp7 protein, which contains two CCHC zinc fingers. The NCp7 protein contains a single tryptophan at position 37 in the 55-residue protein and its fluorescence was used to monitor the nucleic acid interactions. The Trp-37 fluorescence of the NCp7 was significantly quenched after addition of tRNA except with the presence of higher concentrations of $MgCl_2$ (37 m*M*), which stabilized the folded tRNA structure. Sequence-specific interactions between zinc-bound HIV-1 NCp7 protein and ssDNA were observed when 21-residue deoxyoligonucleotides corresponding to the viral Psi-packaging signal and a non-Psi sequence were titrated into the sample. Differential binding to the Psi and non-Psi sequences was more pronounced under high ionic strength conditions (150 m*M* NaCl), under which electrostatic interactions were reduced. Nucleic acid-induced quenching of the Trp-37 fluorescence was indicative of a less polar environment for this residue, which Summers *et al.*[9] interpreted as resulting from an intercalative-type interaction.

Mély *et al.*[11] also investigated the CCHC zinc fingers from HIV-1 NCp7. They examined the interaction between each CCHC finger and $tRNA^{Phe}$; finger 34–51 contained a native sequence Trp-37 and finger 13–30 contained a tryptophan conservatively substituted for Phe-16. The tryptophan fluorescence in both peptides was almost completely quenched when they were bound to $tRNA^{Phe}$, which is consistent with the observation of Summers *et al.*[9] for the single Trp-37 in full-length HIV-1 NCp7. Because the tryptophan fluorescent decay times were nearly invariant at varying concentrations of $tRNA^{Phe}$, Mély *et al.*[11] concluded that the tryptophan quenching was due to a stacking interaction between the indole ring and the $tRNA^{Phe}$ bases.

Conclusion

Fluorescence methods, both time resolved and steady state, are clearly useful in the investigation of zinc finger peptides and proteins. Both extrinsic and intrinsic probes have been used to characterize zinc-binding properties, zinc-induced conformational properties, and nucleic acid interactions. Anisotropy, quenching, and FRET measurements are useful in examining the metal-free and zinc-bound conformations of zinc fingers; time-resolved FRET is particularly useful for determining not only the donor–acceptor distances, but the degree of conformational flexibility present as well. Thus, fluorescence techniques can be used to examine both conformation and dynamics of zinc finger peptides and proteins.

Acknowledgments

P.E. acknowledges the Center for Fluorescence Spectroscopy (NIH RR-08119) and NIH GM35154 for the FRET studies of distance distributions in zinc fingers.

[17] Fluorescence Assays for DNA Cleavage

By S. PAUL LEE and MYUN K. HAN

Introduction

Cleavage and joining reactions of nucleic acids are important processes in cellular events such as replication, recombination, and repair of DNA. Nucleic acids are readily cleaved by a variety of enzymes that recognize DNA sequences either specifically or nonspecifically. Examples of enzymes that recognize specific DNA sequences are the restriction endonucleases, which cleave double-stranded DNA at specific sites within, or adjacent to, their recognition sequences. As a result, discrete fragments of DNA are generated that can be isolated for molecular cloning procedures or used to provide specific landmarks in obtaining physical maps of DNA. Nucleases other than restriction endonucleases (i.e., DNases, RNases, exonucleases, helicases) are also routinely used in molecular biology to effect strand separation, cleavage, and denaturation of nucleic acids. Some enzymes, such as polymerases, ligases, and RNase H, are utilized in numerous processes that serve to amplify and detect small quantities of DNA. These processes include the polymerase chain reaction (PCR), ligase chain reaction (LCR), and catalytic hybridization amplification (CHA).

The efficiency of enzymatic cleavage processes can be determined by numerous methods such as gel electrophoresis,[1] thin-layer chromatogra-

[1] A. A. Yolov, M. N. Vinogradova, E. S. Gromova, A. Rosenthal, D. Cech, V. P. Veiko, V. G. Metelev, V. G. Kosykh, Y. I. Buryanov, A. A. Bayev, and Z. A. Shabarova, *Nucleic Acids Res.* **13,** 8983 (1985).

0076-6879/97 $25

phy,[2] elution of the products from a DEAE-cellulose filter,[3] and ultraviolet (UV) absorbance following high-performance liquid chromatography (HPLC) to monitor the disappearance of substrate or appearance of product.[4–6] Although these assay systems each have distinct advantages, they all have the disadvantage of being both time consuming and laborious. In addition, these methods are not continuous assay systems. Kinetic measurements, therefore, require that aliquots be removed from an overall reaction mixture and reanalyzed individually. This is clearly a drawback in situations where precise quantitation and enzyme kinetic constants must be established. A further disadvantage is that the level of sensitivity often required for these procedures necessitates either that the DNA be radioactively labeled or that a high concentration of substrates be utilized for detection. Therefore, it is desirable to design a precise, continuous, sensitive nonradioisotopic assay for enzymatic cleavage reactions.

A continuous spectroscopic assay for endonucleases based on hyperchromic effects resulting from the turnover of duplex oligonucleotide substrates to single-stranded DNA products has been reported.[7] Although this technique has the advantage of being continuous, it relies on UV absorbance for analytical detection and is therefore limited by its narrow dynamic range and range of potential substrate concentrations to be utilized. More recently, Jeltsch *et al.*[8] described a sensitive nonisotopic enzyme-linked immunosorbent assay (ELISA) for determining the DNA cleavage activity of restriction endonucleases. This assay employs double-stranded DNA substrates that are labeled on the 5′ end of each strand; one end is labeled with biotin, and the other end is labeled with fluorescein or digoxigenin. Although this assay does not depend on radioisotopes, the use of biotin-labeled DNA renders this method discontinuous and necessitates extensive sample handling prior to signal detection.

The development of automated DNA synthesis methods has facilitated the covalent introduction of fluorescent probes at specific positions of any DNA sequence, which, in turn, has facilitated the evolution of a variety of

[2] E. Jay, R. Bambara, R. Padmanabhan, and R. Wu, *Nucleic Acids Res.* **1,** 331 (1974).

[3] L. W. McLaughlin, F. Benseler, E. Grawser, N. Piel, and S. Scholtissek, *Biochemistry* **26,** 7238 (1987).

[4] J. Alves, A. Pingoud, W. Haupt, J. Langowski, F. Peters, G. Maass, and C. Wolff, *Eur. J. Biochem.* **140,** 83 (1984).

[5] P. C. Newman, V. U. Nwosu, D. M. Williams, R. Cosstick, F. Seela, and B. A. Connolly, *Biochemistry* **29,** 9891 (1990).

[6] P. C. Newman, D. M. Williams, R. Cosstick, F. Seela, and B. A. Connolly, *Biochemistry* **29,** 9902 (1990).

[7] T. R. Waters and B. A. Connolly, *Anal. Biochem.* **204,** 204 (1992).

[8] A. Jeltsch, A. Fritz, J. Alves, H. Wolfes, and A. Pingoud, *Anal. Biochem.* **213,** 234 (1993).

steady-state and time-resolved fluorescence techniques for structural studies of DNA and RNA. These studies include the application of fluorescence resonance energy transfer (FRET) to measure distances in tRNA and rRNA,[9-12] to detect hybridization,[13,14] to examine the structure of a synthetic DNA four-way junction,[15-17] and to measure the end-to-end distances of both single- and double-stranded DNA.[18] These studies demonstrate the ability of FRET to be applied as both a qualitative and quantitative tool to measure distances in DNA.

In this chapter, we discuss the use of fluorescence spectroscopic approaches to design fluorescence assay systems monitoring enzyme-catalyzed DNA cleavage reactions. These assays are based on either FRET or a non-FRET quenching mechanism.

Principle of Fluorescence Spectroscopy

As interaction of light with matter can be characterized in terms of energy, probability, and direction, fluorescence can be characterized by the excitation and emission wavelengths, intensity, decay time, and polarization. Fluorescence excitation and emission spectra reflect information on energies of transitions; decay times or quantum yields (the number of quanta emitted/the number of quanta absorbed) provide relative information regarding the probability of fluorescence emission versus other processes that an excited molecule may undergo; emission anisotropy provides the relative orientations of electronic absorption and emission transition dipoles, and the decay of the emission anisotropy can determine detailed information about the principle axes of rotational freedom and the rotational behavior of the chromophore in a macromolecule. Fluorescence

[9] K. Beardsley and C. R. Cantor, *Proc. Natl. Acad. Sci. U.S.A.* **65,** 39 (1970).
[10] C.-H. Yang and D. Soll, *Proc. Natl. Acad. Sci. U.S.A.* **71,** 2838 (1974).
[11] O. W. Odom, Jr., D. L. Robbins, J. Lynch, D. Dottavio-Martis, G. Kramer, and B. Hardesty, *Biochemistry* **19,** 5947 (1980).
[12] D. J. Robbins, O. W. Odom, Jr., J. Lynch, G. Kramer, B. Hardesty, R. Liou, and J. Ofengand, *Biochemistry* **20,** 5301 (1981).
[13] K. M. Parkhurst and L. J. Parkhurst, *Biochemistry* **34,** 285 (1995).
[14] M. S. Urdea, B. D. Warner, J. A. Running, M. Stempien, J. Clyne, and T. Horn, *Nucleic Acids Res.* **16,** 4937 (1988).
[15] R. A. Cardullo, S. Agrawal, C. Flores, P. C. Zamecnik, and D. E. Wolf, *Proc. Natl. Acad. Sci. U.S.A.* **85,** 8790 (1988).
[16] A. I. H. Murchie, R. M. Clegg, E. von Kitzing, D. R. Duckett, S. Diekmann, and D. M. J. Lilly, *Nature (London)* **341,** 763 (1989).
[17] J. P. Cooper and P. J. Hagerman, *Biochemistry* **29,** 9261 (1990).
[18] K. M. Parkhurst and L. J. Parkhurst, *Biochemistry* **34,** 292 (1995).

measurements are influenced not only by microheterogeneity around the ground-state environment of the fluorophore but also by the kinetics of any excited-state interactions that take place on the nanosecond time scale. Numerous excited-state processes such as resonance energy transfer, excimer formation, proton transfer, and Brownian rotational motion are well understood. These spectroscopic properties of fluorescence have been widely used to obtain a variety of structural and functional information regarding the solvent accessibility, microviscosity, concentrations of the probe, the probe environment, and the distance between the probes. Furthermore, conformational changes, complex formation, group reactivity, and structure–function correlations of macromolecules have also been studied.

The focus of this chapter is on applications of fluorescence resonance energy transfer (FRET) to studies of cleavage processes of nucleic acids. Resonance energy transfer has the ability to measure distances between pairs of fluorophores in macromolecules in the range of 10–80 Å.[19,20] FRET is defined as the transfer of electronic excitation energy as a result of dipole–dipole interactions between the donor and acceptor. This does not involve the emission and reabsorption of photons. The rate of energy transfer depends on several factors, such as (1) the extent of overlap between the emission spectrum of the donor and the absorption spectrum of the acceptor, (2) the relative orientation of the donor and acceptor transition dipoles, and (3) the distance between the pair of fluorophores.

The distance between a pair of fluorophores can be estimated from Förster's equation:

$$E = R_0^6/(R_0^6 + r^6) \tag{1}$$

where E is the efficiency of energy transfer (the proportion of photons absorbed by the donor transferred to the acceptor), r is the distance between a donor and an acceptor, and R_0 is the Förster distance at which the transfer rate is equal to the decay rate of the donor in the absence of acceptor. The Förster distance is given as

$$R_0 = 9.79 \times 10^3(\kappa^2 n^{-4} f_d J)^{1/6} \tag{2}$$

where κ^2 is the orientation factor, n is the refractive index of the medium, f_d is the quantum yield of the donor in the absence of acceptor, and J is

[19] T. Forster, *Ann. Phys. (Leipzig)* **2,** 55 (1948).

[20] L. Stryer, *Annu. Rev. Biochem.* **47,** 819 (1978).

the overlap integral that expresses the degree of spectral overlap between the donor emission and acceptor absorption. The overlap integral (J) is given as

$$J = \int_0^\infty F_d(\lambda)\varepsilon_a(\lambda)\lambda^4 \, d\lambda \tag{3}$$

where F_d is the corrected fluorescence intensity of the donor, and ε_a is the extinction coefficient of the acceptor.

A major complication in the determination of the distance by FRET is that the orientation factor, κ^2, cannot be directly measured. Hence, there are two unknowns in FRET: the distance and the orientation factor, κ^2. A κ^2 value of 2/3 is often assumed because of the randomization of the polarization of fluorescent probes. To reduce the error in FRET experiments, multiple pairs of fluorophores are frequently used. An alternative approach is to use fluorescence polarization studies to estimate the range of κ^2.[21,22] Although fluorescence anisotropy measurements can reduce the uncertainties of κ^2 to some degree, there is still substantial error for distance measurements because the fluorescence anisotropy may not be sufficient to provide the exact information regarding the orientation of the fluorophores. Therefore, the orientation factor, κ^2, remains an important and critical issue in determining the donor–acceptor distance using resonance energy transfer.

Another complication associated with FRET distance measurements is that one may encounter distributions of the donor and acceptor that are not separated by discrete distances owing to the flexibilities of the long spacer arms to which the probes are attached to the macromolecule. Probes with long linkers introduced to macromolecules are expected to have a heterogeneity of distances due to contributions from possible conformational fluctuations, the flexibility of the extrinsic probes with linker arms, and the varying orientation factors. Furthermore, if one end of a flexible polymer chain is labeled with an energy donor and the other with an acceptor, the end-to-end distance should be characterized by a distribution among the possible configurations; thus, the observed energy transfer efficiency must reflect this multiplicity in configuration.

To account for this, a concept of distribution analysis of FRET distances was introduced.[23] Many investigations have demonstrated the recovery of distribution functions from steady-state[24] and time-resolved

[21] R. E. Dale, E. Eisinger, and W. E. Blumberg, *Biophys. J.* **26,** 161 (1979).
[22] P. M. Torgerson and M. F. Morales, *Proc. Natl. Acad. Sci. U.S.A.* **81,** 3723 (1984).
[23] F. Grinvald, E. Haas, and I. Z. Steinberg, *Proc. Natl. Acad. Sci. U.S.A.* **69,** 2273 (1972).
[24] I. Gryczynski, W. Wiczk, M. L. Johnson, and J. R. Lakowicz, *Biophys. J.* **54,** 577 (1988).

data.[25–28] The feasibility of using frequency-domain data to recover the distribution of FRET distances from a model system, *N*-dansylundecaoyl-tryptamide, and also for native and denatured troponin I was demonstrated by Lakowicz *et al.*[25,26] In addition, Haas and co-workers determined the distribution of several intramolecular distances in bovine pancreatic trypsin inhibitor (BPTI)[27] and bovine pancreatic RNase.[28] James *et al.*[29] reported the distributions of distances between two specific residues of a staphylococcal nuclease mutant that were recovered from time–domain data as a function of guanidinium concentration. These results indicate the potential applications of the distribution of FRET distances to detect conformational changes in protein and to studies of protein folding/unfolding. It is in light of these facts and DNA being a dynamic molecule in solution that it is essential to test both discrete and distribution models for distance measurements when applying FRET to DNA. This is exemplified in a study by Parkhurst and Parkhurst[18] that implemented distribution analysis methods to measure the end-to-end distance of the probes labeled at both ends of the double-labeled fluorescent oligonucleotides.

Furthermore, the contribution of orientation to the observed "apparent" average distance and distance distribution widths for both simulated and real data has been estimated.[30] Their results indicate that the shape of the "apparent" distribution fit depends on a ratio of Förster distance to average distance when static orientation dominates (a real distance distribution should have no such dependence). The use of donor–acceptor pairs with Förster distances close to or smaller than the average distance produces more reliable results in average distance estimation alone, whereas donor–acceptor pairs with Förster distances close to or larger than the average distances are more suitable for shape information of an "apparent" distance distribution. Therefore, by choosing different donor–acceptor pairs (either $R_0 > r$ or selecting a series of pairs with increasing R_0), it may be possible to better define the nature of the distribution fit. Hence, models that include orientational order contributions can be applied to energy transfer experiments to minimize the error in distance estimation and for better interpretation of the data.

[25] J. R. Lakowicz, M. L. Johnson, W. Wiczk, and R. F. Steiner, *Chem. Phys. Lett.* **138,** 587 (1987).
[26] J. R. Lakowicz, I. Gryczynski, H. C. Cheung, C. K. Wang, and M. L. Johnson, *Biopolymers* **27,** 821 (1988).
[27] D. Amir and E. Haas, *Biochemistry* **26,** 2162 (1987).
[28] E. Haas, C. A. McWheter, and H. A. Scheraga, *Biopolymers* **27,** 1 (1988).
[29] E. James, P. G. Wu, W. Stites, and L. Brand, *Biochemistry* **31,** 10217 (1992).
[30] P. G. Wu and L. Brand, *Biochemistry* **31,** 7939 (1992).

Experimental Methods

Introduction of Fluorophores into DNA

Chemically modified nucleotide analogs have made a significant impact on applications of fluorescence techniques in structural studies of nucleic acids. These analogs can be directly incorporated into oligonucleotides and DNA fragments by using a DNA synthesizer or by enzymatic reactions. For example, phosphoramidites containing aliphatic primary amines or thiol groups can be introduced into the oligonucleotides at desired positions using amino modifiers or thiol modifiers, respectively. The resulting amino or thiol groups can be reacted with a variety of substrates such as biotin,[31,32] fluorescent dyes,[15–18] EDTA,[33] or alkaline phosphatase[34] to produce oligonucleotide conjugates. Alternatively, chemically modified nucleotide analogs can be introduced to DNA fragments using various enzymatic reactions.

Homogeneously purified fluorescent DNA samples are advantageous in quantitative analyses of nonradioactive assays and essential for specific applications such as fluorescence *in situ* hybridization, fluorescence quenching/dequenching, emission anisotropy, and FRET. Inefficient chemical reactions result in incomplete modification of the fluorescent probes to the oligonucleotide samples. For example, the coupling efficiency of Aminolink 2 is approximately 75% (according to Applied Biosystems, Foster City, CA). This results in a significant fraction of oligonucleotides without an aliphatic primary amine group. Furthermore, the subsequent modification of the fluorescent probes to the amine group on the oligonucleotides is generally not 100% efficient. The presence of unlabeled samples will complicate the interpretation of the fluorescence data. Therefore, this potential problem must be avoided by the complete separation and purification of fluorescently labeled DNA from its unlabeled sample.

Selection of the fluorescent probes to be utilized is based on the design of the particular experiment and the fluorescence properties of the probe. Among many fluorecent probes, fluorescein derivatives are probably the most common reagents owing to their high quantum yield, which provides sensitivity of signal detection. Common fluorescence acceptors for the fluorescein in energy transfer measurements are rhodamine, tetramethylrhodamine, and eosin.

[31] B. C. F. Chu and L. E. Orgel, *DNA* **4,** 327 (1985).

[32] A. Chollet and E. H. Kawashima, *Nucleic Acids Res.* **13,** 1529 (1985).

[33] G. B. Dreyer and P. B. Dervan, *Proc. Natl. Acad. Sci. U.S.A.* **82,** 968 (1985).

[34] E. Jablonski, E. W. Moomaw, R. H. Tullis, and J. L. Ruth, *Nucleic Acids Res.* **14,** 6115 (1986).

Preparation of Fluorescent DNA Substrates

Oligonucleotides containing the nucleotide analogs can be derivatized with the appropriate fluorophores [e.g., fluorescein 5-isothiocyanate (FITC) or eosin isothiocyanate] in 100 m*M* $NaHCO_3$–Na_2CO_3 buffer, pH 9.0. The excess dye can be initially removed by filtrating the reaction mixture through a Sephadex G-25 column (DNA grade). The resulting samples can then be electrophoresed on a denaturing (7 *M* urea) 20% (w/v) polyacrylamide gel to purify the oligonucleotides and to remove any residual free dyes. The appropriate labeled oligonucleotides can be electroeluted from the sliced gel using the S&S Elutrap electro-separation system from Schleicher & Schuell (Keene, NH). In general, labeled versus unlabeled oligonucleotides can be separated by denaturing polyacrylamide gel electrophoresis. The purified oligonucleotides can then be annealed with a threefold molar excess of complementary strand to ensure complete annealing of all the donor labeled samples.

Spectroscopic Measurements

Steady-state fluorescence studies can be performed with a spectrophotofluorometer. It is usually preferable to perform the fluorescence emission measurements under "magic angle" emission conditions utilizing polarizers.[35] Most reactions can be performed with 10-mm path length cuvettes. However, a 3-mm path length microcuvette can also be used with smaller amounts of sample. To avoid inner filter effects, the absorbances of all the samples should be less than 0.1 at the wavelengths of excitation. The temperature of the samples must be regulated with a temperature controller and a bath cooler. Time-resolved fluorescence experiments can be performed by either a time-correlated single photon-counting method or a phase-modulation method. Proper functioning of the instrument and correction for the wavelength-dependent transmit time of the photomultiplier can be achieved utilizing the proper fluorescence standards.

Data Analysis

The efficiency of energy transfer, E, can be estimated by three methods: (1) from the quenched emission of the donor, (2) from the quenched quantum yield of the donor, and (3) from the enhanced emission (sensitized) of the acceptor. Among these three methods, steady-state donor quenching is perhaps the most convenient assay for energy transfer. However, the decrease in the donor quantum yield does not always indicate the energy

[35] R. D. Spencer and G. Weber, *J. Chem. Phys.* **52,** 1654 (1970).

transfer. The quenching of the donor fluorescence due to a conformational change induced by the presence of acceptor and a direct radiative absorption of donor emission by the acceptor must be avoided.

As described in a later section of this chapter, two different fluorometric assay systems have been designed to monitor the enzyme-mediated DNA cleavage reaction. These assays are based on (1) resonance energy transfer and (2) quenching of the donor fluorescence by the interaction of the fluorophore attached to the DNA with the nucleotide bases of the double-stranded DNA. Thus, the double-stranded fluorogenic substrates display the quenched donor fluorescence in both cases. On cleavage of the DNA substrates, the cleaved DNA products then exhibit enhancement of the emission fluorescence owing to the reversal of the quenched mechanisms.

Accordingly, the extent of the DNA cleavage can be determined by the ratio of the quantum yields of the cleaved products in the reaction mixture to the quantum yield of the completely cleaved DNA substrate. The extent of DNA cleavage of a fluorogenic substrate can be estimated by using Eq. (4)[36]:

$$[\mathrm{DNA}]_\mathrm{c} = \frac{F_\mathrm{t} - F_0}{F_\infty - F_0}[\mathrm{DNA}]_\mathrm{i} \tag{4}$$

where $[\mathrm{DNA}]_\mathrm{c}$ is the concentration of cleaved DNA, F_t is the fluorescence intensity at time t, F_∞ is the fluorescence intensity of the completely cleaved DNA substrate or the fluorescence intensity of the substrate in the presence of DNase I, F_0 is the initial fluorescence intensity, and $[\mathrm{DNA}]_\mathrm{i}$ is the initial concentration of DNA. As mentioned, these assays require a fluorescence quantum yield of the standard of reference DNA sample representing the fluorescence intensity of the completely cleaved fluorogenic DNA substrate. Single-stranded DNA containing only the donor fluorophore can serve as such a control. Alternatively, DNase I can provide a similar function because DNase I results in the complete digestion of the fluorogenic substrate. The resultant fluorescence intensity represents the reference for the total recovered donor fluorescence. Furthermore, this approach can also rule out any intensity variations due to possible incomplete annealing of the substrate or secondary structure formation by the single strand.

Fluorometric Cleavage Assay Using DNA Substrates Containing Single Fluorophore

Rationale. Previous concerns with FRET measurements of DNA were that effects other than resonance energy transfer may influence FRET

[36] S. P. Lee, D. Porter, J. G. Chirikjian, J. R. Knutson, and M. K. Han, *Anal. Biochem.* **220,** 377 (1994).

measurements in oligonucleotide systems. This concern was in fact substantiated by the observation of donor fluorescence quenching in the absence of the acceptor when the probe-linked strand was annealed to its nonfluorescent complement. The observed fluorescence quenching was attributed to interactions between the dye and double-stranded DNA. This type of interaction has been previously observed with studies of DNA containing the fluorescent analog, 2-aminopurine.[37] The fluorescence of the 2-aminopurine deoxymononucleoside in solution has been characterized by a monoexponential decay of 10 nsec. In contrast, the introduction of 2-aminopurine to double-stranded DNA results in a quenched fluorescence of at least four decay times. This fluorescence quenching was ascribed to a stacking effect of the surrounding helix. The alteration of fluorescence lifetimes owing to stacking effects includes cases of ε-adenosine[38] and aminoacridine[39] quenching by DNA bases and the formation of sandwich stacks of fluorescent molecules.[40] These results in concert with one another indicate that effects other than dipolar energy transfer mechanisms alter the fluorescence of the probe on DNA. Accordingly, these effects must be accounted for to measure reliably distances in DNA molecules by FRET.

To characterize further this quenching behavior, a 14-mer oligonucleotide was synthesized with the primary amino group at the 5′ end from the Aminolink 2. This primary amine was derivatized with FITC as previously described. Figure 1 illustrates the fluorescence quenching of this fluorescein-containing single-stranded oligonucleotide as a result of annealing to its nonfluorescent complementary strand. Increasing concentrations of nonlabeled complementary strand were annealed to a fixed concentration of FITC-labeled oligonucleotide. The fluorescence quantum yield of the fluorescein progressively decreased as the extent of the annealing increased. The addition of excess complementary strand resulted in complete annealing and approximately twofold quenching of the fluorescein intensity. The quenching of the fluorescein fluorescence was accompanied by a blue-shifted emission spectrum, confirming the fluorescence quenching due to probe–DNA interaction as previously suggested.[17] More importantly, the quantum yield of the quenched fluorescence was completely reversed by DNase I cleavage of the double-stranded fluorogenic substrate. The reversibility of this fluorescence quenching process led to the design of a fluorometric assay to monitor the DNA cleavage reaction.

[37] T. M. Nordlund, S. Andersson, L. Nilsson, R. Rigler, A. Grandslund, and L. W. McLaughlin, *Biochemistry* **28,** 9005 (1989).

[38] Y. Kubota, A. Sanjoh, Y. Fujisaki, and R. F. Steiner, *Biophys. Chem.* **18,** 225 (1983).

[39] R. F. Steiner and Y. Kubota, *In* "Excited States Biopolymers" (R. F. Steiner, ed.), pp. 203–254. Plenum, New York, 1983.

[40] J. B. Birks, *In* "Organic Molecular Photophysics" (J. B. Birks, ed.), pp. 409–613. John Wiley & Sons, New York, 1975.

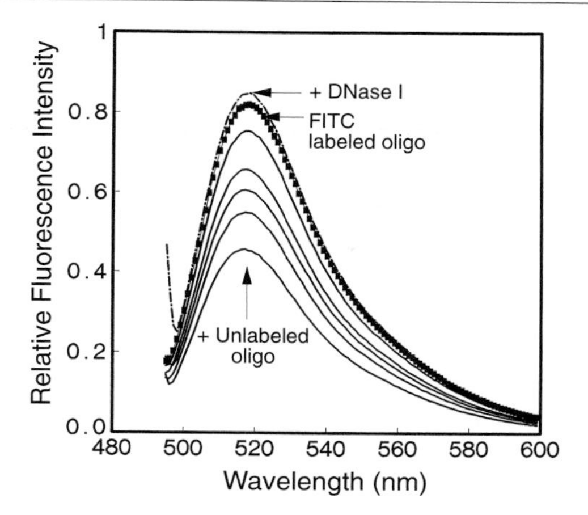

FIG. 1. Fluorescence quenching of a single labeled oligonucleotide. Fluorescence emission spectra of FITC-labeled oligonucleotide were recorded as a fixed concentration of 14-mer FITC-labeled oligonucleotide (0.137 pmol) was annealed to varying concentrations of its unlabeled complementary strand. The concentrations of the unlabeled strand were 0.03, 0.06, 0.09, 0.117, and 0.15 pmol in a 0.5-ml reaction volume. The fluorescence intensities were recorded with an excitation wavelength of 490 nm. (Reprinted with permission from Ref. 36.)

Characterization of BamHI-Mediated DNA Cleavage Reaction. We have reported the design of a fluorometric assay for DNA cleavage reaction using the *Bacillus amyloliquefaciens* HI (*Bam*HI) restriction endonuclease and a single fluorescein-containing oligonucleotide substrate. A 14-mer oligonucleotide substrate for *Bam*HI was synthesized with the Aminolink 2 modifier and derivatized with FITC (isomer I) as shown in Scheme I.

As depicted in Scheme I, *Bam*HI will cleave the 14-mer substrate and generate two DNA fragments under the appropriate reaction conditions. The two resulting short DNA fragments should have relatively low melting temperatures and subsequent temperature-dependent dissociation of the two strands should result in the recovery of the quenched fluorescence

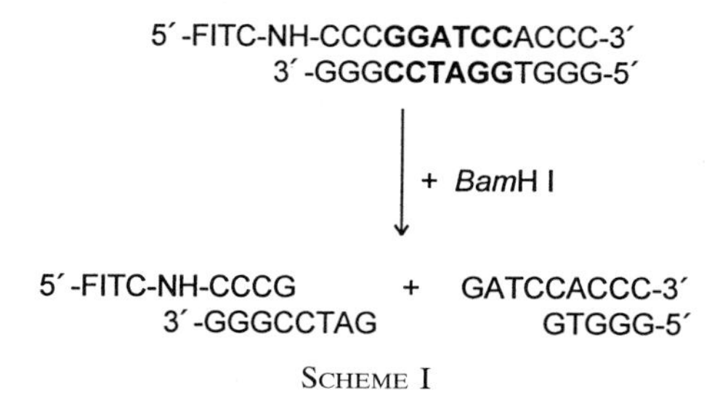

SCHEME I

intensity. In this fluorescence assay system, reaction temperatures become an important factor because fluorescence intensity is sensitive to temperature and the monitored signal corresponding to the cleaved product is also dependent on the reaction temperature. The fluorogenic substrate must be prepared to have a relatively high melting temperature to avoid temperature-dependent dissociation of the substrate in the absence of enzyme. In addition, the assay sample should be preincubated at the reaction temperature for several minutes to achieve thermal equilibration prior to the addition of the enzyme. Addition of a relatively small volume of enzyme is recommended to prevent significant changes in temperature of the reaction mixture.

Figure 2 illustrates the recovery of the fluorescence intensity reflecting the cleavage of the substrate by *Bam*HI restriction endonuclease at 25 and 37°. The spectra were recorded with an excitation of 490 nm after 2 hr of incubation at both temperatures to eliminate the effects of temperature-dependent enzymatic activity. The spectra shown in Fig. 2 represent the intensity peak normalized according to the quantum yields of the uncleaved substrates at the two different temperatures because the quantum yield of the substrate is temperature dependent. An enhancement of fluorescence intensity was observed at both temperatures. However, greater enhancement of fluorescence intensity was observed at 37° than at 25°. This can be accounted for by the incomplete dissociation of the 5-bp products to their single-stranded components at 25°. In fact, the predicted T_m of the 5-bp sequence utilizing the equation (AT × 2° + GC × 4°) is 20°. Therefore, it is not expected that all of the double-stranded DNA products will be

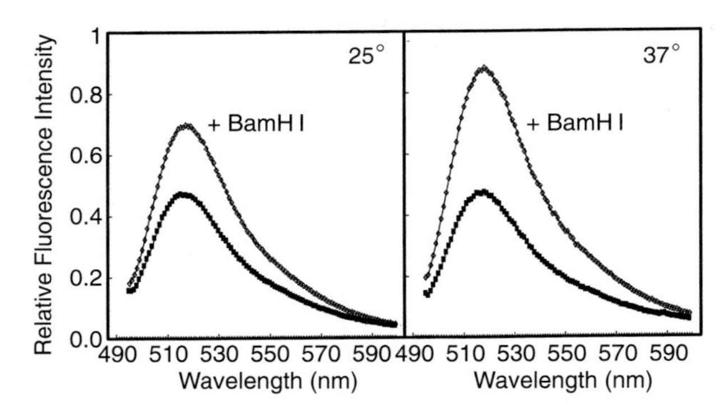

FIG. 2. DNA cleavage reactions monitored by fluorescence. Emission spectra of fluorescein-labeled double-stranded DNA substrate in the presence and absence of *Bam*HI endonuclease at 25 and 37° in 50 m*M* Tris (pH 8.0), 10 m*M* $MgCl_2$, and 0.1 *M* NaCl. Emission spectra were recorded with an excitation wavelength of 490 nm. (Reprinted with permission from Ref. 36.)

dissociated at 25°, whereas all the products will dissociate at 37°. Accordingly, the changes in the fluorescence intensity directly reflect the temperature-dependent dissociation of the cleaved products, thereby making this fluorogenic substrate ideal for enzymatic reactions at the optimum reaction temperature of 37°. To confirm that the enhancement of fluorescence intensity was indeed due to cleavage of the substrate, reactions were performed in the presence of EDTA or after the methylation of the substrate. Under these conditions, no changes in fluorescence intensity were observed.

The fluorometric assay can monitor the activity of an enzyme by two methods using steady-state fluorescence. One method is to take the fluorescence spectrum before and after the addition of the enzyme. The enhancement of fluorescence indicates cleavage of the substrate. This is convenient when trying to determine the activity of enzyme purifications or estimating the enzyme content of various samples. Under these assay conditions, all components of the reaction such as the substrate(s), cofactors, and required activators should be at optimum concentrations. However, kinetic studies require that the initial velocities of the enzymatic reaction be determined. Therefore, a time course of a single wavelength can also be monitored rather than the initial and postincubation spectra. From the linear portion of the kinetic data, the initial velocities can be obtained. Furthermore, the kinetic studies are often performed at subsaturating substrate and effector levels to evaluate the rate–saturation behavior. Accordingly, it is ideal that the assay be designed to fulfill these features.

To further characterize the assay, enzyme concentration-dependent kinetic studies of the *Bam*HI-catalyzed reactions at 37° were performed. The apparent rates of DNA cleavage were monitored as increasing concentrations of enzyme were added to a fixed concentration of substrate (Fig. 3). The observed kinetic rates are the composite of two processes: the cleavage of the substrate and the subsequent dissociation of the cleavage products. The apparent rate-limiting step of the overall process is cleavage by *Bam*HI because the rate of DNA strand dissociation is a rapid process. The advantage of the continuous fluorescence assay is that the initial velocities can be easily obtained. A plot of the initial velocities as a function of enzyme concentration resulted in a linear plot (Fig. 3, inset), suggesting that the rate of cleavage reactions with respect to *Bam*HI concentration was first order, and that the changes in fluorescence signal reflect substrate cleavage by *Bam*HI.

Final confirmation of the validity of the assay was achieved by performing the cleavage reaction as a function of substrate concentration. The DNA cleavage reactions were analyzed by both fluorometric and gel electrophoretic methods. The fractions of cleaved and uncleaved substrate from the fluorometric analysis and the densitometric analysis of the gel are

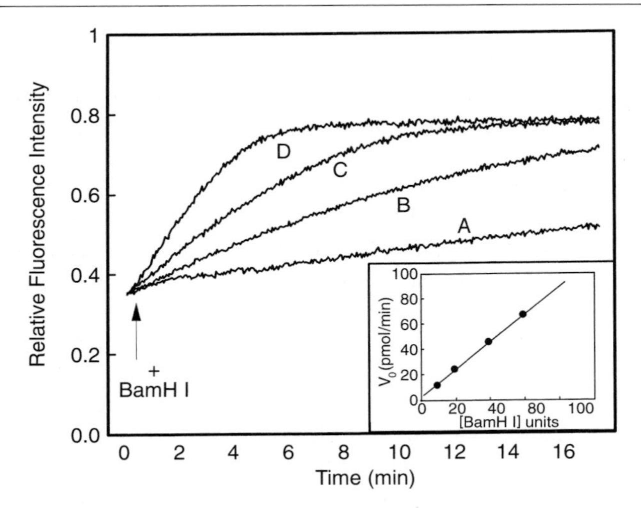

FIG. 3. Enzyme concentration-dependent *Bam*HI cleavage of fluorescein-labeled DNA substrate. Curves A through D represent kinetics of DNA cleavage by 10, 20, 40, and 60 units of *Bam*HI, respectively. The kinetic experiments were performed with 0.5 μM DNA substrate in 420 μl of 50 mM Tris (pH 8.0), 10 mM $MgCl_2$, and 0.1 M NaCl at 37°. Fluorescence intensity was monitored with excitation and emission wavelengths of 490 and 520 nm, respectively. *Inset:* Initial velocities of cleavage reactions determined from the linear portions of the kinetic data and plotted as a function of enzyme concentration. (Reprinted with permission from Ref. 36.)

summarized in Table I. In both analyses, the presence of 25 mM EDTA resulted in the absence of product formation. At a low substrate concentration, the cleavage reaction reached a plateau at approximately 6 min as monitored by fluorescence, indicating that the reaction had reached comple-

TABLE I
PERCENT CLEAVAGE OF FLUORESCEIN ISOTHIOCYANATE-LABELED DNA SUBSTRATES BY *Bam*HI[a]

		Percent cleavage	
Experiment	[DNA] (μM)	Fluorescence assay	PAGE
A	0.21	100	100
B	0.36	91.9	92.3
C	0.72	59.5	60.6
D	1.08	32.2	30.1
E[b]	0.18	0	0

Data from Ref. 36.

[a] Estimated by both fluorometric and PAGE analyses.

[b] Performed in the presence of EDTA.

tion. This was confirmed by gel electrophoretic analysis, which showed 100% conversion of the substrate to product. Increasing substrate concentrations yielded incomplete reactions in the given time frame as measured by both fluorescence and gel electrophoresis. As illustrated in Table I, the percent cleavage estimated from the fluorescence assay and the gel electrophoresis assay are in good agreement. Therefore, the implementation of the quenching/dequenching phenomenon of a single labeled substrate to monitor DNA cleavage reactions is a viable alternative to the previously established assay systems.

Fluorometric Assay for DNA Cleavage Using Fluorescence Resonance Energy Transfer

Rationale. The aforementioned quenching/dequenching phenomenon provides a useful tool for monitoring DNA cleavage reactions. While this assay is both rapid and simple, application of the assay may be limited to enzymes catalyzing efficient DNA cleavage reactions because the extent of quenching is approximately twofold. Enzymes that are inefficient in terms of their cleavage activity will require a more sensitive method for detection of DNA cleavage reactions by maximizing the signal-to-noise ratio. This can be achieved by increasing the efficiency of the donor fluorescence quenching. Large differences in the ratio of fluorescence signal for the cleaved DNA and the undigested substrate will facilitate the detection of DNA cleavage. For example, if the fluorescence intensity of a fluorogenic substrate is 10-fold lower than the intensity of the cleaved substrate (i.e., 90% donor quenching), 10% cleavage of the substrate will yield a twofold increase in fluorescence intensity. This is in contrast to complete cleavage of a substrate with a quantum yield of cleaved products being only two fold higher than the uncleaved, which results in only a twofold increase in fluorescence intensity. Such an enhancement of signal detection can easily be achieved by maximizing the efficiency of FRET because the efficiency of electronic excitation energy transfer by the Förster mechanism is related to the distance between the pair of fluorophores. The selection of a donor and acceptor for FRET is an important consideration because the efficiency of energy transfer (E) is dependent on the distance (R) between the donor and acceptor. However, the efficiency of energy transfer (E) is sensitive to a function of the distance (R) only over the range $0.5R_0 < R < 1.9R_0$ (where R_0 is the Förster distance).[20,41] This range becomes a critical factor for distance estimation from the efficiency of energy transfer. Accordingly, the most reliable distance measurements can be made when R_0 for the chosen donor–acceptor is comparable to the actual distance to be measured.

[41] R. H. Fairclough and C. R. Cantor, *Methods Enzymol.* **48,** 347 (1978).

When the actual distance of the donor–acceptor is less than R_0, the efficiency of energy transfer is less sensitive to the distance, but the maximal transfer of energy can be achieved. Hence, the selection of a donor–acceptor pair that has a considerably longer R_0 than the actual distance between the fluorophores is a poor choice for the distance measurement; however, it is advantageous in designing a fluorometric assay. In fact, maximal donor quenching can be obtained with the minimal distance between the probes. Accordingly, 30 to 40-fold quenching of donor fluorescence has been achieved previously.[42]

It is also important to note that the interpretation of results obtained by FRET relies on the fluorescence quenching to be attributed only to the nonradiative energy transfer process. However, as previously demonstrated, fluorescence quenching can be observed with fluorophores labeled to DNA by both FRET and a mechanism other than FRET. As discussed earlier, this possible complication in fluorescently modified DNA is believed to be the result of interactions between the fluorescence probe and the nucleotide bases.

In general, the fluorescent probes attached to synthetic oligonucleotides via 5′ or 3′ modifiers (e.g., Aminolink 2) usually contain linker arms consisting of chains of fewer than six carbons. These analogs are designed so that the attached chemical group interacts with the double-stranded oligonucleotides. In contrast to these modifiers, 5-amino(12)-2′-dUTP (5′-dimethoxytrityl-5-[*N*-(trifluoroacetylaminohexyl)-3-acrylamido]-2′-deoxyuridine triphosphate)[43] is designed to prevent the labeled chemical groups from interacting with the double-stranded DNA. Hence, the fluorescent modification of an oligonucleotide via a 5-amino(12)-2′-dUTP analog may be an alternative approach to avoid the fluorescence quenching due to the interaction between the attached probe and duplex DNA.

To exploit this possibility, we have characterized the fluorescence properties of the probes covalently attached to DNA via two different modifiers.[36,44] Steady-state fluorescence studies of these fluorogenic substrates indicated that the fluorescence quenching observed from a fluorescein-labeled oligonucleotide using an Aminolink 2 containing a six-carbon linker arm[36] was not observed from a fluorogenic substrate prepared using a 5-amino(12)-2′-dUTP analog.[44] Time-resolved studies further indicate that the fluorescein-labeled single strand using a 5-amino(12)-2′-dUTP analog

[42] E. D. Matayoshi, G. T. Wang, G. A. Krafft, and J. Erickson, *Science* **247,** 954 (1990).

[43] J. A. Brumbaugh, L. R. Middendorf, D. L. Grone, and J. L. Ruth, *Proc. Natl. Acad. Sci. U.S.A.* **85,** 5610 (1988).

[44] S. P. Lee, M. L. Censullo, H. G. Kim, J. R. Knutson, and M. K. Han, *Anal. Biochem.* **227,** 295 (1995).

exhibits a major decay component of 4.1 nsec and a small fraction of 0.7 nsec.[44] No changes in fluorescence lifetimes were observed on annealing this strand to its unmodified complementary strand. However, the fluorescein-labeled single-stranded oligonucleotide using an Aminolink 2 was characterized by a triexponential decay with lifetimes of 0.5 and 2.7 nsec, and a major component of 4.2 nsec, suggesting that the fluorescence intensity was somewhat altered. On annealing to its complementary strand, the 4.2-nsec component was decreased significantly, resulting in twofold quenching of total fluorescence, further confirming the existence of the probe–DNA interaction.[36] Both steady state and time-resolved measurements indicated the absence of fluorescence quenching when the fluorescent probe was introduced to DNA via a 5-amino(12)-2′-dUTP analog containing a longer linker arm. Therefore, the longer linker arm will allow for the donor fluorescence quenching to be entirely attributed to FRET and this further makes it possible to measure the distance between the probes on DNA using FRET.

In addition to the absence of probe–base interactions, the 5-amino(12)-2′-dUTP has additional advantages: (1) This nucleotide analog substitutes for dTTP, which usually resides in the minor groove of DNA. Because DNA-binding proteins usually interact with the major groove of DNA, the analog would have minimal DNA interaction with proteins. Moreover, the steric hindrance due to the bulky fluorescent probe would be reduced owing to the length of the linker arm; and (2) a potential problem with end labeling with other modifiers is that the ends of the oligonucleotides may not be completely annealed, or the degree of annealing could vary depending on the base pairs. Furthermore, these end modifiers have potential limitations in examining long sequences of DNA. The application of the 5-amino(12)-2′-dUTP analog allows fluorophores to be introduced to any position on the DNA either chemically or enzymatically.

Fluorometric Assay for DNA Cleavage Mediated by Human Immunodeficiency Virus Type 1 Integrase

We have prepared a fluorogenic substrate utilizing the 5-amino(12)-2′-dUTP and demonstrated that the DNA cleavage reaction of human immunodeficiency virus type 1 (HIV-1) integrase (IN) can be monitored by FRET. Integrase, which is a key protein responsible for the integration of the viral DNA into the host chromosome, recognizes a specific sequence within both ends of the long-terminal repeat of viral DNA.[45–47] The enzyme

[45] H. E. Varmus and P. O. Brown, Retrovirus. *In* "Mobile DNA" (D. E. Berg and M. M. Howe, eds.), pp. 53–108. American Society for Microbiology, Washington, D.C., 1989.

[46] D. P. Grandgenett and S. R. Mumm, *Cell* **60,** 3 (1990).

[47] R. A. Katz and M. Skalka, *Annu. Rev. Biochem.* **63,** 133 (1994).

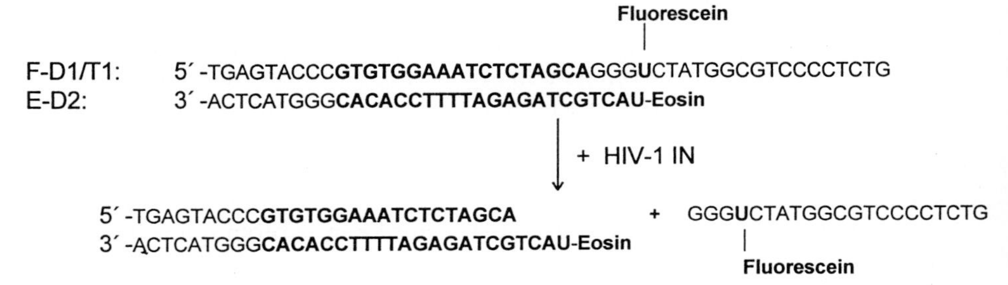

SCHEME II

removes two terminal nucleotides at the 3′ ends (3′-donor processing) in the cytoplasm and also catalyzes a joining reaction of the viral DNA to the 5′ ends of target chromosomal DNA (strand transfer reaction) in the nucleus. In addition, integrase can also catalyze the cleavage of the synthetic donor/target hybrid strand *in vitro*.[44,48] A schematic diagram of such a reaction is shown in Scheme II.

The oligonucleotide substrate containing both specific viral donor and random target sequences was modified to include a fluorescence donor (fluorescein) and acceptor (eosin). The sequence of the substrate and integrase-mediated cleavage reaction of the resulting fluorogenic substrate are shown in Scheme II. Because there is a strong spectral overlap between the emission spectrum of fluorescein and the absorption spectrum of eosin, a severe donor fluorescence quenching is expected from the fluorogenic substrate (the calculated Förster distance of the pair was reported to be 54 Å[49]). The observed intensity ratio of the donor fluorescence in the absence and presence of its acceptor was approximately eightfold.

The recoveries of the quenched donor fluorescence owing to enzyme-mediated DNA cleavage demonstrated with nonspecific DNA cleavage by DNase I and specific cleavage by HIV-1 IN are illustrated in Figs. 4 and 5, respectively. It is apparent that cleavage of the fluorogenic substrate by DNase I was near completion while HIV-1 integrase resulted in a partial cleavage. A comparison of the ratio of recovered donor fluorescence by both enzymes can be used to estimate the efficiency of DNA cleavage due to HIV-1 integrase. The efficiency of DNA cleavage reaction catalyzed by HIV-1 integrase estimated by the fluorescence method and a radioactive assay showed an excellent correlation, confirming the validity of the fluorescence assay system.

[48] S. P. Lee, H. G. Kim, M. L. Censullo, and M. K. Han, *Biochemistry* **34,** 10205 (1995).
[49] K. L. Carraway, III, J. G. Koland, and R. A. Cerione, *J. Biol. Chem.* **264,** 8699 (1989).

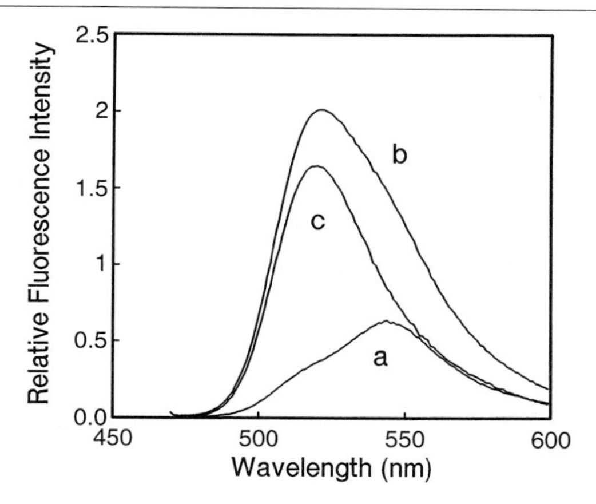

FIG. 4. Steady-state emission spectra of a fluorogenic substrate in the absence (a) and presence (b) of DNase I. Emission spectra of 4 pmol of fluorogenic substrate were recorded with an excitation wavelength of 460 nm in 400 μl of reaction buffer containing 25 m*M* HEPES, (pH 7.5), 50 m*M* NaCl, 2 m*M* dithiothreitol (DTT), 5% (v/v) glycerol, and 7.5 m*M* $MgCl_2$ at 37°. Emission spectrum c denotes the difference between spectra a and b. (Reprinted with permission from Ref. 44.)

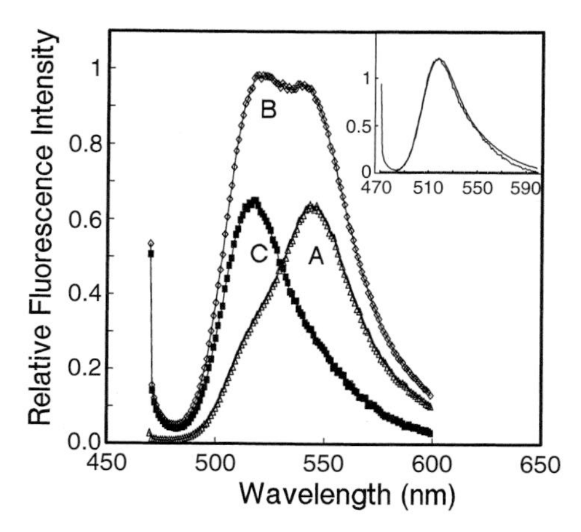

FIG. 5. Steady-state emission spectra of a fluorogenic substrate in the absence (A) and presence (B) of HIV-1 IN. Emission spectra were recorded with 4 pmol of fluorogenic substrate in 400 μl of reaction buffer containing 25 m*M* HEPES (pH 7.5), 50 m*M* NaCl, 2 m*M* DTT, 5% (v/v) glycerol, and 7.5 m*M* $MgCl_2$ at 37°. Spectrum C represents the enhanced fluorescence due to cleavage of the substrate by 40 pmol of HIV-1 IN for 1 hr. Inset: Peak normalized spectra of F-D1/T1 and the difference spectrum C. (Reprinted with permission from Ref. 44.)

As demonstrated earlier, one important feature of the fluorescence method is its ability to monitor reaction rapidly and continuously. This makes it possible to perform precise kinetic analysis of DNA cleavage reactions. An example of kinetic studies using the FRET assay is shown in Fig. 6. A direct verification of the fluorescence kinetics was also provided by comparison studies with a radioactive assay. The fluorogenic substrate was radiolabeled with ^{32}P at the 5′ end of the cleavable strand and a time-dependent cleavage reaction was examined. The reaction products were separated by denaturing polyacrylamide gel electrophoresis (PAGE) followed by autoradiography and quantitation by densitometry. As shown in Fig. 6, both the gel electrophoresis data and fluorescence data displayed similar kinetics. Further kinetic studies of HIV-1 integrase resulted in an apparent K_m and V_{max} of 145 ± 17 nM and 0.24/hr, respectively.[48] The apparent low catalytic turnover may suggest that HIV-1 integrase may remain bound to DNA following the cleavage of the viral DNA. These

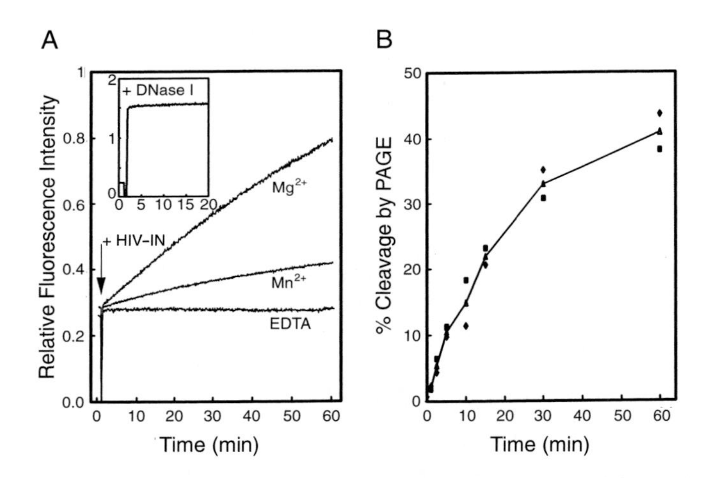

FIG. 6. Kinetics of HIV-1 IN cleavage reaction monitored by FRET and denaturing PAGE. (A) The DNA cleavage reaction was initiated by the addition of HIV-1 IN to a preincubated reaction mixture containing 4 pmol of fluorogenic substrate in the presence of either Mg^{2+} or Mn^{2+} at 37°. The conditions for the reactions are the same as in Fig. 5. Changes in fluorescence intensity were monitored with excitation and emission wavelengths of 460 and 510 nm, respectively. *Inset:* Kinetics of DNase I cleavage reaction of the same substrate. (B) Time course of ^{32}P-radiolabeled fluorogenic substrate by radioactive assay: the reactions were initiated by the addition of 4 pmol of HIV-1 IN to 20 μl of reaction mixture containing 0.15 pmol of the substrate at 37°. The reactions were stopped at 1, 2.5, 5, 10, 30, and 60 min by the addition of an equal volume of stop solution containing EDTA. Reaction mixtures were analyzed by denaturing PAGE. The product bands were quantitated by utilizing a densitometer. The line indicates the mean of the two experiments bounded by the actual values. (Reprinted with permission from Ref. 44.)

results demonstrate that FRET assays can facilitate real-time kinetic studies of enzymatic DNA cleavage reactions.

Conclusion

The results presented here clearly demonstrate that fluorescence can be used as a tool to provide an easy and rapid method for acquiring high data density essential for assaying DNA cleavage reactions. These continuous fluorometric assays will facilitate the kinetic studies of a variety of enzyme-mediated DNA cleavage reactions and the resulting data will play important roles in elucidating the mechanism of enzymes involved in various molecular processes such as replication, recombination, and repair of DNA. In addition, the fluorometric assay can facilitate the screening of massive numbers of potential inhibitors of many enzymes.

[18] Fluorescent Nucleotide Analogs: Synthesis and Applications

By David M. Jameson and John F. Eccleston

Introduction

Nucleoside 5′-triphosphatases are involved in many important cellular processes such as energy transduction in molecular motors, ion transport, and signal transmission, as well as in the topological processing of nucleic acids and the fidelity of protein synthesis. One of the primary reasons for the widespread use of fluorescence to study nucleotide-binding proteins is the inherent high sensitivity of the method—fluorophores with high extinction coefficients, good quantum yields (i.e., >0.1), and absorption bands at longer wavelengths than endogenous backgrounds [e.g., mid-ultraviolet (UV) and longer] can be readily detected at submicromolar concentrations. Fluorescence can provide information about the size and structure of the proteins, allows quantification of the kinetic and equilibrium constants describing the system, and can also shed light on the cellular distribution of the proteins. Intrinsic protein fluorescence, primarily from tryptophan and tyrosine residues, has been used to obtain detailed information about many of these parameters but there are significant advantages in studying nucleotide fluorescence. Specifically, if the nucleotide itself can be spectroscopically isolated, then a unique fluorophore, which can be localized to a defined position, exists in the system.

0076-6879/97 $25

The intrinsic fluorescence of the common nucleotides, at ambient temperatures and neutral pH, is far too low (quantum yields on the order of 10^{-4})[1] to be of general use in the investigation of nucleotide-binding proteins, especially since the longest absorption maxima of nucleotides are near 260 nm, which overlaps with intrinsic protein absorption. Some naturally occurring modified nucleosides, including 4-thiouridine, 7-methylguanosine, N^6-acetylcytidine, and the Wye derivatives (formerly known as the Y bases) are fluorescent[2] and have been useful as probes of tRNA but have found few applications outside of this area.

Considerable effort has been expended on modifying nucleotides to improve their utility as fluorescent probes. Among the first such modified nucleotides were the "etheno" series synthesized by Leonard and collaborators.[2,3] The most commonly used nucleotide in this class is the 1,N^6-ethenoadenosine derivative, εATP (**I** in Fig. 1). Leonard has written an excellent review of the use of etheno-bridged nucleotides in enzyme reactions and protein-binding studies.[4] More conservative changes to the nucleotide base in formycin 5′-triphosphate (**II** in Fig. 1) and 2-aminopurine riboside 5′-triphosphate (**III** in Fig. 1) also produce fluorescent nucleotides that have been used to study the myosin ATPase mechanism.[5] However, many of the rate constants of the nucleotide–protein interaction are altered by such modifications of the purine ring. The situation with guanine nucleotide-binding proteins is even worse because virtually any modification of the purine ring of the guanine nucleotide results in loss of binding activity.[6]

Modifications that do not alter the purine or pyrimidine ring systems offer an alternative approach to fluorescent nucleotide analogs. For example, an ATP analog with an altered phosphoryl structure, namely adenosine-5′-triphosphoro-γ-1-(5-sulfonic acid)naphthylamine [(γ-AmNS)ATP (**IV**) in Fig. 1], which was a good substrate for *Escherichia coli* RNA polymerase, has been described.[7]

Modification of the ribose moiety of nucleotides offers, in most cases, a satisfactory approach in that the resultant analogs mimic more closely the behavior of their parent nucleotides than do pyrimidine or purine

[1] P. Vigny and M. Duquesne, *Photochem. Photobiol.* **20,** 15 (1974).
[2] N. J. Leonard and G. L. Tolman, *Ann. N.Y. Acad. Sci.* **255,** 43 (1975).
[3] J. R. Barrio, J. A. Secrist III, and N. J. Leonard, *Biochem. Biophys. Res. Commun.* **92,** 597 (1972).
[4] N. J. Leonard, *Chemtracts Biochem. Mol. Biol.* **4,** 251 (1993).
[5] C. R. Bagshaw, J. F. Eccleston, D. R. Trentham, D. W. Yates, and R. S. Goody, *Cold Spring Harbor Symp. Quant. Biol.* **37,** 127 (1972).
[6] J. F. Eccleston, T. F. Kanagasabai, D. P. Molloy, S. E. Neal, and M. R. Webb, *In* "Guanine-Nucleotide Binding Proteins" (L. Bosch, B. Kraal, and A. Parmeggiani, eds.), p. 87. Plenum, New York, 1989.
[7] L. R. Yarbrough, J. G. Schlageck, and M. Baughman, *J. Biol. Chem.* **254,** 12069 (1979).

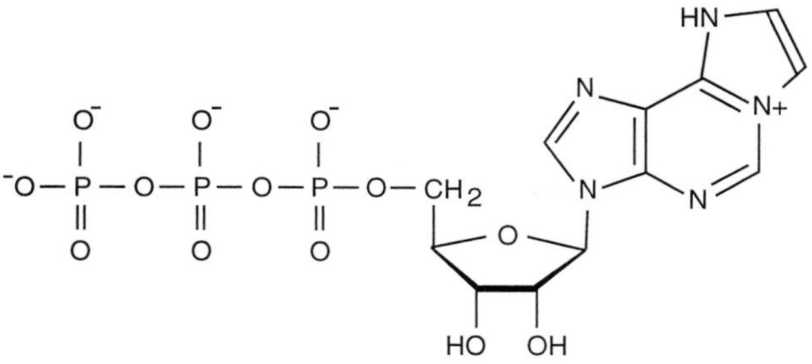

I

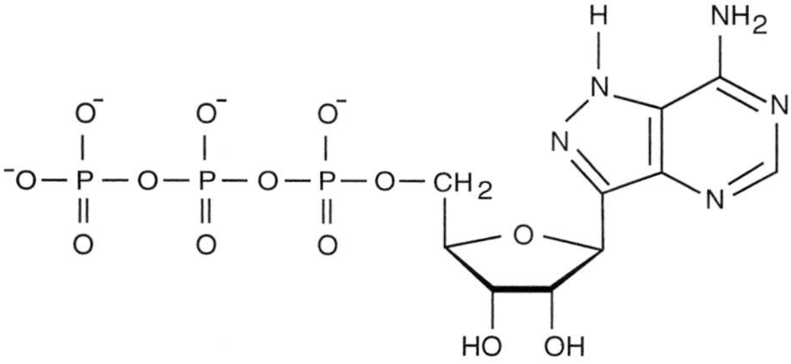

II

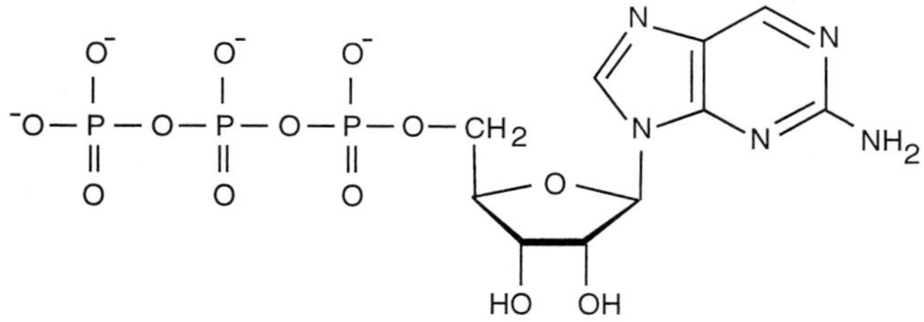

III

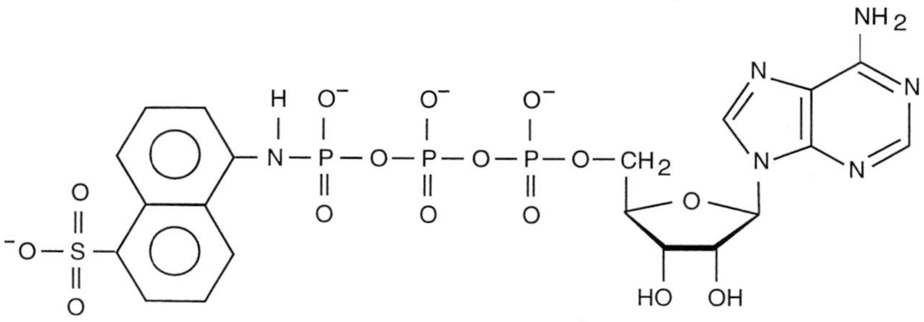

IV

FIG. 1. The structures of etheno-ATP **(I),** formycin 5′-triphosphate **(II),** 2-aminopurine ribose 5′-triphosphate **(III),** and (γ-AmNS)ATP **(IV).**

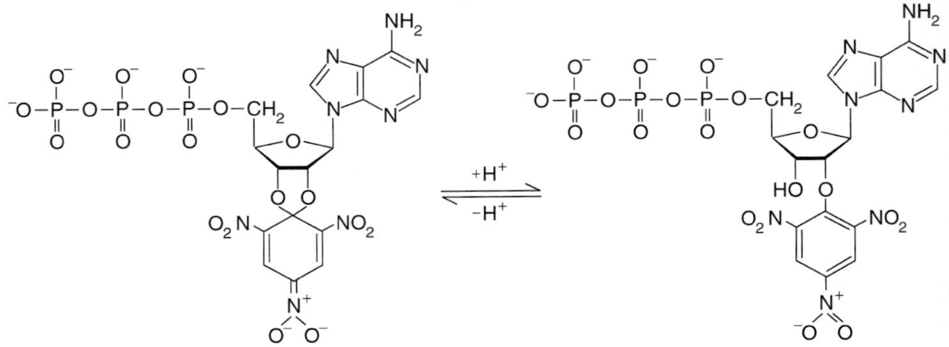

V

FIG. 2. TNP-ATP (**V**) exists as an equilibrium mixture depending on the pH. The right-hand structure represents the TNP group attached to either the 2′ or 3′ ribose hydroxyl.

riboside-modified analogs. We shall focus our attention on these derivatives. The benign nature of such ribose modification can be understood by examination of the structures of nucleoside 5′-triphosphatases such as myosin subfragment 1,[8] elongation factor Tu,[9–11] and p21^{ras},[12] in which cases the 2′,3′-hydroxyl groups of the nucleotide project out of the binding domain. Scheidig *et al.*[13] determined the crystallographic structure of the catalytic domain of p21$^{H\text{-}ras}$ complexed with the nonhydrolyzable fluorescent analog, mantdGppNHp (mant probes are discussed in detail as follows) and found that the mant moiety was located on the surface of the protein.

The first fluorescent ribose-modified nucleotide appears to have been 2′,3′-*O*-(2,4,6-trinitrocyclohexadienylidene) adenosine 5′-triphosphate (TNP-ATP) introduced by Hiratsuka and Uchida,[14] which exists as an equilibrium mixture (**V** in Fig. 2). Hiratsuka[15] then introduced the 2′(3′)-*O*-anthraniloyl (ant) and 2′(3′)-*O*-methylanthraniloyl (mant) derivatives of

[8] A. J. Fischer, C. A. Smith, J. Thoden, R. Smith, K. Sutoh, H. M. Holden, and I. Raymant, *Biophys. J.* **68,** 19s (1995).

[9] F. Jurnak, *Science* **230,** 32 (1985).

[10] M. Kjeldgaard and J. Nyborg, *J. Mol. Biol.* **223,** 721 (1992).

[11] H. Berchtold, L. Reshetnikova, C. O. A. Reiser, N. K. Schirmer, M. Sprinzl, and R. Hilgenfeld, *Nature (London)* **365,** 126 (1993).

[12] S. M. Franken, A. J. Scheidig, U. Krengel, H. Rensland, A. Lautwein, M. Geyer, K. Scheffzek, R. S. Goody, H. R. Kalbitzer, E. F. Pai, and A. Wittinghofer, *Biochemistry* **32,** 8411 (1993).

[13] A. J. Scheidig, S. M. Franken, J. E. T. Corrie, G. P. Reid, A. Wittinghofer, E. F. Pai, and R. S. Goody, *J. Mol. Biol.* **253,** 132 (1995).

[14] T. Hiratsuka and K. Uchida, *Biochim. Biophys. Acta* **320,** 635 (1973).

[15] T. Hiratsuka, *Biochim. Biophys. Acta* **742,** 496 (1983).

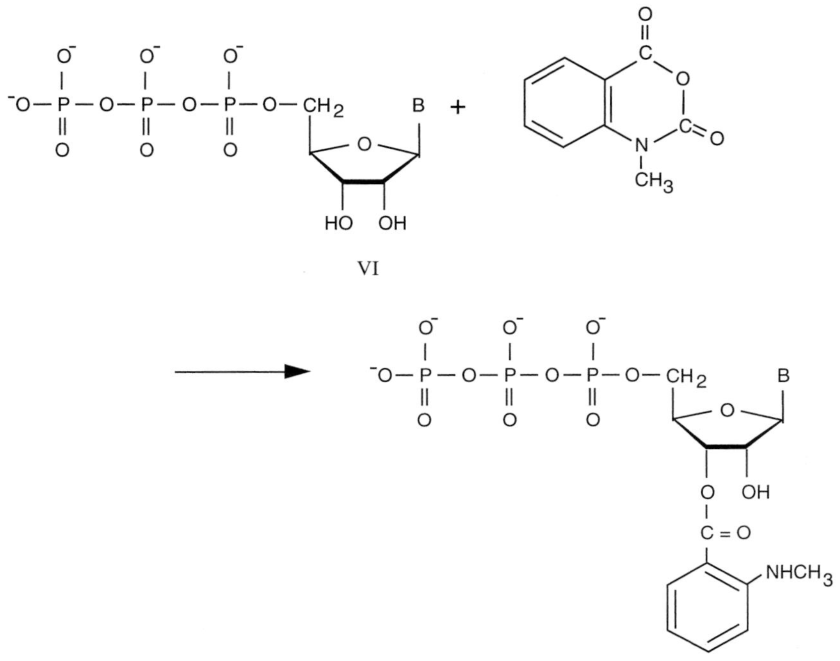

VI

VII

FIG. 3. The synthesis of mant-nucleotides (**VII**) by reaction of nucleotides with methylisatoic anhydride.

adenine and guanine nucleotides (**VII** in Fig. 3). One of the motivations for this choice of fluorophores was their relatively small size, which suggested that perturbation of binding properties would be minimal, and the fact that the fluorescence properties of these probes are environmentally sensitive. These probes have been used extensively in studies of nucleoside 5′-triphosphatases in a wide range of systems and closely mimic the parent nucleotides as regards rate and equilibrium constants, although there are exceptions. However, a more stringent test of their suitability is their ability to promote a particular biological process rather than interaction with a single protein. This concept is shown in Fig. 4, where the effects of ATP, εATP, and mantATP on the mechanical properties of permeabilized muscle fibers are shown. The ability of εATP and mantATP to produce tension in single muscle fibers (in the presence of calcium) and to relax them (in the absence of calcium) was compared to ATP. It can be seen that the εATP resulted in only 78% of the tension obtained with ATP whereas

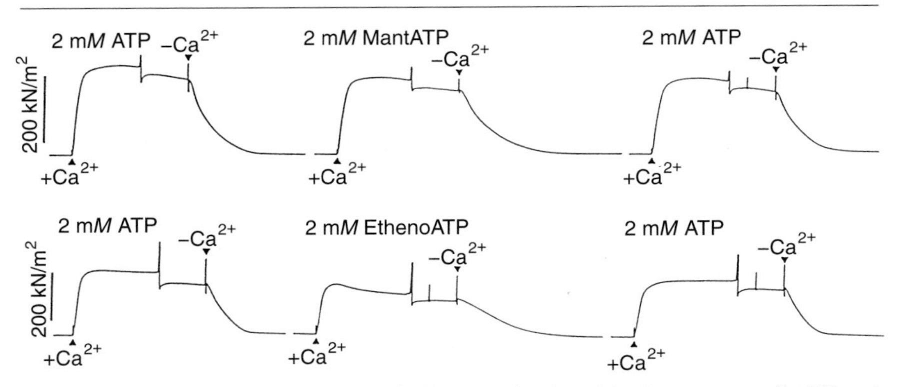

FIG. 4. A single permeabilized muscle fiber was incubated in the presence of ATP and Ca^{2+}, then ATP in the absence of Ca^{2+}, then either in the presence of EATP or mantATP in the presence and then absence of Ca^{2+}. The ATP cycle was then repeated. Muscle fiber tension was recorded with time. [Reprinted with permission from Woodward[71].]

mantATP gave a 95% increase in tension. The halftime of the relaxation process with εATP was 87% slower than with ATP but only 32% slower with mantATP. Similarly, Sowerby *et al.*[16] reported that mantATP supported movement of phalloidin-labeled actin filaments on immobilized rabbit skeletal muscle. On the other hand, Lark and Omoto,[17] studying axonemal dynein ATPases, showed that ant- and mantATP did not support the movement of wild-type *Chlamydomonas* axonemes, although they did reactivate sea urchin sperm axonemes.

Since the introduction of mant-nucleotides, other syntheses of fluorescent nucleotide analogs have been developed that allow a wide range of fluorophores to be introduced onto the ribose moiety. Jeng and Guillory[18] used the imidazolidate of carboxylic acids to acylate the 2′(3′)-hydroxyl of nucleotides (**IX** in Fig. 5) to produce photoaffinity labels and this approach has been extended to fluorescent analogs.[19,20] Cremo *et al.*[21] introduced a new class of analogs in which the fluorophore was attached to the ribose oxygens by a carbamoyl linkage. Hileman *et al.*[22] described the synthesis of fluorescein- and rhodamine-labeled GDP and ADP derivatives in which

[16] A. J. Sowerby, C. K. Seehra, M. Lee, and C. R. Bagshaw, *J. Mol. Biol.* **234,** 114 (1993).
[17] E. Lark and C. K. Omoto, *Cell Motil. Cytoskel.* **27,** 161 (1994).
[18] S. J. Jeng and R. J. Guillory, *J. Supramol. Struct.* **3,** 448 (1975).
[19] G. Onur, G. Schafer, and H. Strotmann, *Z. Naturforsch.* **382,** 49 (1983).
[20] I. Mayer, A. S. Dhams, W. Riezler, and M. Klingenberg, *Biochemistry* **23,** 2430 (1984).
[21] C. R. Cremo, J. M. Neuron, and R. G. Yount, *Biochemistry* **29,** 3309 (1990).
[22] R. E. Hileman, K. M. Parkhurst, N. K. Gupta, and L. J. Parkhurst, *Bioconjugate Chem.* **5,** 436 (1994).

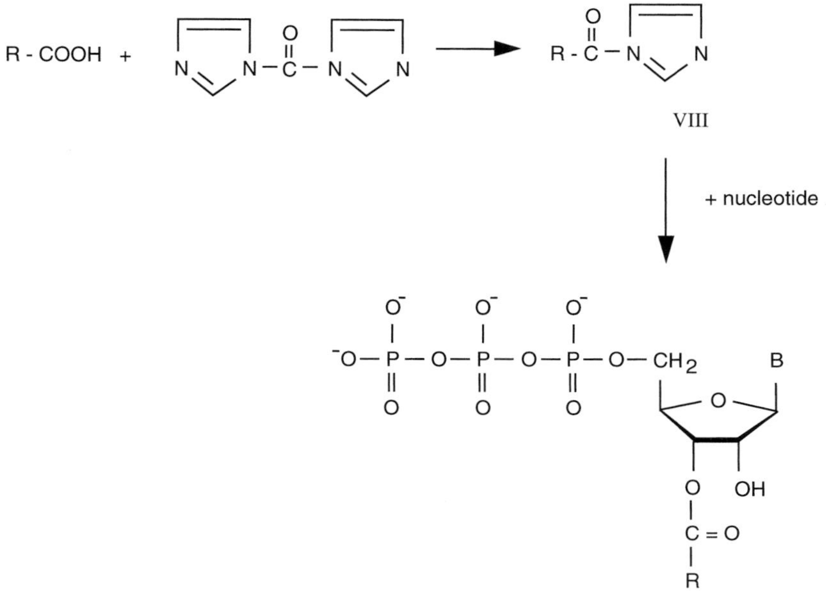

VIII

IX

Fig. 5. The reaction of nucleotides with the imidazolidates of carboxylic acids (**IX**).

the fluorescent groups were initially coupled through amine-containing linker arms to periodate-oxidized nucleotides followed by reduction to yield a six-membered morpholine-like ring. Fluorescent derivatives of 2′-amino-2′-deoxy- and 3′-amino-3′-deoxynucleotides (Fig. 6) offer a potential route to fluorescent nucleotide analogs by modification with fluorescent amine-

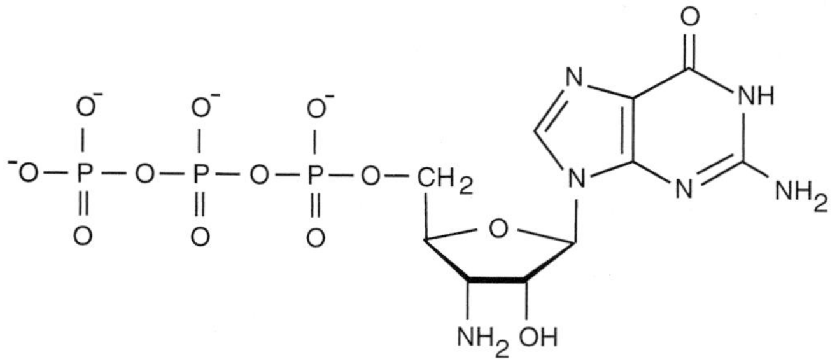

Fig. 6. The structure of 3′-amino-3′-deoxyGTP.

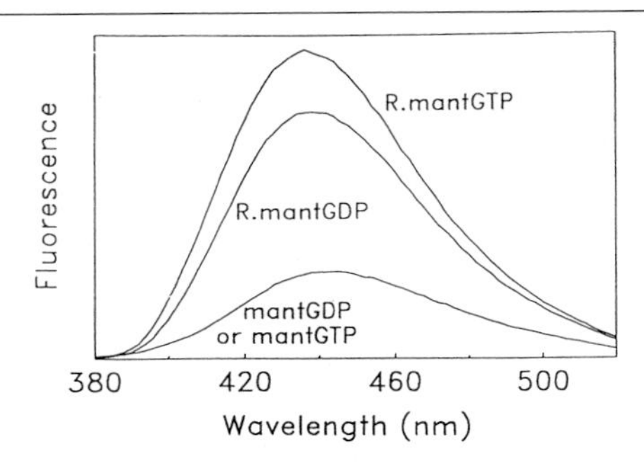

FIG. 7. Emission spectra of equal concentrations of mantGDP, mantGTP, $p21^{N\text{-}ras}$ · mantGDP, and $p21^{N\text{-}ras}$ · mantGTP. Excitation was at 350 nm.

reactive probes but, owing to the difficulties encountered in their synthesis, have so far found few applications (see, for example, Hobbs and Eckstein,[23] for a description of a 2′-amino-2′-deoxyGTP analog).

This chapter describes practical details of the synthesis of several fluorescent ribose-modified analogs together with their purification and characterization. Choices of fluorophore are discussed and examples of their applications to nucleoside 5′-triphosphatase systems are given.

Synthesis of Acyl Derivatives

Anthraniloyl and N-Methylanthraniloyl Derivatives of Adenine and Guanine Nucleotides

The anthraniloyl and *N*-methylanthraniloyl derivatives of nucleotides generally have a two- to threefold enhancement of their fluorescence intensity on binding to proteins (Fig. 7 and see below) with a corresponding increase in their excited-state lifetime. They were first synthesized and characterized by Hiratsuka[15] (Fig. 3). The method is equally applicable to adenine and guanine ribo- and 2′-deoxy-5′-di- and 5′-triphosphates and to the β,γ-imido and γS analogs. The methylanthraniloyl derivatives are more often used because of their longer excitation and emission maxima and the fact that their excitation maxima are near the 366-Hg line of arc lamps and the 364-nm line of argon-ion lasers.

In the synthesis, 0.27 mmol of the nucleotide is dissolved in 4 ml of water at 37°. The pH of the solution is adjusted to pH 9.6 with 1 *M* NaOH.

[23] J. B. Hobbs and F. Eckstein, *J. Org. Chem.* **42,** 714 (1977).

To this solution, with continuous stirring, is added 0.4 mmol of isatoic or methylisatoic anhydride. (These anhydrides are supplied by some manufacturers in a form stabilized with mineral oil but crystalline material gives a faster and cleaner reaction.) The pH of the solution is maintained at pH 9.6 by addition of 1 *M* NaOH and the reaction is considered complete when the pH ceases to fall (except owing to absorption of atmospheric CO_2) and the anhydride has all dissolved. The pH is then reduced to pH 7.6 with HCl before subsequent purification (see below). The mant-nucleotides elute after their parent nucleotides on DEAE-cellulose bicarbonate columns eluted with triethylammonium bicarbonate, pH 7.6. They can be characterized by their absorbance spectrum containing peaks of the base and anthraniloyl (λ_{max} 332 nm) or methylanthraniloyl (λ_{max} 354 nm) groups. Purification, characterization, and isomerization of the analogs are discussed in more detail on page 376.

Other Fluorescent Acyl Derivatives of Nucleotides

The reaction of isatoic and methylisatoic anhydride with nucleotides produces ribose-modified analogs with the fluorophore attached to the ribose by an acyl linkage. A more general approach to the synthesis of such analogs is based on the work of Jeng and Guillory,[18] who used the imidazolidate of carboxylic acids to acylate the 2′(3′)-hydroxyls of ATP to produce photoaffinity labels (Fig. 5). The method is described here by showing the synthesis of a julolidine derivative of GTP, but is applicable to a wide range of fluorescent carboxylic acids.

N,N′-Carbonyldiimidazole (0.17 mmol) is added to 0.05 mmol of 9-(2-carboxy-2-cyanovinyl)julolidine[24] in 0.70 ml of dry dimethylformamide and the solution is stirred for 15 min at room temperature. Eight microliters (0.2 mmol) of absolute methanol is added to remove excess *N,N′*-carbonyldiimidazole. A solution of 0.05 mmol of triethylammonium salt of GTP in 30 μl of water is added dropwise and then left stirring at room temperature overnight. The solution is evaporated to dryness and the remaining solid is washed three times with acetone to remove unreacted carboxylic acid. The nucleotide is then purified on a DE-52 bicarbonate column.

Synthesis of Carbamoyl Derivatives

Ethylenediamine Derivatives

This method for the acylation of nucleotides was first applied by Cremo *et al.*[21] to the synthesis of fluorescent nucleotide analogs. Reaction of nucleo-

[24] T. Iwaki, C. Torigoe, M. Noji, and M. Nakanishi, *Biochemistry* **32,** 7589 (1993).

tide with *N,N'*-carbonyldiimidazole in dry dimethylformamide results in the formation of the 2',3'-cyclic carbonate of the nucleotide in which the γ-phosphate is also converted to the phosphoroimidazolidate (**XI** in Fig. 8). Reaction of this compound with dansylethylenediamine results in attachment of the fluorophore to the 2'(3')-O via a carbamoyl linkage and the phosphoramidate of the γ-phosphate (**XII**). The latter group is then removed by acid hydrolysis to give the product **XIII**. Although the original synthesis[21] was applied to an ethylenediamine derivative, it is applicable to other fluorescent amines. The method is described here for the synthesis of a coumarin derivative of GTP.[25]

The trisodium salt of GTP (0.5 mmol) is converted to the bicarbonate form by applying it to a DE-52 column and washing with 10 m*M* triethylammonium bicarbonate, after which it is eluted with the same buffer at 0.6 *M*. After removal of triethylammonium bicarbonate by rotary evaporation, 0.48 ml (2 mmol) of tributylamine is added and the nucleotide is made anhydrous by repeated (3×) additions and vacuum evaporation of dry dimethylformamide. The nucleotide is then dissolved in 8 ml of dry dimethylformamide to which 405 mg (2.5 mmol) of *N,N'*-carbonyldiimidazole is added. The solution is stirred for 6 hr at 4° and the remaining *N,N'*-carbonyldiimidazole is removed by the addition of 144 μl (3.6 mmol) of absolute methanol. To this solution of the 2'(3')-cyclic carbonate is added a solution of 2.5 mmol of 3-amino-7-dimethylamino-4-methylcoumarin[26] in 5 ml of dry dimethylformamide and the mixture is incubated for 3 days at room temperature. The phosphoramidate of the γ-phosphate is removed by diluting the reaction mixture with 50 ml of water, adjusting to pH 2.5 with HCl, and leaving this mixture at 4° for 16 hr. The pH is then adjusted to pH 7.6 before subsequent purification on a DE-52 bicarbonate column.

2'(3')-O-(2-Aminoethylcarbamoyl)ATP Synthesis

An alternative approach to the synthesis of fluorescent nucleotide analogs involved reacting the 2',3'-cyclic carbonate with ethylenediamine instead of a fluorescent amine.[27] The resulting 2'(3')-*O*-(2-aminoethylcarbamoyl)ATP (**XVI** in Fig. 9) has a primary amine attached to the ribose moiety that can be modified by a wide range of amine-reactive fluorophores

[25] T. L. Hazlett, K. J. M. Moore, P. N. Lowe, D. M. Jameson, and J. F. Eccleston, *Biochemistry* **32,** 13575 (1993).
[26] J. E. T. Corrie, *J. Chem. Soc. Perkins Trans. 1,* 2151 (1990).
[27] S. Braxton and R. G. Yount, *Biophys. J.* **53,** A178 (1988).

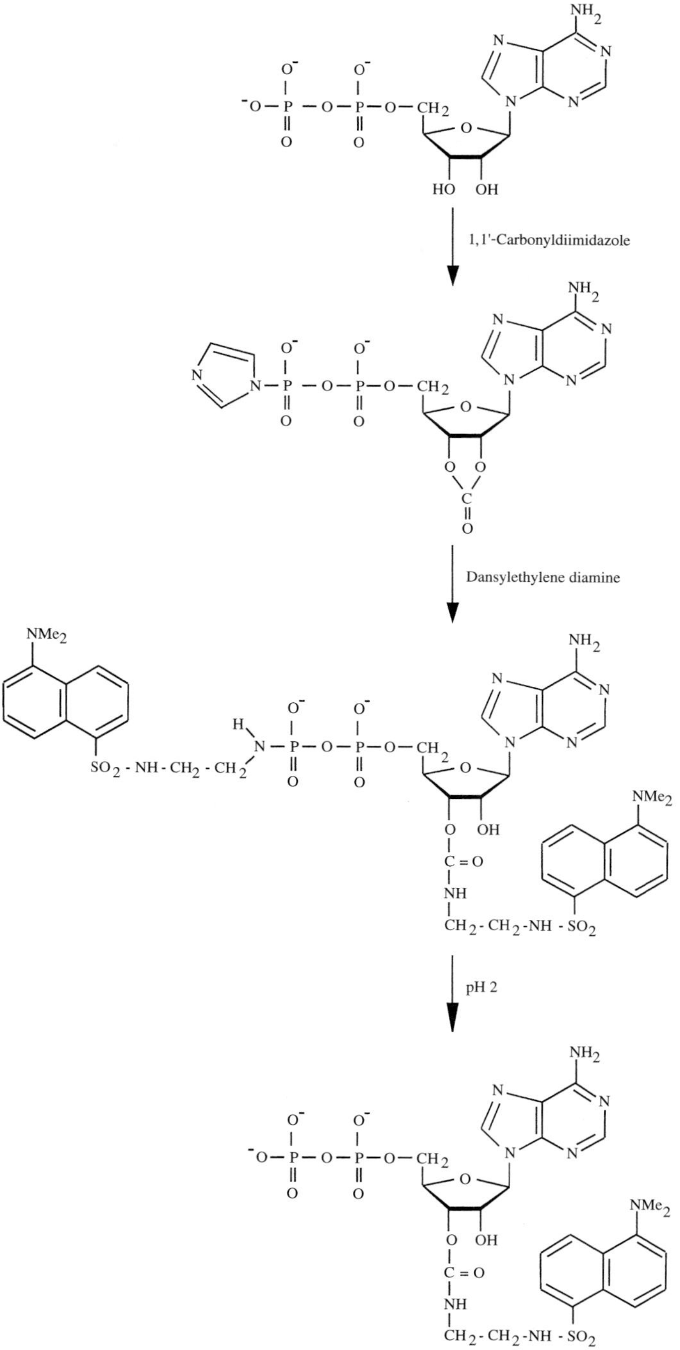

X

1,1'-Carbonyldiimidazole

XI

Dansylethylene diamine

XII

pH 2

XIII

FIG. 8. Synthesis (**X–XIII**) of dansyl-EDA-ADP as described by Cremo *et al.*[21]

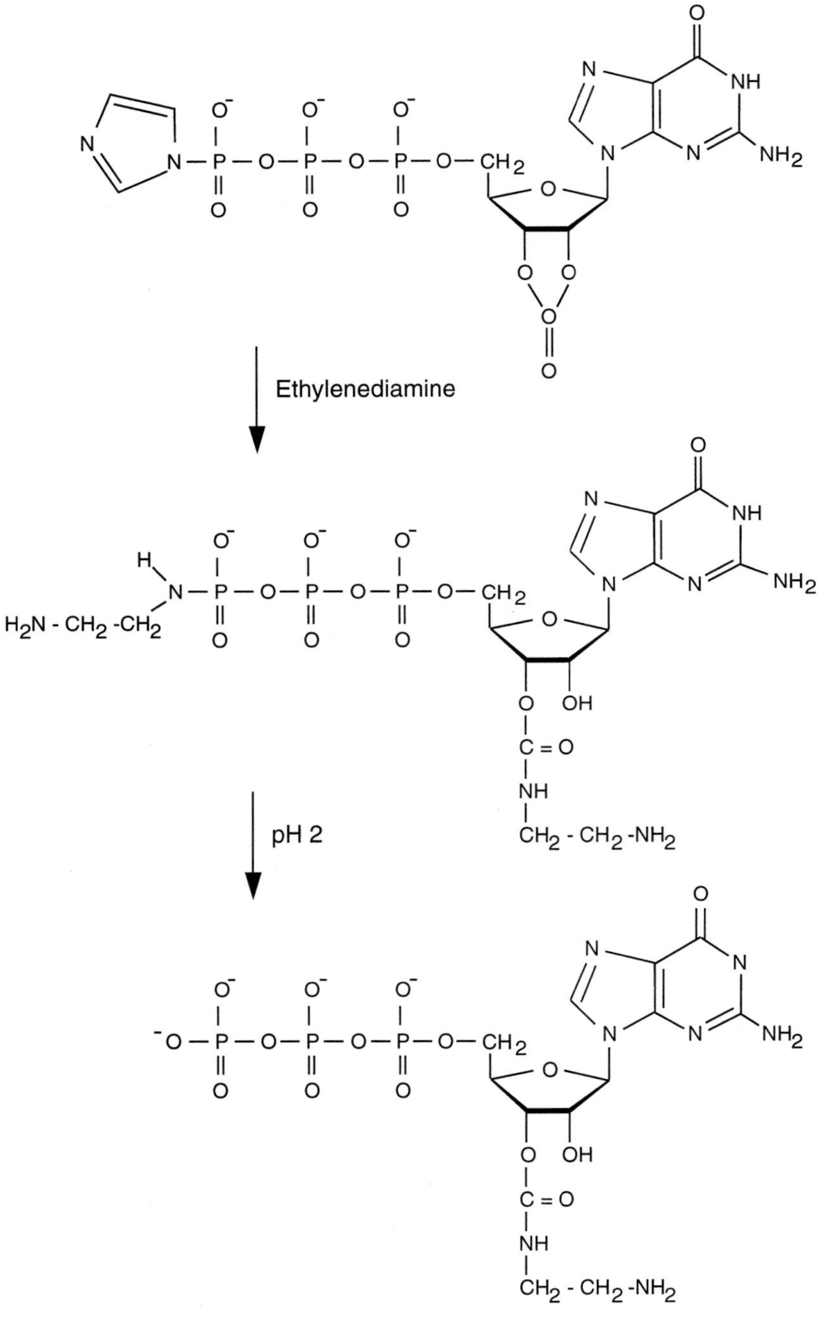

XIV

Ethylenediamine

XV

pH 2

XVI

FIG. 9. Synthesis (**XIV–XVI**) of 2′(3′)-*O*-(2-aminoethylcarbamoyl)ATP.

(see, for example, Refs. 25 and 28–30). To synthesize this compound, 167 μl (2.5 mmol) of ethylenediamine in 5 ml of dry dimethylformamide is added dropwise to a solution of 0.5 mmol of the 2′,3′-cyclic carbonate (prepared as described previously) in 8 ml of dry dimethylformamide. A dense white precipitate is formed, which is recovered by centrifugation (several minutes in a benchtop centrifuge) and washed three times with dimethylformamide. The precipitate is then dissolved in 30 ml of water, adjusted to pH 2.5 with HCl, and left for 16 hr at 4° to remove the phosphoramidate on the γ-phosphate. It is then adjusted to pH 7.6 and purified on a DE-52 bicarbonate column. The presence of the primary amine results in the product eluting before unreacted nucleoside triphosphate. This method can also be used to attach the primary amine to the ribose by a longer spacer arm, for example by using hexamethylenediamine in place of ethylenediamine.

Reactions of 2′(3′)-O-(2-Aminoethylcarbamoyl) Nucleotides

The 2′(3′)-*O*-(2-aminoethylcarbamoyl) derivatives of ATP and GTP can be reacted with any fluorescent reagent containing an amine-reactive group, such as sulfonyl chlorides, isothiocyanates, and succinimidyl esters. Some representative examples are as follows.

1. Rhodamine derivatives of 2′(3′)-*O*-(2-aminoethylcarbamoyl)GTP can be prepared by reaction with Lissamine rhodamine sulfonyl chloride.[28] To a solution of 1 m*M* 2′(3′)-*O*-(2-aminoethylcarbamoyl)GTP in 20 m*M* $NaHCO_3$ is added an equal volume of 2 m*M* Lissamine rhodamine sulfonyl chloride in acetone. After 2 hr at room temperature the acetone is removed by rotary evaporation and the analog purified by DE-52 chromatography. These types of analogs have a sulfonamide linkage between the fluorophore and starting material (Fig. 10a).
2. Fluorescein derivatives of GTP and GDP can be prepared by reaction with fluorescein isothiocyanate.[25] To a solution of 1 m*M* 2′(3′)-*O*-(2-aminoethylcarbamoyl)GTP in 20 m*M* sodium bicarbonate is added an equal volume of 5 m*M* fluorescein isothiocyanate in acetone. The reaction is allowed to proceed for 3 hr at room temperature. The acetone is removed by rotary evaporation before purification of the product. These types of analogs link the fluorophore via a thiourea (Fig. 10b).

[28] B. S. Watson, T. L. Hazlett, J. F. Eccleston, C. Davis, D. M. Jameson, and A. E. Johnson, *Biochemistry* **34,** 7904 (1995).
[29] B. P. Conibear and C. R. Bagshaw, *FEBS Lett.* **380,** 13 (1996).
[30] B. P. Conibear, D. S. Jeffreys, C. K. Seehra, R. J. Eaton, and C. R. Bagshaw, *Biochemistry* **35,** 2299 (1996).

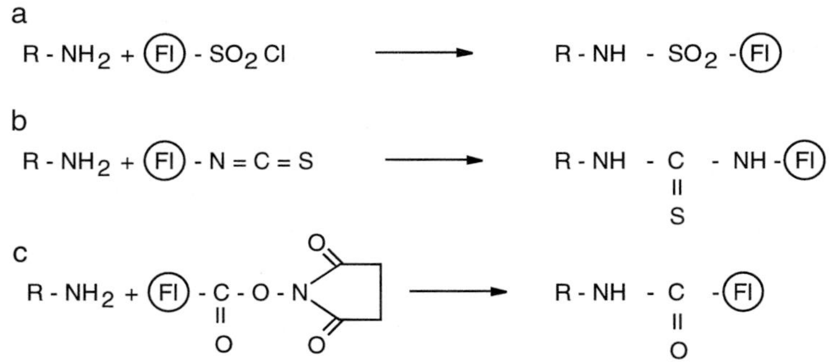

a

R - NH2 + (Fl) - SO2 Cl ⟶ R - NH - SO2 - (Fl)

b

R - NH2 + (Fl) - N = C = S ⟶ R - NH - C(=S) - NH - (Fl)

c

R - NH2 + (Fl) - C(=O) - O - N(succinimide) ⟶ R - NH - C(=O) - (Fl)

FIG. 10. Methods of reacting the amine of 2′(3′)-*O*-(2-aminoethylcarbamoyl) nucleotides or the 2′-amino-2′deoxy- or 3′-amino-3′-deoxynucleotides (RNH_2) with fluorescent amine-reactive probes, where Fl is the fluorophore.

3. The sulfoindocyanine dyes Cy3 and Cy5, developed by Mujumdar *et al.*,[31] can be conjugated to ATP or ADP.[32] The carboxylic acids Cy3.29-OH or Cy5.29-OH are first converted to their succinimidyl esters. Forty milligrams of disuccinimidyl carbonate is dissolved in 600 μl of dry dimethyformamide and 30 μl of dry pyridine. Sixty-five microliters of this solution (containing 16 μmol of disuccinimidyl carbonate) is then added to 3 mg (2.5 μmol) of Cy3.29-OH or Cy5.29-OH under anhydrous conditions and the solution is stirred at 60° under nitrogen for 90 min. After cooling to room temperature, excess disuccinimidyl carbonate is quenched by the addition of 0.48 μl (18.3 μmol) of methanol. The reaction mixture is then added to a solution of 7 m*M* (11.2 μmol) 2′(3′)-*O*-(2-aminoethylcarbamoyl)ATP in 20 m*M* sodium bicarbonate, pH 8.4, and left for 1 hr at room temperature. (Note that in this synthesis, because the fluorophore is expensive, excess nucleotide is used in the reaction rather than excess fluorophore.) The product is purified by DE-52 bicarbonate chromatography. These types of analogs link the fluorophore via an amide linkage (Fig. 10c).

Analysis and Purification of Analogs

Monitoring Synthesis of Nucleotide Analogs by Thin-Layer Chromatography

All reaction mixtures of the above syntheses can be rapidly analyzed for the appearance of product by thin-layer chromatography (TLC). On

[31] R. B. Mujumdar, L. A. Ernst, S. R. Mujumdar, C. J. Lewis, and A. S. Waggoner, *Bioconjugate Chem.* **4,** 105 (1993).

[32] K. Oiwa, G. P. Reid, C. T. Davis, J. F. Eccleston, J. E. T. Corrie, and D. R. Trentham, *Biophys. J.* **68,** A64 (1995).

Merck (Rahway, NJ) silica gel F_{254} TLC plates eluted with propan-2-ol–NH_4OH–H_2O (7 : 1 : 2, v/v/v), the parent nucleotides and their fluorescent derivatives stay at or close to the origin while the fluorescent reagents (and their hydrolysis products) generally elute with an $R_f > 0.4$. Comparison of a TLC plate of the reaction mixture with a control reaction containing fluorescent reagent but no nucleotides shows whether the expected product has been synthesized and the time course of its production. Addition of alkaline phosphatase to a sample of the reaction mixture in 10 m*M* Tris-HCl, pH 8.0, followed by TLC analysis should confirm that the putative product is indeed a fluorescent nucleotide because it is hydrolyzed to the much faster running fluorescent nucleoside. However, more quantitative characterization is required of the purified compound (see as follows).

Purification of Fluorescent Nucleotide Analogs

Reaction mixtures of fluorescent nucleotide analogs can be purified by hydrophobic interaction chromatography using Sephadex LH20 (see Hiratsuka[15] for protocols). However, in the laboratory of one of the authors (J.F.E.), they are generally purified on a DE-52 column in the bicarbonate form and eluted with a gradient of the volatile buffer, triethylammonium bicarbonate. A potential disadvantage of this method is that the pH could rise during rotary evaporation of the buffer, resulting in the hydrolysis of acyl-linked fluorophores. However, if a high vacuum is obtained by a good oil pump and leak-free system this problem has not been found to occur in the purification of many such analogs. DE-52 in the chloride form is first converted to the bicarbonate form by equilibration in 1 *M* ammonium bicarbonate (this is easier to prepare than triethylammonium bicarbonate and suitable for this purpose). The DE-52 is then equilibrated with water. Triethylammonium bicarbonate is prepared from redistilled triethylamine, which is necessary to remove nonvolatile impurities that may form during storage. Redistilled triethylamine (140 ml) is added to 500 ml of distilled water on ice and a stream of carbon dioxide is passed through the solution. This latter procedure is best done by placing dry ice in a stoppered Büchner flask. The tubing from the side arm is attached to a sintered frit that allows a constant stream of CO_2 into the triethylamine–water mixture at 0°. The solution is initially biphasic and strongly alkaline. After a few hours the pH drops to 7.6, but does not go lower. It is then diluted to 1 liter with cold water, which gives a 1 *M* solution of triethylammonium bicarbonate. It should be stored in glass (not plastic) at 4° and used within 2 weeks.

The analog reaction mixture is adjusted to pH 7.6, if necessary, and, after dilution with 10 m*M* triethylammonium bicarbonate, pH 7.6, is applied to the DE-52 bicarbonate column at 4°. For the 0.27-mmol scale preparation described for mant-nucleotides, a 36- × 3-cm column gives good separation.

The column is eluted with a 3l linear gradient of 10 mM to 0.6 M triethylammonium bicarbonate at 100 ml hr^{-1}. The absorbance of the eluant can be monitored at 254 nm and if possible by fluorescence detection. However, fractions containing the fluorophores can also be readily detected using a handheld UV light. The pooled fractions are then rotary evaporated to 3–4 ml. Fifty to 100 ml of methanol is added to the evaporated solution, which is then reevaporated to 3–4 ml. This procedure is repeated twice more and finally the solution is taken to dryness. The sample is then redissolved in a few milliliters of water and stored at −20°.

High-Performance Liquid Chromatography of Ribose-Modified Nucleotide Analogs

It is important that fluorescent nucleotide analogs be fully characterized in terms of structure (see the following section) and purity. The latter should demonstrate the absence of nonmodified nucleotides or nonnucleotide fluorescent species that could seriously affect binding studies. Checks on purity are most easily achieved by high-performance liquid chromatography (HPLC). A mixture of GMP, GDP, and GTP can be separated on a Whatman (Clifton, NJ) SAX (250 × 4.6 mm) ion-exchange column by eluting at 1.5 ml min^{-1} with 0.6 M $(NH_4)_2HPO_4$ adjusted to pH 4.0 with HCl. Typical retention times are 2.8, 3.7, and 7.6 min for GMP, GDP, and GTP, respectively [it should be noted that if the buffer is made with $(NH_4)H_2PO_4$ the retention times are longer because of the lower ionic strength of the solution]. However, under these conditions fluorescent analogs of nucleotides do not elute, presumably resulting from their hydrophobic interaction with the column matrix. If the 0.6 M $(NH_4)_2HPO_4$, pH 4.0, is made with methanol (25% final concentration), then GDP and GTP are eluted at 3.6 and 5.4 min whereas mantGDP and mantGTP are eluted at 3.5 and 5.2 min, respectively. Therefore, this isocratic method gives a good separation of mantGDP and mantGTP but is not appropriate to check for the presence of GTP and GDP in such preparations. To achieve this goal it is first necessary to elute the column isocratically for 10 min with aqueous buffer to elute GDP and GTP and then to apply a 0 to 25% methanol gradient in the same buffer over 30 min to elute mantGDP and mantGTP.

The above ion-exchange HPLC method does not separate the 2′-O and 3′-O isomers of mant-nucleotides (see as follows). Also, derivatives containing larger fluorophores, such as fluorescein, rhodamine, coumarin, and Cy5 groups, are not eluted even at high methanol concentrations. For these separations, reversed-phase HPLC can be used. Eccleston *et al.*[33]

[33] J. F. Eccleston, K. J. M. Moore, G. G. Brownbridge, M. R. Webb, and P. N. Lowe, *Biochem. Soc. Trans.* **19,** 436 (1991).

separated the 2′-O and 3′-O isomers of mantGTP and mantGDP using a Waters C_{18} Novapak column (150 × 3.9 mm) eluting isocratically with 8 m*M* KH_2PO_4, pH 5.5, containing 25% methanol (v/v) at 2 ml min^{-1}. For analogs containing larger fluorophores, it is better to replace the methanol with acetonitrile. For the analysis of Cy5-EDA-ATP Eccleston *et al.*[34] initially used a Waters Novapak C_{18} column (150 × 3.9 mm) initially eluting isocratically with 10 m*M* KH_2PO_4, pH 6.0, followed by a 0 to 30% (v/v) linear gradient of acetonitrile over 30 min. These conditions separated Cy5.29-OH, Cy5-EDA-ADP, and Cy5-EDA-ATP. However, for the separation of the 2′-O and 3′-O isomers of the two nucleotides (see 2′ ⇌ 3′ Isomerization, as follows), separation by isocratic elution was best optimized at 10 m*M* KH_2PO_4, pH 6.8, containing 13% (v/v) acetonitrile. Even then, only partial resolution of the 2′-O and 3′-O isomers of Cy5-EDA-ATP was achieved although baseline resolution of the two isomers of Cy5-EDA-ADP was realized. The exact conditions for separating isomers of other fluorescent nucleotide analogs, on a particular reversed-phase column, need to be determined individually by experiments based on these guidelines.

Characterization of Fluorescent Nucleotide Analogs

Preliminary characterization of analogs can be made on the basis of the absorption spectrum of the purified material, which should be the sum of the spectra of the purine base and the fluorophore. Acyl-modified analogs can be readily hydrolyzed by a few minutes at pH 10 and the resulting reaction mixture analyzed by HPLC to determine that the carboxylic acid and nucleotide are present in stoichiometric ratios. Ideally, particularly for novel analogs, it is preferable to confirm the structure by nuclear magnetic resonance (NMR)[21] and mass spectroscopy. For the latter, fast atom bombardment in the negative ion mode is preferable for nucleotide analogs because of the ease of interpretation of the results.

2′ ⇌ 3′ Isomerization

When 2′(3′)-*O*-methylanthraniloyl adenosine was initially characterized, it was thought to be derivatized only on the 3′-O position because methylation with diazomethane followed by deacylation in 0.1 *M* NaOH resulted only in the formation of 2′-*O*-methyladenosine.[15] However, it has long been known that monoacyl derivatives of *cis*-1,2-diol systems undergo

[34] J. F. Eccleston, K. Oiwa, M. A. Ferenczi, M. Anson, J. E. T. Corrie, A. Yamada, H. Nakayama, and D. R. Trentham, *Biophys. J.* **70,** A159 (1996).

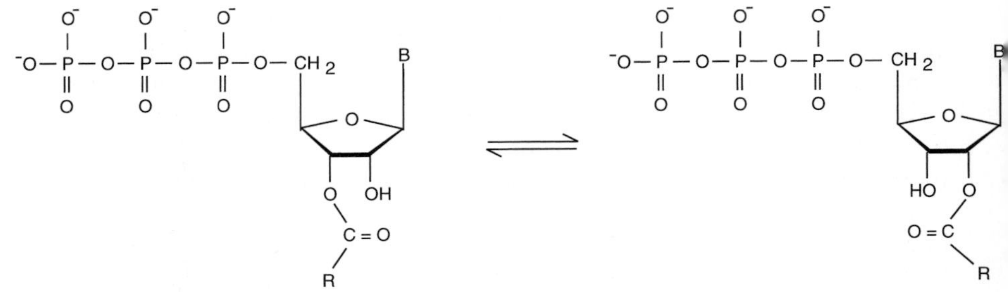

FIG. 11. Isomerization between the 2′- and 3′-oxygen atoms of nucleotide acyl derivatives.

isomerization by acyl migration[35] (Fig. 11). Onur *et al.*[19] showed that naphthoylacyl derivatives of ATP exist as an equilibrium mixture of the two isomers. Cremo *et al.*[21] then showed that this type of equilibrium also exists with the mant derivatives. These demonstrations were achieved by ^{1}H NMR measurements and relied on the fact that the chemical shift of the anomeric H-1′ proton is at lower field for the 2′-derivative than for the 3′-derivative and that the coupling constant, $J_{1'2'}$, for the 3′-substituted compound has a higher value than does $J_{1'2'}$ for the 2′ compound. Integration of the two peaks gives the percentage of each isomer at equilibrium. The result was 70% 3′-isomer and 30% 2′-isomer for the mant analogs. Eccleston *et al.*[32] and Rensland *et al.*[36] separated the two isomers of mantGTP by HPLC and showed that the proportion of the 3′-O isomer is approximately twice that of the 2′-O isomer. They then measured the rate of reequilibration of the isomers. This process was pH dependent with a half-time of 7 min at pH 7.4 and 2 hr at pH 6.0.

This equilibrium mixture presents problems as regards using ribose-modified nucleotides in studies involving their binding to proteins because the two isomers may bind with different affinities or be hydrolyzed at different rates, or the fluorophore could report on different steps of a protein–nucleotide interaction. Also, observed fluorescence changes could result from the isomerization of the fluorophore between the 2′-O and 3′-O groups on the nucleotide either free in solution or bound to the protein. The problem can be overcome by separating the isomers at low pH and using them at low pH if possible. The more usual approach is to derivatize the 2′-deoxy- or 3′-deoxynucleotides so that only a single ribose hydroxyl group is available to react. Ma and Taylor[37] have shown that on mixing

[35] C. B. Reese and D. R. Trentham, *Tetrahedron Lett.* **29,** 2467 (1965).
[36] H. Rensland, A. Lautwein, A. Wittinghofer, and R. S. Goody, *Biochemistry* **30,** 11181 (1991).
[37] Y.-Z. Ma and E. W. Taylor, *Biochemistry* **34,** 13223 (1995).

mantATP with kinesin a biphasic change in fluorescence occurs; an initial fast increase followed by a slower decrease. However, when the experiment was repeated with 3′-deoxyATP only the first, fast increase was observed.[37a] 2′-Deoxynucleotides of both ATP and GTP are freely available commercially as is the 3′-deoxyATP (cordycepin), although this nucleotide is rather expensive. 3′-DeoxyGTP is not, at present, commercially available. Fluorescent analogs bound to 2′-amino-2′-deoxy- and 3′-amino-3′-deoxynucleotides offer one approach to the synthesis of pure 2′- or 3′-modified isomers because migration of the fluorescent group cannot occur.

Although the acyl-linked fluorophores migrate rapidly between the 2′-O and 3′-O positions at neutral pH, the use of carbamate-linked fluorophores offers an alternative approach to the use of pure isomers. Oiwa *et al.*[32] synthesized 2′(3′)-*O*-(*N*-2-acetamidoethylcarbamoyl)ADP and separated the 2′- and 3′-isomers by HPLC, characterized them by ^{1}H NMR, and investigated their equilibration rates. The ratio of 2′- to 3′-substituted nucleotide is 0.7 at equilibrium with a halftime for equilibration of 44 hr at pH 7.1 and 0.3 hr at pH 9.4. These derivatives, therefore, offer the possibility of pure isomers that still contain a ribose hydroxyl. Single turnover studies in which excess myosin subfragment 1 is mixed with either ATP or 2′-*O* or 3′-*O*(*N*-2-acetamidoethylcarbamoyl)/ATP, and the tryptophan fluorescence monitored, showed that the rate constants of binding and turnover were identical in all three cases (J. F. Eccleston, unpublished work, 1996). However, in carbamate-linked fluorescent analogs, the fluorophore may report different events depending on the substitution position.

Factors Influencing Analog Design

Choice of Fluorescent Probe

The choice of fluorophore depends on the type of information being sought. For equilibrium and kinetic experiments involving protein–nucleotide or protein–protein interactions, environmentally sensitive probes are usually necessary (although this caveat need not apply if one uses an anisotropy approach[37b]). The mant-nucleotides are well suited in this regard. Attempts have been made to synthesize an even more environmentally responsive probe by making an acrylodan derivative of GDP.[6] (Acrylodan[38] is a sulfhydryl reactive form of prodan, a fluorophore designed

[37a] Y.-Z. Ma and E. W. Taylor, *J. Biol. Chem.* **272,** 717 (1997).

[37b] D. M. Jameson and W. H. Sawyer, *Methods Enzymol.* **246,** 283 (1995).

[38] F. G. Prendergast, M. Meyers, G. L. Carlson, S. Iida, and D. Potter, *J. Biol. Chem.* **258,** 7541 (1983).

by Weber and Farris[39] to have an exceptionally large excited-state dipole moment, which results in very environmentally sensitive emission properties.) Although this derivative was more sensitive to the environment than mantGDP, as judged by a comparison of its spectral properties in water and ethanol, on binding to elongation factor Tu the changes in spectral properties were no greater than for the mant probe. This reduction in the potential spectral sensitivity is presumably because the fluorophore of the acrylodan derivative is farther away from the nucleotide than in the corresponding mant derivative. Similarly, Molloy[40] synthesized derivatives by reacting 1,5-, 2,5-, and 2,6-dansyl chloride with ED-GTP and ED-GDP and all analogs were less responsive than mantGDP on binding to elongation factor Tu.

For fluorescence energy transfer measurements, the choice of donor and acceptor usually depends on the distances to be measured. Given the extreme distance dependency of nonradiative energy transfer processes (inverse sixth power[41]) the important parameter in this regard is the critical transfer distance, designated R_0. This factor is determined in large part by the extent of overlap between the emission of the donor and the absorption of the acceptor.[42] A compilation of R_0 values for a number of donor–acceptor pairs can be found in Van der Meer *et al.*[43] A complete discussion of energy transfer methods, including the infamous orientation factor, κ^2, is beyond the scope of this chapter and readers are referred to more comprehensive treatments.[43–48] When the intended application of the fluorescent probe involves polarization or anisotropy measurements then considerations such as the limiting polarization (P_0) as well as the lifetime of the probe are often relevant. We should note that polarization and anisotropy are related terms (see, for example, Jameson and Sawyer[37b]) and that both quantities have the same information content. An excellent treatment on

[39] G. Weber and F. J. Farris, *Biochemistry* **18,** 3075 (1979).
[40] D. P. Molloy, Ph.D. thesis. CNAA, United Kingdom (1991).
[41] F. Perrin, *Ann. Phys. Ser. 10* **XVII,** 20 (1932).
[42] Th. Förster, *Ann. Phys.* (*Leipzig*) **2,** 55 (1948).
[43] B. W. Van der Meer, G. Coker, and S.-Y. S. Chen, "Resonance Energy Transfer: Theory and Data." VCH, New York, 1994.
[44] R. F. Dale, J. Eisinger, and W. E. Blumberg, *Biophys. J.* **26,** 161 (1979).
[45] P. Wu and L. Brand, *Biochemistry* **31,** 7939 (1992).
[46] W. Jiskoot, V. Hlady, J. J. Naleway, and J. N. Herron, *In* "Physical Methods to Characterize Pharmaceutical Proteins" (J. N. Herron, W. Jiskoot, and D. J. A. Crommelin, eds.), p. 1. Plenum, New York, 1996.
[47] H. C. Cheung, *In* "Topics in Fluorescence Spectroscopy" (J. R. Lakowicz, ed.), p. 127. Plenum, New York, 1991.
[48] P. Selvin, *Methods Enzymol.* **246,** 300 (1995).

the origins of polarization has been given by Weber.[49] A description of the explicit application of polarization/anisotropy measurements to study biomolecular interactions, including considerations such as the excitation wavelength dependence of P_0 and the relationship between probe lifetime and observed polarization/anisotropy, has been given by Jameson and Sawyer.[37a] We should note that if determination of rotational aspects of the system is the goal then the lifetime of the probe is an important consideration. Certainly the lifetime of the bound probe should be within an order of magnitude of the rotational relaxation time being determined. For example, mant derivatives, with bound lifetimes of about 9 nsec, are well suited for small to medium-size proteins. If the goal is to quantify slow motions, such as for large proteins, protein aggregates, or membrane-associated proteins then one should consider using fluorophores such as pyrene (with intrinsic lifetimes in the range of 100–200 nsec) or even phosphorescence probes.

When fluorescent analogs are to be used for microscopy, the main criteria for choice of fluorophore are high fluorescence intensity (hence high extinction coefficient and quantum yield), excitation and emission maxima best suited to the excitation source and emission detection and stability against photobleaching. We may note here that advances in two-photon microscopy[50,51] suggest a significant role for this technique in such studies because the method greatly simplifies the elimination of scattered light and also reduces photobleaching compared to conventional fluorescence microscopy. Mant-nucleotides are generally not useful for fluorescence microscopy because they lack these properties but rhodamine derivatives have proved more useful.[16] Funatsu *et al.*[52] have used cyanine dye derivatives, which seem to have ideal properties for fluorescence microscopy and have been used for detecting fluorescence from single molecules (see as follows).

Choice of Linker

As discussed previously, for the case of acrylodan, the nature of the linker between ribose and fluorophore may affect the ability of the fluorophore to report on the binding process. For this reason, mant-nucleotides and analogs synthesized by acylation of the ribose by carboxylic acid imida-

[49] G. Weber, *In* "Fluorescence and Phosphorescence Analysis" (D. M. Hercules, ed.), p. 217. John Wiley & Sons, New York, 1966.

[50] W. Denk, J. H. Strickler, and W. W. Webb, *Science* **248,** 73 (1990).

[51] P. T. C. So, T. French, W. M. Yu, K. M. Berland, C. Y. Dong, and E. Gratton, *Bioimaging* **3,** 49 (1995).

[52] T. Funatsu, Y. Harada, M. Tokunaga, K. Saito, and T. Yanagida, *Nature* (*London*) **374,** 555 (1995).

zolidates or derivatives of 2′-amino-2′-deoxynucleotides are preferred, because the fluorophore is closer to the ribose and is more likely, in the case of environmentally sensitive probes, to be perturbed on binding to the protein. However, the acyl linkage is susceptible to base hydrolysis and high pH should be avoided. The derivatization of amino groups of either 2′-amino-2′-deoxynucleotides or ethylenediamine-derivatized nucleotides can be achieved by reaction with isothiocyanates, sulfonyl chlorides, or *N*-hydroxysuccinimidyl esters. The latter two are best used because the sulfonamide and amide linkages are stable compared to the thiourea linkage.

Fluorophore Purity

Many fluorescent reagents such as fluorescein and rhodamine derivatives exist in isomeric forms. Even when a specific isomer is specified, the purity and isomeric composition of many commercially available fluorescent reagents may be questioned. For example, in the original synthesis of Texas Red the product was described as consisting primarily of monosulfonyl chloride derivatives, probably a mixture of the *ortho* and *para* isomers[53] although it is now often assumed to be only the *para* derivative. This consideration may not be a problem for many applications of fluorescent nucleotide analogs but could cause difficulties with heterogeneous kinetic data or when measurements depend on a knowledge of the quantum yield of the fluorophore such as in energy transfer measurements.

Binding Specificity

The presence of hydrophobic groups in the fluorescent nucleotide analogs presents the possibility that they may bind to proteins other than at the nucleotide-binding site. This problem overcomes the advantage of the use of these analogs because they would then not be localized to a known site. The occurrence of such nonspecific binding can be determined by carrying out the relevant experiment in the presence of a large excess of the physiological nucleotide, which should not give rise to changes in the fluorescent signal. Mant-nucleotides show little or no nonspecific binding to ATPases or GTPases. However, larger groups may give problems. For example, the julolidine derivative of ATP (as described previously) shows a large enhancement of its fluorescence yield on binding to subfragment 1 but this effect cannot be reversed by addition of ATP, indicating that it also binds to sites other than the ATPase site (J. F. Eccleston, unpublished results, 1996).

[53] J. A. Titus, R. Haugland, S. O. Sharrow, and D. M. Segal, *J. Immunol. Methods* **50,** 193 (1982).

Applications

Kinetic Studies

The most extensive use of ribose-modified nucleotide analogs to date has been in kinetic and equilibrium measurements of the interactions of nucleotides with ATPases and GTPases. The kinetic methods generally involve the study of fluorescence changes associated with nucleotide binding and dissociation, nucleoside 5′-triphosphate cleavage, and any possible conformational changes. These studies often involve the use of stopped-flow instrumentation for studying pre-steady-state kinetics of the system. A detailed account of such methods is beyond the scope of this article and the reader is referred to an excellent treatment by Gutfreund.[54] Specific examples of the application of fluorescent nucleotide analogs to kinetic and/or equilibrium measurements on systems such as myosin, Ras, elongation factor Tu, kinesin, and others are given as follows.

Several groups have carried out kinetic observations on the interaction of mant-guanine nucleotide derivatives with the p21^{ras} system. These studies include those of Neal *et al.*[55] and Eccleston *et al.*[33] on p21$^{\text{N-}ras}$, the product of the N-*ras* protooncogene and John *et al.*[56] and Rensland *et al.*[36] with p21$^{\text{H-}ras}$, the product of the human c-Ha-*ras* protooncogene. All studies reported an increase in the fluorescence intensity of the mant probes on binding and a decrease in the fluorescence on hydrolysis of the terminal phosphate (see, for example, Fig. 7). These observations have been considered in light of a Ras crystal structure. Specifically, Scheidig,[13] as mentioned above, determined that the mant-nucleotide was on the surface of the protein and that it interacted with the aromatic ring of Tyr-32. They suggested that the enhanced fluorescence observed on binding of the probe was due to relief of quenching interactions between the mant moiety and the guanine base, which exist in solution. They furthermore suggested that the change in fluorescence yield observed on hydrolysis of mantGTP to mantGDP is associated with the movement of this tyrosine residue. Sadhu and Taylor[57] reported transient state kinetic measurements on mantADP and mantATP interactions with kinesin. Lockhart *et al.*[58] studied the release of mantADP from recombinant double-headed non-claret disjunctional (ncd) motor proteins activated by microtubules. The microtubule-activated

[54] H. Gutfreund, "Kinetics for Life Sciences." Cambridge University Press, London, 1995.

[55] S. E. Neal, J. F. Eccleston, and M. R. Webb, *Proc. Natl. Acad. Sci. U.S.A.* **87,** 3562 (1990).

[56] J. John, R. Sohmen, J. Feuerstein, R. Linke, A. Wittinghofer, and R. S. Goody, *Biochemistry* **29,** 6058 (1990).

[57] A. Sadhu and E. W. Taylor, *J. Biol. Chem.* **267,** 11352 (1992).

[58] A. Lockhart, R. A. Cross, and D. F. A. McKillop, *FEBS Lett.* **368,** 531 (1995).

turnover rates for mantATP were also determined. Dissociation kinetics of several nonhydrolyzable mant-labeled guanine nucleotides from recombinant myristoylated G protein α subunit were reported by Remmers and Neubig.[59]

Ligand Binding and Binding Site Environments

Bujalowski and Klonowska[60] studied the binding of mantADP (as well as TNP-ATP, TNP-ADP, and εADP) to the hexameric helicase protein DnaB from *E. coli.* The binding process was followed by the quenching of the intrinsic tryptophan fluorescence of DnaB by bound nucleotide analog. The quenching data were used to generate binding isotherms that were biphasic and interpreted as resulting from negative cooperativity. These authors later described more detailed spectroscopic studies on the fluorescence properties of these various probes, including lifetime and anisotropy results, bound to the protein.[61] Mueser and Parkhurst[62] coupled dansyl-β-alanine to GDP and ATP. By observing changes in fluorescence intensity and anisotropy on binding of the GDP analog to eukaryotic initiation factor 2 they determined a dissociation constant of 33 n*M*, which was in good agreement with the results from filter-binding assays. They also observed binding of the ATP analog to H1 histone. Churchich[63] showed that 2′-deoxy-3′-anthraniloyladenosine-5′-triphosphate (ant-dATP) was a useful environmental probe of the nucleotide-binding site of the molecular chaperone Hsc70. In this system the lanthanide ion complex, Tb^{3+}–ant-dATP, binds to the protein and sensitized luminescence arising from energy transfer from the anthraniloyl group to the terbium is significantly enhanced on protein binding.

Hydrodynamic Studies

Time-resolved fluorescence methods are particularly useful in hydrodynamic studies on biomolecules. This utility arises from the fact that the rotational diffusion of many proteins and ligands occurs on the same time scale as the emission process, namely on the order of nanoseconds. Because the rotational diffusion rate of a molecule is dependent on molecular mass and shape characteristics, methods such as time decay of anisotropy, which can measure rotational relaxation times, give direct information on hydro-

[59] A. E. Remmers and R. R. Neubig, *J. Biol. Chem.* **271,** 4791 (1996).
[60] W. Bujalowski and M. M. Klonowska, *Biochemistry* **32,** 5888 (1993).
[61] W. Bujalowski and M. M. Klonowska, *Biochemistry* **33,** 4682 (1994).
[62] T. C. Mueser and L. J. Parkhurst, *Int. J. Biochem.* **25,** 1689 (1993).
[63] J. E. Churchich, *Eur. J. Biochem.* **231,** 736 (1995).

dynamic aspects. This information can be used to determine size and shape characteristics of biomolecules (see, for example, Brunet *et al.*[64] and Jiskoot *et al.*[46]) or to study binding processes, such as the association of small ligands with macromolecules or macromolecule–macromolecule interactions (see, for example, Jameson and Hazlett,[65] Jameson and Sawyer,[37a] and Jiskoot *et al.*[46]). Some examples of the use of fluorescent nucleotide analogs in hydrodynamic studies are described as follows.

Eccleston *et al.*[66] described the use of a fluorescamine GDP derivative to study the rotational parameters of elongation factor Tu (EF-Tu). 2′-Amino-2′-deoxyGDP, modified on the 2′-amino group by fluorescamine, binds to EF-Tu with an affinity similar to unmodified GDP (i.e., dissociation constants in the nanomolar range). The fluorescence lifetime of the free fluorescamineGDP is 7.7 nsec and increases to 11.0 nsec on binding to EF-Tu. Dynamic polarization methods (the frequency domain equivalent of anisotropy decay) were utilized to determine a rotational relaxation time of 88 nsec for EF-Tu. The data also indicated that the fluorophore experienced little "local" mobility, i.e., the probe did not demonstrate significant motion independent from the overall rotational diffusion of the protein. This result indicated that EF-Tu was nonspherical, which was consistent with the overall shape later determined from crystallographic studies.

Hazlett *et al.*[25] utilized a series of fluorescent nucleotide analogs to investigate the aggregation state of $p21^{N\text{-}ras}$. Specifically, they synthesized three ribose-modified nucleotide analogs, using coumarin, mant, and fluorescein as the fluorophores, and studied the steady state and time-resolved properties of the protein-bound probes. Although the lifetimes of the fluorescein and coumarin probes did not change significantly on binding to the protein (approximately 4.0 and 3.2 nsec for the fluorescein and coumarin derivatives, respectively) the lifetimes of mantGDP, mantGTP, and mantdGTP, at 5°, changed, respectively, from 4.1, 4.2, and 4.1 nsec free in solution to 9.0, 9.3, and 9.2 nsec bound to $p21^{N\text{-}ras}$ (interestingly, the lifetimes of mantGDP bound to $p21^{H\text{-}ras}$ and $P21^{K\text{-}ras}$ were 8.5 and 8.4 nsec, respectively). Using the dynamic polarization method they were able to separate the "global" motion of the protein–probe complex from the "local" mobility of the bound probe (for reviews of this methodology see, for example, Jameson and Hazlett[65] or Gratton *et al.*[67]). Using the global analysis approach (Globals Unlimited; Laboratory for Fluorescence Dynamics, Ur-

[64] J. E. Brunet, V. Vargas, E. Gratton, and D. M. Jameson, *Biophys. J.* **66,** 446 (1994).

[65] D. M. Jameson and T. L. Hazlett, *In* "Biophysical and Biochemical Aspects of Fluorescence Spectroscopy" (T. G. Dewey, ed.), p. 105. Plenum, New York, 1991.

[66] J. F. Eccleston, E. Gratton, and D. M. Jameson, *Biochemistry* **26,** 3902 (1987).

[67] E. Gratton, D. M. Jameson, and R. D. Hall, *Annu. Rev. Biophys. Bioeng.* **13,** 105 (1984).

bana, IL) they were able to analyze simultaneously the data from all three probes, fixing the "global" rotational relaxation time in all data sets and showing that the extent of "local" mobility of each probe was different. This study illustrates the technique of using multiple probes to elucidate hydrodynamic aspects, a method originally introduced by Beechem *et al.*[68]

Energy Transfer Studies

Watson *et al.*[28] used a rhodamineGTP (REDA–GTP) analog to study the arrangement of elongation factor Tu and aminoacyl-tRNA in the ternary complex. Specifically, they prepared EF-Tu · REDA–GTP from EF-Tu · GDP by stripping the nucleotide from the protein and adding REDA–GTP to the nucleotide free EF-Tu. The 4-thiouridine at position 8 of *E. coli* $tRNA^{Phe}$ was modified using 5-(iodoacetamido)fluorescein and the modified tRNA was acylated with phenylalanine. The ternary complex of the Phe-$tRNA^{Phe}$-Fl^8 and EF-Tu · REDA-GTP was then formed and the efficiency of energy transfer between the fluorescein and rhodamine moieties was evaluated using both intensity and lifetime data. After correction for the lengths of the probe attachment tethers, the distance found for the separation of the 2′(3′)-oxygen of the GTP ribose and the sulfur in the s^4U was greater than or equal to 49 Å, which is consistent with the published crystal structure.[69]

Several studies have appeared that take advantage of the possibility of energy transfer between intrinsic protein fluorophores (i.e., tyrosine or tryptophan residues) and anthraniloyl or methylanthraniloyl moieties. Woodward *et al.*,[70,71] for example, showed that the fluorescence excitation spectrum of myosin subfragment 1 complex had a maximum at 294 nm in addition to one at the expected 368 nm. They attributed this peak to energy transfer from tryptophan to mant because it did not appear in the case of free nucleotide. This phenomenon was particularly useful for some of their measurements when experiments had to be carried out with a large excess of the fluorescent nucleotide relative to the protein, namely, because the excitation of free nucleotide is low at 290 nm the bound nucleotide can be preferentially excited and hence significantly increase the signal-to-background ratio relative to excitation at 366 nm. Leonard *et al.*[72] studied the

[68] J. M. Beechem, J. R. Knutson, and L. Brand, *Biochem. Soc. Trans.* **14,** 832 (1986).

[69] P. Nissen, M. Kjeldgaard, S. Thirup, G. Polekhina, L. Reshetnikova, B. F. C. Clark, and J. Nyborg, *Science* **270,** 1464 (1995).

[70] S. K. A. Woodward, J. F. Eccleston, and M. A. Geeves, *Biochemistry* **30,** 422 (1991).

[71] S. K. A. Woodward, Ph.D. thesis. King's College, London, 1990.

[72] D. A. Leonard, T. Evans, M. Hart, R. A. Cerione, and D. Manor, *Biochemistry* **33,** 12323 (1994).

binding of mantGDP to Cdc42Hs, a member of the rho superfamily of low molecular weight GTP-binding proteins. Their data suggest that the single tryptophan residue of Cdc42Hs (W97) can transfer energy to the mant moiety of mantGDP bound to the protein. They further estimate that the distance between the mant moiety and the tryptophan residue is 21 Å (we note that the critical distance for 50% energy transfer between a tryptophan residue and mant will depend on the extent of overlap between the tryptophan emission and the mant absorption, which will vary somewhat from case to case but should be in the range of 20 Å). Remmers *et al.*[73] showed that mantGTP and mantGTPγS displayed a magnesium-dependent increase in fluorescence on binding to bovine brain G_0 and that the increase was much greater on excitation at 280 nm than with 350-nm excitation, suggesting that one or both of the two tryptophans in the α subunit of G_0 can transfer energy to the bound fluorophore. Moore and Lohman[74] found a 20-fold enhancement of the fluorescence of mantATP, on excitation at 290 nm, on binding to the Rep helicase from *E. coli*—this large enhancement, compared to only a 20% increase in mantATP fluorescence on binding observed with 364-nm excitation, demonstrates that one or more of the tryptophan residues of the protein can efficiently transfer excitation energy to bound mant. This finding facilitated stopped-flow kinetics studies on the binding and displacement of mantATP to the Rep helicase. Giovane *et al.*[75] described the use of 2′(3′)-*O*-anthraniloyl-GDP (antGDP) to investigate EF-Tu. They found that antGDP binds to EF-Tu with an affinity even higher than unmodified GDP and their data qualitatively addressed the presence of energy transfer between the intrinsic protein fluorophores (tyrosine and tryptophan) and antGDP. G. Mocz and M. Helms (personal communication, 1996) found that mantATP bound to dynein from sea urchin sperm flagella and that the increase in fluorescence intensity on binding, although minimal on excitation at 360 nm, increased to severalfold on 280-nm excitation. S. Swenson and S. Seifried (personal communication, 1996) noted that mantATP fluorescence increased severalfold on binding to transcription termination factor rho and that excitation at either 360 or 280 nm resulted in similar increases. This result suggests that the single tryptophan residue in each rho subunit is not sufficiently proximate to the mant moiety to transfer excitation energy effectively, as has been observed for other systems.

[73] A. E. Remmers, R. Posner, and R. R. Neubig, *J. Biol. Chem.* **269,** 13771 (1994).
[74] K. J. M. Moore and T. M. Lohman, *Biochemistry* **33,** 14550 (1994).
[75] A. Giovane, C. Balestrieri, M. L. Balestrieri, and L. Servillo, *Eur. J. Biochem.* **227,** 428 (1995).

Microscopy

Several groups have reported on the use of fluorescent nucleotide analogs coupled with microscopy to greatly enhance the sensitivity of the observations. Sowerby *et al.*[16] linked fluorescein to ATP, using the ethylenediamine approach, to form 2′(3′)-*O*-[*N*-2-[3-[5-fluoresceinyl]thioureido-lethyl]carbamoyl]adenine 5′-triphosphate (FEDA-ATP), which they used with epifluorescence light microscopy to look at turnover rates on single myosin filaments from clam red adductor muscles containing several thousand myosin molecules each. Conibear and Bagshaw[29] synthesized the analogous tetramethylrhodamine probe, REDA-ATP, and, using selective observation with total internal reflection fluorescence microscopy, studied the kinetics of its release from isolated myosin filaments after flash photolysis of caged ATP. Funatsu *et al.*[52] were able to detect the association and dissociation of individual Cy3-ATP molecules with a single-headed myosin subfragment in solution using epifluorescence and total internal reflection microscopy.

Acknowledgments

We acknowledge support from National Science Foundation Grant MCB 9506845 (D.M.J.) and the Medical Research Council, U.K. (J.F.E.). We also thank Drs. John Corrie, Michael Helms, Arthur Johnson, Steve Seifried, Colin Davis, and David Trentham for helpful comments.

[19] Fluorescence Approaches to Study of Protein–Nucleic Acid Complexation

By JOHN J. HILL and CATHERINE A. ROYER

Introduction

The central role of specific protein–nucleic acid complexes in the control of growth, differentiation, and development is now well established. At the most fundamental level, cellular DNA is packaged through its interactions with histones and other chromosomal protein components. At the level of transcription, such complexes include those between transcriptional enhancers and repressors and their response elements, those of the general transcription machinery and transcription start sites, and those between the regulatory proteins and the general transcription factors. The interactions between tRNA and the aminoacyl-tRNA synthetases, as well as all of the important protein–RNA interactions involved in ribosomes, and the

0076-6879/97 $25

splicesosome protein–RNA complex, are examples of key protein–nucleic acid interactions involved in translation. Other areas involving protein–nucleic acid complexes include the synthesis and maintenance of telomeres, and the protein–DNA complexes involved in replication and recombination events. It is clear that a deeper understanding of these processes can reveal how malfunctions within particular protein–nucleic acid complexes lead to particular diseased states. Moreover, the elucidation of the structural, energetic, and dynamic characteristics of these complexes will continue to provide cellular targets for therapeutic agents. Given the central importance of protein–nucleic acid complexes, the development of reliable, sensitive, and versatile methodologies for their characterization will greatly enhance the pace at which new discoveries are made in this area.

Limitations of Present Methodologies

Although fluorescence-based methods for DNA sequencing have become widespread, the vast majority of studies involving the characterization of protein–nucleic acid complexes still rely on gel electrophoretic methods coupled with radioactive-based detection. The most common experimental approach, the gel mobility shift assay, while useful, presents a number of drawbacks, the most serious of which is that it is not an equilibrium assay. Complexes often dissociate during electrophoresis in a manner that depends on the pore size in the gel and the conditions of electrophoresis.[1] Moreover, proteins can interact with DNA in a variety of modes and stoichiometries that depend on solution conditions and concentration. Some of these complexes are more stable to electrophoresis than others, and accurate quantitation of the bands can be problematic owing to diffusion. Finally, a true understanding of the forces involved in complexation requires complete binding studies under varying conditions of temperature and salt concentration. This type of investigation is not possible in a gel mobility shift assay.

Other commonly used methodologies for studying protein–DNA interactions include enzymatic or chemical footprinting and nitrocellulose filter-binding assays. Footprinting can provide structural and thermodynamic information about the complexes, and as such is a useful methodology. However, changes in complex affinities and binding modes with temperature or salt concentration are convoluted with the effects of these variables on the enzymatic digestion or chemical modification. Moreover, quantitative footprinting is a rather labor-intensive, cumbersome experiment. Filter binding suffers from the fact that it is not easily applied to systems of moderate affinity and is often operative only under specific solution conditions.

[1] P. J. Czernik, D. S. Shin, and B. K. Hurlburt, *J. Biol. Chem.* **269,** 27869 (1994).

Advantages of Fluorescence-Based Approaches

Despite the overwhelming prevalence of electrophoretic/radioactive-based methodologies in the characterization of protein–nucleic acid complexes, a number of more biophysically oriented research groups have begun to apply fluorescence methods to these systems. The primary advantage of the fluorescence-based assays over gel-based and filter-binding methodologies is that the fluorescence-based assays are carried out in solution, without a need for separation and quantitation of bound and free species. Furthermore, they provide this true equilibrium approach to complexation studies over a wide range of solution conditions. The suitability of fluorescence-based methodologies lies in the fundamental physics of the phenomenon. First, the time scale of the emission of photons extends from picoseconds to hundreds of nanoseconds, depending on the fluorophore and on conditions. Over this time scale many dynamic events take place, including macromolecular rotational diffusion, diffusion of small molecules, solvent reorientation, energy transfer, and domain motion. By choosing the appropriate probes and strategies, fluorescence can provide information on macromolecular size, distances, ligation state, conformational rearrangements, and sample heterogeneity, all of which can be related to the structure, energetics, and dynamics of macromolecular complexes. Another powerful aspect of fluorescence approaches lies in their sensitivity and dynamic range, which extends eight orders of magnitude from millimolar down to near picomolar. This combination of sensitivity and versatility along with relatively low cost and ease of use makes fluorescence ideal for both detailed characterization of complex structure and dynamics, as well as more routine characterization of affinities. Both the broad dynamic range and the enormous sensitivity of fluorescence allow for the study of complexation phenomena under conditions approaching those found *in vivo*, as well as studies carried out under extreme conditions of buffer, temperature, and pressure. Moreover, with appropriate background subtraction, these assays can be used in clinical assays or for drug-screening purposes. Fluorescence approaches are not only ideal for the study of equilibrium phenomena, but are easily extended to kinetic analysis for experiments carried out in real time on a millisecond time scale.

Fluorescence Approaches

Fluorescence Anisotropy

One of the most useful fluorescence observables for monitoring complexation events is fluorescence anisotropy. This measurement makes use

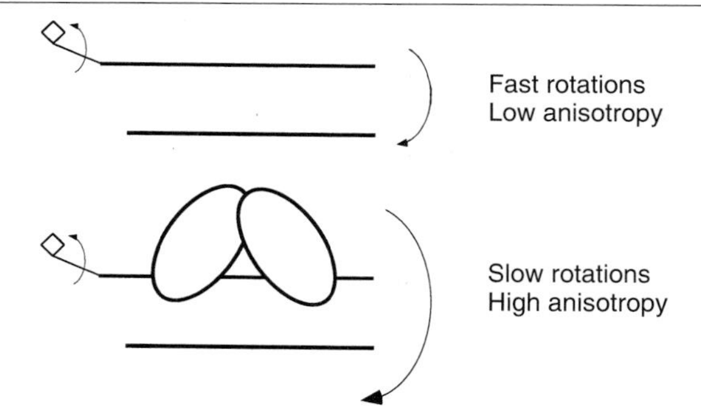

FIG. 1. Schematic representation of the effect of complexation on the rotational diffusion and hence anisotropy of a fluorescently labeled oligonucleotide.

of polarized exciting light and polarized detection to monitor rotational diffusion events (Fig. 1). Large molecular complexes tumble slowly relative to the life time of a fluorophore, depolarize the light only slightly, and exhibit a relatively high anisotropy value. Small complexes tumble rapidly, depolarize the light more readily, and exhibit a low anisotropy value. This relationship of size to anisotropy value can provide information about the number of complexes of different stoichiometries, as well as the extent to which complexation has occurred. The first to apply this approach to measure a binding constant using a fluorescently labeled oligonucleotide and titrating with a DNA-binding protein were Heyduk and Lee.[2] Although the sensitivity of their assay did not allow for determination of highly specific interactions, the general utility of this technique was established. Equilibrium titrations at subnanomolar concentrations[3] have demonstrated that this technique could replace most radioactivity-based assays for measuring subnanomolar to 100 nM affinities and for detecting specific proteins in cell extracts. Anisotropy of labeled proteins[4] and labeled oligonucleotides[2,3,5–10] has been used to monitor complexation between histones and

[2] T. Heyduk and J. C. Lee, *Proc. Natl. Acad. Sci. U.S.A.* **87,** 1744 (1990).

[3] V. LeTilly and C. A. Royer, *Biochemistry* **32,** 7753 (1993).

[4] C. A. Royer, T. Ropp, and S. F. Scarlata, *Biophys. Chem.* **43,** 197 (1992).

[5] C. R. Guest, R. A. Hochstrasser, C. G. Dupuy, D. J. Allen, S. J. Benkovic, and D. P. Millar, *Biochemistry* **30,** 8759 (1991).

[6] T. E. Carver, Jr., R. A. Hochstrasser, and D. P. Millar, *Proc. Natl. Acad. Sci. U.S.A.* **91,** 10670 (1994).

[7] R. P. Kwok, J. R. Lundblad, J. C. Chrivia, J. P. Richards, H. P. Bachinger, R. G. Brennan, S. G. Roberts, M. R. Green, and R. H. Goodman, *Nature (London)* **370,** 223 (1994).

[8] G. M. Perez-Howard, P. A. Weil, and J. M. Beechem, *Biochemistry* **34,** 8005 (1995).

[9] X. Cheng, L. Kovac, and J. C. Lee, *Biochemistry* **34,** 10816 (1995).

[10] M. Bailey, P. Hagmar, D. P. Millar, B. E. Davidson, G. Tong, J. Haralambidis, and W. H. Sawyer, *Biochemistry* **34,** 15802 (1995).

DNA and between specific transcription factors and their target sites on DNA.

Fluorescence Resonance Energy Transfer

Traditional structural biological approaches [X-ray and nuclear magnetic resonance (NMR)] have proven extremely powerful in the field of protein–nucleic acid complexes; a large number of important high-resolution three-dimensional structures have been solved, one exciting example of which is the ternary transcriptional activation complex between the TATA-binding protein, TFIIB, and the TATA box DNA.[11] However, the larger the complex (e.g., eukaryotic transcriptional regulatory complexes contain dozens of polypeptides), the less likely it is that information will be forthcoming from these traditional approaches. Moreover, despite the increasing efficiency of structure determinations, it is still a lengthy process, such that we cannot expect that all of the accessible protein–nucleic acid structures of interest can be solved. Fluorescence resonance energy transfer (FRET) allows for distance mapping between dye molecules at specific sites in macromolecular complexes and can provide three-dimensional structural information where otherwise none is available. This technique has already been applied to a number of nucleic acid structures including the hammerhead ribozyme,[12] DNA and RNA duplexes,[13–16] triplexes,[17–19] and four-way junctions.[20–22] It has also been used to distance map protein–nucleic acid complexes such as the CRP/*lac* promoter bent DNA complex,[23] the aminoacyl-tRNA/EfTu/GTP ternary complex,[24] tRNA/ribosome interac-

[11] D. B. Nikolov, H. Chen, E. D. Halay, A. A. Usheva, K. Hisatake, D. K. Lee, R. G. Roeder, and S. K. Burley, *Nature* (*London*) **377,** 119 (1995).

[12] T. Tuschl, C. Gohlke, T. M. Jovin, E. Westhof, and F. Eckstein, *Science* **266,** 785 (1994).

[13] L. E. Morrison and L. M. Stols, *Biochemistry* **32,** 3095 (1993).

[14] C. Gholke, A. I. Murchie, D. M. Lilley, and R. M. Clegg, *Proc. Natl. Acad. Sci. U.S.A.* **91,** 11660 (1994).

[15] J. Y. Ju, I. Kheterpal, J. R. Scherer, C. C. Ruan, C. W. Fuller, A. N. Glazer, and R. A. Mathies, *Anal. Biochem.* **231,** 131 (1995).

[16] B. P. Maliwal, J. Kusba, and J. R. Lakowicz, *Biopolymers* **35,** 245 (1995).

[17] J. L. Mergny, T. Garestier, M. Rougee, A. V. Lebedev, M. Chassignol, N. T. Thuong, and C. Helene, *Biochemistry* **33,** 15321 (1994).

[18] M. Yang, S. S. Ghosh, and D. P. Millar, *Biochemistry* **33,** 15329 (1994).

[19] P. V. Scaria, S. Will, C. Levenson, and R. H. Shafer, *J. Biol. Chem.* **270,** 7295 (1995).

[20] R. M. Clegg, A. I. Murchie, A. Zechel, C. Carlberg, S. Diekmann, and D. M. Lilley, *Biochemistry* **31,** 4846 (1992).

[21] P. S. Eis and D. P. Millar, *Biochemistry* **32,** 13852 (1993).

[22] R. M. Clegg, A. I. Murchie, and D. M. Lilley, *Biophys. J.* **66,** 99 (1994).

[23] T. Heyduk and J. C. Lee, *Biochemistry* **31,** 5165 (1992).

[24] B. S. Watson, T. L. Hazlett, J. F. Eccleston, C. Davis, D. M. Jameson, and A. E. Johnson, *Biochemistry* **34,** 7904 (1995).

tions,[25] the DNA polymerase III/primer complex,[26] the histone core particle,[27] and the *Eco*RV/DNA and complex.[28] FRET has also been used in kinetic studies of enzymes that function on nucleic acids.[29–31] In this volume Millar and co-workers[31a] cover FRET in greater detail; therefore, we do not discuss it further here.

Fluorescence Intensity

Fluorescence intensity changes can be used in the characterization of protein–nucleic acid complexes for their equilibrium, kinetic, and dynamic characteristics. A large number of studies have relied on changes in the intrinsic fluorescence of proteins on complexation with nucleic acids.[8,32–52]

[25] T. R. Easterwood, F. Major, A. Malhorta, and S. C. Harvey, *Nucleic Acids Res.* **22,** 3779 (1994).

[26] M. A. Griep and C. S. McHenry, *J. Biol. Chem.* **267,** 3052 (1992).

[27] J. Widom, In preparation (1996).

[28] S. G. Erskine and S. E. Halford, *Gene* **157,** 153 (1995).

[29] S. S. Ghosh, P. Eis, K. Blumeyer, K. Fearon, and D. P. Millar, *Nucleic Acids Res.* **22,** 3155 (1994).

[30] K. P. Bjornson, M. Amaratunga, and T. M. Lohman, *Biochemistry* **33,** 14306 (1994).

[31] S. P. Lee, H. G. Kim, M. L. Censullo, and M. K. Han, *Biochemistry* **34,** 10205 (1995).

[31a] M. Yang and D. P. Millar, *Methods Enzymol.* **278,** [20], 1997 (this volume).

[32] J. R. Casas-Finet, A. Kumar, G. Morris, S. H. Wilson, and R. L. Karpel, *J. Biol. Chem.* **266,** 19618 (1991).

[33] L. B. Overman, W. Bujalowski, and T. M. Lohman, *Biochemistry* **27,** 456 (1988).

[34] W. Bujalowski, L. B. Overman, and T. M. Lohman, *J. Biol. Chem.* **263,** 4629 (1988).

[35] W. T. Ruyechan and J. W. Olson, *J. Virol.* **66,** 6273 (1992).

[36] A. P. Stassen, B. J. Harmsen, J. G. Schoenmakers, C. W. Hilbers, and R. N. Konings, *Eur. J. Biochem.* **206,** 605 (1992).

[37] J. B. A. Ross, D. F. Senear, E. Waxman, B. B. Kombo, E. Rusinova, Y. T. Huang, W. R. Laws, and C. A. Hasselbacher, *Proc. Natl. Acad. Sci. U.S.A.* **89,** 12023 (1992).

[38] Y. T. Kim, S. Tabor, J. E. Churchich, and C. C. Richardson, *J. Biol. Chem.* **267,** 15032 (1992).

[39] K. A. Maegley, L. Gonzalez, Jr., D. W. Smith, and N. O. Reich, *J. Biol. Chem.* **267,** 18527 (1992).

[40] J. R. Casas-Finet and R. L. Karpel, *Biochemistry* **32,** 9735 (1993).

[41] M. Crowe and M. G. Fried, *Eur. J. Biochem.* **212,** 539 (1993).

[42] W. Bujalowski and T. M. Lohman, *J. Mol. Biol.* **217,** 63 (1991).

[43] C. A. Gelfand, Q. Wang, S. Randall, and J. E. Jentoft, *J. Biol. Chem.* **268,** 18450 (1993).

[44] C. L. Chan, Z. Wu, T. Ciardelli, A. Eastman, and E. Bresnick, *Arch. Biochem. Biophys.* **300,** 193 (1993).

[45] C. Kim, B. F. Paulus, and M. S. Wold, *Biochemistry* **33,** 14197 (1994).

[46] J. B. Clendenning and J. M. Schurr, *Biophys. Chem.* **52,** 227 (1994).

[47] M. L. Carpenter and G. G. Kneale, *Methods Mol. Biol.* **30,** 313 (1994).

[48] J. R. Wisniewski and E. Schulze, *J. Biol. Chem.* **269,** 10713 (1994).

[49] M. S. Soengas, J. A. Esteban, M. Salas, and C. Gutierrez, *J. Mol. Biol.* **239,** 213 (1994).

[50] P. Thommes, C. L. Farr, R. F. Marton, L. S. Kaguni, and S. Cotterill, *J. Biol. Chem.* **270,** 21137 (1995).

The fluorescence-modified bases such as 2-aminopurine and ethenodeoxyadenosine[53–59] and of extrinsic probes[8,60,61] have also proven to be quite useful and sensitive to protein–nucleic acid complexation events. We contain this discussion to the observation of intensity changes of extrinsically labeled oligonucleotides, because their high quantum yields provide good sensitivity.

In this chapter we present an overview of the fluorescence characteristics of a number of fluorescein-labeled oligonucleotides containing target sequences for site-specific DNA-binding proteins that have been characterized in our laboratory. We discuss how the fluorescence observables of anisotropy and intensity, both steady state and time resolved, change on complexation with protein and changing solution conditions. While others[62] have examined the properties of various probes, in our laboratory at the University of Wisconsin–Madison we have mainly worked with fluorescein-derivatized oligonucleotides and have examined systematically the effects of linker chemistry and DNA structure and sequence on the fluorescence characteristics.

Labeling Strategies

Commercially available strategies for labeling of oligonucleotides are quite numerous and present some advantages for the various assays discussed as follows. Most of the studies we discuss here were carried out using a 5′-labeling strategy in which a fluorescein-labeled phosphoramidite is coupled to the last nucleoside of the oligonucleotide, using automated

[51] T. Lundback, C. Cairns, J.-A. Gustafsson, J. Carlstedt-Duke, and T. Hard, *Biochemistry* **32,** 5074 (1993).
[52] T. Lundback, J. Zilliacus, J.-A. Gustafsson, J. Carlstedt-Duke, and T. Hard, *Biochemistry* **33,** 5955 (1994).
[53] M. Chabbert, H. Lami, and M. Takahashi, *J. Biol. Chem.* **266,** 5395 (1991).
[54] P. Wittung, B. Norden, and M. Takahashi, *Eur. J. Biochem.* **224,** 39 (1994).
[55] D. P. Giedroc, R. Khan, and K. Barnhart, *Biochemistry* **30,** 8230 (1991).
[56] P. Wittung, B. Norden, S. K. Kim, and M. Takahashi, *J. Biol. Chem.* **269,** 5799 (1994).
[57] L. B. Bloom, M. R. Otto, R. Eritja, L. J. Reha-Krantz, M. F. Goodman, and J. M. Beechem, *Biochemistry* **33,** 7576 (1994).
[58] K. D. Raney, L. C. Sowers, D. P. Millar, and S. J. Benkovic, *Proc. Natl. Acad. Sci. U.S.A.* **91,** 6644 (1994).
[59] P. Wittung, M. Funk, B. Jernstrom, B. Norden, and M. Takahashi, *FEBS Lett.* **368,** 64 (1995).
[60] P. Hagmar, B. Norden, D. Baty, M. Chartier, and M. Takahashi, *J. Mol. Biol.* **226,** 1193 (1992).
[61] J. J. Hill, J. Lefstin, J. Miner, K. R. Yamamoto, and C. A. Royer, Submitted (1996).
[62] D. P. Millar, R. M. Hoschstrasser, C. R. Guest, and S.-M. Chen, *In* "SPIE Proceedings 1640, Time-Resolved Spectroscopy in Biochemistry III" (J. R. Lakowicz, ed.), pp. 592–598. SPIE, Bellingham, Washington, 1992.

synthesis. Our work using these 5′-fluorescein phosphoramidites involved coupling of the fluorescein through a rather long six-carbon tether (see Fig. 2), although shorter linkers are also now available. One can also employ 5′-phosphoramidites (Glen Research, Sterling, VA), which couple a free amine or thiol group to the 5′ end of an oligonucleotide through variable-length tethers (Fig. 2). These amino and thiol modifiers can be labeled with amine- or thiol-reactive dyes, allowing one to test the performance of different fluorophores, such as rhodamine-X.[8] Another strategy that we have used and discuss here involves 5′ labeling of oligonucleotides with a more restricted linker by incorporating a reactive thiol group on a 5′-OH using ATP γS and T4 ligase. As previously, the free thiol is then reacted with a thiol-reactive dye. Kits for labeling with fluorescein are available commercially (Promega, Madison, WI). We have used the Promega kit (FluoroAmp) to label some of our target oligonucleotides with fluorescein maleimide (Fig. 2). Oligonucleotides bearing fluorophores internal to the sequence and directed toward the major groove can be generated through the use of modified bases. For example, a fluorescent group tethered to the 5-position of deoxyuridine, through a variety of linker chemistries, will be situated in the major groove. Millar and co-workers have used a thiol-modified deoxyuridine to label oligonucleotides with a naphthalene derivative in fluorescence footprinting experiments.[5,60] a modifiable thymidine, Amino-Modifier C2dT, is available for Glen Research, which has a reactive amine coupled through an eight-atom linker arm to the 5-position of thymine (Fig. 2).

Labeled Oligonucleotide Purification

We have found, not surprisingly, that the purity of the labeled oligonucleotides is key to the success of the fluorescence-based binding assay. This is true for a number of reasons. First, the sensitivity of the assay will be greatest when 100% of the target oligonucleotide molecules are fluorescently labeled. This labeling ratio is akin to the specific activity of a target oligonucleotide in a radioactivity-based assay. Second, oligonucleotide targets that are either labeled or unlabeled and that contain partial sites ($n - 1$, $n - 2$, etc.) will compete more or less effectively for binding. There is a larger proportion of these partially synthesized molecules that are fluorescently labeled in the unpurified mixture of a synthesis based on a 3′-fluorophore. In the few instances in which we have tried to work with 3′-labeled oligonucleotides, the results have been less than ideal. For this reason, we have limited our work to 5′-labeled oligonucleotides, as only completed molecules will fluoresce. Finally, a large amount of contaminating free fluorophore that can result when labeling is carried out postsyn-

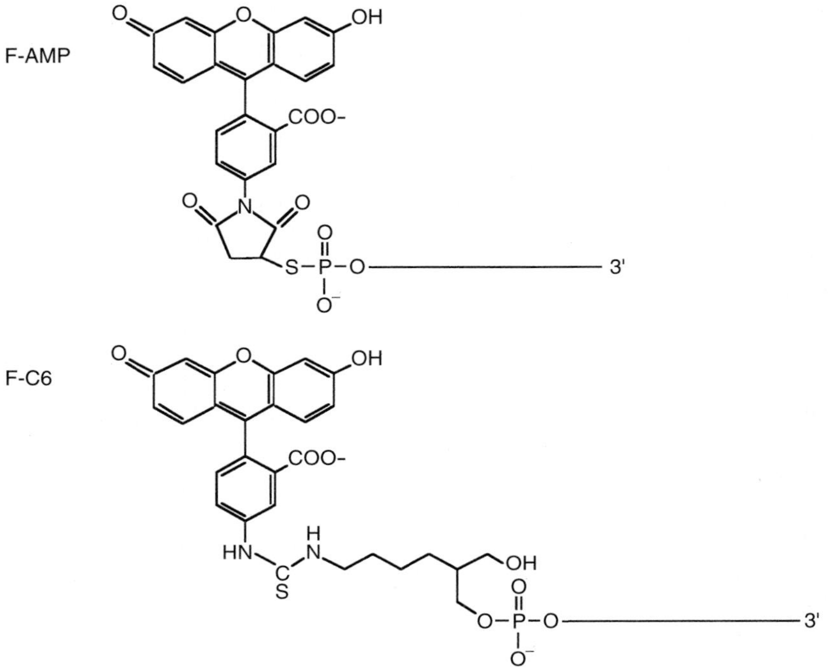

F-AMP

F-C6

Fig. 2. Labeling strategies for obtaining fluorescent oligonucleotides. The F-C6-5′-fluorescein phosphoramidite, the 5′-amino, 5′-thio, and amino C2dT modifiers are available from Glen Research (Sterling, VA). The FluoroAmp labeling kit is available from Promega (Madison, WI).

thesis will contribute large amounts of low anisotropy to the overall signal and will thus severely limit the sensitivity of the assay.

The purification procedure that one chooses, of course, depends on the labeling strategy used in the first place. When using a fluorescently labeled phosphoramidite in automated synthesis, we have explored two procedures, both of which are quite effective. If one uses a fluorescein-labeled phosphoramidite, the full-length fluorescently labeled oligonucleotides can be separated from the abortive, unlabeled synthesis products using anti-fluorescein antibodies (AFABs) (Panvera Corp., Madison, WI) coupled to an agarose gel matrix (AminoLink Plus immobilization kit; Pierce, Rockford, IL). Details for this procedure are provided in Appendix I. Another approach we have found to be efficient is to use the Oligonucleotide Purification Cartridge (Cruachem, Dulles, VA). These cartridges are designed for rapid purification of 5′-trityl on synthetic oligonucleotides. The fluorescein

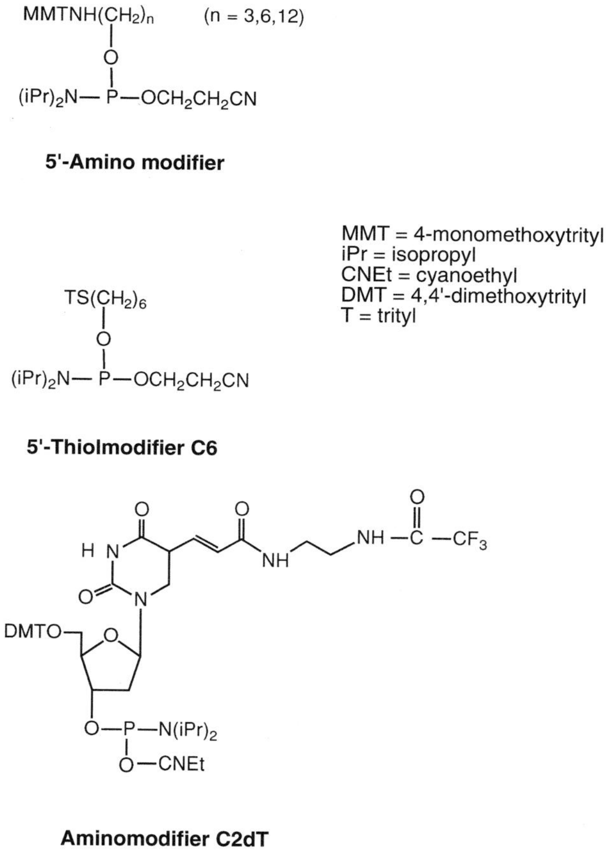

MMTNH(CH$_2$)$_n$ (n = 3,6,12)

O

(iPr)$_2$N— P—OCH$_2$CH$_2$CN

5'-Amino modifier

MMT = 4-monomethoxytrityl
iPr = isopropyl
CNEt = cyanoethyl
DMT = 4,4'-dimethoxytrityl
T = trityl

TS(CH$_2$)$_6$

O

(iPr)$_2$N— P—OCH$_2$CH$_2$CN

5'-Thiolmodifier C6

O O O

H N NH NH —C —CF$_3$

O N

DMTO— O

O—P—N(iPr)$_2$

O—CNEt

Aminomodifier C2dT

FIG. 2. (*continued*)

phosphoramidite, containing the trityl group, interacts with the column matrix and the intact, fluorescinated oligonucleotide will elute under normal trityl off elution conditions. Using nearest neighbor estimates of extinction coefficients, both methods yield labeling ratios near unity for the purified oligonucleotide.

If one chooses to add a label to an oligonucleotide containing a reactive amine or thiol group two purification problems must be solved. The first is the purification of the oligonucleotide itself. Any method typically used for oligonucleotide purification can therefore be used, and these include high-performance liquid chromatography (HPLC), gel electrophoresis, or oligonucleotide purification cartridges. Second, one must separate the labeled oligonucleotide from the excess free dye after the coupling reaction has been carried out. We have found that, contrary to the instructions provided with the Promega FluoroAmp kit, at least a 20-cm (≈20- to 25-ml volume) Sephadex G-10 (Pharmacia, Piscataway, NJ) superfine desalting column is required to remove all excess dye. Details of this purification procedure can be found in Appendix II. Thin-layer chromatography (TLC) characterization of the final product should be performed to ensure free dye is absent.

Theoretical Bases for Fluorescence Assays

Anisotropy Changes

On complexation the rotational diffusion properties of the labeled oligonucleotide change dramatically and these changes can be assessed by measurement of the anisotropy of fluorescence emission. Anisotropy is defined as the difference between the parallel and perpendicular components of the emitted light ($I_{\parallel}$ and $I_{\perp}$), when parallel excitation is used, with respect to the total intensity of the sample:

$$A = I_{\parallel} - I_{\perp}/I_{\parallel} + 2I_{\perp} \tag{1}$$

We note that one can use an alternate quantity, the fluorescence polarization P, which is defined as:

$$P = I_{\parallel} - I_{\perp}/I_{\parallel} + I_{\perp} \tag{2}$$

and is related to the anisotropy

$$A = 2/3[1/P - 1/3]^{1} \tag{3}$$

It is preferable to employ anisotropy rather than polarization because the former is normalized to the total intensity. It is therefore a sum of the anisotropies of the individual species present in solution weighted for their fractional populations and if necessary relative quantum yields.

The measured anisotropy is related to the rotational correlation time τ_c and the fluorophore lifetime τ through the Perrin relation:

$$A_0/A - 1 = \tau/\tau_c \tag{4}$$

where A_0 is the limiting anisotropy, a known photophysical constant. The correlation time is related to the hydrated volume (V_h) of the rotating molecule as follows:

$$\tau_c = \eta V_h/kT \tag{5}$$

where k is the Boltzmann constant, T is the Kelvin temperature, and η is the viscosity. In cases where the fluorophore is covalently bound to a macromolecule, the loss in anisotropy during the excited-state lifetime will reflect both the tumbling of the macromolecule and the local motions of the probe. The relative contributions of these two phenomena to the observable steady state anisotropy must be assessed using time-resolved techniques.

Desired Fluorophore Attributes

In the anisotropy-based assays for monitoring protein–nucleic acid interactions the fluorophore is typically coupled to the nucleic acid, rather than to the protein(s). This allows for the titration of a single target oligonucleotide with the many proteins of interest without synthesis of multiple labeled molecules. Moreover, in many cases pure protein is precious or unavailable, whereas the synthesis and purification of labeled oligonucleotides are relatively straightforward. The attributes of the fluorophore desired for such assays include a high quantum yield and little or no sensitivity of the quantum yield to complexation. As is discussed in this section, this has been established through trial and error. A high quantum yield allows for carrying out equilibrium binding assays using subnanomolar quantities of labeled oligonucleotide. These low concentrations are necessary for the determination of dissociation constants in high-affinity complexes. In addition, it is desirable that the anisotropy signal not be complicated by changes in fluorescence lifetime [see Eq. (2)]. Moreover, the labeling chemistry should be designed to limit the local mobility of the probe such that the majority of fluorophores report on the global tumbling of the oligonucleotide. The larger the fraction of signal that is due to global macromolecular tumbling, the larger the observed anisotropy change will be on an increase in macromolecular size, thus increasing the overall sensitivity of the technique. Trial and error has again led us to the appreciation of the importance of linker chemistries in determining the sensitivity of the anisotropy-based assays.

Intensity Changes

As is discussed in this section, intensity changes often accompany or even obliterate anisotropy changes on oligonucleotide complexation with protein. The quantum yield ϕ of a fluorophore is defined as the ratio of photons emitted to photons absorbed:

$$\phi = \Gamma/(\Gamma + k) \tag{6}$$

where Γ is the rate of photon emission and k is the sum of all of the rates for competing nonradiative processes for return to the ground state. The measured fluorescence lifetime τ of the fluorophore is the ratio of the quantum yield to the radiative rate or

$$\tau = \Gamma/\phi = 1/(\Gamma + k) \tag{7}$$

The processes competing with emission include excited-state electron transfer, intersystem crossing, internal conversion, and external conversion. All of these processes contribute to k and result in a decrease in the fluorescence lifetime, and thus a decrease in the observed intensity of fluorescence. Regardless of the exact process, quenching by any of these mechanisms is termed *dynamic*, as it occurs by encounters of quenching moieties (solvent, external quenching molecules, or intramolecular functional groups) with the fluorophore in the excited state.

Loss of intensity can also result from a ground-state phenomenon. In one case formation of a dark complex that absorbs, but does not emit, light results in static quenching. Alternately, the ground-state complexation can result in a decrease in extinction coefficient. Either phenomenon is characterized by a loss of intensity with no change in the fluorescence lifetime. In the cases we have studied so far, and present in this section, the loss of intensity on complexation arises in large part from static quenching or modification of the extinction coefficient or both. The concentrations at which the quenching occurs are too low for accurate determination of changes in extinction coefficient. In most cases the sites for protein binding are much too distant from the site of the labeling for direct interaction of the protein to be responsible for the quenching process. We have thus concluded that the loss of intensity results from a long-range effect of the protein on the DNA.

Desired Fluorophore Attributes

The most obvious fluorophore attribute for monitoring protein–nucleic acid complexation by decreases (or increases) in intensity is that its quantum yield be sensitive to its environment. The probe should be intimately coupled to the DNA, such that the effects of protein binding are efficiently transferred to the probe. The probe should have a high quantum yield, because in the case of quenching the signal will decrease as the titration progresses. The emission energy should be well removed from any contaminating fluorescence as well, because as the desired signal is quenched any background fluorescence will become increasingly troublesome. In general, the farther to the red the fluorophore emits the better. This is especially true for fluorophores in turbid solutions, because the intensity of scattered light depends on the wavelength (λ) as $1/\lambda^4$. As we discuss, the intensity of fluorescein is highly sensitive to complexation events (not surprisingly because it is a pH probe), and we have used it as a probe of protein–DNA interactions. Similarly, rhodamine-X also exhibits intensity changes on protein binding at distances far removed from the binding site (J. Beechem, personal communication, 1996).

Instrumentation

The key to making these high-sensitivity fluorescence measurements is to use a bright excitation source and an optical module with high throughput. In our laboratory we use a photon-counting ISS Koala (ISS, Champaign, IL) with no excitation or emission monochromator, and a high-pass filter (530 cuton; Corning, Corning, NY) in emission. Excitation is provided by a 3-mW argon-ion laser from Omnichrome (Chino, CA), coupled to the excitation optics of the Koala via a multifiber optical fiber bundle from Oriel (Stratford, CT). The optical fiber provides excitation of both horizontal and vertical polarization owing to the fiber-induced polarization scrambling, allowing accurate G-factor[62a] determination. It also tends to defocus the light such that photobleaching effects are avoided. Typically the laser is used at ≤1.5-mW intensity levels at the 488-nm line for fluorescein. A number of lamp-based instruments are commercially available, including those for fluorescein (Jolley Instruments, Round Lake, IL) and multiple dye excitation (Panvera Corp., Madison, WI), with sensitivity between 50 and 100 p*M*. If the protein solutions are turbid, scatter effects can be corrected for by subtracting the fluorescence of a solution of buffer at the same protein concentration as the solution containing the fluorescent oligonucleotide. This is a particularly important control because scatter will cause an increase in both intensity and anisotropy.

Overview of Results

In this section we present our observations of the fluorescence characteristics of the fluorescein-labeled oligonucleotides that we have investigated to date, and how these characteristics change on binding of protein under various solution conditions. We do not address the thermodynamic basis for binding, which will be the focus of other published work, but simply highlight the consequences of complexation on the observed fluorescence parameters.

[62a] The G-factor is a measure of instrumental response to parallel and perpendicularly polarized emission, and requires horizontal excitation for determination.

GRE$_{s3}$ 5'-F-CTCTCGCCAGAACATCATGTTCTGCGTCGGCCC C
3'-AGAGCGGTCTTGTAGTACAAGACGCAGCCGCC

GRE$_{s4}$ 5'-F-CTCTCGCCAGAACATCGATGTTCTGCGTCGGCCC C
3'-AGAGCGGTCTTGTAGCTACAAGACGCAGCCGCC

plfG 5'-F-ATTCTTGCAGGGCTACTCACAGTATGATTTGTTTTTCTAGAGCGGCCGC-3'
3'-TAAGAACGTCCCGATGAGTGTCATACTAAACAAAAAGATCTCGCCGGCG-F-5'

plf 5'-F-GCGGGCTACTCACAGTATGATTTGTTTT GGGCG-3'
3'-CGCCCGATGAGTGTCATACTAAACAAAA CCGGC-5'

24-HydroxylaseVDRE (DRE I) 5'-F-AGAGAGCGCACCCGCTGAACCCTGGGCCC C
3'-TCTCTCGCGTGGGCGACTTGGGACCCGCC

Ratosteocalcin VDRE 5'-F-GAGCGCGGGTGAATGAGGACAGTAAGTCC C
3'-CTCGCGCCCACTTACTCCTGTCATTCACC

Vitellogenin ERE 5'-F-AGCTTCGAGGAGGTCACAGTGACCTGGAGCGGATC-3'
3'-TCGAAGCTCCTCCAGTGTCACTGGACCTCGCCTAG-5'

TR20 5'-F-CGAACTAGTTAACTAGTACCG T
3'-GCTTGATCAATTGATCATGGA A

TR25 5'-F-TTGCGTACTAGTTAACTAGTTCGAT-3'
3'-AACGCATGATCAATTGATCAAGCTA-5'

"Nonspecific" 5'-F-GAGCGCTCTGTGATGTCTGTGGTAAGTCC C
3'-CTCGCGAGACACTACAGACACCATTCACC

FIG. 3. Sequences of the target oligonucleotides used in our studies. All are 5'-fluorescein labeled, but the linker chemistries (either F-C6 or F-Amp) are not specified here.

Labeled Oligonucleotides

In Fig. 3 are shown the sequences for the various fluorescein-labeled oligonucleotides used to date in this laboratory. They contain target sequences for *trp* repressor, called *operators* (TRops); and glucocorticoid receptor (GRE), estrogen receptor (ERE), and vitamin D receptor (DRE), termed *response elements* (REs). Some were synthesized as single-stranded oligonucleotides that would form base-pairing duplex hairpin structures containing the appropriate double-stranded target sites. This is the case for the TRop 20-mer sequence, the GREs3 and GREs4 targets, the DREs, and the random target. The composite GRE sequences, plf and plfG, as well as the TRop 25-mer, were double-stranded oligonucleotides. In the case of plfG, labeling was carried out on the 5′ end of both the sense and the antisense strands, and double-stranded oligonucleotides were used with only the sense or only the antisense strand labeled and an unlabeled complementary strand. Most of these targets were labeled via automated synthesis with the C_6-fluorescein phosphoramidite linker chemistry (Fig. 2). To date, the F-AMP labeling has been characterized only on the GREs3, plfG, and ERE sequences, which contain specific targets for the glucocorticoid receptor, the AP1 transcription factor, and the estrogen receptor.

Steady-State Fluorescence

In general, the steady-state fluorescence anisotropy of the fluorescein-labeled oligonucleotides is highly dependent on the linker chemistry, and on the identity of the 5′-nucleoside to which the probe is attached. We suspect that the quantum yield and extinction coefficients are also strong functions of the base identity and linker chemistry, as well, although the concentrations at which we work do not allow for accurate determination of these parameters. Quantum yield and extinction coefficient effects have been observed on protein complexation with oligonucleotides labeled with both tetramethylrhodamine and fluorescein.[62] In Table I are given the steady-state anisotropy (A) values and normalized intensity (I) changes for our labeled oligonucleotides in the absence and in the presence of specific or nonspecifically binding proteins under various conditions of salt concentration. Under conditions of low salt, these proteins bind indiscriminately to any DNA target, whereas at 100 mM NaCl, all bind selectively to their target sites.

The fluorescence observations made on binding of the DNA-binding domain of the glucocorticoid receptor (GRDBD) were obtained using two single-stranded hairpin sequences carrying two glucocorticoid response element (GRE) half-sites. The GREs3 (Fig. 3) contains these two half-sites separated by a 3-base pair (bp) spacer, while the GREs4 contains the half-sites separated by a 4-bp spacer. The GREs3 and GREs4 sequences were synthesized with an overhanging deoxycytidine nucleoside and then the 5′ six-carbon (C_6)-linked fluorescein referred to here as F-C6 and shown in Fig. 2. The GREs3 sequence was also synthesized with a 5′-thiol and labeled with fluorescein maleimide, referred to here as F-Amp and shown in Fig.

TABLE I
STEADY-STATE FLUORESCENCE CHARACTERISTICS OF LABELED OLIGONUCLEOTIDES FREE AND COMPLEXED WITH PROTEIN

Target DNA	Protein	[NaCl] (m*M*)	Anisotropy (*A*)		Δ*I* (%)
			Free	Bound	
F-C6-GREs3	GRDBD	10	0.042	0.120	−10
F-Amp-GREs3	GRDBD	100	0.065	0.100	−50
F-C6-GREs3	GRDBD	100	0.040	0.065	−25
F-C6-GREs4	GRDBD	10	0.060	0.120	−10
	GRDBD	100	0.054	0.068	−25
F-C6-plfG (sense)	GRDBD	10	0.110	0.180	−10
	GRDBD	100	0.09	0.110	−50
F-C6-plfG (antisense)	GRDBD	100	0.042	0.079	−55
F-C6-plf	GRDBD	100	0.06	0.07	−60
F-C6-plfG (sense)	c-Jun	100	0.07	0.115	0
F-Amp-plfG	c-Jun	100	0.130	0.265	−10
	GRDBD	100	0.150	0.190	−50
F-C6-TR25	TR	10	0.07	0.113	−10
F-C6-TR20	TR	100	0.067	0.067	−33
F-C6-rose	VDR	10	0.076	0.118	0
	VDR	100	0.078	0.090	−10
	RXR	10	0.076	0.185	0
	RXR	100	0.078	0.100	0
F-C6-DREI	RXR	100	0.060	0.074	0
F-C6-random	RXR	100	0.075	0.090	0
F-Amp-vitERE	ER	100	0.68	0.097	+30

2 as well. The values reported in Table I show that the initial anisotropy of the GREs3 and GREs4 sequences with the F-C6 labeling scheme are significantly different, 0.042 and 0.060. The anisotropy of both labeled sequences increases as a function of protein concentration. Although the total change in the anisotropy on binding of protein under conditions of specific binding (100 m*M* salt) is quite small on both targets, a 25% decrease in total intensity also is observed under specific binding conditions. Conversely, at low salt, where binding is nonspecific, no significant change in intensity is observed, and the total change in anisotropy is much larger, indicating complexation of higher order than the dimer that binds at high salt.

Comparing the results obtained with the C_6 and F-AMP linker chemistries for the GREs3 target, we find that the initial anisotropy is slightly higher for the F-Amp linker and the total change in anisotropy on complexation is also larger (Table I). Moreover, the total change in intensity on specific complexation is also larger, 50% as compared to 25% with the C_6 linker. We assume that the tighter coupling of the probe to the DNA enhances any effects of protein binding on the emission (and absorption) properties of the fluorescein probe.

We have also studied the binding of the GRDBD to a series of double-stranded oligonucleotides derived from the upstream control region of the human proliferin gene. These so-called composite sequences are 49-bp (plfG) or 33-bp (plf) double-stranded targets containing specific sites for both GR (underlined) and AP1 (boxed) transcription factors (Fig. 3). We have studied the plfG sequence with the fluorescein label with a C_6 linker at the 5′ end of either the sense (F-C6-plfGs) or the antisense strand (F-C6-plfGs) and the F-AMP linker on the 5′ end of the sense strand (F-Amp-plfGs). As can be seen in Fig. 3, the fluorescein is coupled to a deoxyadenosine in the plfG sense target and to a deoxyguanosine in the antisense target. The plf sequence was labeled only on the 5′-deoxyguanosine of the sense strand with the C_6 linker (F-C6-plf).

The results obtained with the composite GREs are given in Table I. The uncomplexed F-C6-plfG sense target exhibits initial anisotropies that are significantly higher than those of its antisense counterpart. Small effects of salt are also seen in the initial anisotropy values of the free oligonucleotides. At low salt, under nonspecific binding conditions, binding of GRDBD results in a large change in anisotropy and a minimal decrease in intensity on the F-C6-plfGs target. At high salt the changes in anisotropy are significantly more modest, while a 50% quenching of the intensity is observed on complexation. The F-Amp linkage on the plfGs target results in a significantly larger initial anisotropy than the F-C6 linker (0.15 vs. 0.09), and a significantly larger increase on complexation with either the AP1 factor (c-Jun) or GRDBD. Thus, the shorter, more constrained linker chemistry of the maleimide-labeling scheme results in larger signal changes on complexation with protein. Binding of ER to the F-Amp-labeled vitellogenin ERE (Fig. 3) resulted in a large increase in both anisotropy and intensity (Table I), which we have shown by background protein solution subtraction not to arise from contaminating scattered light.

We have examined the interactions of the *trp* repressor with a double-stranded 25-bp oligonucleotide labeled with fluorescein through the C_6 linker on the 5′ end of the sense strand.[3] Figure 4 is an example of the anisotropy profiles obtained on binding *trp* repressor to the TR25 target at low salt. These profiles, like those for GRDBD–GRE interactions at

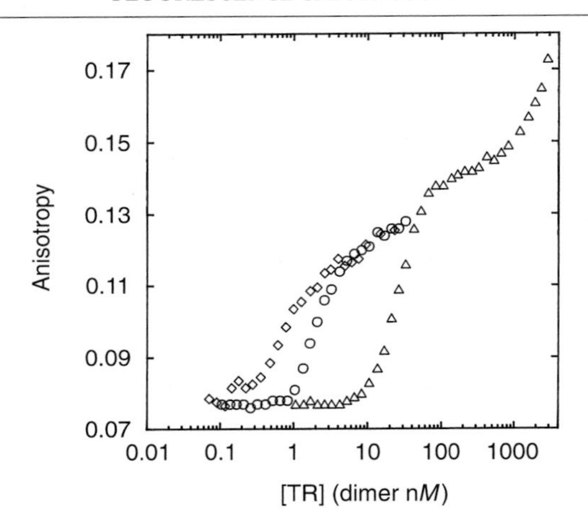

FIG. 4. Anisotropy profiles of *trp* repressor binding to the fluorescein labeled TRop 25-mer (TR25) shown in Fig. 3. The buffer contained 10 m*M* potassium phosphate, and 0.1 m*M* EDTA at pH 7.6. Data for (◇) 0.2 n*M* TR25, (○) 2 n*M* TR25, and (△) 20 n*M* TR25. [Reprinted with permission from V. LeTilly and C. A. Royer, *Biochemistry* **32,** 7753 (1993). Copyright 1993 American Chemical Society.]

low salt, show that large anisotropy changes occur on binding of the protein, but with little change in intensity observed. We have since examined the interaction of *trp* repressor with a 5′-C_6-fluorescein-labeled 44-base single-stranded oligonucleotide that forms a hairpin duplex and to which the protein binds as a single dimer. With this target at high salt no change in anisotropy is observed whatsoever, but a 33% decrease in intensity results from addition of protein (Table I). This intensity decrease provides an alternative signal for monitoring complexation (Fig. 5).

The final series of oligonucleotides we present here are single-stranded hairpin constructs with F-C6 linker chemistries for the fluorescein label on the 5′ end that contain target sequences for the vitamin D receptor (VDR), sequences that are also specifically bound by the retinoid X receptor (RXR). Unlike the GREs3 and GREs4 hairpins, these are 5′-blunt ended constructs, without the single base overhang of the GREs. The rat osteocalcin sequence (r-oscDRE) exhibits a 5′-GC base pair, while the 24-hydroxylase DRE (DREI) has a 5′-AT base pair. We have also examined binding to a nonspecific (random) sequence, with a 5′-GC base pair. Under relatively high salt conditions (100 m*M*, Table I), modest increases in anisotropy and almost no change in intensity are observed for RXR binding to either the r-osc or the 24-hydroxylase (DREI) targets. Retinoid X receptor also binds to the random sequence, resulting in an increase in anisotropy in this case as well, but at significantly higher protein concentrations. The relatively small

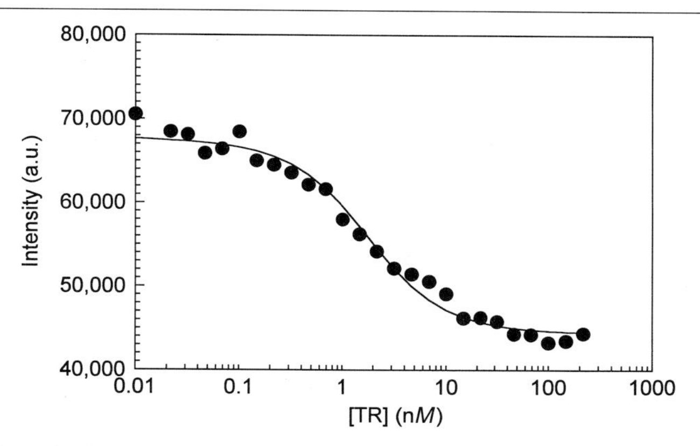

FIG. 5. Intensity loss profile for *trp* repressor binding to the fluorescein labeled TRop 20-mer (TR20), shown in Fig. 3, at 10 m*M* NaCl in 10 m*M* potassium phosphate, 0.1 m*M* EDTA (pH 7.6) and 1 n*M* TR20 at 21°.

increases in anisotropy found at high salt are similar in magnitude to those observed for GRDBD binding to the GREs3 under these conditions.

Time-Resolved Fluorescence

Most of the target oligonucleotides that we have studied using steady-state methods have also been examined using time-resolved fluorescence techniques. Excitation was at 488 nm, either from an argon-ion laser or from an Aladdin (Stoughton, WI) synchrotron light source. Time resolution with the CW laser system was achieved by Pockels cell modulation as previously described.[63] The base frequency of near 4 MHz was achieved at the Aladdin light source using the missing bunch mode.[64] Whenever feasible, both the fluorescence lifetimes and time-resolved anisotropy were determined in the absence and presence of protein. Because time-resolved measurements are much less sensitive than the 10 p*M* limit on the steady-state instrument, target oligonucleotide concentrations were between 10 and 100 n*M*. In all cases, however, similar anisotropy changes were observed on saturation by protein, as those observed in the steady-state assays, indicating that the same complexes were formed.

[63] D. W. Piston, G. Marriott, T. Radiveyovich, R. M. Clegg, T. M. Jovin, and E. Gratton, *Rev. Sci. Instrum.* **60,** 2596 (1989).

[64] E. Gratton, W. W. Mantulin, G. Weber, C. A. Royer, D. M. Jameson, R. Reininger, and R. W. Hansen, *Rev. Sci. Instrum.* In press (1996).

TABLE II
TIME-RESOLVED FLUORESCENCE PARAMETERS OF FLUORESCEIN-LABELED TARGET OLIGONUCLEOTIDES

DNA	[x556] (nM)	α_1	τ (nsec)	α_2	τ_2 (nsec)	β_1	τ_{c1} (nsec)	τ_{c2} (nsec)
F-C6-3S	0	1.0	4.2	—	—	0.15	6.8	0.3
	50	0.84	4.3	0.16	1.1	0.18	12	0.3
F-Amp-3S	0.0	1.0	4.2	—	—	0.16	7.1	0.4
F-C6-plfg (S)	0.0	0.42	5.1	0.58	2.8	0.32	9.5	0.3
F-Amp-plfG (S)	0.0	0.86	4.9	0.14	0.2	0.53	15.1	0.4
F-C6-plfG (S)	900	0.53	5.1	0.47	2.8	0.44	16.5	0.4
F-C6-plfG (AS)	0.0	0.23	5.1	0.77	3.0	0.16	3.8	0.3
	300	0.26	5.1	0.74	3.0	0.21	8.0	0.28
F-C6-TR20	0	0.76	3.5	0.24	0.8	—	—	—
	300	0.72	3.6	0.28	0.5	—	—	—
F-C6-r-ose	0	0.85	3.8	0.15	0.2	0.58	1.7	0.1
F-C6-DRE-I	0	1.0	4.1	—	—	0.44	1.8	0.2
F-C6-random	0	1.0	4.3	—	—	0.55	2.0	0.1
F-Amp-VitERE	0	1.0	4.5	—	—	0.46	3.7	0.2

Table II gives the time-resolved parameters for the GRE-containing targets with the various linker chemistries. We have fit the frequency response curves to either single or double exponential decays of the intensity. Two components were sufficient to obtain good fits of the data, in all cases, which did not improve on increasing the number of components. We have assumed a double exponential decay of the anisotropy to represent the fast local mobility of the probe and a slower composite of the modes of rotation of the oligonucleotide. Perez-Howard and co-workers[8] found this to be an appropriate analysis for a rhodamine-X-labeled oligonucleotide, as the spin and tumble modes of the oligonucleotide rotational diffusion could not be resolved. Where decays were not single exponential, we have not assumed any association between lifetime components and correlation times.

With the F-C6-GREs3 target, while we observe a 25% decrease in intensity for the oligonucleotide on complexation with GRDBD under high-salt conditions, there is only a small change in the lifetime. This indicates that most of the intensity loss is due to static quenching or changes in extinction coefficient, neither of which would affect the lifetime of fluorescence. Examination of the time-resolved characteristics of the fluorescence of the labeled oligonucleotides also provides explanations for many of the observations made with these targets in steady-state mode. For example, the time-resolved anisotropy results for the F-C6-GREs3 target in the absence of protein indicate that only 15% of the molecules undergo rotations that are coupled to the global tumbling of the DNA. The remaining 85% of the probe molecules rotate freely around the flexible C_6 linker with a correlation time near 300 psec. This would account for the relatively low value of the anisotropy of fluorescence of the free target oligonucleotide (0.042). We also find in these data an explanation for why the change in anisotropy on complexation with GRDBD is so small (0.042 to 0.063). An increase in the long correlation time associated with global rotational diffusion of the DNA (τ_{c1}) was observed on addition of GRDBD, consistent with the steady-state anisotropy change. However, in order to reconcile the anisotropy and intensity binding isotherms,[64a] we must assume 90% rather than 25% quenching of the fluorophore. This can be understood if the time-resolved anisotropy data actually reflect an equilibrium between two conformations of the DNA-bound fluorophore, one in which it is free to rotate and the other in which it interacts more intimately with the DNA, perhaps through a stacking interaction on the end of the helix (Fig. 6). By this view, the freely rotating fraction of probe molecules would be largely insensitive to protein binding (no quenching and no anisotropy change), and would contribute a high intensity and low anisotropy signal to the overall measured fluorescence. Only the small fraction of the probe molecules that is tightly coupled to the DNA motions would report on the protein-binding event. Because these are highly quenched, only the remaining 10% of these probe molecules actually give rise to the change in anisotropy, which when averaged with the high intensity, low anisotropy of the majority of freely rotating probe molecules results in a small overall anisotropy change on complexation. Small changes in anisotropy have been reported for a number of protein–DNA complexation studies.[65,66]

In the extreme, the equilibrium between probe conformers and the quenching of the tightly coupled fluorophore moieties can preclude any anisotropy increase on complexation. This phenomenon is observed on binding of the *trp* repressor to its target 20-mer, which is evidenced by a 33% quench. In this case, the fluorophore molecules in the tightly coupled conformation are quenched essentially 100%, and thus no anisotropy change is observed because the probe molecules that would give rise to it do not fluoresce. Despite the large change in intensity observed for this target on complexation by protein, there was no detectable change in lifetime, indicating that the loss in intensity arises owing to static quenching, a change in extinction coefficient, or both.

[64a] J. J. Hill *et al.*, Submitted (1996).

[65] X. Cheng, L. Kovac, and J. C. Lee, *Biochemistry* **34,** 10816 (1995).

[66] P. K. Wittmayer and R. T. Raines, *Biochemistry* **35,** 1076 (1996).

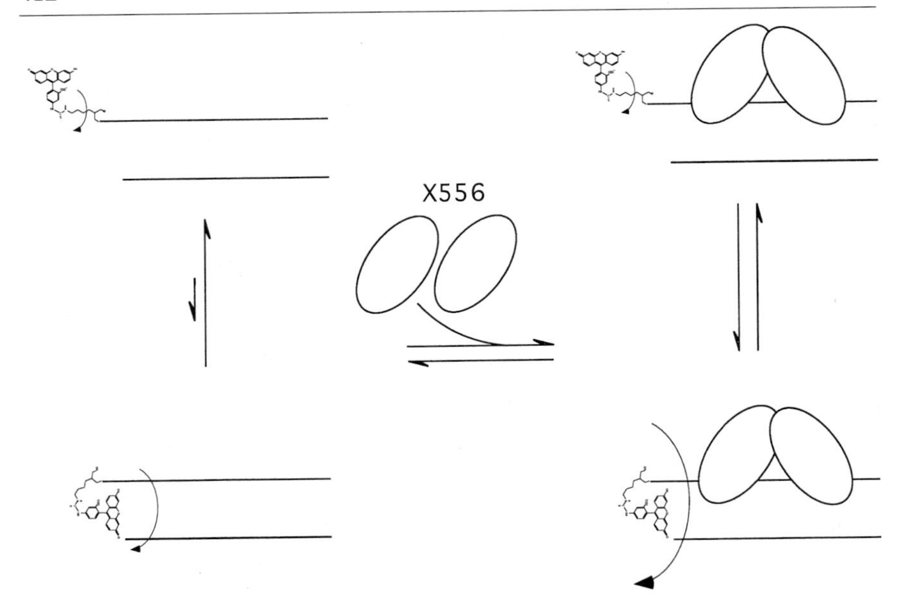

FIG. 6. Schematic representation of the conformational equilibrium of the fluorescein moiety, covalently attached to the 5′ end of the target oligonucleotides. Free rotation about the long linker arm between the 5′-terminal phosphate and the fluorescein renders this conformation insensitive to binding of the target oligonucleotide by X556. The conformation in which the dye interacts more intimately with the DNA leads to quenching of the probe fluorescence and an increase in anisotropy of the residual unquenched dye molecules. We have indicated by the equilibrium arrows that the binding of protein can alter this equilibrium.

Unlike the single-stranded hairpin GREs3, the fluorescein-labeled double-stranded plfG targets exhibit heterogeneous fluorescence decay profiles, even in the absence of protein (Table II), with identical long components (5.1 nsec) and nearly identical shorter decays, 2.8 and 3.0 nsec, respectively. The amplitude of the longer component was nearly twofold larger for the F-C6-plfG sense strand with coupling to a 5′-AT base pair than for the antisense counterpart with coupling to a 5′-GC base pair. Thus, the average lifetime is slightly longer with the fluorescein linked to a deoxyadenosine (3.8 nsec) than to a deoxyguanosine (3.5 nsec). Millar and co-workers[62] have found for similar C_6-fluorescein double-stranded constructs that the average lifetime is significantly longer for a linkage to a thymidine (4.16 nsec) than to a deoxycytidine (0.97 nsec). From the work of Millar and co-workers it would also appear that in a single-stranded context, linkage to a deoxycytidine is quite detrimental to quantum yield. However, our GREs3 hairpin with the overhanging 5′-C-linked fluorescein displayed a monoexponential decay with a rather long lifetime (4.1 nsec). There was little change in the lifetime of the fluorescein-labeled C6-plfG targets on addition of protein. Therefore, the large loss in total intensity observed in the steady-state measurements (Table I) does not arise from dynamic quenching.

Investigation of the time-resolved anisotropy of the plfG targets reveals that the F-C6-plfG sense target exhibits a more intimate coupling of the fluorophore to the DNA, with 32% of the molecules exhibiting a long correlation time, as compared to only 16% for the antisense counterpart. This tighter coupling of the fluorescein to the sense plfG oligonucleotide results in a significantly larger steady-state anisotropy for the sense strand (Table I). It is unclear how the lifetime components may relate to the rotational components, and in the present analysis we assumed no association. For both the sense and antisense plfG targets, the amplitude and the value of the long correlation time increased on binding the protein.

The F-AMP-labeled plfG target also exhibited a nonexponential intensity decay, with a large amplitude for the long lifetime. Only 14% of the molecules decay with a 200-psec lifetime, such that the vast majority of the emission intensity is due to the long-lived component. As we found for the GREs3, the maleimide linker chemistry resulted in a decreased local mobility of the probe, with 53% exhibiting a long (15 nsec) correlation time. This tighter coupling of the probe motions to those of the oligonucleotide result in a larger steady-state anisotropy value for the free oligonucleotide and a larger change in anisotropy observed on binding of c-Jun (Table I) as compared to the C_6 linker. We have also measured the time-resolved intensity and anisotropy properties of an F-Amp-labeled estrogen response element (ERE). Unlike the F-Amp-plfG targets, the intensity decayed as a single exponential with a lifetime slightly over 4 nsec (Table II). Analysis of the differential polarization data revealed that nearly half of the probe molecules exhibit a long correlation time, although its value is rather small, 3.7 nsec, giving rise to a rather low steady-state anisotropy (0.075). It is unclear why the anisotropy properties of the F-Amp-ERE are so different from the F-Amp-plfG target, given that the 5′ bases are both deoxyadenosine.

The time-resolved fluorescence of the fluorescein-labeled DRE targets and the random, nonspecific sequence with the C_6 linkers were quite similar to those observed for the GREs, although the DRE targets, while single-stranded hairpins, did not have a 5′-overhanging base (see Fig. 3). The lifetimes and amplitudes recovered from fits of the frequency response profiles are given in Table II. Only the rat osteocalcin (r-osc) sequence exhibited a multiexponential decay, with 15% of the emitting species exhibited a short lifetime. In this target the fluorescein is coupled to a deoxyguanosine. The fluorescein labels on both the nonspecific sequence and the

24-hydroxylase DRE (DREI) target decayed as single exponentials with lifetimes slightly greater than 4 nsec, typical of fluorescein. We note that in the nonspecific target, the probe is also coupled to a deoxyguanosine nucleoside. No change in the lifetime was observed for any of the targets on complexation by protein. The rotational properties of the C_6-linked fluorescein on these three targetes derived from analysis of differential polarization data are consistent with the relatively low steady-state anisotropy values reported in Table I. Although analysis of the r-osc differential polarization data in terms of two correlation reveals that approximately 50% of the molecules exhibit a long correlation time, its value is only 1.9 nsec (see Table II), consistent with the low steady-state anisotropy values observed both in the absence and in the presence of protein.

Conclusion

We have presented here results from an ongoing effort in our laboratory to develop fluorescence-based methods for studying protein–nucleic acid complexes. To date our efforts have centered on a single fluorophore, fluorescein, mainly because of its high quantum yield. In the context of this dye we have evaluated the effects of the chemistry of the linkage between the dye and the DNA, as well as the DNA sequence and structure. We have also determined the effects of binding by protein, specifically or nonspecifically, to the target oligonucleotides, through changing the target sequence or the salt concentration. What is quite clear from the results presented here, is that the C_6-linker chemistry, while providing adequate data for complexation, is not ideal. The F-Amp linkage is much more efficient, in that the probe is more tightly coupled to the oligonucleotide and thus senses the binding by protein much more acutely. Whether the signal change is an increase in anisotropy or a change in the fluorescence intensity, the tighter coupling increases the amplitude of the change, thereby increasing the sensitivity and precision of the assay.

In general we believe that anisotropy is a more precise measurement of complexation than intensity, as it is less prone to instrumental noise and artifact. From this point of view, fluorescein, owing to the exquisite sensitivity of its quantum yield and extinction coefficient to the environment, is probably not the most appropriate fluorophore. We are currently undertaking a systematic study of a variety of fluorophores and linker chemistries for optimization of the anisotropy assays. Despite the drawbacks of the observed intensity changes of fluorescein on complexation, we find this phenomenon quite interesting. The lifetime and intensity of fluorescein clearly depend on a variety of characteristics of the oligonucleotides that include the linker chemistries and 5′ base pair, but also must include the subtle sequence, structural, dynamic, and energetic properties of each specific target. These properties apparently are altered by binding of protein. In most cases, the distance between the site of protein binding and the 5′ probe is too long for direct contact between the protein and the DNA to account for the observed intensity changes. They must, therefore, reflect long-range effects of protein binding on the environment of the probe. These may be electrostatic, involving cation condensation changes,[67] or structural effects of complexation on the oligonucleotide. In any case they are intriguing and merit further investigation.

Appendix I

Anti-Fluorescein Antibody-Based Purification of Fluorescein-Labeled Oligonucleotides

The oligonucleotide mixture from a 0.25 μM synthesis is resuspended in 500 μl of 10 mM Tris, 1 mM EDTA, 100 mM NaCl, pH. 8.0 buffer (TE_{100}). The 100 mM NaCl is present to minimize nonspecific association of the DNA with the protein–agarose gel. The oligonucleotide mixture is added to approximately 400 μl of immobilized AFAB contained in a compact reaction column (U.S. Biochemicals, Cleveland, OH) at a gravity flow rate. These columns work well for the small volumes used in this purification scheme, and are easily outfitted with Luer-Lock adapters for use with a syringe. Binding can be observed because the green color of the oligonucleotide turns pink as the AFAB quenches the fluorescence of the dye. After binding is complete the mixture is gently washed with approximately 60 ml of TE_{100}, via a 60-cm^3 syringe, in order to elute the unbound DNA. To remove the fluorescein-labeled oligonucleotide from the AFAB a solution of 500 μl of 0.2 N HCl is added to the suspension at a gravity flow rate. The acid will denature the AFAB and release the fluorescein-labeled oligonucleotide with a recovery of the green color. Typically, a significant amount of the oligonucleotide does not elute with the HCl but remains with the column until it is neutralized by the addition of 500 μl of 0.2 N NaOH. Any remaining oligonucleotides can now be further eluted with two 100-μl additions of TE_{100}. There does not appear to be any destruction of either the fluorescein or the oligonucleotide by the addition of either the acid or the base at the concentrations employed in this protocol. The elution fractions containing the labeled oligonucleotide are then concentrated, combined, and ethanol precipitated. Typically this involves resuspending and combining the nearly desiccated pellets in a total volume of 250 μl of TE_{100} and

[67] M. C. Olmsted, J. P. Bond, C. F. Anderson, and M. T. Record, Jr., *Biophys. J.* **68,** 634 (1995).

addition of 700 μl of spectroscopic grade ethanol (McCormick Distilling, Weston, MO). Before addition of the ethanol the fractions are vortexed and centrifuged in a microcentrifuge (12,000 rpm for 5 min) to pellet any insoluble column material that may have eluted with the labeled oligonucleotide. After ethanol precipitation overnight at $-20°$ the DNA pellet can be resuspended in TE buffer.

Appendix II

Separation of Labeled Oligonucleotides from Excess Free Dye

This separation is carried out on a 20-cm Sephadex G-10 Superfine column. Typically, when TE_{100} buffer is used three fluorescent bands can be observed on the column with a handheld UV lamp. The fastest flowing band contains the labeled oligonucleotide, while the slowest band contains the free dye. A middle band, close to the fastest flowing one, appears to contain some unlabeled DNA as well as free dye. Several 500-μl fractions of the eluent are collected and those containing labeled oligonucleotide are then concentrated and ethanol precipitated as described previously. These postsynthesis labeling strategies should be conducted on single-stranded oligonucleotides because a high degree of tight, noncovalent binding of dye to double-stranded DNA will occur. These bound dye molecules cannot be easily separated from the DNA in a normal desalting column, but only under denaturing conditions.

Acknowledgments

The authors acknowledge Theodore Hazlett and other members of the Laboratory for Fluorescence Dynamics (LFD) in Urbana, Illinois for their assistance in obtaining some of the time-resolved data presented here. The LFD is a research resource of the National Institutes of Health. We also acknowledge the staff of the Aladdin Synchrotron Radiation Center at the University of Wisconsin for their help in building and installing the fluorescence beamline (The Aladdin Biofluorescence Center, ABC) where some of the time-resolved data were also acquired. The ABC is jointly funded by the NSF and the Universities of Illinois and Wisconsin. Finally, we acknowledge our collaborators on the many projects for which these labeled oligonucleotides have been synthesized and characterized. These include Dr. Keith Yamamoto (UCSF) and Drs. Jack Gorski and Hector DeLuca (University of Wisconsin). This research was funded by a grant from the Whitaker Foundation to C.A.R.

[20] Fluorescence Resonance Energy Transfer as a Probe of DNA Structure and Function

By MENGSU YANG and DAVID P. MILLAR

Introduction

Fluorescence resonance energy transfer (FRET) is a process by which the excited-state energy of a fluorescent donor molecule is transferred to an unexcited acceptor molecule by means of dipole–dipole coupling. The rate of energy transfer depends on the distance between the donor and acceptor chromophores, as well as the spectral properties of the donor and acceptor, and the relative orientation of the donor and acceptor transition dipoles. The relationship between the efficiency of resonance energy transfer and the donor–acceptor distance was derived by Förster[1] in the late 1940s and the FRET technique has been widely used to determine proximity relationships in biological systems, particularly protein structures.[2–4]

Because of a lack of suitable intrinsic chromophores and difficulties in dye labeling, applications of FRET to nucleic acids have been relatively limited in the past.[5,6] However, advances in automated DNA synthesis and the convenient site-specific labeling of synthetic oligonucleotides with suitable probes have spurred a renewed interest in the application of FRET, both in steady-state and lifetime domain, to structural and functional studies of DNA and RNA. For example, FRET has been used to elucidate the overall geometry of four-way DNA junctions[7,8] and the hammerhead ribozyme[9] and to observe kinking of DNA and RNA helices by bulged nucleotides.[10] In addition, FRET methods have been employed to investigate

[1] T. Förster, *Z. Naturforsch. A: Astrophys. Phys. Phys. Chem.* **4,** 321 (1949).
[2] L. Stryer, *Annu. Rev. Biochem.* **47,** 819 (1978).
[3] R. H. Fairclough and C. R. Cantor, *Methods Enzymol.* **48,** 347 (1978).
[4] P. Wu and L. Brand, *Anal. Biochem.* **218,** 1 (1994).
[5] K. Beardsley and C. R. Cantor, *Proc. Natl. Acad. Sci. U.S.A.* **65,** 39 (1970).
[6] C.-H. Yang and D. Söll, *Proc. Natl. Acad. Sci. U.S.A.* **71,** 2838 (1974).
[7] A. I. H. Murchie, R. M. Clegg, E. von Kitzing, D. R. Duckett, S. Diekmann, and D. M. J. Lilley, *Nature (London)* **341,** 763 (1989).
[8] J. P. Cooper and P. J. Hagerman, *Biochemistry* **29,** 9261 (1990).
[9] T. Tuschl, C. Gohlke, T. M. Jovin, E. Westhof, and F. Eckstein, *Science* **266,** 785 (1994).
[10] C. Gohlke, A. I. H. Murchie, D. M. J. Lilley, and R. M. Clegg, *Proc. Natl. Acad. Sci. U.S.A.* **91,** 11660 (1994).

0076-6879/97 $25

DNA hybridization,[11] DNA triple helix formation,[12] and the kinetics of enzyme-catalyzed DNA cleavage and unwinding.[13,14]

Most applications of the FRET technique to nucleic acids have been based on steady-state measurements of the integrated emission intensity of the donor or acceptor. Such measurements provide qualitative information on the relative proximity of donor and acceptor probes, and can also detect relatively small conformational differences among a series of related molecules. Steady-state FRET measurements also provide a simple tool for monitoring the kinetic processes of DNA–DNA and DNA–protein interactions in real time. In contrast, time-resolved measurements of FRET can be used to obtain detailed information on the structure and solution dynamics of nucleic acids. In time-resolved FRET, the decay of donor fluorescence is measured after picosecond pulse excitation and the decay profile is analyzed with a distance distribution model to obtain information on the mean donor–acceptor distance and the distribution of distances.[15,16] Time-resolved FRET has been used to recover intramolecular distance distributions in proteins,[17,18] peptides,[19] and carbohydrates.[20] More recently, time-resolved FRET methods have been applied to recover the distribution of end-to-end distances in single-stranded and duplex DNA,[16,21,22] and to study the solution dynamics of four-way and three-way DNA junctions.[23,24]

There are a number of monographs and reviews in the literature,[2–4] including two chapters in *Methods in Enzymology*,[25,26] describing the principles and methodology of FRET and its applications to the study of biological

[11] R. A. Cardullo, S. Agrawal, C. Flores, P. C. Zamecnik, and D. E. Wolf, *Proc. Natl. Acad. Sci. U.S.A.* **85,** 8790 (1988).

[12] M. Yang, S. Ghosh, and D. P. Millar, *Biochemistry* **33,** 15329 (1994).

[13] S. Ghosh, P. S. Eis, K. Blumeyer, K. Fearon, and D. P. Millar, *Nucleic Acids Res.* **22,** 3155 (1994).

[14] K. P. Bjorson, M. Amaratunga, K. J. M. Moore, and T. M. Lohman, *Biochemistry* **33,** 14306 (1994).

[15] E. Haas, M. Wilchek, E. Katchalski-Katzir, and I. Z. Steinberg, *Proc. Natl. Acad. Sci. U.S.A.* **72,** 1807 (1975).

[16] R. A. Hochstrasser, S.-M. Chen, and D. P. Millar, *Biophys. Chem.* **45,** 133 (1992).

[17] D. Amir and E. Haas, *Biochemistry* **26,** 2162 (1987).

[18] J. R. Lakowicz, I. Gryczynski, H. C. Cheung, C.-K. Wang, M. L. Johnson, and N. Joshi, *Biochemistry* **27,** 9149 (1988).

[19] P. S. Eis and J. R. Lakowicz, *Biochemistry* **32,** 7981 (1993).

[20] P. Wu, K. G. Rice, Y. C. Lee, and L. Brand, *Proc. Natl. Acad. Sci. U.S.A.* **88,** 9355 (1991).

[21] K. M. Parkhurst and L. J. Parkhurst, *Biochemistry* **34,** 293 (1995).

[22] B. P. Maliwal, J. Kusba, and J. R. Lakowicz, *Biopolymers* **35,** 245 (1995).

[23] P. S. Eis and D. P. Millar, *Biochemistry* **32,** 13852 (1993).

[24] M. Yang and D. P. Millar, *Biochemistry* **35,** 7959 (1996).

[25] R. M. Clegg, *Methods Enzymol.* **211,** 353 (1992).

[26] P. R. Selvin, *Methods Enzymol.* **246,** 301 (1995).

problems. Here, we focus on the most recent applications of steady-state and time-resolved FRET to studies of DNA structure and function. This chapter includes brief descriptions of FRET formalism, strategies for fluorescent labeling of DNA, FRET methodologies, and data analysis. Selected examples of interest to biochemists and/or biologists are discussed in detail to highlight the uniqueness of the FRET methods in investigating a variety of biochemical problems involving nucleic acids. It is hoped that the methods and the examples described here are helpful for the reader, who may not have a background in spectroscopy or in the specific labeling of DNA, to design and carry out FRET experiments involving DNA.

Fluorescence Resonance Energy Transfer Methodologies: Steady-State and Time-Resolved Measurements

Fluorescence resonance energy transfer occurs when an excited fluorescent donor molecule transfers energy nonradiatively to an unexcited acceptor molecule by means of long-range (through-space) resonance coupling between the donor and the acceptor transition dipoles. The energy transfer leads to a decrease in the fluorescence intensity and lifetime of the donor owing to the additional decay pathway in the presence of an acceptor. The acceptor molecule is promoted to its excited state and may become fluorescent depending on its quantum yield. Because the long-range resonance dipole–dipole interactions occur through space, the rate of energy transfer depends on the distance between the donor and acceptor molecules. Förster derived the relationship between the energy transfer rate, k_{ET}, and the donor–acceptor (D–A) distance (R)[1]:

$$k_{\mathrm{ET}} = \frac{1}{\tau_{\mathrm{D}}}\left(\frac{R_0}{R}\right)^6 \tag{1}$$

where τ_{D} is the fluorescence lifetime of the donor in the absence of the acceptor and R_0 is the critical transfer distance at which the energy transfer rate is equal to the intrinsic decay rate.

The critical transfer distance can be determined from the spectral properties of the donor and acceptor chromophores according to Eq. (2):

$$R_0^6 = \left\{\frac{9000(\ln 10)\kappa^2\phi_{\mathrm{D}}}{128\pi^5 N_{\mathrm{A}} n^4}\right\}\int_0^\infty F_{\mathrm{D}}(\lambda)\varepsilon_{\mathrm{A}}(\lambda)\lambda^4\,d\lambda \tag{2}$$

where κ^2 is the factor that describes the orientational dependence of energy transfer, ϕ_{D} is the fluorescence quantum yield of the donor, N_{A} is Avogadro's number, n is the refractive index of the medium separating the donor and acceptor, $F_{\mathrm{D}}(\lambda)$ is the normalized donor fluorescence intensity at wavelength λ, and $\varepsilon_A(\lambda)$ is the extinction coefficient of the acceptor at the same wavelength. The integral in Eq. (2) expresses the degree of spectral overlap between the donor emission and the acceptor absorption and can be calculated as described elsewhere.[4] The determination of κ^2 values is discussed in a later section.

Because the rate of energy transfer, k_{ET}, is difficult to measure accurately, a parameter called the energy transfer efficiency (E) is usually determined. E is defined as the fraction of the excited donor molecules decaying via the energy transfer process.

$$E = \frac{k_{ET}}{k_D + k_{ET}} \tag{3}$$

where k_D is the decay rate of the donor in the absence of the acceptor and contains contributions from all processes that deactivate the excited state other than energy transfer.

There are a number of methods to determine E, including measurements of decreased donor fluorescence intensity and enhanced acceptor fluorescence intensity, decrease in donor lifetime, changes in the fluorescence anisotropy of the donor and acceptor, and changes in photobleachability of the donor. This chapter focuses on steady-state and time-resolved FRET methods based on measurements of donor fluorescence. The reader is referred to earlier reviews for detailed descriptions of the other measurement schemes.[25,26]

Donor–Acceptor Distance Determined by Steady-State Fluorescence Resonance Energy Transfer

The most common method for determining E in the steady state is to measure the decrease in donor fluorescence intensity or quantum yield. The efficiency of energy transfer can be calculated from donor emission in the absence (I_D) and the presence of acceptor (I_{DA}), normalized to the same donor concentration, and related to the donor–acceptor distance as in Eq. (4).

$$E = 1 - (I_{DA}/I_D) = \frac{1}{1 + (R/R_0)^6}; \quad R = R_0(E^{-1} - 1)^{1/6} \tag{4}$$

Donor–Acceptor Distance Determined by Time-Resolved Fluorescence Resonance Energy Transfer

The efficiency of energy transfer can be determined directly from fluorescence lifetime measurements of the donor in the presence and absence of the acceptor. The advantage of such measurements is that they are independent of donor concentration. The fluorescence decay of a donor molecule can be expressed empirically as a sum of exponentials:

$$I(t) = \sum_{i=1}^{N} \alpha_i \exp(-t/\tau_i) \tag{5}$$

where α_i is the fractional amplitude associated with each donor lifetime τ_i, and N is the number of lifetimes. The fluorescence decay of the donor can be measured by time-correlated single-photon counting[27] or by frequency domain methods.[28] In the time-correlated single photon-counting method, the expression in Eq. (5) is convoluted with the instrument response function to account for the finite time resolution of the photomultiplier and detection electronics. The amplitude and lifetime parameters are typically determined by nonlinear least-squares fitting of the donor decay profile.[29] The average fluorescence lifetime of the donor is calculated according to Eq. (6):

$$\bar{\tau} = \sum_{i=1}^{N} \alpha_i \tau_i \tag{6}$$

The efficiency of energy transfer can be calculated from the average lifetimes of the donor in the absence ($\bar{\tau}_D$) and presence ($\bar{\tau}_{DA}$) of the acceptor,

$$E = 1 - (\bar{\tau}_{DA}/\bar{\tau}_D) \tag{7}$$

where E is related to the D–A distance through Eq. (4).

Alternatively, the donor–acceptor distance can be recovered directly from the analysis of fluorescence decay of the molecule labeled with both donor and acceptor. By combining Eqs. (1) and (5), it can be shown that

$$I_{DA}(t) = \sum_i \alpha_i^D \exp(-t/\tau_i^{DA}) = \sum_i \alpha_i^D \exp\{(-t/\tau_i^D)[1 + (R_0/R)^6]\} \tag{8}$$

where the superscripts "DA" and "D" indicate the lifetime components of the donor in the presence and the absence of the acceptor, respectively. The intrinsic donor lifetimes, τ_i^D, and amplitudes, α_i^D, can be obtained from the decay of a suitable donor-only molecule and kept constant in the analysis

[27] D. V. O'Connor and D. Philips, "Time-Correlated Single Photon Counting." Academic Press, London, 1984.

[28] J. R. Lakowicz and I. Gryczynski, Frequency-domain fluorescence spectroscopy. *In* "Topics in Fluorescence Spectroscopy" (J. R. Lakowicz, ed.), Vol. 1, p. 293. Plenum, New York, 1991.

[29] P. R. Bevington, "Data Reduction and Error Analysis for the Physical Sciences." McGraw-Hill, New York, 1969.

of the D–A distance. Strictly speaking, Eq. (8) is valid when only a single distance exists between the donor and acceptor.

Donor–Acceptor Distance Distributions Determined by Time-Resolved Fluorescence Resonance Energy Transfer

One of the most significant characteristics of the donor fluorescence decay is its sensitivity toward the variation of D–A distance. In solution, owing to thermal fluctuations, orientational and conformational flexibilities of macromolecules, and the linker arms used to attach the dye molecules, the D–A distance may adopt a variety of different values on the time scale of the emission process. These lead to multiple D–A energy transfer rates and thus multiexponential decays of the donor fluorescence. Consequently, the fluorescence decay of the donor can be modeled by an apparent distribution of D–A distances [Eq. (9)]:

$$I_{DA}(t) = \int_{R_{min}}^{R_{max}} \sum_i \alpha_i^D \exp\{(-t/\tau_i^D)[1 + (R_0/R)^6]\} P(R)\, dR \tag{9}$$

wherein the probability distribution of D–A distances, $P(R)$, is assumed to be static on the time scale at which energy transfer occurs; R_{min} is the distance of closest approach of the donor and acceptor; and R_{max} is the maximum D–A distance. If the sample is a mixture of molecules labeled with both the donor and acceptor and those labeled with only the donor, the fluorescence decay of the donor in the mixture also consists of the contribution from the donor-only molecules [Eq. (10)]:

$$I_{DA}(t) = (1-f) \int_{R_{min}}^{R_{max}} \sum_i \alpha_i^D \exp\{(-t/\tau_i^D)[1 + (R_0/R)^6]\} P(R)\, dR + f I_D(t) \tag{10}$$

where the first term describes the decay of donors that undergo energy transfer to the acceptor, and the second term accounts for a fraction (f) of donors that do not undergo energy transfer because of incomplete labeling with the acceptor. $I_D(t)$ describes the decay of the donor in the absence of the acceptor [Eq. (5)].

The D–A distance distribution is usually described by a continuous Gaussian distribution of distances in three dimensions[15]:

$$\begin{aligned} P(R) &= 4\pi R^2 c \exp[-a(R-b)^2] \quad \text{for} \quad R_{min} < R < R_{max} \\ &= 0 \quad \text{elsewhere} \end{aligned} \tag{11}$$

where a and b are adjustable parameters that describe the shape of the distribution and c is a normalization constant. Nonlinear least-square methods are used to fit the donor decay in the presence of the acceptor by optimizing the distribution parameters a and b, and the fraction of free donor species (f) if acceptor labeling is incomplete. The resulting D–A distance distributions are calculated from Eq. (11) using the fitted values of a and b.

Rotational Mobility and Orientation of Donor and Acceptor Dyes

The main factors affecting the accuracy of FRET-based distance measurements, both in steady-state and time-resolved methods, are the limits of precision in determining the critical transfer distance R_0. The fluorescence quantum yield of the donor, ϕ_D, may vary for all the samples to be compared. A change in ϕ_D will lead to a change in R_0 in different experiments. This dependence can be corrected by measuring the relative ϕ_D values of a molecule labeled only with donor under different solution conditions. A more important problem is uncertainty in the precise value of κ^2, the factor describing the relative orientation of the donor and acceptor dipoles. In many applications the flexibility of the linker arm is sufficient to provide enough rotational freedom for the dyes so that the conjugated donor and acceptor behave as free entities in solution. As a result, orientational effects are considered to be less significant in affecting the accuracy of the distance measurements and a single value for the orientation factor ($\kappa^2 = 2/3$ for free dye rotation) is used in calculating R_0 [Eq. (2)]. Nonetheless, because the rotation of donors and acceptors conjugated to DNA is usually not completely free and the orientations of the donor and acceptor transition dipoles within the dyes are often unknown, only a distribution of κ^2 values can be assumed in most cases. There is an extensive literature dealing with the orientation factor in FRET measurements.[30–33] Limits can be set on the possible range of values for κ^2 by determining the rotational mobility of the dye molecules by steady-state or time-resolved fluorescence anisotropy measurements.

In the time domain, fluorescence decays measured with the emission polarizer oriented either parallel, $I_{\parallel}(t)$, or perpendicular, $I_{\perp}(t)$, to the excitation polarization can be analyzed according to Eqs. (12) and (13),

$$I_{\parallel}(t) = \{1 + 2r(t)\} \sum_{i=1}^{N} \alpha_i \exp(-t/\tau_i) \tag{12}$$

[30] R. E. Dale and J. Eisinger, *Biopolymers* **13,** 1573 (1974).

[31] R. E. Dale and J. Eisinger, *In* "Biochemical Fluorescence Concepts" (R. F. Chen and H. Edelhoch, eds.), Vol. 1, p. 115. Marcel Dekker, New York, 1975.

[32] R. E. Dale, J. Eisinger, and W. E. Blumberg, *Biophys. J.* **26,** 161 (1979); R. E. Dale, J. Eisinger, and W. E. Blumberg, *Biophys. J.* **30,** 365 (1980).

[33] P. Wu and L. Brand, *Biochemistry* **31,** 7939 (1992).

$$I_{\perp}(t) = \{1 - r(t)\} \sum_{i=1}^{N} \alpha_i \exp(-t/\tau_i) \tag{13}$$

where the time-dependent fluorescence anisotropy, $r(t)$, is represented by Eq. (14):

$$r(t) = \sum_{k=1}^{2} r_{0k} \exp(-t/\phi_k) \tag{14}$$

where r_{01} and ϕ_1 are the limiting anisotropy and decay time associated with local rotation of the dyes, respectively, and r_{02} and ϕ_2 are the corresponding quantities describing overall rotation of the dye-labeled molecules. These parameters can be optimized for a simultaneous best fit to $I_{\parallel}(t)$ and $I_{\perp}(t)$,[34] with the parameters for the donor lifetime kept fixed at the values determined from the isotropic fluorescence decay. The total limiting anisotropy, r_0, is given by $r_0 = r_{01} + r_{02}$. The fractional depolarization arising from local dye rotation is given by $d = (r_{02}/r_0)^{1/2}$.

In dye–DNA conjugates, the rotation of the dyes is usually hindered[16,22,23] and consequently κ^2 is described by a distribution rather than a single value. The minimum (κ_1^2) and maximum (κ_2^2) values for the orientation factor can be estimated based on the fractional depolarization values for both the donor and acceptor dyes.[32,35] If the probability of occurrence of a particular value of κ^2 is constant between these limits and zero elsewhere, Eq. (9) can be generalized as follows[36]:

$$I_{DA}(t) = \frac{1}{(\beta_2 - \beta_1)} \int_{\beta_1}^{\beta_2} \int_{R_{min}}^{R_{max}} \sum_i \alpha_i^D \exp\{(-t/\tau_i^D)[1 + (\beta R_0/R)^6]\} P(R)\, dR\, d\beta \tag{15}$$

where β is defined as $(3/2)\kappa^2$ and is used to scale the value of R_0^6 accordingly; β_1 and β_2 correspond to κ_1^2 and κ_2^2, respectively. If necessary, a free donor term can be added to the right-hand side of Eq. (15) to account for incomplete labeling with the acceptor.

Determination of Binding Parameters by Steady-State Fluorescence Resonance Energy Transfer

In addition to distance determination and distance distribution analysis, FRET may be used to determine thermodynamic and kinetic parameters

[34] A. J. Cross and G. R. Fleming, *Biophys. J.* **46,** 45 (1984).
[35] E. Haas, E. Katchalski-Katzir, and I. Z. Steinberg, *Biochemistry* **17,** 5064 (1978).
[36] S. Albaugh and R. F. Steiner, *J. Phys. Chem.* **93,** 8013 (1989).

describing biomolecular interactions. When a given concentration of a donor-labeled species, D, is incubated with various concentrations of an acceptor-labeled species, A, to form a complex, D–A, the equilibrium binding constant and the total observed fluorescence intensity of the donor emission, I_{obs}, are given by Eqs. (16) and (17):

$$K_A = \frac{1}{K_D} = \frac{[D\text{–}A]_{eq}}{[D]_{eq}[A]_{eq}} \tag{16}$$

$$I_{obs} = I_D + I_{D\text{–}A} \tag{17}$$

where K_A is the association constant, K_D is the dissociation constant, and $[D\text{–}A]_{eq}$, $[D]_{eq}$, and $[A]_{eq}$ are the equilibrium concentrations of the complex, the free donor species, and the free acceptor species, and I_D and $I_{D\text{–}A}$ are the fluorescence intensities of the free donor species and fully bound complex, respectively. The concentration of the complex can be estimated from Eq. (18),

$$[D\text{–}A] = \frac{I_D^0 - I_{obs}}{I_D^0 - I_{D\text{–}A}^{\infty}}[D_0] \tag{18}$$

where I_D^0 is the fluorescence intensity of the donor in the absence of the acceptor, $I_{D\text{–}A}^{\infty}$ is the fluorescence intensity when all the donor is completely bound with the acceptor to form the complex D–A, and $[D_0]$ is the total donor concentration. Combining Eq. (16) with Eq. (18) yields

$$I_{obs} = I_D^0 - \frac{(I_D^0 - I_{D\text{–}A}^{\infty})}{2[D_0]}[b - (b^2 - 4[D_0][A_0])^{1/2}] \tag{19}$$

where $b = [D_0] + [A_0] + K_D$, and $[A_0]$ is the total concentration of the acceptor species. An equilibrium binding isotherm may be obtained from measurements of I_{obs} vs $[A_0]$. These measurements can be carried out using a standard steady-state fluorimeter. The dissociation constant for the complex is obtained by nonlinear least-squares fitting of the fluorescence data using Eq. (19).

The kinetics of the binding process can also be analyzed by means of FRET measurements. In the presence of an excess amount of the acceptor species, and assuming that the association rate is substantially faster than the dissociation rate, the concentration of unbound donor species at time t after mixing with the acceptor species is given by Eq. (20),

$$[D] = [D_0]\exp(-k_{obs}t) = [D_0]\exp(-k_2[A_0]t) \tag{20}$$

where k_{obs} is the observed pseudo-first-order rate constant and k_2 is the second-order association rate constant. The kinetic profile of donor–

acceptor binding can be obtained from the time course of the donor fluorescence and the rate constants can be determined from Eq. (22).

$$[D] = \frac{I_{obs} - I^{\infty}_{D-A}}{I^{0}_{D} - I^{\infty}_{D-A}}[D_0] \tag{21}$$

$$I_{obs} = I^{\infty}_{D-A} + (I^{0}_{D} - I^{\infty}_{D-A})\exp(-k_{obs}t) \tag{22}$$

In principle, Eqs. (19)–(22) can be used to analyze the association between any two species labeled with suitable donors and acceptors. Typical examples involving DNA include triple helix formation, drug–DNA interactions, and DNA–protein interactions. The application of this formalism to obtain thermodynamic and kinetic parameters describing formation of DNA triple helices is considered below.

Preparation of Dye-Labeled DNA Samples

The study of DNA structure and function by means of FRET measurements requires the attachment of fluorescent dye molecules to oligonucleotides, usually at the 5′ or 3′ termini. Most FRET studies of DNA employ fluorescein as the donor and tetramethylrhodamine as the acceptor. This donor–acceptor pair is characterized by a large R_0 value, typically in the 50- to 54-Å range,[23] depending on the sequence of the DNA, which permits measurement of long distances. In addition, the emission peaks of these dyes are sufficiently well separated that the donor emission can be measured without interference from acceptor emission. The procedures described below are the common protocols used in our laboratory for preparing dye-labeled DNA samples. Detailed discussions of dye labeling of DNA can be found elsewhere.[37]

Synthesis and Labeling of Oligonucleotides

Synthesis of oligonucleotides is carried out using standard solid-phase chemistry with β-cyanoethyl phosphoramidite derivatives on an automated DNA synthesizer. Fluorescein or tetramethylrhodamine molecules can be attached to the 5′ or 3′ termini and other positions of an oligonucleotide (1) during the automated chemical synthesis or (2) as a postsynthetic modification of the oligonucleotide. For direct synthesis, 5′-fluorescein-labeled DNA may be prepared by using fluorescein derivatives of phosphoramidite reagents [Fluorescein-ON (Clontech, Palo Alto, CA) or FluorePrime

[37] A. Waggoner, *Methods Enzymol.* **246,** 362 (1995).

(Pharmacia, Piscataway, NJ)] in the last coupling step of the automated synthesis. Similarly, 3′-fluorescein-labeled oligonucleotides may be prepared via automated DNA synthesis using modified controlled pore glass (3′-Fluorescein-ON CPG; Clontech) packed in standard synthesis columns.

For postsynthetic modification, a 5′-hexylamine-linked oligonucleotide is first prepared using hexylamine-modified phosphoramidites [Aminolink 2 (Applied Biosystems, Foster City, CA)] during the final synthesis cycle. Modified controlled pore glass (3′-DMT-C6-Amine-ON CPG; Clontech) can be used to label the 3′ terminus with a desired amino linker. Similarly, amino linkers may be introduced at any position along the phosphate backbone using 2-aminobutyl-1,3-propanediol (Uni-Link AminoModifier; Clontech) or a protected 5-propylaminodeoxyuridine phosphoramidite[38] during the automated synthesis. The donor- or acceptor-labeled oligonucleotides are prepared by reacting the amino-modified oligonucleotides with an excess amount of succinimidyl ester derivatives (Molecular Probes, Eugene, OR) of 5-carboxyfluorescein or 5-carboxytetramethylrhodamine. Unreacted dyes are removed using a Sephadex G-25 column. Dye-labeled and unlabeled oligonucleotides are separated by reversed-phase high-performance liquid chromatography (HPLC) using a 0.1 *M* triethylammonium acetate/acetonitrile eluent system.

For structural studies of DNA based on distance information derived from FRET measurements, it is essential that all dye-labeled oligonucleotide strands be of uniform length. In such cases it is necessary to remove truncated oligonucleotides resulting from synthesis failures. This is best achieved by purifying the crude oligonucleotides by polyacrylamide gel electrophoresis. The full-length product is isolated as a single band on the gel, which is then excised and the purified DNA recovered by electroelution. Amino-linked oligonucleotides purified by this procedure are subsequently used in the dye-labeling and HPLC separation steps previously described.

The methods described typically yield 100% dye-labeled DNA strands of uniform length. However, it is always prudent to determine the stoichiometry of labeling before undertaking detailed fluorescence experiments. This can be achieved by means of absorbance measurements at 260 nm (nucleotides) and at a wavelength corresponding to the peak absorption of the particular dye. One complication of these measurements is that the extinction coefficient of DNA-linked dyes can be considerably different from the values pertaining to the free dyes in aqueous solution. The extinction

[38] K. J. Gibson and S. J. Benkovic, *Nucleic Acids Res.* **15,** 6455 (1987).

coefficient may also vary depending on the base sequence around the point of dye attachment.

Related to this last point is the issue of excited-state quenching of dyes linked to DNA. The fluorescence quantum yield of conjugated dyes is frequently smaller than for the free dye,[23] implying that local interactions between the dye and the DNA structure cause excited-state quenching. Indeed, the average fluorescence liftime of fluorescein conjugated to DNA is highly dependent on the particular base to which the dye is attached,[39] which suggests that the dye–DNA interactions are highly specific. The rotational properties of dyes attached to DNA might also be influenced by dye–DNA interactions, which could cause complications in the interpretation of FRET data. The effects of DNA sequence on the fluorescence and rotational properties of typical dyes used in FRET studies have not yet been studied in detail, but it has been observed that attachment of fluorescein and tetramethylrhodamine via six-carbon linkers to a 5′-thymidine residue results in negligible quenching[39] and imparts considerable rotational freedom to the dyes.[23]

Dye-labeled DNA Samples

The doubly-labeled DNA samples typically required for FRET experiments are usually prepared by annealing suitable combinations of donor- and acceptor-labeled strands. Duplex DNA molecules can be labeled on opposite strands,[8,10,16] or at the 3′ and 5′ ends of the same strand.[21] DNA junctions used for FRET studies are assembled from three or more oligonucleotides, two of which are labeled with either a donor or acceptor.[7,8] In FRET experiments based on measurements of donor quenching, it is important to ensure that all donor strands are bound in complexes containing an acceptor strand as well. For example, in studies of four-way DNA junctions, an excess amount of the acceptor-labeled and unlabeled strands can be added to the sample to ensure that all donor strands are bound in junctions. The free acceptor-labeled and unlabeled strands are invisible in the measurements of the donor fluorescence. Alternatively, the intact, doubly labeled junction complex can be isolated by gel electrophoresis prior to the fluorescence measurements.[40]

[39] D. P. Millar, R. A. Hochstrasser, C. R. Guest, and S.-M. Chen, *Proc. S.P.I.E. Int. Soc. Opt. Eng.* **1640,** 592 (1992).

[40] R. M. Clegg, A. I. H. Murchie, A. Zechel, C. Carlberg, S. Diekmann, and D. M. J. Lilley, *Biochemistry* **31,** 4846 (1992).

Applications of Fluorescence Resonance Energy Transfer

DNA Triple-Helix Formation

Oligonucleotide-directed DNA triple-helix formation has become an active area of research.[41,42] Sequence-specific recognition of double-stranded DNA is primarily achieved through the formation of specific Hoogsteen-type hydrogen bonds (A-T-A, C$^+$-G-C) when pyrimidine oligonucleotides bind to purine tracts in the major groove of DNA parallel to the purine Watson–Crick strand. A variety of techniques have been employed to measure thermodynamic and kinetic parameters of DNA triple-helix formation.[43–46] However, most of the methods are discontinuous and limited in terms of sensitivity and time resolution, and usually require radiolabeling and separation.

Steady-state FRET methods have been used to study the structure and stability of triplex DNA. Mergny *et al.* have used the approach to discriminate between a fully complementary and a mismatched triplex sequence.[47] FRET was observed with concurrent binding of a 5′-acridine (donor)-labeled oligopyrimidine and a 3′-ethidium (acceptor)-labeled oligopyrimidine to a DNA fragment containing two homopurine sites separated by a variable number of base pairs. A single base-pair change in one of the target sequences strongly reduced the efficiency of energy transfer. The FRET technique has also been used to determine the orientation of the third DNA strand relative to the other two strands in DNA triple helices.[48,49]

We have used the FRET technique to measure directly thermodymamic and kinetic parameters of DNA triplex formation in solution.[12] We prepared triple helices with fluorescent donor and acceptor groups attached to the 5′ termini of the duplex and the third strand, respectively (Fig. 1). During triplex formation, the 5′-fluorescein group on the homopurine-containing strand of the duplex DNA and the 5′-tetramethylrhodamine group on the single-stranded oligonucleotide are brought into close proximity, and the efficiency of FRET increases greatly as the acceptor-labeled oligonucleotide binds to the donor-labeled duplex DNA. By analyzing the steady-state

[41] H. E. Moser and P. B. Dervan, *Science* **238,** 645 (1987).
[42] J. S. Sun and C. Hélène, *Curr. Opin. Struct. Biol.* **3,** 345 (1993).
[43] L. E. Xodo, G. Manzini, and F. Quadrifoglio, *Nucleic Acids Res.* **18,** 3557 (1990).
[44] R. W. Roberts and D. M. Crothers, *Proc. Natl. Acad. Sci U.S.A.* **88,** 9397 (1991).
[45] S. F. Singleton and P. B. Dervan, *J. Am. Chem. Soc.* **114,** 6957 (1992).
[46] H. Shindo, H. Torigoe, and S. Akinori, *Biochemistry* **32,** 8963 (1993).
[47] J. L. Mergny, T. Garestier, M. Rougee, A. V. Lebedev, M. Chassignol, T. N. Thuong, and C. Hélène, *Biochemistry* **33,** 15321 (1994).
[48] E. M. Evertsz, K. Rippe, and T. M. Jovin, *Nucleic Acids Res.* **22,** 3293 (1994).
[49] P. V. Scaria, S. Will, C. Levenson, and R. H. Shafer, *J. Biol. Chem.* **270,** 7295 (1995).

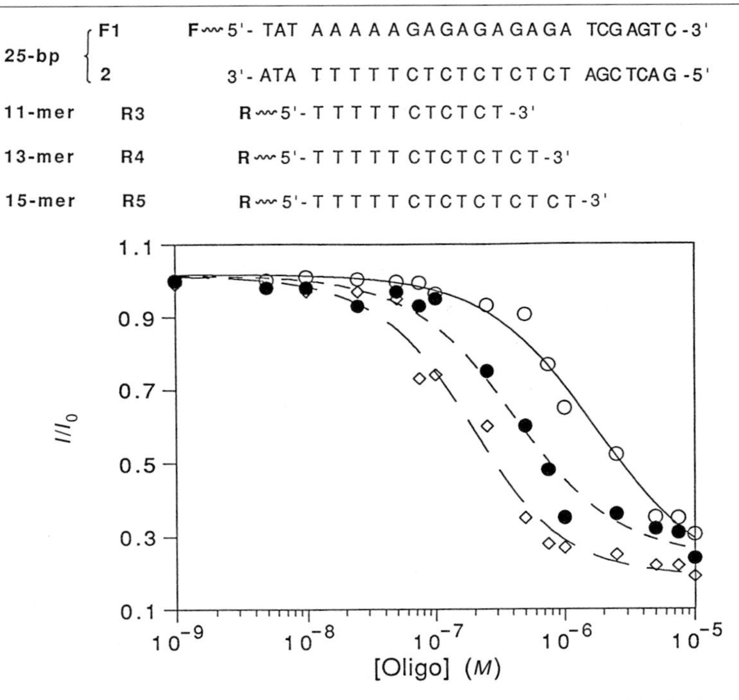

FIG. 1. FRET measurements of DNA triple-helix formation. *Top*: A 25-bp DNA duplex F1–2 containing a 15-bp homopurine–homopyrimidine target site is shown. Triple-helix formation is mediated by Hoogsteen pairing of the homopyrimidine oligonucleotides R3 (11-mer), R4 (13-mer), and R5 (15-mer) with the purine tract of F1 within the F1–2 duplex. Oligonucleotides are labeled with a fluorescein donor (F) and tetramethylrhodamine acceptor (R) where indicated. *Bottom*: Binding isotherms from FRET experiments are shown for the association of acceptor-labeled oligodeoxynucleotides R3 (circles), R4 (filled circles), and R5 (diamonds) with donor-labeled DNA duplex F1–2 (1.2×10^{-7} *M*) in association buffer (50 m*M* Tris–acetate, 0.1 *M* NaCl, 1 m*M* spermine tetrahydrochloride at pH 6.8) at 20°. The data points were normalized to the initial intensity. The solid curves are the nonlinear least-squares best-fit binding isotherms according to Eq. (19). [Reprinted with permission from M. Yang, S. Ghosh, and D. P. Millar, *Biochemistry* **33,** 15329 (1994). Copyright 1994 American Chemical Society.]

fluorescence intensity of the fluorescein donor, it is possible to determine the equilibrium binding constants and the association rates of DNA triplex formation, as well as the thermally induced melting behavior of the triple helix.

The binding constants for three DNA triple helices have been determined by fitting the fluorescence titration data points of I_{obs} vs $[A_0]$ with the theoretical binding isotherm based on Eq. (19) (Fig. 1). The length of the single-stranded oligonucleotide was varied between 11 and 15 nucleotides. Both the K_T (association constant for triplex formation) and the ΔG values obtained, as well as the effect of oligonucleotide length on K_T, were in excellent agreement with those obtained from an affinity cleavage assay.[45] FRET was not observed on the addition of a nonspecific, acceptor-labeled oligonucleotide. When a fluorescein-labeled duplex solution was mixed with an excess amount of tetramethylrhodamine-labeled oligonucleotide, the decrease in donor fluorescence intensity over time was directly related to the kinetics of triplex formation. The recovered pseudo-first-order rate constants, k_{obs}, and the second-order rate constants, k_2, were comparable in magnitude with those determined by a restriction endonuclease assay[50] and a filter-binding assay[46] for triplex DNA species of similar sizes. While the latter assays are discontinuous or indirect, the FRET measurement allows triple-helix formation to be monitored continuously in real time.

For a DNA triplex containing a donor-labeled duplex and an acceptor-labeled oligonucleotide, the efficiency of nonradiative energy transfer decreases as the dissociation of triplex to duplex occurs, resulting in an increase in the fluorescence intensity of the donor group as the temperature increases (Fig. 2a). The melting profile and T_m value determined by the FRET method for the DNA triplex were in excellent agreement with those for the triplex-to-duplex transition obtained from ultraviolet (UV) hypochromicity (Fig. 2b). While the UV triplex-to-duplex transition is usually overlapped with the duplex-to-single strand transition, the triplex melting profile based on FRET shows no interference from the duplex-to-single strand transition. The effects of amino linkers and dye labels on the stability of duplex and triplex DNA were also examined for a series of unmodified and modified duplex and triplex DNAs and the T_m values indicate that the addition of a linker or a linker–dye conjugate has minimal effect on the stability of the triple helix.

Holliday Junctions

Four-arm branched DNA structures, known as Holliday junctions, are the key intermediates formed during the genetic recombination of two duplex DNA molecules.[51] Cellular resolvases cleave the branched DNA structures into either parental DNA duplexes or recombinant products. Because the mode of resolution appears to depend on the three-dimensional structure at the branch point, it is important to characterize the structural features of Holliday junctions. A variety of biophysical and biochemical

[50] L. J. Maher III, P. B. Dervan, and B. J. Wold, *Biochemistry* **29,** 8820 (1990).
[51] R. Holliday, *Genet. Res.* **5,** 282 (1964).

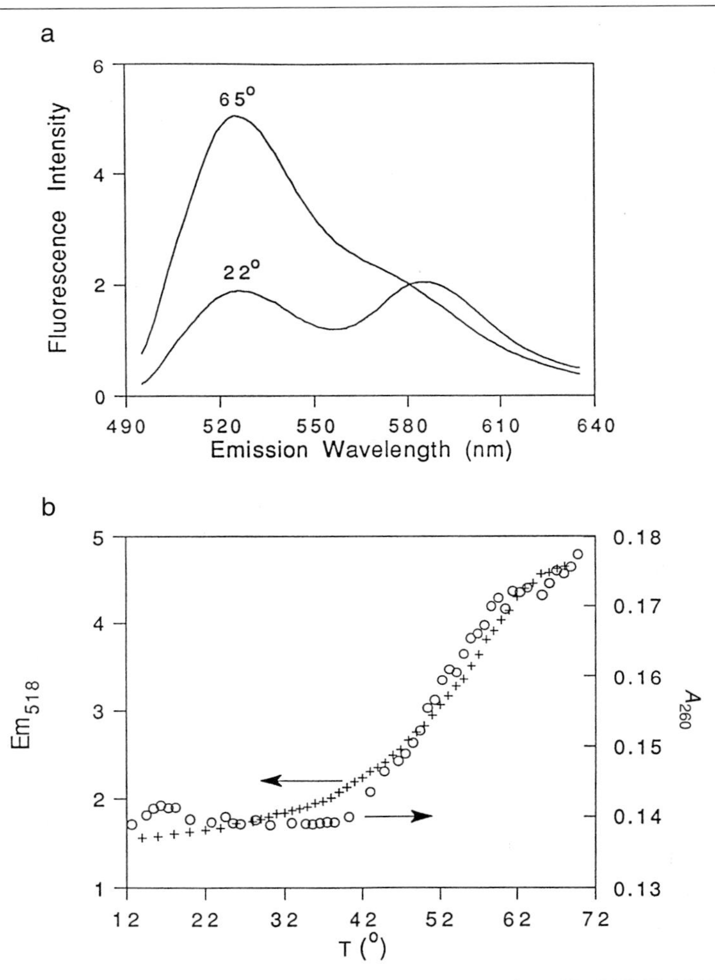

FIG. 2. Thermally induced melting of triple-helical DNA monitored by FRET. (a) Emission spectra for a mixture of the donor-labeled duplex F1–2 (1.0×10^{-6} M) and the acceptor-labeled oligonucleotide R5 (1.0×10^{-6} M) in association buffer equilibrated at 22 and 65°. The excitation wavelength was 485 nm. (b) Dissociation profiles of the DNA triplex F1–2–R5 (1.0×10^{-6} M) as observed by FRET (crosses) and UV absorbance hypochromicity (circles) measurements. The FRET melting curve was directly recorded by monitoring the fluorescence intensity of the fluorescein moiety at 518 nm with an excitation wavelength of 485 nm. The UV melting curve was obtained by subtracting the 260-nm absorbance of the DNA duplex F1–2 from that of a mixture of F1–2 + R5. The temperature was increased from 10 to 80° at a rate of 1°/min. [Reprinted with permission from M. Yang, S. Ghosh, and D. P. Millar, *Biochemistry* **33,** 15329 (1994). Copyright 1994 American Chemical Society.]

techniques have been used to probe the structure of Holliday junctions and these studies have been reviewed.[52,53] Most physical studies have focused on synthetic four-way DNA junctions in which the base sequence is designed to prevent branch migration.[54,55] Steady-state FRET measurements have played an important role in the determination of the global structure of the Holliday junction.[6–8,40,56] Clegg and co-workers have determined the efficiency of FRET between donor (fluorescein) and acceptor (tetramethylrhodamine) molecules attached pairwise in all possible permutations to the 5′ termini of the duplex arms of the four-way structure.[6,7,40] A comparison between a series of identical DNA molecules that were labeled at different positions was used to establish the stereochemical arrangement of the four DNA helices. The FRET results are in excellent agreement with the general conclusion from other biophysical studies that in the presence of cations, the junction folds into an X-shaped structure by pairwise coaxial stacking of duplex arms with antiparallel alignment of the two noncrossed strands.[52,53] The FRET measurements also revealed that at low salt concentrations, the Holliday junction exists as an unstacked, extended, square arrangement of the four duplex arms.[56]

Time-resolved FRET methods have also been used in studies of the Holliday junction. In the following example, we describe the use of the time-resolved FRET technique to estimate the range of distances present between each pair of arms in the Holliday junction and the flexibility of the junction in solution.[23] This information can be obtained by means of time-resolved measurements because the rate of resonance energy transfer between donors and acceptors attached to the junction arms depends strongly on the interarm distance, and a range of interarm distances gives rise to a characteristic nonexponential decay of the donor fluorescence.

We synthesized a series of four-way junctions with fluorescent donor and acceptor dyes attached to the 5′ termini of the duplex arms in all six pairwise combinations (Fig. 3). Care was taken to ensure that the local dye environment was the same at each labeled position, thereby ensuring the constancy of R_0 in each FRET measurement. The fluorescence decay of the donor was measured for all six doubly-labeled species and the D–A distance distributions were recovered according to Eqs. (10) and (11). Typical decay profiles and corresponding D–A distance distributions are shown in Fig. 4. Each distribution represents the probability that a particular

[52] D. M. J. Lilley and R. M. Clegg, *Annu. Rev. Biophys. Biomol. Struct.* **22,** 299 (1993).
[53] N. C. Seeman and N. R. Kallenbach, *Annu. Rev. Biophys. Biomol. Struct.* **23,** 53 (1994).
[54] N. C. Seeman, *J. Theor. Biol.* **99,** 237 (1982).
[55] N. C. Seeman and N. R. Kallenbach, *Biophys. J.* **44,** 201 (1983).
[56] R. M. Clegg, A. I. H. Murchie, and D. M. J. Lilley, *Biophys. J.* **66,** 99 (1994).

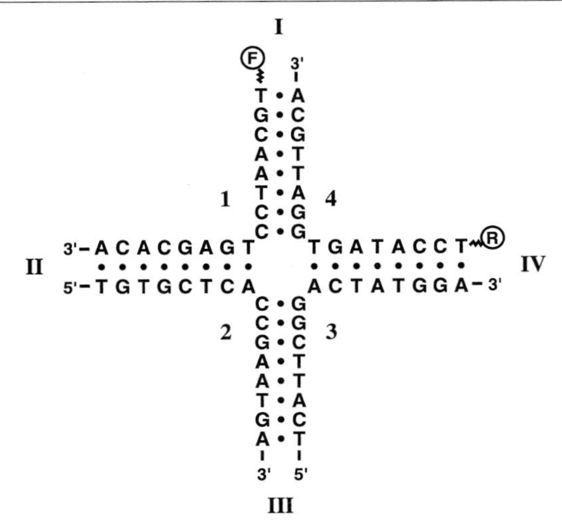

FIG. 3. Base sequence of the Holliday junction used for time-resolved FRET measurements. The junction is composed of four 16-base oligonucleotides (labeled 1–4) that hybridize to form a structure with four duplex arms (designated I–IV). The strand sequences are designed to immobilize the position of the branch point.[55] Two of the strands (strands 1 and 4 in the example shown here) are conjugated at their 5′ ends to either fluorescein (F) or tetramethylrhodamine (R). Each pair of junction arms can be labeled in this manner by hybridizing suitable combinations of labeled and unlabeled strands. A complete set of doubly labeled junctions for FRET measurements consists of six pairwise combinations of donor and acceptor positions. [Reprinted with permission from P. S. Eis and D. P. Millar, *Biochemistry* **32,** 13852 (1993). Copyright 1993 American Chemical Society.]

D–A distance is present in the labeled junction species. The width of the distribution is representative of the range of distances between the 5′ ends of the strands as reported by the donor and acceptor chromophores. The overall spatial agreement of the four junction arms was deduced by comparison of the mean D–A distances in the six doubly labeled junction constructs. The results were consistent with the expected X-shape structure for the four-way junction and further specified the arm-stacking arrangement and the relative orientation of the junction strands (Fig. 5). The interesting new finding revealed by the time-resolved FRET analysis was the observation of broad distributions of distances between certain pairs of junction arms. The end-to-end distances measured along the two continuous stacking domains are described by relatively narrow distributions (that simply reflect the flexibility of the dye linkages), indicating that these domains are relatively rigid, whereas the distances along the sides of the X structure have broad distributions (Fig. 4). These observations can be rationalized in terms of a dynamic model of the Holliday junction, whereby the two helical

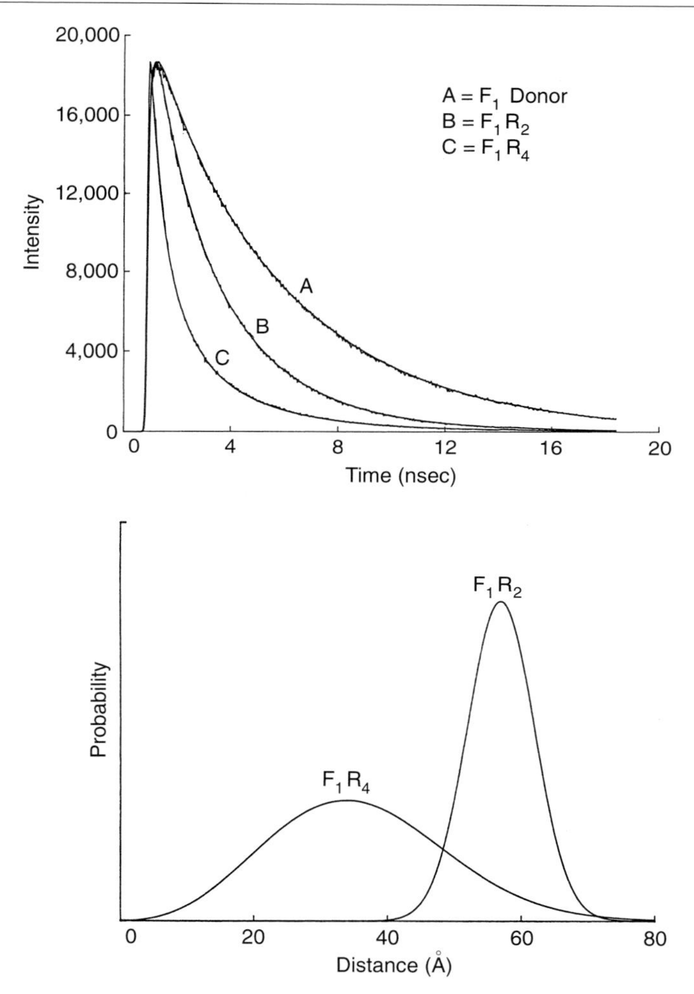

FIG. 4. Donor decay profiles and D–A distance distributions in labeled Holliday junctions. *Top*: Fluorescence decay curves of the fluorescein donor in donor-only junction F1 (curve A), D–A junction F1R2 (curve B), and D–A junction F1R4 (curve C). Labeled junctions are referred to by specifying the labeled strand(s) and dye combinations. Donors were excited at 514.5 nm, and their fluorescence emission was observed at 530 nm. The decay profiles for the donor-only and D–A junctions were analyzed as described in text. *Bottom*: D–A distance distributions in junctions F1R2 and F1R4 recovered from analysis of the corresponding donor decay profiles shown in the upper panel. The distributations are normalized to unit area. [Reprinted with permission from P. S. Eis and D. P. Millar, *Biochemistry* **32,** 13852 (1993). Copyright 1993 American Chemical Society.]

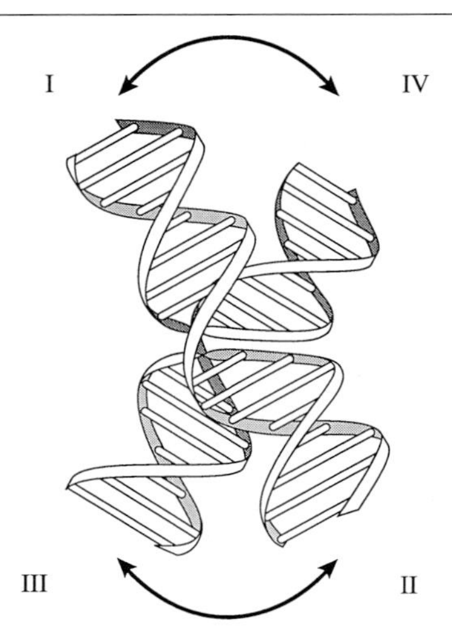

FIG. 5. Simplified ribbon model of the antiparallel stacked-X structure of the Holliday junction, showing the arm-stacking arrangements and modes of internal flexibility identified by time-resolved FRET. The junction arms are labeled as in Fig. 3. A scissors-like rotation of the two rigid stacking domains is indicated by the curved arrows. These rotations produce a range of distances along the short and long sides of the X structure without changing the end-to-end distances along the two double-helical domains. [Reprinted with permission from P. S. Eis and D. P. Millar, *Biochemistry* **32,** 13852 (1993). Copyright 1993 American Chemical Society.]

domains in the stacked-X structure rotate with respect to each other in a manner akin to the opening and closing of a pair of scissors (depicted by the arrows in Fig. 5). The major conclusion of the time-resolved FRET study is that the Holliday junction exhibits considerable conformational flexibility and does not adopt a unique structure in solution.

The analysis of the time-resolved FRET data for the Holliday junction considered alternative models based on a single distance between each pair of arms, as well as two discrete distances. Comparison of the reduced χ^2 values and weighted residuals from curve fitting served to confirm the uniqueness of the Gaussian distribution model. Furthermore, distance measurements made with the dyes in opposite orientations showed that there was no preferential interaction of the dyes with the DNA that might bias the energy transfer measurements. Finally, orientational effects on energy transfer were shown to be negligible because the donor and acceptor dyes both had considerable rotational mobility, as manifested in the time-resolved anisotropy decay of each dye.

Three-Way DNA Junctions

Three-way junctions are the simplest examples of branched nucleic acid structures. Structural and biophysical studies of three-way junctions may therefore provide insight into the interactions that govern the folding and stabilization of branched nucleic acids in general. Three-way junctions are also a common structural motif in single-stranded RNA molecules, such as 5S RNA and the hammerhead ribozyme. Moreover, three-way DNA junctions are involved in certain types of replication and recombination events.[57,58] Biologically active three-way junctions frequently contain one or more unpaired bases at the branch point, raising the question of how unpaired bases affect the folding and stability of these species. Gel electrophoresis and enzymatic ligation experiments[59–61] have shown that the bulged bases perturb the overall geometry and the dynamic structure of three-way DNA junctions. Calorimetric[62] and nuclear magnetic resonance (NMR) studies[63–65] have demonstrated that the addition of two unpaired bases to a three-way junction increases its thermodynamic stability, and that the unpaired bases can either stack into the junction or adopt an extrahelical conformation. We have used time-resolved FRET methods to investigate how bulges affect the overall geometry and conformational flexibility of three-way DNA junctions.[24]

Three-way DNA junctions were prepared using synthetic oligonucleotides and the bulged bases in the junctions were formed by adding two extra thymine, cytosine, adenine, or guanine bases to one junction strand (Fig. 6). The junction arms were labeled in pairwise fashion with donor and acceptor dyes, and the decay of donor fluorescence was measured by time-correlated single-photon counting. The basic strategy is similar to that adopted in the study of four-way DNA junctions. In the junction without added bases, two of the three interarm distances are identical, while the remaining distance is slightly larger, indicating that the overall structure of the junction is somewhat asymmetric (Fig. 7). The three-way junction ap-

[57] T. Minigawa, A. Murakami, Y. Ryo, and H. Yamagashi, *Virology* **126,** 183 (1983).

[58] F. Jensch and B. Kemper, *EMBO J.* **5,** 181 (1986).

[59] J. B. Welch, D. R. Duckett, and D. M. J. Lilley, *Nucleic Acids Res.* **21,** 4548 (1993).

[60] M. Zhong, M. S. Rashes, N. B. Leontis, and N. R. Kallenbach, *Biochemistry* **33,** 3660 (1994).

[61] L. S. Shlyakhtenko, E. Appella, R. E. Harrington, I. Kutyavin, and Y. L. Lyubohenko, *J. Biomol. Struct. Dynam.* **12,** 131 (1994).

[62] N. B. Leontis, W. Kwok, and J. S. Newman, *Nucleic Acids Res.* **19,** 759 (1991).

[63] M. A. Rosen and D. J. Patel, *Biochemistry* **32,** 6563 (1993); M. A. Rosen and D. J. Patel, *Biochemistry* **32,** 6576 (1993).

[64] N. B. Leontis, M. T. Hills, M. Piotto, I. V. Ouporov, A. Malhotra, and D. G. Gorenstein, *Biophys. J.* **68,** 251 (1994).

[65] I. V. Ouporov and N. B. Leontis, *Biophys. J.* **68,** 266 (1995).

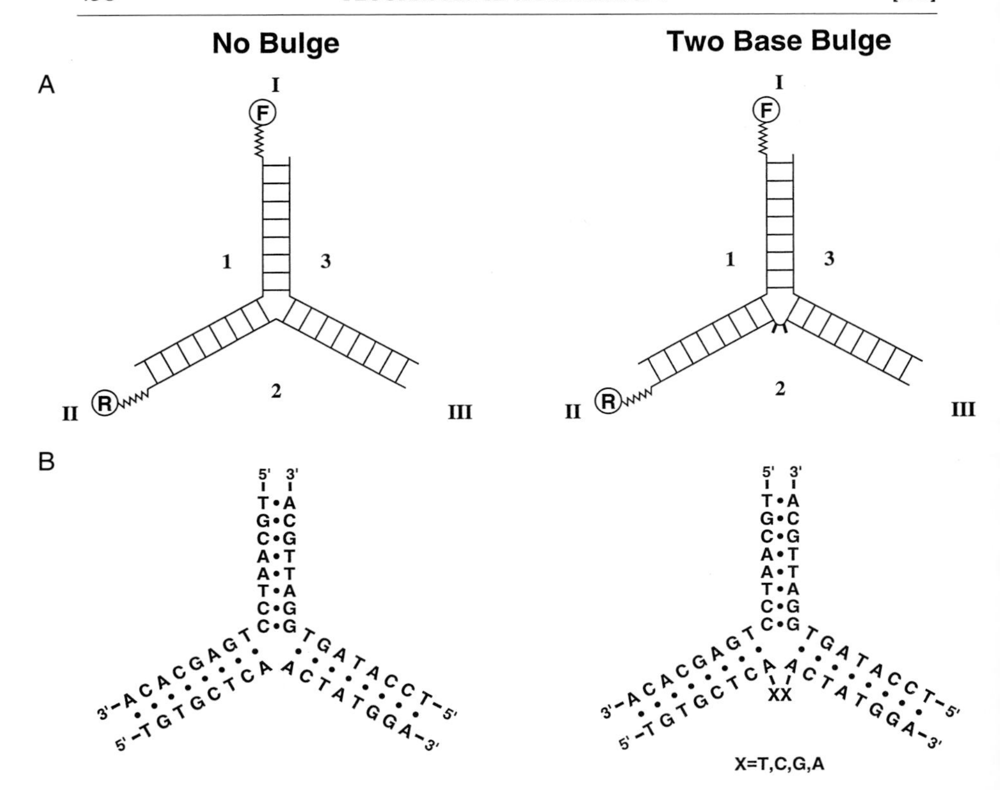

FIG. 6. Three-way DNA junctions used in time-resolved FRET studies. (A) Schematic representation of three-way junctions without added bases (left-hand side) or with various added bases in one strand (right-hand side). In the right-hand diagram, the two extra bases introduce a two-base bulge at the branch point of the junction. Oligonucleotide strands are labeled 1–3. Duplex arms are labeled I–III. The ends of the junction arms are labeled pairwise with fluorescein (F) and tetramethylrhodamine (R) for measurements of interarm distances and distance distributions (arms I and II in the example here). Two other pairwise combinations of the labeled arms can also be prepared. (B) Sequences of the three-way DNA junctions. In the bulged junctions, the bulge composition was varied as shown. [Reprinted with permission from M. Yang and D. P. Millar, *Biochemistry* **35,** 7959 (1996). Copyright 1996 American Chemical Society.]

pears to have a uniformly flexible structure, as judged by the similarity in the widths of the three interarm distance distributions (Fig. 7). Similar to the four-way junction study, the effect of the linker flexibility, and orientation of dye molecules, were considered and shown to have a negligible effect on the recovered D–A distance distributions. The validity of the weighted Gaussian distribution for the D–A distances was also verified.

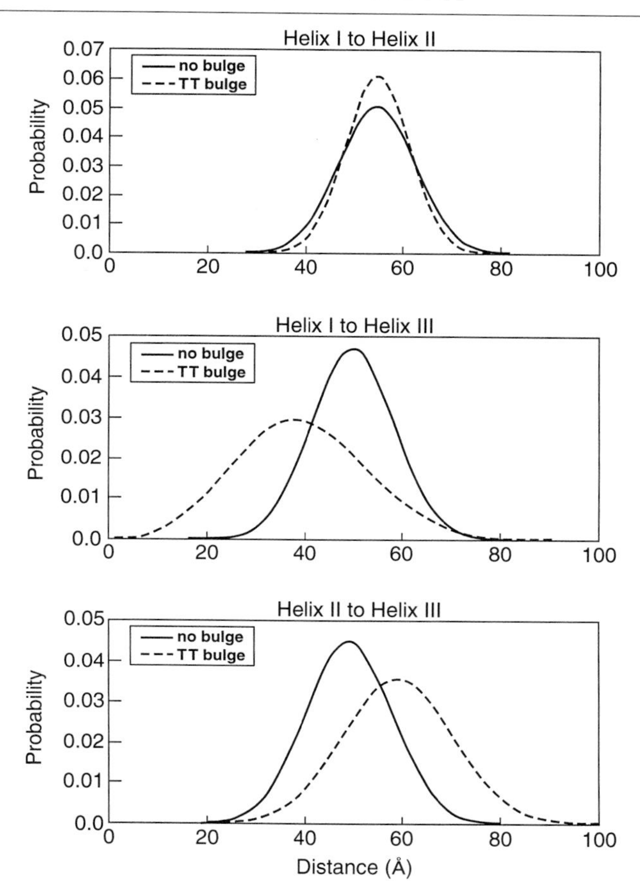

FIG. 7. Interarm distance distributions in three-way DNA junctions. Distributions were recovered from analysis of donor decay profiles according to Eqs. (10) and (11). Distributions are shown for a three-way junction without added bases (solid lines) and for the corresponding junction with two extra thymine bases in strand 2 (dashed lines). See Fig. 6 for the junction sequences. [Reprinted with permission from M. Yang and D. P. Millar, *Biochemistry* **35,** 7959 (1996). Copyright 1996 American Chemical Society.]

The time-resolved FRET analysis for the bulged junctions revealed that the overall geometry and flexibility of the complex are significantly affected by the presence of unpaired bases at the branch point. The interarm distance distributions for the three-way junction containing two added thymine bases are compared with the corresponding distributions for the perfect junction in Fig. 7. Addition of the unpaired thymine bases increases the distance between helices II and III, whereas helices I and III move closer together.

There is also a significant broadening in the range of distance between helix III and these two other helices. In contrast, the distance between helices I and II is hardly affected by the bulged bases, both in terms of the mean distance and the distribution of distances. Similar results were obtained for bulged junctions containing added cytosine and adenine bases. These data indicate that the bulged nucleotides are accommodated by the repositioning of helix III, which also becomes more mobile in the bulged junctions (Fig. 8).

The addition of two unpaired guanine bases was found to have a different effect on the overall structure and flexibility of the complex. The bulged guanines appear to specifically affect helix II, causing it to move closer to

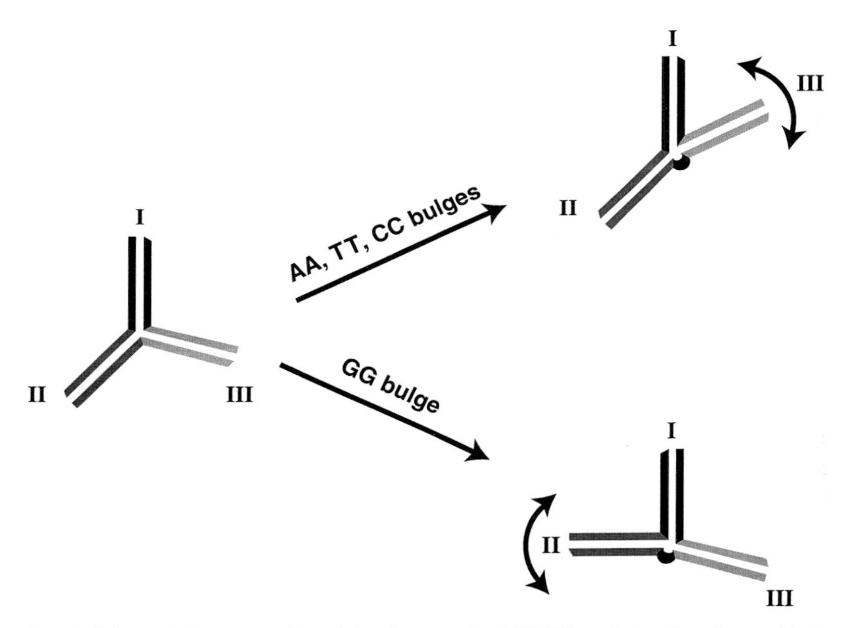

FIG. 8. Schematic interpretation of the time-resolved FRET analysis of perfect and bulged three-way DNA junctions. The overall structure of the perfect junction is asymmetric, with the I–II distance being somewhat larger than the other two interarm distances. Addition of two extra thymine, cytosine, or adenine bases at the branch point of the junction causes helix III to move closer to helix I, and further away from helix II, while having little effect on the distance between helices I and II. Moreover, helix III has greater mobility in the bulged junctions and can sample a wide range of positions, indicated by the curved arrow. In contrast, addition of two extra guanine bases to the three-way junction displaces helix II, causing it to move closer to helix I, while having no effect on the distance between helices I and III. In addition, helix II has greater mobility in the bulged junction, also indicated by the curved arrow. [Reprinted with permission from M. Yang and D. P. Millar, *Biochemistry* **35,** 7959 (1996). Copyright 1996 American Chemical Society.]

helix I, while also increasing its mobility (Fig. 8). In contrast, the spatial arrangement of helices I and III is unaffected by the bulged bases. This strongly suggests that unpaired guanine bases adopt a different conformation in the three-way junction than do the other bulges. The results of this study showed that the overall geometry of bulged three-way junctions is markedly asymmetric, owing to the repositioning of one of the helices flanking the bulge site, and that the mobility of the perturbed helix is also markedly enhanced. Interestingly, the nature of the unpaired bases dictates which of the two helices flanking the bulge site is affected in this manner.

Duplex DNA

One of the most important applications of FRET is in the study of duplex DNA structure. Resonance energy transfer between fluorescein and eosin site-specifically labeled on phosphorothioate-modified DNA backbones was used to estimate the distances between preselected sites.[66] The end-to-end distances of a series of DNA oliogonucleotides (ranging from 8 to 20 bp) labeled with fluorescein and rhodamine were evaluated by measurements of FRET and the helical geometry of double-stranded DNA in solution was observed.[67] FRET methods have also been used to determine helical handedness in hybrid DNA molecules comprised of two helical segments, one of which is in a known helical reference state.[68] By varying the relative number of base pairs in each segment and monitoring the change in the efficiency of FRET between dyes located at the ends of the two segments, it is possible to deduce the helical twist and rise of the unknown segment. FRET has also been used to probe the primary and secondary structures of single-stranded nucleic acids and to reveal the formation of hairpin structures and the translocation of genes between two chromosomes.[69] Hybridization and denaturation of DNA, both *in vitro*[11,69–72] and intracellular,[73] have also been detected by FRET. Nonradia-

[66] H. Ozaki and L. W. McLaughlin, *Nucleic Acids Res.* **20,** 5205 (1992).

[67] R. M. Clegg, A. I. H. Murchie, A. Zechel, and D. M. J. Lilley, *Proc. Natl. Acad. Sci. U.S.A.* **90,** 2994 (1993).

[68] E. A. Jares-Erijman and T. M. Jovin, *J. Mol. Biol.* **257,** 597 (1996).

[69] J. L. Mergny, A. S. Boutorine, T. Garestier, F. Belloc, M. Rougee, N. V. Bulychev, A. A. Koshkin, J. Bourson, A. V. Lebedev, B. Valeur, T. N. Thuong, and C. Hélène, *Nucleic Acids Res.* **22,** 920 (1994).

[70] L. E. Morrison and L. M. Stols, *Biochemistry* **32,** 3095 (1993).

[71] M. Hiyoshi and S. Hosoi, *Anal. Biochem.* **221,** 306 (1994).

[72] K. M. Parkhurst and L. J. Parkhurst, *Biochemistry* **34,** 285 (1995).

[73] S. Sixou, F. C. Szoka, Jr., G. A. Green, B. Giusti, G. Zon, and D. J. Chin, *Nucleic Acids Res.* **22,** 662 (1994).

tive energy transfer has also been employed in the development of fluorescent dye-labeled DNA primers for DNA sequencing and analysis.[74]

The accuracy and limitations of FRET measurements in nucleic acids with respect to the localization of the dyes and the flexibility of the dye–DNA linkages have been studied by time-resolved FRET methods.[16,21] The flexibility of the dye–DNA linkages was reflected in the distribution of donor–acceptor distances between dyes attached at opposite ends of a short DNA helix.[16] Donor–acceptor distance distributions in a double-labeled oligonucleotide, both as a single strand and in duplexes, were determined and, similarly, the flexibility of the polymer was implicated.[21] The fluorescence decays of a variety of linear DNA-bound donor–acceptor pairs were analyzed with frequency-domain spectroscopy and described satisfactorily by the Förster model of energy transfer in one dimension.[22]

The FRET technique has been used to demonstrate the bending of DNA and RNA helices for a series of double-stranded DNA and RNA molecules containing bulge loops of unopposed adenosine nucleotides (A_n, $n = 0$–9).[10] Fluorescein and tetramethylrhodamine were covalently attached to the 5′ termini of the two component strands. The extent of energy transfer within each series increased as the number of bulged nucleotides varied from 1 to 7, indicating a shortening of the end-to-end distance. This is consistent with a bending of DNA and RNA helices that is greater for larger bulges and the ranges of bending angles were estimated from the FRET results. In an earlier report,[75] FRET has been used to determine the distortion of duplex DNA in a protein–DNA complex. Both steady-state and time-resolved FRET measurements showed that in a 26-bp Lac DNA fragment complexed with cyclic AMP receptor protein, the end-to-end distance is about 77 Å, which corresponds to a bending angle between 80 and 100°, depending on the actual contour length between the fluorophores in the free DNA fragment.

DNA Enzymatic Assays

FRET methods are also useful for monitoring the activity of enzymes that act on DNA substrates. The kinetics of *Pae*R7 endonuclease-catalyzed cleavage of fluorophore-labeled oligonucleotide substrates have been examined by steady-state FRET measurements.[13] A series of duplex substrates were synthesized with an internal CTCGAG *Pae*R7 recognition site and donor (fluorescein) and acceptor (rhodamine) dyes conjugated to the opposing 5′ termini. The time-dependent increase in donor fluorescence re-

[74] J. Ju, C. Ruan, C. W. Fuller, A. N. Glazer, and R. A. Mathies, *Proc. Natl. Acad. Sci. U.S.A.* **92,** 4347 (1995).

[75] T. Heyduk and J. C. Lee, *Biochemistry* **31,** (5165) 1992.

sulting from restriction cleavage of these substrates was continuously monitored and the initial rate data was fitted to the Michaelis–Menten equation. The steady-state kinetic parameters for these substrates were in agreement with the rate constants obtained from a gel electrophoresis-based fixed time point assay using radiolabeled substrates. The FRET method provides a rapid continuous assay as well as high sensitivity and reproducibility. Similar FRET-based DNA enzymatic assays have been developed for *Eco*RV endonuclease[76] and human immunodeficiency virus type 1 (HIV-1) integrase.[77] Fluorescence energy transfer methods have also been used to monitor the DNA-unwinding activity of helicases.[14,78,79] One of the most important features of the FRET-based assays is that they allow the DNA cleavage or unwinding reactions to be continuously monitored in real time under homogeneous solution conditions.

RNA Structure

While the emphasis of this chapter is on the application of FRET methods to the study of DNA structure and function, it should be noted that the technique is also applicable to RNA. In addition to the study of kinking of RNA helices by bulged nucleotides,[10] FRET has been used to determine the orientations of transfer RNA in the ribosomal A and P sites[79] and the macromolecular arrangement in the aminoacyl-tRNA-elongation factor Tu-GTP ternary complex.[80] In a more recent study, FRET has been used to establish the relative spatial orientation of the three constituent Watson–Crick base-paired helices constituting the RNA hammerhead ribozyme.[9] Synthetic constructs were labeled with the fluorescence donor (5-carboxyfluorescein) and acceptor (5-carboxytetramethylrhodamine) located at the ends of each helix. The acceptor helix in helix pair I and III, and in II and III, was varied in length from 5 to 11 and 5 to 9 base pairs, respectively, and the FRET efficiencies were determined and correlated with a reference set of labeled RNA duplexes. The FRET efficiencies were predicted on the basis of vector algebra analysis, as a function of the relative helix orientations in the ribozyme constructs, and compared with experimental values. The data were consistent with a Y-shaped arrangement of the ribozyme with helices I and II in close proximity and helix III pointing

[76] S. G. Erskine and S. E. Halford, *Gene (Amsterdam)* **157,** 153 (1995).

[77] S. P. Lee, H. G. Kim, M. L. Censullo, and M. K. Han, *Biochemistry* **34,** 10205 (1995).

[78] P. Houston and T. Kodakek, *Proc. Natl. Acad. Sci. U.S.A.* **91,** 5471 (1994).

[79] T. R. Easterwood, F. Major, A. Malhotra, and S. C. Harvey, *Nucleic Acids Res.* **22,** 3779 (1994).

[80] B. S. Watson, T. L. Hazlett, J. F. Eccleston, C. Davis, D. M. Jameson, and A. E. Johnson, *Biochemistry* **34,** 7904 (1995).

away. The orientational constraints were used for molecular modeling of a three-dimensional structure of the complete ribozyme. The proposed structure based on FRET measurements is in general agreement with that determined by X-ray crystallography.[81]

Acknowledgments

We acknowledge the contributions of Peggy Eis and Shoumitra Ghosh to the FRET studies performed in the laboratory. Work in the authors' laboratory was funded by an operating grant from the National Science Foundation (MCB-9317369).

[81] H. W. Pley, K. M. Flaherty, and D. M. McKay, *Nature* (*London*) **372,** 68 (1994).

[21] Energy Transfer Methods for Detecting Molecular Clusters on Cell Surfaces

By JANOS MATKO and MICHAEL EDIDIN

Introduction

Modern ideas about the organization of cell plasma membranes begin with the so-called "fluid mosaic" model.[1] This model emphasizes the molecular mobility and the autonomy of individual membrane molecules. However, since the model first appeared more than 20 years ago, it has become apparent that there is a high degree of lateral organization of plasma membranes. On the scale of angstroms to nanometers this organization has been detected by immunoprecipitation experiments [examples for insulin receptors and major histocompatibility complex (MHC) molecules reviewed in Ref. 2; other examples reviewed in Ref. 3], or by other immunological techniques, such as cocapping.[4] These methods suffer from serious disadvantages. Immunoprecipitation begins with dissolution of cell membranes and dispersion of their component molecules. Cocapping perturbs native membrane organization with cross-linking antibodies. In contrast,

[1] S. J. Singer and G. Nicolson, *Science* **175,** 720 (1972).

[2] J. Reiland and M. Edidin, *Diabetes* **42,** 619 (1993).

[3] J. Szollosi and S. Damjanovich, Mapping of membrane structure by energy transfer measurements. *In* "Mobility and Proximity in Biological Membranes" (S. Damjanovich, M. Edidin, J. Szollosi, and L. Tron, eds.), p. 49. CRC Press, Boca Raton, Florida, 1994.

[4] U. Dianzani, M. Bragardo, D. Buonfiglio, V. Redoglia, A. Funaro, P. Portoles, J. Rojo, F. Malavasi, and A. Pileri, *Eur. J. Immunol.* **25,** 1306 (1995).

0076-6879/97 $25

resonance energy transfer (RET) methods can be used to detect molecular clusters with little or no perturbation of membrane structure. Furthermore, with sufficient signal, the methods can resolve changes in this organization consequent to changes in cell physiology.[5]

Study of Molecular Clusters

A fluid mosaic membrane allows interacting molecules to meet and form clusters through lateral diffusion. Some of these molecular clusters, for example, acetylcholine receptors at neuromuscular junctions[6] and antigen–receptor complexes of B and T lymphocytes,[7,8] are stable on a time scale of hours to days. Such clusters are functional units for receiving and transmitting signals. In other cases clusters may form as a consequence of stimuli from peptide hormones[9] or antibodies,[10] or from receptors on other cells.[11] Another class of clusters reports the arrival of newly synthesized molecules of some proteins at the cell surface.[12] Still other molecular clusters, clusters of MHC molecules may be important for antigen presentation.[13,14]

Molecular clusters in native cell membranes may be detected and characterized by RET methods. In this chapter we take our examples from studies of the clustering of class I MHC. Our approach to cell RET to detect molecular clustering is illustrated within the terms of this problem.

Physical Basis of Resonance Energy Transfer

Classic Förster-Type Fluorescence Resonance Energy Transfer

The history of the fluorescence resonance energy transfer (FRET) theory has been reviewed by Clegg.[15] Here we restrict ourselves to a brief

[5] R. S. Mittler, S. J. Goldman, G. L. Spitalny, and S. J. Burakoff, *Proc. Natl. Acad. Sci. U.S.A.* **86,** 8531 (1989).

[6] M. Gesemann, A. J. Denzer, and M. A. Ruegg, *J. Cell Biol.* **128,** 625 (1995).

[7] C. M. Pleiman, D. D'Ambrosio, and J. C. Cambier, *Immunol. Today* **15,** 393 (1994).

[8] H. Clevers, B. Alarcon, T. Wileman, and C. Terhorst, *Annu. Rev. Immunol.* **6,** 629 (1988).

[9] A. Ullrich and J. Schlessinger, *Cell* **16,** 203 (1990).

[10] H. Metzger, G. Alcaraz, R. Hohman, J.-P. Kinet, V. Pribluda, and R. Quarto, *Annu. Rev. Immunol.* **4,** 419 (1986).

[11] C. A. Janeway, Jr., *Annu. Rev. Immunol.* **10,** 645 (1992).

[12] L. A. Hannan, M. P. Lisanti, E. Rodriguez-Boulan, and M. Edidin, *J. Cell Biol.* **120,** 353 (1993).

[13] J. Matkó, Y. Bushkin, T. Wei, and M. Edidin, *J. Immunol.* **152,** 3353 (1994).

[14] G. G. Capps, B. E. Robinson, K. D. Lewis, and M. C. Zuniga, *J. Immunol.* **151,** 159 (1993).

[15] R. M. Clegg, Fluorescence resonance energy transfer (FRET). *In* "Fluorescence Imaging Spectroscopy and Microscopy" (X. F. Wang and B. Herman, eds.), 1996 (in press).

description of the basic phenomenon. In classic FRET energy is transferred nonradiatively from an excited donor molecule to an acceptor molecule. The process takes place through a long-range dipole–dipole coupling.[16] The rate of energy transfer is inversely proportional to the sixth power of the distance between the donor and acceptor; hence, FRET efficiency is extremely sensitive to dipole–dipole distances. FRET is most efficient when this distance is 1–10 nm. The method thus becomes an excellent tool for mapping molecular proximity and molecular interactions at the cell surface.

Let us consider a system with two different fluorophores, where the molecule with higher energy (shorter wavelength) absorption is defined as the donor (D) and the one with lower energy (longer wavelength) absorption as acceptor (A). If the donor is in the excited state, it will lose energy by internal conversion until it reaches the lowest vibrational level of the first excited state. If the donor emission energies overlap with the acceptor absorption energies, the following resonance can occur by means of weak coupling between the dipoles:

$$D^* + A \rightleftharpoons D + A^*$$

where D and A denote the donor and the acceptor molecules in ground state, while D* and A* denote the first excited states of the fluorophores. The rate of the forward process is k_T, while the rate of the inverse process is k_{-T}. Because vibrational relaxation converts the excited acceptor into the ground vibrational level, the inverse process is highly unlikely to occur. As a result the donor molecules are quenched, while the acceptor molecules are excited and, under favorable conditions, emit photons (sensitized emission).

According to the theory of Förster, the rate (k_T) and efficiency (E) of the resonance energy transfer can be derived as

$$k_T = \text{const } J k_F n^{-4} \kappa^2 R^{-6} \tag{1}$$

and

$$E = k_T/(k_T + k_F + k_{ND}) \tag{2}$$

where k_F is the rate constant of fluorescence emission of the donor (in the absence of acceptor) and k_{ND} is the sum of the rate constants of all other nonradiative deexcitation processes for the donor. R is the separation distance between the donor and acceptor dipoles, κ^2 is an orientation factor that is a function of the relative spatial orientation of the donor emission dipole and the acceptor absorption dipole, n is the refractive index of the intervening medium, and J is the spectral overlap integral, which is the

[16] T. Förster, *Discuss. Faraday Soc.* **27,** 7 (1959).

measure of the overlap between the emission spectrum of the donor and the absorption spectrum of the acceptor:

$$J = \int_0^{\infty} [\varepsilon_A(\lambda)\lambda^4] F_D(\lambda) d\lambda \tag{3}$$

where $F_D(\lambda)$ is the fluorescence intensity of the donor at wavelength λ, and $\varepsilon_A(\lambda)$ is the molar extinction coefficient of the acceptor.

For dipole–dipole resonance energy transfer the orientation factor can be derived as

$$\kappa^2 = (\cos\alpha - 3\cos\beta\cos\gamma)^2 \tag{4}$$

where α is the angle between the transition moments of the donor and the acceptor, and β and γ are the angles between the line connecting donor to acceptor and their transition moments.

Uncertainties in the value of κ^2 can cause the greatest error in distance determination by enery transfer. [Fortunately R depends on $(\kappa^2)^{1/6}$.] Its direct measurement is impossible; however, fluorescence anisotropy measurements may give estimates for its range of values.[17,18] From theoretical considerations κ^2 varies between 0 and 4. If a large number of random orientations of the donor and the acceptor dipoles are sampled during the singlet lifetime, κ^2 becomes 2/3, a statistical average. Changes of κ^2 between 1 and 4 may result in at most a 25% error in distance estimates but an unfavorable (perpendicular) orientation of dipoles ($\kappa^2 = 0$) may cause serious errors in FRET measurement.

The efficiency of the resonance energy transfer (E) is related to the dipole–dipole distance as

$$E = R_0^6/(R^6 + R_0^6) \tag{5}$$

where R_0 is the so-called critical Förster distance, at which the efficiency of the energy transfer is 50%:

$$R_0 = 9.7 \times 10^2 (J\Phi_D\kappa^2 n^{-4})^{1/6} \quad \text{(nm)} \tag{6}$$

where Φ_D is the fluorescence quantum yield of the donor in the absence of acceptor.

Selection of Proper Donor–Acceptor Pairs. The choice of a proper donor–acceptor pair for an effective energy transfer should generally meet a number of theoretical and practical requirements.

1. The emission spectrum of the donor and the excitation spectrum of the acceptor should overlap significantly; R_0 should be large [see Eqs.

[17] R. E. Dale and J. Eisinger, *Biopolymers* **13,** 1573 (1974).
[18] R. E. Dale, J. Eisinger, and W. Blumberg, *Biophys. J.* **26,** 161 (1979).

(1), (3), and (6)]. An excellent technical review[19] includes a useful and comprehensive list of R_0 values for a large number of donor–acceptor pairs. For mapping intermolecular proximities at the cell surface the most frequently used dye pairs are fluorescein (donor)–tetramethylrhodamine (acceptor), R_0 49–54 Å[19] and fluorescein (donor)–Texas Red (acceptor), R_0 39 Å.[20] The fluorescein (donor)–phycoerythrin (PE) (acceptor) pair is also useful for probing molecular clusters at the cell surface. The large R_0 for this pair, ~100 Å, and the extremely high molar absorption coefficient of PE[21] make it especially sensitive for FRET.

2. The quantum yield of donor (Φ_D) and the molar absorption coefficient of the acceptor (ε_A) should be sufficiently high (e.g., $\Phi_D > 0.05$ and $\varepsilon_A > 10{,}000$).

3. The absorption maximum of donor should be well matched with the emission maximum of the light source (lamp or laser).

4. There should be little or no direct excitation of the acceptor at the excitation wavelength of the donor.

5. A large Stokes shift between the absorption and emission spectrum of the donor helps to avoid donor–donor self-transfer.

6. For labeling of proteins (especially antibodies) the selected dyes should have a relatively small size to minimize conformational perturbation of the host biomolecule and they should exhibit a sufficiently low hydrophobicity to avoid nonspecific, adsorptive interactions.

7. If possible, the donor and acceptor molecules should have a high degree of rotational freedom, so that κ^2 is averaged. This is often the case for fluorophores attached to biomolecules through molecular spacers. If this is not the case then it may be possible to use rare earth chelates as donors. These have a low emisison anisotropy and hence FRET data using these donors are insensitive to κ^2.[20]

Determination of Transfer Efficiency. The efficiency of the energy transfer, E [see Eqs. (2) and (5)], can be determined experimentally in a number of different ways. Because energy is transferred from the excited donor to the acceptor, the lifetime, quantum efficiency, or the fluorescence intensity of the donor decrease, if the acceptor is present:

$$E = 1 - \tau_D^A/\tau_D = 1 - \Phi_D^A/\Phi_D = 1 - F_D^A/F_D \tag{7}$$

As a consequence the fluorescence intensity of the acceptor (excited in the excitation band of the donor) increases if the donor is also present:

[19] P. G. Wu and L. Brand, *Anal. Biochem.* **218,** 1 (1994).

[20] J. Matko, A. Jenei, T. Wei, and M. Edidin, *Cytometry* **19,** 191 (1995).

[21] G. Szabo, Jr., J. L. Weaver, P. S. Pine, P. E. Rao, and A. Aszalos, *Biophys. J.* **68,** 1170 (1995).

$$F_A^D/F_A = 1 + (\varepsilon_D C_D/\varepsilon_A C_A)E \tag{8}$$

In Eqs. (7) and (8) the sub- and superscripts refer to the donor (D) or acceptor (A). For example, F_D^A refers to donor fluorescence in the presence of acceptor. C_D and C_A are the molar concentrations of the donor and the acceptor, respectively. Detailed descriptions of evaluation of energy transfer data can also be found in several reviews.[19,22–24]

Calculation of intermolecular distances from energy transfer efficiencies is relatively easy in the case of a single donor–single acceptor system, if the localization and relative orientation of the fluorophores are known. However, when investigating cell membrane components, a two-dimensional restriction applies for the labeled molecules. There are some excellent theoretical papers describing in detail analytical solutions for randomly distributed donor and acceptor molecules and numerical solutions for nonrandom distribution.[24–29] However, it is usually enough to estimate molecular proximity in terms of E, the transfer efficiency, rather than in terms of absolute distance. Note that to differentiate between random and nonrandom distributions, energy transfer efficiency, E, must be determined at different acceptor densities.[30]

Detecting Molecular Clusters: Long-Range Electron Transfer. The long-range electron transfer (LRET) method involves an electron transfer reaction between an electron–donor and an electron–acceptor molecule located within a "sphere of action." The rate and distance interval of quenching strongly depend on the redox potential of the donor and acceptor molecules, and on the nature of the intervening medium[31,32]:

$$k_{ET} = A \exp\{-\Delta G^*/RT\} \tag{9}$$

here k_{ET} is the rate of the electron transfer, A is a preexponential factor, ΔG^* is the free energy of activation, which is related to the free energy of the

[22] S. Damjanovich, M. Edidin, J. Szollosi, and L. Tron, "Mobility and Proximity in Biological Membranes." CRC Press, Boca Raton, Florida, 1994.

[23] J. Matko, L. Matyus, J. Szollosi, L. Bene, A. Jenei, P. Nagy, A. Bodnar, and S. Damjanovich, *J. Fluoresc.* **4,** 303 (1994).

[24] J. Szollosi, S. Damjanovich, S. A. Mulhern, and L. Tron, *Prog. Biophys. Mol. Biol.* **49,** 65 (1987).

[25] K. Wolber and B. S. Hudson, *Biophys. J.* **28,** 197 (1979).

[26] B. Snyder and E. Freire, *Biophys. J.* **40,** 137 (1982).

[27] N. Estep and T. E. Thompson, *Biophys. J.* **26,** 195 (1979).

[28] T. G. Dewey and G. G. Hammes, *Biophys. J.* **32,** 1023 (1980).

[29] C. Gutierrez-Merino, *Biophys. Chem.* **14,** 259 (1981).

[30] B. Kwok-Keung Fung and L. Stryer, *Biochemistry* **17,** 5241 (1978).

[31] R. A. Marcus, *J. Chem. Phys.* **24,** 966 (1956).

[32] R. A. Marcus and N. Sutin, *Biochim. Biophys. Acta* **811,** 265 (1985).

redox reaction, ΔG°, and the total (internal + solvent shell) reorganization energy, λ, as follows:

$$\Delta G^{*} = (\lambda/4)(1 + \Delta G^{0})^{2} \tag{10}$$

Besides recognizing this unusual quadratic dependence of the activation free energy on the free energy of redox reaction, Marcus[31] and Dexter[33] predicted an exponential dependence of the electron transfer rate on the donor–acceptor separation distance:

$$k_{\mathrm{ET}}(R) = k_0 \exp\{-\beta(R - R_0)\} \tag{11}$$

where k_0 is the rate at the van der Waals contact distance, R_0, β characterizes the steepness of distance dependence, and R is the actual separation distance between the donor and the acceptor. The effective range of ET distances is often greatly increased by the opening of effective electron-tunneling pathways, for example through polypeptide backbones or polyhydroxy(carbohydrate) moieties of cell surface macromolecules.[34]

The predictions of Marcus and Dexter have been verified experimentally.[35–38] Long-range electron transfer (LRET) appears to be a useful molecular ruler in the range 0.5–2.5 nm. Application of LRET to cells has three advantages. First, only quenching of donor fluorescence is measured; hence, cells are rapidly screened and data analysis is simplified. Second, the small size and the hydrophilic character of the nitroxide quencher used in LRET reduce perturbation of antibody conformation by the label and also reduce nonspecific binding of antibodies to the cells.[20] The third and most important advantage of the method over FRET is the narrow range of effective ET distances as well as the reduced sensitivity to the spatial orientation of the donor and acceptor molecules, which make the approach extremely useful in detecting physical associations, clusters of proteins at the cell surface, with great certainty. A difficulty of LRET is that the stoichiometry of the applied unlabeled fluorescein isothiocyanate (FITC)-labeled, and nitroxide radical-labeled antibodies must be accurately determined and controlled. The equal competition of the labeled antibodies should also be tested. The method requires determination of the population means of FITC–monoclonal antibody (MAb) fluorescence at varying surface densities of nitroxide–

[33] D. L. Dexter, *J. Chem. Phys.* **21,** 836 (1953).

[34] J. J. Hopfield, *Proc. Natl. Acad. Sci. U.S.A.* **71,** 3640 (1974).

[35] P. Maróti, *J. Photochem. Photobiol. B: Biol.* **19,** 235 (1993).

[36] J. N. Betts, D. N. Beratan, and J. N. Onuchic, *J. Am. Chem. Soc.* **114,** 4043 (1992).

[37] J. Matkó, K. Ohki, and M. Edidin, *Biochemistry* **31,** 703 (1992).

[38] S. A. Green, D. J. Simpson, G. Zhou, P. S. Ho, and N. V. Blough, *J. Am. Chem. Soc.* **112,** 7337 (1990).

MAb (electron acceptor). The surface concentration of spin label must be measured in an electron paramagnetic resonance (EPR) spectrometer in order to determine the donor: quencher ratio accurately. However, an alternative approach can also be used in which donor fluorescence emission is compared in fully labeled samples and in samples treated with sodium ascorbate at a concentration sufficient to inactivate completely the nitroxide radicals (quenchers). The donor emission intensity of this sample is regarded then as the unquenched donor fluorescence. The emission of the same sample in the absence of ascorbate will provide the quenched fluorescence intensity.[20,23]

We have described LRET measurements on living cells.[20] This approach has been successfully applied to monitor homo-associations of class I MHC molecules on a number of different types of human lymphoid cells.[13,20,23,39] The results are in good agreement with FRET data on the same types of cells.

Limitations of Resonance Energy Transfer Methods for Cells

An ideal RET experiment would probe a transparent solution of labeled molecules. We expect that the population of molecules is pure and that it can be maintained unchanged in a cuvette for the time of the experiment. RET can be determined in such a sample using steady-state or lifetime fluorimeters. RET measurements on cells depart substantially from this ideal. To begin with, cells are relatively large and scatter light strongly. Hence, measurements of cell suspensions in fluorimeter cuvettes require great care in correcting for cell settling and for the effects of light scattering.[22,40] Another possible error source is the contribution from unbound fluorophores and cell debris to the specific fluorescence. These are difficult, if not impossible, to control, especially if the fluorescent label has a low binding constant. Multiple washings decrease the contribution of free fluorophores to the fluorescence intensity, but unavoidably increase the amount of cell debris. Despite these problems some early measurements of molecular clustering were made using cell suspensions in cuvettes.[41] In other experiments scattering was substantially reduced by using a microscope to make RET measurements on single cells.[42]

[39] A. Chakrabarti, J. Matko, N. A. Rahman, B. G. Barisas, and M. Edidin, *Biochemistry* **31,** 7182 (1992).

[40] J. Szollosi, L. Tron, S. Damjanovich, S. Helliwell, D. J. Arndt-Jovin, and T. M. Jovin, *Cytometry* **5,** 210 (1984).

[41] R. E. Dale, J. Novros, S. Roth, M. Edidin, and L. Brand, *In* "Fluorescent Probes" (G. S. Beddard and M. A. West, eds.), pp. 159–189. Academic Press, London, 1981.

[42] S. M. Fernandez and R. D. Berlin, *Nature (London)* **264,** 411 (1976).

Other difficulties in RET measurements on cells arise from their biology. Cell populations are heterogeneous, so measurements on individual cells are preferable to measurements that give population averages. Cell surfaces are also differentiated and heterogeneous; RET measured on one area may not be the same as for another area of the same cell. Finally, even within a single area, molecules of interest may be constrained in their motions so that the orientations of donor and acceptor are not averaged. Although it may be possible to detect molecular proximity, calculation of average distance between donor and acceptor may be impossible under these conditions.

Instrumentation for Cell Resonance Energy Transfer

Once the limitations of cell RET are recognized, the technique may be used to probe cell membranes. Cell RET measurements are made either by flow cytometry or by quantitative microscopy. The flow cytometer was developed, and is best used, for cells in suspension, particularly lymphocytes. It sacrifices spatial information on each cell for large samples, 5000–10,000 cells or more per measurement. Information about cell-to-cell variation is not lost. RET can be calculated for each cell in a sample.[22,40,43]

RET microscopy can resolve details of lateral heterogeneity in the surface of a single cell, and can also detect cell-to-cell variations in molecular clustering. A good example of this approach can be found in Ref. 44. On the other hand, compared to flow cytometry, which can sample 5000 cells in a few seconds, microscopy can sample relatively few cells—at best, several per minute.

Most cell RET experiments using either flow cytometry or microscopy measure fluorescence intensities rather than fluorescence lifetimes. All of the techniques that we describe here are based on changes in fluorescence intensity of either donor or acceptor molecules. However, there has been some work on time-resolved detection of fluorescence by flow cytometry[45–47] and a number of approaches have been made to time-resolved fluorescence measurements in digital microscopy.[48,49] Measuring lifetimes

[43] L. Tron, J. Szollosi, S. Damjanovich, S. Helliwell, D. J. Arndt-Jovin, and T. M. Jovin, *Biophys. J.* **45,** 939 (1984).

[44] T. W. Gadella, Jr. and T. M. Jovin, *J. Cell Biol.* **129,** 1543 (1995).

[45] B. G. Pinsky, J. J. Ladasky, J. R. Lakowicz, K. Berndt, and R. A. Hoffman, *Cytometry* **14,** 123 (1993).

[46] C. Deka, L. A. Sklar, and J. A. Steinkamp, *Cytometry* **17,** 94 (1994).

[47] J. A. Steinkamp, *Methods Cell Biol.* **42B,** 627 (1994).

[48] T. Oida, Y. Sako, and A. Kusumi, *Biophys. J.* **64,** 676 (1993).

[49] H. Szmacinski, J. R. Lakowicz, and M. L. Johnson, *Methods Enzymol.* **240,** 723 (1994).

of fluorescence instead of intensities avoids uncertainties due to errors in measuring labeling ratios of antibodies and donor : acceptor stoichiometry. It is also insensitive to light-scattering artifacts. Further development of time-resolved techniques applicable to single cells could lead to more reliable and sensitive measurements of RET.

Principles of Flow Cytometric Energy Transfer Measurements

We have relied mainly on flow cytometry for our RET measurements. We believe that flow cytometry offers a good compromise for measuring energy transfer on cell surfaces, combining some of the advantages of the spectrofluorimetric and microscopic methods. Dynamic processes induced by receptor clustering or changes of macromolecular interactions can be followed in real time. Because the volume of illumination is close to the cell volume, unbound fluorophores make little contribution to the specific signal, and therefore receptors with low binding affinities can also be studied. The interfering effect of dead cells and cell debris can also be eliminated by triggering with scattered light. Light scatter, which is a major source of error in spectrofluorimetric measurements, has negligible effect in flow cytometers. Flow cytometry reveals heterogeneities in cell populations. If high-speed computers are used for data analysis, and real-time energy transfer values are calculated, even cell sorting based on energy transfer might be possible.[24,50]

Two types of dual-wavelength excitation configuration can be used for flow cytometry energy transfer (FCET) measurements. One is a flow cytometer equipped with a single laser only. In this case one can generate a second excitation beam (e.g., the 514-nm line) from the primary 488-nm line by introducing dispersive elements into the input optics, spatially separating the two lines of different wavelength.[22,51] Here the laser beams are displaced by about 0.5 mm at the so-called intersection point, where the cells are illuminated. Another possible configuration is to use a second laser (preferentially a tunable dye–laser pumped by an argon-ion laser) separated spatially and optical elements to focus both beams onto the sample stream.[13,20,39]

One can also use a flow cytometer equipped with only a single laser to estimate RET efficiency. In this case instead of dual-laser excitation emission readings are taken from at least three different spectral regions. This, however, requires narrow bandwidths to assure the specificity of the contribution of donor or acceptor to the measured intensity. The narrow band-

[50] J. Matko, J. Szollosi, L. Tron, and S. Damjanovich, *Q. Rev. Biophys.* **21,** 479 (1988).

[51] J. Szollosi, L. Matyus, L. Tron, M. Balazs, I. Ember, M. J. Fulwyler, and S. Damjanovich, *Cytometry* **8,** 120 (1987).

width results in a diminished signal, reducing the precision of FRET measurements. The third fluorescence intensity detectable with dual-beam excitation in this case could be substituted by a precisely determined acceptor-to-donor ratio. This ratio can be determined in separate measurements, using single-labeled cells and the 488- and 514-nm single lines of the laser (after tuning the laser). Of course the measurements must be corrected for the difference in sensitivity of the detector system to green and red photons, as well as for the differences in the number of binding sites and dye/protein labeling ratios of the donor- and acceptor-carrying molecules, respectively. A data acquisition and evaluation protocol for such measurements was elaborated by Szollosi *et al.*[51]

Resonance Energy Transfer Labels for Cell Surface Antigens

The problem in cell biology that we use as an example here involves cell surface antigens. Fluorescent labels for these antigens are prepared from specific immunoglobulin G (IgG) antibodies, either polyclonal or monoclonal. Once purified, the intact IgG may be conjugated with donor or acceptor fluorophore, or may be cleaved to Fab fragments. Fab fragments are the smallest Ig fragments that bind efficiently to cell surfaces, although the affinity of a typical Fab is usually one to two orders of magnitude lower than that of the IgG from which it was prepared. Fab labels are more likely to report molecular proximity accurately, and less likely to induce proximity by cross-linking than are IgG molecules. However, in most instances orientation of donor and acceptor are critical to RET efficiency, and the assumption of random orientation that is used so effectively in solution RET almost certainly does not apply to rather rigid Fab bound to membrane proteins. Hence in some cases, while the membrane proteins themselves may be within RET distance of one another, the orientation of donor and acceptor Fab labels may not allow detectable RET, while the size and flexibility of intact IgG molecules may allow it (Fig. 1). On the other hand, labels at the ends of IgG molecules may be too far apart for RET even though the IgGs themselves are bound to surface membrane proteins that are in close proximity. IgG labels may also be required if the affinity of Fab is too low for good labeling, or if a particular MAb proves resistant to digestion by papain.

Under standard conditions mouse IgG is rapidly cleaved to Fab and peptides by papain. Hence our protocol uses brief digestion at 37°[13,20]; as an alternative it may be convenient to run the digestion outlined as follows at suboptimal pH, pH 7, for longer times.

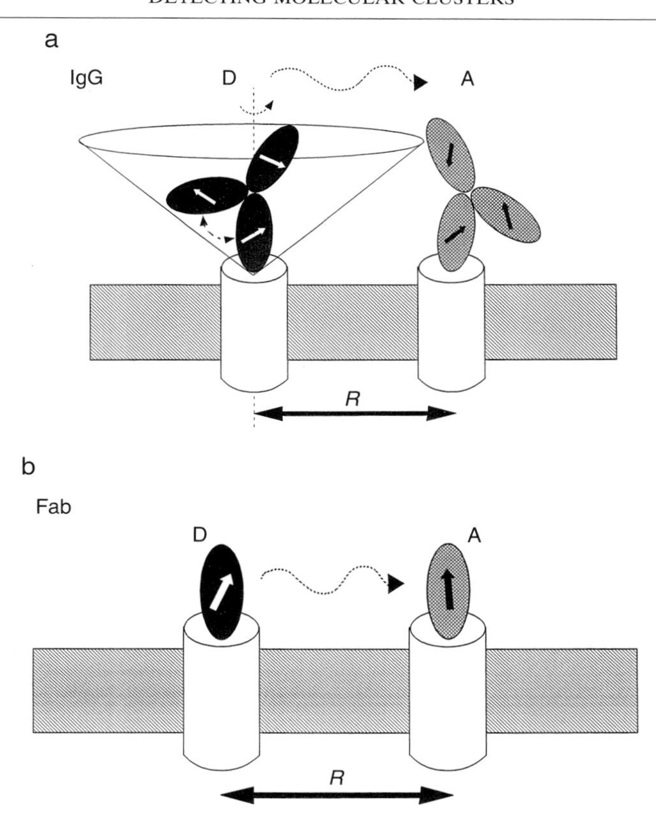

FIG. 1. Schematic views of IgG and Fab labeling of membrane proteins for RET. (a) Monovalently bound IgG with donor and acceptor fluorophores. The cone indicates the volume explored by the flexible IgG molecule, and the dashed arrows suggest some of the possible motions (wobbling, bending) of an IgG. This flexibility may allow RET between donor (D) and acceptor (A). IgG even when R, the distance between the centers of the two membrane proteins, is significantly larger than R_0, and may also be helpful in orientational averaging of donors and acceptors. (b) Monovalently bound Fab with donor and acceptor fluorophores. Motions of the rather rigid Fab molecules are limited and hence the pair may not detect RET (in case of unfavorable donor–acceptor orientations) in instances where IgG would detect it.

Fab Fragments of Mouse IgG

Purified protein should be concentrated to at least 1 mg/ml; higher concentrations, 3–4 mg/ml, are recommended. The protein solution is dialyzed overnight against digestion buffer—0.15 *M* NaCl, 1 m*M* EDTA, 0.1 *M* sodium phosphate adjusted to pH 8. Papain stock solution (usually 1000s of units per milliliter) is diluted to 240 units/ml in 0.1 *M* sodium acetate, pH 5.2, 1 mM EDTA, 10 mM cysteine. This stock is diluted ~1/3 in the IgG solution (final enzyme concentration is ~80 units/ml) and the solution incubated at 37°. The reaction is stopped after 10 min by adding 10 mg of iodoacetamide per milliliter of solution.

After digestion each preparation is fractionated on a sizing column (Sephadex G-100 fine). This isolates a peak of Fab and removes both the bulk of the undigested IgG and all of the lower molecular weight peptides that are particularly abundant in digests of mouse IgG where the Fc is usually almost completing degraded. The Fab peak from this fractionation is passed over protein A–Sepharose to remove traces of IgG. The run-through material is concentrated for conjugation with fluorophore.

In our view, fluorescein remains the best fluorophore for use as an RET donor. Both tetramethylrhodamine and sulforhodamine (Texas Red; Molecular Probes, Inc., Eugene, OR) are useful RET acceptors for fluorescein. Texas Red is usually the better choice if appropriate light sources are available, because it is more water soluble, and less likely to stick noncovalently to proteins than is tetramethylrhodamine. Proteins are conjugated with reactive isothiocyanates of fluorescein and tetramethylrhodamine and with the sulforhodamine sulfonyl chloride in the usual way, at alkaline pH, between pH 8 and 9. We find that ratios of approximately 15 mg of fluorescein isothiocyanate per milligram of protein and 40 mg of tetramethylrhodamine or Texas Red per milligram of protein give adequate labeling of most Fab or IgG. After conjugation, free dye is removed by passing the reaction mixture over a sizing column (Sephadex G-50). This is sufficient purification for fluorescein and Texas Red conjugates. Noncovalently bound tetramethylrhodamine is stripped from labeled proteins by passing these over Bio-Beads (Bio-Rad, Hercules, CA) or other beads designed for removing traces of hydrophobic molecules from solution.[52] Because these columns also adsorb some protein it is best to keep the volume of column bed smaller than that of the applied protein solution. We usually use a 1-ml column (in a Pasteur pipette) to purify 1–2 mg of labeled protein in a volume of 1–5 ml.

Dye : protein ratios are estimated in the usual manner, from absorption of the conjugates at 280 nm and at or near the absorption maximum for a particular fluorophore. We take the extinction coefficients for the dyes we commonly use as 72,000 M^{-1} cm^{-1} for fluorescein at 495 nm and 85,000 M^{-1} cm^{-1} for sulforhodamine (Texas Red) at 595 nm. These values are based on the absorption of free dye. More accurate standards may be made from conjugates of each dye with lysine,[41] but only small errors are introduced using the published extinction coefficients. A more serious error

[52] E. Spack, Jr., B. Packard, M. L. Wier, and M. Edidin, *Anal. Biochem.* **158,** 233 (1986).

arises from the fact that the fluorophores used absorb significantly at 280 nm. Hence, estimates of Ig or Fab protein will be high unless corrected for this absorbance.

For LRET measurements IgG or Fab has been spin-labeled using the isothiocyanate ITC-EECP (3-[[2-(2-isothiocyanatoethoxy)ethyl]carbamoyl]PROXYL). Conjugation and purification follow the steps outlined for fluorescent labels. Spin-label:protein ratios are estimated from absorbance at 280 nm and by EPR spectrometry. [Unfortunately, ITC-EECP is no longer available from Sigma Chemical Company (St. Louis, MO) but it is still available at Janssen Chimica (Geel, Belgium) or at Fisher Scientific/Acros (Pittsburgh, PA)].

IgG and Fab aggregate to some extent during their handling and conjugation. Freezing and thawing conjugates also enhances aggregate formation. These aggregates are known to affect measurements of membrane dynamics[53] and they may also distort RET measurements. They may be removed by centrifuging conjugates just before use for 1–4 $\times$ 10^6 $g \cdot$min (by air-centrifuge) at 4–20°. The centrifuged conjugates can be stored for 1–2 weeks in ice (0°) in a cold room. They are far more stable when stored in this way than if stored in the refrigerator (4°).

Cell Labeling and Flow Cytometry

In our experiments FRET or LRET is usually measured in terms of quenching of donor, fluorescein, fluorescence, or in terms of sensitized acceptor emission. Cell suspensions are labeled with (1) donor-labeled protein (Fab or IgG) plus unlabeled protein, (2) donor-labeled Fab or IgG plus acceptor-labeled protein, and (3) unlabeled protein plus acceptor-labeled protein. Suspension 1 serves as the standard for unquenched donor fluorescence, we expect to detect energy transfer in suspension 2, and suspension 3 serves as a standard for bound acceptor. If sensitized acceptor emission is measured, suspension 3 also serves as a control for fluorescence excited directly by 488-nm light. The affinity of each species of labeling protein should be measured either directly (for labeled Fab or IgG) or in terms of competition with a labeled protein (for unlabeled or nitroxide-labeled Fab or Ig). If even relative affinities of Fab or IgG are known then cells can be labeled with controlled ratios of donor to acceptor labels. This is crucial when trying to detect clusters of like molecules, for example the clusters of MHC molecules that we have studied.[13] In that case donor, unlabeled, and acceptor ligands are all competing for the same binding

[53] B. E. Bierer, S. H. Herrmann, C. S. Brown, S. J. Burakoff, and D. E. Golan, *J. Cell Biol.* **105,** 1147 (1987).

site, and displacement of donor by acceptor may give a false indication of RET. However, unless the affinities of the labels differ by an order of magnitude or more, an accurate estimate of RET may be obtained by labeling cells with mixtures of the fluorescent conjugates at saturating (10–20 K_d) concentrations.[54]

In a typical experiment, using conjugated Fab whose estimated K_d values were within threefold of each other, cells of the human B lymphoma line, JY, were resuspended in a small volume of Fab solution at a concentration of $\sim 5 \times 10^6$/ml. A constant amount of donor, Fl-Fab was used, 75 pmol of Fab/10^6 cells, with varying amounts of unlabeled or acceptor labeled Fab to give a total of 150 pmol of Fab/10^6 cells. Cells were incubated in Fab for 45 min in ice. They were then washed twice in HEPES-buffered Hanks' balanced salt solution (BSS) and kept in ice until used. These samples can be fixed in 1–2% paraformaldehyde without affecting observed RET or LRET.[13,20]

All of our RET measurements are made on a dual-laser flow cytometer [Coulter (Hialeah, FL) Epics 752]. Fluorescein is excited with the 488-nm line of an argon-ion laser and Texas Red is excited with the 595-nm line of a dye laser pumped by the argon-ion laser. Mirrors and dyes of the dye laser can be changed so that tetramethylrhodamine can be used as acceptor. Our RET data are evaluated in terms of mean intensities for a populations of cells. A typical histogram showing unquenched and quenched donor fluorescence is shown in Fig. 2.

Calculating Resonance Energy Transfer

In this section we describe the calculation of RET efficiency and give a numerical example using the complete data set from the experiment illustrated in Fig. 2. In this experiment we were interested in the proximity of mouse class I MHC molecules, H-2L^d, on the surface of mouse lymphoma cells. We tested for both donor quenching (the phenomenon shown in Fig. 2) and acceptor sensitization, to avoid the possibility that donor quenching was due to other factors than RET. An accurate determination of antibody (protein) concentration and of the dye : protein ratio is essential for these calculations. A reasonably accurate estimate of E, energy transfer efficiency, can be derived from population averages only if the emission spectra of the donor and acceptor dyes do not overlap significantly, and if there is little direct excitation of the acceptor by the wavelength optimum for exciting the

[54] J. Szollosi, S. Damjanovich, M. Balazs, P. Nagy, L. Tron, M. J. Fulwyler, and F. M. Brodsky, *J. Immunol.* **143,** 208 (1989).

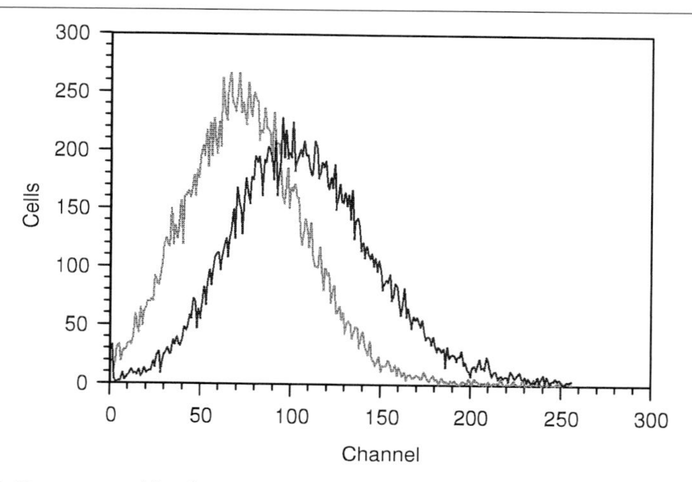

FIG. 2. Donor quenching by acceptor, one indication of energy transfer between labels on mouse class I MHC molecules. The histograms of number of cells versus intensity are displayed with a 256-channel, 3 log abscissa.

donor. In this experiment these requirements are met by the fluorescein donor/Texas Red acceptor pair.[20,39]

Resonance Energy Transfer Efficiency Estimated from Acceptor Sensitization

Correction factors, Z, for spillover of donor fluorescence into the acceptor fluorescence and for direct excitation of acceptor fluorescence are first derived from cells labeled with a single fluorophore (either by donor or by acceptor).

$$Z_1 = I_D\,(488/630)/I_D\,(488/525), \qquad Z_2 = I_A\,(488/630)/I_A\,(595/630)$$

where I is the mean fluorescence intensity of cells and the numbers in parentheses are the wavelengths of excitation and detection of fluorescence. [We assume no acceptor emission at donor (fluorescein) emission wavelengths.] The acceptor sensitization can then be estimated from the following two fluorescence intensities (histogram means):

$$F_A^D\,(488/630) = I\,(488/630) - Z_1 I\,(488/520)$$
$$F_A\,(488/630) = Z_2 I\,(595/630)$$

where I denotes the mean fluorescence intensity of double-labeled cells.

Using these intensities and a correction factor, β, one can derive a parameter B that is proportional to the efficiency of energy transfer.

$$B = [F_A^D\ (488/630) - F_A\ (488/630)]/\beta I\ (488/525)$$

where β equals $(M_A L_D \varepsilon_D/(M_D L_A \varepsilon_A)$. Here M_A and M_D are the population means of FITC- and Texas Red-labeled cells (at saturation), and L_D and L_A are the dye/protein labeling ratios for the donor and acceptor Fabs, respectively.

From B the efficiency of energy transfer is obtained as

$$E = B/(1 + B)$$

Resonance Energy Transfer Efficiency Estimated from Donor Quenching

The efficiency of RET (E) can also be estimated from the quenching of donor fluorescence according to Eq. (7). An accurate estimation may occasionally require two correction factors:

$$g = C_{FD}/C_{FDA}$$

correcting for a different fractional quantity of FITC–Fab in the labeling cocktail for single- and double-labeled cells (in the case of a D : A ratio different from 1 : 1), and

$$\delta = [I_D\ (488/570) - I_{UL}\ (488/570)]/[I_{DA}\ (488/570) - I_{UL}\ (488/570)]$$

which corrects for the different percent contribution of cellular autofluorescence to the detected fluorescein emission signal of cells in the presence or absence of acceptor, respectively.

RET efficiency, E, is then calculated as

$$E = 1 - (F_{DA}g)/(F_D\delta)$$

From a complete data set including that shown in Fig. 2, we obtain the following results. For RET measured in terms of acceptor sensitization:

$$Z_1 = 0.752, \qquad Z_2 = 0.741$$

Checking assumption: $I\ (488/525)_A = 7.3$ and $I\ (488/525)_{UL} = 6.5$, so there is no acceptor emission to the donor emission band.

$$\beta = [86.34(1.7)72{,}000]/[78.86(1.2)85{,}000] = 1.31$$
$$F_A^D\ (488/630) = 66.15 - [(0.752)(21.9)] = 49.66$$
$$F_A\ (488/630) = (0.741)(54.05) = 40.05$$
$$B = (49.66 - 40.05)/(1.31)(21.9) = 0.334$$
$$E = 0.334/1.334 = 0.25$$

For RET calculated in terms of donor quenching:

$$g = 1.66; \quad \delta = 0.595$$
$$F_{\mathrm{D}} = 78.86; \quad F_{\mathrm{DA}} = 21.9$$
$$E = 1 - [(21.9)(1.66)/(78.86)(0.595)] = 0.22$$

There is a reasonably good agreement between the energy transfer efficiencies calculated by the two approaches.

Resonance Energy Transfer Efficiency Estimated on Single Cells

An ideal presentation of flow cytometric RET data involves calculations of RET efficiency for each cell of the 5000–10,000 constituting a typical sample. Programs have been written to do this calculation. These transform so-called list mode data, lists of all parameters recorded for each cell, into histograms of the distribution of RET efficiencies in the cell sample. Software for this approach has been developed, and examples of its application may be found in Refs. 3, 5, 40, 43, and 54. Figure 3 also shows an example of this application, comparing the distribution of RET efficiencies in a population of human B lymphoblasts and T lymphoblasts whose ICAM-1 molecules were labeled with donor- and acceptor-conjugated MAb

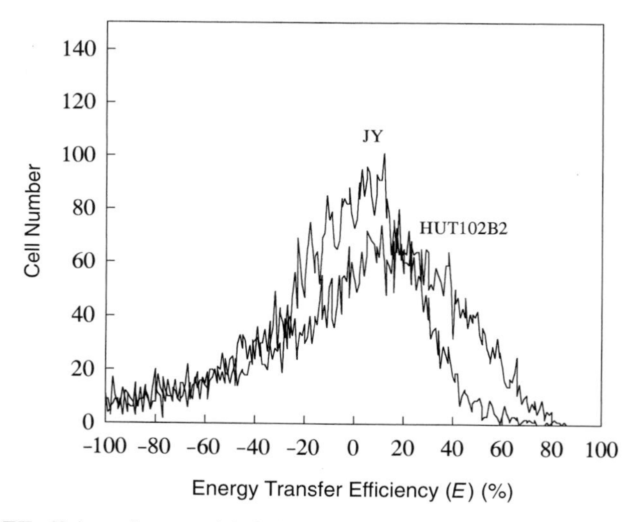

FIG. 3. RET efficiency between labeled ICAM-1 molecules on JY human B lymphoblasts compared to that between the same molecules on HUT102B human T lymphoblasts, The ordinate is as in Fig. 2, number of cells, but the abscissa is scaled directly in terms of percent RET efficiency (E). Notice that the distribution of transfer efficiencies for JY cells is centered close to 0, while that for HUT102B2 is centered at 10–15%.

(IgG).[55] It is apparent that ICAM-1 molecules self-associate in the T lymphoblast membranes, but not in the B lymphoblast membranes. The histograms are relatively broad and noisy because the level of ICAM-1 expression is low on both cell types, but they clearly show a difference in the surface clustering of ICAM-1 in the two different cell types.

The original RET programs performing cell-by-cell analysis were run on minicomputers. However, a DOS program for calculating RET efficiency, TRANSFER, is currently available and a new extended version, running under Microsoft Windows, FLOWIN 2.0, is in late stages of development. Information on both programs is available by e-mail at *Balkay@atomki.hu*.

Conclusion

Clustering, trapping, and confinement of plasma membrane proteins have become increasingly apparent.[56] The scale of this lateral organization varies considerably. At the smallest scale, molecular clustering, the techniques that we have outlined here will serve well to characterize lateral heterogeneity. The approaches that we have described are effectively based on the reagents and methods of immunofluorescence and so offer the possibility of quantitative assay of molecular clustering using well-characterized reagents and cells. As techniques develop for characterizing RET on a cell-by-cell basis, it may also be possible to use energy transfer as part of a multiparameter approach to characterizing cells in health and disease.

[55] L. Bene, M. Balazs, J. Matko, J. Most, M. P. Dierich, J. Szollosi, and S. Damjanovich, *Eur. J. Immunol.* **24,** 2115 (1994).
[56] M. Edidin, *Curr. Topics Membr.* in press (1996).

[22] Distribution Analysis of Depth-Dependent Fluorescence Quenching in Membranes: A Practical Guide

By ALEXEY S. LADOKHIN

Determination of membrane organization and dynamics is one of the most challenging problems of structural biology because many of the high-resolution (or even low-resolution) methods developed for water-soluble systems are not directly applicable to membranes. By far the most popular schematic representation of the membrane is of a "lipid sea with icebergs

0076-6879/97 $25

of proteins" in it. Inevitably this picture leads to a grossly oversimplified view of a membrane bilayer as a two-dimensional hydrocarbon fluid. This view has been seriously challenged in a series of X-ray and neutron diffraction studies demonstrating a great complexity of the membrane transverse organization.[1] The depth-dependent fluorescence-quenching technique, which utilizes lipids with bromine atoms or spin labels selectively attached to certain positions along their acyl chains, is a perfect tool with which to explore the membrane structure along the depth coordinate. This chapter focuses on the basic principles for and the nature of the information available from the method of distribution analysis (DA)[2] of depth-dependent fluorescence quenching.[3-7] It also is intended to provide practical guidance on the implementation of this method (see Appendix).

Considerations for Quantitative Analysis

The single most important experimental observation in the quantitation of depth-dependent fluorescence quenching (DFQ) is that the density distribution of the lipid-attached moiety can be adequately described by a Gaussian function of depth.[8] First, it implies that the Stern–Volmer formalism based on the Smoluchowski equation for free diffusion would no longer be applicable because the quencher is not uniformly distributed over the corresponding space coordinate (depth). This will distinguish the DFQ from the more familiar and much more explored cases of the three-dimensional diffusion of water-soluble quencher or two-dimensional lateral diffusion in the membrane. Nevertheless, it was demonstrated that the transverse fluctuations of the lipid-attached quencher result in a temperature-activated, dynamic quenching.[6] Second, the simple parametrized description

[1] S. H. White and M. C. Wiener, *In* "Permeability and Stability of Lipid Bilayers" (E. A. Disalvo and S. A. Simon, eds.), p. 1. CRC Press, Boca Raton, Florida, 1995; and references therein.

[2] DA, Distribution analysis; DFQ, depth-dependent fluorescence quenching; DFQP, depth-dependent fluorescence quenching profile; DPEP, depth probability/exposure profile. POPC, palmitoyloleoylphosphatidylcholine; 4,5-, 6,7-, 9,10-, 11,12-, or 15,16-BRPC, 1-palmitoyl-2-(dibromostearoyl)phosphatidylcholine with bromine atoms at the 4,5-, 6,7-, 9,10-, 11,12-, or 15,16-position, respectively; TOE, tryptophan octyl ester; $G(S, \sigma, h_m)$, Gaussian distribution and its parameters: area, dispersion, position of the maximum.

[3] A. S. Ladokhin, *Biophys. J.* **64,** A290 (1993).

[4] A. S. Ladokhin, P. W. Holloway, and E. G. Kostrzhevska, *J. Fluoresc.* **3,** 195 (1993).

[5] V. G. Tretyachenko-Ladokhina, A. S. Ladokhin, L. Wang, A. W. Steggles, and P. W. Holloway, *Biochim. Biophys. Acta* **1153,** 163 (1993).

[6] A. S. Ladokhin and P. W. Holloway, *Biophys. J.* **69,** 506 (1995).

[7] A. S. Ladokhin and P. W. Holloway, *Ukrainian Biochem. J.* **67,** 34 (1995).

[8] M. C. Wiener and S. H. White, *Biochemistry* **30,** 6997 (1991).

of the depth distribution of both the fluorescent probe and the quencher allows the design of a simple fitting function to analyze DFQ. It also justifies some of the assumptions that necessarily are made for quantitative analysis.[9]

The following assumptions are made for the DA method [Eq. (1) discussed in the next section]: (1) the probability of quenching is proportional to the distance separating probe and quencher along the depth coordinate; (2) the distribution of separating distances is set by a Gaussian function; and (3) the dynamics of the quencher on the relevant time scale (set by the excitation lifetime of the probe) are independent of its depth.

Although not intuitively obvious, the third assumption is validated for the nanosecond time window of fluorescence by nuclear magnetic resonance (NMR) measurements and molecular dynamics simulations for a broad range of depth.[10,11]

Distribution Analysis Formalism and Information It Provides

The depth-dependent fluorescence quenching profile (DFQP) is defined as a depth (h) function of the logarithm of the ratio of fluorescence without the quencher (F_0) to the fluorescence in the presence of the quencher [$F(h)$]. Defined this way the DFQP was suggested to be approximated by a Gaussian function, $G(h - h_{\rm m}, \sigma, S)$[3]:

$$\ln\left[\frac{F_0}{F(h)}\right] = G(h - h_{\rm m}, \sigma, S) = \frac{S}{\sigma\sqrt{2\pi}} \exp\left[-\frac{(h-h_{\rm m})^2}{2\sigma^2}\right] \tag{1}$$

characterized by three parameters: mean position, dispersion, and area.

The position of the maximum, $h_{\rm m}$, is the most probable location of the fluorescent probe. It is important to remember that DFQ, like any other fluorescence technique, is capable of registering only emitting molecules. Consider a heterogeneous population of molecules that in the absence of lipid-attached quencher have various quantum yields. In this case the DFQP will be weighted by the depth dependence of the quantum yield. Therefore, $h_{\rm m}$ will not necessarily coincide with mean depth derived from, for example, diffraction experiments (ignoring the effects of variation in hydration and other differences between the two experimental systems).

The dispersion, σ, arises from the several broadening terms, such as finite size of the probe and the quencher and fluctuational distribution of their depth owing to the thermal motion. Therefore, even when the DFQP

[9] More discussion is presented in Ref. 6.

[10] M. F. Brown, A. A. Ribeiro, and G. D. Williams, *Proc. Natl. Acad. Sci. U.S.A.* **80,** 4325 (1983).

[11] R. W. Pastor, R. M. Venable, M. Karplus, and A. Szabo, *J. Chem. Phys.* **89,** 1128 (1988).

appears to be broader than the bilayer thickness, in reality the much narrower distribution of the position of the probe will be well within boundaries of the membrane. Assuming that the broadening caused by the finite sizes of probe and quencher can also be approximated by a Gaussian function[8] it is possible to present the dispersion of DFQP as the following sum:

$$\sigma^2 = \sigma_{PC}^2 + \sigma_{PS}^2 + \sigma_{QC}^2 + \sigma_{QS}^2 \tag{2}$$

where individual dispersions represent four constituting Gaussian functions. The first index denotes the species, probe (P) or quencher (Q), and the second index denotes the origin of broadening, distribution of the center (C) or size of species (S). What we are interested in is the value of σ_{PC} that is indicative of the probe transverse heterogeneity. When brominated lipids are utilized, the broadening due to the quencher could be estimated from diffraction experiments[8]: $(\sigma_{BrC}^2 + \sigma_{BrS}^2)^{1/2} = 3.5$ Å. When the tryptophan fluorescence was analyzed the value of 2.5 Å was suggested for σ_{PS}.[6] Using those numbers and experimentally obtained DFQP one can, for example, estimate the conformational freedom of tryptophan residues in membrane proteins and/or the conformational heterogeneity of the protein itself (see Extended Distribution Analysis for examples).

The area under the DFQP, S, is proportional to the quenchibility of the probe, γ, determined by the nature of the quenching mechanism, excited-state lifetime in the absence of quenching, τ, the degree to which probe is exposed to a lipid phase, ω, and the concentration of the quencher, C[12]:

$$S = \gamma\omega\tau C \tag{3}$$

The variation of the exposure could arise from either incomplete binding to the membrane or, for example, from shielding of tryptophan side chains by the protein moiety. In the latter case the relative exposure, Ω, could be estimated as the ratio of absolute exposures of indole chromophore in tryptophan residue in a protein, ω_P, to that in model compound, ω_M, such as tryptophan octyl ester (TOE) (the quenching mechanism is assumed to be the same, $\gamma_P = \gamma_M$):

$$\Omega = \frac{\omega_P}{\omega_M} = \frac{S_P \tau_M}{S_M \tau_P} \tag{4}$$

If the lifetime measurements are not available, the ratio of τ values can be approximated with the ratio of quantum yields of a protein and a model compound in a nonquenching lipid membrane. The S parameter, available from the DA, will allow us to estimate the lipid exposure of tryptophan

[12] Usually expressed either as a molar fraction or as a molar fraction per area per lipid.

side chains of the integral proteins in the same way the Stern–Volmer constant allows us to estimate the exposure of tryptophanyl to aqueous solvent for water-soluble proteins (see Extended Distribution Analysis for examples). Note that the water-soluble quencher will not be able to differentiate tryptophanyl in a lipid environment from a tryptophanyl buried in the protein interior. Another potential application of this technique would be to follow the changes of accessibility of probe to quencher owing to membrane domain formation.

Applications

Simulations

Consider a hypothetical probe that will partition into the membrane but will stay only on that half of the bilayer to which it was added. Consider the hypothetical continuous set of quenchers that cover all possible depth. The depth is defined as a distance from the middle of the bilayer so that the outer leaflet will have positive depth and the inner leaflet will have negative depth. The DFQP of the probe in the outer leaflet (thick solid line in Fig. 1) has the following parameters: h_m = 8 Å (A) or 5 Å (B); σ = 4 Å; S = 18. The DFQP for the probe in the inner leaflet (dotted line in Fig. 1) is a mirror image of the former. The sum of the two (thin solid line in Fig. 1) is the full DFQP. In the standard quenching experiment both leaflets contain the quenching lipid and, regardless of the distribution of the probe between the leaflets, only the positive region of the full DFQP can be examined.[13] Nevertheless, the contribution from the quenching of the probe by lipid from the opposite leaflet can be substantial. Therefore, in general the complete function containing contributions of both symmetrical terms should be used to describe a DFQP:

$$\ln\left[\frac{F_0}{F(h)}\right] = G(h - h_m, \sigma, S) + G(h + h_m, \sigma, S) \tag{5}$$

In the real experiment, however, only several points along the positive part of the depth coordinate could be sampled. For example, squares in Fig. 1 correspond to quenching with the set of lipids brominated at the 4,5-, 6,7-, 9,10-, 11,12-, or 15,16-position of the *sn*-2 acyl chain. Their depth

[13] A rare example of the use of the bilayers with asymmetrical distribution of quenchers in leaflets is presented in an elegant set of experiments that determine the topology of cytochrome b_5 binding by J. Everett, A. Zlotnick, J. Tennyson, and P. W. Holloway, *J. Biol. Chem.* **261,** 6725 (1986).

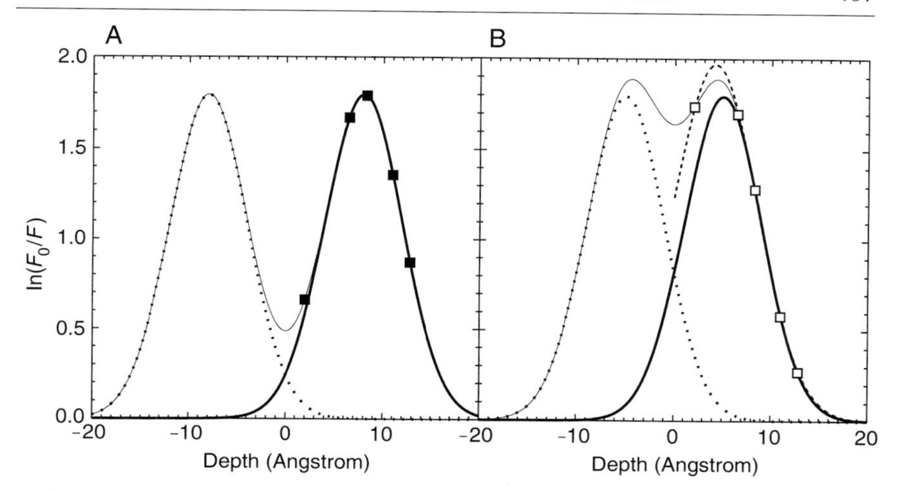

FIG. 1. Depth-dependent fluorescence quenching profiles (DFQPs) of the hypothetical probe in a membrane. The DFQP of the probe in the outer leaflet (thick solid line) has the following parameters: $h_m = 8$ Å (A) or 5 Å (B); $\sigma = 4$ Å; $S = 18$. The DFQP for the probe in the inner leaflet (dotted line) is a mirror image of the former. The sum of the two (narrow solid line) is the full DFQP. In the standard quenching experiment, when quenchers are distributed in both leaflets, only several points along the positive part of the depth coordinate could be sampled (squares correspond to the quenching with the set of five bromolipids). When ignored, the action of the quencher located in the opposite leaflet from a probe can introduce a significant error in the analysis. For example, an incorrectly applied Eq. (1) [instead of Eq. (5)] to fit the simulated data (open squares) results in a DFQP with parameters that significantly differ from the inputted [$h_m = 4.2$ Å; $\sigma = 4.4$ Å; $S = 21.5$ and dashed line in (B)].

positions calibrated from the X-ray diffraction experiments[14] are, correspondingly: 12.8, 11.0, 8.3, 6.5, and 2.0 Å. Alternatively, the set of spin-labeled lipids can be used.[15] It is clear that when h_m is larger than 2σ (Fig. 1A), the contribution of the second term in Eq. (5) is negligible at all experimentally sampled depths. For the distribution of the same width, but with smaller h_m (Fig. 1B), this is not so. When ignored the action of the quencher located in the opposite leaflet from a probe can introduce a significant error in the analysis. For example, an incorrectly applied Eq. (1) [instead of Eq. (5)] to fit the simulated data (open squares in Fig. 1) results in a DFQP with parameters that significantly differ from those input ($h_m = 4.2$ Å; $\sigma = 4.4$ Å; $S = 21.5$ and dashed line in Fig. 1B).

[14] T. J. McIntosh and P. W. Holloway, *Biochemistry* **26,** 1783 (1987).

[15] A. Chattopadhyay and E. London, *Biochemistry* **26,** 39 (1987).

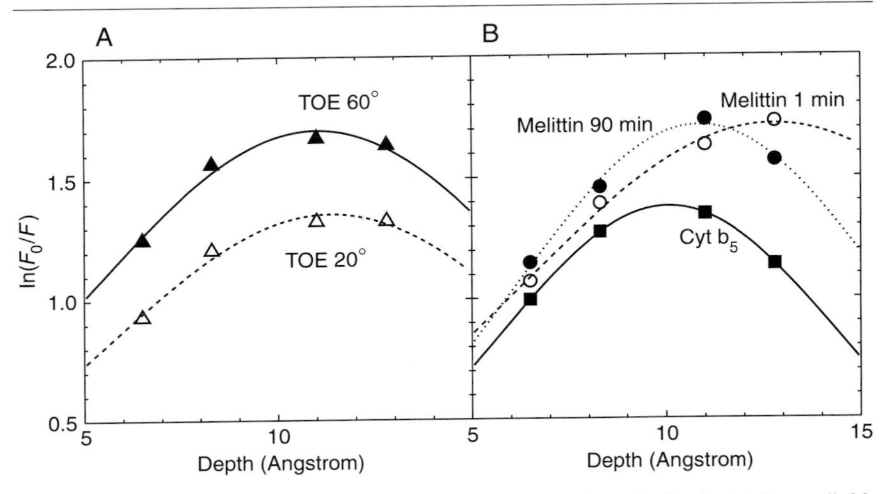

FIG. 2. DFQPs (lines) and the fluorescence quenching data obtained with bromolipids (symbols) for the model compound TOE (A) and for single tryptophan-containing proteins (B). DFQPs have the following parameters as recovered by Eq. (5): TOE at 20°: h_m = 11.3 Å; σ = 5.5 Å; S = 18.9; TOE at 60°: h_m = 11.0 Å; σ = 5.7 Å; S = 24.6. Cytochrome b_5 mutant (W108) at 20°: h_m = 10.1 Å; σ = 4.4 Å; S = 15.3. Melittin (W19) at 20°: 1 min of incubation: h_m = 12.5 Å; σ = 6.2 Å; S = 26.6; 90 min of incubation: h_m = 10.9 Å; σ = 4.9 Å; S = 21.0. [Original analysis was presented by A. S. Ladokhin and P. W. Holloway, *Biophys. J.* **69,** 506 (1995).]

Analysis of Experimental Data

A few examples of application of DA to real experimental data are presented in Fig. 2. The set of four brominated lipids was used to quench tryptophan fluorescence of the model compound TOE (Fig. 2A) or that of the single tryptophan-containing proteins, melittin and mutant cytochrome b_5 (Fig. 2B). Fluorescence of each in palmitoyloleoylphosphatidylcholine (POPC) was used as a corresponding F_0 value. Equation (5) was used for fitting, which for these particular data gave the same result as Eq. (1), used when the original calculations were presented.[6] A detailed discussion of the result is presented elsewhere[6,7]; however, it is worth emphasizing several points that provide an additional empirical justification for DA: (1) fit is statistically adequate (on the other hand, application of the alternative parallax method,[15,16] which operates with only two parameters, resulted in a poor fit[6]); (2) a significant variation in the S and σ parameters is observed (in the parallax method[15,16] the area and width of the DFQP are related to a single parameter R_C,[6] which according to the assumptions should be

[16] F. S. Abrams and E. London, *Biochemistry* **31,** 5312 (1992).

constant for the same fluorophore/quencher pair); and (3) for the same system S increases with the temperature, confirming the dynamic nature of the quenching (the parallax method[15,16] assumes a static mechanism of quenching).

Extended Distribution Analysis

In the previous sections we established the way in which to analyze DFQ data with the help of DA. As a result we have recovered three parameters that describe the DFQP and are related to certain properties of both probe and quencher. Next, we extract from these results and from results of some additional experiments the information that is relevant only to the probe and its interaction with the bilayer. This extended DA procedure includes (1) calculation of the width of the transverse distribution of the geometric center of the probe [according to Eq. (2)] and (2) determination of the quenching efficiency [or relative exposure; see Eqs. (3) and (4)]. It utilizes an experimentally determined DFQP, measurements of the fluorescence lifetime in the absence of the quencher, and an independently determined contribution of quencher into the broadening of the DFQP (as discussed in Distribution Analysis Formalism and Information It Provides).

The extended DA results in the depth probability/exposure profiles (DPEPs) shown in Fig. 3. The DPEP is a Gaussian function with the maximum at h_m, dispersion of σ_{PC} (in this case $\sigma_{PC}^2 \equiv \sigma^2 - \sigma_{PS}^2 - \sigma_{QC}^2 - \sigma_{QS}^2 = \sigma^2 - 18.5$ Å^2), and area of S/τ as derived from the corresponding DFQP (Fig. 2). This way the shape of the curve corresponds to the distribution of depth of the center of the fluorophore while the area is proportional to the efficiency of quenching. The following τ values used for the calculation are either amplitude averages for experimentally measured decay or are the result of approximation using both lifetime and quantum yield measurements: 2.9 nsec (TOE 20°)[6]; 1.8 nsec (TOE 60°)[6]; 2.3 nsec (melittin at all times)[7]; 3.3 nsec (cytochrome b_5 mutant).[17] For simplicity of presentation the areas of DPEPs in Fig. 3 were also normalized to the area for TOE at 20°.

The dramatic increase in area for the DPEP of TOE with the temperature increase (Fig. 3A) is related to a diffusion-controlled increase in quenching efficiency, γ. The width of the distribution is rather high, but the limits are within the dimensions of the bilayer. The DPEP for cytochrome b_5 (Fig. 3B, solid line) is much narrower, suggesting a significant reduction in the conformational freedom, and also has a smaller area. The exposure of

[17] A. S. Ladokhin, H. Malak, M. L. Johnson, J. R. Lakowicz, L. Wang, A. W. Steggles, and P. W. Holloway, *Proc. S.P.I.E.* **1640,** 562 (1992).

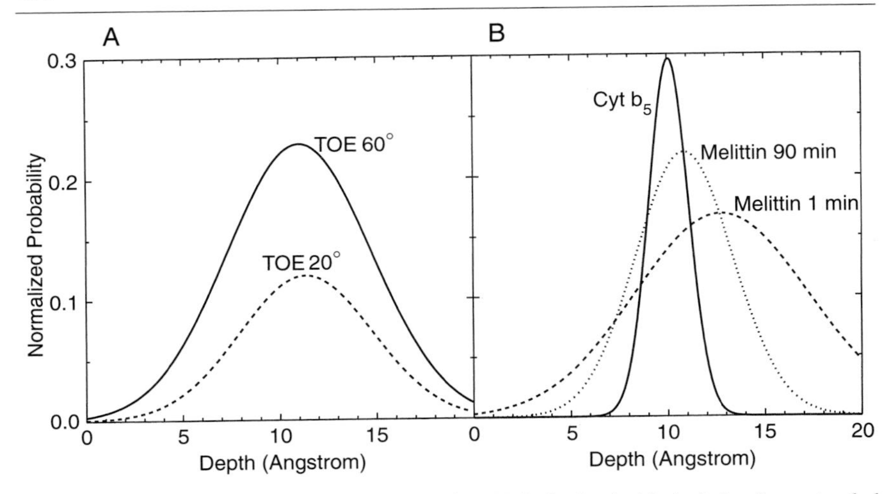

FIG. 3. Depth probability/exposure profiles (DPEPs) obtained with the help of an extended DA procedure from the corresponding DFQPs (Fig. 2) and the lifetime measurements as described in text. The shape of the curve represents a probability distribution of depth position of the geometric center of the fluorophore for the TOE (A) and single tryptophan-containing proteins (B). The areas are proportional to the quenchibility of the fluorophore and depend on temperature (A) and degree of exposure of tryptophan residue into the lipid phase (B). A broad distribution that narrows with time of incubation exhibited by melittin is attributed to the apparent effects of coexistence of several subpopulations of the protein immersed at a different depth and the time-dependent transition between them. Cytochrome b_5, on the other hand, has a tight distribution suggesting the lack of conformational freedom.

W108 in cytochrome b_5 relative to that of TOE at the same temperature is determined by the ratio of the corresponding areas [Eq. (4)] and is equal to $\Omega = 0.66$.

The apparent exposure of W19 in melittin appears to be higher than one and the width of the DPEP is also high (Fig. 3B). At initial incubation time the distribution even appears to be "falling" out of the bilayer (dashed line). The explanation of such behavior lies in the multiple conformations of melittin.[18,19] Consider that initial binding occurs on the interface and that the tryptophan is located at a shallow depth (perhaps h_m = 14–17 Å). However, this conformation is metastable and another, with the tryptophan located deeper (perhaps h_m = 9–12 Å), is populated with time. At any moment in time both populations exist, bringing the number of parameters required to describe this simplest case up to six. The limited number of

[18] A. S. Ladokhin, E. G. Kostrzhevska, and A. P. Demchenko, *Proc. Acad. Sci. Ukrainian SSR Ser. B.* **11,** 65 (1988). [In Ukrainian]

[19] J. P. Bradshaw, C. E. Dempsey, and A. Watts, *Mol. Membr. Biol.* **11,** 79 (1994).

experimentally available data points along the depth coordinate prevents the direct resolution of those populations. Instead, the effective distribution, with artificially large width and area, is recovered.

Practical Comments

Number of Quenchers to Use

The more quenchers used the better, as long as they differ only by depth; at least three should be used. For bromolipids the best minimal choice would be to use 6,7-, 9,10-, and 11,12-BRPCs.[2] Two other commonly used lipids (4,5- and 15,16-BRPC) can be added to increase accuracy and reliability. Note that the phase transition for 4,5-BRPC is higher (about 10°) than for the rest of the set (about 0°), suggesting somewhat different packing.[20] On the other hand 15,16-BRPC forms somewhat thicker bilayers[14] and the set that includes this lipid also does not completely satisfy the third assumption for the quantitative analysis discussed above. Therefore the choice between Scylla and Charybdis should be made on an individual basis, depending on the expected average depth of the probe under study.

Spin Labels or Bromolipids

Using spin labels does not require any particular variation from the standard DA procedure. The full extended DA, however, will require estimation of the contribution of the spin labels into the broadening of the DFQP. Also, the complete model data are required if one wishes to determine the relative exposure with the help of spin labels. More labels at different depths are available for bromolipids and they are much better calibrated by diffraction measurements. However, the quality of currently available commercial bromolipids is not always sufficient for quantitative analysis.[21] As a precautionary measure, to check the homogeneity of bromination, a mass spectroscopy examination of lipid samples is recommended. Similar problem exists with spin labels, which were found to disattach from lipid during such procedures as sonication.[22]

When Fractions of Different Quenching Lipids Are Not Equal

All the examples presented so far utilized bilayers formed entirely of brominated lipids, so the fraction of the quenching lipid was independent

[20] P. W. Holloway, Personal communication, 1994.
[21] S. H. White, Personal communication, 1996.
[22] E. London, Personal communication, 1992.

of depth and was simply equal to one. Although convenient for many studies this may not always be practical, especially when spin labels are used. To account for the possible difference in quencher concentration the normalization procedure should be utilized. Although a rigorous procedure would have involved the complicated task of modeling lateral diffusion, an empirical observation that the logarithm of fluorescence intensity linearly depends on the fraction of quencher[23] yields a solution. Before the data are subjected to DA the intensities should be corrected as follows: $F_{corr}(h) = F_{obsrvd}(h)/\exp[C(h)]$, where $C(h)$ is the mole fraction of the particular quenching lipid.

Intensities or Lifetimes

Those few who would consider this question seriously should note that DA could operate on both owing to the predominantly dynamic nature of quenching. Lifetime measurements could be beneficial because they eliminate the possible error in determination of probe concentration. Also, they allow an easy correction for the contribution of unbound probe (provided the lifetimes change with membrane partitioning). Those theoretical considerations will become practical advantages mainly when the fluorescence decay is monoexponential in the absence of quenching. However, even then quenching is expected to introduce a nonexponentiality in the decay. When the decay is heterogeneous to begin with, such as for tryptophan residue, much care should be taken in interpreting the lifetime data.[9] In any case, whether measuring intensity or lifetime, the use of polarizers configured at "magic" angles is a must.

Corrections of Intensity Measurements

When working with membranes the correction of intensity measurements for scattering (of both excitation and emission light) is absolutely essential. However, a dilution procedure,[24] traditionally used for such a correction, suffers from the possibility of artifacts caused by changes in the fraction of membrane-bound probe/protein owing to a decrease in lipid concentration. Instead, it is recommended that one use a standard fluorescence compound that will not interact with the membrane at all, so that the only changes in intensity on addition of membrane would be those that are due to scattering and dilution. To correct the data one would have to divide the intensities of the probe in different lipids (and in buffer if needed) by the corresponding intensities of a standard. The standard, of course,

[23] L. A. Chung, J. D. Lear, and W. F. DeGrado, *Biochemistry* **31,** 6608 (1992).
[24] J. Eisinger and J. Flores, *Biophys. J.* **48,** 77 (1985).

should have an emission spectrum similar to that of a probe (and ideally a similar quantum yield) and the samples should be of the same optical density. For example, for studies of intrinsic fluorescence of membrane proteins, a tryptophan zwitterion has been used as such a standard.[25] Also, because the quenching is associated with notable spectral shifts,[6] the entire area under the spectrum, rather than the intensity at a certain wavelength, should be used when possible.

Appendix

Presented here is an example of the application of DA. Data for cytochrome b_5 mutant[26] were fitted to the following equation:

$$F(h)/F_0 = \exp - [G(h - h_m, \sigma, S) + G(h + h_m, \sigma, S)] \quad (A.1)$$

where G is a Gaussian function. Data were weighted by the corresponding errors. The nonlinear least-squares minimization routine, NONLIN,[27] was used. Several commercially available programs (Sigmaplot, Origin) gave results that differed only in the confidence limits but not in the mean values of the parameters. The subroutines are available from the author on request: ladokhin@uci.edu.

Input	Output
Data (F/F_0; error; h in angstroms)	Variance of fit = 0.00178
0.37; 0.01; 6.5	F(NDF, NDF, 1 − P) = 3.206
0.28; 0.01; 8.3	Correlation matrix
0.26; 0.01; 11.0	S 1
0.32; 0.01; 12.8	σ 0.958 1
	h_m 0.482 0.454 1
Confidence probability 67.0%	Parameter values and confidence limits
Fraction change in variance 0.0001	S = 15.27 (15.23; 15.32)
	σ = 4.42 (4.40; 4.44)
	h_m = 10.07 (10.06; 10.08)

Acknowledgments

I am grateful to Drs. S. H. White and W. C. Wimley for helpful discussions and for reading the manuscript, and to Dr. M. L. Johnson for making his NONLIN program available for the calculations. It is my pleasant duty to acknowledge Dr. P. W. Holloway for helpful advice, for reading the manuscript, and, above all, for introducing me to the field of membranology.

[25] A. S. Ladokhin, M. E. Selsted, and S. H. White, *Biophys. J.* **72,** 794 (1997).
[26] A. S. Ladokhin, L. Wang, A. W. Steggles, and P. W. Holloway, *Biochemistry* **30,** 10200 (1991).
[27] M. L. Johnson and S. G. Frasier, *Methods Enzymol.* **117,** 301 (1985).

[23] Mechanism of Leakage of Contents of Membrane Vesicles Determined by Fluorescence Requenching

By ALEXEY S. LADOKHIN, WILLIAM C. WIMLEY, KALINA HRISTOVA, and STEPHEN H. WHITE

Introduction

A commonly used method for studying the permeabilization of membranes by peptides or proteins utilizes vesicle-encapsulated fluorescent dyes and quenchers that change fluorescence intensity on release. Self-quenching dyes such as carboxyfluorescein or calcein and dye/quencher pairs such as ANTS/DPX (8-amino-napthalene-1,3,6 trisulfonic acid/p-xylene-bis-pyridinium bromide) and terbium/dipiccolinic acid are the most frequently used marker systems. There are several mechanisms by which leakage can occur and distinguishing among them is of fundamental importance. For the most part, leakage can be *graded,* in which all of the vesicles lose some of their contents, or it can be *all-or-none,* in which vesicles either lose all of their contents or none (Fig. 1). These two mechanisms are commonly distinguished by measuring the degree of quenching inside of vesicles that have been physically separated from released contents.[1–6] The requenching method we describe here utilizes the dye/quencher pair ANTS/DPX to give a quantitative description of the mechanism of release without physical separation of the vesicles from entrapped material. This is accomplished through measurements of the intensity changes that occur when the quencher DPX is titrated into a suspension of vesicles that have leaked some of their contents.

We first present the theory of the requenching method and show how it distinguishes between leakage mechanisms. We then describe in detail the experimental implementation, discuss the potential sources of errors, and present several examples in which the requenching method is used to determine the exact release mechanism of several antimicrobial peptides.

[1] G. Schwarz and A. Arbuzova, *Biochim. Biophys. Acta* **1239,** 51 (1995).
[2] J. N. Weinstein, R. D. Klausner, T. Innerarity, E. Ralston, and R. Blumenthal, *Biochim. Biophys. Acta* **647,** 270 (1981).
[3] K. Matsuzaki, O. Murase, H. Tokuda, S. Funakoshi, N. Fujii, and K. Miyajima, *Biochemistry* **33,** 3342 (1994).
[4] T. Benachir and M. Lafleur, *Biochim. Biophys. Acta* **1235,** 452 (1995).
[5] R. A. Parente, S. Nir, and F. Szoka, *Biochemistry* **29,** 8720 (1990).
[6] H. Ostolaza, B. Bartolomé, I. Ortiz de Zárate, F. de la Cruz, and F. M. Goñi, *Biochim. Biophys. Acta* **1147,** 81 (1993).

0076-6879/97 $25

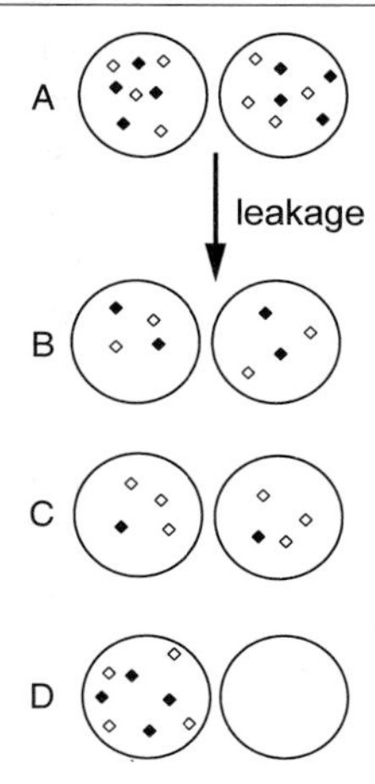

FIG. 1. Mechanisms of peptide-induced release of bilayer vesicle contents. (◇) Fluorescent dye; (◆) quencher. In this schematic representation, vesicles are initially loaded (A) with a self-quenching dye or a binary dye–quencher pair. Graded nonpreferential release (B) leads to partial release of dye and quencher from all of the vesicles while graded preferential release (C) occurs when dye and quencher are different molecules that leak out unequally. All-or-none leakage (D) can occur as a result of peptide-induced pores in the membranes or peptide-induced lysis. Examples of each type of leakage can be found in the literature.[1–8,10,12] Graded release can be distinguished from all-or-none release by means of the fluorescence requenching method[7,8] described in text.

The examples demonstrate that the leakage mechanism can be affected by the changes in the physical state of a peptide or by changes in lipid composition. Finally, we show through an example how the kinetic measurements of fluorescence intensities can be converted into true effluxes of the encapsulated dye and quencher.

Theory

Principal Definitions

The general idea of the requenching method[7,8] is to determine the degree to which ANTS dye molecules remaining inside vesicles are quenched by the remaining DPX after partial leakage has occurred. This is done by titrating the vesicle suspension with DPX at some chosen time after the addition of a leakage-inducing agent.

[7] W. C. Wimley, M. E. Selsted, and S. H. White, *Protein Sci.* **3,** 1362 (1994).
[8] A. S. Ladokhin, W. C. Wimley, and S. H. White, *Biophys. J.* **69,** 1964 (1995).

The total fluorescence observed at any time will be $F = F_o + F_i$, where F_o is the fluorescence originating from outside the vesicles and F_i that from inside. If there were no quenching, the observed total fluorescence from ANTS inside and outside the vesicles would have the maximal value $F^{max} = F_o^{max} + F_i^{max}$. When quencher is present in the vesicles, the addition of Triton X-100 causes lysis of the vesicles and dilution of the DPX to a negligible concentration so that the fluorescence observed in that case will essentially be F^{max}. We define quenching outside (Q_{out}) and inside (Q_{in}) the vesicles and the total quenching (Q_{total}) as follows:

$$Q_{out} = F_o/F_o^{max} \tag{1a}$$
$$Q_{in} = F_i/F_i^{max} \tag{1b}$$
$$Q_{total} = F/F^{max} \tag{1c}$$

Defined in this way, the Q parameters have a value of 1 when there is no quenching and 0 when there is complete quenching. The total fluorescence F is given by $F = Q_{total}F_{max} = Q_{out}F_o^{max} + Q_{in}F_i^{max}$. In the absence of quenching, the ratio of fluorescence coming from inside the vesicles to that coming from the outside equals the molar ratio of dye molecules inside and outside. Therefore, the fractions of ANTS outside and inside the vesicles are $f_{out} = F_o/F^{max}$ and $f_{in} = F_i/F^{max}$ and it must be true that $f_{out} + f_{in} = 1$. The total quenching now can be expressed as

$$Q_{total} = Q_{out}f_{out} + Q_{in}(1 - f_{out}) \tag{2}$$

The goal of the requenching experiment is to determine the behavior of Q_{in} with the change in f_{out} because it reflects the state of the vesicle contents. To accomplish this, one measures Q_{total} as a function of Q_{out} to obtain Q_{in} and f_{out} by fitting the data with Eq. (2). This is described in detail under Experimental Procedures later in this chapter.

Leakage Mechanism

The dependence of Q_{in} on f_{out} reveals the mechanism of leakage: For all-or-none leakage, Q_{in} is constant, and equals the initial value, while for graded leakage Q_{in} increases with f_{out} in a way that depends on the selectivity of the leakage pathway for quencher and dye. To account for this selectivity, a parameter of preferential release, α, can be defined that relates the fractions of quencher and dye inside or outside of the vesicle.[8] Keeping in mind that f_{in} and f_{out} refer to the fraction of ANTS inside and outside, we define α through the relations

$$f_{in}^{DPX} = (f_{in})^{\alpha} \tag{3a}$$
$$1 - f_{out}^{DPX} = (1 - f_{out})^{\alpha} \tag{3b}$$

One can then obtain an expression[8] for the dependence of Q_{in} on f_{out} for graded release that contains only two unknown parameters, the initial concentration of encapsulated quencher, $[DPX]_0$, and the parameter of preferential release α:

$$Q_{in} = [\{1 + K_d[DPX]_0(1 - f_{out})^{\alpha}\}\{1 + K_a[DPX]_0(1 - f_{out})^{\alpha}\}]^{-1} \quad (4)$$

where $K_d = 50\ M^{-1}$ is the dynamic quenching constant and $K_a = 490\ M^{-1}$ is the association constant for static quenching.[9]

Equation (4) is the basic equation of the requenching technique and may be used to distinguish all-or-none from graded release and to estimate the parameter α.

Simulations

Simulations using Eq. (4) allow one to explore the effect of $[DPX]_0$ and α on the Q_{in} dependence of f_{out}. The results are shown in Fig. 2. The solid lines in Fig. 2A correspond to equal release of dye and quencher in a graded manner, while the dashed lines correspond to all-or-none release. Experimentally, one is limited to certain values of f_{out} (not more than 0.9) because the reduction in the fraction of molecules remaining inside increases the experimental error. The ability to distinguish the two mechanisms is affected by the choice of $[DPX]_0$ because the two curves become more similar at lower f_{out} values as $[DPX]_0$ increases. The optimal conditions are met when moderate concentrations of 4–8 m*M* DPX are encapsulated in vesicles.

The variation in α has a dramatic effect on the shape of the $Q_{in}(f_{out})$ curves and consequently on the determination of the leakage mechanism, as shown in Fig. 2B. While preferential release of DPX ($\alpha > 1$) increases the difference between all-or-none and graded leakage, the preferential release of ANTS diminishes it. In the limiting case of complete absence of DPX release ($\alpha = 0$), the graded release of ANTS becomes indistinguishable from the all-or-none release of ANTS and DPX.

[9] The interpretation of the requenching experiment depends little on the exact form of the dependence of fluorescence intensity, F, on DPX concentration {in our case $F_o/F = (1 + K_d[DPX])(1 + K_a[DPX])$}. Therefore one can use any empirical formula as long as it describes the value of F when [DPX] is changed between zero and $[DPX]_0$. The reported values of K_d and K_a were measured in 50 m*M* KCl with 10 m*M* HEPES buffer (pH 7.0) at 25° as described by Ladokhin *et al.*[8] These values, however, might depend on pH and buffer ionic strength and, therefore, they need to be determined separately for each set of experimental conditions. Note that the domination of static quenching prevents the direct resolution of leakage mechanism with lifetime measurements.

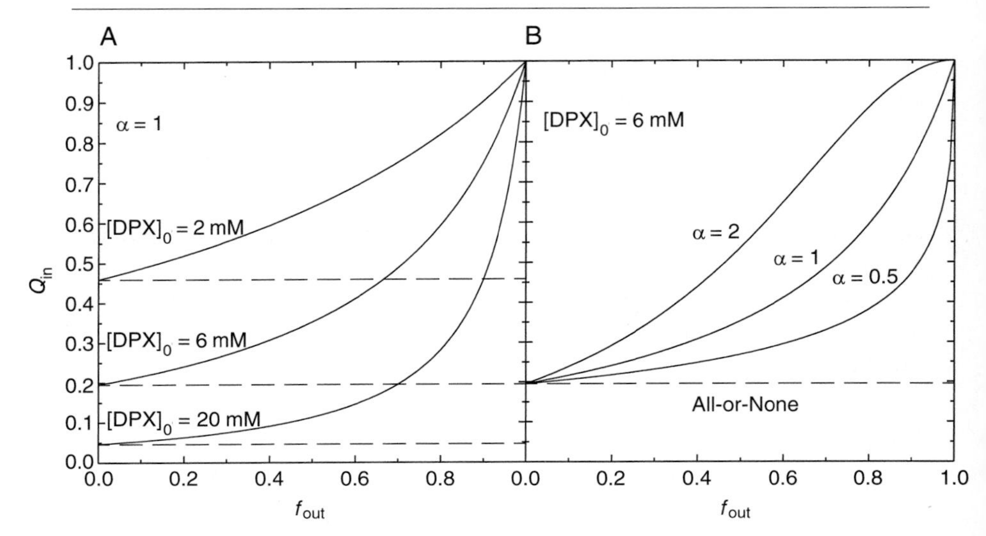

FIG. 2. Simulations of internal quenching (Q_{in}) of ANTS inside vesicles as a function of the ANTS released (f_{out}). Dashed lines correspond to all-or-none release and solid lines to graded release simulated using Eq. (4). (A) The effect of initial DPX concentration for a nonpreferential release ($\alpha = 1$). (B) The effect of preferential release for a fixed initial DPX concentration. Because one is experimentally limited to $f_{out} < 0.9$, both $[DPX]_o$ and α affect the ability to distinguish all-or-none from graded release in a requenching experiment. (Reprinted from Ladokhin *et al.*,[8] with permission.)

Experimental Procedures

Overview

To determine how Q_{in} varies with f_{out}, one first establishes a calibration curve by disrupting untreated vesicle-encapsulated ANTS/DPX preparations with Triton X-100, making incremental additions of DPX, and measuring the normalized fluorescence $F_{DPX}/F^{max} = Q_{out}$ following each addition so that Q_{out} for a particular DPX addition can be determined. One then carries out the leakage experiment. The peptide (or other agent that induces leakage) is added to an ANTS/DPX-containing vesicle solution and the system incubated long enough to reach a plateau level of fluorescence. F_{total} as a function of DPX concentration for the preparation is then measured after each incremental addition of DPX. Following the last addition of DPX, Triton X-100 is added to determine F^{max} from Eq. (1c). Using those measurements and the Q_{out} calibration curve, one plots Q_{total} versus Q_{out} and thereby obtains a linear curve with slope f_{out} and intercept $Q_{in}(1 - f_{out})$ by means of standard linear least-squares fitting procedures.

TABLE I
TYPICAL EXPERIMENTAL RESULTS OBTAINED USING REQUENCHING PROTOCOL[a]

Sample	$(F - F_{bckgrnd}) \times$ volume	$F_{normlzd}$	Q_{total}	Q_{out}
A				
POPC + Triton	1602	3552		1.000
+DPX	1019	2259		0.636
+DPX	737	1634		0.460
+DPX	573	1271		0.358
+DPX	444	984		0.277
+Buffer	451	1000		
B				
POPC + peptide	1279	2812	0.792	
+DPX	839	1844	0.519	
+DPX	611	1343	0.378	
+DPX	478	1051	0.296	
+DPX	368	810	0.228	
+Triton	455	1000		
C				
POPC + buffer	182	404	0.114	
+DPX	127	283	0.080	
+DPX	99	221	0.062	
+DPX	84	186	0.052	
+DPX	71	157	0.044	
+Triton	450	1000		

[a] See explanations in text.

Example Protocol

The following is a description of a requenching experiment[10] intended to establish the mode of leakage caused by the antimicrobial peptide indolicidin.[11] The detailed protocol, which is repeated over a range of concentrations of leakage-inducing peptide, is presented in Table I. A 25-μl aliquot of a stock solution of 25 m*M* POPC vesicles preloaded with ANTS/DPX was added to a cuvette containing a solution (1.025 ml) of 0.24% Triton X-100 (sample A, in Table I), leakage-inducing peptide (in this case, 30 μM indolicidin, sample B), or just buffer (sample C). After the fluorescence had stabilized, each sample was titrated with 45 m*M* DPX solution in four 25-μl additions. Fluorescence intensities were subsequently corrected for background and dilution (Table I). Then, 25 μl of 10% Triton is added to samples B and C and 25 μl of buffer to sample A. The intensities resulting

[10] A. S. Ladokhin, M. E. Selsted, and S. H. White, *Biophys. J.* **72,** 794 (1997).
[11] M. E. Selsted, M. J. Novotny, W. L. Morris, Y.-Q. Tang, W. Smith, and J. S. Cullor, *J. Biol. Chem.* **267,** 4292 (1992).

from these final samples were used for normalization. By dividing the normalized intensities by the initial intensity of sample A (3552), Q_{total} is obtained from samples B and C, and Q_{out} from sample A. A fit of Q_{total} against Q_{out} with Eq. (2) is used to determine Q_{in} and f_{out}. The following values were obtained: For sample B, $Q_{in} = 0.079$, $f_{out} = 0.778$ and for sample C, $Q_{in} = 0.019, f_{out} = 0.097$. The observed values of f_{out} are corrected for the amount of unencapsulated ANTS (sample C) as described under Incomplete Entrapment. f_{out} is found to be $(0.778 - 0.097)/(1 - 0.097) = 0.754$ for sample B and, similarly, 0 for sample C. These points are among the data plotted in Fig. 4B, which were used in further analysis with Eq. (4).

Common Sources of Errors

There are a number of factors that can seriously complicate the implementation and analysis of requenching experiments. In our work on leakage-inducing peptides, we have observed each of these complexities. Here we briefly describe some of them.

Back-Leakage of DPX

The requenching technique is based on the assumption that additions of DPX affect only Q_{out} but not Q_{in} or f_{out}. For graded release, however, the quenching of ANTS molecules remaining inside might be affected by externally added DPX, due to the leakage of DPX back into the vesicles. This possibility is supported by experimental evidence.[8,12] Under these circumstances, the acquisition time for the requenching experiment should be as short as possible to reduce DPX backflow effects. When the DPX backflow is particularly fast, the titration of the same sample with consecutive additions of DPX is not recommended. Instead, experiments on several identical samples, one for each concentration of added DPX, should be performed, and the size of the initial drop in fluorescence should be used as F_{total}. However, the very fact of a strong back-leakage of DPX in the requenching experiment indicates that the leakage-inducing agent causes graded leakage with the preference for DPX.[8]

Incomplete Entrapment

The removal of ANTS from outside the vesicles during sample preparation is sometimes incomplete. For this reason, it is essential that the fraction of ANTS outside the vesicles be assessed before the requenching experiment is performed. This is accomplished through a requenching analysis

[12] K. Hristova, M. E. Selsted, and S. H. White, *Biochemistry* **35,** 11888 (1996).

of the vesicles that have not been treated with peptide (see previous text). In such cases, $f_{out} = f_o$ is the fraction of ANTS that is outside and Q_{in} is the quenching inside the vesicles at the beginning of the experiment. If f_o is significant, then the experimentally observed values of f_{out} should be corrected: $f_{corr} = (f_{obs} - f_o)/(1 - f_o)$. Some vesicle preparations will spontaneously leak a fraction of their contents during prolonged storage of stock solution and we therefore recommend repeating the determination of f_o each time the leakage experiments are conducted.

Nonreleasable ANTS/DPX

The presence of multilamellar or oligolamellar vesicles can lead to the entrapment of ANTS/DPX that cannot be released by peptides acting on the vesicle surfaces. However, this nonreleasable fraction will be released by Triton. A nonreleasable fraction can be included in the analysis of requenching experiments by adding an extra term to Eq. (2):

$$Q_{total} = Q_{out} f_{out} + Q_{in}(1 - f_{out} - f_{NR}) + Q_{NR} f_{NR} \tag{5}$$

where f_{NR} and Q_{NR} are the fraction of ANTS and the quenching of the nonreleasing fraction, respectively. These parameters can either be estimated from independent experiments or used as parameters to fit Q_{in} to f_{out}. If ignored, a significant nonreleasable fraction leads to a bimodal dependence of Q_{in} on f_{out} in the case of graded release where Q_{in} initially increases with f_{out}, and then decreases as the nonreleasable fraction begins to dominate the apparent Q_{in}. Obviously, the best way of dealing with this problem is to make sure that the stock vesicle preparation is homogeneous.

Complex Leakage Mechanism

The previous analysis and comments assume that a single mechanism is responsible for leakage of a vesicles contents. However, both graded and all-or-none release can be present simultaneously in a system; this seriously complicates the analysis. Consider a peptide that is capable of releasing f^{graded} and $f^{a\text{-}o\text{-}n}$ fractions of the dye in graded and all-or-none manners, respectively. However, the effect of graded leakage is limited to only the portion of vesicles that have not already been completely emptied via an all-or-none mechanism. Therefore, the total fraction of dye released is

$$f_{out}^{total} = f^{a\text{-}o\text{-}n} + (1 - f^{a\text{-}o\text{-}n}) f^{graded} \tag{6}$$

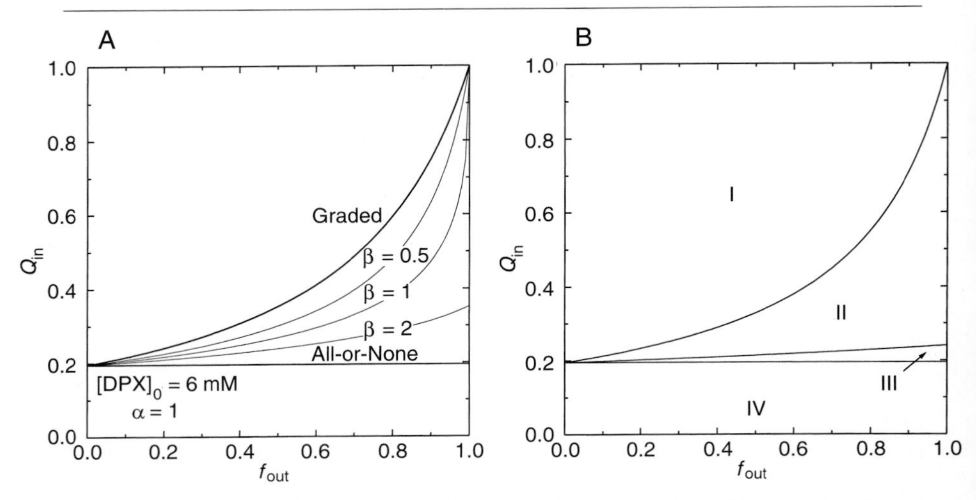

FIG. 3. Simulations of internal quenching (Q_{in}) of ANTS inside vesicles as a function of the ANTS released (f_{out}) for a combination of graded and all-or-none release ($[DPX]_o = 6$ mM, $\alpha = 1$) using Eqs. (4), (6), and (7). (A) The effect of various relative contributions of all-or-none and graded mechanisms: $\beta = 0.5$ (domination by graded release), $\beta = 1$ (equal contributions), $\beta = 2$ (domination by all-or-none release). (B) The effect of complex leakage mechanism on the determination of mode of leakage with the requenching analysis (see text for details). The curves that divide the figure into zones I–IV correspond, from top to bottom, to $\beta = 0.2$, 5, and ∞. See text for description and use of zones.

For the special case that all-or-none and graded leakage are proportional to each other, preference for all-or-none release, β, is given by

$$\beta = \frac{f^{\text{a-o-n}}}{f^{\text{graded}}} \tag{7}$$

After solving Eqs. (6) and (7) for f^{graded} and substituting it into Eq. (4) for f_{out}, one can easily obtain the dependence of Q_{in} on f_{out}^{total} for this combination of two mechanisms.[13]

We have examined the predictions of this simplified model for combinations of graded nonpreferential and all-or-none mechanisms (Fig. 3A). Depending on β, the curves could appear close to pure graded (low β) or all-or-none (high β) mechanisms. Note that for intermediate values of β, the curves appear to be similar to those for graded release with the preference for ANTS, and, for example, the curve for $\beta = 1$ coincides with the one for $\alpha = 0.5$ from Fig. 2B.

[13] A somewhat different approach to complex leakage has been described by Schwarz and Arbuzova.[1]

This inability to differentiate certain mechanisms is a limitation of the requenching technique and cannot be overcome by further data analysis. Nevertheless, there can be value in understanding what models may be consistent with the data. This can be done by dividing the Q_{in} vs f_{out} plane into four different zones, as shown in Fig. 3B.

Zone I: Here the mechanism is quite well defined: Leakage is graded, with the preference for DPX ranging from high preference to no preference.

Zone II: The mechanism is poorly defined. Multiple possibilities exist, including graded preferential leakage of ANTS or a combination of graded and all-or-none release. These possibilities may be distinguished in some cases through additional independent experiments. For example, a size-dependent leakage of large dextrans can indicate a contribution of an all-or-none mechanism.[7]

Zone III: The mechanism is reasonably well defined as mostly all-or-none. Be sure, however, to go to the highest achievable values of f_{out} to differentiate from graded leakage with strong preference for ANTS.

Zone IV: Data in this region are not consistent with any of the described models. Check calculations, measurements, materials, etc., for errors.

Applications

Using the requenching method, we have determined the mechanism of leakage from vesicles of different compositions induced by a variety of natural and synthetic peptides.[7,8,10,12] We present here two specific examples that demonstrate that leakage can be sensitive to the physical state of the leakage-inducing peptide and to lipid composition of the vesicles. Finally, we show how the information obtained in the requenching experiments can be used to recover the efflux kinetics of dye and quencher by means of the observed fluorescence intensities.

Leakage Induction by a Human Neutrophil Peptide (Defensin)

The human neutrophil peptide (defensin) HNP-2 is one of a potent class of antimicrobial peptides found in the dense granules of neutrophils isolated from several species.[14] Neutrophil defensins are cationic arginine-rich peptides of 30 or so residues that form a triple-stranded β-sheet structure stabilized by three disulfide bonds.[15–17] Unlike known neutrophil defen-

[14] R. I. Lehrer, T. Ganz, and M. E. Selsted, *Cell* **64,** 229 (1991).

[15] S. H. White, W. C. Wimley, and M. E. Selsted, *Curr. Opin. Struct. Biol.* **5,** 521 (1995).

[16] A. Pardi, X. L. Zhang, M. E. Selsted, J. J. Skalicky, and P. F. Yip, *Biochemistry* **31,** 11357 (1992).

[17] C. P. Hill, J. Yee, M. E. Selsted, and D. Eisenberg, *Science* **251,** 1481 (1991).

sins from other species, human defensins form extremely stable dimers that may be related to their ability to form large-diameter multimeric pores.[7] Fluorescence requenching studies of leakage induction in anionic palmitoyl oleoylphosphatidylglycerol (POPG) vesicles by the native and reduced (denatured) forms of HNP-2 provide excellent examples of, respectively, all-or-none and graded leakage.[7] The results are summarized in Fig. 4A. As can be seen, the leakage induced by reduced HNP-2 is in zone I and therefore unequivocally graded. Furthermore, the leakage is preferential for leakage of DPX with $\alpha = 1.7$. The data for native HNP-2 lie in zone

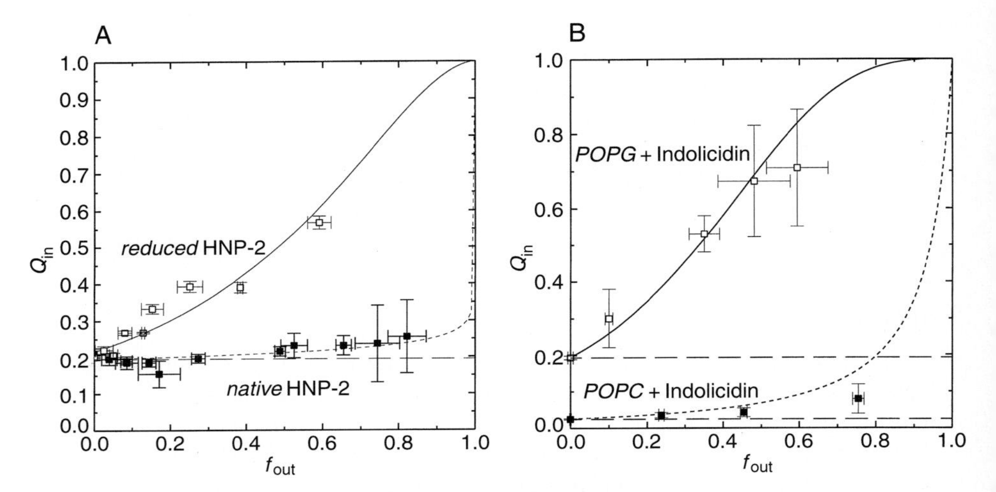

FIG. 4. Examples of results of fluorescence requenching experiments that demonstrate both graded and all-or-none leakage of the fluorophore–quencher pair ANTS/DPX. (A) Requenching of ANTS/DPX released from POPG vesicles by native (■) and reduced (□) forms of HNP-2. (Data replotted from Wimley *et al.*[7]) Reduced and unfolded HNP-2 induces graded leakage of ANTS/DPX from POPG vesicles (solid line, $\alpha = 1.7$), Native HNP-2 induces either all-or-none leakage (dashed line) or highly preferential leakage of ANTS (dotted line, $\alpha = 0.12$). Wimley *et al.*[7] distinguished between these two possibilities by examining the leakage of fluorescent dextrans of different sizes. They found that native HNP-2 appears to cause all-or-none leakage through pores approximately 25 Å in diameter. These data demonstrate that the requenching method is sensitive to changes in the mechanism of release resulting from changes in the state of the peptide. (B) The release of ANTS/DPX from POPG (□) and POPC (■) vesicles induced by indolicidin. (Data replotted from Ladokhin *et al.*[10]) Indolicidin induces graded leakage of ANTS/DPX from POPG in a manner that is highly preferential for DPX ($\alpha = 3.1$, $[\mathrm{DPX}]_o = 6.1$ mM, solid line). The mechanism of leakage from POPC vesicles is different, but not well defined by requenching experiments alone. Determination of the mechanism requires additional experiments with large dextrans (see text). The dotted line describes graded nonpreferential release ($\alpha = 1$, $[\mathrm{DPX}]_o = 30$ mM) and the dashed line describes all-or-none release. These results demonstrate that the requenching method is sensitive to changes in mechanism that accompany changes in lipid.

III and are mostly consistent with all-or-none leakage, although graded release with a low value of α is feasible (dotted line, Fig. 4A). The later possibility has been ruled out because native HNP-2 induces leakage of large dextrans.[7]

Leakage Induction by Antimicrobial Peptide Indolicidin

Indolicidin is a cationic 13-residue peptide amide with potent antimicrobial activity isolated from the dense granules of bovine neutrophils. Its five tryptophans and its four positive charges render it both hydrophobic and cationic. As a result, unlike neutrophil defensins, it will bind strongly to neutral (zwitterionic) palmitoyloleoylphosphatidylcholine (POPC) vesicles as well as POPG vesicles. The results,[10] summarized in Fig. 4B, show that the leakage-induction activity is different for the two lipids. Leakage data for POPG vesicles fall in zone I with $\alpha = 3.1$. Thus, the leakage is graded with a remarkably strong preference for DPX. For POPC vesicles, the leakage data fall in zone II so that the mechanism of leakage is not well defined. The POPC data are consistent with either graded leakage with some preference for ANTS or a combination of graded and all-or-none. Independent results on size-dependent leakage of large dextrans indicate

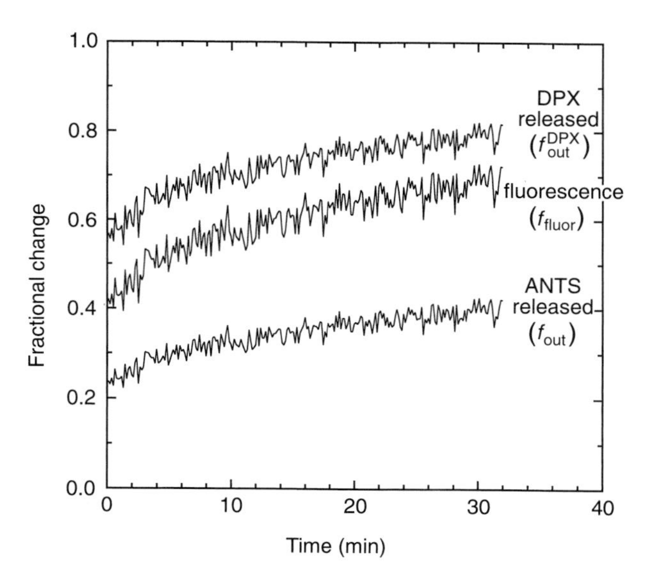

FIG. 5. Comparison of fractional changes in fluorescence (f_{fluor}) with the fractions of ANTS (f_{out}) and DPX (f_{out}^{DPX}) released as a function of time. The fractions are calculated as described in text.

a strong contribution of all-or-none leakage (A. S. Ladokhin and S. H. White, unpublished data).

Conversion of Fluorescence Intensity Changes to Fractional Release of Loaded Material

The fluorescence changes observed during the graded release of quenched fluorophores, unlike those observed for simple all-or-none release, do not coincide directly with the amount of material released.[1,8] Unless appropriate corrections are made, the observed fluorescence overestimates both the amount of dye released and the rate of release. These erroneous estimates can be corrected using the same basic equations developed for the requenching procedure. In the absence of the externally added DPX and infinite dilution of the DPX that has leaked out, $Q_{out} = 1$. Equation (2) may therefore be rewritten in combination with Eq. (1c) to give

$$F = F^{max}[f_{out}(1 - Q_{in}) + Q_{in}] \quad (8)$$

For all-or-none release, Q_{in} is constant so that the observed changes in F are proportional to f_{out}. For graded release, however, Q_{in} depends strongly on f_{out} so that F is not a direct and simple measure of f_{out}. This is true for self-quenching dyes as well.[1,8] We have developed an analytical procedure for the analysis of efflux of dye or quencher using the parameterized dependence of f_{out} on the time elapsed after the vesicles were mixed with leakage-inducing agent.[8,10] The same result can be achieved numerically as follows. (1) Using the values of $[DPX]_0$ and α that were determined in the requenching experiment, generate a table of F/F^{max} values as a function of f_{out} values, using Eq. (8); (2) obtain from the table by interpolation the value of f_{out} that corresponds to a given experimental value of F/F^{max}. The corresponding f_{out}^{DPX} is calculated with Eq. (3b). In Fig. 5 we show the time courses of the fractional releases of ANTS (f_{out}) and DPX (f_{out}^{DPX}) for $[DPX]_0 = 6.1$ mM and $\alpha = 3.1$. The fractional change in fluorescence intensity $f_{fluor} = [F(f_{out}) - F(0)]/[F^{max} - F(0)]$ has also been plotted in Fig. 5 for comparison. The preferential nature of the release is revealed by the relative values of f_{out} and f_{out}^{DPX}.

Acknowledgments

Research supported by NIH grant GM-46823 awarded to S. White and NIH Grant AI-22931 awarded to Dr. Michael Selsted.

[24] Fluorescence Probes for Studying Membrane Heterogeneity

By LESLEY DAVENPORT

Introduction

This chapter discusses the application of fluorescence probes for studying structural and dynamic heterogeneity of lipid packing in bilayer membranes. Although large-scale phase separations, or *domains,* arising from nonideal mixing of chemically heterogeneous lipid mixtures in membrane systems may be directly visualized by microscopy tools,[1–2] structural "gel–fluid" lipid clusters (composed of several hundred lipid molecules) and their associated dynamics can be studied only via indirect approaches. Fluorescence spectroscopy provides an excellent tool for unraveling, and in some cases, quantifying such complex microheterogeneous systems. Using extrinsic fluorescence membrane probes, which partition into the different coexisting lipid phases, heterogeneous fluorescence signals can often be resolved and associated with the different environments. Measurable probe parameters include fluorescence intensity (FL), excitation [$FL(\lambda_{ex})$] and/or emission [$FL(\lambda_{em})$] spectra, fluorescence lifetime (τ_{FL}), rotational correlation times (ϕ), and absorption spectra [$A(\lambda)$].

Because several excellent reviews on fluorescent probes for studying membrane structure are already available,[3–5] the goal here is to (1) outline desirable properties for a membrane heterogeneity probe by identifying specific spectroscopic properties of commonly used fluorescence probes, (2) highlight fluorescence spectroscopic approaches for targeting structural and dynamic lipid microheterogeneity, and (3) indicate some of the pitfalls for these kinds of studies. As a result of the enormous variety of available fluorescence probes, the focus here is restricted predominantly to those probes that reside within the hydrophobic region of the bilayer.

[1] S. C. Juo and M. P. Sheetz, *Trends Cell Biol.* **2,** 116 (1992).
[2] X. F. Wang, K. Florine-Casteel, J. J. Lemasters, and B. Herman, *J. Fluoresc.* **5,** 71 (1995).
[3] L. M. Loew (ed.), "Spectroscopic Membrane Probes," Vols. 1–3. CRC Press, Boca Raton, Florida, 1988.
[4] B. R. Lentz (ed.), *J. Fluoresc.* **5,** 1995.
[5] C. D. Stubbs and B. W. Williams, *in* "Topics in Fluorescence Spectroscopy." (J. R. Lakowicz, ed.), Vol. 3, p. 231. Plenum, New York, 1991.

0076-6879/97 $25

Nature of Membrane Microheterogeneity

The extent and nature of detected bilayer heterogeneity depends on the biophysical method employed. Spectroscopic studies (magnetic resonance or optical) are sensitive to "local" (microscopic) environmental effects and are thus ideal for microheterogeneity studies. In contrast, differential scanning calorimetry (DSC) provides information on larger scale (macroscopic or cooperative) lipid packing.[6] In addition, the extent of bilayer ordering observed is also dependent on the experimental "time window" over which the study is conducted.[7] For example, on the nanosecond (fluorescence) time scale, the bilayer system may clearly demonstrate gel–fluid lipid heterogeneity. This pattern, however, is continually shifting and can be blurred to homogeneity within microseconds as detected via nuclear magnetic resonance (NMR) methods.

The following discusses thermally controlled gel–fluid lipid microheterogeneity. Larger and more persistent long-term lipid heterogeneity (or phase separations), as induced by intrinsic (e.g., cholesterol) or extrinsic (e.g., Ca^{2+}) chemical effectors, can also be successfully assayed using fluorescent probe techniques.[8] Indeed, these are the basis for many membrane fusion assays (for a review, see Ref. 9).

Choosing a Fluorescent Probe

A probe is generally selected because of specific attractive spectroscopic characteristics (e.g., rotational correlation time, fluorescence lifetime, or fluorescence intensity) that demonstrate sensitivity to the physical (gel–fluid) state (or lipid packing) of the membrane (the terms *gel* and *fluid* are used here to indicate a wide range of acyl-chain order). This sensitivity can be assessed by preparing and labeling model membrane systems composed of phospholipids of varying lipid-phase transition temperatures (T_c) where "gel" or "fluid" lipid contributions can be thermally controlled. The most frequently employed heterogeneity probes are hydrophobic and readily partition within the apolar region of the bilayer system of interest. These may be "free" or attached to charged or polar organic groups, or a biological anchor (e.g., fatty-acyl chain). The members of a second class of commonly adopted probe molecules are water soluble (neutral or charged), again

[6] R. Biltonen, *J. Chem. Thermodynam.* **22,** 1 (1990).

[7] J. F. Holzwarth, V. Eck, and A. Genz, *in* "Spectroscopy and the Dynamics of Molecular Biological Systems" (P. M. Bayley and R. E. Dale, eds.), p. 351. Academic Press, London, 1985.

[8] L. Davenport, J. R. Knutson, and L. Brand, *Subcell. Biochem.* **14,** 145 (1989).

[9] N. Düzgünes (ed.), "Methods in Enzymology," Vol. 220 Academic Press, San Diego, 1993.

either "free" or covalently attached to intrinsic membrane components, and are located at the polar head group region of the bilayer. Spectral and temporal characteristics of some commonly employed structural and dynamic membrane heterogeneity fluorescence probes are summarized in Table I. For convenience the probes are arranged according to their specific spectroscopic sensitivities in bilayer studies.

Probes with high quantum yield (Φ) values and extinction coefficients (ε) are preferred when working with highly scattering vesicle samples, or with cell membrane systems that exhibit strong autofluorescence, as they provide easily measurable fluorescence intensity signals and are only required at low (nanomolar) concentrations. This minimizes concerns of possible bilayer perturbation effects arising from the presence of the reporter fluorophore (as discussed later in this chapter).

General Considerations

Confidence in the interpretation of measured heterogeneous fluorescence signals from bilayer-embedded extrinsic probe data via heterogeneous membrane models requires knowledge of the position, partitioning, and perturbation properties of the probe.

Probe Position. Simple estimation of internal or surface bilayer location of the probe can be addressed using fluorescence-quenching experiments (for a review, see Eftink[10]), employing membrane-permeable [e.g., acrylamide,[11] oxygen,[12] and Cu(II) ions[13]] or membrane-impermeant (e.g., iodide[14] and cesium anions[15]) quencher (Q) molecules. In addition, positioning of the fluorescence quencher at fixed locations across the bilayer width, e.g., brominated phospholipids,[16] or spin-labeled lipids,[13,17] can provide a more precise depth estimate of fluorophore location within the bilayer (see [22] in this volume[18]).

A more complex question of probe position arises when using lipid-soluble "free" molecules. It is critical that the fluorescence properties of

[10] M. R. Eftink, *In* "Topics in Fluorescence Spectroscopy" (J. R. Lakowicz, ed.), Vol. 2, p. 53. Plenum, New York, 1991.

[11] D. B. Chalpin and A. M. Kleinfeld, *Biochim. Biophys. Acta* **731,** 465 (1983).

[12] J. R. Lakowicz and J. R. Knutson, *Biochemistry* **19,** 905 (1980).

[13] E. A. Haigh, K. R. Thurlborn, and W. H. Sawyer, *Biochemistry* **18,** 3535 (1979).

[14] G. W. Stubbs, B. J. Litman, and Y. Barenholz, *Biochemistry* **15,** 2766 (1976).

[15] M. Shinitzky and B. Rivnay, *Biochemistry* **16,** 982 (1977).

[16] T. Markello, A. Zlotnick, E. James, J. Tennyson, and P. W. Holloway, *Biochemistry* **24,** 2895 (1985).

[17] R. C. Chatelier, P. J. Rogers, K. P. Ghiggino, and W. H. Sawyer, *Biochim. Biophys. Acta* **776,** 75 (1984).

[18] A. Ladokhin, *Methods Enzymol.* **278,** [22], 1997 (this volume).

TABLE I
PROPERTIES OF SOME COMMON FLUORESCENCE MEMBRANE HETEROGENEITY PROBES[a]

Structure	Properties
A. Probes for Structural Heterogeneity	
$(CH_3)_2N$–naphthalene–$C(=O)-CH_2-CH_3$ Prodan[b]	Neutral (spectral sensitivity) 6-propionyl-2-dimethylaminonaphthylene λ_{ex} 360 nm (G)/360 nm (F); λ_{em} 460 nm (G)/520 nm (F) ε_{max} 18,000 $M^{-1} \cdot cm^{-1}$ (methanol)
$H_3C-(CH_2)_{14}-C(=O)$–naphthalene–$N(CH_3)-(CH_2)_2-\overset{\oplus}{N}(CH_3)_3$ Patman[c]	Cationic (spectral sensitivity) 6-palmitoyl-2-[(2-trimethylammonium-ethyl)methyl]aminonaphthylene λ_{ex} 357 nm (G)/357 nm (F); λ_{em} 425 nm (G)/465 nm (F) ε_{max} 18,000 $M^{-1} \cdot cm^{-1}$ (methanol/H_2O)
Benzoxazole(N–$(CH_2)_3$–$SO_3^{\ominus}$)$=CH-CH=CH-CH=$(pyrimidine-dione, N,N′-C_4H_9) Merocyanine 540[d]	Anionic (intensity sensitivity) λ_{ex} 550 nm; λ_{em} 590 nm ε_{max} 144,000 $M^{-1} \cdot cm^{-1}$ (methanol)

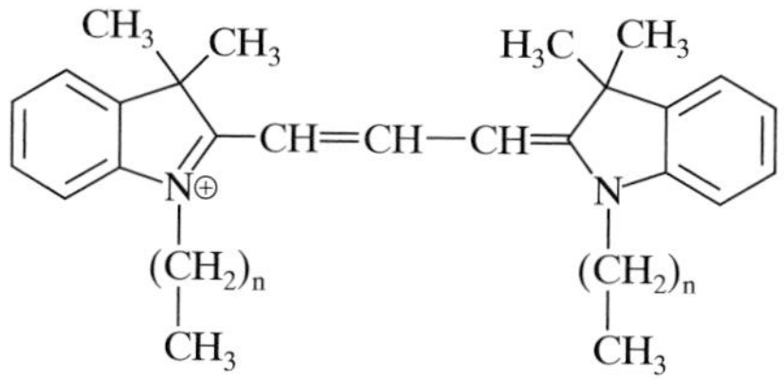

Probe	Properties
1,1′-Dialkyl-3,3,3′,3′-tetramethylindocarbocyanine [DiIC$_n$(3)][e,f]	Cationic (intensity/lifetime sensitivity) (n = 11, 15, 17, 21) λ_{ex} ~550 nm; λ_{em} ~568 nm ε_{max} 110,000–128,000 $M^{-1} \cdot cm^{-1}$ (methanol) $\langle \tau_G \rangle$ ~(0.78–0.81 nsec); $\langle \tau_F \rangle$ ~(0.89–1.87 nsec)
DPH[g]	Neutral (lifetime/rotational sensitivity) 1,6-Diphenyl-1,3,5-hexatriene λ_{ex} 340/355/380 nm; λ_{em} 430 nm ε_{max} 80,000 $M^{-1} \cdot cm^{-1}$ (methanol) $\langle \tau_G \rangle$ ~(10–11 nsec); $\langle \tau_F \rangle$ ~(6–9 nsec); r_0 = 0.395
TMA-DPH[h]	Cationic (lifetime/rotational sensitivity) 1-(4-Trimethylammoniumphenyl)-6-phenyl-1,3,5-hexatriene λ_{ex} 353 nm; λ_{em} 426 nm ε_{max} 53,000 $M^{-1} \cdot cm^{-1}$ (methanol) $\langle \tau_G \rangle$ ~(6.7 nsec); $\langle \tau_F \rangle$ ~(4–7 nsec); r_0 = 0.39
n-(9-Anthroyloxy) fatty acids[i,j]	Anionic [rotational (anisotropic) sensitivity] λ_{ex} ~330/350/380 nm; λ_{em} ~445 nm ε_{max} 7300–7900 $M^{-1} \cdot cm^{-1}$ (methanol) $\langle \tau_G \rangle$ ~(9–12 nsec); $\langle \tau_F \rangle$ ~(5.5–10 nsec) r_0^{316} = 0.1; r_0^{380} = 0.34 where: x = 0; y = 15 x = 7; y = 8 x = 1; y = 14 x = 8; y = 7 x = 4; y = 11 x = 10; y = 5 x = 5; y = 10

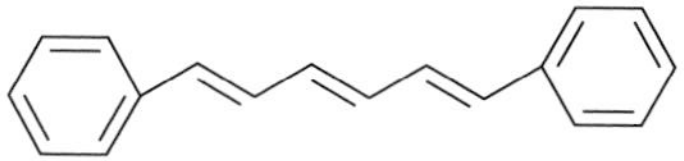

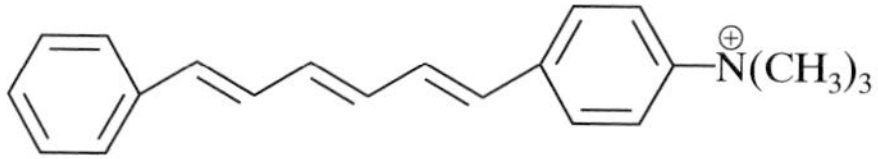

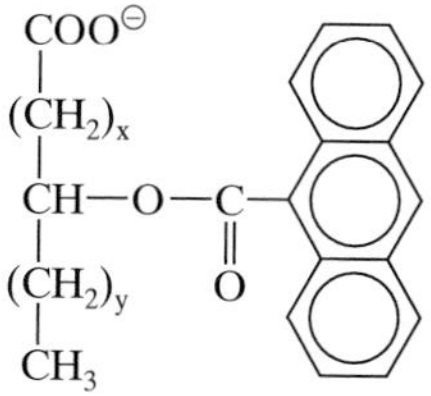

(*continues*)

TABLE I (*continued*)

Structure	Properties
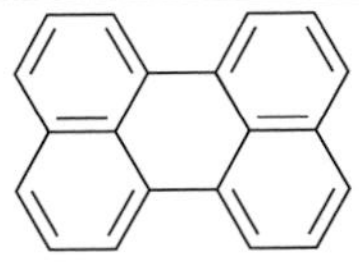 Perylene[k,l,m]	Neutral [rotational (anisotropic) sensitivity] λ_{ex} 256/410/430 nm; λ_{em} 440/475/500 nm ε_{max} 40,000 $M^{-1} \cdot cm^{-1}$ (methane) $\langle\tau_G\rangle$ ~(5.5 nsec); $\langle\tau_F\rangle$ ~(4.5–5.5 nsec) $r_0^{256} = -0.14$; $r_0^{315} = 0.1$; $r_0^{410} = 0.314/0.34$
B. Probes for Structural and Dynamic Heterogeneity	
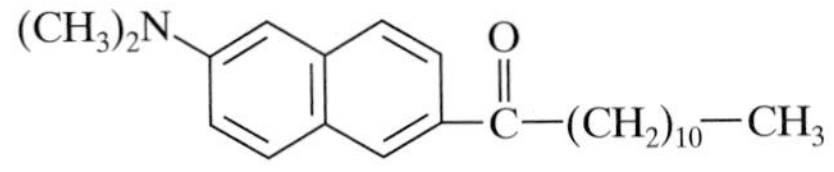 $(CH_3)_2N$–naphthalene–$C(=O)$–$(CH_2)_{10}$–CH_3 Laurdan[n]	Neutral (spectral sensitivity) 6-Dodecanoyl-2-dimethylaminonaphthylene λ_{ex} 410 nm (G)/340 nm (F); λ_{em} 440 nm (G)/490 nm (F) ε_{max} 19,000 $M^{-1} \cdot cm^{-1}$ (methanol)
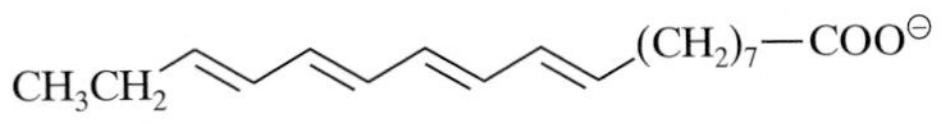 CH_3CH_2–(CH=CH)$_4$–$(CH_2)_7$–$COO^{\ominus}$ *trans*-Parinaric acid[o]	Anionic (rotational/lifetime sensitivity) λ_{ex} ~305/320 nm; λ_{em} ~445 nm ε_{max} 84,000 $M^{-1} \cdot cm^{-1}$ (methanol) $\langle\tau_G\rangle$ ~(30–50 nsec); $\langle\tau_F\rangle$ ~(3 nsec); $r_0 = 0.391$

C. Probe for Dynamic Heterogeneity

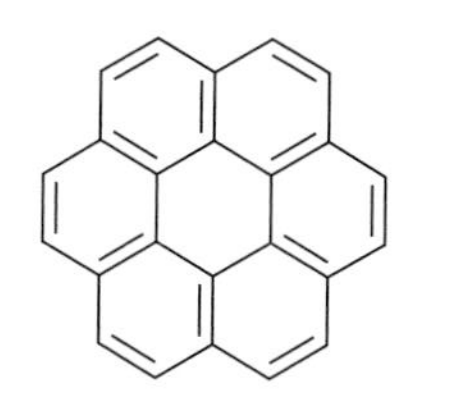

Coronene[p]

Neutral [rotational (out of plane) sensitivity]
λ_{ex} 305/340 nm; λ_{em} 448 nm
$\langle\tau_G\rangle$ ~(246 nsec); $\langle\tau_F\rangle$ ~(166 nsec); $r_0 = 0.07$

---

[a] Extinction coefficients were obtained from R. Haughland, *In* "Handbook of Fluorescent Probes and Research Chemicals" (K. D. Larison, ed.). Molecular Probes, Eugene, Oregon, sixth edition (1996).

[b] G. Weber and F. J. Farris, *Biochemistry* **18,** 3075 (1979).

[c] J. R. Lakowicz, D. R. Bevan, B. P. Maliwal, H. Cherek, and A. Balter, *Biochemistry* **18(22),** 5714 (1983).

[d] P. Williamson, K. Mattocks, and R. A. Schlegel, *Biochim. Biophys. Acta* **732,** 387 (1983).

[e] R. D. Klausner and D. E. Wolf, *Biochemistry* **19,** 6199 (1980).

[f] B. S. Packard and D. E. Wolf, *Biochemistry* **24,** 5176 (1985).

[g] B. R. Lentz, *In* "Spectroscopic Membrane Probes" (L. M. Loew, ed.), Vol. 1, p. 13. CRC Press, Boca Raton, Florida, 1988.

[h] P. Herman, I. Konopasek, J. Plasek, and J. Svobodova, *Biochim. Biophys. Acta* **1190,** 1 (1994).

[i] M. Vincent, B. DeForesta, J. Gallay, and A. Alfsen, *Biochemistry* **21,** 708 (1982).

[j] K. R. Thurlborn, L. M. Tilley, W. H. Sawyer, and F. E. Treloar, *Biochim. Biophys. Acta* **558,** 166 (1979).

[k] P. L.-G. Chong, B. Wieb van der Meer, and T. E. Thompson, *Biochim. Biophys. Acta* **813,** 253 (1985).

[l] U. Cogan, M. Schnitzky, G. Weber, and T. Nishida, *Biochemistry* **12,** 521 (1973).

[m] L. Brand, J. R. Knutson, L. Davenport, J. M. Beechem, R. E. Dale, D. G. Walbridge, and A. A. Kowalczyk, *In* "Spectroscopy and the Dynamics of Molecular Biological Systems" (P. M. Bayley and R. E. Dale, eds.), p. 259. Academic Press, London, 1985.

[n] T. Parasassi, G. De Stasio, G. Ravagnan, R. M. Rusch, and E. Gratton, *Biophys. J.* **60,** 179 (1991).

[o] A. Ruggiero and B. Hudson, *Biophys. J.* **55,** 1111 (1989).

[p] L. Davenport and P. Targowski, *Biophys. J.* **71,** 1837 (1996).

the probe accurately reflect those of its surrounding lipid environment; probe (or positional) heterogeneity should be avoided. Experimentally, it is difficult to resolve probe versus environmental microheterogeneity unequivocally. Studies attempting to address this point have been primarily focused on bilayers labeled with 1,6-diphenyl-1,3,5-hexatriene (DPH) (although the orientations of membrane probes in stretched polymer films and liquid crystals have been examined[19,20]). Angle-resolved measurements of oriented bilayer systems by Levine *et al.*[19,21] and detailed time-dependent polarized fluorescence investigations,[22–24] suggest at least two separate DPH probe populations: aligned parallel with the fatty-acyl chains of the bilayer, or lying between the bilayer leaflets and parallel with the membrane surface. Two orientational distributions for DPH have also been reported using polarized fluorescence imaging microscopy studies of labeled large unilamellar vesicles (LUVs).[25] Potential heterogeneity of "free" probe location can be excluded by attachment of charged or polar groups (e.g., trimethylammonium-DPH; TMA-DPH), or intrinsic membrane components, such as fatty acids, phospholipids, or steroids,[26] to neutral hydrophobic probes, which serve also to locate the probe at a defined location and depth within the bilayer interior. In addition, when studying cell systems, probe attachment minimizes permeation of the fluorophore to other intracellular membranes; an important consideration for fluorescence microscopy and imaging studies.

Probe Partitioning. A fluorescence probe can (1) distribute nonpreferentially into all lipid subfractions, with observable spectroscopic parameters that can be associated with a specific lipid phase, or (2) reside almost exclusively within a particular lipid phase. The gel–fluid partition coefficient ($K_P^{g/f}$) is written

$$K_P^{g/f} = \frac{(P_G/X_G)}{(P_F/X_F)} \quad \text{and} \quad X_G + X_F = 1 \tag{1}$$

[19] M. Von Gurp, H. Van Langen, G. Van Ginkel, and Y. K. Levine, *In* "Polarized Spectroscopy of Ordered Systems" (B. Samorì and E. W. Thurlstrup, eds.), NATO ASI Series. Kluwer, Amsterdam, The Netherlands, 1987.

[20] A. Arcioni and C. Zannoni, *Chem. Phys.* **88,** 113 (1984).

[21] U. A. van der Heide, G. Van Ginkel, and Y. K. Levine, *Chem. Phys. Lett.* **253,** 118 (1996).

[22] W. Van der Meer, H. Pottel, W. Herreman, and M. Ameloot, *Biophys. J.* **46,** 515 (1984).

[23] S. Wang, J. M. Beechem, E. Gratton, and M. Glaser, *Biochemistry* **30,** 5565 (1991).

[24] D. Toptygin and L. Brand, *J. Fluoresc.* **5,** 39 (1995).

[25] K. Florine-Casteel, *Biophys. J.* **57,** 1199 (1990).

[26] R. Haughland, *In* "Handbook of Fluorescent Probes and Research Chemicals" (K. D. Larison, ed.). Molecular Probes, Inc., Eugene, Oregon, 1992–1994.

where P_G and P_F refer to mole fractions of the probe in gel and fluid lipid environments, respectively, corrected for unequal mole fractions of gel (X_G) and fluid (X_F) phases present in the heterogeneous lipid mixture. Quantitative estimation of P_G and P_F for a probe embedded within a binary lipid system (with thermally controlled X_G and X_F contributions) is possible from measured quantum yield, lifetime, and/or steady-state emission anisotropy (EA) measurements. Excellent descriptions of the experimental determination of gel–fluid partition coefficients ($K_P^{g/f}$) have been described in detail elsewhere.[27,28] Values of $K_P^{g/f}$ for a particular probe can vary depending on the nature of the gel phase under investigation.[29]

For a fluorescent probe that demonstrates a strong partitioning preference in one lipid phase, the observed (average) fluorescence parameters FL_{obs} will be weighted in favor of the signal from that phase (assuming that the lifetime of the lipid phase persists beyond that of the fluorescence lifetime of the probe). For the simple case where $\phi_G = \phi_F$ and $\varepsilon_G = \varepsilon_F$:

$$FL_{obs} = FL_G P_G + FL_F P_F \tag{2}$$

where F_G and F_F are the values for the fluorescence parameter in gel and fluid environments, respectively. Evidence for preferential partitioning of a probe (at nonperturbing probe-to-lipid molar labeling ratios) can be demonstrated from fluorescence intensity versus temperature plots,[30] or via temperature shifts in lipid "melt" profiles [defined here as fluorescence emission anisotropy (EA) versus temperature plots].[27,28] For a probe with equal ($K_P^{g/f} = 1$) partitioning between coexisting gel- and fluid-like lipid phases (e.g., DPH embedded within a binary mixture composed of 50/50 mol% dimyristoylphosphatidylcholine (DMPC) and dipalmitoylphosphatidylcholine (DPPC) vesicles), measured temperature versus EA profiles accurately reflected the expected average lipid-phase transition temperature for these two phospholipids (expected average T_c values can vary depending on the membrane radius of curvature for the bilayer system employed[31]). For probes with $K_P^{g/f} < 1$, profiles are shifted to apparent lower temperatures, as a result of preferentially partitioning into fluid-phase lipid (e.g., *cis*-parinaric acid) on increasing the temperature; for $K_P^{g/f} > 1$ (e.g., *trans*-parinaric acid), apparent T_c values are shifted to higher temperatures. This effect is summarized in the simulated data shown in

[27] B. R. Lentz, Y. Barenholz, and T. E. Thompson, *Biochemistry* **15,** 4529 (1976).
[28] L. A. Sklar, G. P. Miljanich, and E. A. Dratz, *Biochemistry* **18,** 1707 (1979).
[29] N.-N. Huang, K. Florine-Casteel, G. W. Feigenson, and C. Spink, *Biochim. Biophys. Acta* **939,** 124 (1988).
[30] R. D. Klausner and D. E. Wolf, *Biochemistry* **19,** 6199 (1980).
[31] B. R. Lentz, Y. Barenholz, and T. E. Thompson, *Biochemistry* **15,** 4521 (1976).

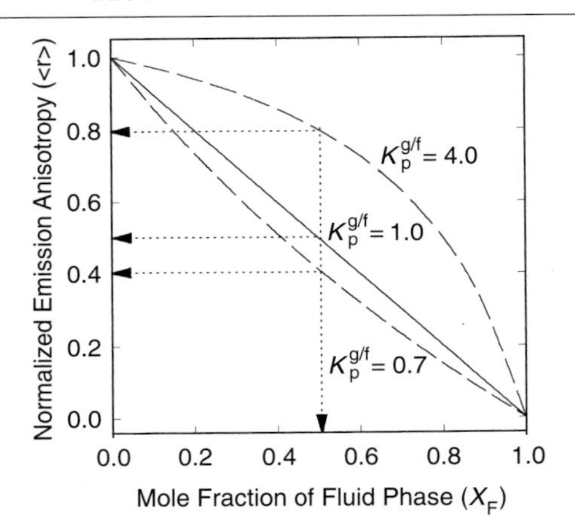

FIG. 1. Dependence of normalized fluorescence emission anisotropy values on mole fraction of fluid-phase phospholipid (X_F), for differing gel–fluid partition coefficients ($K_P^{g/f}$), where $K_P^{g/f} < 1$ indicates preferential partitioning of a probe into fluid-phase lipid. For a 50/50 gel–fluid mixture, EA values are significantly weighted toward higher values for probes exhibiting $K_P^{g/f} > 1$, and corresponding "melt" profiles will thus be shifted to higher apparent T_c values. Curves are calculated from Eqs. (1) and (2) as shown in text, with [$P_G = 1 - P_F$], essentially according to L. A. Sklar, G. P. Miljanich, and E. A. Dratz, *Biochemistry* **18,** 1707 (1979).[28]

Fig. 1, which demonstrates the effect of varying $K_P^{g/f}$ on normalized EA values, versus mole fraction of fluid-phase lipid (X_F). Clearly for a 50/50 mol% gel–fluid lipid comixture, EA values can vary significantly with partitioning properties of the chosen membrane probe. Preferential partitioning can thus provide a sensitive marker (or assay) for the presence of a particular lipid phase (e.g., indocarbocyanine dyes[30] and merocyanine 540 for fluid-phase lipid[32]). A summary of $K_P^{g/f}$ values for selected bilayer probes is shown in Table II.

Probe Perturbation. Probe-induced "local" perturbation effects are difficult to estimate because of the low molar labeling ratios used for fluorescence studies (typically 0.05 to 0.50 mol%). This is not surprising, because a bilayer labeling ratio of 0.20 mol% corresponds on average to only 5 molecules of dye per vesicle (assuming 2500 phospholipid molecules per SUV[32a]). At these concentrations, probes have little detectable effect on indicators of perturbation, such as the width and position of differential scanning calorimetry (DSC) enthalpy profiles, and/or morphological differences in electron micrographs of dye-labeled membrane systems. Bilayer

[32] P. Williamson, K. Mattocks, and R. A. Schlegel, *Biochim. Biophys. Acta* **732,** 387 (1983).
[32a] G. C. Newman and C.-H. Huang, *Biochemistry* **14,** 3363 (1975).

TABLE II
GEL–FLUID PARTITION COEFFICIENTS ($K_P^{g/f}$) FOR SOME COMMON FLUORESCENCE MEMBRANE PROBES

Probe name	$K_P^{g/f}$
DPH	0.97 ± 0.04; 1.04 ± 0.04[a]
TMA-DPH	0.31[b]
DPH-PC	0.30[b]
trans-Parinaric acid	4 ± 1[c]; 4–9[c]
cis-Parinaric acid	0.7 ± 0.2[d]
C_nDiI	
$n = 12$	0.8 ± 0.1[e,f]
$n = 16$	0.5 ± 0.1[e,f]
$n = 18$	0.35 ± 0.05[e,f]
$n = 20$	0.25 ± 0.05[e,f]
$n = 22$	0.7 ± 0.1[e,f]
16-(9-Anthroyloxy)palmitate	0.67 ± 0.5[f,g]
6-(9-Anthroyloxy)stearate	0.25 ± 1[g]
12-(9-Anthroyloxy)stearate	0.25 ± 1[g]
11-(9-Anthroyloxy)undeconate	0.25 ± 1[g]
3-(9-Anthroyloxy)stearate	0.40[g]

[a] B. R. Lentz, Y. Barenholz, and T. E. Thompson, *Biochemistry* **15,** 4529 (1976).
[b] R. A. Parente and B. R. Lentz, *Biochemistry* **24,** 6178 (1985).
[c] B. Hudson and S. A. Cavalier, *In* "Spectroscopic Membrane Probes" (L. M. Loew, ed.), Vol. 1, p. 43. CRC Press, Boca Raton, Florida, 1988.
[d] L. A. Sklar, G. P. Miljanich, and E. A. Dratz, *Biochemistry* **18,** 1707 (1979).
[e] C. H. Spink, M. D. Yeager, and G. W. Feigensen, *Biochim. Biophys. Acta* **1023,** 25 (1990).
[f] Gel–fluid partition coefficients determined from fluorescence quenching in DPPC model membranes.
[g] N.-N. Huang, K. Florine-Casteel, G. W. Feigensen and C. Spink, *Biochim. Biophys. Acta* **939,** 124 (1988).

disturbances have been reported using X-ray,[33] DSC,[34] and fluorescence studies[30,35] at atypically high probe-to-lipid ratios (20–25 mol%), which often results in probe aggregation (and in some cases probe-induced lipid-phase separation[30]) within the bilayer. This condition is often recognized

[33] W. Lesslauer, J. E. Cain, and J. K. Blaise, *Proc. Natl. Acad. Sci. U.S.A.* **69,** 1499 (1972).
[34] D. A. Cadenhead, B. M. J. Kellner, K. Jacobson, and D. Papahadjopoulos, *Biochemistry* **16,** 5386 (1977).
[35] R. G. Ashcroft, K. R. Thurlborn, J. R. Smith, H. G. Coster, and W. H. Sawyer, *Biochim. Biophys. Acta* **602,** 299 (1980).

spectroscopically with increasing probe-labeling ratios, as nonlinear plots of fluorescence intensity, altered absorption spectra and/or lifetime values, variation of EA values (all unaffected by sample dilution), and "melt" curves where lipid T_c values are lower than expected.[35] Static fluorescence quenching, fluorescence energy homotransfer mechanisms (particularly for dyes with small Stokes shifts), and excited-state interactions are the most common explanations. Possible specific chemical interactions between lipid and probe can be of concern when working with unsaturated lipid systems (including steroids, or added membrane effectors), where adduct or singlet oxygen production ($^1O_2$) arising from embedded bilayer probes can be photochemically induced, resulting in facilitated lipid peroxidation and severe membrane damage.[36]

Labeling of Bilayer Membranes with Fluorescence Probes

The choice of molar labeling ratio must balance the need for a good signal-to-noise ratio and the desire to minimize any potential perturbing effects. In all cases the possibility of lipid oxidation and/or photodegradation of the organic dye can be minimized by carrying out labeling procedures in the absence of ultraviolet and under an inert atmosphere (polyaromatic probes with long fluorescence lifetimes are sensitive to dissolved oxygen concentrations[37] and it is advisable to maintain labeled membrane samples either fully aerated or sealed under nitrogen for lifetime reproducibility). In general, probes are added either to preformed vesicles or membrane preparations, or incorporated at the time the vesicles are prepared.

Direct Organic Solvent Injection. Direct organic solvent injection is generally employed for lipid-soluble hydrophobic probes, which readily partition into bilayer systems, or for asymmetric labeling of the outer bilayer leaflet with fluorescent lipid analogs. Addition of probes to vesicle preparations (0.2 to 1.0 m*M* phosphorus concentration) is accomplished by direct injection of microliter aliquots from a concentrated stock solution [for example, 1–5 μM in dimethyl sulfoxide, dimethylformamide, ethanol, acetone, tetrahydrofuran (THF, although less desirable due to peroxide contamination), or other organic–aqueous miscible solvent], into a vortexing suspension (~3 ml) of vesicles or membranes. Because bilayer partitioning of the organic carrier solvent can cause severe disruption of the membrane architecture (e.g., as low as 50 m*M* for ethanol,[38] equivalent to ~9 μl of ethanol for 3 ml of membrane preparation), the final volume of organic

[36] J. E. Valinsky, T. G. Easton, and E. Reich, *Cell. Vol.* **13,** 487 (1978).

[37] H. Szmacinski and J. R. Lakowicz, *In* "Topics in Fluorescence Spectroscopy" (J. R. Lakowicz, ed.), Vol. 4, p. 295, Plenum, New York, 1994.

[38] S. J. Slater, C. Ho, F. J. Taddeo, M. B. Kelly, and C. D. Stubbs, *Biochemistry* **32,** 3714 (1993).

solvent added should be minimized. Membrane permeation of dye varies with probe or membrane type, and may take up to several hours.[39,40] For hydrophobic probes, which exhibit relatively low quantum yield values in polar solvents, incorporation can be followed directly by the accompanying increase in fluorescence signal, or alternatively via increased EA values. Lentz *et al.*[31] have noted that the efficiency of probe incorporation is strongly dependent on temperature and is maximal above the appropriate lipid T_c. Potential probe aggregates (polyaromatic molecules are most problematic), membrane-adsorbed dye, and micellar structures (from fluorescent lipid adducts) can be stripped from labeled vesicles by passing the preparation over a prepacked PD-10 Sephadex G-25M column (9-ml bed volume; Pharmacia-LKB, Piscataway, NJ). Labeled vesicles elute with the void volume, and can be detected by simple scattering and/or by using an ultraviolet hand lamp. For labeled cell membranes, brief centrifugation will leave unincorporated dye in the supernatant, allowing pelleted fluorescently labeled membranes to be washed with buffer. Confirmation of the desired labeling ratio can be achieved by back-extraction of the dye into excess THF or other suitable membrane-dissolving, water-miscible solvent [e.g., sodium dodecyl sulfate (SDS)[41]] into which the probe will partition. From the extinction or fluorescence of the dye in the resultant disrupting solvent, the concentration of embedded probe can be determined.

Dilution Method. The dilution method is useful for labeling cell membranes where sensitivity to organic solvents is a problem. Incorporation of water-soluble fluorescent probes can also be carried out by this approach. A small (microliter) volume of concentrated stock solution of probe in organic solvent is first diluted into an excess of an appropriate buffer. This is further diluted with an equal volume of the membrane suspension, to give the desired molar labeling ratio. The general aqueous insolubility of many fluorescence probes often leads to aggregate formation. Therefore, careful separation steps must be taken to ensure that after appropriate incubation, unabsorbed dye or micelles are removed by centrifugation or by column chromatography as described above.

Cosonication/Coextrusion. Cosonication/coextrusion can be used for introducing fluorescent probes covalently attached to integral membrane components (such as lipids or steroids). Organic solvent preparations of the fluorescent probes are first comixed with the matrix phospholipid in the correct molar ratio. Removal of the organic solvent is achieved by evaporation using nitrogen followed by vacuum desiccation for several

[39] M. Shinitzky and Y. Barenholz, *J. Biol. Chem.* **249,** 2652 (1974).
[40] C. Ho, B. W. Williams, and C. D. Stubbs, *Biochim. Biophys. Acta* **1104,** 273 (1992).
[41] J. Eisinger, J. Flores, and W. P. Petersen, *Biophys. J.* **49,** 987 (1986).

hours to ensure removal of residual solvent. On resuspension in the buffer of choice, the resultant (milky) fluorescently labeled multilamellar suspension can be used to form bilayer vesicles.[42] Such vesicles, when incubated with cell systems, often facilitate introduction of fluorescently labeled lipid analogs (in a relatively nonevasive fashion) into cell membrane systems either directly through membrane fusion or spontaneous lipid exchange, or (alternatively) indirectly by use of lipid transfer proteins.[43]

Scattering Effects

Rayleigh scattering effects of bilayer membrane vesicle samples (λ^{-4} wavelength dependency) are of concern when performing polarized fluorescence measurements[44] and/or when the fluorescence intensity yield of the desired embedded probe is low. The quantitative effect of scattering on measured EA values has been discussed elsewhere.[45] Simple light scattering can be minimized experimentally by using double monochromators, especially in the excitation optical path (although compromising sensitivity), and/or by insertion of a cutoff filter in the emission train (where the wavelength range for cutoff is greater than the wavelength of excitation, but less than that for emission). For time-dependent fluorescence studies of membrane or vesicle samples, it is often necessary to introduce a scattering component into the analysis of the decay data using a delta impulse response function ($\tau \rightarrow 0$). Potential inner filter effects are minimized by ensuring that absorbance values measured at the wavelength of excitation (scatter plus absorption) do not exceed values of about 0.1. Preparation of appropriate blank samples is generally achieved by matching the scatter values for the blank and sample measured at longer wavelengths where absorption of the probe is zero.

Interpretation of Fluorescence Probe Data

Structural Lipid Heterogeneity

Under conditions where the fluorescence lifetime of the probe is shorter than the rate of gel–fluid lipid exchange, useful information regarding structural or "static" lipid heterogeneity can be extracted:

[42] G. Gregoriadis (ed.), "Liposome Technology," Vols. 1–3. CRC Press, Boca Raton, Florida, 1993.
[43] J. W. Nichols and R. E. Pagano, *J. Biol. Chem.* **258,** 5368 (1983).
[44] M. Shinitzky, A. C. Dianoux, C. Gitler, and G. Weber, *Biochemistry* **12,** 2106 (1971).
[45] J. R. Lakowicz, *In* "Principles of Fluorescence Spectroscopy" Chapter 2, p. 19. Plenum, New York, 1983.

$$\langle \tau_{\mathrm{GEL}} \rangle \gg \langle \tau_{\mathrm{FL}} \rangle \tag{3}$$

Most common fluorescence probes used for membrane heterogeneity studies fall within this experimental regime, exhibiting fluorescence lifetime values of 20 nsec or less (Table I). In the simplest case, the membrane system may be approximated using a simple compartmental model,[46] where gel (G) and fluid (F) lipid regions within the bilayer coexist in equilibrium and fluorescence probes reside within these lipid subfractions. If the fraction of probe residing in these specific lipid phases (P_G and P_F, respectively) can be directly determined from measured spectroscopic parameters, and the gel/fluid partition coefficient for the probe is known, then quantification of gel-phase (X_G) or fluid-phase (X_F) lipid fractions is possible [see Eq. (1)].

Spectral Heterogeneity. Here discrete spectral [$F(\lambda)$] signatures (excitation or emission) may be associated with gel or fluid regions of the bilayer. Examples of the more popular spectrally sensitive fluorescence probes include PATMAN,[47] prodan,[48] and laurdan.[49] Estimation of the contribution from gel- and fluid-embedded probe subensembles to the observed (average) emission spectrum can be obtained from an intensity ratio measured at two characteristic emission wavelengths, or alternatively from determination of the center of spectral mass ($\langle \nu_g \rangle = \int \nu FL(\nu)\, d\nu / \int FL(\nu)\, d\nu$). For example, the emission spectrum of PATMAN shifts about 40 nm from 425 nm when embedded within the gel phase, to 465 nm when in liquid–crystalline or fluid phase. The characteristic red shift in the emission spectrum observed for these fluorescence probes is presumed to arise from the relaxation of the membrane environment around the excited state dipole of the fluorophore. For laurdan, which resides deeper in the bilayer owing to its dodecanoic acid side chain (Table I), the fluorescence excitation spectrum also demonstrates sensitivity to gel–fluid lipid packing and it is possible to selectively excite fluorescence probes located within the different lipid phases. Parassasi *et al.*[49] have developed a general polarization factor parameter (GP) for interpreting the average fluorescence signals from fluid- and gel-embedded laurdan, which has been further extended to provide quantitative estimates of gel-phase (X_G) and fluid-phase (X_F) lipid.[50]

[46] G. L. Atkins, "Multicomponent Models in Biological Systems." Methuen, London, 1969.
[47] J. R. Lakowicz, D. R. Bevan, B. P. Maliwal, H. Cherek, and A. Balter, *Biochemistry* **18(22),** 5714 (1983).
[48] G. Weber and F. J. Farris, *Biochemistry* **18,** 3075 (1979).
[49] T. Parasassi, G. De Stasio, A. D'Ubaldo, and E. Gratton, *Biophys. J.* **57,** 1179 (1990).
[50] T. Parasassi, G. De Stasio, G. Ravagnan, R. M. Rusch, and E. Gratton, *Biophys. J.* **60,** 179 (1991).

Lifetime Heterogeneity. Many popular fluorescence probes employed for studies of membrane heterogeneity, e.g., DPH,[51] perylene,[52] and coronene,[53] exhibit multiexponential fluorescence lifetime decay behavior when embedded in lipid bilayer systems. It is tempting to interpret such heterogeneous decay profiles [$s(t)$] by directly mapping discrete fluorescence lifetime values with a particular lipid (gel or fluid) phase:

$$s(t)_{\text{obs}} = \alpha_{\text{G}} e^{-t/\tau_{\text{G}}} + \alpha_{\text{F}} e^{-t/\tau_{\text{F}}} \tag{4}$$

Under such conditions, associated preexponential terms (α_i) are expected to provide estimates of the fractional contribution of gel- or fluid-embedded probe subfractions (P_{G} and P_{F}) to the intensity decay (given that $\sum_{i=n}\alpha_i = 1$, where n is the number of compartments), and demonstrate significant temperature sensitivity. For *trans*-parinaric acid, a long-lifetime component (10–40 nsec) can be unambiguously assigned to gel regions of the bilayer.[54] Because α_{G} is directly proportional to the fraction of probe localized in the gel phase (P_{G}), and if $K_{\text{P}}^{\text{g/f}}$ is known, a phase diagram for DMPC/DPPC membrane mixtures can be successfully constructed.[55] In the case of indocarbocyanine dyes (C_n diI), which exhibit complex biexponential behavior when embedded in homogeneous gel- or fluid-phase lipid systems, Packard and Wolf[56] have found that variation in the average fluorescence lifetime [$\langle\tau\rangle = (\sum_{i=1}^{n}\alpha_i\tau_i^2)/(\sum_{i=1}^{n}\alpha_i\tau_i)$] is a more useful parameter for quantifying the lateral organization of binary lipid mixtures. For DPH, fluorescence lifetime parameters are only weakly sensitive to gel (10–11 nsec) or fluid (6–9 nsec) lipid environments [the persistent minor (1%) lifetime component (2–3 nsec) appears to be independent of the physical state of the lipid and possible origins have been discussed elsewhere[57]]. Nevertheless, DPH and its analogs have been successfully employed for the investigation of lateral organization in bilayers and membranes.[40,58,59]

In an alternate approach, observed variances in measured lifetime values with temperature for DPH have been rationalized as transverse environmental heterogeneity. Here the probe is expected to distribute across the

[51] L. A. Chen, R. E. Dale, S. Roth, and L. Brand, *J. Biol. Chem.* **252,** 2163 (1977).

[52] P. L.-G. Chong, B. Wieb van der Meer, and T. E. Thompson, *Biochim. Biophys. Acta* **813,** 253 (1985).

[53] L. Davenport and P. Targowski, *Biophys. J.* **71,** 1837 (1996).

[54] A. Ruggiero and B. Hudson, *Biophys. J.* **55,** 1111 (1989).

[55] C. R. Mateo, J.-C. Brochon, M. P. Lillo, and A. U. Acuña, *Biophys. J.* **65,** 2237 (1993).

[56] B. S. Packard and D. E. Wolf, *Biochemistry* **24,** 5176 (1985).

[57] T. Parasassi, F. Conti, M. Glaser, and E. Gratton, *J. Biol. Chem.* **259,** 14011 (1984).

[58] D. A. Barrow and B. R. Lentz, *Biophys. J.* **48,** 221 (1985).

[59] G. Ferretti, A. Tangorra, G. Zolese, and G. Curatola, *Membrane Biol.* **10,** 17 (1993).

bilayer width[60,61] and lifetime variations now reflect potential dielectric[62,63] gradients. Decay data can be analyzed using distributional analyses,[62,64] where the width (w) of the distribution reflects the heterogeneity of probe location. Toptygin and Brand have demonstrated that transverse bilayer refractive index gradients can also give rise to multiexponential intensity decay profiles for DPH.[65]

For other probes (e.g., coronene[53]), associated preexponential terms (α_i) exhibit little sensitivity to temperature. Definitive assignment of fluorescence lifetimes to gel and fluid regions within the bilayer is therefore not possible for these probes.

Rotational Heterogeneity. A wide variety of hydrophobic membrane probes, including rod-shaped (e.g., DPH,[51] *trans*- and *cis*-parinaric acids[66]) and planar disklike molecules [e.g., *n*-(9-anthroyloxy)-fatty acids,[31,67] perylene,[12,52] and coronene[53]] can exhibit rotational sensitivity to the immediate lipid environment during their excited-state lifetime. Steady-state EA values for probes of cylindrical symmetry, with partition coefficients close to unity (Table II), can directly quantify the contribution of gel (X_G) and fluid (X_F) lipid subfractions (Fig. 1). However, only qualitative information regarding membrane dynamics and ordering is possible, and values represent the average rotational properties of all contributing probe subensembles. The steady-state EA ($\langle r \rangle$) is defined as the difference ($\langle d \rangle$) over the sum ($\langle s \rangle$) of polarized emission intensities:

$$\langle r \rangle = \frac{GI_{VV} - I_{VH}}{GI_{VV} + 2I_{VH}} \equiv \frac{\langle d \rangle}{\langle s \rangle} \tag{5}$$

where $G = (I_{HH}/I_{HV})$ is the grating factor and represents a correction factor for the inequality of the detection system sensitivity to horizontally and vertically polarized emissions.[68] The first subscript, V or H, refers, respectively, to the vertical or horizontal orientation of the dielectric vector of the excitation and the second to those for emission.

Membrane probes often reveal multicomponent polarized exponential decay profiles, superimposed on a residual anisotropy term (r_∞) when embedded in lipid bilayers[51,52,66,69]:

[60] L. Davenport, R. E. Dale, R. H. Bisby, and R. B. Cundall, *Biochemistry* **24,** 4097 (1985).

[61] E. E. Pebay-Peyroula, E. J. Dufourq, and A. G. Szabo, *Biophys. Chem.* **53,** 45 (1994).

[62] E. Gratton and T. Parasassi, *J. Fluoresc.* **5,** 51 (1995).

[63] C. D. Stubbs, C. Ho, and S. J. Slater, *J. Fluoresc.* **5,** 19 (1995).

[64] J. R. Alcala, E. Gratton, and F. G. Prendergast, *Biophys. J.* **51,** 587 (1987).

[65] D. Toptygin and L. Brand, *Biophys. Chem.* **48,** 205 (1993).

[66] A. Ruggiero and B. Hudson, *Biophys. J.* **55,** 1125 (1989).

[67] M. Vincent, B. DeForesta, J. Gallay, and A. Alfsen, *Biochemistry* **21,** 708 (1982).

[68] R. F. Chen and R. L. Bowman, *Science* **147,** 729 (1965).

[69] S. Kawato, K. Kinosita, and A. Ikegami, *Biochemistry* **16,** 2319 (1977).

$$r(t) = (r_0 - r_\infty) \sum_j \beta_j e^{-t/\phi_j} + r_\infty \tag{6}$$

For probes approximating cylindrical symmetry, the limiting anisotropy term (r_∞) is related to the coefficient of the second Legendre polynomial ($\langle P_2 \rangle$) and to the order parameter $S = \sqrt{r_\infty / r_0} = \langle P_2 \rangle$, which indirectly characterizes the ordering of the acyl-chain region of the bilayer.[70] Possible origins of the complex decay are (1) anisotropic or hindered rotational behavior of the probe (*homogeneous* rotational model),[21,69,71,72] (2) microheterogeneity of probe location (*heterogeneous* rotational model),[55,73,74] and (3) a combination of (1) and (2). This latter condition implies (at minimum) a double-hindered rotational model, and has been explored by Calafut *et al.*[75]

Attempts to conclusively demonstrate rotational microheterogeneity of probe location focus on possible linkage of a particular rotational correlation time (ϕ) with a particular fluorescence lifetime value (i.e., $\tau \longleftrightarrow \phi$ linkage). The difference decay for the EA [Eq. (5)] is now described via an associative model [Eq. (7a)].[74] For a simple gel–fluid two–phase system:

$$d(t)_{\text{obs}} = (\alpha_G e^{-t/\tau_G} \beta_G e^{-t/\phi_G}) + (\alpha_F e^{-t/\tau_F} \beta_F e^{-t/\phi_F}) \tag{7a}$$

$$d(t)_{\text{obs}} = \alpha_{\text{OBS}} e^{-t/\tau_{\text{OBS}}} (\beta_G e^{-t/\phi_G} + \beta_F e^{-t/\phi_F}) \tag{7b}$$

Alternatively, probes may demonstrate rotational sensitivity, but have no detectable (associated) lifetime change when embedded in gel or fluid lipid environment.[76] Under such conditions, polarized decays are best described using the nonassociative expression shown in Eq. (7b). In the limiting case, where rotational motions for the probe embedded in gel-phase lipid are restricted on the nanosecond time scale ($\phi_G \geq \tau_{FL}$), the term $[\beta_G \exp(-t/\phi_G)] \rightarrow r_{\infty G}$.

Only a few membrane probes have provided clear evidence for associative behavior. *trans*-Parinaric acid, however, is an excellent example. For this probe, the long fluorescence lifetime, previously assigned to gel-embedded probes, is fully associated with the r_∞ term. As such, experimental time-dependent EA profiles observed for this probe are signatory for associative behavior, exhibiting upward curvature at long times after excitation.[77]

[70] M. P. Heyn, *FEBS Lett.* **108,** 359 (1979).

[71] G. Lipari and A. Szabo, *Biophys. J.* **30,** 489 (1980).

[72] C. Zannoni, *Mol. Phys.* **42,** 1303 (1981).

[73] L. Davenport, J. R. Knutson, and L. Brand, *Biochemistry* **25,** 1811 (1986).

[74] L. Brand, J. R. Knutson, L. Davenport, J. M. Beechem, R. E. Dale, D. G. Walbridge, and A. A. Kowalczyk, *In* "Spectroscopy and the Dynamics of Molecular Biological Systems" (P. M. Bayley and R. E. Dale, eds.), p. 259. Academic Press, London, 1985.

[75] T. M. Calafut, J. A. Dix, and A. S. Verkman, *Biochemistry* **28,** 5051 (1989).

[76] D. Toptygin and L. Brand, *J. Fluoresc.* **5,** 39 (1995).

[77] P. K. Wolber and B. S. Hudson, *Biochemistry* **20,** 2800 (1981).

Structural heterogeneity studies of lipid bilayers using planar disklike molecules including perylene,[52] and n-(anthroyloxy) lipids,[67] are complicated by hindered anisotropic rotational motions of the embedded probes. Their motional properties can be characterized by in-plane and out-of-plane rotations. Evaluation of the hindrances imposed on each type of rotational motion can be approached by selective excitation [absorption and emission oscillators for perylene and the n-(anthroyloxy) lipids lie within the plane of the molecule], taking into account the fundamental anisotropy values, r_0. For $r_0 = 0.1$, exclusive estimation of the out-of-plane rotational modes is expected. At $r_0 = 0.324$, in-plane and out-of-plane modes or rotations contribute equally to the depolarization. For a series of n-(9-anthroyloxy)-fatty acids embedded in bilayers, Vincent *et al.*[67] have demonstrated, using selective excitation approaches, unrestricted out-of-plane rotational motions and varying degrees of restricted in-plane rotational motions, dependent on the depth of the fluorophore within the bilayer. A combination of both hindered in-plane and out-of-plane rotational rates for "free" perylene incorporated within DMPC SUVs have been suggested by Brand *et al.*[74] However, while evaluation of hindrances imposed on each type of rotational motion of an anisotropic fluorophore is expected to provide more insight into the anisotropy of the membrane structure, the influences of lipid microheterogeneity have yet to be fully considered.

Associative Fluorescence Spectroscopy. Associative fluorescence spectroscopy, a multidimensional approach (which assumes ground- or excited-state heterogeneity of the fluorophore), can be used for spectrally resolving complex mixtures of embedded fluorophores by associating a characteristic emission spectral envelope for a given probe subfraction with a particular fluorescence decay component [τ and/or ϕ; decay associated spectra (DAS) and anisotropy decay-associated spectra (ADAS), respectively]. Component spectra for the probe subfractions may exhibit only weak environmental sensitivity, with a high degree of spectral overlap. Thus simple spectral resolution as discussed previously is not a practical option.

The theory of associative spectroscopy has been described in detail elsewhere by Knutson *et al.*, and will not be elaborated on here.[74,78,79] Experimentally, "magic angle" [$s(t)$] or orthogonally polarized [$I_{VV}(t)$ and $I_{VH}(t)$] decay curves are collected as a function of emission wavelength, stepping the monochromator through the wavelength region of interest [with steps less than or equal to 10 nm, with appropriate dwell times such that the number of photons counted per channel (not less than 70 counts/

[78] J. R. Knutson, D. G. Walbridge, and L. Brand, *Biochemistry* **21,** 4671 (1982).

[79] J. R. Knutson, L. Davenport, and L. Brand, *Biochemistry* **25,** 1805 (1986).

channel) at each wavelength easily satisfy Gaussian statistics]. The resultant three-dimensional data decay surface (wavelength, intensity, time) is analyzed using "global" procedures (for a review, see Ref. 80) with linkage of fluorescence lifetimes and/or rotational correlation times over the emission wavelength axis. Decay-associated spectra or DAS $[a_i(\lambda)]$[74,78] thus represent the wavelength dependency of the preexponential terms associated with a particular lifetime [compare Eq. (4)]:

$$s(\lambda, t) = \sum_{j}^{n} a_i(\lambda)e^{-t/\tau_i} \tag{8}$$

DAS methodologies have been successfully applied in the investigation of cholesterol-induced heterogeneity of lipid packing, using a pyrene-cholesterol adduct (pyrenemethyl–cholesterol, PMC) as the probe.[81]

Anisotropy decay-associated spectra (ADAS) can be employed for spectrally resolving complex rotationally heterogeneous populations of fluorophores in lipid bilayers where there is no detectable $\tau \longleftrightarrow \phi$ linkages. Extraction of ADAS $[b_i(\lambda)]$ from experimentally measured polarized emission decay intensities can be achieved via two "global" analysis procedures: (1) direct vector analysis of $I_{VV}(t)$ and $I_{VH}(t)$ versus wavelength, or (2) indirect analysis of difference decay $d(t)$ profiles [Eq. (7)] versus wavelength, constructed from experimental polarized decay data $[d(\lambda, t) = GI_{VV}(\lambda, t) - I_{VH}(\lambda, t)]$, linking appropriate τ_{FL} and ϕ terms across the wavelength region of interest. In analogy to Eq. (7), for associative and nonassociative cases, respectively, the difference decay wavelength surface can be written

$$d(\lambda, t) = [b_G(\lambda)e^{-t/\tau_G}r_{0_G}e^{-t/\phi_G}] + [b_F(\lambda)e^{-t/\tau_F}r_{0_F}e^{-t/\phi_F}] \tag{9a}$$

$$d(\lambda, t) = e^{-t/\tau_{obs}}[b_G(\lambda)r_{0_G}e^{-t/\phi_G} + b_F(\lambda)r_{0_F}e^{-t/\phi_F}] \tag{9b}$$

The difference spectrum or ADAS $[b_i(\lambda)]$ of each component (n) of the probe subensemble is associated with the proper total intensity and/or anisotropy decay functions through the index i.

Alternatively, as discussed by Knutson *et al.*,[74,78,79] where lifetime or rotational decay parameters for the mixture are sufficiently disparate, simple "time gating" of the decay profiles collected over the emission spectral envelope (achieved using the multichannel analyzer of the time-correlated single photon-counting equipment) allows storage of decay contributions

[80] J. M. Beechem, E. Gratton, M. Ameloot, J. R. Knutson, and L. Brand, *In* "Topics in Fluorescence Spectroscopy: Principles" (J. R. Lakowicz, ed.), Vol. 2, p. 241. Plenum, New York, 1991.

[81] L. Davenport, J. R. Knutson, and L. Brand, *Faraday Discuss. Chem. Soc.* **81,** 81 (1986).

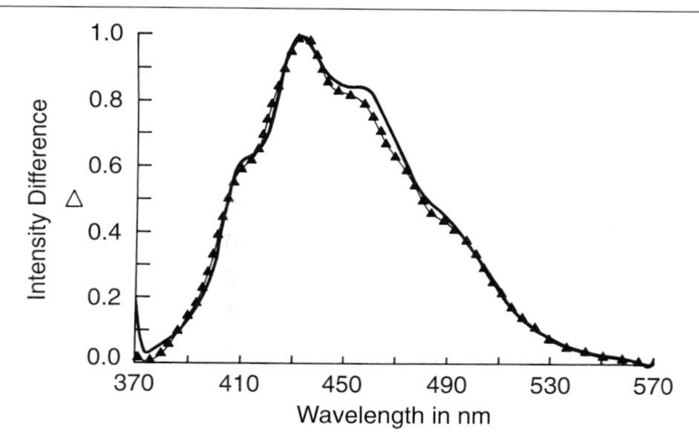

FIG. 2. Anisotropy decay-associated spectrum for DPH-labeled DMPC vesicle samples at 25°. The late difference spectrum (▲) directly represents the ADAS of the immobile probe fraction in this system. The early difference (—) includes contributions from both mobile and immobile probes. Detailed conditions used in this experiment are described more fully in Ref. 83. [Reprinted with permission from L. Davenport, J. R. Knutson, and L. Brand, *Biochemistry* **25,** 1811 (1986). Copyright 1986 American Chemical Society.]

arising from longer lived polarizations within a "late" time window. This approach minimizes the acquisition of large data arrays, and has been successfully used for lifetime gating in time-resolved microscopy studies.[82] ADAS have resolved multiple rotational environments for DPH embedded in pure DMPC SUVs at T_c, where gel and fluid lipid phases are expected to coexist (Fig. 1).[83] It is advantageous to choose an appropriate time window such that contributions from immobile probes are equal in both time windows. Under these conditions, because ADAS from the early time window correspond to decay contributions from a mixture of "mobile" and "immobile" probe signals and the late window contains contributions exclusively from immobile probes, ADAS for mobile probes can now be determined by simple subtraction of the late time window from the mixed early time window. ADAS corresponding to mobile and immobile DPH probe fractions (Fig. 2) are nonsuperimposable, indicative of rotational heterogeneity.

Dynamic Lipid Heterogeneity

For fluorescence probes with lifetimes of several tens of nanoseconds (e.g., pyrene, coronene, *trans*-parinaric acid) the experimental averaging regime is extended:

[82] D. Phillips, *Analyst* **119,** 543 (1994).

[83] L. Davenport, J. R. Knutson, and L. Brand, *Biochemistry* **25,** 1811 (1986).

$$\langle \tau_{FL} \rangle \sim \langle \tau_{GEL} \rangle \tag{10}$$

Under such conditions, submicrosecond gel–fluid lipid dynamics or fluctuations (k_{FG}) can be successfully targeted:

$$G \underset{k_{GF}}{\overset{k_{FG}}{\rightleftharpoons}} F \tag{11}$$

where F and G define the average concentrations of the fluid and gel lipid fractions, respectively. Evolution of the lipid environment surrounding the fluorophore is now reflected in the measured fluorescence lifetime and/or rotational behavior of the embedded probe. Experimental fluorescence approaches to the investigation of lipid fluctuations have been limited by the availability of suitable long-lived fluorescence probes (for a review, see Davenport and Targowski[84]). However, rotational sensitivity has been observed using planar probes with long-lived fluorescence lifetimes, such as coronene ($\langle \tau_{FL} \rangle \sim 200$ nsec) and pyrenemethyl-cholesterol (PMC) ($\langle \tau_{FL} \rangle \sim 80$ nsec).[85] Depolarizing motions for coronene (a planar molecule with D_{6h} symmetry; $r_0 = 0.1$) are influenced exclusively by out-of-plane rotations ($\langle r_{op} \rangle$).[53,86] Steady-state EA versus temperature, for coronene embedded within DPPC SUVs, reveals broad and low temperature-shifted melt profiles (heavy line, Fig. 3). However, rather than preferential partitioning or lipid perturbations effects arising from the probe, these melt profiles are suggestive of submicrosecond lipid dynamics occurring at temperatures well below the normal lipid T_c value, and are detected by virtue of the long fluorescence lifetime of the probe. This condition is illustrated in Fig. 3. By simulation of steady-state EA versus temperature profiles (via the Perrin equation $[r_0/\langle r \rangle = 1 + (1\langle \tau_{FL} \rangle / \langle \phi \rangle)]$[51]) for a probe with coronene-like rotational motions but short lifetime values ($10 \text{ nsec} \leq \langle \tau_{FL} \rangle \leq 200$ nsec), we see that melt profiles become more "DPH-like." Hence unusual melt profiles are not unexpected for long-lived probes, which demonstrate sensitivity to submicrosecond lipid dynamics. (Other peculiarities of employing long-lived probes, e.g., depolarization arising from whole vesicle rotation or, alternatively, from two-dimensional lateral diffusion effects, have been discussed elsewhere.[53])

Appropriate analyses of complex multiexponential EA profiles for coronene embedded in DPPC bilayers have allowed extraction of gel–fluid lipid

[84] L. Davenport and P. Targowski, *J. Fluoresc.* **5,** 9 (1995).

[85] L. Davenport, J. Z. Wang, and J. R. Knutson, *In* "Biological and Synthetic Membranes" (A. Butterfield, ed.), p. 97. Alan R. Liss, New York, 1989.

[86] L. Davenport, J. R. Knutson, and L. Brand, *In* "Time Resolved Laser Spectroscopy in Biochemistry" (J. Lakowicz, ed.), p. 909. SPIE, Bellingham, Washington, 1988.

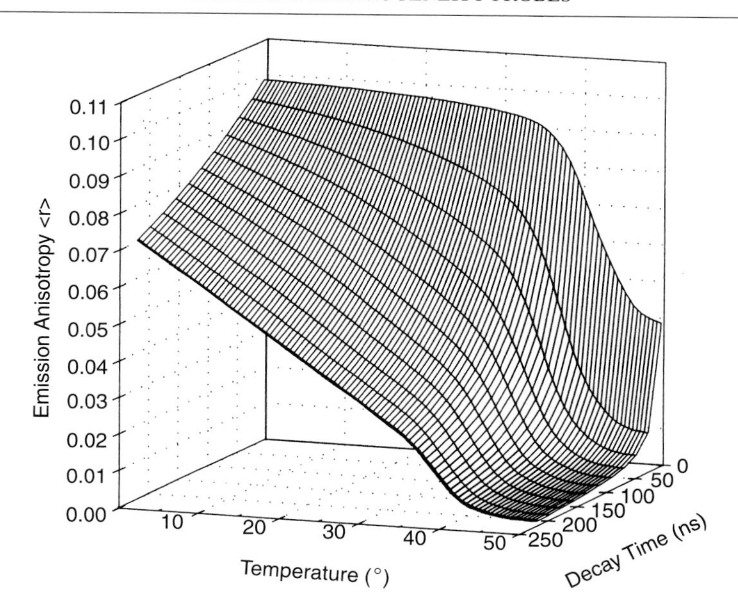

FIG. 3. Surface plot of synthesized steady state emission anisotropy as a function of temperature and lifetime of the fluorescence probe. From experimental time-resolved data for coronene-labeled DPPC SUVs at 5° ($\alpha_1 = 0.75$, $\tau_1 = 256$ nsec; $\alpha_2 = 0.09$, $\tau_2 = 37$ nsec; $\alpha_3 = 0.16$, $\tau_3 = 2$ nsec) and the experimental $\langle r \rangle (T)$ profile, the average lifetime and rotational correlation time (from the Perrin equation) are determined. The average fluorescence lifetime is then reduced from 200 to 5 nsec, while the rotational parameters remain unchanged. The heavy line indicates the true, experimentally obtained steady state melt curve for coronene, which is characteristically broad and shifted to colder temperatures. [Reproduced from L. Davenport and P. Targowski, *J. Fluoresc.* **5,** 9 (1995), with permission.]

exchange rates (k_{FG}). Two approaches have been proposed (discussed in detail elsewhere[53]). In the first, a compartmental model, three distinct lipid fractions are defined (F, fluid; G, gel; N, nonexchangeable gel), in which the probe is randomly distributed. G and F are in equilibrium on the submicrosecond time scale [Eq. (11)]. N represents refractory gel lipid and is excluded from this equilibrium. The decay of the EA may thus be expressed as:

$$r(t) = \beta_F e^{-t/\phi_F} + (\beta_G e^{-k_{FG}t} + \beta_N) e^{-t/\phi_N} \quad \text{where} \quad k_{FG} \equiv \frac{1}{\phi_{EFF}} \tag{12a}$$

and

$$\frac{\beta_F}{r_0} = \frac{F}{(F+G+N)}; \quad \frac{\beta_G}{r_0} = \frac{G}{(F+G+N)}; \quad \frac{\beta_N}{r_0} = \frac{N}{(F+G+N)} \tag{12b}$$

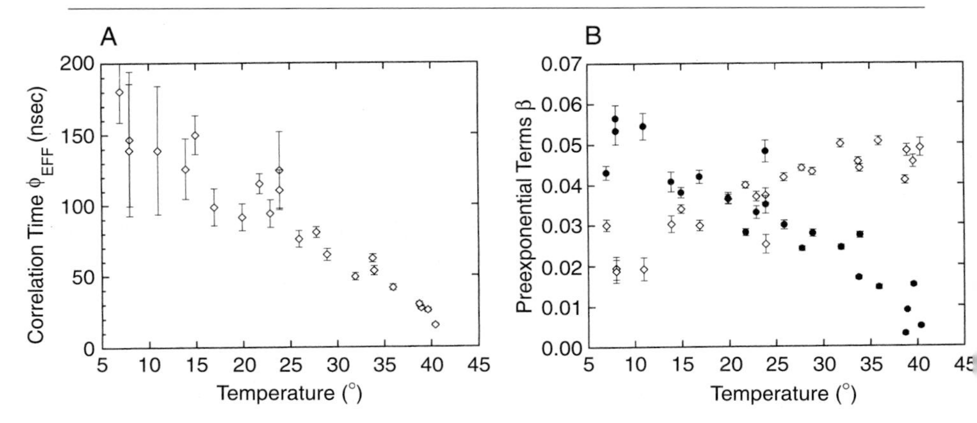

FIG. 4. Summary of rotational decay parameters for coronene-labeled DPPC SUVs, as a function of temperature, analyzed according to the compartmental model. (A) Rotational correlation time, $\phi_{EFF} \equiv 1/k_{FG}$. (B) Preexponential factors: β_G (◇) and β_N (●). The lifetime decay parameters used for these analyses, for a given temperature, are summarized in Fig. 3. Excitation was accomplished using a 340-nm interference filter (bandwidth = 12 nm) and emission was recorded at 448 nm, using a 16-nm slit width. [Reproduced from L. Davenport and P. Targowski, *Biophys. J.* **71,** 1837 (1996), with permission.]

The rotational correlation times, ϕ_F and ϕ_N, are assigned to probes residing in fluid or nonexchanging gel lipid phases (F or N, respectively). An additional (effective) rotational correlation time (ϕ_{EFF}) for coronene, linked with the melting of its surrounding lipid environment, provides an estimate of the gel–fluid exchange rate (k_{FG}), where $\phi_F \ll k_{FG} \ll \tau_{FL} \ll \phi_N$. Plots of experimentally obtained preexponential terms for coronene (β_G and β_N) as a function of increasing temperature are indicative of this exchange process (Fig. 4). Values for ϕ_{EFF} range from 150 ± 40 nsec at temperatures well below T_c to around 20 nsec at T_c for coronene embedded in DPPC SUVs (Fig. 4) and compare favorably with probe-independent approaches, e.g., ultrasound.[87]

An alternative approach invokes "gated" lipid fluctuations. Here a distribution of membrane lipid ordering ($0 \leq S_i \leq 1$) rather than discrete gel and fluid regions is assumed, with a corresponding distribution of lipid melting rates $d(S, T)$ and hence effective probe rotational correlation times $[\phi(S, T)]$:

$$r(t) = r_0 \int_0^1 P(S, T) e^{-d(S,T)t}\, dS \quad \text{where} \quad d(S, T) = \frac{1}{\phi(S, T)} \tag{13}$$

[87] S. Mitaku, A. Ikegami, and S. Sakanishi, *Biophys. Chem.* **8,** 295 (1978).

$P(S, T)$ defines the probability of a phospholipid molecule achieving the appropriate activation required for fluidizing, and is proportional to the difference in the free energy of fluid- and gel-phase lipid.[88] Effective rotational correlation times $[\phi(S, T)]$ for the embedded fluorescence probe can be associated with this same activation energy barrier. Hence known thermodynamic quantities for the lipid may be linked to measured dynamic spectroscopic quantities. It is expected that some multiple (or gating factor; γ) of the activation energy barrier (i.e., simultaneous melting of several surrounding shells of lipid molecules) actually permits probe rotation. Analyses of time-resolved EA profiles for coronene embedded in DPPC SUVs provides estimates of the gating factor that range from around 10 at 20°, to around 5 close to T_c. This corresponds to an estimated critical lipid shell diameter on the order of 25–40 Å. From time-resolved emission spectral shifts of laurdan fluorescence, Parasassi *et al.*[89] have estimated lipid fluctuation rates of 25 nsec with lipid cluster sizes of 20–50 Å in diameter. Furthermore, Ruggerio and Hudson,[54] using *trans*-parinaric acid, have demonstrated that plots of the amplitude (α_G) associated with the long-lifetime component for this probe reveal a temperature dependence $[(T - T_c)/T_c]$ that is characteristic of classic critical fluctuation behavior.[88] Estimates for the lifetime of the lipid fluctuations range from 10 to 40 nsec on approaching T_c from temperatures greater than T_c.

Conclusions

In conclusion, fluorescence membrane probes, as discussed here, can provide sensitive tools for the examination of structural and dynamic microheterogeneity of lipid packing. For successful interpretation of heterogeneous fluorescence probe signals, characterization of the position, perturbation, and partitioning properties of the fluorophore is required. Studies using short-lived probes, such as DPH, provide a static "snapshot" of structural lipid packing, whereas longer lived fluorescence probes provide insights into the dynamics or shifting of the lipid heterogeneity. Because evidence for coexisting gel–fluid lipid clearly exists within ±10–15° of the lipid T_c in model bilayer systems, heterogeneity of lipid packing must be considered when modeling complex lipid probe fluorescence signals. With the development of new near-infrared (IR) fluorescence probes,[26,90] (remote from contaminating autofluorescence and scatter), combined with

[88] F. Jähnig, *Biophys. J.* **36,** 329 (1981).

[89] T. Parasassi, G. Ravagnan, R. M. Rusch, and E. Gratton, *Photochem. Photobiol.* **57,** 403 (1993).

[90] D. J. S. Birch and G. Hungerford, *In* "Topics in Fluorescence Spectroscopy" (J. R. Lakowicz, ed.), Vol. 4, p. 377. Plenum, New York, 1994.

two-photon spectroscopic methods[91] and new near-IR diode lasers, high-resolution heterogeneity mapping of large-scale domains of cellular membranes is now an exciting future possibility.

Acknowledgments

I thank Professor Ludwig Brand, Dr. Robert E. Dale, Dr. Jay Knutson, and Dr. Piotr Targowski for their continued support and encouragement. They were also involved in some of the studies cited here. I thank Drs. Jay R. Knutson, Piotr Targowski, and Michael Straher for critical reading of the manuscript. I am grateful for the patience and enthusiasm of both Professor Ludwig Brand and Ms. Shirley Light of Academic Press. Also, my thanks to Dr. Michael Straher for assistance with table and figure preparations, and to Ms. Hazel Ward for manuscript preparation. ADAS studies were performed in the laboratory of Professor Ludwig Brand and supported by NIH Grant No. GM 11632.

Research was supported in part by the American Heart Association, NYC Affiliate; NSF Award No. DMB-9006044; PSC-CUNY Internal Award Program; and Petroleum Research Fund of the ACS. The author was an Investigator of the American Heart Association, NYC Affiliate.

[91] K. M. Berland, P. T. C. So, and E. Gratton, *Biophys. J.* **68,** 694 (1995).

[25] Preparation of Bifluorescent-Labeled Glycopeptides for Glycoamidase Assay

By KYUNG BOK LEE and YUAN CHUAN LEE

Introduction

The oligosaccharide chains of glycoproteins have many biological functions.[1,2] To investigate the biological effects of carbohydrate components in glycoproteins, it is often desirable to release intact oligosaccharide chains from glycoproteins. Peptide-N^4-(*N*-acetyl-β-D-glucosaminyl)asparagine amidase (EC 3.5.1.52), such as peptide:*N*-glycosidase F (PNGase F) from *Flavobacterium meningosepticum* and glycopeptidase A (PNGase A) from sweet almond, has been used to remove oligosaccharide from glycopeptides or glycoproteins. These enzymes are also called glycoamidases.[3] The activity

[1] R. Dwek, *FASEB. J.* **7,** 1330 (1993).

[2] R. L. Schnaar, *Glycobiology* **1,** 477 (1991).

[3] N. Takahashi, *In* "Handbook of Endoglycosidases and Glycoaminidases" (N. Takahashi and T. Muramatsu, eds.), p. 183. CRC Press, Boca Raton, Florida, 1992.

0076-6879/97 $25

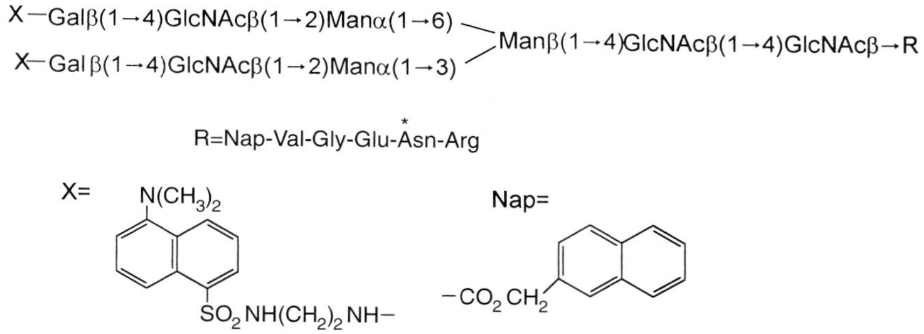

X—Galβ(1→4)GlcNAcβ(1→2)Manα(1→6) ⟩ Manβ(1→4)GlcNAcβ(1→4)GlcNAcβ→R
X—Galβ(1→4)GlcNAcβ(1→2)Manα(1→3)

R=Nap-Val-Gly-Glu-Asn*-Arg

X= [dansyl: $N(CH_3)_2$ … $SO_2NH(CH_2)_2NH-$] Nap= $-CO_2CH_2$-(2-naphthyl)

FIG. 1. Structure of the substrates for glycoamidases. X, 2-(Dansylamido)ethylamine, conjugated to the C-6 of galactosyl residues; Nap, 2-naphthylacetyl.

of the glycoamidases has been measured, usually by measuring the release of fluorescence-labeled or radiolabeled peptide.[4,5]

The fluorescent energy transfer assay has been used for measuring endoprotease activities.[6–8] However, this principle has not been applied to the measurement of glycoamidases activities until recently.[9] Here, we describe the preparation of fluorescence energy transfer substrates (Fig. 1) and the continuous fluorometric assay for glycoamidases using this substrate.

Materials and Methods

Bovine fibrinogen (95% clottable) is from Miles, Inc. (Kankakee, IL). Trypsin and chymotrypsin are from Worthington Enzymes (Freehold, NJ). Glycopeptidase F (PNGase F) is from New England BioLabs (NEB, Beverly, MA). One NEB unit is defined as the amount of enzyme required to remove all of the carbohydrate from 10 μg of denatured RNase B at 37° in 1 hr (1 mIU is equal to 500 NEB units). Glycopeptidase A is from Seikagaku America, Inc. (Rockville, MD). One milliunit of this enzyme is defined as the amount of enzyme capable of hydrolyzing 1 μmol of ovalbumin glycopeptide per minute at 37°, pH 5.0. Neuraminidase from *Arthobacter ureafaciens* is from Boehringer Mannheim (Indianapolis, IN). Dithi-

[4] T. H. Plummer, J. H. Elder, S. Alexander, A. W. Phelan, and A. L. Tarentino, *J. Biol. Chem.* **259,** 10700 (1984).
[5] A. L. Tarentino and T. H. Plummer, *Methods Enzymol.* **138,** 770 (1987).
[6] L. L. Maggiora, C. W. Smith, and Z. Zhong-Yin, *J. Med. Chem.* **35,** 3727 (1992).
[7] E. D. Matayoshi, G. T. Wang, G. A. Krafft, and J. Erickson, *Science* **247,** 954 (1990).
[8] G. T. Wang, C. C. Chung, T. F. Holzman, and G. A. Krafft, *Anal. Biochem.* **210,** 351 (1993).
[9] K. B. Lee, M. Koji, S. Nishimura, and Y. C. Lee, *Anal. Biochem.* **230,** 31 (1995).

othreitol, iodoacetamide, dicyclohexylcarbodiimide, galactose oxidase (EC 1.1.3.9), and catalase (EC 1.11.1.6) are from Sigma Chemical Co. (St. Louis, MO). Sodium cyanoborohydride, borane–pyridine complex, 1-hydroxybenzotriazole, and 2-naphthylacetic acid (Nap) are from Aldrich Chemical Co. (Milwaukee, WI). 2-(Dansylamido)ethylamine (Dan) is from Molecular Probes, Inc. (Eugene, OR). Reversed-phase high-performance liquid chromatography (RP-HPLC) columns (Spherisorb S5C8 and S5ODS2) are from Phase Separation (Norwalk, CT). An HPLC system, equipped with two Gilson model 302 pumps, a Rheodyne 7125 injector, a Fiatron TC 50 column oven, and an ISCO V4 ultraviolet (UV) detector, is used for all HPLC operations. Fluorescence detection of HPLC eluate is performed with a Perkin-Elmer (Norwalk, CT) LS40 scanning fluorescence detector.

A Dionex (Sunnyvale, CA) Bio-LC system equipped with a CarboPac PA-1 column (0.46 × 25 cm) and a pulsed amperometric detector (PAD-II) is used for monosaccharide compositional analysis.[10] Amino acid analysis by PITC (phenylisothiocyanate)-derivatization is as described.[11]

Preparation of Glycopeptides from Bovine Fibrinogen

Reduced and alkylated bovine fibrinogen (5 g)[12] is digested with 25 mg of trypsin and 25 mg of chymotrypsin in 100 ml of Tris buffer, pH 8.0, at 37° for 16 hr, and freeze-dried. The digest, dissolved in 20 ml of pyridine–acetate buffer [2.5% (v/v), pH 4.7], is fractionated on a Sephadex G-50 column (5 × 200 cm), equilibrated and eluted with the same buffer. The glycopeptide fractions, detected by the phenol–sulfuric acid method,[13] are pooled, freeze dried, and further digested with 5 U of neuraminidase in 5 ml of ammonium acetate buffer, pH 5.5, at 37° for 16 hr (or in 0.1 *M* H_2SO_4 at 80° for 1 hr). The reaction mixture (50 μmol of glycopeptides) is purified on the same Sephadex G-50 column as described previously. Fractions containing carbohydrates (asialoglycopeptides) are combined and freeze dried.

Preparation of Bifluorescence-Labeled Glycopeptides: NapGpDan

2-Naphthylacetylation of the N terminus of the asialoglycopeptides is performed as described in [15] in this volume.[14] The product (Nap-Gp) is purified on a C_8 RP-HPLC column (0.46 × 25 cm) and eluted at a flow

[10] J. Q. Fan, Y. Namiki, K. Matsuoka, and Y. C. Lee, *Anal. Biochem.* **219,** 375 (1994).
[11] R. F. Ebert, *Anal. Biochem.* **154,** 431 (1986).
[12] K. G. Rice, N. B. N. Rao, and Y. C. Lee, *Anal. Biochem.* **184,** 249 (1990).
[13] J. F. McKelvy and Y. C. Lee, *Arch. Biochem. Biophys.* **132,** 99 (1969).
[14] P. Wu and L. Brand, *Methods Enzymol.* **278,** [15], 1997 (this volume).

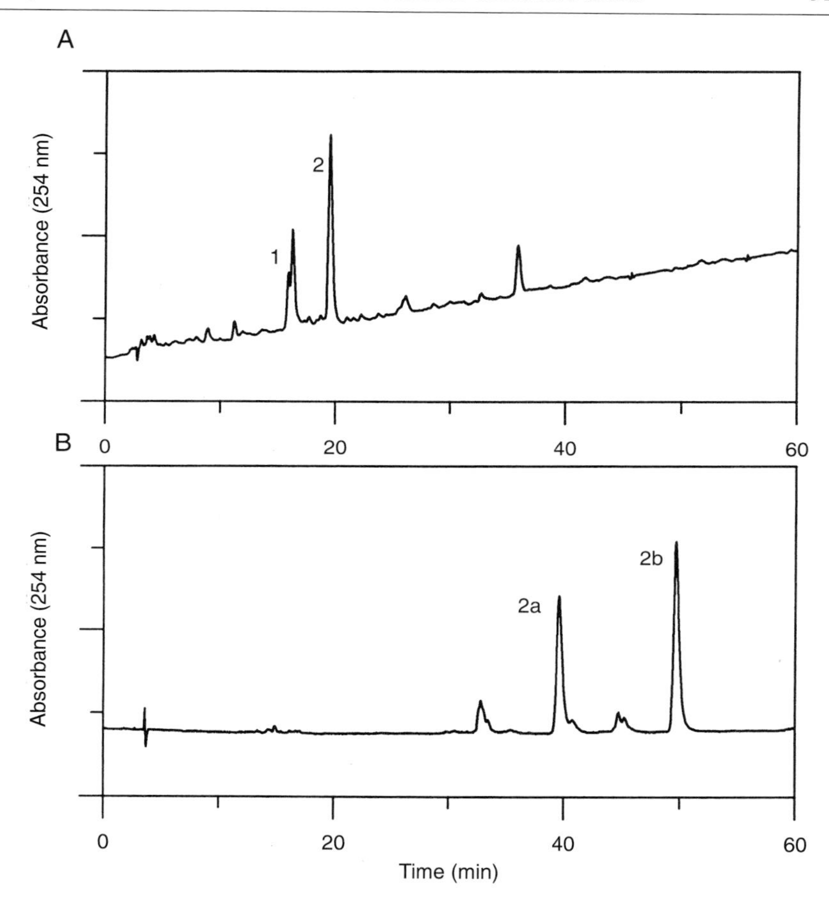

FIG. 2. (A) RP-HPLC analysis of Nap-Gp. The mixture of Nap-Gp (20 nmol) is chromatographed on a C_8 RP-HPLC column. (B) RP-HPLC analysis of the doubly fluorescence-labeled glycopeptides (NapGpDan) after reductive amination with 2-(dansylamido)ethylamine. The column is eluted with a gradient formed by 50 m*M* ammonium acetate (eluant 1) and 50 m*M* ammonium acetate containing 50% acetonitrile (eluant 2), varying eluant 2 from 15 to 80% over 55 min at a flow rate of 1 ml/min. The effluent is monitored by $A_{254\text{ nm}}$ at 2.0 AUFS. Peak 2a is the monodansylated biantennary glycopeptide derivative and peak 2b is the didansylated biantennary glycopeptide derivative.

rate of 1 ml/min with a gradient formed by 50 m*M* ammonium acetate, pH 6.5 (eluant 1), and 50% acetonitrile in 50 m*M* ammonium acetate (eluant 2), increasing eluant 2 from 15 to 80% over 55 min (Fig. 2A). The effluent is monitored by $A_{254\text{ nm}}$ at 2 AUFS (absorbance units full scale). The *N*-2-naphthylacetylated glycopeptides (Nap-Gp) are separated on a C_8

RP-HPLC column (Fig. 2A) into two major peaks, in a combined yield of about 75% (43 μmol of biantennary type glycopeptides). Peak 1 contains valine, glutamate, asparagine, and lysine at a ratio of 1 : 1.9 : 1 : 0.9, and peak 2 contains valine, glycine, glutamate, asparagine, and arginine at a ratio of 1 : 0.8 : 0.7 : 1 : 0.9.[15,16] The unreacted glycopeptides are collected and reused.

Nap-Gp (9 μmol, peak 2 in Fig. 2A) in 570 μl of 100 m*M* phosphate buffer, pH 7.0, containing catalase (50 μg/ml) is treated with galactose oxidase (60 U in 400 μl of 100 m*M* phosphate buffer, pH 7.0) at 37° for 18 hr. The reaction mixture is fractionated on a Sephadex G-10 column (1.6 × 30 cm) in water, monitoring $A_{210\ nm}$. The void volume peak is collected and freeze dried and the residue is dissolved in 1 ml of sodium phosphate, pH 7.0, mixed with 100 μmol of 2-(dansylamido)ethylamine in 200 μl of ethanol and 50 μl of borane–pyridine complex (495 μmol), and kept at 37° overnight. The product is purified on the Sephadex G-10 column as previously described, freeze dried, and further fractionated on the C_8 RP-HPLC column (0.46 × 25 cm; Fig. 2B) as described. The products are separated on a C_8 RP-HPLC column (Fig. 2B) to obtain two dansyl-labeled glycopeptides (peaks 2a and 2b). Peak 2a is the monodansylated biantennary glycopeptide derivative and peak 2b is the didansylated product (Fig. 2B). peak 2a is separated into two peaks of isomers on a C_{18} RP-HPLC column (data not shown).

High-Performance Liquid Chromatography Analysis of Digestion of NapGpDan by PNGase F and PNGase A

The NapGpDan (10 nmol) is digested with 5 μl of PNGase F (40 μU) in 45 μl of 10 m*M* ammonium acetate buffer, pH 7.5, or with PNGase A (50 μU), in 10 m*M* sodium acetate, pH 5.0, at 37° overnight, and the products are analyzed on a C_{18} RP-HPLC column (0.46 × 25 cm) at a flow rate of 1 ml/min with a linear gradient, varying eluant 2 from 40 to 70% in 55 min. The effluent is monitored by $A_{254\ nm}$ at 0.02 AUFS as well as by fluorescence (λ_{ex} 280 nm, λ_{em} 340 nm for naphthyl and λ_{ex} 334 nm, λ_{em} 520 nm for dansyl).

The NapGpDan (peak 2a in Fig. 2) can be completely digested by PNGase F, yielding two new fluorescent peaks in HPLC (data not shown). The peak eluted earlier showed λ_{ex} 280 nm and λ_{em} 337 nm and the latter showed λ_{ex} 340 nm and λ_{em} 520 nm, indicating the presence of naphthyl and dansyl groups, respectively. Peak 2b in Fig. 2B is also digested with

[15] L. V. Medved, T. N. Platonova, S. V. Litvinovich, and N. I. Lukinova, *FEBS Lett.* **232,** 56 (1988).

[16] W. M. Brown, K. M. Dziegielewska, R. C. Foreman, and N. R. Saunders, *Nucleic Acids Res.* **17,** 6397 (1989).

PNGase F and the reaction products are analyzed on a C_{18} RP-HPLC column. The earlier eluting peak matched the peptide peak from the digest of peak 2a, while the later eluting peak is much more retarded, indicative of its being the didansyl derivative of the oligosaccharide.

Fluorometric Assay for Glycoamidase Activities Using NapGpDan

To reduce the inner filtering effect, assays are performed in a microcuvette (3 × 3 mm) and nonoptimal excitation (295 nm) and emission (340 nm) wavelengths are used for the initial rate determination. A mixture of the positional isomers of monodansyl-labeled Nap-Gp is used without further separation. While the fluorescence quantum yield of didansylated substrate is approximately 1.6-fold higher than that of the monodansylated substrate, the former is not suitable as substrate for kinetic studies, because of the greater inner filtering effect. The cuvette containing the substrate (final concentration 1.3 μM in 65 μl of 10 mM ammonium acetate buffer, pH 7.5, for PNGase F; or 10 mM sodium acetate buffer, pH 5.0, for PNGase A) is kept at 37°, and various amounts of glycoamidases are added to the

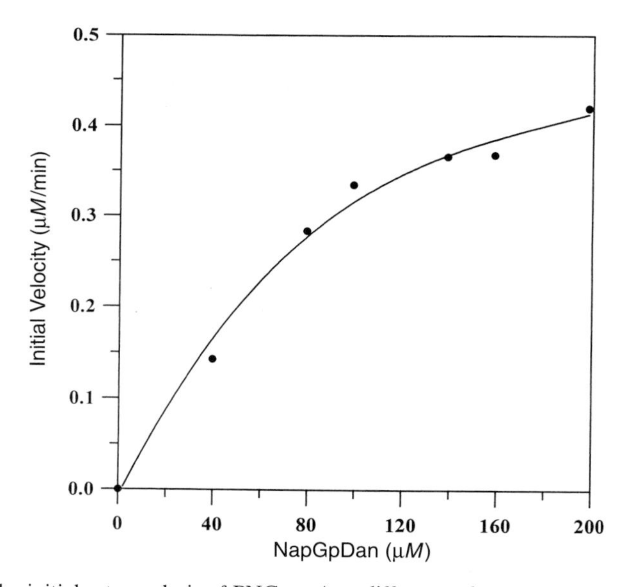

FIG. 3. The initial rate analysis of PNGase A at different substrate concentrations. Initial rates are determined at substrate concentrations from 1 to 200 μM. The reaction mixture contains 0.2 mU of PNGase A in 100 μl of 10 mM sodium acetate buffer (pH 5.0) and is incubated for 10 min at 37° are used. The excitation is at 295 nm and emission is at 340 nm. The slit bandwidths for both are set to 5 nm.

mixture. The increase in fluorescence intensity is continually recorded as a function of time (λ_{ex} 273 nm; bandwidth, 15 nm; λ_{em} 335 nm; bandwidth, 20 nm). Analysis of the data in Fig. 3 reveals that PNGase A has a K_m of 114 μM ± 3 μM and a V_{max} of 0.663 ± 0.08 μM/min by the integrated Michaelis–Menten equation, using EZ-FIT (Perrella Scientific Inc., Springfield, PA).

As an alternative approach a fixed-time 5-min incubation assay has been developed to obtain initial rates at various levels of enzymes. The reaction is stopped by heating the mixture at 100° for 30 sec. The reaction mixture is transferred to a cuvette and fluorescence is measured. The initial rates are found to be linear with respect to the glycoamidase added (Fig.

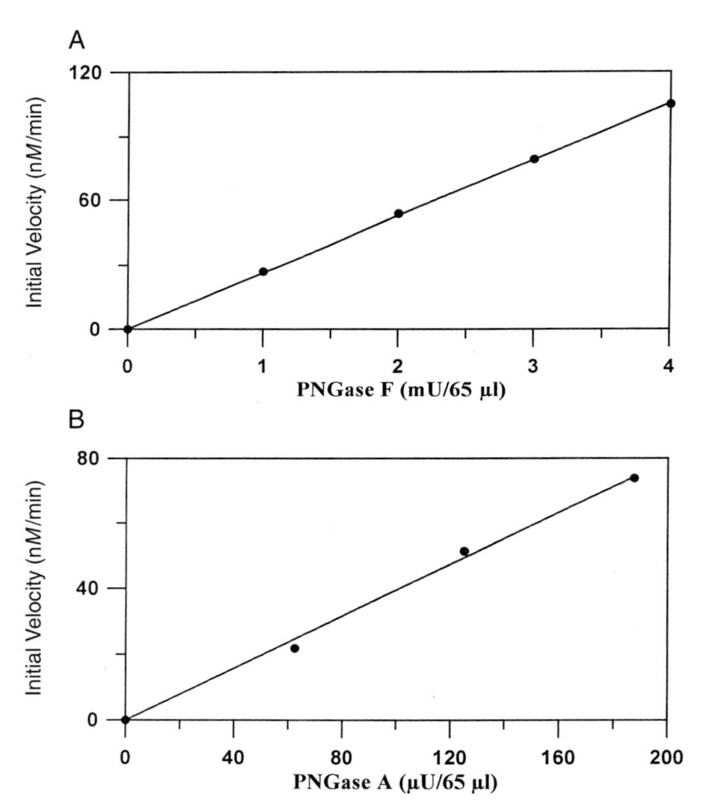

FIG. 4. Initial rates of hydrolysis by PNGase F (A) or PNGase A (B). The data were obtained at a final concentration of 1.3 μM in 65 μl of 10 mM ammonium acetate buffer, pH 7.5, for PNGase F, or 10 mM sodium acetate buffer, pH 5.0, for PNGase A (λ_{ex} 273 nm; bandwidth, 15 nm; λ_{em} 335 nm; bandwidth, 20 nm).

4A and B). A control experiment with the nondansylated substrate (i.e., Nap-Gp) shows no increase in fluorescence of naphthyl emission when digested with glycoamidase.

Thus fluorescence energy transfer substrates such as those described here allow sensitive assays for glycoaminidases, requiring only submicromolar substrate concentration and allowing continuous monitoring of the enzyme activity without product separation.

Acknowledgments

We are indebted to Prof. Ludwig Brand and Drs. Reiko T. Lee and Michael S. Quesenberry for valuable suggestions and discussions.

[26] Preparation of Fluorescence-Labeled Neoglycolipids for Ceramide Glycanase Assays

By KOJI MATSUOKA, SHIN-ICHIRO NISHIMURA, and YUAN C. LEE

Introduction

Fluorescence energy transfer, most suitable for the distance range of 10–60 Å, has been applied to proteins, oligonucleotides, and lipids.[1,2] Rice *et al.*[3–5] have utilized fluorescence energy transfer to determine solution conformations of complex-type glycopeptides. Another useful application is in the design of substrates for the assay of endo-type hydrolases. Many such applications for endopeptidases[6] have appeared, but few applications of the same principle to the endoglycosidases, glycoamidases,[6a] and ceramide glycanases have appeared.

Ceramide glycanase (CGase) is an enzyme that cleaves the glycosidic linkage between oligosaccharide and ceramide in the ceramide glycosides (glycosphingolipids). A number of ceramide glycanases have been iso-

[1] R. Fairclough and C. R. Cantor, *Methods Enzymol.* **48,** 347 (1977).
[2] L. Stryer, *Annu. Rev. Biochem.* **47,** 819 (1978), and references cited therein.
[3] K. G. Rice, P. Wu, L. Brand, and Y. C. Lee, *Biochemistry* **30,** 6646 (1991).
[4] K. G. Rice, P. Wu, L. Brand, and Y. C. Lee, *Biochemistry* **32,** 7264 (1993).
[5] P. Wu, K. G. Rice, L. Brand, and Y. C. Lee, *Proc. Natl. Acad. Sci. U.S.A.* **88,** 9355 (1991).
[6] For example, see E. D. Matayoshi, T. G. Wang, G. A. Krafft, J. Erickson, *Science* **247,** 954 (1990).
[6a] K. B. Lee and Y. C. Lee, *Methods Enzymol.* **278,** [25], 1997 (this volume).

0076-6879/97 $25

lated.[7–13] Unlike the assays for exoglycosidases, for which chromogenic or fluorogenic substrates are available, the assay for CGases (and other endo-type carbohydrates) is by means of separation of the products followed by quantification (e.g., Zhou *et al.*[8]). The hitherto reported methods usually require considerable amounts of both the enzyme and the substrate, and are not convenient for a large number of samples; nor are they amenable to continuous monitoring of the enzyme activity. We describe here the synthesis of the bifluorescence-labeled lactoside as a simple glycolipid mimetic substrate and its usefulness as substrate for some ceramide glycanases.[14–16]

Synthesis of Bifluorescence-Labeled Lactoside

General Procedures. Unless otherwise stated, all commercially available solvents and reagents are used without further purification. Chloroform, dimethylformamide (DMF), tetrahydrofuran (THF), and methanol are stored over molecular sieves (3-Å pore size) (MS 3Å) before use. Pyridine and triethylamine are stored over NaOH pellets. Pulverized molecular sieves 4 Å (MS 4 Å) is dried *in vacuo* at ca. 100° overnight before use.

Melting points (uncorrected) are determined with a Fisher–Johns apparatus. ^{1}H nuclear magnetic resonance (NMR) and proton-decoupled carbon NMR spectra are recorded at 300 and 75.5 MHz, respectively, with a Bruker AMX-300 spectrometer in [^{2}H] chloroform or [^{2}H] methanol using tetramethylsilane (TMS) or methanol (3.3 ppm for ^{1}H or 49.0 ppm for ^{13}C) as internal standard. Assignment of the ring protons is made by first-order analysis of the spectra, and are confirmed by homonuclear decoupling experiments. Samples are dried (ca. 24 hr) *in vacuo* (50°, 0.1 Torr) over NaOH pellets before elemental analyses (Galbraith Laboratories, Inc., Knoxville, TN).

[7] S.-C. Li, R. DeGasperi, J. E. Muldrey, and Y.-T. Li, *Biochem. Biophys. Res. Commun.* **141,** 346 (1986).
[8] B. Zhou, S.-C. Li, R. A. Laine, R. T. C. Huang, and Y.-T. Li, *J. Biol. Chem.* **264,** 12272 (1989).
[9] Y.-T. Li, B. Z. Carter, B. N. N. Rao, H. Schweingruber, and S.-C. Li, *J. Biol. Chem.* **266,** 10723 (1991).
[10] Y.-T. Li, Y. Ishikawa, and S.-C. Li, *Biochem. Biophys. Res. Commun.* **149,** 167 (1987).
[11] M. Ito and T. Yamagata, *J. Biol. Chem.* **261,** 14278 (1986).
[12] M. Ito and T. Yamagata, *J. Biol. Chem.* **264,** 9510 (1989).
[13] H. Ashida, Y. Tsuji, K. Yamamoto, H. Kumagai, and T. Tochikura, *Arch. Biochem. Biophys.* **305,** 559 (1993).
[14] K. Matsuoka, S.-I. Nishimura, and Y. C. Lee, *Tetrahedron: Asymmetry* **5,** 2335 (1994).
[15] K. Matsuoka, S.-I. Nishimura, and Y. C. Lee, *Carbohydr. Res.* **276,** 31 (1995).
[16] K. B. Lee, K. Matsuoka, S.-I. Nishimura, and Y. C. Lee, *Anal. Biochem.* **230,** 31 (1995).

Synthetic and hydrolytic reactions are monitored by thin-layer chromatography (TLC) on precoated plates of silica gel 60 F_{254} (layer thickness, 0.25 mm; E. Merck AG, Darmstadt, Germany). The solvent systems used are as follows: (A) 5:1 (v/v) toluene–ethyl acetate, (B) 9:1, (C) 5:1 (v/v) chloroform–methanol, (D) 65:25:4 (v/v/v) chloroform–methanol–water. For detection of the components, TLC plates are sprayed with a solution of 85:10:5 (v/v/v) methanol–concentrated sulfuric acid–*p*-anisaldehyde, and heated for a few minutes (for carbohydrate) or with an aqueous solution of 5% (w/v) potassium permanganate and heated similarly (for double bonds). Visualization of aldehydic components is by heating the plate after spraying it with a 2,4-dinitrophenylhydrazine (DNPH) solution which is made by dissolving 0.4 g of DNPH in 100 ml of 2 *M* HCl and 100 ml of 95% (v/v) ethanol. An amino group is visualized by heating plates sprayed with 0.2% (w/v) ninhydrin in 95% (v/v) ethanol. Column chromatography is performed on silica gel (Silica Gel 60; 0.015–0.040 mm, E. Merck AG). A 4W F4T5/CW, Cool White ultraviolet (UV) lamp is used in the Michael addition reaction.

n-Pentenyl-β-lactoside (**3**) is prepared by the method previously reported.[17] 2-Naphthaldehyde, borane–trimethylamine complex, sodium cholate, and dansyl chloride are purchased from Aldrich Chemical Co. (Milwaukee, WI). G_{M1} as a standard solution of 5 mg/ml in 2:1 (v/v) chloroform–methanol is purchased from Matreya, Inc. (Pleasant Gap, PA). Ceramide glycanase (CGase) from leech (*Macrobdella decora*) is purchased from V-Labs, Inc. (Covington, LA). The CGase as supplied usually contains 0.29 units/10 ml and 21.8 mg of protein/10 ml. One unit of CGase is defined as the amount of the enzyme that hydrolyzes 1 nmol of G_{M1} per minute at pH 5.0, 37°. The stock sodium cholate solution (1.0 g/10.0 ml) is prepared in deionized water. A 1 m*M* methanolic solution of the 2-naphthylmethylated lactoside with ω-dansyl aglycon (NLD) is prepared and stored at −15° in the dark. Fluorescence measurement is carried out with a Perkin-Elmer luminescence spectrometer LS50B (Perkin-Elmer Corp., Rockville, MD), and the data are processed with OBEY fluorescence data manager software (Perkin-Elmer).

A summary of reagents and conditions is presented in Scheme I.

Preparation of n-Pentenyl-β-4′,6′-(2-naphthylmethylidine)lactoside

2-Naphthaldehyde is readily converted to an acetal derivative in an excellent yield in the presence of 10-camphorsulfonic acid (CSA), isobutanol, and powdered Drierite as dehydrating agent. When an *n*-pentenyllacto-

[17] K. Matsuoka and S.-I. Nishimura, *Macromolecules* **28,** 2961 (1995).

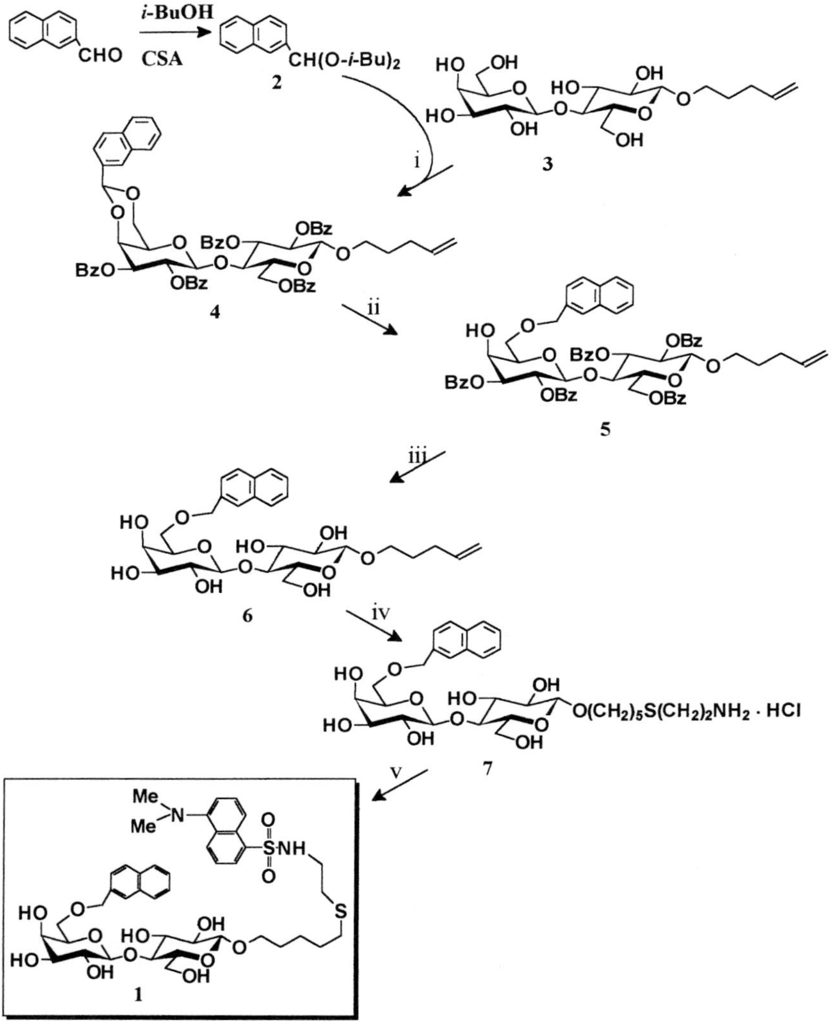

SCHEME I. Reagents and conditions: (i) Step 1: **2**, CSA, DMF, reduced pressure, 55°, 2 hr; step 2: Bz-Cl, pyridine, 0° to room temperature, 3 hr; (ii) borane–triethylamine complex, MS 4 Å, $AlCl_3$, THF, room temperature, 11 hr; (iii) NaOMe, methanol; (iv) $HSCH_2CH_2NH_2 \cdot HCl$, methanol, UV irradiation, room temperature, 3 days; (v) triethylamine, dansyl chloride, methanol, room temperature, 30 min.

side is treated with 2-naphthaldehyde di(isobutyl)acetal (**2**) with continual removal of isobutanol produced in the presence of CSA as an acid catalyst, the acetal exchange reaction proceeds smoothly. The cyclic acetal derived from the lactoside is purified with a Sephadex LH-20 column, and the acetal is successively benzoylated to give **4** in 55% yield.

2-Naphthaldehyde Di(isobutyl)acetal (**2**). To a solution of 2-naphthaldehyde (2.00 g, 12.8 mmol) in isobutanol (20 ml) containing powdered Drierite (3.0 g) is added 10-camphorsulfonic acid (3.0 mg, 12.8 μmol) with stirring, and the mixture is continuously stirred for 2 days at room temperature. The reaction mixture is diluted with chloroform (50 ml) and filtered. The filtrate is washed with cold saturated sodium hydrogen bicarbonate solution and brine, then dried over anhydrous sodium sulfate. After filtration, the chloroform solution is concentrated *in vacuo* to give syrupy **2** (3.29 g, 89.6%) containing a trace of starting material: R_f 0.81 (solvent toluene), and the compound is used for the next step without further purification.

n-Pentenyl-O-[2,3-di-O-benzoyl-4,6-O-(2-naphthylmethylidene)-β-D-galactopyranosyl]-(1 → 4)-2,3,6-tri-O-benzoyl-β-D-glucopyranoside (**4**). To a solution of *n*-pentenyl-*O*-(β-D-galactopyranosyl)-(1 → 4)-β-D-glucopyranoside (**3**)[17] (2.04 g, 4.97 mmol) and **2** (2.85 g, 9.94 mmol) in dry DMF (20 ml) at 0° is added 10-camphorsulfonic acid (0.93 g, 4.00 mmol), and the mixture is heated at 55° for 2 hr while removing the liberated isobutanol with a water aspirator. The reaction mixture is neutralized with triethylamine (0.56 ml, 4.02 mmol) and vacuum evaporated to remove DMF. The residual syrup is fractionated on a column of Sephadex LH-20 (5 × 200 cm), using 95% (v/v) ethanol as eluant. The fractions, containing both carbohydrate and aldehyde when analyzed by TLC, are concentrated. The syrup is dissolved in pyridine (30 ml), to which benzoyl chloride (5.83 ml, 49.7 mmol) is added with stirring at 0°. After 3 hr at room temperature, methanol (0.5 ml) is added dropwise to the mixture, the reaction mixture is diluted with chloroform, and the solution is successively washed with cold 0.5 *M* sulfuric acid, cold saturated sodium hydrogen bicarbonate, and brine. The extract is dried over anhydrous sodium sulfate, filtered, and evaporated. The residue is chromatographed on a column of silica gel eluted first with 20:1 (v/v), then 10:1 (v/v), toluene–ethyl acetate to give pure **4** (2.90 g, 54.6%): R_f 0.45 (solvent A); mp 243–245° (from methanol); ^{1}H NMR δ ($CDCl_3$) 1.53 (m, 2H, CH_2CH=), 2.35 (m, 2H, OCH_2CH_2), 3.02 (s, 1H, H-5′), 3.6 (m, 2H, OCH_2), 3.63 (dd, 1H, $J_{6'a,6'b}$ 10.7 Hz, H-6′a), 3.8 (m, 2H, H-5, 6′b), 4.23 (t, 1H, $J_{4,5}$ 9.0 Hz, H-4), 4.37 (d, 1H, $J_{4',5'} \sim 0$ Hz, H-4′), 4.44 (dd, 1H, $J_{5,6a}$ 4.2 Hz and $J_{6a,6b}$ 12.6 Hz, H-6a), 4.62 (dd, 1H, $J_{5,6b}$ 2.0 Hz, H-6b), 4.67 (d, 1H, $J_{1,2}$ 7.8 Hz, H-1), 4.76 (m, 2 H, CH_2=CH), 4.86 (d, 1H, $J_{1',2'}$ 7.9 Hz, H-1′), 5.20 (dd, 1H, $J_{3'4'}$ 3.6 Hz, H-3′), 5.33 (dd, 1H, $J_{2,3}$

9.5 Hz, H-2), 5.44 [s, 1H, CH-(2-naphthyl)], 5.57 (m, 1H, C*H*=CH_2), 5.83 (dd, 1H, $J_{2',3'}$ 10.4 Hz, H-2′), 5.86 (t, 1H, $J_{3,4}$ 9.5 Hz, H-3), 7.59 (m, 32H, 5 Bz, 2-naphthyl).

Anal. Calcd. for $C_{63}H_{56}O_{16} \cdot H_2O$: C, 69.61; H, 5.37. Found C, 69.74; H, 5.55.

Preparation of n-Pentenyl-β-6′-(2-naphthylmethyl)lactoside

The regioselective ring-opening reaction of a cyclic acetal reported by Garegg[18] requires that the substituent adjacent to the cyclic acetal be a bulky group (e.g., benzyl, benzoyl, or a sugar). Because the 2-naphthylmethyl group is susceptible to hydrogenolysis, the benzyl group is excluded as the neighboring substituent. The benzoyl group, however, meets our requirement, and yields the desired product successfully. When the reductive opening of a cyclic acetal is carried out with a small excess of the reagents, the reaction is sometimes incomplete. Therefore the reductive cleavage of the cyclic acetal (**4**) is carried out with 7 mol equivalent each of borane–trimethylamine complex and aluminum chloride in the presence of dried and powdered MS 4 Å in THF at room temperature. After 11 hr, the reaction is complete as judged by TLC, and 2-naphthylmethylated lactoside (**5**) is obtained in 95% yield. The lactosyl derivative is de-O-benzoylated by the Zemplén method to provide **6**, which shows fluorescence emission at 334 nm for 2-naphthyl group when excited at 260 nm in 50 m*M* sodium acetate buffer, pH 5.0, including sodium cholate (200 mg/3.0 ml) at ambient temperature.

*n-Pentenyl-O-[2,3-di-O-benzoyl-6-O-(2-naphthylmethyl)-β-D-galactopyranosyl]-(1 → 4)-2,3,6-tri-O-benzoyl-β-D-glucopyranoside (**5**).* A mixture of **4** (703 mg, 0.66 mmol), powdered MS 4 Å (1.5 g), and borane–trimethylamine complex (346 mg, 4.61 mmol) in 10 ml of dry THF is stirred for 40 min at room temperature, to which is added anhydrous aluminum chloride (614 mg, 4.61 mmol) with stirring, and the stirring is continued for 11 hr at room temperature. When TLC shows complete disappearance of the acetal group, the mixture is filtered, and the powdered MS 4 Å is thoroughly washed with chloroform. The filtrate is successively washed with cold 0.5 *M* sulfuric acid, cold saturated sodium hydrogen bicarbonate, and brine, dried over anhydrous sodium sulfate, filtered, and concentrated. The syrup is separated on a column of silica gel eluting with 10 : 1 (v/v) toluene–ethyl acetate to give pure **5** (668 mg, 94.7%): R_f 0.42 (solvent A); mp 171–173° (from methanol); ^{1}H NMR δ ($CDCL_3$) 1.52 (m, 2H, C*H*$_2$CH=), 1.87 (m, 2H, OCH_2C*H*$_2$), 2.26 (d, 1H, OH-4′), 3.05 (m, 2H,

[18] P. J. Garegg, *Pure Appl. Chem.* **56,** 845 (1984).

H-6′a, 6′b), 3.46 (t, 1H, $J_{5,6'}$ 5.9 and 6.2 Hz, H-5′), 3.58 (m, 2H, OCH_2), 3.78 (m, 1H, H-5), 4.16 (t, 1H, $J_{4,5}$ 9.2 Hz, H-4), 4.19 (br t, 1H, $J_{4',OH}$ 2.3 Hz, H-4′), 4.38 [m, 2H, OCH_2-(2-naphthyl)], 4.39 (dd, 1H, $J_{5,6a}$ 4.6 Hz and $J_{6a,6b}$ 11.8 Hz, H-6a), 4.57 (dd, 1H, $J_{5,6b}$ ~ 1 Hz, H-6b), 4.62 (d, 1H, $J_{1,2}$ 7.8 Hz, H-1), 4.74 (m, 2H, CH_2=CH), 4.77 (d, 1H, $J_{1',2'}$ 7.8 Hz, H-1′), 5.11 (dd, 1H, $J_{3',4'}$ 3.1 Hz, H-3′), 5.37 (dd, 1H, $J_{2,3}$ 9.5 Hz, H-2), 5.56 (m, 1H, CH=CH_2), 5.69 (dd, 1H, $J_{2',3'}$ 10.3 Hz, H-2′), 5.71 (t, 1H, $J_{3,4}$ 9.0 Hz, H-3), 7.58 (m, 32H, 5Bz, 2-naphthyl).

Anal. Calcd. for $C_{63}H_{58}O_{16}$: C, 70.65; H, 5.45. Found C, 70.76; H, 5.65.

n-Pentenyl-O-[6-O-(2-naphthylmethyl)-β-D-galactopyranosyl]-(1 → 4)-β-D-glucopyranoside (**6**). To a solution of **5** (667.6 mg, 0.623 mmol) in dry methanol (5 ml) and dry THF (5 ml) is added sodium methoxide (20.2 mg, 0.374 mmol), and the solution is stirred for 64 hr at room temperature. The reaction mixture is neutralized by adding Dowex 50W X-8 (H^+) resin. The suspension is filtered, and the filtrate is evaporated *in vacuo*. The residue is purified by passing it through a column of Sephadex LH-20 (2.5 × 50 cm), eluting with 95% ethanol, to give homogeneous **6** (260.8 mg, 76.0%): R_f 0.18 (solvent B); ^{1}H NMR δ (CD_3OD) 1.66 (m, 2H, CH_2CH=), 2.10 (m, 2H, OCH_2CH_2), 3.25 (m, 1H, H-2′), 3.51 (m, 1H, H-3′), 3.53 (m, 1H, H-2), 4.24 (d, 1H, $J_{1'2'}$ 7.8 Hz, H-1′), 4.32 (d, 1H, $J_{1,2}$ 7.4 Hz, H-1), 4.68 [m, 2H, CH_2-(2-naphthyl)], 4.95 (m, 2H, CH_2=CH), 5.79 (m, 1H, CH=CH_2), 7.62 (m, 10H, 2-naphthyl); ^{13}C NMR δ (CD_3OD) 30.06 (OCH_2CH_2), 31.21 (CH_2CH=), 62.05 (C-6), 70.24, 70.37, 70.55, 72.38, 74.48, 74.68, 74.79, 75.30, 76.27, 76.45, 81.53 [C-2, -3, -4, -5, -2′, -3′, -4′, -5′, -6′, OCH_2CH_2, CH_2-(2-naphthyl)], 104.22 (C-1), 105.19 (C-1′), 115.19 (CH_2=CH), 126.92, 127.06, 127.10, 127.73, 128.64, 128.98, 129.11, 134.49, 134.74, 136.95 (10C, attributable to 2-naphthyl), 139.47 (CH=CH_2).

Preparation of Bifluorescence-Labeled Lactoside

Examples of Michael addition of thiol (cysteamine) to a C=C to provide thioether products in an anti-Markovnikov manner have been reported.[19,20] Using an analogous reaction, the lactosyl derivative **6** is extended with cysteamine by the Michael addition reaction to the ω-alkene group to afford **7** in 66% yield. Dansylation of the amino group is carried out in dry methanol at room temperature to give **8** (NLD) in 91% yield.

5-(2-Aminoethanethio)pentyl-O-[6-O-(2-naphthylmethyl)-β-D-galactopyranosyl]-(1 → 4)-β-D-glucopyranoside Hydrochloride Salt (**7**). A solution of **6** (40.4 mg, 73.4 μmol) in dry methanol (0.5 ml) is allowed to react with

[19] R. T. Lee and Y. C. Lee, *Carbohydr. Res.* **37,** 193 (1974).

[20] R. Roy and F. D. Tropper, *J. Chem. Soc. Chem. Commun.* 1058 (1988).

2-aminoethanethiol hydrochloride (85.1 mg, 0.734 mmol) under a UV lamp for 3 days at room temperature. The reaction mixture is evaporated *in vacuo*, and the residue is dissolved in a small amount of water. The aqueous solution is fractionated on a column of Sephadex G-10 (1.5 × 14 cm), first eluted with water then with 0.5 *M* aqueous acetic acid, to give homogeneous **7** (32.1 mg, 65.9%): R_f 0.13 (solvent D).

*5-[2-(5-Dimethylamino-1-naphthalenesulfonamido)ethanethio]-pentyl-O-[6-O-(2-naphthylmethyl)-β-D-galactopyranosyl]-(1 → 4)-β-D-glucopyranoside (**8**).* To a solution of **7** (32.1 mg, 48.3 μmol) and dry triethylamine (13.4 μl, 96.7 μmol) in dry methanol (2.0 ml) is added dansyl chloride (52.1 mg, 193 μmol) at room temperature. After 30 min, the mixture is directly applied to a column of Sephadex LH-20 (2.5 × 50 cm) eluted with 95% ethanol to give crude **8** (37.8 mg, 90.9%). An analytical sample is chromatographed on silica gel with 10:1 (v/v) chloroform–methanol as eluent to give pure **8** (NLD): R_f 0.33 (solvent C); ^{1}H NMR δ (CD_3OD) 1.36 (m, 4H, 2 CH_2), 1.55 (m, 2H, OCH_2CH_2), 2.26 (m, 2H, $CH_2SCH_2CH_2NH$), 2.39 (m, 2H, SCH_2CH_2NH), 2.98 (m, 2H, CH_2NH), 4.28 (d, 1H, $J_{1',2'}$ 7.8 Hz, H-1′), 4.37 (d, 1H, $J_{1,2}$ 7.3 Hz, H-1), 4.73 [m, 2H, OCH_2-(2-naphthyl)], 7.25–8.57 (m, 13H, dansyl, 2-naphthyl).

Time Course of Enzymatic Reaction Using 2-Naphthylmethylated Lactoside with ω-Dansyl Aglycon (8) as Substrate

The UV-absorption spectrum of NLD shows the presence of peaks at 260 and 335 nm, as expected from the presence of the 2-naphthyl group and the dansyl group. The NLD having these donor and acceptor fluorescence probes at each end of the molecule is tested as a function of excitation wavelength from 260 to 300 nm and the emission spectra of a 5 μ*M* solution of NLD. Those emission spectra include two peaks at around 335 nm (due to the 2-naphthyl group) and at 540 nm (due to the dansyl group) because the fluorescence energy transfer is observed from 2-naphthyl as the energy donor to dansyl as the acceptor. The relative fluorescence intensities of the acceptor emission peaks when the donor is excited at 260 nm are found to be most effectively enhanced. Therefore, we have selected this excitation wavelength for all subsequent enzyme reactions.

Digestion of NLD monitored by fluorescence energy transfer is carried out in a cuvette at ambient temperature, using the following conditions: A 3.0-ml solution of 10 μ*M* NLD containing 200 μg of sodium cholate is made fresh. To the solution (3.0 ml) placed in the cuvette is added CGase (40 μl, 1.14U), which is mixed quickly. An emission spectrum is taken immediately after the enzyme addition and this is regarded as the 0-min spectrum. The spectra taken at preselected intervals (up to 240 min) are shown in Fig. 1.

The emission spectra of the reaction mixture shown in Fig. 1 are taken by excitation at 260 nm, which also shows a double scattered peak at around 520 nm. As expected, the enzymatic digestion causes the decrease in the dansyl emission by diminishing the energy transfer from the 2-naphthyl group. It also causes an increase in the 2-naphthyl emission by removing the quenching by the dansyl group. The 2-naphthyl emission at 334.5 nm gradually increases up to 90 min, after which the reaction tapers off until it has almost stopped at 210 min. The decreases in the dansyl emission at 540.3 nm as the reaction progresses are also measured. The inset in Fig. 1 shows the time course of the formation of the product (lower curve), 2-naphthymethylated lactose (NL), determined from the increase in 2-naphthyl emission at 334.5 nm, and that of hydrolysis of NLD (upper curve) determined by the dansyl emission at 540.3 nm. The concentrations of the NL are calibrated from a standard curve of the emission of **6** (2-naphthyl-

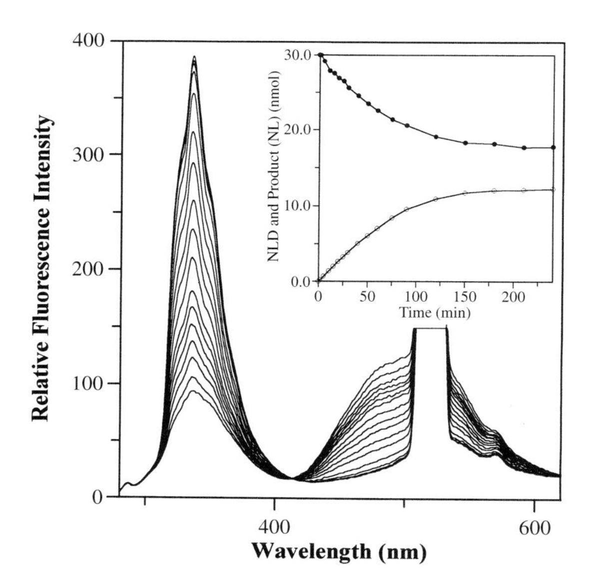

FIG. 1. The time course (1.5, 5, 10.5, 15, 20, 25, 30, 40, 50, 60, 75, 90, 120, 150, 180, 210, and 240 min) of the relative fluorescence emission of NLD during hydrolysis with CGase. The reaction conditions were as follows: ambient temperature, 3.0 ml of 50 mM sodium acetate, pH 5.0, containing 200 μg of sodium cholate, a 10 μM concentration of the NLD (30 nmol/3.0 ml), and 1.14 U of the CGase. A slit size of 5.0 nm was used for both the excitation and emission. The spectra were scanned at 500 nm/min, exciting at 260 nm for the 2-naphthyl group. *Inset*: the result of the time course expressed as the concentrations of the product NL (○) and the remaining NLD (●).

Lac-Pent or NLP) under the same spectrometric conditions. Analysis of the data in the inset in Fig. 1 gives 7.7 μM for K_m and 0.16 nmol/min for V_{max}.

Continuous Assay of Ceramide Glycanase with NLD by Intramolecular Fluorescence Quenching. Fluorescence measurement of emission spectra and excitation spectra is carried out at ambient temperature, in a Teflon-stoppered cuvette (12.5 × 45 mm) containing a 3.0-ml sample. The slit width of the excitation and the emission is 5.0 nm, and the scan speed is 500 nm/min.

A typical enzymatic reaction is carried out at room temperature as follows: Appropriate amounts of 1 mM NLD solution and the sodium cholate solution in a screw-capped vial are evaporated to dryness. To the vial is added 4.0 ml of 50 mM sodium acetate buffer, pH 5.0, and the vial is sonicated for 1 min. A 3.0-ml portion is pipetted into a Teflon-stoppered cuvette, and an appropriate amount of the enzyme (10–40 μl, 0.29–1.14 U) is added and quickly mixed by inversion. The initial spectrum at 0 min is taken immediately after the enzyme is added. The quantification of NLD by measurement of dansyl emission is estimated by subtracting the dansyl emission (540.3 nm) at 240 min from the value at 0 min. The amount of NLD hydrolyzed during this period is regarded to be equal to the 6′-(2-naphthylmethyl)lactose (NL) formed in the same reaction period. Linear proportionation of the decrease of the dansyl emission yields the quantity of the hydrolyzed NLD. These emission spectra show an isoemissive point at 414 nm.

[27] Applications of Fluorescence Resonance Energy Transfer to Structure and Mechanism of Chloroplast ATP Synthase

By RICHARD E. MCCARTY

Introduction

The lack of detailed structural information hampers the investigation of complex proteins. With 9 different polypeptides and a total of about 20 chains,[1] the chloroplast ATP synthase is certainly complex. Although the task of elucidating mechanistic and structural features of the enzyme and interaction among its subunits seems daunting, applications of biochemical techniques and of fluorescence resonance energy transfer (FRET) have

[1] R. E. McCarty, *J. Exp. Biol.* **172,** 431 (1992).

0076-6879/97 $25

resulted in some remarkable insights. This chapter outlines some results of these procedures that may be useful to others who investigate complex proteins.

The chloroplast ATP synthase catalyzes ATP synthesis coupled to the flow of protons down their electrochemical potential established by electron transfer. The enzyme resembles its counterparts in the coupling membranes of many bacteria and mitochondria in that it is composed of two parts, F_1 and F_0 (CF_1 and CF_0, for chloroplast F_1 and F_0).[1] CF_1, an extrinsic membrane protein, is composed of five polypeptides labeled α–ε in order of decreasing molecular mass (55 to 14.7 kDa). CF_1, which once released from the membrane is very water soluble, contains the active sites of the enzyme, and has the unusual subunit stoichiometry of $\alpha_3\beta_3\gamma\delta\varepsilon$. The subunit composition of CF_1 shows that it is structurally asymmetric, an intriguing fact that has mechanistic implications and has permitted an unusual application of FRET. CF_0 is the complex of four different polypeptides of the synthase that is an integral part of the thylakoid (green) membrane of chloroplasts. CF_0 functions to anchor CF_1 to the membrane and contains a rapid proton conductance mechanism. To date, there has been relatively little FRET work with CF_0.

Physical chemists tend, for quite rational reasons, to shy away from complex systems. Gordon G. Hammes, then at Cornell University, was, in contrast, a biophysical chemist intrigued by complex enzymes. The initial FRET measurements and cross-linking studies on CF_1 were performed in his laboratory. This earlier work is reviewed in Ref. 2. In the author's laboratory, then also at Cornell, we had gained extensive experience in CF_1 subunit depletion and reconstitution as well as site-specific chemical modification. The two laboratories entered into a productive and stimulating collaboration lasting for several years.

Structural Mapping by Fluorescence Resonance Energy Transfer

One of the earliest applications of fluorescence spectroscopy to the study of CF_1 was structural mapping by FRET. The Hammes laboratory worked out procedures for labeling of nucleotide-binding sites on CF_1 with modified nucleotides,[3] including 2′(3′)-*O*-trinitrophenyl-ATP or -ADP (TNP-ATP/TNP-ADP). A complete description of the nucleotide-binding sites of CF_1 is beyond the scope of this chapter; a brief outline of these sites is, however, necessary. There are likely to be six nucleotide-binding

[2] G. G. Hammes, *In* "Protein–Protein Interactions" (C. Frieden and L. W. Nichol, eds.), p. 257. Wiley Interscience, New York, 1981.

[3] M. F. Bruist and G. G. Hammes, *Biochemistry* **20,** 6298 (1981).

sites,[4] at least two of which do not participate in catalysis. There are two sites that bind Mg^{2+}-ATP so tightly that it is not lost even after extensive turnover; these sites are clearly noncatalytic. Two other sites bind ATP or ADP and are operationally defined as "tight sites" because the rate of release of nucleotide bound to these sites is so slow that even extensive dialysis or gel filtration fails to remove significant amounts of bound nucleotide. Despite the slow release of nucleotide bound to these sites, exchange of bound nucleotide with medium nucleotide can be fast, especially under conditions that favor catalysis. These two sites are, thus, referred to as "tight, exchangeable" sites. Finally, there is at least one site with broad specificity for nucleoside di- and triphosphates that is characterized by relatively weak association (K_D about 2 μM) with CF_1 and relatively rapid dissociation.

The α and β polypeptides of CF_1 contain the nucleotide-binding sites near α–β subunit interfaces. Chemically, the three α subunits are identical; there is a single gene for α in the plastid genome and there is no evidence for posttranslational modification of α. These facts apply to the β subunit as well. The α and β polypeptides form a heterohexamer in which the two proteins alternate. The $\alpha_3\beta_3$ structure is symmetric, but the nucleotide-binding properties of CF_1 are not. Asymmetric interactions among $\alpha_3\beta_3$ and the centrally located γ subunit are likely to induce the marked differences in the nucleotide-binding site properties.

These differences in properties allowed the selective filling of nucleotide-binding sites. For example, incubation of CF_1 with TNP-ADP, followed by removal of free and loosely bound TNP-ADP by gel filtration, allows the remaining TNP-ADP to specifically occupy a tight, exchangeable site. Dissociable, weaker sites could be titrated directly with TNP-ADP. The noncatalytic site(s) could also be selectively filled.[3] These remarkable properties of CF_1 made the initial FRET mapping of nucleotide sites possible.[5]

Chemical modification studies of membrane-bound CF_1 were initiated in the author's laboratory during the early 1970s. Subunit depletion experiments were initiated about 10 years later. Why and how we started these lines of investigation are not important; their results are. Over the years we developed methods to label covalently and specifically a number of different residues of the protein. Much of our success was with the modification of various cysteine residues of CF_1 by N-substituted maleimides, including fluorescent maleimides.[6] A certain amount of good fortune combined with hard work resulted in our ability to modify, with reasonable stoichiom-

[4] A. B. Shapiro and R. E. McCarty, *J. Biol. Chem.* **260,** 17276 (1991).
[5] R. A. Cerione and G. G. Hammes, *Biochemistry* **21,** 745 (1983).
[6] C. M. Nalin, R. Béliveau, and R. E. McCarty, *J. Biol. Chem.* **258,** 3376 (1983).

etry and excellent specificity, the sole cysteine residue of ε (Cys-6),[7] Cys-89 and Cys-322 of the γ subunit, as well as Cys-199 and -205, partners in the γ disulfide bond.

There are a total of just 11 cysteine residues in CF_1, 1 each in the 3 α and β subunits, 4 in γ, and only 1 in ε. Unless the α and β chains are unfolded, the cysteine residues they contain react slowly with –SH alkylating reagents. In contrast, Cys-322 of γ and Cys-6 of ε react rapidly with maleimides. Cysteine at position 89 is modified only under specific conditions.[8]

The methods used to obtain specific chemical modification of each group are too detailed to be considered in depth here. An illustration of how subunit depletion and chemical modification may be used to achieve specific labeling suffices. Suppose the specific labeling of Cys-322 of the γ subunit by a fluorescent maleimide is desired. We know that both this residue and Cys-6 of ε react relatively rapidly with alkylating reagents. To achieve specific labeling of γ-Cys-322, we can label the intact enzyme and remove the labeled ε or remove the ε subunit first and then incubate with the maleimide. Under these conditions, only γ-Cys-322 labels. The ε subunit may be reconstituted fully with CF_1 deficient in ε, a property that permits its specific labeling and incorporation into the CF_1 complex.[7]

During the early 1980s, William J. Patrie, a graduate student, suggested to Carlo M. Nalin, a postdoctoral associate in the laboratory, and the author that the reagent Lucifer Yellow VS (where VS stands for "vinyl sulfone") was a stable fluorophore that might label –SH groups. The reagent could (and did) react with the accessible thiols of CF_1 but, surprisingly, reacted far more rapidly with α than with either γ or ε.[9] α incorporated Lucifer Yellow VS much more rapidly than the smaller subunits and the rapid phase of incorporation plateaued at about one Lucifer Yellow per CF_1.

It seemed possible that only one of the three α subunits reacts rapidly with Lucifer Yellow VS. To confirm this, structural mapping by FRET was used to show that the Lucifer Yellow occupied a unique position, fixed in space by six independent distance measurements. Lysine at position 378 of one of the α subunits is abnormally reactive with the vinyl sulfone function of the Lucifer Yellow.[10] This high reactivity is probably a result of the structural asymmetry of the enzyme. Asymmetric interactions among αβ pairs and γ cause local changes in the region of one of three α subunits that enhance the rate of reaction of this α-Lys-378 with Lucifer Yellow VS. Nucleotide-binding site occupancy could also affect the reactivity of α-Lys-378.

[7] M. L. Richter, B. Snyder, R. E. McCarty, and G. G. Hammes, *Biochemistry* **24,** 5755 (1985).
[8] R. E. McCarty and J. Fagan, *Biochemistry* **12,** 1503 (1973).
[9] C. M. Nalin, B. Snyder, and R. E. McCarty, *Biochemistry* **24,** 2318 (1985).
[10] A. B. Shapiro, Ph.D. thesis. Cornell University, Ithaca, New York, 1991.

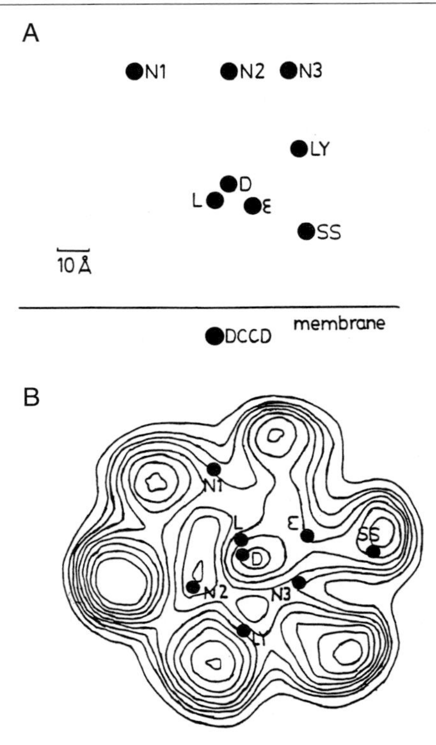

FIG. 1. A view of CF_1 by FRET mapping. N1, N2, and N3 refer to three distinct nucleotide-binding sites on CF_1. LY, Lucifer Yellow covalently attached to Lys-378 of one of the three α subunits; L and D, Cys-89 and Cys-322, respectively, of γ; SS, Cys-199 and -205 of γ; ε, Cys-6 of the ε subunit. In (A), the "side" view of the map is given, with the phase of the membrane surface perpendicular with respect to that of the page. DCCD, The site of covalent attachment of carbodiimides to subunit III of CF_0 (Glu-65). In (B), a top view is shown in which the sites mapped are superimposed on an image of CF_1 devised by single-molecule averaging of electron micrographs of CF_1. [Adapted with permission from A. B. Shapiro, K. D. Gibson, H. A. Scheraga, and R. E. McCarty, *J. Biol. Chem.* **266,** 17276 (1991).]

In addition, specific site labeling of CF_1 in CF_1–CF_0, followed by incorporation of the intact synthase into phospholipid vesicles, permitted the distance between CF_1 sites and the surface of the membrane to be measured. About 35 individual FRET measurements have been made from 8 different specific sites on CF_1. A possible interpretation of the FRET structural mapping is given in Fig. 1. This figure is based on a structure refinement performed by a minimization of a least-squares function.[11] The agreement between the measured and fitted distances was within $\pm 5\%$.

[11] A. B. Shapiro, K. D. Gibson, H. A. Scheraga, and R. E. McCarty, *J. Biol. Chem.* **266,** 17276 (1991).

Important features of the FRET structural map include the observations that (1) the nucleotide-binding sites so far depicted are 80–90 Å away from the membrane; (2) Cys-322 (D, Fig. 1) of the γ subunit and Cys-89 (L, Fig. 1) of the γ subunit occupy a central position in the $\alpha\beta$ heterohexamer. A central position of at least part of the γ subunit was also indicated by analysis of subunit-depleted CF_1 by single-molecule averaging of negatively stained images of CF_1[12]; (3) the structural asymmetry of the enzyme is clearly evident; (4) Cys-89 and Cys-322 of γ are close together, as predicted by the ability of the two cysteine residues to form a disulfide bond under some conditions; (5) Cys-89 alkylation by *N*-ethylmaleimide inhibits ATP synthesis and hydrolysis by the enzyme despite the large ($\geq$50 Å) distances between this residue and the mapped nucleotide sites. More astonishing, perhaps, is the fact that alkylation of Cys-89 greatly enhances the rate of exchange of bound nucleotide with medium nucleotide.[13]

There has been but one report on the application of FRET to the mapping of the entire ATP synthase.[14] Subunit III of CF_0, a hydrophobic 8-kDa protein, contains a glutamate residue at position 65 that reacts with hydrophobic carbodiimides such as *N,N'*-dicyclohexylcarbodiimide. The covalent modification of Glu-65 of subunit III blocks proton conductance by CF_0 and ATP synthesis. *N*-Cyclohexyl-*N'* · (1-pyrenyl)carbodiimide (NCP) also reacts specifically with subunit III of CF_1–CF_0 in thylakoids and CF_1–CF_0 was purified. The NCP-labeled enzyme was also prepared with *N*-[7-(diethylamino)-4-methylcoumarin-3-yl]maleimide (CPM) at γ-Cys-89 or the γ disulfide thiols (Cys-199 and -205). Energy transfer from NCP to CPM was determined by lifetime measurements after incorporation of the modified CF_1–CF_0 into asolectin (crude soybean lipids) vesicles. A distance of 41 Å from NCP to CPM on Cys-89 was calculated, whereas that from NCP to the Cys-199/205 of γ was 50 Å.

The quenching of NCP fluorescence by spin-labeled fatty acids in the reconstituted membranes and energy transfer measurements placed Glu-65 about 6–10 Å from the surface of the CF_1 side of the membrane. At first glance, the distance of 41 Å between γ-Cys-89 and Glu-65 of subunit III would seem at odds with the distance (45 Å) between the membrane surface and Cys-89 of γ. The distance to the membrane is a maximum value because the line between Cys-89 and the membrane does not have to be perpendicular. Assuming that the NCP is directly below Cys-89 of γ, the perpendicular distance between Cys-89 of γ and the membrane would be just 35–41 Å. Because Cys-322 of γ is close to Cys-89, the C terminus of γ must also be relatively close to the membrane.

[12] E. J. Boekema, J.-P. Xiao, and R. E. McCarty, *Biochim. Biophys. Acta* **1020,** 49 (1990).
[13] P. Soteropoulos, A. M. Ong, and R. E. McCarty, *J. Biol. Chem.* **269,** 19810 (1994).
[14] B. Mitra and G. G. Hammes, *Biochemistry* **28,** 3063 (1989).

Fluorescence resonance energy transfer has also been used effectively for distance mapping of purified CF_1 β subunits in solution.[15] β-Cys-63 was labeled with CPM and the β nucleotide-binding site was titrated with trinitrophenyl-ADP. A distance of 41 Å was calculated.

A partial structure at 2.85 Å resolution of bovine heart mitochondrial F_1 is available.[16] A wealth of detail on the major parts of the nucleotide-binding $\alpha\beta$ core of the enzyme was provided. However, only a part of the γ subunit and neither of the δ and ε subunits was revealed. Some of the FRET measurements with CF_1 are not in accord with assignments made to the MF_1 structure for γ. First, the FRET measurements clearly indicate that Cys-322, the penultimate amino acid in γ, is located >40 Å from any of the mapped nucleotide-binding sites. Second, the MF_1 crystal structure predicts that γ-Cys-89 and γ-Cys-322 should be about 70 Å apart. Yet, FRET mapping gives a strong indication that the two residues are close together.

Several of the elements of the FRET map have been independently confirmed by biochemical evidence. For example, cleavage of γ by trypsin is strongly prevented by the presence of the ε subunit.[7] Once γ is cleaved, ε binds poorly, if at all.[17] The structural mapping studies indicate that ε and γ are close to each other. Under appropriate conditions, Cys-89 and Cys-322 of γ may be cross-linked. The least-squares minimization of the FRET structure reveals that the two γ cysteine residues are ~8 Å apart.

To date, no FRET mapping data for δ have been published. The δ subunit is devoid of cysteine and, until recently, it has been difficult to obtain δ that was labeled specifically. By subcloning site-directed mutagenesis, expression, and folding of δ, it is now possible to obtain δ with cysteine substituted for specific serine residues. Structural mapping of the δ subunit by FRET is in progress.[18]

Fluorescence resonance energy transfer may be combined with site-directed mutagenesis to obtain more detailed maps. For example, serine-to-cysteine mutations could be introduced into specific sites on the ε, δ, and γ subunits of CF_1. These proteins may be overexpressed and folded to active forms. Thus, specific labeling of the mutagenized polypeptides and their reconstitution with appropriately labeled α/β cores can be used to enhance the resolution of the FRET map.

[15] K. Colvert, D. A. Mills, and M. L. Richter, *Biochemistry* **31,** 3930 (1992).

[16] J. B. Abrahams, A. G. W. Leslie, R. Lotter, and J. E. Walker, *Nature* (*London*) **230,** 621 (1994).

[17] P. Soteropoulos, K.-H. Süss, and R. E. McCarty, *J. Biol. Chem.* **267,** 10348 (1992).

[18] S. Engelbrecht, Personal communication (1995).

There are drawbacks to FRET mapping of proteins that extend beyond the necessity to label sites with great specificity. A problem that is often construed as more serious than it can be, is the so-called orientation factor. The efficiency of FRET depends on the orientation of the fluorescent donor and the acceptor. This is the κ^2 term, which in structural mapping studies of CF_1 was assumed to be the average value of 2/3. In some cases, at least, the fluorescence of probes bound to specific sites on CF_1 is strongly polarized, making the average value of κ^2 unlikely. Fortunately, as long as distances in the range of $\geq$20 Å are to be determined, the error in the distances calculated by assuming a κ^2 of 2/3 is small. A second problem at one time was the lack of ready availability of donor and acceptor molecules of high quality that covered a broad range in useful distance measurements. Donor/acceptor molecules that form pairs with R_0 values (the distance between the donor and the acceptor at which the efficiency of energy transfer is 50%) in excess of 50 Å may now be purchased.

Applications of Fluorescence Resonance Energy Transfer to Mechanistic Studies

The asymmetry of CF_1 is also the basis for the asymmetry of the properties of its nucleotide-binding sites. At least two of the six nucleotide-binding sites of CF_1 are catalytic and likely interact in an unusual way.[19] The sites are negatively cooperative with respect to binding affinity, but positively cooperative with respect to catalysis. In essence, the catalytic sites alternate their properties. A "loose site," characterized by a K_D on the order of the micromolar range, binds substrate and promotes catalysis in a site to which substrate is "tightly bound." Concurrently, the site to which medium substrate had bound (the "loose site") becomes a "tight site" (slow release) and product in the former tight site is released into the medium as a consequence of a major decrease in substrate/product binding affinity. Simply put, two sites in CF_1 (or more) are likely to switch their properties as a result of nucleotide (substrate) presence in the medium.

It is difficult to obtain evidence for switching of the properties of the nucleotide-binding sites of CF_1 as a result of either catalysis (ATP hydrolysis) or substrate (ATP) binding. A novel use of FRET, in combination with fortuitous locations of two of the nucleotide-binding sites, provided an approach. Previously, it was mentioned that Lucifer Yellow VS reacts rapidly with Lys-378 of one of the three α subunits. Thus, we had a fixed point of reference on the α/β heterohexamer. The Lucifer Yellow is 64 Å

[19] P. D. Boyer, *FASEB J.* **3,** 2164 (1989).

from a site that binds TNP-ADP tightly, but just 36 Å from a site that binds TNP-ATP in a freely dissociable manner (K_D, about 2 μM). When there is 1 mol of bound TNP-ADP per mole of sites, the energy transfer efficiency from Lucifer Yellow to TNP-ATP in the tight site would be only about 2–3%, whereas that between Lucifer Yellow and TNP-ADP at the dissociable site would be about 40%. These values hold true for CF_1 that had not been exposed to ATP for several hours prior to FRET assay.

Remarkably, incubation of Lucifer Yellow-labeled CF_1 with Mg^{2+}-ATP prior to TNP-ATP loading of the tight site significantly increased the efficiency of energy transfer between the Lucifer Yellow and tightly bound TNP-ATP.[20] The degree of energy transfer between TNP-ATP bound to the dissociable site was decreased. Thus, as a result of exposure of CF_1 to Mg^{2+}-ATP, at least two of the substrate-binding sites switch their properties. This result is in accordance with the idea that there is catalytic site negative/positive cooperativity. That is, sites switch their properties when exposed to substrate in the medium.

Use of Fluorescence Resonance Energy Transfer to Estimate Binding Affinity of CF_1 for Thioredoxin

Thioredoxins are small (11 kDa) proteins that have several cellular functions, including the conversion of ribonucleotides to deoxyribonucleotides. In chloroplasts, there are multiple forms of thioredoxin. A major role of thioredoxins in chloroplasts is to mediate the reduction of critical disulfide bonds in several enzymes.[21] Thioredoxin is reduced in light by electrons from the chloroplast electron transport chain and in turn reduces the enzymes. CF_1 is one of the enzymes reduced in light by thioredoxin. Reduction enhances the rate of ATP synthesis by the enzyme, especially at physiological values of the electrochemical proton potential.

In Fig. 2, an example of the use of FRET to examine thioredoxin–CF_1 interactions is illustrated. CF_1 was labeled on Cys-322 of its γ subunit and thioredoxin (ThR) was labeled on unspecified $-NH_2$ groups by eosin isothiocyanate. A fixed concentration of CF_1 (0.1 μM) was titrated with the concentrations of ThR shown in Fig. 2. The concentration of the ThR–CF_1 complex was calculated from the relationship Q/Q_{max} times the total CF_1 concentration, where Q stands for "quenching" and Q_{max} is the quenching at infinite labeled ThR concentration. The line through the data points is a fit to the quadratic solution of the equilibrium binding equation. The good fit ($R = 0.997$) indicates that CF_1 and ThR form a one-to-one complex

[20] A. B. Shapiro and R. E. McCarty, *J. Biol. Chem.* **265,** 4340 (1990).

[21] B. B. Buchanan, *Annu. Rev. Plant Physiol.* **31,** 341 (1980).

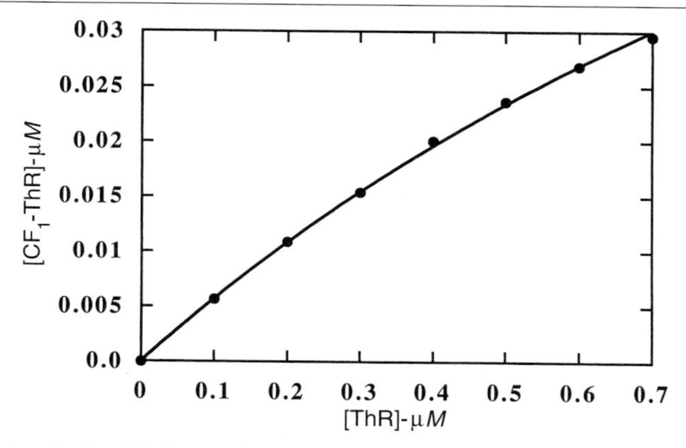

FIG. 2. Thioredoxin–CF_1 interactions quantitated by FRET. As detailed in Ref. 12, thioredoxin was labeled with eosin isothiocyanate and the γ subunit was labeled at Cys-322 of γ with fluoresceinylmaleimide. The R_0 for this donor (fluoresceinylmaleimide) and acceptor pair is 53 Å. Labeled CF_1 deficient in its ε subunit (0.1 μM) was titrated with ThR modified with eosin isothiocyanate. The amount of ThR bound to CF_1 as a function of ThR concentration was determined as outlined in text. The line through the data points is a fit ($R = 0.997$) of the data to the quadratic solution of the equilibrium equation in terms of the total CF_1 and ThR concentrations. the K_D value is 0.9 μM. (The data are previously unpublished data obtained by M. S. Dann.)

with a K_D close to 1 μM (0.9 ± 0.002). Thus, ThR is only weakly associated with CF_1 and its absence from the isolated, purified enzyme is explained.

A combination of FRET analysis with subunit depletion studies can be revealing. The accessibility of a particular region (residues 199–205) of the γ subunit[7] to either reducing agents or to selective proteolysis is markedly enhanced by removal of the ε subunit. The formation of a transmembrane electrochemical potential has similar effects. The rate of reduction of the γ disulfide bond is, for example, increased when an electrochemical proton potential is formed across the thylakoid membrane. There are several lines of evidence to suggest that the γ and ε subunits interact strongly.

Removal of the ε subunit of CF_1 (CF_1-ε) enhanced the efficiency of energy transfer at all labeled ThR concentrations. At all ThR concentrations, the efficiency of energy transfer was doubled by removal of ε. The binding constant for ThR to CF_1-ε was not significantly different from that to CF_1. Thus, the increase in energy transfer between fluorescein on Cys-322 of γ and eosin attached to ThR resulted from a closer approach of the ThR to the γ subunit. This simple FRET result explains why removal of ε promotes reduction of the γ disulfide. ε can hinder the approach of ThR.

Dynamic versus Static Structural Analysis

Although crystallographic analysis of the structure of the F_1 part of ATP synthases has revealed partial, static structures, an attempt was made[16] to interpret the structure with respect to a dynamic mechanism. This interpretation is speculative. Fluorescence spectroscopy has great promise in resolving kinetic events surrounding the regulation and mechanism of the ATP synthase. For example, the ε and γ subunits likely move with respect to each other during the conversion by the electrochemical proton gradient of the chloroplast ATP synthase to an active form.[22] The kinetics and properties of these energy-dependent changes could be monitored by phase-modulated FRET. These changes take place in the first several milliseconds after the initiation of electron transport. Also, stopped-flow fluorescence measurements may be used to monitor substrate-induced movements between parts of the γ subunit and the $\alpha_3\beta_3$ core. There are thus applications of FRET to the dynamic structure of the chloroplast ATP synthase.

[22] M. S. Dann and R. E. McCarty, *Plant Physiol.* **99,** 153 (1992).

[28] Intrinsic Fluorescence of Hemoglobins and Myoglobins

By ZYGMUNT GRYCZYNSKI, JACEK LUBKOWSKI, and ENRICO BUCCI

Introduction

In proteins, an excited luminophore may lose its electronic excitation energy either by emitting radiation or by nonradiative deexcitation. In the presence of other chromophores it may transfer its excitation energy to a suitable acceptor chromophore by a nonradiative resonance process.

Almost half a century ago Förster developed an exact quantum mechanical theory of resonance excitation energy transfer for weak dipole–dipole interactions.[1,2] This technique is suitable for the determination of macromo-

[1] Th. Förster, *Ann. Phys.* **2,** 55 (1948).

[2] Th. Förster, Delocalized excitation and excitation transfer. *In* "Modern Quantum Chemistry" (O. Sinanoglu, ed.), Part III, pp. 93–137. Academic Press, New York, 1965.

0076-6879/97 $25

lecular dimensions and for investigating conformational dynamics.[3–10] In practice, radiationless interaction affects fluorescence intensities and lifetimes of both donor and acceptor chromophores.

In hemoglobin and myoglobin radiationless excitation energy transfer from tryptophan residues to nonfluorescent hemes can be monitored by observing the emission characteristics of the donor tryptophans. These hemoproteins are highly quenched systems where the transfer reduces the intensity of tryptophan emission to levels approachable only with modern instrumentation. The time resolution of these weak signals has been made possible by the availability of laser excitation sources, ultrafast detectors, and data collection systems.[11–14]

At present, knowledge of the transition moment orientations for tryptophan emission and heme absorption, which are responsible for excitation energy transfer, and the availability of the atomic coordinates for many hemoglobin and myoglobin systems allow explicit analysis of the parameters responsible for radiationless tryptophan–heme interaction. Therefore we can compute the expected residual lifetimes that identify the various molecular conformers of hemoglobin and myoglobin. In this way we can now monitor conformational attitudes and their changes in myoglobin and hemoglobin systems previously beyond the reach of experimental approaches.

Deactivation of Tryptophan Excited State in Hemoglobins and Myoglobins

In a typical experiment the tryptophan chromophore is excited by ultraviolet (UV) light (290–300 nm), usually from a frequency doubler coupled

[3] L. Stryer and R. Haugland, *Biochemistry* **58,** 719 (1967).

[4] R. Haugland, J. Yguerabide, and L. Stryer, *Chemistry* **63,** 23 (1969).

[5] I. Z. Steinberg, *Annu. Rev. Biochem.* **40,** 161 (1971).

[6] R. E. Dale and J. Eisinger, *Biopolymers* **13,** 1573 (1974).

[7] E. Katchalski-Katzir and I. Z. Steinberg, *Ann. N.Y. Acad. Sci.* **366,** 44 (1981).

[8] P. W. Schiller, *In* "Biochemical Fluorescence: Concepts" (R. F. Chen and H. Edelhoch, eds.), Vol. 1, p. 285. Marcel Dekker, New York, 1975.

[9] P. Wu and L. Brandt, *Biochemistry* **31,** 7939 (1992).

[10] H. C. Cheung, *In* "Topics in Fluorescence Spectroscopy: Principles" (J. R. Lakowicz, ed), Vol. 2. Plenum, New York, 1994.

[11] R. F. Stainer, G. Holtom, and Y. Kubota, Laser spectroscopy of nucleic acid complexes. *In* "Lasers in Polymer Science and Technology: Applications" (J. P. Fouassier and J. F. Rabel, eds.), 4th Ed. CRC Press, Boca Raton, Florida, 1989.

[12] G. R. Fleming, "Chemical Application of Ultrafast Spectroscopy." University Press, New York, 1986.

[13] G. Laczko, I. Gryczynski, Z. Gryczynski, W. Wiczk, H. Malak and J. R. Lakowicz, *Sci. Instrum.* **61,** 2331 (1990).

[14] B. Barbieri, F. De Piccoli, M. Van de Van, and E. Gratton, Time-resolved laser spectroscopy. *In* "Biochemistry II," Vol. 1204, pp. 158–170. SPIE, Bellingham, Washington, 1990.

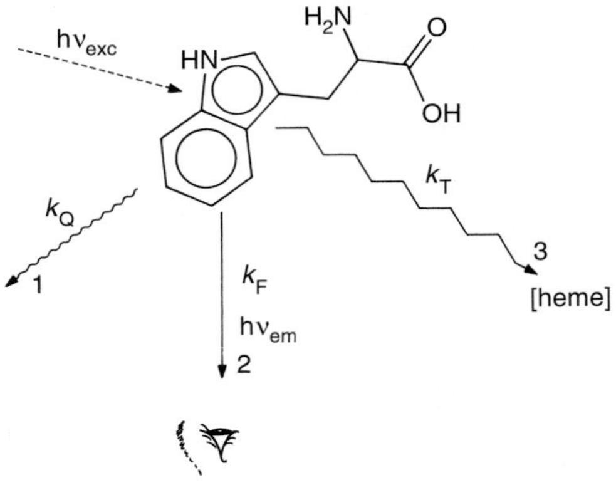

FIG. 1. Possible deactivation processes affecting excited state of donor (tryptophan) in the presence of energy transfer to the heme. k_F, Decay rate of fluorescence; k_Q, rate of internal and external conversion processes; k_T, rate of radiationless excitation energy transfer process.

to a rhodamine laser. The acquired excitation energy is deactivated in three principal decay processes. In a simplified form they are depicted in Fig. 1. Pathway 1 with rate k_Q is a nonradiative form of thermal deactivation (internal conversion plus external conversion processes). Pathway 2, with rate k_F, is the radiative form detectable as fluorescence emission. The third energy pathway with rate k_T is the nonradiative excitation energy transfer to a suitable acceptor, such as a heme for hemoglobins and myoglobins.

The decay in time of the observable emission intensity, I, is characterized by its lifetime τ, which regulates the exponential form of the decay as in[15]

$$I_t = I_0 \exp(-t/\tau) \tag{1}$$

It should be stressed, however, that the decay may be not exponential, and most often can be analyzed assuming a sum of exponentials

$$I_t = I_0 \sum_i \alpha_i \exp(-t/\tau_i) \tag{2}$$

where α_i is the preexponential factor representing the fractional contributions of the particles emitting with the lifetime τ_i.

In turn, the lifetime is related to the three deactivation rates shown in Fig. 1 by

$$\tau = \frac{1}{k_Q + k_F + k_T} \tag{3}$$

[15] J. R. Lakowicz, "Principles of Fluorescence Spectroscopy." Plenum, New York, 1983.

The efficiency of the radiationless energy transfer process[3–10,15] E can be defined as

$$E = \frac{k_T}{k_Q + k_F + k_T} \tag{4}$$

The transfer efficiency E regulates the relative fluorescence yield observed in the presence (ϕ_T) and absence of energy transfer (ϕ_0)

$$\phi_T = \phi_0(1 - E) \tag{5}$$

For donor–acceptor systems with high energy transfer efficiency the relative fluorescence yield ϕ_T is small and the observed fluorescence intensity is correspondingly weak.

Hemoglobin and myoglobin are highly quenched systems where the efficiency of excitation energy transfer from tryptophan to heme is $E > 0.97$, which implies that more than 97% of the excitation energy of tryptophans is transferred to hemes. This is the reason for their weak fluorescence and short (picosecond) lifetimes.

Fluorescence Lifetime in Presence of Excitation Energy Transfer

The Förster theory for weak dipole–dipole interaction between donor and acceptor establishes a precise relationship between the rate of energy transfer, k_T; the spectroscopic properties of both fluorophores; and their separation, R, and the lifetime of donor emission in the absence of acceptor, τ_D, as in

$$k_T = \frac{1}{\tau_D}\left(\frac{R_0}{R}\right)^6 \tag{6}$$

where R_0 is the classic Förster distance at which transfer depopulates the donor excited state at the same rate as the sum of all other deactivation processes ($k_T = k_F + k_Q$); i.e., when the transfer efficiency is $E = 0.5$.

The constant R_0 is a characteristic of the energy transfer system, which can be computed from[1–10]

$$R_0^6 = 8.785 \times 10^{-25} \kappa^2 n^{-4} \varphi_0 J \tag{7}$$

where n is the refractive index of the solvent, ϕ_0 is the quantum yield of the donor in the absence of acceptors, κ^2 is the orientation factor of the donor–acceptor transition moments, and J is the overlap integral between the normalized emission spectrum of the donor and the absorption spectrum of the acceptor.

From Eqs. (3) and (6) it is possible to infer that the residual lifetime of the system, τ_c, after energy transfer, is

$$\tau_c = \frac{\tau_D}{1 + \sum_i (R_0/R_i)^6} \tag{8}$$

The ratios $(R_0/R_i)^6$ are the transfer rate factors between the donor and each of the molecules that can accept transfer from it. In myoglobin each tryptophan has only one heme that accepts transfer. In tetrameric hemoglobin each tryptophan has four hemes that accept transfer.

For computing τ_c it is necessary to evaluate the spectroscopic parameters involved in Eqs. (7) and (8). The quantum yield and the lifetime of the donor in the absence of energy transfer can be estimated from previous steady-state and time-resolved fluorescence studies on different systems.[15–18] Donor–acceptor separation, the overlap integral J, and the orientation factor κ^2 are specific for each system and must be evaluated.

Overlap Integral in Hemoglobin and Myoglobin

The overlap integral is defined by

$$J = \int_0^\infty \frac{I(\nu)\varepsilon(\nu)\,d\nu}{\nu^4} \tag{9}$$

where $I(\nu)$ is the normalized donor emission spectrum, $\varepsilon(\nu)$ is the extinction coefficient of the acceptor, and ν is the wave number.

Figure 2 shows the absorption spectra of met-, CO-, and deoxymyoglobin together with the corrected emission spectrum of heme-free globin. It shows an evident overlap of tryptophan emission with the near-UV region of absorption, between 300 and 400 nm. Deconvolution in terms of Gaussian components of the absorption spectra of either free hemes (namely CO-heme and Mg-heme) or of several hemoglobin and myoglobin derivatives is shown in Fig. 3. Deconvolved spectra show the presence of three Gaussian components in the 380 to 450-nm region, and only one component in the 300 to 380-nm region.[19] As shown in Figs. 2 and 3 the presence of ligands does not greatly influence the heme absorption spectrum in the region where it overlaps with the emission of tryptophan.

Table I presents values of overlap integrals computed for several deriva-

[16] J. M. Beechem and L. Brand, *Annu. Rev. Biochem.* **54,** 43 (1985).

[17] A. P. Demchenko, "Ultraviolet Spectroscopy of Proteins." Springer-Verlag, Berlin, 1986.

[18] M. Eftink, *In* "Topics in Fluorescence Spectroscopy: Principles" (J. R. Lakowicz, ed.), Vol. 2. Plenum Press, New York, p. 53, 1994.

[19] Z. Gryczynski, E. Bucci, and J. Kusba, *Photochem. Photobiol.* **58,** 492 (1993).

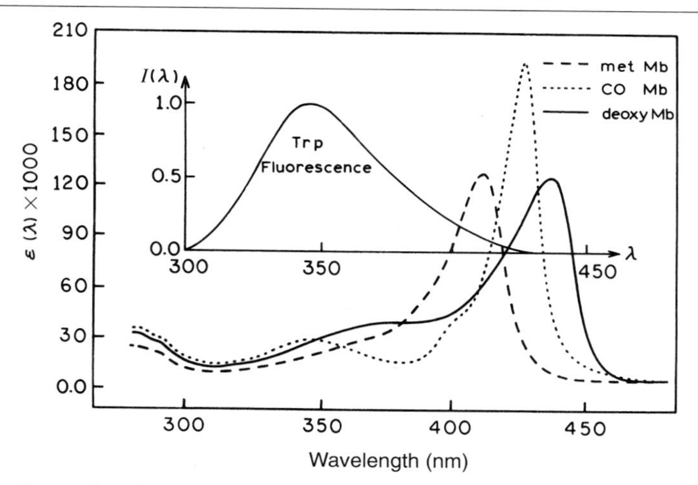

FIG. 2. Comparison between the emission spectrum of tryptophan (apomyoglobin) and the absorption spectra of met-, CO-, and deoxymyoglobin, showing the extensive overlap.

tives of hemoglobin and myoglobin.[20,21] They are similar in spite of the substantial difference of the absorption above 380 nm (the Soret band) for the various derivatives. This confirms that the energy transfer process is almost completely regulated by the absorption band of the heme in the near-UV region, between 300 and 380 nm. It should be noted that this absorption is produced by a single absorption band. This fact is extremely relevant because it implies that a single transition moment of the heme is the acceptor of radiationless energy transfer from tryptophan. This allows a precise determination of the orientation factor for tryptophan and heme in hemoglobin and myoglobin.

Evaluation of Orientation Parameter κ^2 in Hemoglobin and Myoglobin

According to the Förster theory the orientation factor κ^2 for dipole–dipole radiationless interaction is a function of the angle (α_{DA}) as defined by the direction of the donor transition moments, D, and acceptor, A, and of the angles (α_D, α_A), that the two vectors form, respectively, with the translation vector **T** connecting the centers of the two oscillators[4–7,20]:

$$\kappa^2 = (\cos \alpha_{DA} - \cos \alpha_D \cos \alpha_D)^2 \tag{10}$$

[20] Z. Gryczynski, T. Tenenholtz, and E. Bucci, *Biophys. J.* **63,** 648 (1992).
[21] Z. Gryczynski, C. Fronticelli, T. Tenenholtz, and E. Bucci, *Biophys. J.* **65,** 1951 (1993).

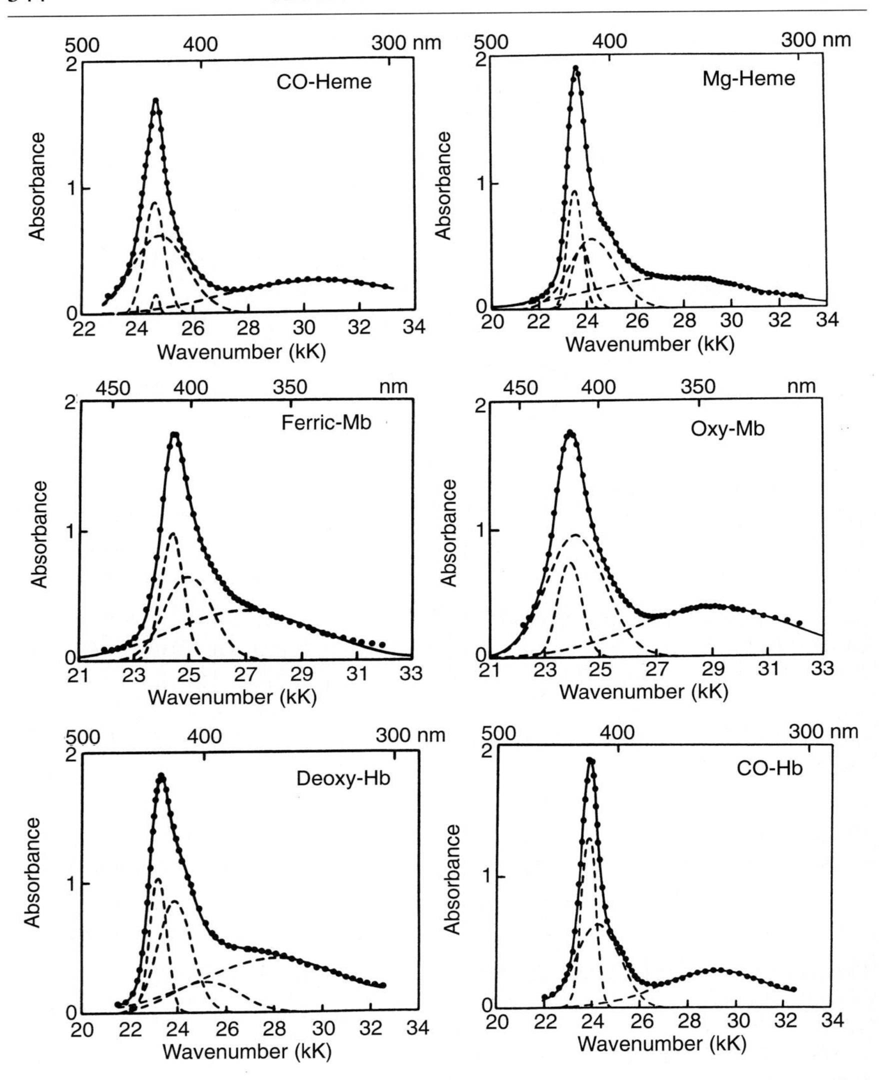

FIG. 3. Absorption spectra in the near-UV and Soret range of CO-heme and Mg-heme in isotropic polyvinyl alcohol (PVA) film, myoglobin and hemoglobin derivatives in 0.05 phosphate buffer, pH 7.0, at room temperature. The measured values are given by filled circles and the Gaussian components by dashed lines.

TABLE I
OVERLAP INTEGRALS FOR SEVERAL DERIVATIVES OF SPERM WHALE MYOGLOBIN AND HUMAN HEMOGLOBIN

Protein	Form	Overlap integral (cm/M)
Myoglobin (sperm whale)	Met	5.16×10^{14}
	Deoxy	5.28×10^{14}
	CO	4.70×10^{14}
Hemoglobin A_0 (human)	Oxy	5.75×10^{14}
	Deoxy	5.89×10^{14}
	CO	5.56×10^{14}

In flexible systems total averaging of the reciprocal orientation of donor and acceptor chromophores occurs on a time scale shorter than the transfer time, allowing the use of $\kappa^2 = 2/3$. In hemoglobin and myoglobin the quenched lifetimes are in the picosecond time range, and are too short for allowing dynamic averaging of the relative donor–acceptor positions, whose motions are also strongly restricted by the rigidity and stereochemistry of the protein matrix. In the picosecond time scale, these systems must be considered rigid structures. Therefore the orientation factor must be computed with the greatest possible accuracy. This requires a precise estimation of the reciprocal orientation of donor–tryptophan and acceptor–heme transition moments.

Transition Moment Orientation of Tryptophan

Extensive polarized spectroscopic studies of indole derivatives have been made both because of its intrinsic interest and because of the importance of tryptophan in protein fluorescence. There is general agreement that the near-UV absorption of indole, tryptophan, and their derivatives arises from two overlapping $\pi\pi^*$ transitions,[22–28] ${}^1L_a \leftarrow {}^1A_1$ and ${}^1L_b \leftarrow {}^1A_1$. Studies on crystals of 3-indolylacetic acid indicated that the direction of the ${}^1L_a$ and ${}^1L_b$ transition moments are at $-38°$ and $+54°$, respectively,

[22] Y. Yamamoto and J. Tanaka, *Bull. Chem. Soc. Jpn.* **65,** 1362 (1972).
[23] B. Valeur and G. Weber, *Photochem. Photobiol.* **25,** 441 (1977).
[24] J. E. Hensen, J. Rosenthal, and G. R. Fleming, *J. Phys. Chem.* **96,** 3034 (1992).
[25] A. J. Ruggiero, D. Todd, and G. R. Fleming, *J. Am. Chem. Soc.* **112,** 1003 (1990).
[26] A. Suwaiyan and R. Zwarich, *Spectrochem. Acta* **43A,** 605 (1987).
[27] B. Albinsson, M. Kubista, B. Norden, and E. Thulstrup, *J. Phys. Chem.* **93,** 6646 (1989).
[28] B. Albinsson and B. Norden, *J. Phys. Chem.* **96,** 6204 (1992).

from the major axis of tryptophan ring.[22] Studies of indole derivatives in stretched polymer films[26–28] confirm this finding and show only minor perturbations of the transition moment orientations by different substituents.[28]

Subpicosecond fluorescence depolarization studies of tryptophan and tryptophanyl residues of proteins[24,25] demonstrate that interconversion between ^1L_a and ^1L_b occurs in the first few picoseconds of excited-state decay. Our fluorescence polarization studies on hemoglobin systems[29] also confirm the presence of an ultrafast anisotropy decay component in the range of 2–5 psec.

The most often used excitation wavelength for tryptophan fluorescence is close to 300 nm, which preferentially populates the ^1L_a state. Also, the fast decay of the ^1L_b excited state in practice limits the interactions of tryptophan with heme to the ^1L_a excited state. For these reasons it can be safely assumed that the orientation of the donor ^1L_a transition moment is in the plane of tryptophan at $-38°$ from the major axis of the tryptophan ring.[20–28]

Orientation of Transition Moments of Heme

Orientation of the transition moments of the near-UV heme absorption band is a critical parameter for evaluating the orientation parameter κ^2 of the tryptophan–heme (donor–acceptor) system. The spectroscopic properties of porphyrins and metalloporphyrins have been extensively investigated, both theoretically and experimentally. The classic free electron or LCAO (linear combination of atomic orbitals) molecular orbital model[30–33] properly describes the frequencies and intensities of their π-electronic transitions. These theoretical calculations show that in porphyrin with fourfold symmetry, the allowed transitions are polarized parallel to the symmetry axes in the plane of the molecule. Experimental studies of fluorescence polarization,[34] fluorescence in Shpolski-type matrixes,[35–37] and Fourier-transform spectroscopy in noble gas matrixes[38] clearly indicate that the

[29] E. Bucci, Z. Gryczynski, C. Fronticelli, I. Gryczynski, and J. R. Lakowicz, *J. Fluoresc.* **2,** 29 (1992).

[30] W. T. Simpson, *J. Chem. Phys.* **17,** 1218 (1949).

[31] M. Gouterman, *J. Mol. Spectrosc.* **6,** 138 (1961).

[32] C. Weiss, H. Kobayashi, and M. Gouterman, *J. Mol. Spectrosc.* **16,** 415 (1965).

[33] J. V. Knop and A. Knop, *Z. Naturforsch.* **25,** 1720 (1970).

[34] M. Gouterman and L. Stryer, *J. Chem. Phys.* **37,** 2260 (1962).

[35] S. Volker and R. M. Macfarlane, *J. Chem. Phys.* **73,** 4476 (1980).

[36] L. L. Gladkow, A. T. Gradyushko, A. M. Shulga, K. N. Solovyov, and S. Starukhin, *J. Mol. Struct.* **47,** 463 (1978).

[37] U. Even, J. Jortner, and Z. Berkovitch-Yellin, *Can. J. Chem.* **63,** 2073 (1985).

[38] J. G. Radziszewski, J. Waluk, and J. Michl, *J. Mol. Spectrosc.* **140,** 373 (1990).

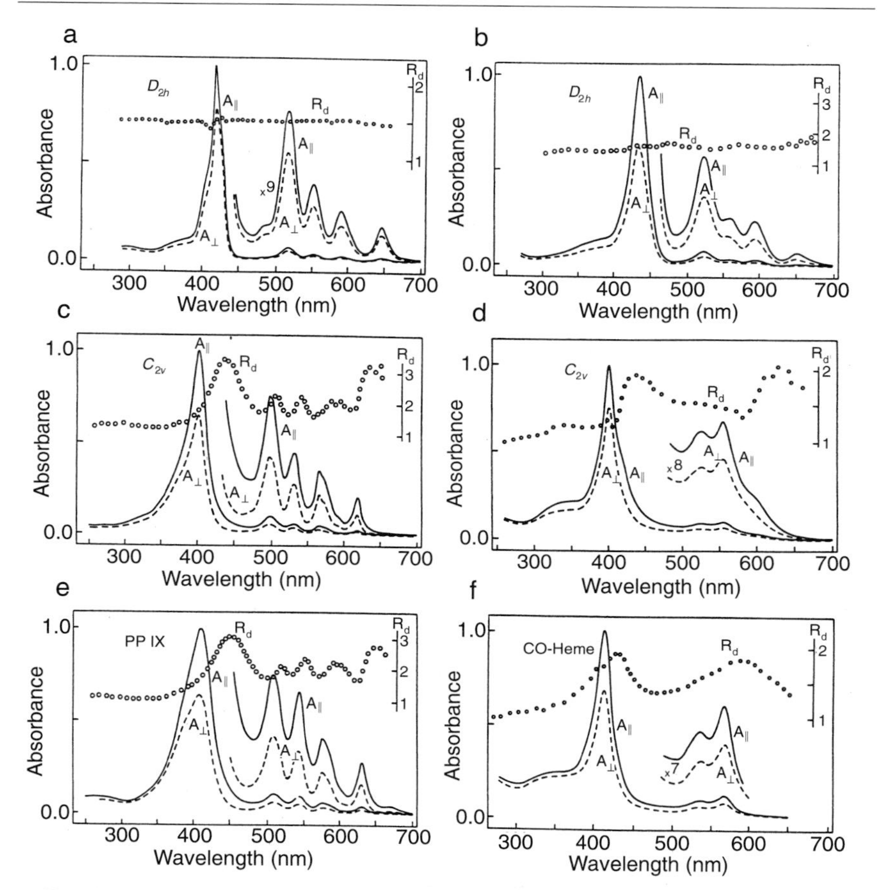

FIG. 4. Absorbance components, $A_\parallel/A_\perp$, and dichroic ratio $R_d = A_\parallel/A_\perp$, as a function of the wavelength of various protoporphyrin derivatives in stretched PVA films.

electronic transitions in porphyrin derivatives are linearly polarized in the plane of the molecule. We have studied polarized absorption spectra of different porphyrin derivatives to determine whether metal substitutions in the center of the porphyrin ring would alter the electronic structure of the transition moments.[39] The linear dichroism of several porphyrin derivatives with different symmetries was investigated in stretched PVA films. In Fig. 4 we present the linear dichroism spectra of fully symmetric (D_{4h})[39a]

[39] Z. Gryczynski, R. Paolesse, K. M. Smith, and E. Bucci, *J. Phys. Chem.* **98,** 8813 (1994).
[39a] The point symmetry of porphyrin is D_{2h}; however, in stretched PVA films the apparent symmetry is in this case equivalent to D_{4h} symmetry. This was already noted by Guterman.[31]

$\alpha,\beta,\gamma,\delta$-tetraphenylsulfonic porphyrin, and $\alpha,\beta,\gamma,\delta$-tetrakis(1-methyl-4-pyridyl-21*H*,23*H*-porphyrin) tetra-*p*-tosylate salt; of deuteroprotoporphyrin III, protrohemin III with C_{2v} symmetry; and of protoporphyrin IX (PP IX) and heme (no symmetry axis). As anticipated, the observed linear dichroism (Fig. 4) of fully symmetric $\alpha,\beta,\gamma,\delta$-tetraphylsulfonic porphyrin and $\alpha,\beta,\gamma,\delta$-tetrakis(1-methyl-4-pyridyl-21*H*,23*H*-porphyrin) tetra-*p*-tosylate salt are wavelength independent. Deuteroprotoporphyrin III, with C_{2v} molecular symmetry, has a linear dichroism spectrum strongly wavelength dependent, indicating the energetic split of the polarization of the transition moments of the porphyrin into two orthogonal and nonequivalent vectors parallel and perpendicular, respectively, to the main symmetry axis. Notably, addition of iron in the center of C_{2v} symmetry protoporphyrins (proto- and deuterohemin III) does not change the characteristics of their linear dichroism. The linear dichroism of protoporphyrin IX (no symmetry axis) is similar to that of a C_{2v} symmetry porphyrin. Metal substitutions at the center of the ring (Fe) and the presence of ligands (CO-heme) do not introduce substantial changes in the linear dichroism. Also, similar behavior of the dichroism was observed for Mg- and Zn-hemes.[19] This indicates that the metal substitutions may introduce only small out-of-plane components to the electronic transitions. Thus it appears that in porphyrins and metalloporphyrins the electronic transition moments lay in the plane of the porphyrin ring. It should be stressed that in all porphyrin derivatives the linear dichroism between 300 and 380 nm is independent of wavelength, consistent with the observation that in this spectral region optical absorption is produced by a single absorption band, with a single transition moment, positioned essentially in the plane of the porphyrin. As discussed above, this transition moment is the acceptor of the excitation energy transfer from tryptophans. The information that it lays on the plane of the porphyrin or close to it is important for the evaluation of transfer efficiency. The orientation of this transition moment relative to the α–γ-meso axis of the porphyrin ring (angle Θ) was initially evaluated using the Saupe parameters computed from the linear dichroism of the metal porphyrins reported above.[19] That estimation has been refined using the linear dichroism of the metal-free porphyrins to be in the interval $\Theta = 50$–$60°$. The low solubility of these porphyrin derivatives does not, at present, allow a more precise estimation. In the simulations presented here we have used average values of κ^2 obtained every 2° in the 50-to-60° interval. This relatively wide angle for κ^2 averaging may take into account possible wobbling of both tryptophan and heme around their standard positions in the crystal.

Computation of Orientation Parameter κ^2

Once the orientation of the transition moments of tryptophan and heme on the respective molecular frames has been evaluated, we can use the atomic coordinates of the protein for computing the orientation parameter κ^2. Also, from the atomic coordinates we can evaluate the tryptophan–heme separation necessary to calculate the rates of radiationless energy transfer between the two chromophores. We have developed an algorithm that reads Protein DataBank (PDB) files of atomic coordinates, selects the indicated tryptophan residues, computes their frame position relative to the frames of all of the hemes in the structure (proximal or distal position), creates pseudocoordinates for the transition moments on the tryptophan planes and on the porphyrin rings, computes the necessary angles and distances, and prints the resulting values of κ^2. To observe the variability of the orientation factor in various systems, the algorithm computes the values of orientation factors for a fixed position of the ^1L_a transition moment of the tryptophan, while the position of the transition moment of the heme, angle Θ, over the α–γ-meso axis of the porphyrin ring shown in Fig. 5, varies from 0° to +90°. Also, in order to mimic the presence of an inverted (disordered) heme (i.e., rotated 180° around the α–γ-meso axis of the porphyrin),[21,40–43] values of the orientation factor κ^2 are calculated for Θ varying from 0° to −90°. Some examples of this variability are shown in Fig. 6. Owing to the shape of the curves in the plots, we call these "fish plots." These plots indicated the relevance of the position of the heme (normal or disordered) to the values of the orientation factor, and consequently the value of R_0, used for computing the transfer efficiency. It should also be pointed out that within the range Θ = 50–60° (the shaded areas of the plots in Fig. 6), the value of κ^2 usually differs from 2/3, the value for a random orientation of donor and acceptor.

In these simulations we did not take into account possible fluctuations of the tryptophan residue whose transition moment for emission was fixed at 38° from the main axis of the indole ring. This assumption was supported by the observation of Hochstrasser and Negus[44] that the degree of freedom of tryptophan is limited by steric hindrances, so that high energy was necessary to obtain a reposition of tryptophan that would substantially affect the energy transfer to the heme.

Orientation Factor in Sperm Whale and Horse Heart Myoglobins

Figure 6 presents examples of the dependence of κ^2 on Θ for normal (N) and disordered (D) hemes in sperm whale (SW) ferric myoglobin

[40] G. N. La Mar, K. Smith, K. Gersonde, H. Sick, and M. Overkamp, *J. Biol. Chem.* **165,** 255 (1980).

[41] G. N. La Mar, N. L. Davis, D. W. Parish, and K. M. Smith, *J. Mol. Biol.* **168,** 887 (1983).

[42] Y. Yamamoto and G. N. La Mar, *Biochemistry* **25,** 5288 (1986).

[43] H. S. Aojula, M. T. Wilson, and A. Drake, *Biochem. J.* **237,** 613 (1986).

[44] R. M. Hochstrasser and D. K. Negus, *Proc. Natl. Acad. Sci. U.S.A.* **81,** 4399 (1984).

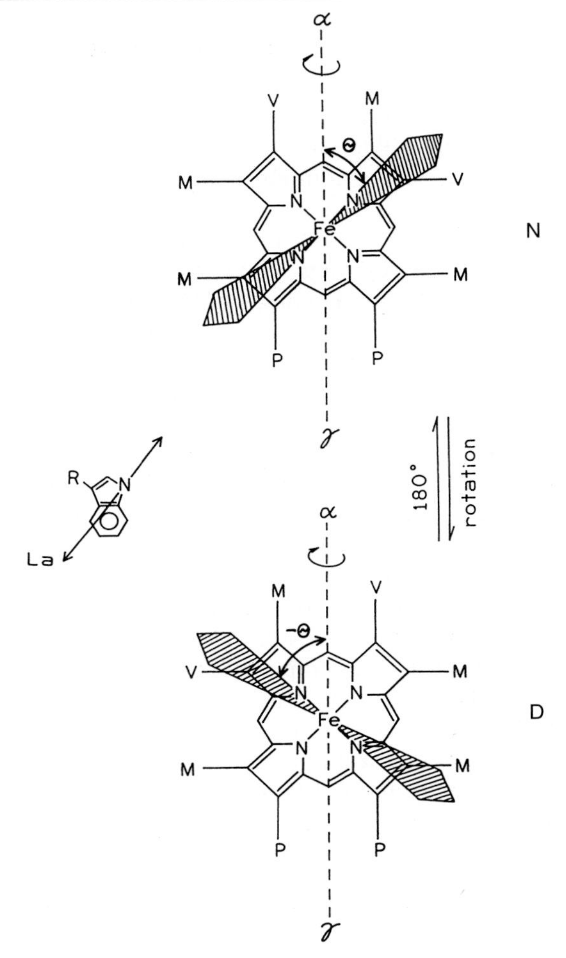

FIG. 5. Schematics of heme before and after 180° rotation around its α–γ-meso axis. Tryptophan moiety and hemes are arbitrarily positioned for emphasizing the drastic change of angular orientation of the respective transition moments. The proximal histidine is above the plane of the figure. N, Normal; D, disordered.

and deoxymyoglobin,[45–47] and horse heart[48] (HH) ferric myoglobins. Both tryptophans in position 7 (dashed lines) and position 14 (solid lines) show a strong dependence of κ^2 on the angle Θ. Consequently the values of the

[45] T. Takano, *J. Mol. Biol.* **110**(3), 569 (1977).
[46] S. E. Phillips, *Nature (London)* **273,** 247 (1978).
[47] S. E. Phillips, *J. Mol. Biol.* **142**(4), 531 (1980).
[48] S. V. Evans and G. D. Brayer, *J. Mol. Biol.* **213,** 885 (1990).

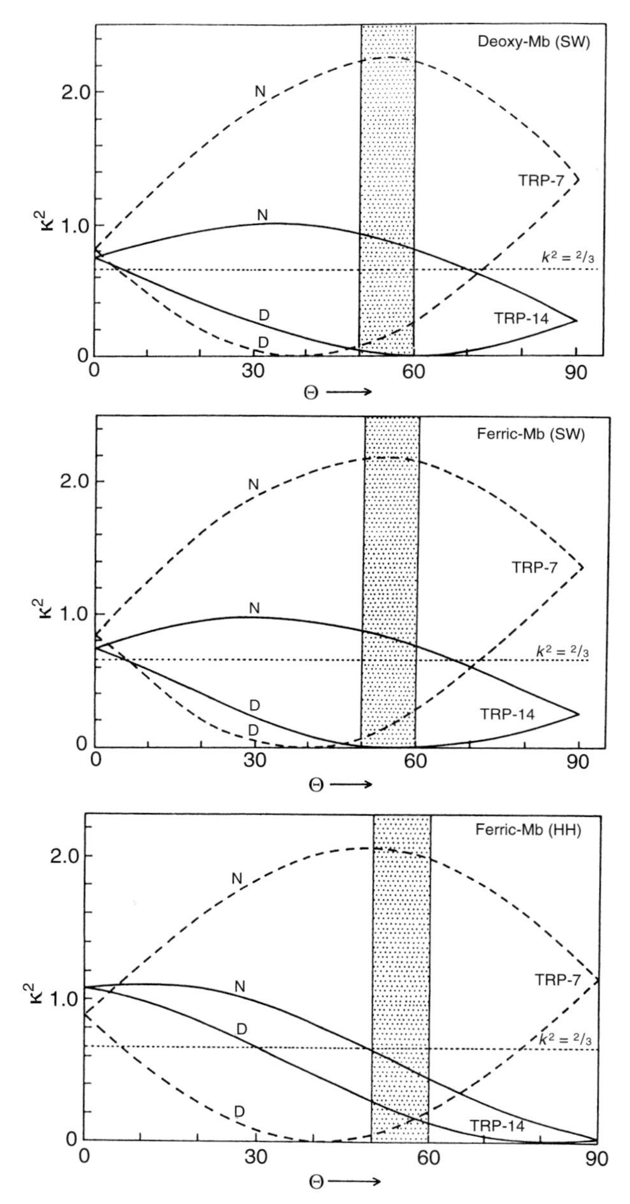

FIG. 6. Dependence of the orientation factor κ^2 on the angle Θ between the direction of the transition moment of the heme absorption and the α–γ-meso axis of the porphyrin ring for different myoglobin systems. Solid lines are for Trp-14 and dashed lines are for Trp-7. N, Normal heme orientation; D, disordered heme orientation. The horizontal dashed lines are for $\langle\kappa^2\rangle = 2/3$, and correspond to the isotropic distribution of interacting dipoles. The shaded areas show the range estimated for the angle Θ by linear dichroism measurements.

orientation factor are distinctly different for normal and disordered heme throughout all of the Θ range investigated. Computed values of orientation factors for several myoglobins are shown in Table II.

It is interesting to note that in spite of a similar tertiary structure, with tryptophans in identical positions, the fish plots for SW and HH myoglobin are different. This is due to a small displacement of about 15° of Trp-14 in HH myoglobin in comparison to the same residue in SW myoglobin. This displacement is illustrated in Fig. 7, where the hemes of the two structures are superimposed. This points out the sensitivity of this technology to conformational changes in these systems.

In Fig. 8 we present the fish plots for the interactions of Trp-14 α_1 of human hemoglobin with all of the hemes in the tetramer. We used the coordinates of Fermi *et al.* for deoxyhemoglobin and Shaanan for oxyhemo-

TABLE II
DISTANCES AND ORIENTATION FACTORS COMPUTED FOR Trp-7 AND Trp-14 FOR NORMAL AND DISORDERED HEME ORIENTATIONS

Mb Source[a]	Trp	Distance Trp–heme (Å)	Heme orientation	Orientation factor $\langle\kappa^2\rangle$
Met SW	7	21.40	N	0.479
			D	2.21
	14	15.00	N	0.813
			D	0.0097
Deoxy SW	7	21.30	N	0.495
			D	2.19
	14	14.83	N	0.813
			D	0.0098
CO SW	7	21.20	N	0.529
			D	2.50
	14	14.87	N	0.760
			D	0.0039
Met HH	7	21.41	N	2.0
			D	0.119
	14	15.36	N	0.549
			D	0.207
YFT	14	15.57	N	0.685
			D	0.068

[a] Sperm whale (SW), horse heart (HH), and yellow fin tuna (YFT) myoglobins.

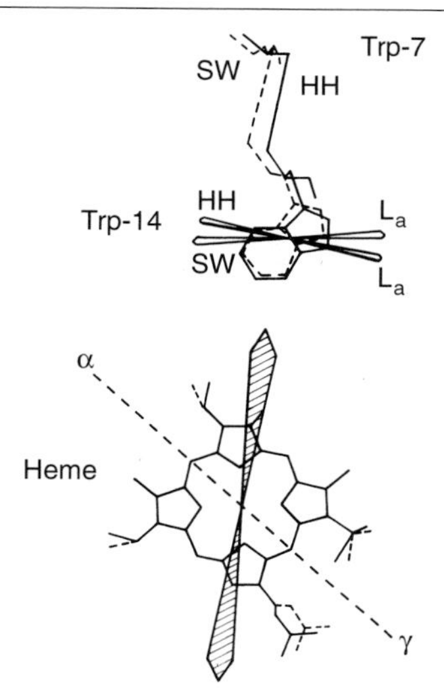

FIG. 7. Structures of hemes and tryptophans extracted from atomic coordinates of horse heart (HH) myoglobin (continuous lines) and sperm whale (SW) myoglobin (dashed lines). Heme rings from HH and SW myoglobins are superimposed and parallel to the plane of the figure.

globin.[49,50] In Table III we report the Trp–heme separations and the average values of orientation factors $\langle\kappa^2\rangle$, of the inter- and intrasubunit interactions of each tryptophan of deoxyhemoglobin with each heme. Intrasubunit tryptophan–heme separations are much smaller than those at intersubunit distances. For this reason the normal or disordered conformation of the heme has little relevance in terms of the intersubunit transfer rate. Therefore the intersubunit interactions were computed only for hemes in normal position.[50a] From Table III it can be seen that the tryptophans most sensitive to disordered hemes are those in positions α-14 and β-15. This is due to their higher proximity to the heme and to the quasi-parallel orientation of their planes to that of the intrasubunit heme.

[49] G. Fermi, M. F. Perutz, B. Shaanan, and R. Forume, *J. Mol. Biol.* **175,** 159 (1984).
[50] B. Shaanan, *J. Mol. Biol.* **171,** 31 (1981).
[50a] In this case intersubunit radiationless energy transfer in the dimer was computed assuming a heme in normal position in the partner subunit. In fact, the energy transfer to the partner subunit heme is regulated mostly by the distance. Also, the low fraction of disordered hemes makes the probability of finding a dimer with both hemes disordered remote.

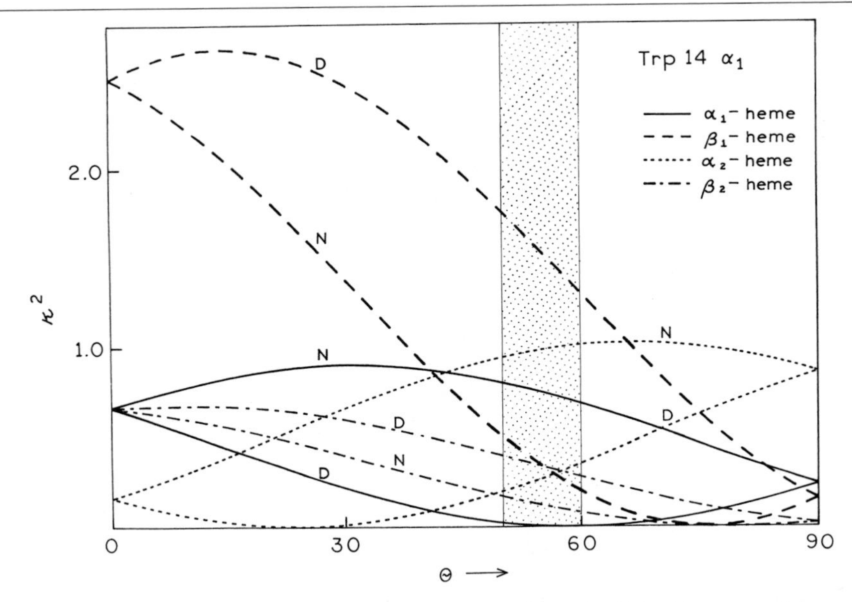

FIG. 8. Dependence of the orientation factor κ^2 on the angle Θ for Trp-14 α_1 of human hemoglobin in interaction with all four hemes in tetrameric hemoglobin.

Experimental Techniques

As discussed above, the emission intensity of hemoglobin and myoglobin is weak. Therefore artifacts that are negligible for the intensity of normal emitters become relevant and may significantly distort the signals. These include purity of the samples and all the optical imperfections such as reflections, stray light, and quartz or paint fluorescence.[51,52]

Purity of Samples

It should be stressed that, because of the weak emission of the hemoproteins, a 1% impurity of nonheme polypeptides containing tryptophans will result in long fluorescence lifetimes with steady state intensities comparable to those of the samples under consideration. Purification can be achieved by a number of chromatographic techniques. We successfully use DEAE ion-exchange HPLC techniques. Another useful technique is preparative free-phase isoelectrofocusing (Rotofor; marketed by Bio-Rad, Richmond,

[51] G. R. Holtom, Time-resolved laser spectroscopy. *In* "Biochemistry II," Vol. 1204, p. 2. SPIE, Bellingham, Washington, 1990.

[52] Z. Gryczynski and E. Bucci, *Biophys. Chem.* **48,** 31 (1993).

TABLE III
DISTANCES AND ORIENTATION FACTORS COMPUTED FOR TRYPTOPHANS α_1-14, β_1-15, AND β_1-37 FOR NORMAL AND DISORDERED HEME ORIENTATIONS[a]

Trp	Heme	Distance Trp–heme (Å)	Heme orientation[b]	Orientation factor $\langle\kappa^2\rangle$
α_1 − 14	α_1	15.24	N	0.718
			D	0.033
	α_2	36.77	N	1.225
			D	0.177
	β_1	34.63	N	1.598
			D	0.510
	β_2	38.66	N	0.277
			D	0.270
β_1 − 15	α_1	33.84	N	1.030
			D	0.511
	α_2	37.57	N	0.477
			D	0.125
	β_1	16.12	N	0.946
			D	0.015
	β_2	40.83	N	1.138
			D	0.123
β_1 − 37	α_1	25.91	N	0.497
			D	0.856
	α_2	14.59	N	1.020
			D	0.304
	β_1	16.31	N	1.944
			D	0.040
	β_2	34.66	N	0.124
			D	1.020

[a] From human deoxyhemoglobin.
[b] N, Normal; D, disordered.

CA). After focusing, the ampholyte can be removed from the protein by repetitive ultrafiltration with the buffer of choice. The ferric forms of hemoglobin and myoglobin are difficult to purify by either technique, even in the presence of ligands such as cyanide. In fact, they tend to release their ligands and, most important, their hemes to either the column matrix or to the buffer in the electric field. Even a small release of 1% hemes, difficult to detect in physicochemical and functional tests, would produce long lifetimes resulting in distorted interpretations of the data.

Square (Right-Angle) Geometry

Square (right-angle) geometry is the commonly used optical beam geometry, in which the excitation and observation channels are positioned at a

90° angle. Square cuvettes are used with variable pathlengths of 3–10 mm. In view of the Soret and near-UV absorption of hemoglobin and myoglobin, protein concentrations must be kept in the range 0.1–0.5 mg/ml, so as to avoid inner filter effects. With this optical arrangement, internal reflection from the walls of the square cuvette may distort the fluorescence signal.[51,52] Holtom proposes to use a 3-mm cuvette with two darkened walls,[51] one facing the incident light beam and the other opposite to the observation window. This helps with the reflections and allows the use of protein concentrations up to 2 mg/ml. The narrow cuvette requires small sample volumes but prevents efficient stirring. These cuvettes are commercially available.

We prefer to use what we have described as the "shielded" cuvette.[52] This is a regular 1-cm^2 cuvette, with black Kodak paper used to mask the windows in front of the excitation source and in front of the photomultiplier tube (PMT), respectively, leaving small apertures (1 mm^2 in excitation and 2×3 mm in emission) positioned near the edge of the cuvette (Fig. 9a). The position of the mask in emission can be adjusted at various distances from the edge, according to protein concentration (higher concentration, shorter distance). In place of the dark walls, as in the Holtom cuvette, we use two neutral-density filters positioned in a slanted way, opposite to the two small apertures (Fig. 9a), so as to absorb and reflect all stray light toward the top of the cuvette. On the bottom of the cuvette there is still space for a small magnetic stirrer. Also, the optical absorption of the solution can be observed on the same cuvette by sending the light beam above the filters. The position of the masks in the excitation and observation windows allows the use of protein concentrations up to 2 mg/ml.

In square geometry, care must be taken to filter out the Raman scattering of the buffer, whose intensity, in these systems, can be comparable to that of the emitted light.[52]

Front-Face Geometry

Front-face geometry technology is useful for studying the fluorescence response of samples with high optical density, which would not allow penetration of the incident beam and would have a strong inner filter effect. It is common to use a square geometry beams configuration by tilting a square cuvette[53,54] in the optical path of the fluorometers or by using triangular cuvettes.[55] This technique reduces the Raman scattering because of the low penetration of the incident beam inside the solution. However, the Raman

[53] J. Eisinger and J. Flores, *Anal. Biochem.* **94,** 15 (1979).

[54] R. E. Hirsch, *Methods Enzymol.* **232,** 231 (1994).

[55] E. Bucci, H. Malak, C. Fronticelli, I. Gryczynski, G. Laczko, and J. R. Lakowicz, *Biophys. Chem.* **32,** 187 (1988).

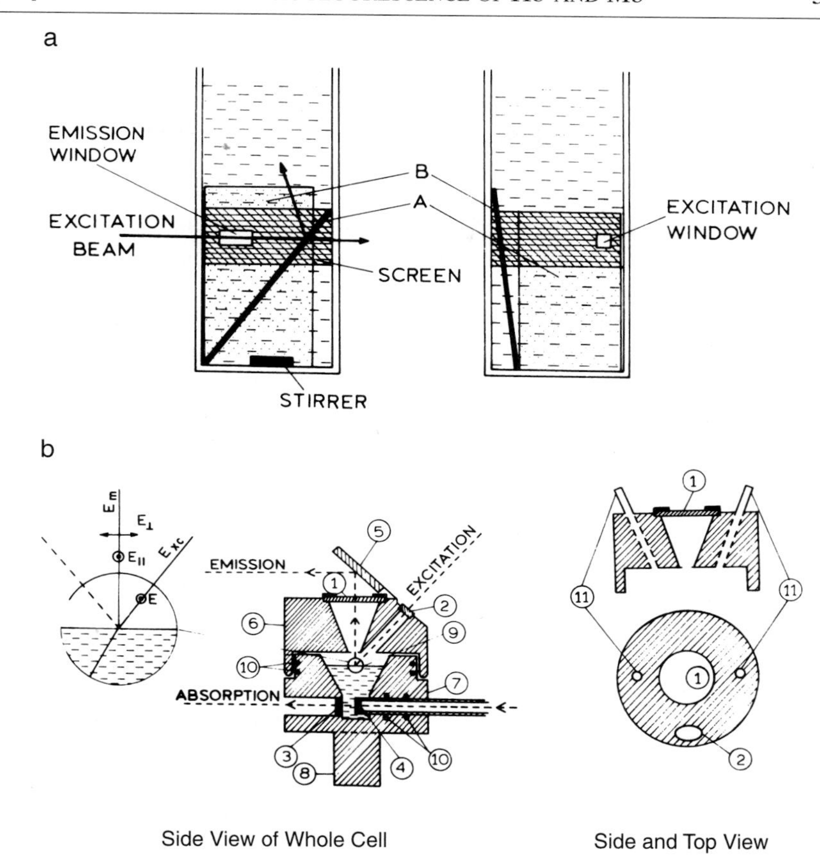

FIG. 9. (a) Emission and excitation sides of the "shielded" cuvette. A and B are slanted neutral density filters inserted into the cuvette for absorbing the residual excitation light and the back reflection of sample fluorescence. (b) Schematic of the GB cell. (1) Emission channel; (2) excitation channel; (3) window; (4) window on sliding sleave; (5) mirror; (6) body of the cover; (7) body of the cuvette; (8) supporting stem; (9) liquid surface; (10) O rings; (11) vent for flushing gases.

scattering is still at its maximum efficiency for a 90° (square geometry) observation. Also, the slanted orientation of the cuvette walls relative to the light beam prevents anisotropy measurements. Moreover, high-energy laser illumination promotes protein denaturation and deposition of degraded protein at the quartz–protein interface on the cuvette wall. Horiuchi and Asai[56] have designed a rhomboidal optical cell with thick windows where the inside compartment is tilted, while the outside walls are in square

[56] K. Horiuchi and H. Asai, *Biochem. Biophys. Res. Commun.* **97,** 811 (1980).

position. This improves anisotropy measurements, because both excitation and emission beams are perpendicular to the external walls. Depolarization on internal walls and other problems mentioned above remain.

We have designed an optical cell that works in a true front-face light beam configuration and operates on a free liquid surface.[52] Figure 9b shows a schematic of this optical cell (GB cell). It is essentially a conical cup that presents a horizontal liquid surface to the excitation beam. A cover slides over the cuvette, through which both the excitation and emission beams are channeled to avoid spilling reflections. The excitation channel makes a 34° angle with the normal to the liquid surface, which is the direction of the observation. This decreases the intensity of Raman scattering about 10-fold with respect to 90° scattering. The conical shape of the emission channel and an additional diaphragm positioned on top of this channel either direct the reflections away from the detector or trap them. All channels are covered with quartz windows. Two vents mounted on the cover of the cuvette allow sample equilibration with the desired gases (e.g., nitrogen for deoxygenation). At the bottom of the cell there are two opposing windows. One of them is mounted on a sliding sleeve, for adjusting the optical path as necessary for absorption measurements and for regulating the height of the liquid level in the cell. The proper directions of the excitation and emission beams are obtained by using metallic mirrors. For anisotropy measurements the excitation beam must be polarized with the electric vector parallel to the surface of the liquid, and perpendicular to the plane described by excitation and emission beams. For fluorescence intensity and lifetime measurements the "magic angle" condition is obtained with a 36° rotation of the polarizer in emission.

Instrumentation

Both time-correlated and frequency domain instruments can be used for these measurements. As presented later, hemoglobin and myoglobin are endowed with ultrafast (picosecond) lifetimes and relatively long lifetimes in the nanosecond time range, which have a small amplitude, below 0.01. This wide, specific distribution of lifetime components implies that it is necessary to use a wide range of frequencies in frequency-domain measurements and a large number of channels in time-correlated measurements. In our studies we successfully used frequency-domain instrumentation available in the Center for Fluorescence Spectroscopy at the University of Maryland (Baltimore, MD) and in the Laboratory for Fluorescence Dynamics at the University of Illinois (Urbana, IL). For time-domain measurements we used instrumentation available in the Regional Laser Laboratory at the University of Pennsylvania (Philadelphia, PA).

With hemoglobin and myoglobin a proper filters allows the use of the same sample for monitoring the emission of fluorescence and the scattering of the excitation light as a reference. The choice of filters may vary in different instruments. In emission the main problem is to eliminate the Raman scattering, especially in square geometry optics. We successfully used a combination of either a 340-nm (10-nm bandpass) interference filter (or an interference filter centered at 335 nm with a bandpass of 30 nm) together with a 0–54 Corning (Corning, NY) cutoff filter and a 7–60 broad-band filter. This combination allowed a 30% transmission at 340 nm. For the reference, which is light scattering from protein, we used an interference filter centered at 289 nm (10-nm bandpass) and a neutral-density filter. The optical delay of the two sets of filters was tested to be equal.[52] This is an important requirement when picosecond lifetimes are measured in the emission.

Sensitivity of Time-Resolved Fluorescence to Small Fractions with Longer Lifetimes

It should be stressed that when a protein exists in several conformations of which one is predominant, the minor fractions are sensitive to small, hardly detectable changes of the main fraction. For example, if there are two components whose fractions are in a 97:3 ratio, a change of the major fraction from 97 to 94% doubles the minor fraction to a ratio of 94:6. The decrease in the major fraction is hardly detectable, while the minor fraction is doubled. In myoglobin and hemoglobin these minor fractions are molecules with disordered or reversibly dissociated hemes, where heme quenching is drastically reduced or eliminated. Therefore their higher quantum yields and longer lifetime make their fractional change easy to detect. In Fig. 10 we report a simulation in frequency domain in which a lifetime of 25 psec is mixed with a lifetime of 2 nsec. The relative amplitudes vary from 99:1 to 95:5. The longer lifetime is responsible for the inflection of the curves at the lower frequencies. The simulation clearly shows the sensitivity of the measurements to variations of less than 1% in the fraction (amplitude) of the minor component with long lifetimes.

Informational Content of Lifetimes of Myoglobin and Hemoglobin

In hemoglobin and myoglobin there are at least three molecular species with different heme–protein relationships. They are the protein with hemes in their normal position as seen in the crystals, the species in which the heme is rotated 180° around its α–γ-meso axis (inverted

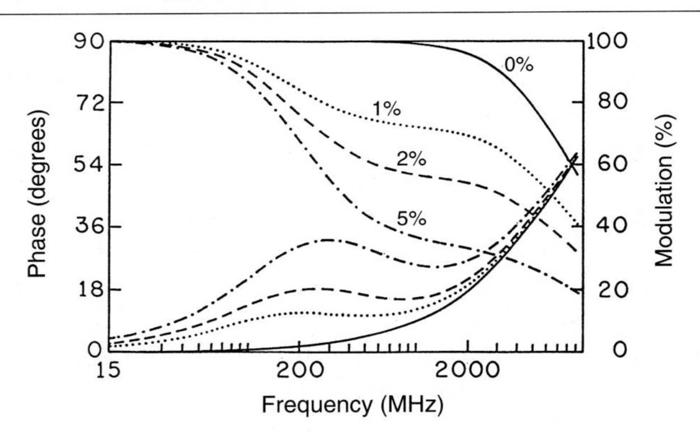

FIG. 10. Sensitivity of the phase shift and demodulation of time-dependent emission to changes in the amplitudes of the long-lifetime components. The computer simulation assumed two lifetime components with fixed lifetimes of 25 and 1000 psec, respectively. The amplitude of the longer component varied from 0 to 5%.

or disordered heme),[41–44] and the reversibly heme-dissociated species (RHD).[57,58] The three species have different lifetimes. The fish plots in Figs. 6 and 8 have shown that for Θ between 50 and 60° hemes in normal position produce high values of the orientation parameter κ^2, yielding very efficient energy transfers and short residual lifetimes in the picosecond range. Instead, disordered hemes produce much lower κ^2 values, resulting in longer lifetimes. The RHD species have fewer hemes, therefore their lifetime will be the longest. In single-subunit hemoprotein it will be like that of free tryptophan. The species with normal hemes are by far the largest fraction—more than 95% of the total. They are characterized by the most efficient radiationless excitation energy transfer between tryptophan and heme, resulting in the lowest quantum yield and shortest tryptophan lifetime. This implies that the other species, in smaller fraction but with longer lifetimes, produce fractional intensities comparable to that of the major species. This makes them detectable with great accuracy.

Calculated lifetimes for various myoglobin and hemoglobin species are drastically different in length. Therefore, they can identify the different species present in the sample. Their respective amplitudes reveal the respective relative fractions of the identified species. This allows a complete description of the conformational distribution of these systems into their heme–protein interaction components under different experimental condi-

[57] M. L. Smith, J. Paul, P. I. Ohlsson, and K. G. Paul, *Proc. Natl. Acad. Sci. U.S.A.* **88,** 882 (1991).
[58] Z. Gryczynski, J. Lubkowski, and E. Bucci, *J. Biol. Chem.* **270,** 19232 (1995).

tions (pH, temperature, concentration, ligation states). Modifications of the lengths and amplitudes can be used to monitor conformational changes of the system, which modify the relative positions and orientations of tryptophans and hemes.

Lifetimes of Myoglobin Systems

The multiplicity of sperm whale myoglobin lifetimes was first recognized by Hochstrasser and collaborators.[40,59] Those data have been confirmed by other authors.[60,61] We have published an interpretation of the origin of those lifetimes, based on the fish plots shown in Fig. 6, for the two tryptophans at positions 7 and 14, respectively.[20]

We have studied in detail the lifetimes of horse heart myoglobin at neutral pH and exposed to acid, using frequency-domain technology.[58] At all pH values we observed the presence of four lifetime components justified by a two- to fourfold decrease in the χ^2 value going from three- to four-component analyses. The data are shown in Table IV, which includes singular analyses and simultaneous, global analyses, of all the data at different pH values. The data reported for global analyses were obtained when the four lifetime components were linked to identical values across the pH titration. This allowed precise recovery of the pH-dependent amplitudes. From Table IV it appears that the different frequency responses at various pH values, shown in Fig. 11, were produced solely by the increasing amplitudes of the lifetimes in the nanosecond range near 1.3 and 4.8 nsec, respectively.

The expected lifetimes were computed from the atomic coordinates with the assumption that, in the absence of heme, the lifetime of both tryptophans in myoglobin (at positions 7 and 14, respectively) was 4.8 nsec (Table V). Calculations anticipate three lifetime components: one near 60 psec, due to Trp-14 in the presence of normal hemes, one near 130 psec, due to both Trp-14 in the presence of inverted hemes and Trp-7 in the presence of normal hemes, and one near 1.8 nsec, due to Trp-7 in the presence of inverted hemes. Therefore, we assigned the 40-psec lifetime to Trp-14 in the presence of normal heme, and the 116-psec lifetime to Trp-7, also in the presence of normal hemes. The 1.3-nsec lifetime was assigned to Trp-7 in the presence of inverted hemes. The longest lifetime of 4.8 nsec

[59] S. M. Janes, G. Holtom, P. Ascenzi, M. Brunori, and R. M. Hochstrasser, *Biophys. J.* **51,** 653 (1987).

[60] K. J. Willis, A. G. Szabo, and D. T. Krajcarski, *Photochem. Photobiol.* **51,** 375 (1990).

[61] K. J. Willis, A. G. Szabo, M. Zuker, J. M. Ridgeway, and B. Alpert, *Biochemistry* **29,** 5270 (1990).

TABLE IV

DISCRETE LIFETIMES, FRACTIONAL INTENSITIES, AND RELATIVE AMPLITUDES FOR INTENSITY DECAY OF HORSE HEART MYOGLOBIN[a]

pH	τ_1	f_1	α_1	τ_2	f_2	α_2	τ_3	f_3	α_3	τ_4	f_4	α_4
7	35	0.142	0.550	130	0.407	0.425	1491	0.198	0.018	4894	0.253	0.007
	40	0.149	0.509	116	0.394	0.463	1363	0.210	0.021	4822	0.247	0.007
4.95	43	0.188	0.603	127	0.342	0.371	1505	0.207	0.019	5167	0.262	0.007
	40	0.160	0.544	116	0.365	0.428	1363	0.191	0.019	4822	0.284	0.008
4.72	47	0.133	0.514	121	0.299	0.449	1599	0.220	0.025	5254	0.347	0.012
	40	0.103	0.455	116	0.335	0.508	1363	0.178	0.023	4822	0.384	0.014
4.62	56	0.122	0.519	113	0.200	0.420	1304	0.209	0.038	4842	0.469	0.023
	40	0.075	0.435	116	0.256	0.510	1363	0.189	0.032	4822	0.480	0.023
4.42	46	0.086	0.522	126	0.184	0.411	1679	0.245	0.041	5236	0.485	0.026
	40	0.065	0.450	116	0.203	0.480	1363	0.188	0.038	4822	0.544	0.031

[a] At each pH, the parameters were recovered by either individual analyses (upper rows) or global analyses (lower rows). In the global analyses the lifetimes were linked to the same value across the pH titration. Notably, the sum of α_1 and α_2, which is the fractional amount of the native protein, with normal hemes, decreases with pH, while α_3 and α_4, which originate from disordered heme and heme-free myoglobin, respectively, increase. The amplitudes and fractional intensities were recalculated using $f_i = \alpha_i\tau_i/\Sigma_i \alpha_i\tau_i$.

was assigned to reversibly heme-dissociated (RHD) myoglobin. For this reason this value was used as that of nonquenched tryptophans in the system.

The lifetime assignment formulated above allows the distinction of three conformations in the myoglobin system. One is the native conformation of crystalline myoglobin with normally oriented hemes yielding the two lifetimes in the picosecond range. The second conformation is myoglobin with

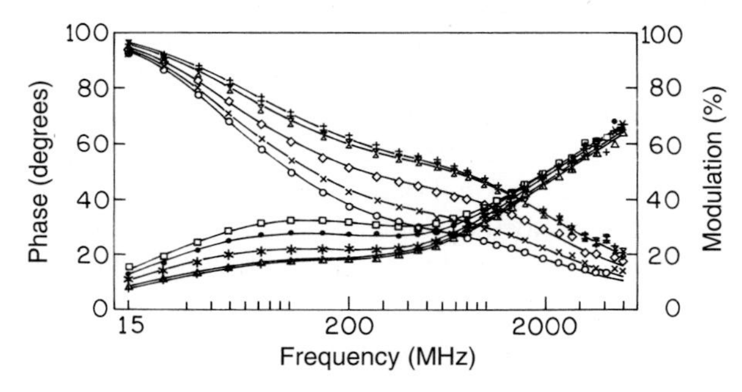

FIG. 11. Frequency dependence of phase shift and demodulation of the intensity decay of horse heart myoglobin at various pH values: (△, +) −7.20; (‡, ⧖) −4.95; (*, ◇) −4.72; (●, ×) −4.62; (□, ○) −4.42.

TABLE V
COMPUTED LIFETIMES FOR TRYPTOPHANS IN HORSE HEART MYOGLOBIN[a]

Trp-14		Trp-7	
τ_N(psec)	τ_D (psec)	τ_N (psec)	τ_D (psec)
62	156	126	1811

[a] The tryptophans were assumed to be in the presence of either normal hemes, τ_N, or inverted hemes, τ_D. The lifetime of nonquenched tryptophans was assumed to be 4.8 nsec. For the orientation parameters, we assumed an average κ^2 value obtained in the interval $\theta = 52$ to $58°$.

disordered hemes, which produced the lifetime near 1.3 nsec. The third conformation is the reversibly dissociated heme-free myoglobin, whose lifetime is near 4.8 nsec. The relative amplitudes of these components indicate that exposure to acid increases the amount of disordered heme and of RHD myoglobin. Notably the lifetimes remain constant throughout the titration in acid, indicating that the protein was not unfolded down to pH 4.5.

The distribution of the three molecular species is described by the lifetime amplitudes in Table IV. The sum of the amplitudes $\alpha_1 + \alpha_2 - \alpha_3 = \alpha_N$ is the fraction of myoglobin with normal hemes, Mb_N. Only one tryptophan (Trp-7) is responsible for the 1.3-nsec lifetime, therefore $2\alpha_3 = \alpha_D$ is the fraction of myoglobin with disordered hemes, Mb_D. The amplitude of the lifetime at 4.8 nsec, $\alpha_R = \alpha_4$, is the fraction of reversibly dissociated heme-free myoglobin. After doubling the amplitude α_3, all fractions are normalized to 1. Also, it can be safely assumed that in the system there are equimolar amounts of Mb_R and free heme.

On these assumptions, the equilibrium between the three conformations can be described by Scheme I.

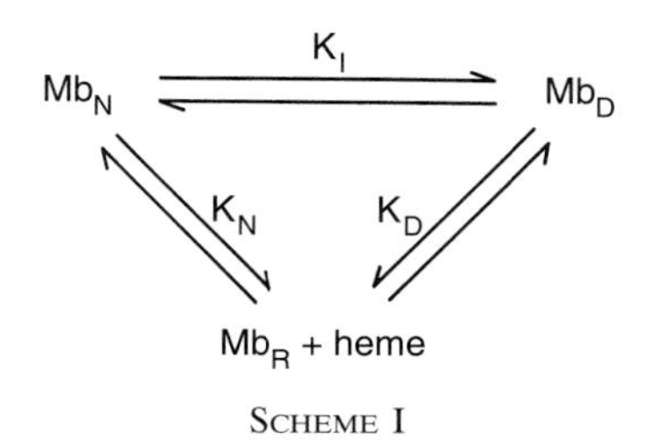

SCHEME I

In Scheme I

$$K_I = \frac{Mb_D}{Mb_N} = \frac{\alpha_D}{\alpha_N}$$

$$K_N = \frac{(Mb_R)^2}{Mb_N} = \frac{(\alpha_R)^2}{\alpha_N} C \qquad (11)$$

$$K_D = \frac{(Mb_R)^2}{Mb_D} = \frac{(\alpha_R)^2}{\alpha_D} C$$

C is the protein concentration at which the measurements were performed ($C = 1.25 \times 10^{-5}$ M).

The computed constants are listed in Table VI. Time-resolved fluorescence decay is probably the only technology that allows direct monitoring of those small fractions of myoglobin conformers and relative equilibrium constants.

Additional information can be obtained by examining the Lorentzian distribution of the lifetime components. These distributions at the various pH values are illustrated in Fig. 12. They show that up to pH 4.5 the longest lifetime in the nanosecond range has a sharp distribution. This indicates a homogeneous protein structure for RHD myoglobin. Only at pH 4.42 did their distributions become broader, indicating an initial unfolding of the protein. In contrast, the distributions of the lifetimes generated by disordered hemes are broad, indicating non homogeneous conformation of the heme pocket around disordered hemes. It is interesting to observe that when the distributions of the lifetime for RHB myoglobin become broader, suggesting an initial unfolding of the protein, that produced by disordered hemes remains unmodified. This suggests that the unfolding in acid starts with the RHD form of the protein. This observation is relevant, and it points out the importance of these minor fractions to the thermodynamics of the folding–unfolding of the system.

TABLE VI
EQUILIBRIUM CONSTANTS FOR CONFORMATIONAL STATES OF HORSE HEART MYOGLOBIN[a]

pH	K_N (M)	K_D (M)	K_I
7.20	5.9×10^{-10}	1.4×10^{-8}	4.4×10^{-2}
4.95	7.8×10^{-10}	2.0×10^{-8}	3.9×10^{-2}
4.72	24.1×10^{-10}	4.9×10^{-8}	4.9×10^{-2}
4.62	67.2×10^{-10}	9.5×10^{-8}	7.0×10^{-2}
4.42	124.7×10^{-10}	14.6×10^{-8}	8.5×10^{-2}

[a] As defined in Eq. (11).

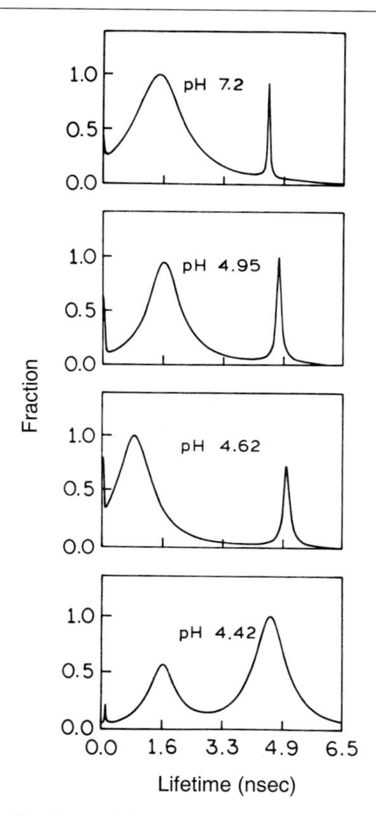

FIG. 12. Lorentzian distributions of the nanosecond lifetimes of horse heart myoglobin at various pH values. The distribution where computed using the equation

$$I(t) = I_0 \sum_i \alpha_i(\tau)e^{t/\tau} \quad \text{where} \quad \alpha_i(\tau) = \frac{1}{\pi}\frac{\Gamma/2}{(\tau-\tau_i)^2(\Gamma/2)^2}$$

τ_i is the center value of the lifetime distribution and Γ is the full width at half-maximum.

Lifetimes of Isolated α Subunits of Human Hemoglobin

The lifetimes of the isolated α subunits of human hemoglobin[62] in their oxy form were measured at two different protein concentrations, 0.5 and 30 mg/ml, respectively, in 0.05 *M* phosphate buffer at pH 7.0, at room temperature. These two concentrations were chosen because this system is essentially monomeric at low concentrations, while it is largely dimeric

[62] Z. Gryczynski, C. Fronticelli, E. Gratton, J. Lubkowski, and E. Bucci, Time-resolved laser spectroscopy. *In* "Biochemistry III," Vol. 2137, p. 129. SPIE, Bellingham, Washington, 1994.

TABLE VII
RECOVERED LIFETIMES AND AMPLITUDES FOR α SUBUNITS OF HEMOGLOBIN[a]

Protein concentration (mg/ml)	τ_1 (psec)	Amplitude	τ_2 (psec)	Amplitude
0.5	42	0.976	1050	0.024
30.0	38	0.971	651	0.029
Monomeric[b]	32 (N)		1132 (D)	
Dimeric[b]	31 (N)		746 (D)	

[a] At high and low protein concentration. The computed lifetimes for monomeric and dimeric α subunits are also reported. N and D, Normal and disordered intrasubunit heme position, respectively.
[b] Calculated.

at higher concentrations.[63] The numerical values of the recovered lifetime parameters are presented in Table VII.

Using the atomic coordinates of human oxyhemoglobin[49] we calculated the lifetimes for the single tryptophan at position 14 of monomeric and dimeric α subunits (Table VII). Calculations anticipate the presence of two lifetime components in the system. One is short, near 30 psec, due to the quenching within the monomer by normal hemes. The other, longer component is due to the disordered heme in the monomer, whose transition moment is not properly oriented and accepts transfer at a much reduced rate, resulting in a lifetime of 1132 psec. This longer lifetime is shortened to 746 psec in the dimer by the additional quenching of the heme at an intersubunit distance. The correspondence between the measured lifetimes (Table VII) and those computed for monomers and dimers, respectively, is excellent. We were not able to detect in this system lifetimes near 5 nsec, as expected from RHD forms. It suggests that the fraction representing this species was extremely small, below 1 part per thousand (ppt).

As anticipated by the simulations, the experimental lifetimes proved to be concentration dependent. It is important to stress that the relative amplitudes of the long-lifetime components are similar at low and high concentrations, 2.4 and 2.9%, respectively, proving that the amount of disordered hemes is not dependent on the polymerization of the system. It is also important to note that these amplitudes are in good agreement with the 2–3% fraction of disordered hemes detected in the α subunits of hemoglobin by NMR spectroscopy in the laboratory of La Mar.[43]

[63] R. Valdes and G. K. Ackers, *J. Biol. Chem.* **252,** 74 (1977).

Our computations assumed that the dimeric form of the isolated α subunits is identical to the α_1–α_2 dimer present in hemoglobin. While it is reasonable to hypothesize that the interface between the two subunits is the same as that in hemoglobin, its rigidity may be "softer" than in the tetramer. Even a slight wobbling of the two subunits with respect to each other would modify both the distance and angular orientations between tryptophans and heme present in the hemoglobin tetramer, thereby becoming detectable. These measurements suggest that the isolated α subunits of hemoglobin have a similar structure and α_1–α_2 interface as in the liganded hemoglobin tetramer. This shows applicability of this technology for studying the polymerization of the α subunits at high protein concentrations, difficult to approach by common physicochemical techniques.

Lifetimes of Human Hemoglobin

Table VIII shows the lifetimes computed from crystallographic structures[49,50] for human deoxy-, oxy-, and CO-hemoglobin for monomeric, dimeric, and tetrameric forms. We have conducted measurements at two different concentrations of carbomonoxy hemoglobin; 30 mg/ml (using the GB cuvette and front-face technology) and 0.2 mg/ml (using the shielded cuvette in square geometry optics). As shown in Table IX, a lifetime is found in the nanosecond range only at low concentration. Comparing these data with the lifetimes computed from the atomic coordinates of liganded and unliganded hemoglobin (Table IX), we can assign the lifetimes at 28–35 psec to Trp-14 α_1 and to Trp-15 β_1 when the heme is in normal position in the respective subunits. Components at 600–1100 psec are produced by Trp-14 α_1 when heme in that subunit is either dissociated or in a disordered position. The shortest component of 2–6 psec is the overall product of two processes. One process is probably the interconversion of the ^1L_a and ^1L_b states within the excited molecule of tryptophan. The other is due to Trp-37 β, which is simultaneously quenched by two hemes; the one in its own subunit and the other in the α subunit of the opposite dimer. It should be stressed that this tryptophan will always produce lifetimes in the short picosecond range because it is statistically impossible to find a hemoglobin molecule with those two hemes simultaneously either inverted or dissociated. The RDH species deserve special attention. Because of intersubunit quenching, they contribute lifetimes in the 600- to 1100-psec range both in the dimeric and tetrameric species of hemoglobin. Therefore the lifetime at 4 nsec, detected in hemoglobin at low concentration, must be referred to the RHD form of a monomeric species. The relative amplitude of this lifetime implies a concentration of 1 ppt for this species. It was not detected at high concentrations because the equilibrium condition decreased that

TABLE VIII

COMPUTED LIFETIMES OF TRYPTOPHANS α-14, β-15, AND β-37 FOR HUMAN DEOXY-, CO-, AND OXY HEMOGLOBIN[a]

Hb	Trp	Heme	τ_m (psec)	τ_d (psec)	τ_t (psec)
Deoxy	α-14	N	45	44	44
		D	830	639	559
		RD	5000	1789	1277
	β-15	N	47	47	46
		D	1865	1248	1039
		RD	5000	2148	1597
	β-37	N	25	24	12
		D	987	607	23
		RD	5000	1196	24
CO	α-14	N	40	39	39
		D	2269	1414	1044
		RD	5000	2144	1395
	β-15	N	47	47	47
		D	1245	976	859
		RD	5000	2374	1783
	β-37	N	8	8	8
		D	312	277	250
		RD	5000	1664	1001
Oxy	α-14	N	74	72	71
		D	2918	1493	1117
		RD	5000	1896	1404
	β-15	N	42	42	42
		D	1654	824	794
		RD	5000	2153	1668
	β-37	N	14	14	12
		D	1381	862	60
		RD	5000	1457	62

[a] Lifetime of 5 nsec for nonquenched tryptophan and average Förster distance R_0 33 Å for tryptophan heme interaction were assumed. The subscripts m, d, and t refer to the monomer, dimer, and tetramer form, respectively.

TABLE IX

MEASURED DISCRETE LIFETIMES, FRACTIONAL INTENSITIES, AND RELATIVE AMPLITUDES FOR INTENSITY DECAY OF HUMAN CO-HEMOGLOBIN[a]

HbA_0	τ_1 (psec)	f_1	α_1	τ_2 (psec)	f_2	α_2	τ_3 (psec)	f_3	α_3	τ_4 (psec)	f_4	α_4
CO L	12	0.217	0.671	31	0.272	0.326	486	0.393	0.003	4405	0.119	0.001
Deoxy H	2	0.250	0.852	28	0.660	0.147	820	0.090	0.001	—	—	—
Oxy H	3	0.230	0.813	35	0.540	0.183	670	0.230	0.004	—	—	—
CO H	6	0.310	0.725	35	0.540	0.272	1100	0.150	0.003	—	—	—

[a] For different concentrations. L, 0.2 mg/ml; H, 30 mg/ml.

fraction below 1 ppt (i.e., below the limit of detection). Experiments in progress indicate that this fraction reversibly increases under hyperbaric conditions. Therefore it is part of the system and not an impurity.

As reported above, also in the isolated α subunits, this component was not detected. This suggests that this long lifetime, in hemoglobin, originates from the RDH species of monomeric β subunits. It is argued that heme exchange with either albumin or apomyoglobin primarily occurs from the β subunits in the dimeric form of hemoglobin. Our data suggest that monomeric forms of β subunits also exist, and that they are in equilibrium with dissociated hemes, therefore contributing to the exchange. This again emphasizes the relevance to the thermodynamics of the system of small components in the parts per thousand range, and the necessity of studying their behavior.

Conclusion

The identification of the transition moment in which hemes accept radationless energy transfer from the excited state of tryptophan is a breakthrough, which allows description of the conformational distribution of heme–protein interaction species in hemoglobin and myoglobin systems. Under equilibrium conditions, as observed here, the species with disordered and dissociated hemes are only minute fractions. Nevertheless they may be important to the thermodynamics of the system if they are in the path of conformational transitions. The data available so far propose that in myoblogin, the RDH species plays a significant role in the unfolding of the system. These data also suggest that in hemoglobin, monomeric forms of β subunits are important for heme release, therefore potentially relevant for the folding of the protein. Additional data are necessary to clarify these phenomena.

[29] Multiple-Domain Fluorescence Lifetime Data Analysis

By MICHAEL L. JOHNSON

It is now common practice to analyze fluorescence lifetime experiments by nonlinear least-squares methods.[1–5] For frequency-domain fluorescence lifetime experiments the phase shift and modulation of the emitted light (the dependent variables) are least-squares fitted as a function of the modulation frequency of the excitation light (the independent variable). Time-domain fluorescence lifetime experiments are analyzed by least-squares fitting the fluorescence intensity (the dependent variable) as a function of time (the independent variable). The objective of these analyses is to obtain parameters of the fitting equations (e.g., fluorescence lifetimes) with the highest probability of being correct. Other equally important goals are to evaluate the precision of these estimated parameters[1–6] and to evaluate the actual quality, or goodness, of the least-squares fit.[1–5,7]

Frequently least-squares parameter estimations are performed globally with several sets of experimental results.[8–10] The global approach simultaneously analyzes several related sets of data. An example is collisional quenching data from a series of different quencher concentrations.[5,11] For the traditional global approach the sets of data are all generated by the same experimental method, i.e., either time- or frequency-domain experiments.

[1] M. L. Johnson and Susan G. Frasier, *Methods Enzymol.* **117,** 301 (1985).
[2] M. L. Johnson and L. M. Faunt, *Methods Enzymol.* **210,** 1 (1992).
[3] M. L. Johnson, *Methods Enzymol.* **240,** 1 (1994).
[4] D. G. Watts, *Methods Enzymol.* **240,** 23 (1994).
[5] M. Straume, S. G. Frasier-Cadoret, and M. L. Johnson, *Topics Fluoresc. Spectrosc.* **2,** 177 (1991).
[6] M. Straume and M. L. Johnson, *Methods Enzymol.* **210,** 117 (1992).
[7] M. Straume and M. L. Johnson, *Methods Enzymol.* **210,** 87 (1992).
[8] G. K. Ackers, M. L. Johnson, F. C. Mills, H. R. Halvorson, and S. Shapiro, *Biochemistry* **14,** 5128 (1975).
[9] J. M. Beechem, *Methods Enzymol.* **210,** 37 (1992).
[10] J. M. Beechem, E. Gratton, M. Ameloot, J. R. Knutson, and L. Brand, *Topics Fluoresc. Spectrosc.* **2,** 241 (1991).
[11] I. Gryczynski, H. Szmacinski, G. Laczko, W. Wiczk, M. L. Johnson, J. Kusba, and J. R. Lakowicz, *J. Fluoresc.* **1,** 163 (1991).

0076-6879/97 $25

A logical generalization of this global approach is also to analyze simultaneously time- and frequency-domain experiments. This chapter discusses the theory, methodology, and pitfalls of this simultaneous approach.

Intensity Decay Laws

The method presented in this chapter is independent of the molecular mechanism that is the origin of the fluorescence intensity decay, $I(t)$. In this chapter the intensity decay law is expressed as $I(t)$ rather than as a specific decay law.

The most commonly used, and abused, intensity decay law is simply the sum of multiple exponentials, i.e.,

$$I(t) = \sum_{i=1}^{N} A_i e^{-t/\tau_i} \tag{1}$$

where t represents time, τ_i is the ith fluorescence lifetime, and A_i is the amplitude of the ith fluorescence lifetime. In this case the least-squares procedure will estimate the A_i and τ_i parameters.

For some complex decays it is sometimes useful to describe the intensity decay as a distribution function of decays. In this case the A_i in Eq. (1) is replaced by a particular probability distribution, such as the Gaussian distribution shown in Eq. (2).

$$I(t) = \int_{\tau=0}^{\infty} P(\tau) e^{-t/\tau}\, d\tau$$

$$P(\tau) = \frac{1}{\sigma_\tau (2\pi)^{1/2}} e^{-\frac{1}{2}\left(\frac{\tau - \tau_{\text{ave}}}{\sigma_\tau}\right)^2} \tag{2}$$

$$\text{Half-width} = 2.354\sigma_\tau$$

where τ_{ave} is the mean of a Gaussian distribution of lifetimes with a standard deviation of σ_τ. Here the least-squares procedure will estimate the τ_{ave} and half-width parameters. Lorentzian, binomial, and rectangular probability distributions are also commonly used.

Please note that the actual form of the actual intensity decay law is a complex function of the chemical and physical mechanism of the fluores-

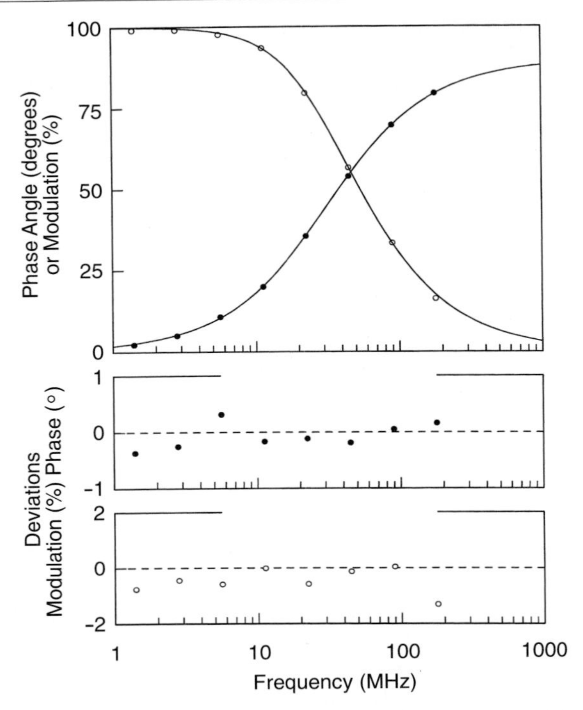

FIG. 1. Typical simulated frequency-domain data set. *Top:* A simulated frequency-domain data set, where the closed circles correspond to the phase shift and the open circles correspond to the modulation. These data were simulated with a Gaussian distribution of fluorescence lifetimes with a mean of 5 nsec and a half-width of 2.5 nsec. The simulated experimental uncertainties were Gaussian distributed with standard deviations of 0.2° for the phase and 0.005 for the modulation. The solid line corresponds to the fit of these data to Eqs. (2) and (3) (see Table I for parameter values). *Middle and bottom:* Weighted residuals of this fit: *middle,* corresponding to the phase residuals; *bottom,* corresponding to the modulation residuals.

cence decay. For example, quenching[12] and resonance energy transfer[13] have intensity decay laws that differ significantly from Eqs. (1) and (2).

Frequency-Domain Analysis

Frequency-domain experiments have two dependent variables: phase and modulation. Figure 1 is an example of a frequency-domain data set. The top portion of Fig. 1 presents a simulated frequency-domain data set

[12] M. R. Eftink, *Topics Fluoresc. Spectrosc.* **2,** 53 (1991).
[13] H. C. Cheung, *Topics Fluoresc. Spectrosc.* **2,** 128 (1991).

in which the closed circles correspond to the phase shift and the open circles correspond to the modulation.

Two fitting equations are utilized because there are two dependent variables:

$$\text{Modulation}(\omega) = \{[D(\omega)]^2 + [N(\omega)]^2\}^{1/2}$$

$$\text{Phase}(\omega) = \tan^{-1}\left[\frac{N(\omega)}{D(\omega)}\right]$$

$$N(\omega) = \frac{\int_0^\infty I(t)\sin[\omega t]\,dt}{\int_0^\infty I(t)\,dt} \tag{3}$$

$$D(\omega) = \frac{\int_0^\infty I(t)\cos[\omega t]\,dt}{\int_0^\infty I(t)\,dt}$$

ω is the angular frequency of the excitation in radians. N and D are both normalized functions and consequently the Phase and Modulation functions are also both normalized. This means that the intensity decay law always has an unknown scaling factor. Thus, if the decay law is the sum of multiple exponentials [i.e., Eq. (1)], all of the amplitude terms (i.e., A_i terms) sum to an unknown (i.e., arbitrary) scaling factor. A consequence of this is that one of the A_i terms must be specified (i.e., not estimated by the least-squares procedure) and the values of the other A_i terms are relative to the specified A_i for frequency-domain data.

Time-Domain Analysis

For time-domain experiments the observed experimental observations, Intensity(t), is a convolution integral of the intensity decay law, $I(t)$, and the instrument response function, $L(t)$.

$$\text{Intensity}(t) = \int_0^t L(z)\,I(t-z)\,dz = L(t) \otimes I(t) \tag{4}$$

Figure 2 is an example of a time-domain data set. The instrument response function is to the left and the observed intensity is to the right.

The instrument response function is sometimes incorrectly called the lamp function. The instrument response function includes both the lamp profile and the response function of all other parts of the instrument. The instrument response function is normally empirically measured concomitantly with the experimental observations. Thus, time-domain experiments

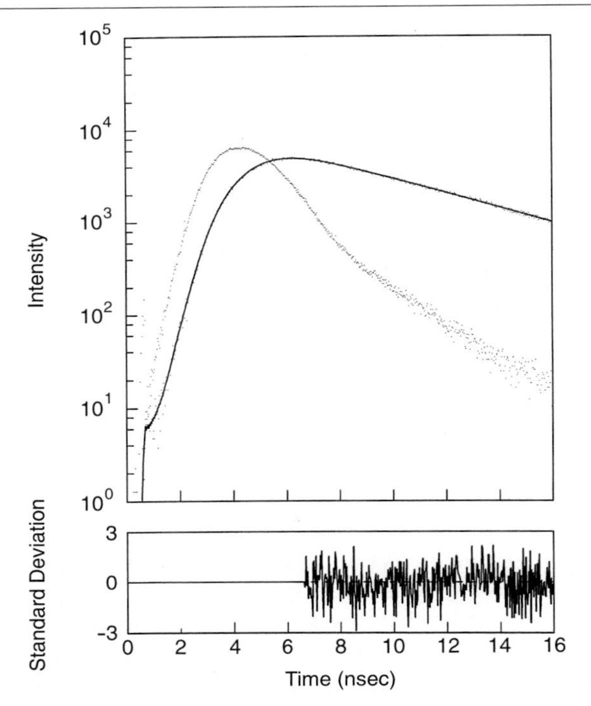

FIG. 2. Typical simulated time-domain data set. *Top:* Simulated time-domain data set. The instrument response function is to the left and the observed time-dependent intensities are to the right. These data were simulated with a Gaussian distribution of fluorescence lifetimes with a mean of 5 nsec and a half-width of 2.5 nsec. The simulated experimental uncertainties were Gaussian distributed with variance equal to the observed intensity. The solid line corresponds to the fit of these data to Eqs. (2) and (4) (see Table II for parameter values). *Bottom:* Weighted residuals of this fit. The region utilized for the analysis is that region where residuals exist.

have only one dependent variable but have two independent variables: time and the instrument response function that is also a function of time.

Other Domains of Analysis

The methods outlined in this chapter are equally applicable to experiments other than time- and frequency-domain lifetime measurements. For example, the steady-state intensity can be used as a third domain (i.e., type of data) for simultaneous analysis.

Least-Squares Parameter Estimation Procedures

The three objectives of a least-squares parameter estimation are (1) to evaluate the parameter values with the highest probability of being correct, (2) to evaluate the precision with which these parameter values have been estimated, and (3) to test the goodness of fit.

There are many nonlinear least-squares parameter estimation algorithms.[1–5] The actual choice of the algorithm is somewhat arbitrary because all should yield equivalent results. For this work the Nelder–Mead simplex algorithm was used.[5,14] The common feature of all least-squares fitting algorithms is that they adjust the parameter values to minimize the sum of the weighted squares of the difference between the data values and the fitted function:

$$\sum_i \left[\frac{\text{Data}_i - \text{Function}_i}{\text{SEM data}_i}\right]^2 = \sum_i [\text{Weighted Residual}_i]^2 \tag{5}$$

In Eq. (5) the summation is over all of the experimental observations. Data_i is the ith data point. This data point could be either a phase value or a modulation value for frequency-domain data. It is a time-dependent intensity for time-domain data. Or it could be virtually any other type of experimental observation, such as a steady-state intensity. SEM data_i corresponds to the standard error of the mean value of the ith data point. Function_i is the fitting function evaluated at the ith data point and with the maximum likelihood parameter values.

The most commonly used estimate of the precision to which the parameter values have been estimated is the asymptotic standard error.[1–6] However, for nonlinear least-squares problems, such as the fluorescence data analysis in this work, the asymptotic standard errors provide a lower limit of the actual precision of the estimated parameters. The actual uncertainty of the estimated parameters is typically roughly five times the asymptotic standard error. Note that this factor of five is only an extremely rough estimate and can vary from one to infinity depending on the fitting equation, the data, the experimental uncertainties within the data, etc. If the confidence intervals are underestimated (i.e., too small), the researcher will assume that the parameters are known to better precision than they actually are. This can lead to incorrect conclusions about the overall significance of the results. This work utilizes the Monte-Carlo and support plane methods for the evaluation of the uncertainties of the parameters.

Least-squares parameter estimation procedures can also be utilized to test hypotheses about molecular mechanisms. For example, the experimen-

[14] J. A. Nelder and R. Mead, *Comput. J.* **7,** 308 (1965).

tal observations can be fit to the appropriate mathematical form for resonance energy transfer.[13] If the least-squares parameter estimation with a particular mathematical form provides a good description of the experimental observations then it can be concluded that the data are consistent with a resonance energy transfer mechanism. Conversely, if the least-squares parameter estimation with a particular mathematical form does not provide a good description of the experimental observations then it can be concluded that the data are inconsistent with the particular resonance energy transfer mechanism. The statistical procedures for this comparison are known as goodness-of-fit criteria.[1–5,7] If the least-squares procedure provides a good description of the experimental data then the weighted residuals, as defined in Eq. (5), will be random and Gaussian distributed. Thus, the goodness-of-fit criteria test that the residuals follow a Gaussian, or normal, distribution. The run's test and the Kolmogorov–Smirnov test are recommended.[5]

Global Parameter Estimations

In its simplest terms global analysis is defined as the simultaneous analysis of multiple sets of related data. The global estimated parameters will be the same for multiple sets of data. Nonglobal parameters are unique to specific data sets. For example, this could be either time- or frequency-domain data at a series of quencher concentrations. Or, as in the present example, it could be the simultaneous analysis of time- and frequency-domain data. If the time- and frequency-domain experiments are done under the same conditions then all of the frequency-domain parameters will be global parameters of the time-domain data. However, the time-domain data will have one additional parameter corresponding to the amplitude eliminated from the frequency-domain data because of the normalization process discussed above. The time-domain data may also have instrument specific constants, such as dark current and time shifts, that do not apply in the frequency domain. This dual-domain concept can be easily extended by adding other types of data such as steady-state intensity data. It can also be easily extended to include variable experimental conditions such as changes in temperature, wavelength, and quencher concentrations.

Doing global fits, and multidomain global fits, simply involves a bookkeeping problem. The program, and thus the programmer, simply needs to keep track of which sets of equations apply to which set of data and keep track of which parameters and constants pertain to which set of equations and/or data sets. Thus for each term of the summation in Eq. (5) the program must select which parameters and equations apply to that specific data point, then evaluate the function and accumulate the summation.

Confidence Intervals by Monte-Carlo Method

The first step in the creation of Monte-Carlo[6] confidence intervals is to do the least-squares parameter estimation as outlined above. A set of noise-free data is then simulated on the basis of the fitting equation, the estimated parameters with the highest probability of being correct, the values of the independent variables (e.g., time, instrument response function, and/or frequency) of the original set of data, and the weighting factors of the original data. If the original least-squares parameter estimation provided a good description of the experimental data, according to the goodness-of-fit criteria, then this noise-free data will be a good simulation of the data without experimental noise.

The next step is to create several hundred or more simulated data sets that include simulated experimental uncertainties. The current application used 500 sets of simulated data. Each of these simulated noisy data sets is simply the noise-free simulated data set with pseudorandom noise added. The distribution and magnitude of the added noise must be a close approximation of the actual noise on the original experimental data. If the original least-squares parameter estimation provided a good description of the experimental data, according to the goodness-of-fit criteria, then the simulated noise should follow a Gaussian distribution and have a weighted variance equal to the original weighted variance of fit. Gaussian distributed pseudorandom deviates with a variance of 1 were created by adding 12 pseudorandom deviates with an even distribution between ± 0.5 that were created with the RAN3 procedure.[15] These were then scaled to the desired variance before use.

A new least-squares parameter estimation is then done for each of these noisy simulated data sets. This yields several hundred estimates, 500 for the present case, of each parameter on the basis of the simulated data. These estimates are presented as a histogram of frequency vs the value for each of the estimated parameters. This histogram provides an approximation of the complete probability distribution for each parameter.

Confidence Intervals by Support Plane Method

The support plane method[5] simply creates graphs of the apparent weighted variance of fit, or weighted sum of squared residuals, as a function

[15] W. H. Press, B. P. Flannery, S. A. Teukolsky, and W. T. Vetterling, "Numerical Recipes: The Art of Scientific Computing," p. 199. Cambridge University Press, Cambridge, 1986.

of each of the parameter values. The creation of the graph for a particular parameter requires that a series of values for that parameter be specified. For each of these specified parameter values the least-squares parameter estimation procedure is repeated to reevaluate all of the parameter values except the particular parameter being plotted. For linear fitting equations these graphs are parabolas increasing to infinity on both sides of the maximum likelihood value. However, for the nonlinear fitting equations utilized here these graphs are not required to be parabolas; they can show multiple minimum values, and they can have distinctly nonparabolic shapes. Statistically significant increases in the weighted variance of fit, and thus significant changes in the particular parameter value, are then evaluated by an F test.

Weighting Functions

A further consideration in doing global fits with different types, i.e., domains, of data should be made in determining the relative weights of one type of experimental data as compared to other types of data. For example, time-domain data sets will typically contain about 1000 data points while frequency-domain data sets may contain only 20 data points. Should the time-domain data be given greater or lesser statistical weight because of the large difference in the number of data points? No! The theoretically, and statistically, correct weighting is done on a per data point basis with the standard error of the mean for that particular data point,[15] as in Eq. (5).

Consider a time-domain experiment with 2000 data points. The standard error of the mean of each observation is approximately the square root of the number of photons found in each time window. Now consider the consequences of binning the adjacent data points to generate a data set with only 1000 data points. The precision of a least-squares parameter estimation is roughly proportional to the square root of the number of data points, so it is expected that a fit to this condensed data set will yield results that are worse by approximately the square root of two. However, each of these data points in the new set contains approximately twice as many counts as a data point in the previous set, and thus its uncertainty (SEM) is improved by approximately the square root of two. Improving the precision of the data points will generally improve the results in direct proportion to the improved precision. Thus, these two results of binning the data points approximately cancel each other. Half as many data points with precision improved by the square root of two will, to a first approximation, not affect the resulting fit.

Typical Analyses

Table I presents the analysis of the simulated frequency-domain data shown in Fig. 1. This analysis was for an intensity decay law with a Gaussian

TABLE I

ANALYSIS OF SIMULATED FREQUENCY-DOMAIN DATA SHOWN IN FIGURE 1 BY GAUSSIAN DISTRIBUTION OF LIFETIMES

Variable	Value	Confidence interval
τ_{ave}	4.969	(4.942–4.991)[a]
Half-width	2.700	(2.481–2.904)
SSR[b]	22.933	

[a] Confidence intervals corresponding to ±1 SEM as evaluated by the Monte-Carlo method.
[b] Weighted sum of the squared residuals of the fit.

distribution of fluorescence lifetimes [Eq. (2)]. Please note that the confidence intervals evaluated by the Monte-Carlo method are not exactly symmetrical. Also note that the confidence interval for the half-width corresponds to ~±8%. The complete probability distribution, as determined by the Monte-Carlo method, from which this confidence interval was determined is shown in Fig. 3 (top). Figure 4 (top) presents the analogous

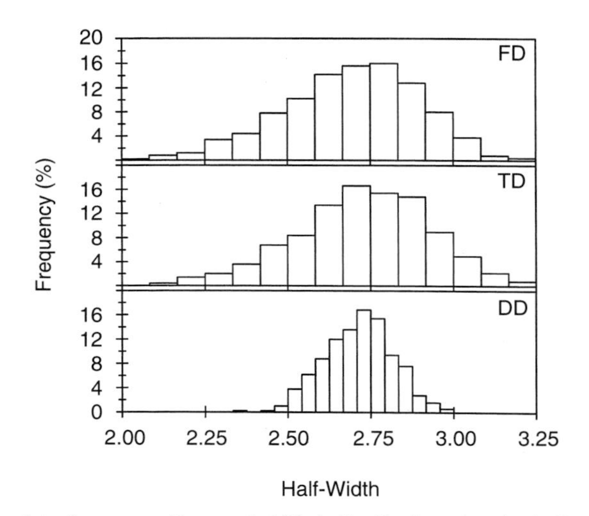

FIG. 3. Complete frequency (i.e., probability) distributions for the half-width evaluated by the Monte-Carlo method. The frequency distribution is presented as a percentage of the Monte-Carlo simulations that yielded a value within the particular bin. *Top:* Corresponds to the frequency-domain analysis presented in Fig. 1 and Table I. *Middle:* Corresponds to the time-domain analysis shown in Fig. 2 and Table II. *Bottom:* Corresponds to the dual-domain analysis presented in Table III.

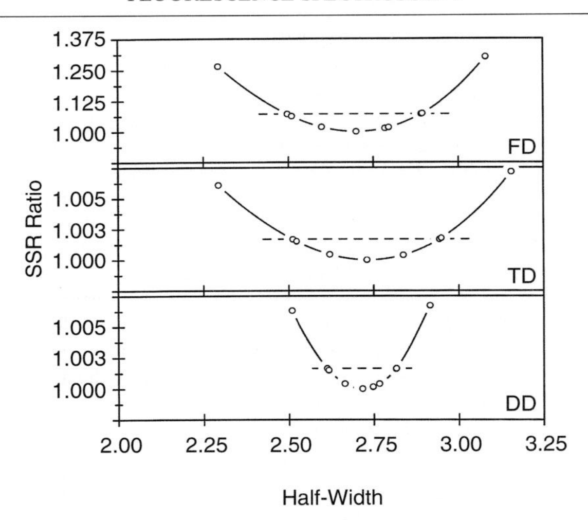

FIG. 4. Confidence intervals as evaluated by the support plane method. The SSR ratio is the SSR at a particular value of half-width divided by the SSR corresponding to the maximum likelihood value of the half-width. The circles correspond to the points where the additional least-squares parameter estimations were performed. The horizontal dashed lines correspond to a ±1 SEM increase in the variance. The projection of the intersections points with the smooth curve through the evaluated points defines the confidence intervals of the half-width. *Top:* Corresponds to the frequency-domain analysis. *Middle:* Corresponds to the time-domain analysis. *Bottom:* Corresponds to the dual-domain analysis.

information as determined by the support plane method. The solid lines and the residuals shown in Fig. 1 were calculated from this analysis.

Table II presents the analysis of the simulated time-domain data shown in Fig. 2. This analysis was for an intensity decay law with a Gaussian

TABLE II
ANALYSIS OF SIMULATED TIME-DOMAIN DATA SHOWN IN FIGURE 2 BY GAUSSIAN DISTRIBUTION OF LIFETIMES

Variable	Value	Confidence interval
τ_{ave}	4.972	(4.954–4.990)[a]
Half-width	2.735	(2.507–2.924)
A	0.399	(0.3966–0.4013)
SSR[b]	342.084	

[a] Confidence intervals corresponding to ±1 SEM as evaluated by the Monte-Carlo method.
[b] Weighted sum of the squared residuals of the fit.

TABLE III
SIMULTANEOUS ANALYSIS OF SIMULATED TIME- AND FREQUENCY-DOMAIN DATA SHOWN IN FIGURES 1 AND 2 BY GAUSSIAN DISTRIBUTION OF LIFETIMES

Variable	Value	Confidence interval
τ_{ave}	4.972	(4.960–4.983)[a]
Half-width	2.715	(2.606–2.817)
A	0.399	(0.3976–0.4003)
SSR[b]	365.098	

[a] Confidence intervals corresponding to ±1 SEM as evaluated by the Monte-Carlo method.
[b] Weighted sum of the squared residuals of the fit.

distribution of fluorescence lifetimes [Eq. (2)]. Please note that the confidence intervals evaluated by the Monte-Carlo method are even more asymmetrical. Also note that the confidence interval for the half-width corresponds to ~±8%. The complete probability distribution, as determined by the Monte-Carlo method, from which this confidence interval was determined is shown in Fig. 3 (middle). Figure 4 (middle) presents the analogous information as determined by the support plane method. The solid lines and the residuals shown in Fig. 2 were calculated from this analysis.

The two data sets were carefully selected such that the precision of the results of the analysis would be roughly equivalent. It is, however, interesting that the time-domain data contained 351 data points and the frequency-domain data contained only 16 observations. Nevertheless, it cannot be concluded from this example that either domain is preferable.

Table III presents the simultaneous global analysis of the frequency-domain data shown in Fig. 1 and the time-domain data shown in Fig. 2. The weighted sum of the squares residuals, SSR, for this global dual-domain fit is only 0.02% higher than the sum of the SSR values from the individual fits. A slight increase in the SSR is expected because the fit to each of the separate data sets will be slightly compromised to obtain the best simultaneous maximum likelihood parameter values.

The other interesting point is that the uncertainty in the half-width corresponds to ~±4%. This is a factor of 2 improvement over the individual least-squares parameter estimations. The addition of a second set of time-domain data to the time-domain analysis is only expected to improve the precision by the square root of two. This is because the improvement in the precision is approximately equal to the square root of the number of data points. The same can be said in the frequency domain. The complete

TABLE IV
COMPARISON OF METHODS FOR EVALUATION OF CONFIDENCE INTERVALS OF HALF-WIDTH FOR DUAL-DOMAIN ANALYSIS

Method	Confidence interval
Support plane[a]	2.613–2.816
Monte-Carlo[a]	2.606–2.817
Monte-Carlo[b]	2.516–2.923

[a] Confidence intervals corresponding to ±1 SEM.
[b] Confidence intervals corresponding to ±2 SEM.

probability distribution, as determined by the Monte-Carlo method, from which this confidence interval was determined is shown in Fig. 3 (bottom). Figure 4 (bottom) presents the analogous information as determined by the support plane method.

Table IV presents the confidence intervals of the half-width as evaluated by the support plane method and the Monte-Carlo method. These two methods yield approximately equivalent results.

Conclusion

The Monte-Carlo method for the evaluation of confidence intervals is generally considered one of the best methods. It requires few assumptions and is easy to implement. However, it requires a minimum of 500-fold more computer time than is required by the original least-squares fit. This time requirement may not be practical for routine analyses. An alternative is the support plane method, which requires only about a 10-fold increase in computer time. This single example cannot be used to demonstrate that the results of the two methods are equivalent. However, it and many other simulations (not shown) have indicated that the two methods yield equivalent results.

The dual-domain (and by analogy multidomain) analysis provides a substantial improvement in precision over simply adding more data in any one domain alone. It is expected that this improved precision will be most notable in cases where the parameters are highly correlated.

An added advantage of the multidomain analysis is that the systematic biases and uncertainties of each method are different. Thus, a simultaneous analysis of data obtained by multiple methods minimizes the effects of the systematic biases of the individual methods.

Acknowledgments

The author acknowledges the support of the University of Maryland at Baltimore Center for Fluorescence Spectroscopy (NIH RR-08119), National Institutes of Health Grant GM-35154, the Clinical Research Center at the University of Virginia (NIH RR-00847), and the National Science Foundation Science and Technology Center for Biological Timing at the University of Virginia.

Author Index

Numbers in parentheses are footnote reference numbers and indicate that an author's work is referred to although the name is not cited in the text.

A

C

D

E

F

G

H

I

J

K

L

M

N

O

P

Q

R

S

T

U

V

W

X

Y

Z

Subject Index

A

B

G

H

I

K

L

M

Q

V

W

Z

ISBN 0-12-182179-X